CALCULUS

Early Transcendentals

Fourth Edition

by Dennis G. Zill and Warren S. Wright

Jeffrey M. Gervasi, EdD

Porterville College

JONES AND BARTLETT PUBLISHERS

Sudbury, Massachusetts

BOSTON   TORONTO   LONDON   SINGAPORE

World Headquarters
Jones and Bartlett Publishers
40 Tall Pine Drive
Sudbury, MA 01776
978-443-5000
info@jbpub.com
www.jbpub.com

Jones and Bartlett
 Publishers Canada
6339 Ormindale Way
Mississauga, Ontario L5V 1J2
Canada

Jones and Bartlett
 Publishers International
Barb House, Barb Mews
London W6 7PA
United Kingdom

Jones and Bartlett's books and products are available through most bookstores and online booksellers. To contact Jones and Bartlett Publishers directly, call 800-832-0034, fax 978-443-8000, or visit our website, www.jbpub.com.

Substantial discounts on bulk quantities of Jones and Bartlett's publications are available to corporations, professional associations, and other qualified organizations. For details and specific discount information, contact the special sales department at Jones and Bartlett via the above contact information or send an email to specialsales@jbpub.com.

Production Credits
Publisher: David Pallai
Senior Acquisitions Editor: Timothy Anderson
Editorial Assistant: Melissa Potter
Production Director: Amy Rose
Associate Production Editor: Tiffany Sliter
Senior Marketing Manager: Andrea DeFronzo
V.P., Manufacturing and Inventory Control: Therese Connell
Cover and Title Page Design: Kristin E. Parker
Composition: Lapiz Online
Cover Image: © NASA, ESA, and The Hubble Heritage Team (STScI/AURA)
Printing and Binding: Courier Kendallville
Cover Printing: Courier Kendallville

ISBN- 10: 0-7637-7353-0
ISBN- 13: 978-0-7637-7353-3

6048
Printed in the United States of America
14 13 12 11 10   10 9 8 7 6 5 4 3 2 1

Preface

To the Instructor

This Student Resource Manual has been developed to accompany the **fourth edition** of *Calculus: Early Transcendentals* by Dennis G. Zill and Warren S. Wright. It consists of two parts: a review of precalculus topics and solutions to the odd-numbered exercises in the text. The precalculus review is a short reference which contains brief discussions of topics students are likely to encounter as they proceed through any calculus course. Students are highly encouraged to review these topics! I have been teaching the calculus sequence for nearly 16 years and have seen students with varying degrees of mathematical knowledge. The precalculus review will help to strengthen their understanding, however, it is by no means a complete exposition of the subject. For greater depth and details into those topics covered here, see the **fourth edition** of *Precalculus with Calculus Previews* by Dennis G. Zill and Jacqueline M. Dewar.

To the Student

You are embarking on a wonderful journey into the world of higher mathematics, one that begins with the study of calculus! Calculus is an amazing tool that can help us solve problems in such fields as physics, engineering, and economics, to name just a few. Indeed calculus is the gateway to some of these disciplines, therefore passing your calculus course successfully is quite necessary. To that end, this Student Resource Manual has been prepared with you in mind. You see, your instructor most likely assumes you possess certain mathematical competencies. These competencies include familiarity with the structure of the real and complex number systems, the ability to algebraically manipulate equations and inequalities, a basic knowledge of elementary analytic geometry, and familiarity with the characteristics and graphs of certain elementary functions. If you are rusty in any or all of these areas, the precalculus review in Part 1 of this manual will be a good reference for you. Consult it often when necessary. Part 2 of this manual, the solutions to the odd-numbered exercises from your text, is meant to be a guide for you, but you must take care not to become too dependent on it. If you get stuck, resist the temptation to rush to the solutions guide right away! Think about the problem, try a different approach, step away from it for a time, or consult your classmates. Only after you have spent some time on the problems will their solutions hold any value for you. Best of luck to you on your journey!

Acknowledgments

My deepest thanks to John David Dionisio, Loyola Marymount University, and Brian and Melanie Fulton, High Point University, for providing the solutions for this Student Resource Manual. Thanks, also, to Roger Cooke, University or Vermont, and Fred S. Roberts, Rutgers University, for contributing the excellent calculus essays.

I would also like to thank the staff at Jones and Bartlett Publishers, Tim Anderson, Senior Acquisitions Editor, Tiffany Sliter, Associate Production Editor, and Melissa Potter, Editorial Assistant.

Contents

Topics in Precalculus

1. Basic Mathematics

1.1 Sets

A set A is a collection of objects called **elements** of A. The symbolism $a \in A$ is used to denote that a is an element of a set A. A set B is a **subset** of set A, written $B \subset A$, if every element of B is also an element of A. Sets A and B are equal, $A = B$, if they contain the same elements. The **union** $A \cup B$ of two sets A and B is the set containing the elements that are either in A or in B. The **intersection** $A \cap B$ is the set of elements that are elements in A and in B. The intersection of sets A and B is the set of elements that are common to both A and B. The set with no elements is called the **empty set** and is denoted with the symbol $\emptyset$. If $A \cap B = \emptyset$, then we say the sets A and B are **disjoint**.

1.2 Laws of Exponents

If n is a positive integer, then

$$x^{-n} = \frac{1}{x^n}, \quad x \neq 0$$
$$x^0 = 1, \quad x \neq 0$$

If m and n are integers, then

$$(i) \qquad x^m x^n = x^{m+n}$$
$$(ii) \qquad \frac{x^m}{x^n} = x^{m-n}, \quad x \neq 0$$
$$(iii) \qquad (x^m)^n = x^{mn}$$
$$(iv) \qquad (xy)^n = x^n y^n$$
$$(v) \qquad \left(\frac{x}{y}\right)^n = \frac{x^n}{y^n}, \quad y \neq 0$$

1.3 Rational Exponents and Radicals

If m and n are positive integers,

$$x^{m/n} = (x^m)^{1/n} = (x^{1/n})^m$$
$$x^{1/n} = \sqrt[n]{x}$$

provided that all expressions represent real numbers. The laws of exponents hold for rational numbers.

EXAMPLE 1

(a) $16^{3/4} = (16^{1/4})^3 = 2^3 = 8$

(b) $(-27)^{2/3} = ((-27)^{1/3})^2 = (-3)^2 = 9$

1.4 Binomial Theorem

If n is a positive integer, the **Binomial Theorem** is

$$(a+b)^n = a^n + \frac{n}{1!}a^{n-1}b + \frac{n(n-1)}{2!}a^{n-2}b^2 + \cdots$$
$$+ \frac{n(n-1)\cdots(n-r+1)}{r!}a^{n-r}b^r + \cdots + b^n$$

where $r! = 1 \cdot 2 \cdot 3 \cdots (r-1)r$. For example,

$$(a+b)^2 = a^2 + 2ab + b^2$$
$$(a+b)^3 = a^3 + 3a^2b + 3ab^2 + b^3$$

EXAMPLE 2

Expand $(2x+4)^3$

Solution First we identify $a = 2x$, $b = 4$, and $n = 3$. Then the Binomial Theorem gives

$$(2x+4)^3 = (2x)^3 + 3(2x)^2 4 + 3(2x)(4)^2 + (4)^3$$
$$= 8x^3 + 48x^2 + 96x + 64$$

1.5 Special Factors

Difference of two squares: $x^2 - y^2 = (x+y)(x-y)$

Difference of two cubes: $x^3 - y^3 = (x-y)(x^2+xy+y^2)$

Sum of two cubes: $x^3 + y^3 = (x+y)(x^2-xy+y^2)$

EXAMPLE 3

(a) $x^3 - 1 = x^3 - 1^3 = (x-1)(x^2+x+1)$

(b) $x^3 + 8 = x^3 + 2^3 = (x+2)(x^2-2x+4)$

1.6 Equations

The equation $ax + b = 0$ is called a **first-degree** or **linear** equation. A linear equation has the unique solution $x = -b/a$, $a \neq 0$. A **second-degree** or **quadratic equation** is of the form $ax^2 + bx + c = 0$, $a \neq 0$. Two methods for finding the solutions of a quadratic equation are factoring and the quadratic formula,

$$x = \frac{-b \pm \sqrt{b^2 - 4ac}}{2a}$$

Recall that $b^2 - 4ac$ is the **discriminant** and it determines what type of number the solutions of a quadratic equation are. When $b^2 - 4ac > 0$, the equation has two distinct real solutions. For example, we see from the quadratic formula that the solutions of $x^2 + x - 1 = 0$ are the irrational numbers $(-1 - \sqrt{5})/2$ and $(-1 + \sqrt{5})/2$. When $b^2 - 4ac = 0$, the solutions of a quadratic equation are **equal** (or have *multiplicity* 2). When $b^2 - 4ac < 0$, the solutions of a quadratic equation are **complex numbers**.

1.7 Complex Numbers

A complex number is any expression of the form

$$z = a + bi \text{ where } i^2 = -1$$

The real numbers a and b are called the real and imaginary parts of z, respectively. In practice, the symbol i is written as $i = \sqrt{-1}$. The **complex conjugate** of z is defined as $\bar{z} = a - bi$.

EXAMPLE 4

If $z_1 = 4 + 5i$ and $z_2 = 3 - 2i$ are complex numbers, then their complex conjugates are, respectively, $\bar{z}_1 = 4 - 5i$ and $\bar{z}_2 = 3 - (-2)i = 3 + 2i$ ∎

1.8 Sum, Difference, and Product

The sum, difference, and product of two complex numbers $z_1 = a_1 + b_1 i$ and $z_2 = a_2 + b_2 i$ are defined as follows:

$$
\begin{aligned}
(i) &\quad z_1 + z_2 = (a_1 + a_2) + (b_1 + b_2)i \\
(ii) &\quad z_1 - z_2 = (a_1 - a_2) + (b_1 - b_2)i \\
(iii) &\quad z_1 z_2 = (a_1 a_2 - b_1 b_2) + (a_1 b_2 + b_1 a_2)i
\end{aligned}
$$

In other words, to add or subtract two complex numbers we simply add or subtract their corresponding real and imaginary parts. To multiply two complex numbers we use the distributive law and the fact that $i^2 = -1$

EXAMPLE 5

If $z_1 = 4 + 5i$ and $z_2 = 3 - 2i$, then

$$
\begin{aligned}
z_1 + z_2 &= (4 + 3) + (5 + (-2))i = 7 + 3i \\
z_1 - z_2 &= (4 - 3) + (5 - (-2))i = 1 + 7i \\
z_1 z_2 &= (4 + 5i)(3 - 2i) \\
&= (4 + 5i)3 + (4 + 5i)(-2i) \\
&= 12 + 15i - 8i - 10i^2 \\
&= (12 + 10) + (15 - 8)i = 22 + 7i
\end{aligned}
$$

The product of a complex number $z = a + bi$ and its conjugate is the real number $a^2 + b^2$. That is,

$$z\bar{z} = (a + bi)(a - bi) = a^2 + b^2$$

1.9 Quotients

The quotient of two complex numbers is found by multiplying the numerator and denominator of the expression by the conjugate of the denominator.

$$\frac{4+5i}{3-2i} = \frac{(4+5i)(3+2i)}{(3+2i)(3-2i)} = \frac{2+23i}{3^2+2^2} = \frac{2}{13} + \frac{23}{13}i$$

■

If $a > 0$, then the square root of the negative number $-a$ is defined as $\sqrt{-a} = \sqrt{a(-1)} = \sqrt{a}\sqrt{-1} = \sqrt{a}i$. For example, $\sqrt{-25} = \sqrt{25}\sqrt{-1} = 5i$.

By the quadratic formula, the solutions of $x^2 - 4x + 20 = 0$ are

$$x = \frac{-(-4) \pm \sqrt{(-4)^2 - 4(1)(20)}}{2(1)} = \frac{4 \pm \sqrt{64}i}{2} = \frac{4 \pm 8i}{2} = 2 \pm 4i$$

Therefore $x_1 = 2 + 4i$ and $x_2 = 2 - 4i$.

■

Notice that the complex solutions in Example 7 are conjugates. In general, the complex solutions of a quadratic equation with real coefficients are conjugates. Note also that every real number can be considered a complex number by taking $b = 0$ in $z = a + bi$.

The relationship between real and complex numbers is shown in Figure 1.9.1.

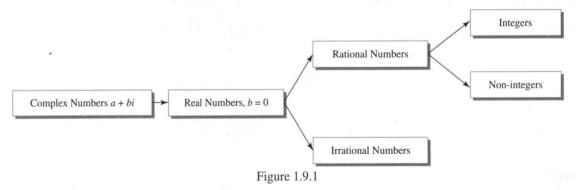

Figure 1.9.1

1.10 Similar Triangles

Two triangles are **similar** if three angles of one are congruent (have equal measures) to three angles of the other. The lengths of corresponding sides of similar triangles are proportional. Figure 1.10.1 represents two similar triangles:

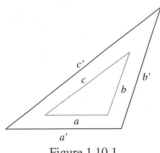

Figure 1.10.1

The ratios of the lengths of the corresponding sides of similar triangles are equal:

$$\frac{a}{a'} = \frac{b}{b'} = \frac{c}{c'}$$

1.11   Right Triangle Trigonometry

Consider the two right triangles shown in Figure 1.11.1 and having the same acute angle θ.

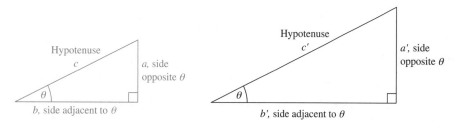

Figure 1.11.1

The side opposite the right angle is the hypotenuse. The other two sides are referred to as the legs of the triangle. Both triangles are similar (why?) therefore ratios of the lengths of corresponding sides are equal. Let **opp** represent the length of the side opposite the angle θ, **adj** represent the length of the side adjacent to angle θ, and **hyp** represent the length of the hypotenuse. Then we have the following equivalent ratios:

$$\frac{a}{c} = \frac{\text{opp}}{\text{hyp}} = \frac{a'}{c'}, \quad \frac{b}{c} = \frac{\text{adj}}{\text{hyp}} = \frac{b'}{c'}, \quad \frac{a}{b} = \frac{\text{opp}}{\text{adj}} = \frac{a'}{b'}$$

Note that these ratios depend only on the acute angle θ and not on the size of the triangles. From these we define the trigonometric functions **sine**, **cosine**, and **tangent**, abbreviated as **sin**, **cos**, and **tan** respectively. That is,

$$\sin \theta = \frac{\text{opp}}{\text{hyp}}, \quad \cos \theta = \frac{\text{adj}}{\text{hyp}}, \quad \tan \theta = \frac{\text{opp}}{\text{adj}}$$

The ratios of lengths of other pairs of sides are also equivalent, namely

$$\frac{c}{a} = \frac{\text{hyp}}{\text{opp}} = \frac{c'}{a'}, \quad \frac{c}{b} = \frac{\text{hyp}}{\text{adj}} = \frac{c'}{b'}, \quad \frac{b}{a} = \frac{\text{adj}}{\text{opp}} = \frac{b'}{a'}$$

We thus have three more trigonometric functions referred to as the **cosecant**, **secant**, and **cotangent**, and abbreviated as **csc**, **sec**, and **cot** respectively. That is,

$$\csc \theta = \frac{\text{hyp}}{\text{opp}}, \quad \sec \theta = \frac{\text{hyp}}{\text{adj}}, \quad \cot \theta = \frac{\text{adj}}{\text{opp}}$$

The sine and cosecant are reciprocals of each other. With a little algebraic manipulation we see that

$$\csc \theta = \frac{\text{hyp}}{\text{opp}} = \frac{1}{\text{opp/hyp}} = \frac{1}{\sin \theta}$$

In the same way it can be shown that

$$\sec \theta = \frac{1}{\cos \theta}, \quad \cot \theta = \frac{1}{\tan \theta}$$

The functions $\csc \theta$, $\sec \theta$, and $\cot \theta$ are typically referred to as the **reciprocal functions**.

EXAMPLE 8

Find the values of the six trigonometric functions given the acute angle $\theta = 45°$.

Solution   The six trigonometric functions depend only on θ and not the size of the right triangle. If one of the acute angles is $45°$ then the other must also be $45°$ giving us an isosceles triangle. This special triangle is sometimes referred to as a **45-45-90 right triangle**.

The legs will have the same length. For the sake of simplicity let's choose their lengths to be 1. By the Pythagorean Theorem, the length of the hypotenuse is $\sqrt{1^2 + 1^2} = \sqrt{2}$. See Figure 1.11.2.

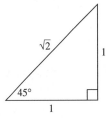

Figure 1.11.2

Here we have opp = adj = 1 and hyp = $\sqrt{2}$. Therefore

$$\sin 45° = \tfrac{1}{\sqrt{2}}, \quad \cos 45° = \tfrac{1}{\sqrt{2}}, \quad \tan 45° = \tfrac{1}{1}$$
$$\csc 45° = \tfrac{\sqrt{2}}{1}, \quad \sec 45° = \tfrac{\sqrt{2}}{1}, \quad \cot 45° = \tfrac{1}{1}$$

■

Be sure you understand that the values obtained above for the angle 45° are the same no matter the size of the right triangle. For example if we assign the lengths of the legs of the 45-45-90 triangle to be 6, then by the Pythagorean Theorem, the length of the hypotenuse is $\sqrt{6^2 + 6^2} = \sqrt{2 \cdot 36} = 6\sqrt{2}$. See Figure 1.11.3.

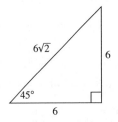

Figure 1.11.3

We then have $\sin 45° = \tfrac{6}{6\sqrt{2}} = \tfrac{1}{\sqrt{2}}$, $\cos 45° = \tfrac{6}{6\sqrt{2}} = \tfrac{1}{\sqrt{2}}$, $\tan 45° = \tfrac{6}{6} = 1$, the same values obtained before. What are the values of the remaining three trigonometric functions?

Motivated by our use of triangles to determine the trigonometric functions, consider now an equilateral triangle with sides of length $2a$, where $a > 0$, as shown in Figure 1.11.4(a). All three angles have measure 60° (why?).

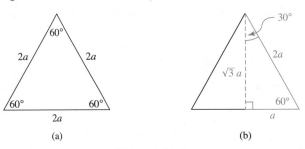

(a)   (b)

Figure 1.11.4

If we drop a perpendicular bisector from the top vertex to the base, we get two congruent right triangles with acute angles 30° and 60°, as shown in Figure 1.11.4(b). This special triangle is referred to as a **30-60-90 right triangle**. The side opposite the 30° angle has length a and the side opposite the angle 60° has length $\sqrt{3}\,a$ (determined by using the Pythagorean Theorem). We then have

$$\sin 30° = \frac{a}{2a} = \frac{1}{2}, \quad \cos 30° = \frac{\sqrt{3}a}{2a} = \frac{\sqrt{3}}{2}, \quad \tan 30° = \frac{a}{\sqrt{3}a} = \frac{1}{\sqrt{3}}$$

$$\sin 60° = \frac{\sqrt{3}a}{2a} = \frac{\sqrt{3}}{2}, \quad \cos 60° = \frac{a}{2a} = \frac{1}{2}, \quad \tan 60° = \frac{\sqrt{3}a}{a} = \frac{\sqrt{3}}{1}$$

The decision to use the length $2a$ in the equilateral triangle was merely one of convenience. In truth any positive value would do. Try it! In your study of calculus and physics, you will find that the sine, cosine, and tangent of the angles $30°$, $45°$, and $60°$ occur so frequently that it will be worth your time to memorize them rather than to repeatedly construct the 45-45-90 and 30-60-90 triangles from which they were derived.

1.12   Unit-Circle Trigonometry

In section 1.11, the sine and cosine functions were defined as the quotient of lengths of sides of a right triangle but recall from trigonometry that they may also be interpreted as the x- and y-coordinates of a point on the unit circle. See Figure 1.12.1(a). More on this in a moment but first a few reminders.

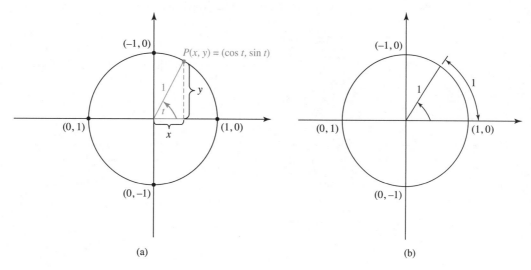

Figure 1.12.1

Angles can be measured in either degrees or radians. In your study of calculus you'll find that radians are used almost exclusively. An angle of **one radian** subtends (intersects) an arc of length 1 unit on the circumference of a unit circle.   See Figure 1.12.1(b). A central angle of the unit circle (in fact any circle) is an angle whose vertex lies at the center of the circle. The central angle is in **standard position** if its vertex is at the origin of the xy-coordinate system and its initial side coincides with the positive x-axis. Think of the angle as being formed by a rotation of the initial side from the positive x-axis to the terminal side. If the angle is formed by a counterclockwise rotation, we say the angle is **positive** whereas if the angle is formed by a clockwise rotation we say the angle is **negative**. Two angles in standard position are **coterminal** if they share the same terminal side. Verify for example that $\pi/3$ is coterminal with $-5\pi/3$ and $19\pi/3$. Note that $\pi/3 - (-5\pi/3) = 2\pi$ and $\pi/3 - 19\pi/3 = -6\pi = -3(2\pi)$. In general, two coterminal angles measured in radians will differ by an integer multiple of 2π. If the initial side of the angle sweeps out one counterclockwise rotation, its measure is 2π radians. This is the equivalent of $360°$ therefore

$$2\pi \text{ radians} = 360° \quad \Rightarrow \quad \pi \text{ radians} = 180°$$

the latter of which may be used in either of the two forms

$$1 \text{ radian} = \left(\frac{180}{\pi}\right)^{\circ} \text{ or } 1° = \frac{\pi}{180} \text{ radians}$$

to convert from radians to degrees or from degrees to radians.

EXAMPLE 9

(a) $30° = 30 \cdot \dfrac{\pi}{180} = \dfrac{\pi}{6}$ radians   (b) $45° = 45 \cdot \dfrac{\pi}{180} = \dfrac{\pi}{4}$ radians

(c) $60° = 60 \cdot \dfrac{\pi}{180} = \dfrac{\pi}{3}$ radians   (d) $\dfrac{\pi}{2}$ radians $= \dfrac{\pi}{2} \cdot \left(\dfrac{180}{\pi}\right)° = 90°$ ■

The following table contains the degree and radian measures of angles which occur frequently.

Degrees	0°	30°	45°	60°	90°	120°	135°	150°	180°	270°	360°
Radians	0	$\frac{\pi}{6}$	$\frac{\pi}{4}$	$\frac{\pi}{3}$	$\frac{\pi}{2}$	$\frac{2\pi}{3}$	$\frac{3\pi}{4}$	$\frac{5\pi}{6}$	π	$\frac{3\pi}{2}$	2π

While these angles occur frequently to be sure, they are by no means the *only* angles. In fact for *any* real number t there corresponds an angle of t radians in standard position. The terminal side of this angle will intersect the unit circle at the point $P(x,y)$ as shown in Figure 1.12.2(a) There's more to the story however! Consider Figure 1.12.2(b) where θ is an acute angle in standard position.

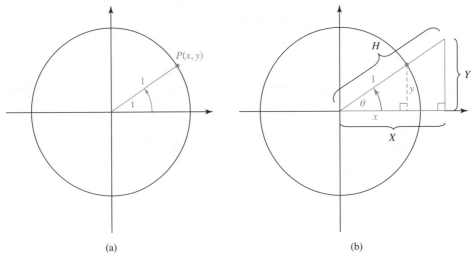

(a)   (b)

Figure 1.12.2

The terminal side intersects the unit circle at the point $P(x,y)$. Earlier we indicated in Figure 1.12.1(a) that $\cos t = x$ and $\sin t = y$ but why and what do those definitions have to do with the definitions of the sine and cosine in section 1.11? Well, the two right triangles shown in Figure 1.12.2(b) share the same angle θ therefore they are similar. By our work in section 1.11, we know that

$$\sin \theta = \frac{Y}{H} = \frac{y}{1} = y \quad \text{and} \quad \cos \theta = \frac{X}{H} = \frac{x}{1} = x$$

We usually measure θ in radians and so if we let the angle $\theta = t$ radians, then we have

$$\sin t = y \quad \text{and} \quad \cos t = x$$

As you can see, unit-circle trigonometry and right-triangle trigonometry are intimately related. Each approach has its advantages depending on the context of the problem you're trying to solve. You should be sure to understand that for each real number t there corresponds a unique point on the unit circle whose coordinates are $(\cos t, \sin t)$. That is, there is only one value of the cosine of t, namely $\cos t$, and one value of the sine of t, namely $\sin t$. By definition then, the sine and cosine are *functions* each with the set of all real numbers for their domains and we typically write

$$y = \sin x \quad \text{and} \quad y = \cos x$$

EXAMPLE 10

Find the exact values of $\sin t$ and $\cos t$ for $t = \pi/6$.

Solution   Draw an angle of $\pi/6$ radians (30°) in standard position as shown below with intersection point $P(x,y)$. The triangle is a 30-60-90 right triangle with $a = 1/2$ (compare to Figure 1.11.4(b)). Therefore the point P has coordinates $(\sqrt{3}/2, 1/2)$. From this section we know that $\sin(\pi/6) = y = 1/2$ and $\cos(\pi/6) = x = \sqrt{3}/2$. ■

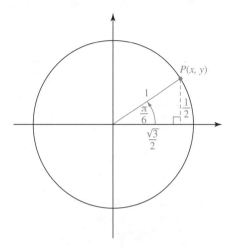

1.13   Law of Sines and Law of Cosines

Consider the triangle shown:

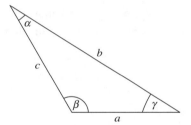

If we know either the length of one side and two angles or the lengths of two sides and the angle opposite one of them, then the remaining parts of the triangle can be found from the **law of sines:**

$$\frac{\sin \alpha}{a} = \frac{\sin \beta}{b} = \frac{\sin \gamma}{c}$$

Furthermore, if we know either the lengths of three sides or the length of two sides and the angle between them, then the remaining parts of the triangle can be found from the **law of cosines:**

$$a^2 = b^2 + c^2 - 2bc \cos \alpha$$
$$b^2 = a^2 + c^2 - 2ac \cos \beta$$
$$c^2 = a^2 + b^2 - 2ab \cos \gamma$$

2.   Real Numbers and Inequalities

Real numbers are classified as either **rational** or **irrational**. A rational number can be expressed as a quotient a/b, where a and b are integers and $b \neq 0$. A number that is *not* rational is said to be irrational. For example, $7/8, 1/2, -6$, and $\sqrt{4} = 2$, are rational numbers, whereas $\sqrt{2}, \sqrt{3}/2$, and π are irrational. The sum, difference, and product of two real numbers are real numbers. The quotient of two real numbers is a real number provided the denominator is not zero. The set of real numbers is denoted by the symbol R; the symbols Q and H are commonly used to denote the set of rational numbers and the set of irrational numbers, respectively. In terms of the union of two sets, we have $R = Q \cup H$. The fact that Q and H have no common elements is summarized by $Q \cap H = \emptyset$, where $\emptyset$ is the empty set.

2.1 Number Line

The set of real numbers can be put into a one-to-one correspondence with the points on a horizontal line, which is called the **number line** or **real line**. A number a associated with a point P on the number line is called the coordinate of P. The point chosen to represent 0 is called the **origin**. As shown in the figure, **positive numbers** are placed to the right of the origin and **negative numbers** are placed to the left of the origin. The number 0 is neither positive nor negative. The arrowhead on the number line points in the **positive direction**. As a rule, the words "point a" and "number a" are used interchangeably with "point P with coordinate a".

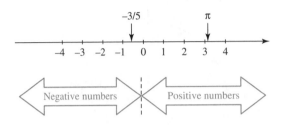

2.2 Inequalities

The set R of real numbers is an **ordered** set. A real number a is **less than** a real number b, written $a < b$, if the difference $b - a$ is positive. For example, $-2 < 5$, since $5 - (-2) = 7$ is a positive number. On the number line $a < b$ means that the number a lies to the left of the number b as shown:

$$a < b$$

Figure 2.2.1

The statement "a is less than b" is equivalent to saying "b is greater than a" and is written $b > a$. Thus, a number a is positive if $a > 0$ and negative if $a < 0$. Expressions such as $a < b$ and $b > a$ are called **inequalities**.

2.3 Properties of Inequalities

The following is a list of some **properties of inequalities**:

$\quad(i)\quad$ If $a < b$, then $a + c < b + c$ for any real number c

$\quad(ii)\quad$ If $a < b$ and $c > 0$, then $ac < bc$

$\quad(iii)\quad$ If $a < b$ and $c < 0$, then $ac > bc$

$\quad(iv)\quad$ If $a < b$ and $b < c$, then $a < c$

Property (iii) indicates that if an inequality is multiplied by a negative number, then the sense of the inequality is *reversed*. The symbolism $a \leq b$ is read "a is **less than or equal** to b" and means $a < b$ or $a = b$. The preceding four properties hold when $<$ and $>$ are replaced by $\leq$ and $\geq$, respectively. If $a \geq 0$, the number a is said to be **nonnegative**.

EXAMPLE 1

Solve the inequality $2x < 5x + 9$

Solution First we use (i) to add $-5x$ to both sides of the inequality:

$$2x + (-5x) < 5x + 9 + (-5x)$$
$$-3x < 9$$

Then when we multiply both sides of the last inequality by $-1/3$, property (iii) implies $x > -3$. ∎

2.4 Intervals

If $a < b$, the set of real numbers x that are *simultaneously* less than b and greater than a is written $\{x \mid a < x < b\}$ which we may simply write as $a < x < b$. This set is called an open interval and is denoted by (a, b). The numbers a and b are the endpoints of the interval. Various other intervals are listed here:

$$\text{Closed interval } [a,b] = \{x \mid a \leq x \leq b\}$$
$$\text{Open interval } (a,b) = \{x \mid a < x < b\}$$
$$\text{Half-open interval } (a,b] = \{x \mid a < x \leq b\}$$
$$\text{Half-open interval } [a,b) = \{x \mid a \leq x < b\}$$
$$\text{Infinite intervals } (a,\infty) = \{x \mid x > a\}$$
$$\text{or } [a,\infty) = \{x \mid x \geq a\}$$
$$\text{or } (-\infty,b) = \{x \mid x < b\}$$
$$\text{or } (-\infty,b] = \{x \mid x \leq b\}$$
$$\text{or } (-\infty,\infty) = \{x \mid -\infty < x < \infty\}$$

The symbols ∞ and $-\infty$ are read "infinity" and "negative infinity," respectively. These symbols do not represent real numbers. Their use is simply shorthand for writing an *unbounded* interval.

EXAMPLE 2

(a) The set $\{x \mid -3 < x < 8\}$ is $(-3,8)$ in interval notation

(b) The set $\{x \mid 5 < x \leq 6\}$ is $(5,6]$ in interval notation

(c) The set $\{x \mid x \leq -1\}$ is $(-\infty, -1]$ in interval notation

■

An interval that is a subset of another interval I is said to be a **subinterval** of I. For example, $[1,2]$, $[2,2.5]$, and $(3,6)$ are each subintervals of $[1,6]$.

EXAMPLE 3

Solve the inequality $4 < 2x - 3 < 8$.

Solution   From the properties of inequalities it follows that

$$4 < 2x - 3 < 8$$
$$4 + 3 < 2x - 3 + 3 < 8 + 3$$
$$\frac{7}{2} < x < \frac{11}{2} \text{ or } (7/2, 11/2) \text{ in interval notation.}$$

■

EXAMPLE 4

Solve the inequality $(x+3)(x-4) < 0$.

Solution   We begin by putting the two roots, -3 and 4, of the quadratic equation $(x+3)(x-4) = 0$ on a number line. By ascertaining the algebraic sign of each factor on the subintervals $(-\infty, -3)$, $(-3,4)$, and $(4,\infty)$ of $(-\infty,\infty)$, we can determine the algebraic sign of the product. For example, by replacing x by any number in $(-\infty,3)$, we see that both factors $x+3$ and $x-4$ are negative, and so the product $(x+3)(x-4)$ must be positive. In this case, the original inequality is not satisfied. Inspection of the sign chart shown indicates that the solution of the problem is the interval $(-3,4)$.

Interval	$(-\infty, -3)$	$(-3, 4)$	$(4, \infty)$
Sign of $x+3$	$-$	$+$	$+$
Sign of $x-4$	$-$	$-$	$+$
Sign of $(x+3)(x-4)$	$+$	$-$	$+$

∎

2.5   Absolute Value

If a is a real number, then its absolute value is defined as

$$|a| = \begin{cases} a, & a \geq 0 \\ -a, & a < 0 \end{cases}$$

On the number line, $|a|$ is the distance between the origin and the number a. In general, the distance between two numbers a and b is either $b-a$ if $a < b$ or $a-b$ if $b < a$. In terms of absolute value, the distance between a and b is $|b-a|$. See Figure 2.5.1. Note also that $|b-a| = |a-b|$.

Figure 2.5.1

EXAMPLE 5
_____

(a) $|-5| = -(-5) = 5$

(b) $|2x-1| = \begin{cases} 2x-1, & 2x-1 \geq 0 \\ -(2x-1), & 2x-1 < 0 \end{cases}$

That is, $|2x-1| = \begin{cases} 2x-1, & x \geq \frac{1}{2} \\ -2x+1, & x < \frac{1}{2} \end{cases}$

∎

EXAMPLE 6
_____

The distance between -5 and 7 is the same as the distance between 7 and -5.

$$|7-(-5)| = |12| = 12 \text{ units and } |-5-7| = |-12| = 12 \text{ units}$$

∎

2.6   Absolute Values and Inequalities

Since $|x|$ gives the distance between a number x and the origin, the solution of the inequality $|x| < b$, $b > 0$, is the set of real numbers x that are _less_ than b units away from the origin. What does this look like? Consider Figure 2.6.1:

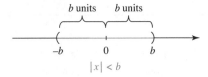

Figure 2.6.1

This figure should convince you that

$$|x| < b \text{ if and only if } -b < x < b \tag{2.1}$$

On the other hand, the solution of the inequality $|x| > b$ is the set of real numbers that are *more* than b units away from the origin. This set is shown Figure 2.6.2:

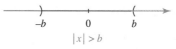

$$|x| > b$$

Figure 2.6.2

Once again you should be convinced that

$$|x| > b \text{ if and only if } x > b \text{ or } x < -b \tag{2.2}$$

The results in (2.1) and (2.2) hold when $<$ and $>$ are replaced with $\leq$ and $\geq$, respectively, and when x is replaced by $x - a$. For example, $|x - a| \leq b, b > 0$, represents the set of real numbers x such that the distance from x to a is less than or equal to b units. This implies that x is a number in the closed interval $[a - b, a + b]$, as shown Figure 2.6.3:

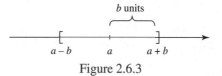

Figure 2.6.3

EXAMPLE 7
_____

Solve the inequality $|x - 1| < 3$.

Solution   Using (2.1), we first rewrite the inequality as

$$-3 < x - 1 < 3$$
$$-3 + 1 < x - 1 + 1 < 3 + 1$$
$$-2 < x < 4$$

The solution is the open interval $(-2, 4)$. ∎

EXAMPLE 8
_____

Solve the inequality $|x| > 2$.

Solution   From (2.2) we have immediately

$$x > 2 \quad \text{or} \quad x < -2$$

The solution is a union of infinite intervals $(-\infty, -2) \cup (2, \infty)$. ∎

EXAMPLE 9
_____

Solve the inequality $0 < |x - 2| \leq 7$.

Solution   The given inequality means $0 < |x - 2|$ *and* $|x - 2| \leq 7$. In the first case, $0 < |x - 2|$ is true for any real number except $x = 2$. In the second case, we have

$$-7 \leq x - 2 \leq 7 \quad \text{or} \quad -5 \leq x \leq 9 \quad \text{or} \quad [-5, 9]$$

The set of real numbers x that satisfies both inequalities consists of all numbers in $[-5, 9]$ except 2. Written as a union of intervals, the solution is $[-5, 2) \cup (2, 9]$. ∎

2.7 Triangle Inequality

In conclusion, we show the proof of the so-called triangle inequality

$$|a+b| \leq |a| + |b| \tag{2.3}$$

First convince yourself of the validity of the two inequalities $-|a| \leq a \leq |a|$ and $-|b| \leq b \leq |b|$. Now adding these two inequalities gives us the following:

$$-|a| + |b| \leq a + b \leq |a| + |b|$$

Comparing this to (2.1), we see that we may rewrite the last inequality as $|a+b| \leq |a| + |b|$.

2.8 Remarks

(*i*) Every real number has a nonterminating decimal representation. The rational numbers are characterized as *repeating* decimals; for example, by long division, we see that

$$\frac{7}{11} = 0.636363\ldots \quad \text{and} \quad \frac{3}{4} = 0.750000\ldots$$

In the first, the digits 63 repeat whereas in the second it is the 0 that repeats itself. On the other hand, irrational numbers are *nonrepeating* decimal numbers. For example, the number π. This number has no repeating decimal:

$$\pi = 3.141592\ldots$$

(*ii*) In Example 8 there is no way of writing the solution of the inequality $|x| > 2$ as a single interval. The statement $x > 2$ or $x < -2$ is *not* equivalent to $2 < x < -2$. There are no real numbers that are simultaneously greater than positive 2 and less than negative 2.

   (*iii*) Finally a brief word about the square root of a nonnegative number is in order:

The square root of $x \geq 0$, written $\sqrt{x}$, is always nonnegative

For example, $\sqrt{64} = 8$, not ± 8. In fact, the square root is related to the absolute value:

$$\sqrt{x^2} = |x|$$

Thus, $\sqrt{x^2} = x$ if $x \geq 0$ and $\sqrt{x^2} = -x$ if $x < 0$. For example, $\sqrt{5^2} = 5$ since $5 > 0$ and $\sqrt{(-5)^2} = -(-5)$ since $-5 < 0$.

3　The Cartesian Plane

In general we typically display data in a coordinate plane formed by the intersection of two perpendicular number lines. In mathematics such a *coordinate plane* is called the **Cartesian plane**. The point of intersection of these number lines, corresponding to the number 0 on both lines, is called the **origin** and is denoted by O. The horizontal number line is called the **x-axis** and the vertical number line is called the **y-axis**. Numbers to the right of the origin on the x-axis are positive; numbers to the left of O are negative. On the y-axis, numbers above O are positive; numbers below O are negative. Because the horizontal and vertical number lines are labeled with the letters x and y, a Cartesian plane is often simply called an **xy-plane**. If P denotes a point in a Cartesian plane, we can draw perpendicular lines from P to both the x- and y-axes.

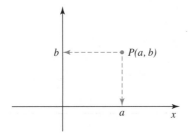

Figure 3.1

   As shown in Figure 3.1, this determines a number a on the x-axis and a number b on the y-axis. Conversely, we see that specified numbers a and b on the x- and y-axes determine a unique point P in the plane. In this manner a one-to-one correspondence between points in a Cartesian plane and ordered pairs of real numbers (a,b) is established. We call a the **x-coordinate** or **abscissa** of the point P; b is called the **y-coordinate** or **ordinate** of the point. The axes are also called **coordinate axes** and P is said to have **coordinates** (a,b).

3.1   Quadrants

The coordinate axes divide the Cartesian plane into four regions known as quadrants. Algebraic signs of the x-coordinate and y-coordinate of any point (x,y) located in each of the four quadrants are indicated in the following figure:

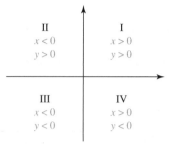

Figure 3.1.1

Points on a coordinate axis, for example $(2,0)$ and $(0,4)$, are considered to be not in any quadrant. This method of describing points in a plane is called a **rectangular** or **Cartesian coordinate system**. In this system, two points (a,b) and (c,d) are equal if and only if $a = c$ and $b = d$.

3.2   Distance Formula

We can determine the distance $d(P_1, P_2)$ between two points $P_1(x_1, y_1)$ and $P_2(x_2, y_2)$ from the Pythagorean Theorem. As shown in Figure 3.2.1, the three points P_1, P_2, and P_3 are the vertices of a right triangle with a hypotenuse of length d and sides of lengths $|x_2 - x_1|$ and $|y_2 - y_1|$.

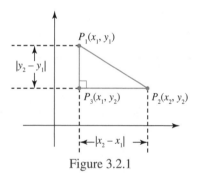

Figure 3.2.1

Thus, $d^2 = |x_2 - x_1|^2 + |y_2 - y_1|^2$ and from this we get the distance formula

$$d(P_1, P_2) = \sqrt{(x_2 - x_1)^2 + (y_2 - y_1)^2} \tag{3.1}$$

EXAMPLE 1

Find the distance between the points $(-2, 3)$ and $(4, 5)$.

Solution   By identifying P_1 as $(-2, 3)$ and P_2 as $(4, 5)$, we obtain from (3.1)

$$d(P_1, P_2) = \sqrt{(4 - (-2))^2 + (5 - 3)^2} = \sqrt{6^2 + 2^2} = \sqrt{40} = 2\sqrt{10}$$

Since $(x_2 - x_1)^2 = (x_1 - x_2)^2$ and $(y_2 - y_1)^2 = (y_1 - y_2)^2$, it doesn't matter which points are designated P_1 and P_2. In other words, $d(P_1, P_2) = d(P_2, P_1)$.

3.3 Midpoint of a Line Segment

The distance formula can be used to prove that the coordinates of the midpoint of a line segment joining the points $P_1(x_1, y_1)$ and $P_2(x_2, y_2)$ are

$$\left(\frac{x_1 + x_2}{2}, \frac{y_1 + y_2}{2} \right) \tag{3.2}$$

So if M denotes a point on the segment $\overline{P_1 P_2}$ with coordinates given in (3.2), then M is the midpoint of $\overline{P_1 P_2}$ provided

$$d(P_1, M) = d(M, P_2) \text{ and } d(P_1, P_2) = d(P_1, M) + d(M, P_2) \tag{3.3}$$

EXAMPLE 2

Find the coordinates of the midpoint of the line segment joining $P_1(-2, 2)$ and $P_2(4, 5)$.

Solution From (3.2) we have

$$x = \frac{-2 + 4}{2} = 1 \text{ and } y = \frac{2 + 5}{2} = \frac{7}{2}$$

The midpoint of the line segment is therefore $(1, 7/2)$. ∎

3.4 Graphs

A graph is any set of points (x, y) in the Cartesian plane. The graph of an equation is the set of points (x, y) in the Cartesian plane that are solutions of the equation. An ordered pair (x, y) is a solution of an equation if substitution of x and y into the equation reduces it to an identity.

EXAMPLE 3

The point $(-2, 2)$ is on the graph of $y = 1 - \frac{1}{8}x^3$, since

$$2 = 1 - \frac{1}{8}(-2)^3 \quad \text{is equivalent to} \quad 2 = 1 + \frac{8}{8} \quad \text{or} \quad 2 = 2$$

3.5 Point Plotting

One way of sketching the graph of an equation, often done in elementary courses, is to plot points and then connect these points with a smooth curve. To obtain points on the graph we assign values to either x or y and then solve the equation for the corresponding values of y or x. Most of the equations you will encounter in calculus are those that define functions. When graphing an equation by plotting points, we need to plot enough of them so that the shape, or pattern, of the graph is evident. Sometimes it's useful to consider the *local* behavior of the graph such as intercepts and symmetry. As you will see in your calculus course, there are a number of tools we can use to refine and analyze graphs beyond point-plotting. For example you will learn how to determine where the graph of a function is increasing and decreasing, where it is concave up or concave down, and if it has any asymptotes (vertical, horizontal, or slanted).

3.6 Symmetry

Before plotting points, it is often useful to determine whether a graph of an equation possess symmetry. Consider Figure 3.6.1. We say that a graph is

  (i) **symmetric with respect to the y-axis** if, whenever (x, y) is a point on the graph, $(-x, y)$ is also a point on the graph

  (ii) **symmetric with respect to the x-axis** if, whenever (x, y) is a point on the graph, $(x, -y)$ is also a point on the graph; and

(iii) **symmetric with respect to the origin** if, whenever (x, y) is a point on the graph, $(-x, -y)$ is also a point on the graph.

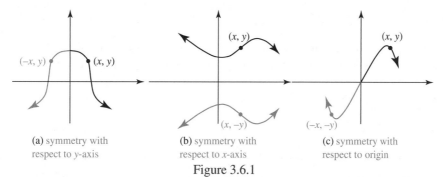

(a) symmetry with
respect to y-axis

(b) symmetry with
respect to x-axis

(c) symmetry with
respect to origin

Figure 3.6.1

3.7 Tests for Symmetry

For an equation, *(i)*, *(ii)*, and *(iii)* yield the following three **tests for symmetry**. A graph of an equation is symmetric with respect to

(i) the y-axis if replacing x by $-x$ results in an equivalent equation;

(ii) the x-axis if replacing y by $-y$ results in an equivalent equation; and

(iii) the origin if replacing both x with $-x$ and y with $-y$ results in an equivalent equation.

EXAMPLE 4

Determine whether the graph of $x = |y| - 2$ possesses any symmetry.

Solution

Test *(i)*: $-x = |y| - 2$ or $x = -|y| + 2$ is not equivalent to the original equation.

Test *(ii)*: $x = |-y| - 2$ or $x = |y| - 2$ which is equivalent to the original equation.

Test *(iii)*: $-x = |-y| - 2$ or $x = -|y| + 2$ is not equivalent to the original equation.

Therefore the graph of $x = |y| - 2$ is symmetric with respect to the x-axis.

Detecting symmetry before plotting points can often save time and effort when graphing. For example, if the graph of an equation is shown to be symmetric with respect to the y-axis, then it is sufficient to plot points with x-coordinates that satisfy $x \geq 0$. We can then find points in the second and third quadrants by taking the mirror images, through the y-axis, of the points in the first and fourth quadrants.

3.8 Intercepts

When you sketch the graph of an equation, it is always a good idea to determine whether the graph has any **intercepts**. The x-coordinate of a point where the graph crosses the x-axis is called an ***x*-intercept**. The y-coordinate of a point where the graph crosses the y-axis is called a ***y*-intercept**. Figure 3.8.1 shows a graph with three x-intercepts and one y-intercept:

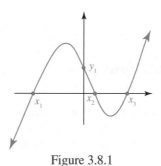

Figure 3.8.1

Since $y = 0$ for any point on the x-axis and $x = 0$ for any point on the y-axis, we can determine the intercepts of the graph of an equation in the following manner:

$$
\begin{array}{ll}
x\text{-intercepts:} & \text{Set } y = 0 \text{ in the equation and solve for } x \\
y\text{-intercepts:} & \text{Set } x = 0 \text{ in the equation and solve for } y
\end{array}
\tag{3.4}
$$

EXAMPLE 5

Find the intercepts for the graph of (a) $y = 4x - 3$ and (b) $y = \frac{x^2+1}{x^2+5}$.

Solution

(a) Setting $y = 0$ gives us $0 = 4x - 3$ or $x = \frac{3}{4}$. Setting $x = 0$ gives $y = -3$. The x-intercept is $(3/4, 0)$ and the y-intercept is $(0, -3)$.

(b) Here $y = 0$ if the numerator $x^2 + 1 = 0$ and the denominator $x^2 + 5 \neq 0$. Since $x^2 + 5 \neq 0$ for all real numbers, we have $x^2 + 1 = 0$ or $x^2 = -1$ which is not satisfied for any real number. Therefore the graph has no x-intercepts. Setting $x = 0$ gives us $y = \frac{1}{5}$. The y-intercept is $(0, 1/5)$.

Recall that a **zero** of a function f is a number c for which $f(c) = 0$. Equivalently, the point $(c, 0)$ is an x-intercept for the graph of the function f.

EXAMPLE 6

Find the zeros of the function $f(x) = x^3 - 2x^2 - x + 2$.

Solution   To find the zeros, we set the function equal to 0 and solve for x

$$
\begin{aligned}
x^3 - 2x^2 - x + 2 &= 0 \leftarrow \text{factor by grouping} \\
x^2(x - 2) - (x - 2) &= 0 \\
(x - 2)(x^2 - 1) &= 0 \\
(x - 2)(x + 1)(x - 1) &= 0
\end{aligned}
$$

From the last equation we see that the zeros are $x = 2$, $x = -1$, and $x = 1$. Therefore the x-intercepts are $(2, 0)$, $(-1, 0)$, and $(1, 0)$.

3.9   Real Zeros of Polynomial Functions

In calculus you will spend a lot of time finding the zeros of functions, especially polynomial functions. Depending on the function, this may be easy or difficult. Fortunately there are a few tools you may recall that we can use to aid us, namely synthetic division, the Factor Theorem, and the Rational Zeros Theorem. Let's revisit each one in turn. When the divisor of a polynomial function has the form $x - c$, for any real number c, long division may be simplified by using synthetic division. Study the next example carefully.

EXAMPLE 7

Use synthetic division to divide $f(x) = 3x^3 + 5x^2 - 4x + 2$ by $x + 2$.

Solution   Here we have $c = -2$.

$$\underline{-2}\ |\quad 3\quad 5\quad -4\quad 2\qquad \leftarrow\ \text{Write the } -2 \text{ from the divisor and the coeffecients of } f$$

$$3\qquad\qquad\qquad \leftarrow\ \text{Bring down the first coefficient}$$

$$\begin{array}{r|rrrr} -2 & 3 & 5 & -4 & 2 \\ & & -6 & & \\ \hline & 3 & -1 & & \end{array}\qquad \leftarrow\ \text{Multiply 3 by } -2.\ \text{Record } -6.\ \text{Add down.}$$

$$\begin{array}{r|rrrr} -2 & 3 & 5 & -4 & 2 \\ & & -6 & 2 & \\ \hline & 3 & -1 & -2 & \end{array}\qquad \leftarrow\ \text{Multiply } -1 \text{ by } -2.\ \text{Record } +2.\ \text{Add down.}$$

$$\begin{array}{r|rrrr} -2 & 3 & 5 & -4 & 2 \\ & & -6 & 2 & 4 \\ \hline & 3 & -1 & -2 & \underline{6} \end{array}\qquad \begin{array}{l} \leftarrow\ \text{Multiply } -2 \text{ by } -2.\ \text{Record } +4.\ \text{Add down.} \\ \leftarrow\ 6 \text{ is the remainder.} \end{array}$$

The coefficients of the quotient, call it $q(x)$, are the first three numbers in the third row, so $q(x) = 3x^2 - 1x - 2$. The remainder is the last entry in the third row, $r = 6$. ■

EXAMPLE 8

Use synthetic division to divide $f(x) = x^3 - 3x^2 + 4$ by $x - 2$.

Solution   Here we have $c = 2$ and we proceed as in the previous example:

$$\begin{array}{r|rrrr} 2 & 1 & -3 & 0 & 4 \\ & & 2 & -2 & -4 \\ \hline & 1 & -1 & -2 & \underline{0} = r \end{array}$$

Therefore the quotient is $q(x) = x^2 - x - 2$. Note that a 0 was added for the coefficient of x. The remainder is 0 so $x - 2$ divides evenly into $f(x)$. That is, $x - 2$ is a *factor* of the polynomial function $f(x)$ and so we may write

$$f(x) = (x - 2)(x^2 - x - 2)$$

■

Next recall the Factor Theorem:

> The number c is a zero of a polynomial function f if and only if $x - c$ is a factor of $f(x)$

EXAMPLE 9

As we found in Example 8, $x - 2$ is a factor of the function $f(x) = x^3 - 3x^2 + 4$. By the Factor Theorem, the number 2 is a zero of $f(x)$ which we verify:

$$\begin{aligned} f(2) &= (2)^3 - 3(2)^2 + 4 \\ &= 8 - 12 + 4 \\ &= 0 \end{aligned}$$

Therefore $f(x) = x^3 - 3x^2 + 4 = (x-2)(x^2 - x - 2)$. Note that we may factor further to get

$$f(x) = (x-2)(x-2)(x+1)$$

from which we see immediately that $x = -1$ is also a zero of the function.   ■

Finally recall the Rational Zeros Theorem:

> If p/s is a rational number (in reduced form) and a zero of the polynomial function
> $$f(x) = a_n x^n + a_{n-1} x^{n-1} + \cdots + a_2 x^2 + a_1 x + a_0$$
> where the coefficients are integers, $a_n \neq 0$, and $a_0 \neq 0$, then p is an integer factor of the constant term a_0 and s is an integer factor of the leading coefficient a_n.

Be sure to understand that the theorem *does not* say a polynomial function with integer coefficients *must* have a rational zero! It does however tells us that *if* a polynomial function with integer coefficients has a rational zero p/s, *then* p is an integer factor of a_0 and s is an integer factor of a_n.

EXAMPLE 10

Find all rational zeros of $f(x) = 3x^4 - 10x^3 - 3x^2 + 8x - 2$.

Solution   Here the constant term $a_0 = -2$ and the leading coefficient $a_4 = 3$. List all the integer factors of a_0 and a_4 respectively:

$$p : \pm 1, \pm 2$$
$$s : \pm 1, \pm 3$$

The list of possible rational zeros p/s are then

$$\frac{\pm 1}{\pm 1}, \quad \frac{\pm 2}{\pm 1}, \quad \frac{\pm 1}{\pm 3}, \text{ and } \quad \frac{\pm 2}{\pm 3}$$

The possible rational zeros are therefore

$$-1, +1, -2, +2, -\frac{1}{3}, +\frac{1}{3}, -\frac{2}{3}, +\frac{2}{3}$$

By direct substitution into $f(x)$ we find that -1 is a zero. By the Factor Theorem, $x + 1$ is a factor of $f(x)$ and so we may write $f(x) = (x+1)q(x)$. To find $q(x)$, use synthetic division to divide $f(x)$ by $x + 1$

$$\begin{array}{r|rrrrr} -1 & 3 & -10 & -3 & 8 & -2 \\ & & -3 & 13 & -10 & 2 \\ \hline & 3 & -13 & 10 & -2 & \underline{0} = r \end{array}$$

Therefore $q(x) = 3x^3 - 13x^2 + 10x - 2$ and so $f(x) = (x+1)(3x^3 - 13x^2 + 10x - 2)$. Any other rational zero of f must be a zero of the quotient $q(x) = 3x^3 - 13x^2 + 10x - 2$. Continuing with our list of possible zeros we find that $q(1/3) = 0$. By the Factor Theorem, $x - \frac{1}{3}$ is a factor of $q(x) = 3x^3 - 13x^2 + 10x - 2$. Verify using synthetic division that $q(x) = (x - \frac{1}{3})(3x^2 - 12x + 6)$ and so therefore

$$f(x) = (x+1)(x - \tfrac{1}{3})(3x^2 - 12x + 6)$$

The remaining real zeros will come from the quadratic polynomial $3x^2 - 12x + 6$ and may be found using the quadratic formula. They are the irrational numbers $2 + \sqrt{2}$ and $2 - \sqrt{2}$.   ■

3.10   Circles

The distance formula (3.1) enables us to find an equation for a very familiar plane curve. A **circle** is the set of all points (x, y) in the Cartesian plane that are equidistant from a fixed point $C(h, k)$, referred to as the center. If r is the fixed distance, then the point $P(x, y)$ is on the circle if and only if $d(C, P) = \sqrt{(x-h)^2 + (y-k)^2} = r$. See Figure 3.9.1(a). Equivalently, we have the **standard form** for the equation of a circle with center $C(h, k)$ and radius r:

$$(x - h)^2 + (y - k)^2 = r^2 \tag{3.5}$$

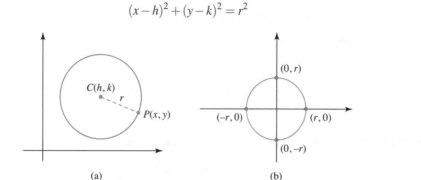

(a)   (b)

Figure 3.10.1

If we set $h = 0$ and $k = 0$ in (3.5), we obtain the standard form for the equation of a circle centered at the origin (Figure 3.10.1(b))

$$x^2 + y^2 = r^2$$

EXAMPLE 11

Find an equation of the circle with center $(-5, 4)$ and radius 2.

Solution   Substituting $h = -5$, $k = 4$, and $r = 2$ into (3.5), we get

$$(x - (-5))^2 + (y - 4)^2 = (2)^2 \text{ or } (x+5)^2 + (y-4)^2 = 4$$

∎

By squaring out the terms in (3.5), we see that every circle has an alternative equation of the form

$$Ax^2 + Ay^2 + Cx + Dy + E = 0, \quad A \neq 0$$

However, the converse is not necessarily true; that is, not every equation of the form $Ax^2 + Ay^2 + Cx + Dy + E = 0$ is a circle.

EXAMPLE 12

Show that $x^2 + y^2 - 4x + 8y + 2 = 0$ is an equation of a circle. Find its center and radius.

Solution   We want to write the given equation in the form (3.5). To do this we must complete the square in both x and y. We begin by grouping the x-terms and the y-terms together and taking the constant term to the right side of the equation:

$$(x^2 - 4x) + (y^2 + 8y) = -2$$

Inside each set of parentheses we add the square of one-half the coefficient of the first-degree term; that is, we add $(-4/2)^2 = 4$ in the first and $(8/2)^2 = 16$ in the second. But to retain the equality we must also add these numbers to the right side:

$$(x^2 - 4x + 4) + (y^2 + 8y + 16) = -2 + 4 + 6$$
$$(x - 2)^2 + (y + 4)^2 = 18$$

The last equation is the standard form for the equation of a circle centered at (h, k). Identifying $h = 2$, $k = -4$, and $r^2 = 18$, we conclude that the circle has center $(2, -4)$ and radius $\sqrt{18} = 3\sqrt{2}$. ∎

EXAMPLE 13

Determine whether $x^2 + y^2 - 12x + 8y + 52 = 0$ is an equation of a circle. If so, give its center and radius.

Solution   As in Example 7, complete the square in both x and y

$$(x^2 - 12x + 36) + (y^2 + 8y + 16) = -52 + 36 + 16$$
$$(x - 6)^2 + (y + 4)^2 = 0$$

The last equation can only be true when $x = 6$ and $y = -4$ therefore the equation represents the point $(6, -4)$ and not a circle. ∎

3.11   Conic Sections

The circle is just one member of a class of curves known as conic sections. An equation of a conic section can always be expressed in the form $Ax^2 + By^2 + Cx + Dy + E = 0$, where A and B are not both zero.

EXAMPLE 14

Graph $9x^2 + 16y^2 = 25$.

Solution   First, observe that the equation satisfies the three tests for symmetry and so its graph is symmetric with respect to the y-axis, the x-axis, and the origin. Now $9(1)^2 + 16(1)^2 = 25$ indicates that the point $(1, 1)$ is on the graph and therefore by symmetry so are the points $(-1, 1)$, $(1, -1)$, and $(-1, -1)$. Next we try to determine intercepts:

$$y = 0 \text{ implies } 9x^2 = 25 \text{ or } x = \pm\frac{5}{3}$$
$$x = 0 \text{ implies } 16y^2 = 25 \text{ or } y = \pm\frac{5}{4}$$

The x-intercepts are $-\frac{5}{3}$ and $\frac{5}{3}$; the y-intercepts are $-\frac{5}{4}$ and $\frac{5}{4}$. By plotting all of these points and connecting them with a smooth curve, we obtain the graph shown in Figure 3.11.1(b).

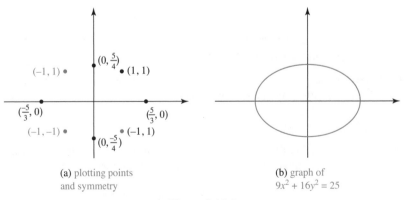

(a) plotting points and symmetry

(b) graph of $9x^2 + 16y^2 = 25$

Figure 3.11.1

3.12   The Ellipse

When written in the equivalent form

$$\frac{x^2}{(5/3)^2} + \frac{y^2}{(5/4)^2} = 1$$

the last equation in Example 14 is a special case of the standard form of an ellipse

$$\frac{(x-h)^2}{a^2} + \frac{(y-k)^2}{b^2} = 1$$

For $a = b$, the graph of this equation is a circle of radius a. If $a \neq b$, then the graph of the equation is called an **ellipse with center** (h, k). Thus, the graph of $9x^2 + 16y^2 = 25$ is an ellipse centered at the origin.

4.   Lines

The notion of a line plays an important role in the study of differential calculus. Before discussing lines however, it is convenient to introduce two special symbols.

4.1   Increments

If $P_1(x_1, y_1)$ and $P_2(x_2, y_2)$ are any two points in the plane, we define the **x-increment** to be the difference in x-coordinates

$$\Delta x = x_2 - x_1$$

and the **y-increment** to be the difference in y-coordinates

$$\Delta y = y_2 - y_1$$

We read the symbols "Δx" and "Δy" as "delta x" and "delta y", respectively.

EXAMPLE 1

Find the x- and y-increments for (a) $P_1(4, 6)$, $P_2(9, 7)$; (b) $P_1(10, -2)$, $P_2(3, 5)$; and (c) $P_1(8, 14)$, $P_2(8, -1)$.

Solution

(a) $\Delta x = 9 - 4 = 5$; $\Delta y = 7 - 6 = 1$

(b) $\Delta x = 3 - 10 = -7$; $\Delta y = 5 - (-2) = 7$

(c) $\Delta x = 8 - 8 = 0$; $\Delta y = -1 - 14 = -15$

Example 1 shows that an increment can be positive, negative, or zero.

4.2   Slope

Suppose L denotes a nonvertical line in the Cartesian plane. Associated with such a line there is a number called the **slope** of the line. If $P_1(x_1, y_1)$ and $P_2(x_2, y_2)$ are distinct points on L, then the slope m of the line is defined to be the quotient

$$m = \frac{\Delta y}{\Delta x} = \frac{y_2 - y_1}{x_2 - x_1} \qquad (4.1)$$

The increment $\Delta x = x_2 - x_1$ is said to be the *change in x* or **run** of the line; the corresponding *change in y* or **rise** of the line is $\Delta y = y_2 - y_1$. Thus,

$$m = \frac{\text{change in } y}{\text{change in } x} = \frac{\text{rise}}{\text{run}}$$

In the following figures, we compare the graphs of lines with positive, negative, zero, and undefined slopes:

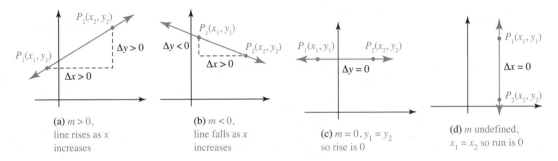

(a) $m > 0$,
line rises as x
increases

(b) $m < 0$,
line falls as x
increases

(c) $m = 0$, $y_1 = y_2$
so rise is 0

(d) m undefined,
$x_1 = x_2$ so run is 0

EXAMPLE 2

Find the slope of the line that passes through the points (a) $(6,3)$, $(-2,5)$; (b) $(7,-2)$, $(4,-2)$; and (c) $(1,5)$, $(1,-3)$.

Solution   Using (4.1) in the first two cases results in

(a) $m = \dfrac{5-3}{-2-6} = \dfrac{2}{-8} = -\dfrac{1}{4}$ and so the line is falling

(b) $m = \dfrac{-2-(-2)}{4-7} = 0$ which means the line through $(7,-2)$ and $(4,-2)$ is horizontal

(c) $\Delta x = 1 - 1 = 0$ which means the slope of the line through $(1,5)$ and $(1,-3)$ is undefined therefore the line is vertical. ∎

Any pair of distinct points on a line will determine the same slope. Consider the two similar right triangles in the figure.

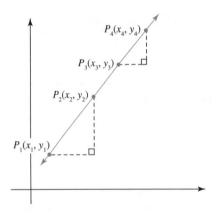

Since the ratios of corresponding sides of similar triangles are equal, we have

$$\frac{y_2 - y_1}{x_2 - x_1} = \frac{y_4 - y_3}{x_4 - x_3}$$

Therefore the slope of a line is independent of the choice of points on the line.

4.3   Parallel and Perpendicular Lines

Two lines L_1 and L_2 with slopes m_1 and m_2 are **parallel** if and only if $m_1 = m_2$. Of course, two vertical lines—that is, lines parallel to the y-axis – are also parallel but have undefined slopes. The lines are **perpendicular** if and only if $m_1 m_2 = -1$. In other words, lines with slopes are perpendicular when one slope is the *negative reciprocal* of the other:

$$m_1 = -\frac{1}{m_2} \quad \text{and} \quad m_2 = -\frac{1}{m_1}$$

4.4 Equations of Lines

The concept of slope enables us to find equations of lines. Through a point $P_1(x_1, y_1)$ there passes only one line L with specified slope m. To find an equation of L, suppose that $P(x, y)$ denotes any other point on L for which $x \neq x_1$. Then by (4.1) we have

$$\frac{y - y_1}{x - x_1} = m \text{ from which we obtain } y - y_1 = m(x - x_1) \tag{4.2}$$

Since the coordinates of all points $P(x, y)$ on the line, including $P_1(x_1, y_1)$, satisfy (4.2), we conclude that it is an equation of L. This particular equation is called the **point—slope form for the equation of a line**.

EXAMPLE 3

Find an equation for the line through $(6, -2)$ with slope 4.

Solution   From the point—slope form (4.2) we have

$$y - (-2) = 4(x - 6)$$

Equivalently,

$$y + 2 = 4x - 24 \text{ or } y = 4x - 26$$

∎

The point—slope form (4.2) yields two other important forms for equations of lines with slope. If the line L passes through the y-axis at $(0, b)$ then (4.2) gives us

$$y - b = mx \text{ or } y = mx + b$$

The number b is called the **y-intercept** of the line. Furthermore, if L is a horizontal line through $P_1(x_1, y_1)$ then setting $m = 0$ in (4.2) yields $y - y_1 = 0(x - x_1)$ or $y = y_1$. To summarize:

The Slope-Intercept Form for the Equation of a Line
$$y = mx + b \tag{4.3}$$

Equation of a Horizontal Line Through $P_1(x_1, y_1)$
$$y = y_1 \tag{4.4}$$

Two distinct points $P_1(x_1, y_1)$ and $P_2(x_2, y_2)$ also determine a unique line. If the line has slope, then an equation can be found from (4.2) by computing $m = (y_2 - y_1)/(x_2 - x_1)$ and using the coordinates of either P_1 or P_2. Finally, if the line through P_1 and P_2 is vertical, then any pair of points on the line have the same x-coordinate. Thus, if $P(x, y)$ is on the vertical line through $P_1(x_1, y_1)$, we must have $x = x_1$. We summarize this last case:

Equation of a Vertical Line Through $P_1(x_1, y_1)$
$$x = x_1 \tag{4.5}$$

EXAMPLE 4

Find an equation for the line through $(2, -3)$ and $(-4, 1)$.

Solution   By designating the first point as P_1, it follows from (4.1) that the slope of the line through the given points is $-\frac{2}{3}$. Using the point—slope form (4.2), we get

$$y - (-3) = -\frac{2}{3}(x - 2) \text{ or } y = -\frac{2}{3}x - \frac{5}{3}$$

As an alternative solution, we could have chosen the point $(-4, 1)$ as P_1 getting

$$y - 1 = -\frac{2}{3}(x - (-4)) \text{ or } y = -\frac{2}{3}x - \frac{5}{3}$$

as before.

We could also use the slope—intercept form (4.3) and write $y = -\frac{2}{3}x + b$. Substituting $x = 2$ and $y = -3$ in the last equation gives $-3 = -\frac{4}{3} + b$, so $b = -\frac{5}{3}$. As before, $y = -\frac{2}{3}x - \frac{5}{3}$. ∎

EXAMPLE 5

(a) An equation of a horizontal line through $(4, 9)$ is $y = 9$.

(b) An equation of a vertical line through $(-1, -2)$ is $x = -1$.

∎

4.5 Linear Equation

Any equation of the form

$$ax + by + c = 0 \tag{4.6}$$

in which both x and y appear to the first power and a, b, and c are constants, is a **linear equation**. As the name suggests, the graph of an equation of form (4.6) is a straight line. The following summarizes three special cases of (4.6):

$$
\begin{array}{lll}
\text{(i)} & a = 0,\ b \neq 0, & \text{line is horizontal with equation } y = -\dfrac{c}{b} \\[3mm]
\text{(ii)} & a \neq 0,\ b = 0, & \text{line is vertical with equation } x = -\dfrac{c}{a} \\[3mm]
\text{(iii)} & a \neq 0,\ b \neq 0, & \text{line may be rewritten as } y = -\dfrac{a}{b}x - \dfrac{c}{b}
\end{array}
\tag{4.7}
$$

From the last result we have $m = -a/b$ and y-intercept $-c/b$.

EXAMPLE 6

Find an equation of the line through $(-1, 5)$ perpendicular to the line with equation $2x + y + 4 = 0$.

Solution   Writing the given equation as $y = -2x - 4$, we see that the slope is -2. Thus the line through $(-1, 5)$ perpendicular to $2x + y + 4 = 0$ has slope $\frac{1}{2}$. Therefore from (4.2) we get

$$y - 5 = \frac{1}{2}(x - (-1)) \text{ or } y = \frac{1}{2}x + \frac{11}{2} \text{ or } x - 2y + 11 = 0$$

∎

4.6 Graphs

To graph an equation of a line, we need determine only two points whose coordinates satisfy the equation. When $a \neq 0$ and $b \neq 0$ in (4.6), the line must cross both coordinate axes. The **x-intercept** is the x-coordinate of the point where the line crosses the x-axis; the **y-intercept** is the y-coordinate of the point where the line crosses the y-axis. Since a point on the x-axis has coordinate $(x, 0)$, the x-intercept is found by setting $y = 0$ in the equation and solving for x. Similarly, we get the y-intercept by setting $x = 0$ and solving for y.

EXAMPLE 7

Graph the line with equation $2x - 3y + 12 = 0$.

Solution   By setting $y = 0$ in the equation, we get

$$2x + 12 = 0 \text{ or } x = -6$$

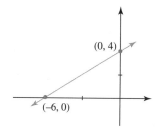

The x-intercept is -6. Now when $x = 0$, we get

$$-3y + 12 = 0 \text{ or } y = 4$$

The y-intercept is 4. The figure shows the line through the intercepts $(-6, 0)$ and $(0, 4)$. ∎

5. Matrices and Determinants

5.1 Matrices

A matrix is any rectangular array of numbers:

$$\begin{bmatrix} a_{11} & a_{12} & \cdots & a_{1n} \\ a_{21} & a_{22} & \cdots & a_{2n} \\ \vdots & \vdots & \ddots & \vdots \\ a_{m1} & a_{m2} & \cdots & a_{mn} \end{bmatrix}$$

The numbers in the array are called **entries** or **elements** of the matrix. If a matrix has m rows and n columns, we say that its size is m by n (written $m \times n$). An $n \times n$ matrix is called a **square** matrix or a matrix of **order n**. A 1×1 matrix is simply a constant. A matrix is usually denoted by a capital boldface letter such as **A**, **B**, **C**, or **X**. The entry in the ith row and jth column of an $m \times n$ matrix **A** is written a_{ij}. An $m \times n$ matrix **A** is then abbreviated as $A = [a_{ij}]_{m \times n}$.

EXAMPLE 1

(a) $\mathbf{A} = \begin{bmatrix} 1 & 2 & 3 \\ 0 & 5 & -6 \end{bmatrix}$ is a 2×3 matrix.

(b) $\mathbf{B} = \begin{bmatrix} 9 & 7 & 0 & 8 \\ \frac{1}{2} & -2 & 6 & 1 \\ 0 & 0 & -1 & 6 \\ 5 & \sqrt{3} & \pi & -4 \end{bmatrix}$ is a 4×4 square matrix of order 4.

∎

5.2 Vectors

A matrix

$$\begin{bmatrix} a_1 \\ a_2 \\ \vdots \\ a_n \end{bmatrix}$$

with n rows and one column is called a **column vector**.

A matrix

$$\begin{bmatrix} a_1 & a_2 & \cdots & a_n \end{bmatrix}$$

with one row and n columns is called a **row vector**.

5.3 Equality

Two $m \times n$ matrices $\mathbf{A}$ and $\mathbf{B}$ are equal if $a_{ij} = b_{ij}$ for each i and j.

5.4 Scalar Multiple

In the discussion of matrices, real numbers are often referred to as **scalars**. If k is a real number, the scalar multiple of a matrix $\mathbf{A}$ is defined as

$$k\mathbf{A} = \begin{bmatrix} ka_{11} & ka_{12} & \cdots & ka_{1n} \\ ka_{21} & ka_{22} & \cdots & ka_{2n} \\ \vdots & \vdots & \ddots & \vdots \\ ka_{m1} & ka_{m2} & \cdots & ka_{mn} \end{bmatrix} = [ka_{ij}]_{m \times n} \tag{5.1}$$

EXAMPLE 2

From (5.1) we have

$$5 \begin{bmatrix} 2 & -3 \\ 4 & -1 \\ \frac{1}{5} & 6 \end{bmatrix} = \begin{bmatrix} 5(2) & 5(-3) \\ 5(4) & 5(-1) \\ 5(\frac{1}{5}) & 5(6) \end{bmatrix} = \begin{bmatrix} 10 & -15 \\ 20 & -5 \\ 1 & 30 \end{bmatrix}$$

■

5.5 Matrix Addition

When two matrices $\mathbf{A}$ and $\mathbf{B}$ are the same size, we can add them by adding their corresponding entries. The sum of two $m \times n$ matrices is the matrix

$$\mathbf{A} + \mathbf{B} = [a_{ij} + b_{ij}]_{m \times n} \tag{5.2}$$

EXAMPLE 3

It follows from (5.2) that the sum of

$$\mathbf{A} = \begin{bmatrix} 2 & -1 & 3 \\ 0 & 4 & 6 \\ -6 & 10 & -5 \end{bmatrix} \text{ and } \mathbf{B} = \begin{bmatrix} 4 & 7 & -8 \\ 9 & 3 & 5 \\ 1 & -1 & 2 \end{bmatrix}$$

is

$$\mathbf{A} + \mathbf{B} = \begin{bmatrix} 2+4 & -1+7 & 3+(-8) \\ 0+9 & 4+3 & 6+5 \\ -6+1 & 10+(-1) & -5+2 \end{bmatrix} = \begin{bmatrix} 6 & 6 & -5 \\ 9 & 7 & 11 \\ -5 & 9 & -3 \end{bmatrix}$$

■

The **difference** of two $m \times n$ matrices is defined as

$$\mathbf{A} - \mathbf{B} = \mathbf{A} + (-\mathbf{B}) \text{ where } -\mathbf{B} = (-1)\mathbf{B}$$

5.6 Matrix Multiplication

Under some circumstances two matrices can be multiplied together. If $\mathbf{A}$ is a matrix with m rows and p columns, and $\mathbf{B}$ is a matrix with p rows and n columns, then the product $\mathbf{AB}$ is the $m \times n$ matrix

$$\mathbf{AB} = \begin{bmatrix} a_{11} & a_{12} & \cdots & a_{1p} \\ a_{21} & a_{22} & \cdots & a_{2p} \\ \vdots & \vdots & \ddots & \vdots \\ a_{m1} & a_{m2} & \cdots & a_{mp} \end{bmatrix} \begin{bmatrix} a_{11} & a_{12} & \cdots & a_{1n} \\ a_{21} & a_{22} & \cdots & a_{2n} \\ \vdots & \vdots & \ddots & \vdots \\ a_{p1} & a_{p2} & \cdots & a_{pn} \end{bmatrix}$$

$$= \begin{bmatrix} a_{11}b_{11}+a_{12}b_{21}+\cdots+a_{1p}b_{p1} & \cdots & a_{11}b_{1n}+a_{12}b_{2n}+\cdots+a_{1p}b_{pn} \\ a_{21}b_{11}+a_{22}b_{21}+\cdots+a_{2p}b_{p1} & \cdots & a_{21}b_{1n}+a_{22}b_{2n}+\cdots+a_{2p}b_{pn} \\ \vdots & \ddots & \vdots \\ a_{m1}b_{11}+a_{m2}b_{21}+\cdots+a_{mp}b_{p1} & \cdots & a_{m1}b_{1n}+a_{m2}b_{2n}+\cdots+a_{mp}b_{pn} \end{bmatrix}$$

(5.3)

Note carefully in (5.3) that the product $\mathbf{C} = \mathbf{AB}$ is defined only when the number of columns in $\mathbf{A}$ is the same as the number of rows in $\mathbf{B}$. The size of the product can be determined from

$$\mathbf{A}_{m \times p}\mathbf{B}_{p \times n} = \mathbf{C}_{m \times n}$$

Also, you might recognize that the entries in, say, the ith row of the final matrix $\mathbf{C} = \mathbf{AB}$ are formed by using the component definition of the inner or dot product of the ith row (vector) of $\mathbf{A}$ with each of the columns (vectors) of $\mathbf{B}$.

EXAMPLE 4

Find the product of the following matrices.

(a) $\mathbf{A} = \begin{bmatrix} 4 & 7 \\ 3 & 5 \end{bmatrix}$ and $\mathbf{B} = \begin{bmatrix} 9 & -2 \\ 6 & 8 \end{bmatrix}$ (b) $\mathbf{A} = \begin{bmatrix} 5 & 8 \\ 1 & 0 \\ 2 & 7 \end{bmatrix}$ and $\mathbf{B} = \begin{bmatrix} -4 & -3 \\ 2 & 0 \end{bmatrix}$

Solution

(a) From (5.3) we have

$$\mathbf{AB} = \begin{bmatrix} 4 \cdot 9 + 7 \cdot 6 & 4 \cdot (-2) + 7 \cdot 8 \\ 3 \cdot 9 + 5 \cdot 6 & 3 \cdot (-2) + 5 \cdot 8 \end{bmatrix} = \begin{bmatrix} 78 & 48 \\ 57 & 34 \end{bmatrix}$$

(b) $\mathbf{AB} = \begin{bmatrix} 5 \cdot (-4) + 8 \cdot 2 & 5 \cdot (-3) + 8 \cdot 0 \\ 1 \cdot (-4) + 0 \cdot 2 & 1 \cdot (-3) + 0 \cdot 0 \\ 2 \cdot (-4) + 7 \cdot 2 & 2 \cdot (-3) + 7 \cdot 0 \end{bmatrix} = \begin{bmatrix} -4 & -15 \\ -4 & -3 \\ 6 & -6 \end{bmatrix}$

∎

In general, matrix multiplication is not commutative; that is, $\mathbf{BA} \neq \mathbf{AB}$. Observe that in part (a) of Example 4, the product $\mathbf{BA}$ is

$$\mathbf{BA} = \begin{bmatrix} 30 & 53 \\ 48 & 82 \end{bmatrix}$$

whereas in part (b) the product $\mathbf{BA}$ is not defined, since the first matrix (in this case $\mathbf{B}$) does not have the same number of columns as the second matrix has rows.

5.7 Associative Law

Although we shall not prove it, matrix multiplication is **associative**. If $\mathbf{A}$ is an $m \times p$ matrix, $\mathbf{B}$ a $p \times r$ matrix, and $\mathbf{C}$ $r \times n$ matrix, then the product

$$\mathbf{A(BC)} = \mathbf{(AB)C}$$

is an $m \times n$ matrix.

5.8 Distributive Law

If **B** and **C** are $r \times n$ matrices and **A** is an $m \times n$ matrix, then the distributive law is

$$\mathbf{A}(\mathbf{B}+\mathbf{C}) = \mathbf{AB}+\mathbf{AC}$$

Furthermore, if the product $(\mathbf{B}+\mathbf{C})\mathbf{A}$ is defined, then

$$(\mathbf{B}+\mathbf{C})\mathbf{A} = \mathbf{BA}+\mathbf{CA}$$

5.9 Systems of Linear Equations

Recall that any equation of the form $ax+by+c=0$, where a, b, and c are real numbers, is said to be a linear equation in the variables x and y. The graph of a linear equation in two variables is a straight line. For real numbers a, b, c, and d, $ax+by+cz=d$ is a linear equation in the variables x, y, and z and is the equation of a plane. In general, an equation of the form

$$a_1x_1 + a_2x_2 + \cdots a_nx_n = b_n$$

where $a_1, a_2, \ldots, a_n$ and b_n are real numbers, is a linear equation in the n variables $x_1, x_2, \ldots, x_n$.

Matrices can be used to solve systems of linear equations. Systems of linear equations are also called linear systems. A general system of m linear equations in n unknowns has the form

$$\begin{cases} a_{11}x_1 + a_{12}x_2 + \cdots + a_{1n}x_n &= b_1 \\ a_{21}x_1 + a_{22}x_2 + \cdots + a_{2n}x_n &= b_2 \\ &\vdots \\ a_{m1}x_1 + a_{m2}x_2 + \cdots + a_{mn}x_n &= b_m \end{cases} \tag{5.4}$$

5.10 Augmented Matrix

The solution of a linear system does not depend on what symbols are used as variables. Thus, the systems

$$\begin{cases} 2x+6y+z=7 \\ x+2y-z=-1 \\ 5x+7y-4z=9 \end{cases} \quad \text{and} \quad \begin{cases} 2u+6v+w=7 \\ u+2v-w=-1 \\ 5u+7v-4w=9 \end{cases}$$

have the same solution $(10, -3, 5)$. In other words, when solving a linear system, the variables are immaterial; it is the coefficients of the variables and the constants that determine the solution of the system. In fact, we solve a system of form (5.4) by dropping the variables entirely and performing operations on the rows of the array of coefficients and constants:

$$\left[\begin{array}{cccc|c} a_{11} & a_{12} & \cdots & a_{1n} & b_1 \\ a_{12} & a_{22} & \cdots & a_{2n} & b_2 \\ \vdots & \vdots & \ddots & \vdots & \vdots \\ a_{m1} & a_{m2} & \cdots & a_{mn} & b_m \end{array} \right]$$

This array is called the augmented matrix of the system (5.4).

EXAMPLE 5

(a) The augmented matrix $\left[\begin{array}{ccc|c} 1 & -3 & 5 & 2 \\ 4 & 7 & -1 & 8 \end{array} \right]$ represents the linear system

$$\begin{cases} x_1 - 3x_2 + 5x_3 = 2 \\ 4x_1 + 7x_2 - x_3 = 8 \end{cases}$$

(b) The linear system

$$\begin{cases} x_1 - 5x_3 = -1 \\ 2x_1 + 8x_2 = 7 \\ x_2 + 9x_3 = 1 \end{cases} \text{ is the same as } \begin{cases} x_1 + 0x_2 - 5x_3 = -1 \\ 2x_1 + 8x_2 + 0x_3 = 7 \\ 0x_1 + x_2 + 9x_3 = 1 \end{cases}$$

Thus the augmented matrix of the system is $\begin{bmatrix} 1 & 0 & -5 & -1 \\ 2 & 8 & 0 & 7 \\ 0 & 1 & 9 & 1 \end{bmatrix}$ ■

5.11 Elementary Row Operations

Two linear systems are **equivalent** if they have exactly the same solutions. Since the rows of an augmented matrix represent equations in a linear system, we can obtain equivalent linear systems by performing the following **elementary row operations** on an augmented matrix:

(i) Multiply a row by a nonzero constant.

(ii) Interchange any two rows.

(iii) Add a nonzero multiple of one row to any other row.

Of course, when we add a multiple of one row to another, we add the corresponding entries in the rows.

5.12 Gauss–Jordan Elimination

To solve a system such as (5.4) using an augmented matrix we shall use the **Gauss–Jordan elimination method**. This means we carry out a succession of elementary row operations until we arrive at an augmented matrix in **reduced row–echelon form**:

(i) The first nonzero entry in a row is a 1.

(ii) The remaining entries in a column containing a first entry 1 are all zeros.

(iii) In consecutive nonzero rows, the first entry 1 in the lower row appears to the right of the 1 in the higher row.

(iv) Rows consisting of all zeros are at the bottom of the matrix.

EXAMPLE 6

(a) The augmented matrices

$$\begin{bmatrix} 1 & 0 & 0 & 2 \\ 0 & 1 & 0 & 8 \\ 0 & 0 & 0 & 0 \end{bmatrix} \text{ and } \begin{bmatrix} 0 & 0 & 1 & -6 & 0 & 2 \\ 0 & 0 & 0 & 0 & 1 & 4 \end{bmatrix}$$

are in reduced row–echelon form. You should verify that the criteria for this form are satisfied.

(b) The augmented matrices

$$\begin{bmatrix} 1 & -7 & 2 \\ 0 & 0 & 0 \\ 1 & 0 & 0 \end{bmatrix} \text{ and } \begin{bmatrix} 2 & 6 & 0 & 5 & 0 \\ 0 & 1 & 0 & 0 & 0 \\ 0 & 0 & 7 & 3 & -2 \\ 0 & 0 & 0 & 1 & 4 \end{bmatrix}$$

are not in reduced row–echelon form. You should determine which of the four criteria listed above are not satisfied.

■

Once an augmented matrix in reduced row–echelon form has been attained the solution of the system will be apparent by inspection. In terms of the equations of the original system, our goal is simply to make the coefficient of x_1 in the first equation equal to 1 and then use multiples of that equation to eliminate x_1 from the remaining equations. The process is repeated on the other variables.

To keep track of the row operations on an augmented matrix let's use the following notation:

Symbol	Meaning
R_{ij}	Interchange the rows i and j
cR_i	Multiply the ith row by the nonzero constant c
$cR_i + R_j$	Multiply the ith row by c and add to the jth row

EXAMPLE 7

Solve the linear system $\begin{cases} 2x_1 + 6x_2 + x_3 = 7 \\ x_1 + 2x_2 - x_3 = -1 \\ 5x_1 + 7x_2 - 4x_3 = 9 \end{cases}$ using Gauss–Jordan elimination.

Solution   Using row operations on the augmented matrix of the system, we obtain

$$
\begin{bmatrix} 2 & 6 & 1 & 7 \\ 1 & 2 & -1 & -1 \\ 5 & 7 & -4 & 9 \end{bmatrix}
\overset{R_{12}}{\Rightarrow}
\begin{bmatrix} 1 & 2 & -1 & -1 \\ 2 & 6 & 1 & 7 \\ 5 & 7 & -4 & 9 \end{bmatrix}
\qquad
\overset{-2R_1+R_2}{\underset{-5R_1+R_3}{\Rightarrow}}
\begin{bmatrix} 1 & 2 & -1 & -1 \\ 0 & 2 & 3 & 9 \\ 0 & -3 & 1 & 14 \end{bmatrix}
$$

$$
\overset{\frac{1}{2}R_2}{\Rightarrow}
\begin{bmatrix} 1 & 2 & -1 & -1 \\ 0 & 1 & \frac{3}{2} & \frac{9}{2} \\ 0 & -3 & 1 & 14 \end{bmatrix}
\qquad
\overset{-2R_2+R_1}{\underset{-3R_2+R_3}{\Rightarrow}}
\begin{bmatrix} 1 & 0 & -4 & -10 \\ 0 & 1 & \frac{3}{2} & \frac{9}{2} \\ 0 & 0 & \frac{11}{2} & \frac{55}{2} \end{bmatrix}
$$

$$
\overset{\frac{2}{11}R_3}{\Rightarrow}
\begin{bmatrix} 1 & 0 & -4 & -10 \\ 0 & 1 & \frac{3}{2} & \frac{9}{2} \\ 0 & 0 & 1 & 5 \end{bmatrix}
\qquad
\overset{4R_3+R_1}{\underset{(-3/2)R_3+R_2}{\Rightarrow}}
\begin{bmatrix} 1 & 0 & 0 & 10 \\ 0 & 1 & 0 & -3 \\ 0 & 0 & 1 & 5 \end{bmatrix}
$$

The last matrix is in reduced row—echelon form and represents the system

$$
\begin{cases} x_1 + 0x_2 + 0x_3 = 10 \\ 0x_1 + x_2 + 0x_3 = -3 \\ 0x_1 + 0x_2 + x_3 = 5 \end{cases}
$$

Therefore it is evident that the solution of the system is $x_1 = 10$, $x_2 = -3$, and $x_3 = 5$.   ∎

5.13 Determinants

Suppose $\mathbf{A}$ is an $n \times n$ matrix. Associated with $\mathbf{A}$ is a number called the determinant of $\mathbf{A}$ which is denoted by $\det \mathbf{A}$. Symbolically we distinguish a matrix $\mathbf{A}$ and the determinant of $\mathbf{A}$ by writing

$$
\mathbf{A} = \begin{bmatrix} a_{11} & a_{12} & \cdots & a_{1n} \\ a_{21} & a_{22} & \cdots & a_{2n} \\ \vdots & \vdots & \ddots & \vdots \\ a_{n1} & a_{n2} & \cdots & a_{nn} \end{bmatrix} \quad \text{and} \quad \det \mathbf{A} = \begin{vmatrix} a_{11} & a_{12} & \cdots & a_{1n} \\ a_{21} & a_{22} & \cdots & a_{2n} \\ \vdots & \vdots & \ddots & \vdots \\ a_{n1} & a_{n2} & \cdots & a_{nn} \end{vmatrix}
$$

A **determinant of order 2** is the number

$$
\begin{vmatrix} a_{11} & a_{12} \\ a_{21} & a_{22} \end{vmatrix} = a_{11}a_{22} - a_{12}a_{21} \tag{5.5}
$$

EXAMPLE 8

By (5.5), $\begin{vmatrix} 4 & -3 \\ 2 & 9 \end{vmatrix} = 4 \cdot 9 - (-3)(2) = 42$

A **determinant of order 3** can be evaluated by **expanding the determinant by cofactors of the first row**:

$$\begin{vmatrix} a_{11} & a_{12} & a_{13} \\ a_{21} & a_{22} & a_{23} \\ a_{31} & a_{32} & a_{33} \end{vmatrix} = a_{11} \begin{vmatrix} a_{22} & a_{23} \\ a_{32} & a_{33} \end{vmatrix} - a_{12} \begin{vmatrix} a_{21} & a_{23} \\ a_{31} & a_{33} \end{vmatrix} + a_{13} \begin{vmatrix} a_{21} & a_{22} \\ a_{31} & a_{32} \end{vmatrix} \qquad (5.6)$$

EXAMPLE 9

By (5.6),

$$\begin{vmatrix} 8 & 5 & 4 \\ 2 & 4 & 6 \\ -1 & 2 & 3 \end{vmatrix} = 8 \begin{vmatrix} 4 & 6 \\ 2 & 3 \end{vmatrix} - 5 \begin{vmatrix} 2 & 6 \\ -1 & 3 \end{vmatrix} + 4 \begin{vmatrix} 2 & 4 \\ -1 & 2 \end{vmatrix}$$
$$= 8(0) - 5(12) + 4(8) = -28$$

■

5.14 Cofactors

In general, the **cofactor** of an entry in the ith row and jth column of a determinant is $(-1)^{i+j}$ times that determinant formed by deleting the ith row and jth column. Thus, the cofactors of a_{11}, a_{12}, and a_{13} are, respectively,

$$\begin{vmatrix} a_{22} & a_{23} \\ a_{32} & a_{33} \end{vmatrix}, \quad -\begin{vmatrix} a_{21} & a_{23} \\ a_{31} & a_{33} \end{vmatrix}, \quad \begin{vmatrix} a_{21} & a_{22} \\ a_{31} & a_{32} \end{vmatrix}$$

The cofactor of, say a_{23} is $-\begin{vmatrix} a_{11} & a_{12} \\ a_{31} & a_{32} \end{vmatrix}$.

5.15 Properties of Determinants

(i) If every entry in a row (or column) is zero, then the value of the determinant is zero

(ii) If two rows (or columns) are equal, then the value of the determinant is zero

(iii) Interchanging two rows (or columns) results in the negative of the value of the original determinant

(iv) A determinant can be expanded by the cofactors of any row (or column).

EXAMPLE 10

(a) From (i), $\begin{vmatrix} 1 & 6 & 0 \\ 3 & 2 & 0 \\ 4 & 5 & 0 \end{vmatrix} = 0$

(b) From (ii), $\begin{vmatrix} 1 & 6 & 6 \\ 3 & 2 & 2 \\ 4 & 5 & 5 \end{vmatrix} = 0$

(c) Since $\begin{vmatrix} 2 & 3 \\ 4 & 5 \end{vmatrix} = -2$, it follows from (*iii*) that $\begin{vmatrix} 4 & 5 \\ 2 & 3 \end{vmatrix} = 2$

■

EXAMPLE 11

Expand the determinant $\begin{vmatrix} 3 & 4 & 3 \\ 2 & 0 & 0 \\ 7 & 2 & 1 \end{vmatrix}$.

Solution Taking advantage of the zeros in the second row, we expand the determinant by the cofactors of the second row:

$$\begin{vmatrix} 3 & 4 & 3 \\ 2 & 0 & 0 \\ 7 & 1 & 2 \end{vmatrix} = -2\begin{vmatrix} 4 & 3 \\ 1 & 2 \end{vmatrix} + 0\begin{vmatrix} 3 & 3 \\ 7 & 2 \end{vmatrix} - 0\begin{vmatrix} 3 & 4 \\ 7 & 1 \end{vmatrix} = -2(8-3) = -10$$

■

5.16 Cramer's Rule

Systems of n linear equations in n unknowns can sometimes be solved by means of determinants. The solution of the system

$$\begin{cases} a_{11}x + a_{12}y = b_1 \\ a_{21}x + a_{22}y = b_2 \end{cases}$$

is

$$x = \frac{\begin{vmatrix} b_1 & a_{12} \\ b_2 & a_{22} \end{vmatrix}}{\begin{vmatrix} a_{11} & a_{12} \\ a_{21} & a_{22} \end{vmatrix}} \quad \text{and} \quad y = \frac{\begin{vmatrix} a_{11} & b_1 \\ a_{21} & b_2 \end{vmatrix}}{\begin{vmatrix} a_{11} & a_{12} \\ a_{21} & a_{22} \end{vmatrix}}$$

provided that the determinant of the coefficient matrix $\begin{bmatrix} a_{11} & a_{12} \\ a_{21} & a_{22} \end{bmatrix}$ is not zero. This last condition also guarantees that there is only one solution. Similarly, the system of three equations in three unknowns

$$\begin{cases} a_{11}x + a_{12}y + a_{13}z = b_1 \\ a_{21}x + a_{22}y + a_{23}z = b_2 \\ a_{31}x + a_{32}y + a_{33}z = b_3 \end{cases}$$

has the unique solution

$$x = \frac{\begin{vmatrix} b_1 & a_{12} & a_{13} \\ b_2 & a_{22} & a_{23} \\ b_3 & a_{32} & a_{33} \end{vmatrix}}{\begin{vmatrix} a_{11} & a_{12} & a_{13} \\ a_{21} & a_{22} & a_{23} \\ a_{31} & a_{32} & a_{33} \end{vmatrix}}, \quad y = \frac{\begin{vmatrix} a_{11} & b_1 & a_{13} \\ a_{21} & b_2 & a_{23} \\ a_{31} & b_3 & a_{33} \end{vmatrix}}{\begin{vmatrix} a_{11} & a_{12} & a_{13} \\ a_{21} & a_{22} & a_{23} \\ a_{31} & a_{32} & a_{33} \end{vmatrix}}, \quad z = \frac{\begin{vmatrix} a_{11} & a_{12} & b_1 \\ a_{21} & a_{22} & b_2 \\ a_{31} & a_{32} & b_3 \end{vmatrix}}{\begin{vmatrix} a_{11} & a_{12} & a_{13} \\ a_{21} & a_{22} & a_{23} \\ a_{31} & a_{32} & a_{33} \end{vmatrix}} \qquad (5.7)$$

provided that the determinant of the coefficient matrix is not zero. The next example illustrates **Cramer's Rule**.

EXAMPLE 12

Solve the system of equations $\begin{cases} x+2y+z=1 \\ x+0y+3z=4 \\ 4x+2y+z=5 \end{cases}$

Solution   Since $\begin{vmatrix} 1 & 2 & 1 \\ 1 & 0 & 3 \\ 4 & 2 & 1 \end{vmatrix} = 18 \neq 0$, the system can be solved by Cramer's Rule (5.7):

$$x = \frac{\begin{vmatrix} 1 & 2 & 1 \\ 4 & 0 & 3 \\ 5 & 2 & 1 \end{vmatrix}}{\begin{vmatrix} 1 & 2 & 1 \\ 1 & 0 & 3 \\ 4 & 2 & 1 \end{vmatrix}} = \frac{24}{18} = \frac{4}{3}, \quad y = \frac{\begin{vmatrix} 1 & 1 & 1 \\ 1 & 4 & 3 \\ 4 & 5 & 1 \end{vmatrix}}{\begin{vmatrix} 1 & 2 & 1 \\ 1 & 0 & 3 \\ 4 & 2 & 1 \end{vmatrix}} = -\frac{11}{18}, \quad z = \frac{\begin{vmatrix} 1 & 2 & 1 \\ 1 & 0 & 4 \\ 4 & 2 & 5 \end{vmatrix}}{\begin{vmatrix} 1 & 2 & 1 \\ 1 & 0 & 3 \\ 4 & 2 & 1 \end{vmatrix}} = \frac{16}{18} = \frac{8}{9}$$

■

Essay

Calculus and Mathematical Modeling

by Fred S. Roberts
Department of Mathematics and Center for Operations Research
Rutgers University
New Brunswick, NJ 08903

We are all familiar with building models. Most everyone enjoys working with clay, which is a wonderful way to build models. We have all encountered road maps at one time or another, and while not everyone enjoys the experience, road maps provide excellent models of transportation systems. In much the same way, scientists use calculus to build models of a different sort—called **mathematical models**.

In building any model, we are usually trying to understand some phenomenon. In building a mathematical model, we are using the tools of mathematics to help us do so. Mathematical modelers are using the tools of calculus to make decisions about health care, map the human genome, build new and highly efficient communication and transportation systems, understand the threats to our global environment, ...

Mathematics has been called "the language of science" and it is the most precise language invented by man. Because it is so precise, we can use it to reason and draw conclusions; we have developed many complex tools for reasoning in this language. Yet, mathematical modeling is as much an art as a science; we learn it by doing and there are often no specific rules (like those for differentiating and integrating) that will help us to build a model or determine just how good it is.

I would like to introduce the "art" of mathematical modeling by giving a variety of examples. These examples make three basic points: (1) Mathematical modelers can afford to be lazy; (2) Mathematical modelers are invariably simple-minded; (3) Mathematical modelers are fickle.

One of the best things about mathematical modeling is that it often allows us to be lazy. Once we have formulated a problem in mathematical terms, we can forget about the interpretation of these terms and concentrate on reasoning in the mathematical language. In this way, we only need to do the reasoning once. If a totally different problem is formulated in much the same mathematical way, we can simply use the solution we had previously derived. Let me illustrate this point with the following examples. Consider the investment of D_0 dollars at p percent interest, compounded annually. If $D(t)$ is the number of dollars after t years and $q = p/100$, then $D(t) = D_0(1+q)^t$. This simple mathematical model allows us to calculate how much money we will have any number of years after the initial investment. If we want to know how long it will take to double our initial investment, we simply try to solve the equation $D(t) = 2D_0$, or $(1+q)^t = 2$. We can solve this by taking logs of both sides, noting that $t\log_{10}(1+q) = \log_{10} 2 = 0{\cdot}30103$. Thus, $t = 0.30103/log_{10}(1+q)$.

Now suppose we are interested in population growth, and are concerned with the growing population in certain countries of the world. We note that a given country with P_0 people has a growth rate of r percent per year. We want to predict the population after t years. Letting $s = r/100$ and $P(t) = the$ population after t years, we find that $P(t) = P_0(1+s)^t$. Now if we simply replace P_0 by D_0, $P(t)$ by $D(t)$, and s by q, we don't have to make new calculations. We simply note that the mathematical model representing our problem is exactly the same, and there is no reason to make calculations over again. For example, if we want to know how many years before the population doubles, we do not have to repeat our previous reasoning. We conclude immediately that the solution is given by $t = 0.30103/log_{10}(1+s)$. In short, we can be lazy.

Let us consider a second example. In the year 1202, Leonardo of Pisa, better known as Fibonacci, posed a problem about breeding of rabbits. Assume that a newborn rabbit becomes an adult two months after it is born. Let us start with one pair of adult rabbits (of opposite sex), and assume that every adult pair of rabbits produce a pair of young (of opposite sex) each month. (Rabbits are very prolific.) Suppose that F_k counts the number of rabbit pairs present at the beginning of month k. Then $F_1 = 1$, $F_2 = 2$, and for

$k \geq 3$, it can be seen that $F_k = F_{k-1} + F_{k-2}$.[1] This equation is an example of a **recurrence**. The number F_k is called the kth **Fibonacci number**. If one calculates the ratios F_k/F_{k-1}, one sees that these seem to be approaching a limit, and in fact one can show that this limit is $\tau = \frac{1}{2}(1 + \sqrt{5}) = 1.618034\ldots$ (The number τ is called the **golden ratio** or **divine proportion** and it is surprising how often rectangles whose sides are in the ratio $\tau : 1$ occur in art; such rectangles are called **golden rectangles**.) It is possible to show that the kth Fibonacci number is given by the equation

$$F_k = \frac{\tau^{k+1} - (1-\tau)^{k+1}}{\sqrt{5}},$$

which gives us a way to calculate the number of rabbit pairs at the beginning of the kth month. It is remarkable how many important applications the Fibonacci numbers have. For instance, in **phyllotaxis** (the study of arrangement of leaves), these numbers appear as the number of rotations of leaves around stems in a branch; and they can be used in understanding the whorls on a pineapple and scales on a fir cone. In numerical analysis, in searching for the maximum value of a function on an interval, the Fibonacci numbers are used to determine where to evaluate the function in getting better and better estimates of the location of the maximum.

Now, consider a completely different problem that arises in telecommunications. A model of a communication channel involves a coding alphabet, some "codewords" defined from that alphabet, and assumptions about how these codewords are transmitted over the channel. If N_t is the number of codewords that can be transmitted in exactly t units of time, Claude Shannon, one of the pioneers of the theory of telecommunications, introduced a notion of the **capacity** C of the transmission channel, through the formula

$$C = \lim_{t \to \infty} \frac{\log_2 N_t}{t}.$$

To use a simple example, suppose that there are only two signals (say 0 and 1), and a codeword is any sequence from the alphabet $\{0, 1\}$. Suppose that 0 takes one unit of time to transmit while 1 takes two units of time. Then, a codeword 001 takes four units of time to transmit. It can be seen that $N_1 = 1$, $N_2 = 2$, and for $t \geq 3$, $N_t = N_{t-1} + N_{t-2}$. If we want to calculate N_t, we can do so with little effort. We can observe that the exact same recurrence occurs here as occurred in our model of the rabbit population. We simply replace N by F and t by k. We therefore conclude that

$$N_t = \frac{\tau^{t+1} - (1-\tau)^{t+1}}{\sqrt{5}}$$

It can be shown from this that the (Shannon) capacity of this channel is given by

$$C = \log_2 \left(\frac{1 + \sqrt{5}}{2} \right)$$

It is remarkable that the problem of computing the (Shannon) capacity remains unsolved even for some very simple channels. However, for the particular channel we have described, we have computed it without much work.

I offer one final example of the so-called "lazy" aspect of mathematical modeling. In physics, it is known that light travels in a path that minimizes time travelled. What does this say about the point where light is reflected in a mirror? We shall answer this question under the assumption that light travels in straight lines and under the assumption that it travels at constant velocity. If we wish to find the point where the light hits the mirror, it is sufficient to find the angles α and β. Using the Pythagorean Theorem, we find that if $d = x + y$ and $S = p + q$, then

$$S = \sqrt{a^2 + x^2} + \sqrt{(d-x)^2 + b^2}$$

With some manipulation, one can show from this that $\frac{dS}{dx} = \cos \alpha - \cos \beta$. Since the path that light takes in reflection is determined by minimizing $S = p + q$, we try to find a point where $\frac{dS}{dx} = 0$. This occurs when $\alpha = \beta$ (since α and β are clearly both less than 90 degrees). It is not hard to check that this is a minimum rather than a maximum. Thus, light travels in such a way as to make the two angles equal. Put another way, "the angle of incidence equals the angle of reflection." This principle is called **Fermat's Principle**.

Now let us turn to a different problem. The methods of calculus are widely used in the field called **operations research**, which is concerned with finding the most efficient, least expensive, or generally optimal ways to solve problems of industry and government. A typical problem in operations research is the problem of locating facilities. Suppose that two towns located on the same side of a river would like to locate a sewage treatment station somewhere in between them. The towns would like to minimize the lengths of the pipelines needed. Where should the treatment station be located? We wish to locate the treatment plant at the point R where $p + q$ is minimized. By the analysis we have already made, we conclude that we should locate the treatment plant at the point R for which the two angles α and β are equal.

---

[1] You do not have to understand why this is so, though it is a good excersise for you to do so. You do not have to check all of the technical points to understand the general message in this essay.

The second point I would like to make is that mathematical modelers are invariably simple-minded. In building a mathematical model, we try to describe in mathematical language the phenomenon being studied, so that we can take advantage of our tools for reasoning in this language. Invariably, our mathematical models involve some oversimplifications. To better understand these oversimplifications, let us consider some of the examples we have discussed so far. In the case of the path of light, we have made quite explicit the assumptions that light travels in straight lines and at constant velocity. Of course, this depends upon the medium through which it is traveling, and so our analysis has to change if for example the mirror is under water but the viewer is not. In the case of the sewage treatment plant, we have assumed that the river bank is straight. It should be clear that the pipelines must be straight in order to minimize length, and we have assumed that they are. However, what if there are various obstacles such as buildings or mountains and it is necessary to go around them? Then the pipelines could take various forms. Our model of the proliferation of rabbit populations leaves out a very important fact: Rabbits die! It also assumes that all adult rabbits are equally fertile and that all rabbits become adults in exactly the same amount of time. Clearly, these are extremely strong oversimplifications. So, too, are the assumptions in the description of our transmission channel. Transmission times may not always be exactly the same, depending upon noise in the system, variation in the senders, etc. There might be errors in transmission that require signals to be sent again. We are disregarding such complications. In the case of interest on our investments, we disregard changes in interest rates over time, different types of compounding (interest is sometimes compounded semi-annually instead of annually, sometimes quarterly, sometimes daily, sometimes "continuously"). Similarly, populations don't always increase at the same percentage. The percentages vary over time.

The third point I would like to make is that mathematical modelers are fickle. Models are only useful if they can be tested. If these tests fail, the modelers give up their models. There may be an attempt to rescue them by making small changes or improvements, but more likely the models will be abandoned if they lead to faulty conclusions. Often, model builders will go through stage after stage to make a model more sophisticated, guided by failed predictions from earlier versions. To illustrate this point, let us consider the case of the growth of populations. A very simple model of population growth is that the rate at which the population grows is proportional to the size of the population. This model can be described by the equation $\frac{dP}{dt} = \lambda P$, where $P(t)$ is the population at time t and λ is a positive constant called the **birth rate**. This model leads to the conclusion that $P(t) = P(0)e^{\lambda t}$. The graph of $P(t)$ *vs.* t is an exponential curve. How well does this model do under a test? For human populations, it is not very good. However, what about other populations? For large populations under "ideal conditions," for example populations that grow by simple cell division, it is a pretty good model. Of course, it leaves out death. We can modify our model to include a **death rate** μ, another positive constant, and get the model $\frac{dP}{dt} = \lambda P - \mu P$. This model leads to the conclusion $P(t) = P(0)e^{(\lambda - \mu)t}$. The corresponding graph of $P(t)$ *vs.* t is either exponential growth without bound (if $\lambda > \mu$), exponential decay to an asymptote of the horizontal axis (if $\lambda < \mu$), or a straight line (if $\lambda = \mu$). Both models disregard immigration and emigration. If we bring in a constant immigration rate α and disregard both death and emigration, we get a model like $\frac{dP}{dt} = \lambda P + \alpha$. The conclusion here is that $P(t) = P(0)e^{\lambda t} + (\alpha/\lambda)(e^{\lambda t} - 1)$. We still get rapid growth without bound. Thus, all of these models we have discussed so far either predict growth without bound, decay to zero, or constant populations. Real populations usually do not exhibit these patterns.

No organism or population grows indefinitely. There are limits set by food supply, shelter, space, physical conditions, etc. To make our models more sophisticated, it is useful to consider a fixed upper bound B on the size of the population, often called the **saturation level** or **carrying capacity**. A quite different and somewhat more sophisticated model of population growth is to assume that populations grow at a rate proportional to the difference between the population P and the carrying capacity B. We thus have the model $\frac{dP}{dt} = k(B - P)$, for k a positive constant. This model says that populations grow rapidly when far from the carrying capacity, and slowly when close to this capacity. This model predicts that $P(t) = B - Ce^{-kt}$ for C a positive constant. This model fits some populations. However, more often the curve is not a bad fit to real data for larger t and is not too accurate for smaller t, where the growth might be closer to exponential. Thus, we are led to reject this model again. However, our observation suggests that we might want to modify the model to allow some combination of unrestricted growth as in the model $\frac{dP}{dt} = \lambda P$ and restricted growth as in the model $\frac{dP}{dt} = k(B - P)$. We are led to consider the still more sophisticated model $\frac{dP}{dt} = \lambda P(B - P)$. Here, if P is small, the right hand side is like $B\lambda P$, and if P is large, *i.e.*, close to B, it is like $\lambda B(B - P)$. This model leads to the prediction that

$$P(t) = \frac{B}{1 + Ke^{-\lambda Bt}}$$

for K constant. This function is called the **logistic function** and the curve of $P(t)$ *vs.* t is S-shaped. It was introduced in 1838 by the Belgian mathematician P.F. Verhulst and is a pretty good model for the growth of populations of protozoa, bacteria, and the like. Still, it predicts steady growth of a population. Real populations have fluctuations: They sometimes go up, sometimes go down. Thus, data about real populations does not often fit this model either.

There are now various ways to proceed. One is to introduce the idea of survivorship (*i.e.*, some offspring fail to survive to maturity). A second is to note that populations don't exist in isolation, but often there are competing populations. Fluctuations are often brought about by having two competing species. A simple model involving two species considers the populations $P_1(t)$ and $P_2(t)$ of the first and second species at time t. It is then assumed that the rate of growth of each population depends upon its rate of growth in the absence of interaction plus its rate of growth due to the interaction. It is natural at first to build a model like our very first one to explain what happens in the absence of interaction, *i.e.*, that the rate of growth is proportional to the size of the population. It is also natural to assume that the rate of growth due to the interaction is proportional to the size of the competing population (perhaps with a negative proportionality constant). Then we get a pair of **simultaneous differential equations**

$$\frac{dP_1}{dt} = aP_1 + bP_2$$
$$\frac{dP_2}{dt} = \alpha P_2 + \beta P_1.$$

One can show that under certain conditions (specifically where the **characteristic equation** $\lambda^2 - (a + \alpha)\lambda + (a\alpha - b\beta) = 0$ has two distinct real roots), this model predicts that

$$P_1(t) = ce^{\lambda_1 t} + c'e^{\lambda_2 t}$$
$$P_2(t) = ke^{\lambda_1 t} + k'e^{\lambda_2 t},$$

for some constants c, c', k, k', λ_1, λ_2. Depending upon the signs of λ_1, λ_2, we either conclude that both species explode in population or both die out. We do not see fluctuations. However, under other circumstances (in particular where the characteristic equation has complex roots), the model does predict oscillations in populations.

A still more sophisticated model involves the pair of simultaneous differential equations

$$\frac{dP_1}{dt} = aP_1 + bP_1P_2$$
$$\frac{dP_2}{dt} = \alpha P_2 + \beta P_1P_2.$$

Here, the rate of growth due to the interaction of the populations is proportional to the product of these two populations. If the first species is a prey population like rabbits and the second is a predator population like foxes, then one can argue that a and β should be positive and b and α negative. (What this says is that in the absence of foxes, rabbits increase, while in the absence of their prey, rabbits, foxes decrease. The fox population also increases at a rate proportional to the number of meetings between rabbits and foxes, namely $\beta P_1 P_2$, and the rabbit population decreases at a rate proportional to the number of such meetings.) This predator-prey model is called the **Lotka-Volterra Model** of population growth. It was first developed by the American biophysicist Alfred Lotka and, independently, by the Italian mathematician Vito Volterra, both around 1925. There is no explicit solution of this system of differential equations. However, for specific values of the parameters a, b, α, β, we may use numerical methods to plot the graphs of P_1 and P_2 vs. t, and a typical situation. Note how both populations fluctuate, but there are times when both are increasing, times when one is increasing, and times when neither is increasing. In spite of the number of steps we have taken to reach the Lotka-Volterra model and the number of models we have abandoned, this model is only a starting point for modern **ecology**, the branch of biology that deals with the relations between living organisms and their environment.

Mathematical modeling can get more and more complicated as it gets more and more realistic. There is a need for increasingly sophisticated mathematical models in many areas of our modern society. Often, dealing with these more realistic models becomes rather difficult. However, with the help of today's high speed computers, the chances are good that mathematical models, and in particular mathematical models using methods of the calculus, will play an important role in solving some of society's problems.

Questions to Think About

1. a). To estimate the cost of living in a city, three economists estimate the cost of renting a two bedroom apartment. They come up with the numbers \$1000, \$1200, and \$1300. To obtain one estimate from several, there are various methods called **consensus methods**. A common model of consensus is the **method of least squares**. This method uses x as an estimate, where x is that number so that the sum of the squares of the differences between it and the observations is as small as possible. That is, if N_i is the estimate by the ith economist, we use as an estimate that number x so that

$$S(x) = (N_1 - x)^2 + (N_2 - x)^2 + (N_3 - x)^2$$

is as small as possible. To find the estimate x, we differentiate $S(x)$ with respect to x, set the derivative equal to 0, and check the second derivative to see that we indeed get a minimum. You should check that the estimate you get by this method is \$1166.67.

b). If there are p economists each making an estimate, derive a formula for the least squares consensus estimate.

c). Ten ecologists estimate the population of fish in a pond, coming up with the estimates 1000, 1100, 1200, 1300, 1400, 1500, 1600, 1700, 1800, and 1900. If they want to find a consensus estimate, what is one way to do so and what is the estimate? (Do they have to do all the work all over again?)

2. a). Suppose $N(t)$ is the number of atoms of a radioactive substance present at time t. Using differential equations, one can show that $N(t) = N_0 e^{-\lambda t}$, where N_0 is the number of atoms present at time 0 and λ is a constant called the **rate of decay**. We are often interested in the **half-life** h, the amount of time required for half the atoms to disappear. (This is terribly important, for example, in carbon dating of fossils and in determining how to deal with radioactive waste.) One can calculate h by picking two different times t_1 and t_2 with $h = t_2 - t_1$, and showing that since $N(t_2) = (1/2)N(t_1)$, then $1/2 = e^{-\lambda h}$. (How do you show this?) It follows that $\ln(1/2) = -\lambda h$ and hence that

$$h = 0.69315/\lambda.$$

Note that as the decay rate decreases, the half-life increases.

b). Light reaching into the water of a lake is significant for the health of the lake because it is critical in the photosythesis carried on by green plants which live in or under the water. These plants produce the oxygen which makes it possible for the lake to maintain fish life. The light near the surface of a lake might be bright, but further down under the surface, the light is rapidly absorbed. The **Bouguer-Lambert Law** states that if $I(x)$ is the intensity of light at a depth x meters below the surface, and I_0 is the intensity of light at the surface, then $I(x) = I_0 e^{-ux}$, where u is a constant called the **absorption coefficient**. This constant differs for different media, and is approximately $1.4 m^{-1}$ for water. Calculate the depth at which the light intensity is only 50% of that at the surface of the lake. How can one avoid doing extra work? In murky water, where not as much light can reach the bottom, is the absorption coefficient higher or lower? What is the effect of a doubling in the absorption coefficient on the depth at which light intensity is 50% that on the surface?

3. a). Let us model an artery through which blood flows as a cylinder of radius r and length L. In such an "idealized" artery, the force F required to push a red blood corpuscle through at any point is inversely proportional to the area A of a cross-section of the cylinder at that point: $F = c(1/A)$. The total amount of work required to move the red corpuscle through the artery can be calculated from the physical model that work with constant force is given by force times distance. Hence, we reach the conclusion that the work W required to move a red blood corpuscle through this artery is given by $W = cL/(\pi r^2)$. (Why?) In particular, if $r = 1$ and $L = 1$, the work is c/π.

b). If there is calcification of the artery (a build-up of hard material on the walls that can lead to heart attacks), then the artery cannot be modeled using a cylinder; the artery might have differing radius at different points. For a second round of modeling, let us suppose that $r(x)$ is the radius at a point x cm. from the beginning of the artery. A simple case is where the calcification takes a parabolic shape. For the sake of illustration, suppose that $r(x) = (1/100)(10 - x)^2$. (Then L must be less than 10, for otherwise the artery would be completely blocked at the point $x = L$. At its widest point, that where $x = 0$, this artery has radius 1, the same as the normal artery of part 3a.) In this case, the area of a cross-section of the artery at point x is given by

$$A(x) = \pi[r(x)]^2 = (\pi/10,000)(10 - x)^4$$

and the force required at point x is given by

$$F(x) = c[1/A(x)] = [10,0000c/\pi]\frac{1}{(10 - x)^4}.$$

(If x were 10, what would you conclude about this force?) The work done moving an object at varying force is now modeled using an integral:

$$W = \int_0^L F(x)dx.$$

(You should apply this integral formula to part 3a to check the result given there.) It can be shown from this integral formula, using the substitution $u = 10 - x$ to work out the integral, that if $L = 1$, then

$$W = \frac{10,000c}{3\pi}[\frac{1}{9^3} - \frac{1}{10^3}].$$

(You should check this.) Compare this to the result with a normal artery with $r = 1$ and $L = 1$. You should conclude that the work done with the calcified artery is almost 25% more than what it was with the normal artery.

c). We might like to investigate the effect of changes in radius of the artery described in part 3b without changing its shape. For instance, suppose that the radius of this artery is doubled everywhere. We might ask what happens to the work in moving the blood corpuscle through it. Specifically, what is the ratio of the work done with the narrower artery to that done with the wider artery?

d). In general, for *any shape* artery, if the radius at every point doubles, what can you conclude about the ratio of work done with the narrower artery to that done with the wider artery? (Can you learn from having done the calculation in part 3c?)

e). What happens to the amount of work required if we double the length of the artery described in part 3b and leave the radius the same everywhere? You should investigate this with a calculator for different values of L. To do so, you should first work out the answer in part 3b for arbitrary L.

Essay

The Story of Calculus

by Roger Cooke
University of Vermont

Calculus is generally considered to be a creation of the seventeenth-century European mathematicians, with the main work having been done by Isaac Newton (1642–1727) and Gottfried Wilhelm Leibniz (1646–1711). This traditional view is correct in broad outline. Any large-scale theory, however, is a mosaic whose tiles were laid over a long period of time; and in any living theory new tiles are continually being laid. The strongest statement the historian dares to make is that a pattern became apparent at a certain time and place. Such is the case with calculus. We can say with some confidence that the main outlines of the subject appeared in the seventeenth century and that the pattern was made much clearer by the work of Newton and Leibniz. However, many of the essential principles of calculus were discovered as early as the time of Archimedes (287–211 BCE), and some of these same discoveries were made independently in China and Japan. Moreover if you dig deeply into the problems and methods of calculus, you will soon find yourself pursuing problems that lead into the modern areas of analytic function theory, differential geometry, and functions of a real variable. To change the metaphor from art to transportation, we can think of calculus as a large railroad station, where passengers arriving from many different places all come together for a brief time before setting out again for a variety of destinations. In the present essay we shall try to look in both directions from that station, to the sources, and to the destinations. Let us begin by describing the station itself.

Isaac Newton

Gottfried Leibniz

What Is Calculus?

Calculus is traditionally divided into two parts, called *differential calculus* and *integral calculus*. Differential calculus investigates the properties of the comparative rates of change of variables that are linked by equations. For example, a fundamental result of differential calculus is that if $y = x^n$, then the rate of change of y with respect to x is nx^{n-1}. It turns out that when we think intuitively about certain phenomena—the motion of bodies, changes in temperature, growth of populations, and many others—we are led to postulate certain relations between these variables and their rates of change. These relations are written down in a form known as *differential equations*. Thus the primary purpose of studying differential calculus is to understand what rates of change are and how to write down differential equations. Integral calculus provides methods of recovering the original variables knowing their rates of change. The technique for doing so is called *integration*, and the primary purpose of studying integral calculus is to learn how to *solve* the differential equations that are provided by differential calculus.

These goals are often masked in calculus books, where differential calculus is used to find the maximum and minimum values of certain variables, and integral calculus is used to compute lengths, areas, and volumes. There are two reasons for emphasizing these applications in a textbook. First, the full use of calculus involving differential equations involves some rather elaborate theory that must be introduced gradually; meanwhile, the student must be shown *some* use for the techniques that are being put forth. Second, such problems were the source of the ideas that led to calculus; the uses we now make of the subject arose only after it was discovered.

In describing the problems that led to calculus and the problems that can be solved using calculus, we still have not pointed out the fundamental techniques that make calculus so much more powerful a tool of analysis than mere algebra and geometry. These techniques involve the use of what was once called *infinitesimal analysis*. The constructions and formulas of high school geometry and algebra all have a finite character. For example, to construct the tangent to a circle or to bisect an angle, you perform a finite number of operations with straightedge and compass. Although Euclid knew considerably more geometry than is found in modern high school courses, he, too, confined himself mostly to finite processes. Only in the limited context of the theory of proportion does he allow the infinite into his geometry, and even there it is surrounded by so much logical caution that the proofs involved are extraordinarily cumbersome and hard to read. This same situation occurs in algebra. In order to solve a polynomial equation, you perform a finite number of operations of addition, subtraction, multiplication, division, and root extraction. When the equation can be solved, the solution is expressed as a finite formula involving the coefficients.

These finite techniques, however, have a limited range of applicability. One cannot find the areas of most curved figures by a finite number of operations with straightedge and compass, nor can one solve most polynomial equations of degree five or higher using a finite number of algebraic operations. It was the desire to escape from the limitations of finite methods that led to the creation of calculus. We shall now look at some of the early attempts to develop techniques for handling the more difficult problems of geometry, after which we shall summarize the process by which calculus was worked out and finally exhibit some of the harvest it has provided.

The Geometric Sources of the Calculus

One of the oldest mathematical problems is that of squaring the circle; that is, constructing a square equal in area to a given circle. It is now known that this problem cannot be solved by use of a finite number of applications of compass and straightedge. However, Archimedes discovered that if one could draw a spiral starting at the center of a circle that makes exactly one revolution before reaching the circle, then the tangent to that spiral at its point of intersection with the circle would form the hypotenuse of a right triangle with area exactly equal to the circle (see Figure 1). Thus if one could draw this spiral and its tangent, one could square the circle. Archimedes, however, was silent on the question of how one might draw this tangent.

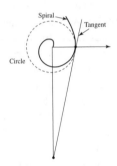

Figure 1 *The spiral of Archimedes.* The tangent at the end of the first turn and the two axes form a triangle with area equal to the circle about the origin through the point of tangency.

We see here that one of the classical mathematical problems could be solved if only we could draw a certain curve and a tangent to it. This problem, and others like it, caused the purely mathematical problem of finding the tangent to a curve to become important. This problem is the main source of differential calculus. The "infinitesimal" trick that allows the problem to be solved is to think of the tangent as the line determined by two points on the curve "infinitely close" together. Another way of saying the same thing is that an "infinitely short" piece of any curve is straight. The trouble is that it is hard to be precise about the meaning of the phrases "infinitely close" and "infinitely short."

Little progress was made on this problem until the invention of analytic geometry in the seventeenth century by Pierre de Fermat (1601–1665) and René Descartes (1596–1650). Once a curve could be represented by an equation, it became possible to say with more confidence what was meant by "infinitely close" points, at least for polynomial equations such as $y = x^2$. With algebraic symbolism to represent points on the curve, it was possible to consider two points on the curve with x-coordinates x_0 and x_1, so that $x_1 - x_0$ is the distance between the x-coordinates. When the equation of the curve was written at each of these points and one of the two equations subtracted from the other, one side of the resulting equation contained the factor $x_1 - x_0$, which could therefore be divided out. Thus if $y_0 - x_0^2$ and $y_1 - x_1^2$, then $y_1 - y_0 = x_1^2 - x_0^2 = (x_1 - x_0)(x_1 - x_0)$, and so $\dfrac{y_1 - y_0}{x_1 - x_0} = x_1 + x_0$. When $x_1 = x_0$, it follows that $y_1 = y_0$, and the expression $\dfrac{y_1 - y_0}{x_1 - x_0}$ has no meaning. However, the expression $x_1 + x_0$ has the perfectly definite value $2x_0$. Thus we can think of $2x_0$ as the ratio of the infinitely small difference in y, namely $y_1 - y_0$, to the infinitely small difference in x, namely $x_1 - x_0$, when the point (x_1, y_1) is infinitely close to the point (y_1, y_0) on the curve $y = x^2$. As you will learn in your study of calculus, this ratio gives enough information to draw the tangent line to the curve $y = x^2$.

The preceding argument is, except for small changes in notation, exactly the way Fermat found the tangent to a parabola. It was open to one logical objection, however: At one stage we divided both sides of an equation by $x_1 - x_0$, then at a later stage we decided that $x_1 - x_0 = 0$. Since division by zero is an illegal operation, we seem to be trying to eat our cake and have it, too. It took some time to find a convincing answer to this objection.

We have just seen that Archimedes was unable to solve the fundamental problem of differential calculus, drawing the tangent to a curve. Archimedes *was* able to solve some of the fundamental problems of integral calculus, however. In fact he found the volume of a sphere in an extremely ingenious way. He considered a cylinder containing a cone and a sphere and imagined this figure cut into infinitely thin slices. By looking at the areas of these sections of the cone, sphere, and cylinder, he was able to show how the cylinder would balance the cone and sphere if the figures were hung on opposite sides of a fulcrum. This balancing gave one relation among the three figures, and Archimedes already knew the volumes of the cone and cylinder; hence he was able to compute the volume of the sphere.

This argument illustrates the second infinitesimal technique that lies at the foundation of calculus: A volume can be regarded as a stack of plane figures, and an area can be regarded as a stack of line segments, in the sense that if every horizontal section of one region equals the same horizontal section of another region, then the two regions are equal. During the European Renaissance this principle came to be widely used under the name of the *method of indivisibles* for finding the areas and volumes of many figures. It is nowadays called *Cavalieri's Principle* after Bonaventura Cavalieri (1598–1647), who used it to prove many of the elementary formulas that now make up integral calculus. Cavalieri's Principle was also discovered in other lands where Euclid's work had never gone. The fifth-century Chinese mathematicians Zu Chongzhi and his son Zu Geng, for example, found the volume of a sphere using a technique very similar to Archimedes' method.

Thus we find mathematicians anticipating the integral calculus by using infinitesimal methods to find areas and volumes at a very early stage of geometry in both ancient Greece and China. Like the infinitesimal method of drawing tangents, however, this method of finding areas and volumes was open to objection. For example, the volume of each plane section of a figure is zero; how can a collection of zeros be put together to yield something that is not zero? Also, why doesn't the method work in one dimension? Consider the sections of a right triangle parallel to one of its legs. Each section intersects the hypotenuse and the other leg in congruent figures, namely one point each. Yet the hypotenuse and the other leg are not the same length. Objections like these were worrisome. The results obtained using these methods were so spectacular, however, that mathematicians preferred to take them on faith, continue to use them, and try to build the foundation under them later, just as a tree grows both roots and branches at the same time.

The Invention of the Calculus

By the middle of the seventeenth century, many of the elementary techniques and facts of calculus were known, including methods for finding the tangents to simple curves and formulas for areas bounded by these curves. In other words, many of the formulas you will find in the early chapters of any calculus textbook were already known before Newton and Leibniz began their work. What was lacking until late in the seventeenth century, however, was the realization that these two kinds of problems are related to each other.

To see how the relation came to be discovered, we need to say a bit more about tangents. We mentioned above that to draw a tangent to a curve at a given point one needs to know how to find a second point on the line. In the early days of analytic geometry this second point was usually taken as the point at which the tangent intersects the x-axis. The projection onto the x-axis of the portion of the tangent between the point of tangency and the intersection with the axis was called the *subtangent*. In the study of tangents one very natural problem arose: *to reconstruct a curve, given the length of its subtangent at every point.* Through the study of this problem it came to be noticed that the ordinates of any curve are proportional to the area under a second curve whose ordinates are the lengths of the subtangents to the original curve. That result is the fundamental theorem of calculus. The honor of explicitly recognizing this relation goes to Isaac Barrow (1630–1677), who pointed it out in a book called *Lectiones Geometricae* in 1670. Barrow stated several theorems resembling the fundamental theorem of calculus. One of these is the following: *If a curve is drawn so that the ratio of its ordinate to its subtangent* [this ratio is precisely what is now called the derivative] *is proportional to the ordinate of a second curve, then the area under the second curve is proportional to the ordinate of the first.*

These relations provided a unifying principle for the large number of particular results on tangents and areas that had been obtained by the method of indivisibles in the early seventeenth century: To find the area under a curve, find a second curve for which the ratio of the ordinate to the subtangent equals the ordinate of the given curve. Then the ordinate of that second curve will give the area under the first curve.

At this point the calculus was ready to be born. It needed only someone to give systematic methods of computing tangents (actually subtangents) and inverting that process in order to find areas. This is the work actually performed by Newton and Leibniz. These two giants of mathematical creativity took quite different paths to their discoveries.

Newton's approach was algebraic and developed out of the problem of finding an efficient method of extracting the roots of a number. Although he had barely begun to study algebra in 1662, by 1665 Newton's reflections on the problem of extracting roots led him to the discovery of the infinite series now known as the binomial theorem; that is, the relation

$$(1+x)^r = 1 + rx + \frac{r(r-1)}{2}x^2 + \frac{r(r-1)(r-2)}{1 \cdot 2 \cdot 3}r^3 + \ldots$$

By combining the binomial theorem with infinitesimal techniques, Newton was able to derive the basic formulas of differential and integral calculus. Central to Newton's approach was the use of infinite series to express the variables in question, and the fundamental problem that Newton did not solve was to establish that such series could be manipulated just like finite sums. Thus in a sense Newton drove the infinite away from one entrance to its burrow, only to find it showing its face in another.

Since he thought of variables as being physical quantities that change their value with time, Newton invented names for variables and their rates of change that reflected this intuition. According to Newton a *fluent* (x) is a moving or flowing quantity; its *fluxion* (x) is its rate of flow, what we now call its velocity or *derivative*. Newton expounded his results in 1671 in a treatise called *Fluxions* written in Latin; but this work was not published until an English version appeared in 1736. (The original Latin version was first published in 1742.) Despite his notation and his arguments, which seem crude and inefficient today, the tremendous power of calculus shines through Newton's *Fluxions* in the solution of such difficult problems as finding the arc length of a curve. This "rectification" of a curve had been thought impossible, but Newton showed that one could find an infinite number of curves whose length could be expressed in finite terms.

Newton's approach to the calculus was algebraic, as we have just seen, and he inherited the fundamental theorem from Barrow. Leibniz, on the other hand, worked out the fundamental result on his own during the 1670s, and his approach was different from Newton's. Leibniz is considered the earliest pioneer of symbolic logic, and he had a much better appreciation than Newton of the importance of good symbolic notation. He invented the notation dx and dy that we still use today. For him dx was an abbreviation for "difference in x" and represented the difference between two infinitely close values of x. In other words, it expressed exactly what we had in mind above when we considered the infinitely small change $x_1 - x_0$. Leibniz thought of dx as an "infinitesimal" number, a number not zero, yet so small that no multiple of it could exceed any ordinary number. Not being zero, it could serve as the denominator in a fraction, and so dy/dx was the quotient of two infinitely small quantities. In this way he hoped to avoid the objections to the newly established method of finding tangents.

In the controversial technique of finding areas by adding up the sections, Leibniz also made a major contribution. Instead of thinking of an area [for example, the area under a curve $y = f(x)$] as a collection of line segments, he regarded it as the sum of the areas of "infinitely thin" rectangles of height $y = f(x)$ and infinitesimal base dx. Hence the difference between the area up to the point $x + dx$ and the area up to the point x was the infinitesimal difference in area $dA = f(x)dx$, and the total area was found by summing up these infinitesimal differences in area. Leibniz invented the elongated S (the integral sign $\int$) that is now universally used for expressing this summation process. Thus he would express the area under the curve $y = f(x)$ as $A = \int dA = \int f(x)dx$, and each part of this symbol expressed a simple and clear geometric idea.

With Leibniz's notation, Barrow's fundamental theorem of calculus merely says that the pair of equations

$$A = \int f(x)dx, \quad dA = f(x)dx$$

are equivalent to each other. Because of what was just stated above, this equivalence is nearly obvious.

Both Newton and Leibniz had made huge advances in mathematics, and there was plenty of credit for both of them. It is unfortunate that the near coincidence of their work led to an acrimonious dispute over priority between their followers.

Some parts of the calculus, involving infinite series, had been invented in India in the fourteenth and fifteenth centuries. The late fifteenth-century Indian mathematician Jyesthadeva gave the series

$$\theta = r\left(\frac{\sin\theta}{\cos\theta} - \frac{\sin^3\theta}{3\cos^3\theta} + \frac{\sin^5\theta}{5\cos^5\theta} - \cdots\right)$$

for the length of an arc of a circle, proved this result, and explicitly stated that this series will converge only if θ is not larger than $45°$. If we write $\theta = \arctan x$, and use the fact that $\dfrac{\sin\theta}{\cos\theta} = \tan\theta = x$, this series becomes the standard series for $\arctan x$.

Likewise some infinite series were developed in Japan independently about the same time as in Europe. The Japanese mathematician Katahiro Takebe (1664-1739) found a series expansion equivalent to the series for the square of the arcsine function. He was considering the square of half the arc at height h in a circle of diameter d; this works out to be the function $f(h) = \left(\dfrac{d}{2}\arcsin\dfrac{h}{d}\right)^2$. Katahiro Takebe had no notation for the general term of a series, but he discovered patterns in the coefficients by computing the function geometrically at the particular value of $h = 0.000001$, $d = 10$ to a very large number of decimal places— more than fifty—and then using this extraordinary accuracy to refine the approximation by successively adding corrective terms. By proceeding in this way he was able to discern a pattern in the successive approximations, from which by extrapolation he was able to state the general term of the series:

$$f(h) = dh\left[1 + \sum_{n=1}^{\infty} \frac{2^{2n+1}(n!)^2}{(2n+2)!}\left(\frac{h}{d}\right)^n\right]$$

After Newton and Leibniz, there remained the problem of putting flesh on the skeleton these two geniuses had created. The majority of this work was completed by the Continental mathematicians, notably the circle around the Swiss mathematicians

James and John Bernoulli ((1655–1705) and (1667–1748), respectively) and John Bernoulli's student the Marquis de l'Hôpital (1661–1704). These mathematicians and others worked out the familiar formulas for the derivatives and integrals of elementary functions that are found in textbooks today. The essential techniques of calculus were known by the early eighteenth century, and an eighteenth-century textbook such as Euler's *Introduction in analysin infinitorum* (1748), if translated into English, would look very much like a modern textbook.

The Legacy of the Calculus

Having looked at the sources of calculus and the procedure by which it was constructed, let us now examine briefly the results it produced.

The calculus scored an amazing number of triumphs in its first two centuries. Dozens of previously obscure physical phenomena involving heat, fluid flow, celestial mechanics, elasticity, light, electricity, and magnetism turned out to have measurable properties whose relations could be described as differential equations. Physics was forever committed to speaking the language of calculus.

By no means were all of the mathematical problems arising from physics solved, however. For example, the area under a curve whose equation involved the square root of a cubic polynomial could not be found in terms of familiar elementary functions. Such integrals arose frequently in both geometry and physics, and came to be known as *elliptic integrals* because the problem of finding the length could be understood only when the real variable x is replaced by a complex variable $z = x + iy$. The reworking of the calculus in terms of complex variables led to many new and fascinating discoveries, which eventually came to be codified as a new branch of mathematics called analytic function theory.

The proper definition of integration remained a problem for some time. Integrals arose out of the use of infinitesimal processes to find areas and volumes. Should the integral be defined as a "sum of infinitesimal differences," or should it be defined as the reverse of differentiation? What functions can be integrated? Many definitions of integral were proposed in the nineteenth century, and the elaboration of these ideas has led to the subject now known as real analysis.

While the applications of calculus have moved on to more and more triumphs in an unending stream for the last three hundred years, its foundations lay in an unsatisfactory state for the first half of this period. The root of the difficulty was the meaning to be attached to Leibniz's dx. What was this quantity? How could it be neither positive nor zero? If zero, it could not be used as a denominator; if positive, then the equations in which it occurred were not truly equations. Leibniz believed that infinitesimals were real things, that areas and volumes could be synthesized by "adding up" their sections, as Zu Chongzhi, Archimedes, and others had done. Newton was less confident of the validity of infinitesimal methods and tried to justify his arguments in ways that would meet Euclidean standards of rigor. In his *Principia Mathematica* he wrote:

> These Lemmas are premised to avoid the tediousness of deducing involved demonstrations *ad absurdum*, according to the method of the ancient geometers. For demonstrations are shorter by the method of indivisibles; but because the hypothesis of indivisibles seems somewhat harsh, and therefore that method is reckoned less geometrical, I chose rather to reduce the demonstrations of the following Propositions to the first and last sums and ratios of evanescent quantities, that is, to the limits of those sums and rations ... Therefore if hereafter I should happen to consider quantities as made up of particles, or should use little curved lines for right [straight] ones, I would not be understood to mean indivisibles, but evanescent divisible quantities ...

> ... For those ultimate ratios with which quantities vanish are not truly the ratios of ultimate quantities, but limits towards which the ratios of quantities decreasing without limit do always converge; and to which they approach nearer than by any given difference, but never go beyond, nor in effect attain to, till the quantities are diminished *in infinitum*.

In this passage Newton was claiming that the lack of rigor involved in using infinitesimal arguments could be compensated for by the use of limits. His formulation of this concept in the passage just quoted is not so clear as one might wish, however. This lack of clarity led the philosopher Berkeley to refer contemptuously to fluxions as "ghosts of departed quantities." The advances achieved in physics using calculus, however, were so outstanding that for more than a century no one bothered to supply the extra rigor Newton alluded to (and physicists still don't bother with it!). A completely rigorous and systematic presentation of the calculus came only in the nineteenth century.

After the work of Augustin-Louis Cauchy (1789–1856) and Karl Weierstrass (1815–1896), the received view was that infinitesimals are merely heuristic in nature, and students were subjected to a rigorous "epsilon-delta" approach to limits. Somewhat surprisingly, however, in the twentieth century it was shown by Abraham Robinson (1918–1974) that a logically consistent model of the real numbers can be developed in which there are actual infinitesimals, just as Leibniz had believed. This new approach, called "nonstandard analysis," does not seem to be supplanting the now-traditional presentation of calculus, however.

Exercises

1. The kind of spiral considered by Archimedes in now named after him. An Archimedean spiral is the locus of a point moving at a constant speed along a ray rotating with constant angular speed about a fixed point. If the linear speed along the ray (the *radial* component of its velocity) is v, the point will be at a distance vt from the center of rotation (assuming that is where it starts) at time t. Suppose the angular speed of rotation of the ray is ω (radians per unit time). Given a circle of radius R, and a radial speed of v, what must ω be in order for the spiral to reach the circle at the end of its first turn? *Ans.* $\left(\frac{2\pi v}{R}\right)$

   The point will have a *circumferential* velocity $r\omega = vt\omega$. According to a principle enunciated in Aristotle's *Mechanics*, the actual velocity of the particle will be directed along the diagonal of a parallelogram (a rectangle in this case) having the two components as sides. Use this principle to show how to construct the tangent to the spiral (it will be the line containing the diagonal of this rectangle). Verify that sides of this rectangle are in the ratio $1 : 2\pi$. See Figure 1.

2. Figure 2 illustrates how Archimedes found the relation between the volumes of a sphere, cone, and cylinder. The diameter AB is doubled, making $BC = AB$. Note also that $AD = AG = AB$. When the figure is revolved about this line, the circle generates a sphere, the triangle DBG generates a cone, and the rectangle $DEFG$ generates a cylinder. Prove the following facts.

   (a) If B is used as a fulcrum, the cylinder has the center K of the circle as its center of grav- ity, and therefore could all be concentrated there without changing the torque about B.

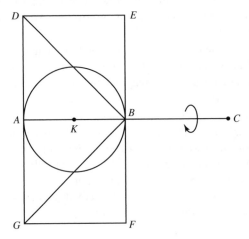

Figure 2 Section of Archimedes' sphere, cone, and cylinder.

   (b) Each section of the cylinder perpendicular to the line AB, remaining in its present position, would exactly balance the same section of the cone plus the section of the sphere if both of the latter were moved to the point C.

   (c) Hence the cylinder concentrated at K would balance the cone and sphere concen- trated at C.

   (d) Therefore the cylinder equals twice the sum of the cone and the sphere.

   (e) Since the cone is known to be one-third of the cylinder, it follows that the sphere must be one-sixth of it.

   (f) Since the volume of the cylinder is $8\pi r^3$ the volume of the sphere is $\frac{4}{3}\pi r^3$.

3. The method by which Zu Chongzhi and Zu Geng found the volume of a sphere is as follows: Imagine the sphere as a ball tightly stuck inside the intersection of two cylinders at right angles to each other. The solid formed by the intersection of the two cylinders (called a *double umbrella* in Chinese) and containing the ball is then tightly fitted inside a cube whose edge equals the diameter of the sphere.

   From this description draw a section of the sphere within the double umbrella within the cube. Imagine this section being made parallel to the plane formed by the axes of the two cylinders and at a distance h below this plane. Verify the following facts.

   (a) If the radius of the sphere is r, the circular section of it has diameter $2\sqrt{r^2 - h^2}$.

   (b) Hence the square formed by the section of the double umbrella has area $4(r^2 - h^2)$, and so the area between the section of the cube and the section of the double umbrella is

$$4r^2 - 4(r^2 - h^2) = 4h^2.$$

(c) The corresponding section of a pyramid whose base is the bottom of the cube and whose vertex is at the center of the sphere (or cube) would also have area $4h^2$. Hence the volume between the double umbrella and the cube is exactly the volume of such a pyramid plus its mirror image above the central plane. Conclude that the region between the double umbrella and the cube is one-third of the cube.

(d) Therefore the double umbrella occupies two-thirds of the volume of the cube; that is, its volume is $\dfrac{16}{3}r^3$.

(e) Each circular section of the sphere is inscribed in the corresponding square section of the double umbrella. Hence the circular section is $\dfrac{\pi}{4}$ of the section of the double umbrella.

(f) Therefore the volume of the sphere is $\dfrac{\pi}{4}$ of the volume of the double umbrella; that is, $\dfrac{4}{3}\pi r^3$.

4. Give an "infinitesimal" argument that the area of a sphere is three times its volume divided by its radius by imagining the sphere to be a collection of "infinitely thin" pyramids with vertices all stuck together at the origin. [*Hint*: Use the fact that the volume of a pyramid is one-third the area of its base times its altitude. Archimedes says that this is the reasoning that led him to discover the area of a sphere.]

Answers to the Essay's exercise questions can be found on page 421.

Use of a Calculator

1. Introduction

This part is not intended to be a comprehensive manual on the use of a graphing calculator in a precalculus course. While much of the material discussed will be pertinent to any graphing calculator, the references in this manual will be to the TI-84 family of calculators. A number of students entering college already have a TI-83 calculator that they used in their high school mathematics classes. The TI-83 is similar to the TI-84 and most of what is discussed here applies to the TI-83 as well. After a few brief comments on the use of the TI-84 calculator, the focus in this manual will be on how to use the calculator to assist in the solution of a number of the problems in the text. While the TI-84 calculator contains many sophisticated routines that could be used, the focus in this manual will be on using techniques that require a minimum amount of expertise with the calculator. Before deciding on the extent to which you use a calculator in your precalculus course, it is very important that you keep two things in mind:

1. Your instructor may not allow the use of calculators on quizzes or tests. If this is the case you should definitely avoid becoming dependent on the calculator as a source of solutions to problems. At most, you should use the calculator as a way to check the plausibility of your answer to a problem.

2. Even if you are allowed free use of a calculator it is important to keep in mind that many of the problems in the text are frequently designed to have "nice" outcomes. That is, the answers will generally involve integers or relatively simple fractions. In actual practice, this will not generally be the case, and your calculator will give only an approximate answer. You may still need to generate the exact answer (for example, in a calculus course following this precalculus course). A simple example involves the equation $x^2 - 3 = 0$. The exact solution is $x = \sqrt{3}$ or $x = -\sqrt{3}$, but the calculator will generate the solutions $x = 1.7320508075\ldots$ and $x = -1.732050807\ldots$, which are very close, but still not exact.

2. Mode Settings

When you first turn on the calculator it is a good idea to check the settings in the calculator. To do this press the $\boxed{\text{MODE}}$ key to see something like

Generally, you will want to use the default settings, which are highlighted in the screenshot shown above.

3. Clearing the Calculator Screen

To clear the screen when you are not in a graphics mode simply position the cursor on a blank line and press the $\boxed{\text{CLEAR}}$ key. If the cursor is not on a blank line, pressing the $\boxed{\text{CLEAR}}$ key will generally erase only that line. In this case, if there are still entries on the screen and you want to erase them, press the $\boxed{\text{CLEAR}}$ key again.

When in the $\boxed{Y=}$ window, to clear a function definition, position the cursor at the end of the function and press the $\boxed{\text{CLEAR}}$ key.

Suppose a graphics object, like a line connecting $(1,3)$ and $(-5,-4)$, has been entered using the DRAW menu. Then, later, you want to graph the function $y = x^2$. In this case you are likely to see the following screen:

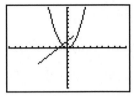

You probably didn't expect to see the line in the graph. To clear the line, press the $\boxed{\text{2ND}}$ $\boxed{\text{DRAW}}$ keys and select the first option, 1 : ClrDraw

When the $\boxed{\text{CLEAR}}$ key doesn't work, try using the $\boxed{\text{2ND}}$ $\boxed{\text{QUIT}}$ keys to clear a graphics screen.

4.   Basic Calculations and Memory

In the TI-84 calculator, expressions like $(2\pi - 3)(5 + \sqrt{2})^3$ can be computed as shown:

If the result of this operation is to be used again, *do not* write the number 866.4143622 on a piece of paper and then key it in (or even worse, a rounded version of it like 866.41) at the appropriate spot. For example, if you later want to find the reciprocal of the square root of this number, then you should store it in a memory register and call it up when needed. This process is demonstrated here.

The fourth line in the window, which stores the result of the calculation in the variable B, is obtained by pressing the $\boxed{\text{STO}}$ $\boxed{\text{ALPHA}}$ $\boxed{\text{B}}$ keys. To clear a value assigned to a variable, simply store the number 0 in that variable.

5.   Functions

To enter a function that is to be evaluated or graphed, press the $\boxed{Y=}$ key and enter the function (or functions) that is to be used. In the screenshot below, the function $y = x^3 - 2x + 4$ is entered.

5.1 Function Evaluation

If a function has been entered as Y1 then it can be evaluated at any number, say $x = -3.2$, using the $\boxed{\text{2ND}}$ $\boxed{\text{TBLSET}}$ and $\boxed{\text{2ND}}$ $\boxed{\text{TABLE}}$ keys. The following sequence of screenshots demonstrates how to do this.

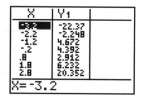

In this case, we see that when $x = -3.2$, $x^3 - 2x + 4$ equals -22.37. When entering a negative number like -3.2 be sure to use the $\boxed{(\text{-})}$ key in the bottom row on the calculator, as opposed to the subtraction key $\boxed{-}$ just to the right of the $\boxed{6}$ key. This is a very common mistake, and should generally be the first possibility you consider whenever the calculator gives you an error message. Also, once you have input the number in the TABLESETUP screen, be sure to press the $\boxed{\text{ENTER}}$ key.

5.2 Graphing a Function

To graph Y1 use the $\boxed{\text{WINDOW}}$ key to set the viewing rectangle. For example, if you set Xmin=-10, Xmax=10, Ymin=-10, and Ymax=10, and then press the $\boxed{\text{GRAPH}}$ key, you obtain the following two screens:

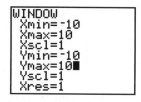
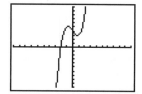

To change the viewing rectangle simply press the $\boxed{\text{WINDOW}}$ key and reset the boundaries of the window. Based on the screen shown above, you might want to reset Xmin to be -4 and Xmax to be 3. In this case the following screen is obtained:

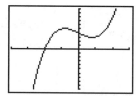

5.3 An Alternative Method of Function Evaluation

An alternative way to evaluate a function that has been entered in Y1 uses the $\boxed{\text{2ND}}$ $\boxed{\text{CALC}}$ key combination. Suppose you want to evaluate $f(-3.4)$ when

$$f(x) = \frac{\sqrt{x^2 - 3}}{x + 4}.$$

First, enter this function as Y1 in the $\boxed{\text{Y=}}$ window. Next, be sure that the graphing window includes the value -3.4. Then, in the $\boxed{\text{2ND}}$ $\boxed{\text{CALC}}$ window, choose 1 : value. Enter the value for x, in this case -3.4, and press $\boxed{\text{ENTER}}$. (Did you remember to use the $\boxed{(\text{-})}$ key instead of the subtraction key?) In the lower right hand portion of the screen you will see the value of $f(x)$, in this case $Y = 4.8762463$. You should see something like the following sequence of screens:

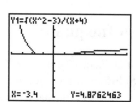

6. Solution of an Equation

To find the solution or solutions of an equation such as $x^3 - x^2 - 12x + 7 = 0$, first graph the function to get a sense of how many solutions there are and approximately where they are. The graph of $Y2 = x^3 - x^2 - 12x + 7 = 0$ is shown below with $Xmin = -10$, $Xmax = 10$, $Ymin = -10$, and $Ymax = 10$.

From the graph above, we see that the equation probably has three solutions near $x = -3$, $x = 1$, and $x = 4$. The word "probably" is used because, without knowledge of the behavior of cubic functions, it is possible that the graph comes back down (or up on the left) and crosses the x-axis outside the window of the calculator. Now, to find the solutions of the equation, use the EQUATION SOLVER in the the MATH menu. This can be accessed by pressing the $\boxed{\text{MATH}}$ key followed by the $\boxed{0}$ key. Where the screen reads eqn : 0 = type in the equation. (If the screen does not say EQUATION SOLVER at the top, press the up cursor.) You should see the following screen:

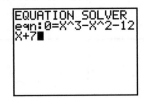

Now press $\boxed{\text{ENTER}}$ and assign a value to X that is close to a solution. For example, to find the smallest positive solution of the equation in this example you could assign x = 1 followed by $\boxed{\text{ALPHA}}$ $\boxed{\text{SOLVE}}$. You should see the following sequence of two screens:

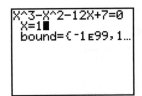 

The smallest positive solution is $x = .57166826799$ rounded to 10 decimal places. To find the other positive solution press $\boxed{4}$ followed by $\boxed{\text{SOLVE}}$.

This shows that the next positive root is $x = 3.7199797055$.

7. Solution of an Inequality

To find the solution set of an inequality such as $x^2 + 3x - 10 \geq 0$, first graph $Y1 = x^2 + 3x - 10$ and then solve the equation $x^2 + 3x - 10 = 0$ as described in **Solution of an Equation** on the previous page. This gives $x = -5$ and $x = 2$. Now, pressing the $\boxed{\text{GRAPH}}$ key again shows that $x^2 + 3x - 10 \geq 0$ (that is, above the x-axis) when $x \leq -5$ and $x \geq 2$. You should see the following sequence of calculator screens:

While the above technique could be used to approximate the solution of an inequality, it should only be used to check the answer that you arrived at using the sign chart technique described in the text.

8. Solution of an Inequality Involving an Absolute Value

To find an approximation to the solution of an inequality involving an absolute value such as $|2x - 8| \geq 12$, first write the inequality as $|2x - 8| - 12 \geq 0$. Then graph $\mathtt{Y1} = \mathtt{abs}(2x - 8) - 12$ and find the solutions of $|2x - 8| = 12$ by estimating where the graph intersects the x-axis. The solution set of the inequality will then be the union of the intervals where the graph is above the x-axis.

9. Graphing a Circle

Option 9 in the DRAW menu is used to draw a circle. In this case, three inputs are needed— the first two are the coordinates of the center and the third is the radius of the circle. For example, to draw a circle centered at $(1, 2)$ with radius 3 use the following sequence of keystrokes: 2ND DRAW 9 1 2 3 . You will probably see a screen that looks something like the following:

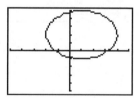

To make this circle actually look like a circle rather than an oval, use the ZOOM key followed by 5 : ZSquare. To see the resulting circle you may need to press the CLEAR key followed by the ENTER key. The screen should now show the following:

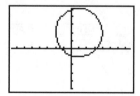

10. Limit of a Function

While you should generally not use your calculator to find the limit of a function, it can be useful to check that the limit you have obtained using hand computations is plausible. For example, suppose you want find

$$\lim_{x \to -1} \frac{2x + 2}{x^2 - x - 2}.$$

By hand, you compute

$$\frac{2x + 2}{x^2 - x - 2} = \frac{2(x + 1)}{(x - 2)(x + 1)} = \frac{2}{x - 2}, \quad x \neq -1,$$

so that

$$\lim_{x \to -1} \frac{2x+2}{x^2-x-2} = \lim_{x \to -1} \frac{2}{x-2} = \frac{2}{-1-2} = -\frac{2}{3}.$$

There are two ways you can use your calculator to check the plausibility of this result. First, graph the function

$$Y1 = \frac{2x+2}{x^2-x-2},$$

and then use the $\boxed{\text{TRACE}}$ key to position the cursor over $x = -1$ (or as close as you can get to $x = -1$). The corresponding y-coordinate will be displayed in the lower right-hand portion of the screen, as shown here:

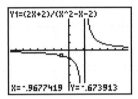

Since $Y = -.673913$ is close to $-\frac{2}{3}$, you have reason to believe that your hand computations are correct. Alternatively, you could evaluate the function (as discussed in Section 5.1 of this part of the manual)

$$Y1 = \frac{2x+2}{x^2-x-2},$$

at a value very close to $x = -1$, say, at $x = -1.01$. In this case, you will see a screen like the one here:

In this case, when $x = -1.01$ the corresponding y value is $-.6645$. Since this number is close to $-\frac{2}{3}$, you have reason to believe that your hand computations are correct.

11. Factorials

Factorials play a very important role in probability, which explains its location in the calculator. To find, say 6!, press $\boxed{6}$ $\boxed{\text{MATH}}$. You will see the following screen:

Now press the right cursor three times to highlight PRB followed by $\boxed{4}$ $\boxed{\text{ENTER}}$. This will give the value of 6!.

To test your ability with factorials on the calculator, you might want to find values of n for which the calculator switches to scientific notation, and values for which it overflows. You should also determine what the calculator thinks 0! is, and what it thinks of $(-4)!$. [In the latter case, be sure to include the parentheses around the -4 before trying to find the factorial.]

12. Graphing Quadratic Functions

As discussed in Section 2.4 of the text, the graph of a quadratic function is always a parabola opening upward or downward. When the coefficients of the function are integers or simple fractions you should be able to sketch a graph of the function by hand. In this case, a calculator may be used to confirm the results of your work. However, when the function does not have simple coefficients, it can be very tedious to find the vertex and the intercepts by hand.

Suppose, for example, that you want to approximate the intercepts and vertex of the parabola determined by $f(x) = 2.83x^2 - 4.32x - 31.09$. To get a sense of the intercepts and vertex of the parabola you could graph the function. To do this, enter the function as Y1 using the $\boxed{Y=}$ key and make sure that all other functions, Y1, Y2, Y3, . . . , are clear. Then press the $\boxed{\text{GRAPH}}$ key. Depending on the settings in the $\boxed{\text{WINDOW}}$ screen you will see a screen like

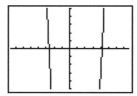

Knowing that the graph is a parabola, you can zoom in on the x-intercepts to determine their locations with reasonable accuracy. To do this, first press the $\boxed{\text{TRACE}}$ key and then use the left and right cursor keys, $\boxed{\blacktriangleleft}$ and $\boxed{\blacktriangleright}$, to place the cursor near an x-intercept. (Initially, the cursor may be off the screen, but its x- and y-coordinates are shown at the bottom of the screen, so you can move it into the viewing window.) Having done this, you should see a screen like the following:

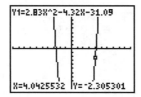

Now, press the $\boxed{\text{ZOOM}}$ $\boxed{2}$ $\boxed{\text{ENTER}}$ key sequence to zoom in on the part of the graph around the cursor. At this point, the cursor is no longer attached to the graph and can be moved using any of the four cursor keys, making it easier to position the cursor more directly on top of the x-intercept. After doing this, press the $\boxed{\text{ZOOM}}$ $\boxed{2}$ $\boxed{\text{ENTER}}$ key sequence to zoom in again. Successively repositioning the cursor and noting the x-values (the y-value should remain 0), you soon see that, to two decimal places, this x-intercept is 4.16.

To zoom back out to find the other x-intercept you can either use the $\boxed{\text{ZOOM}}$ $\boxed{3}$ $\boxed{\text{ENTER}}$ key sequence, or you can reset Xmin, Xmax, Ymin, and Ymax in the $\boxed{\text{WINDOW}}$ screen. For example, if you set Xmin $= -10$, Xmax $= 10$, Ymin $= -10$, and Ymax $= 10$ and then press the $\boxed{\text{GRAPH}}$ key you will see the following screen:

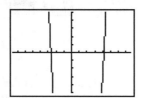

Repeating the process above to find the other x-intercept, you should determine that it is approximately $x = 2.637$. (If in the above process you lose the y-axis from the screen, it can always be recovered by resetting the Ymin and Ymax values in the $\boxed{\text{WINDOW}}$ menu.)

Notice in the picture above (where Xmin $= -10$, Xmax $= 10$, Ymin $= -10$, and Ymax $= 10$) that both the vertex and the y-intercept of the parabola are below the bottom of the calculator screen. To get a more complete view of the parabola, you could change Ymin using the $\boxed{\text{WINDOW}}$ menu to, say, Ymin $= -20$. This is better, but still not enough, so try again. Eventually, Ymin $= -40$ will work and you should see the following screen:

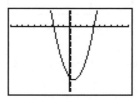

At this point, you can see the y-intercept and could approximate its coordinates using the $\boxed{\text{TRACE}}$ and $\boxed{\text{ZOOM}}$ keys as above when approximating the x-coordinates, but this is an overuse of the graphing capabilities of the calculator. Instead, simply use one of the techniques described earlier in Section 5.1 or 5.3 to find $f(0)$. In this case, it is easier to use Section 5.3, so in the $\boxed{\text{2ND}}$ $\boxed{\text{CALC}}$ window choose 1 : value. Enter the value for x, 0 for the y-intercept, and press $\boxed{\text{ENTER}}$. In the lower right hand portion of the screen you will see the value of $f(x)$, in this case Y $= -31.09$. You should see something like the following screen:

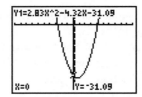

Next, to find the vertex of the parabola, you can use the ⌐TRACE⌐ and ⌐ZOOM⌐ keys as described above for finding the *x*-intercepts. You should find that the vertex is approximately at $(0.76, -32.74)$.

13. Constructing a Table

The table building capability of the calculator can be used to generate the data in a table. We illustrate this by considering part (d) of Problem 51 in Section 2.4 of the text. We begin by entering the function $R(D) = kD(P - D)$ as Y1 in the ⌐Y=⌐ window. To do this we let $k = 0.00003$, $P = 10,000$, and identify D with X. You should see the following screen:

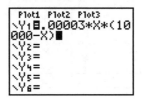

Next, use the ⌐2ND⌐ ⌐TBLSET⌐ key combination to change the Indpnt : setting from Auto to Ask. (The TblStart and ΔTbl setting do not matter.) You should now see the following window with Ask flashing after you press the ⌐ENTER⌐ key:

Now press the ⌐2ND⌐ ⌐TABLE⌐ key combination and enter 125 after X = at the bottom of the window, followed by the ⌐ENTER⌐ key. You should see 125 in the X column of the table and 37.031 in the Y1 column of the table. Round 37.031 to the nearest integer since it represents a number of persons, add it to 125, and enter the result 162 after X =. Press the ⌐ENTER⌐ key, and you should see

X	Y1	
125	37.031	
162	47.813	

X=162

Continuing in this fashion, you will generate the following table containing the number of infected individuals in the Y1 column:

X	Y1	
125	37.031	
162	47.813	
210	61.677	
272	79.38	
351	101.6	
453	129.74	

X=

14. Converting a Decimal to a Fraction

Selecting 1:►Frac in the MATH menu will sometimes convert a decimal to a fraction. The following screens illustrate this:

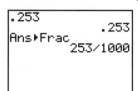 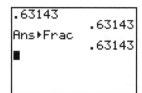 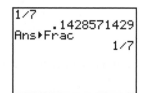 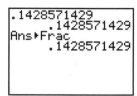

Notes: (1) After the decimal number is entered, press the MATH key, followed by 1   ENTER to obtain the fraction.

(2) As the second screen shows, not every decimal can be converted to a fraction. You can try experimenting with various decimals to see if you can determine those that will be converted to fractions.

(3) If the decimal arises as the quotient of two integers, as shown above in the third screen, the decimal result *may (but not always)* be converted back to the fraction. On the other hand, when the same decimal is entered directly, it may not be converted to a fraction. See the fourth screen above.

15. Arithmetic of Complex Numbers

It is a relatively simple task to have the calculator perform the four basic arithmetic operations of addition, subtraction, multiplication, and division on complex numbers. To do this, you first need to put the calculator in rectangular complex mode. Press the MODE key, scroll down to REAL, and then scroll right to $a + bi$. Then press the ENTER key.

To find $(2 - 4i) - (1 + 3i)$, first clear the calculator screen (see Section 3 in this manual); then the following sequence of keystrokes will cause the calculator to display the difference of the two complex numbers: (2   −   4   2ND   i   )   −   (1   +   3   2ND   i   )   ENTER .

```
(2-4i)-(1+3i)
            1-7i
```

A quotient of complex numbers with integer real and imaginary parts will usually result in a complex number with fractional (meaning noninteger) real and imaginary parts. The calculator will display the resulting quotient as a complex number with decimal real and imaginary parts. In this case, decimal to fraction conversion (see Section 14 of this manual) can be used. The following screen illustrates this in the computation of $(5 + 7i)/(4 - 2i)$:

```
(5+7i)/(4-2i)
          .3+1.9i
Ans►Frac
        3/10+19/10i
```

16. Rational Zeros of a Polynomial With Integer Coefficients

We want to find the rational zeros of a polynomial $f(x) = a_n x^n + a_{n-1} x^{n-1} + \cdots + a_1 x + a_0$, where $n \geq 1$ and a_i is an integer for $0 \leq i \leq n$. As discussed in the text, all rational zeros of $f(x)$ must have the form p/s where p is an integer factor of a_0 and s is an integer factor of a_n. With the calculator it is a simple matter to directly check each possible rational zero. Using the Y= key enter the function as Y1. In the screen below we are trying to find the rational zeros of $f(x) = 6x^4 + 11x^3 + 14x^2 - 7x - 6$.

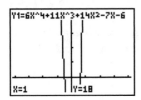

Now press the TRACE key. The set of possible rational zeros is ± 1, $\pm\frac{1}{2}$, $\pm\frac{1}{3}$, $\pm\frac{1}{6}$, ± 2, $\pm\frac{2}{3}$, ± 3, $\pm\frac{3}{2}$, and ± 6. To test -1 simply press $\boxed{(-)}$ $\boxed{1}$ $\boxed{\text{ENTER}}$ and read $Y = 10$. Thus, $f(1) = 10$, so 1 is not a zero of $f(x)$.

Next, press $\boxed{1}$ $\boxed{\text{ENTER}}$ and read $Y = 18$. Continuing in this manner we find that $-\frac{1}{2}$ and $\frac{2}{3}$ are the rational zeros of $f(x)$. Some sample screens are shown here:

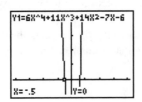 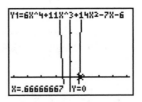 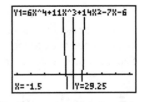 

You should be aware that the technique described above is very mechanical and will easily find the rational zeros of $f(x)$. It will not, however, find any irrational zeros, nor will it find any nonlinear factors of $f(x)$. Most significantly, the technique will not contribute to your understanding of the concepts involved in the theory of rational zeros of polynomials.

17. Calculation of the Other Trigonometric Functions

The keyboard of the calculator contains no keys for directly computing the cotangent, secant, cosecant, inverse cotangent, inverse secant, and inverse cosecant functions. To compute the cot, sec, and csc functions, we use the identities

$$\cot x = \frac{1}{\tan x}, \quad \sec x = \frac{1}{\cos x}, \quad \text{and} \quad \csc x = \frac{1}{\sin x}.$$

For example, to find $\cot 137°$, first be sure that the calculator is set to degree mode by pressing $\boxed{\text{MODE}}$, then scroll down to degree and press $\boxed{\text{ENTER}}$ to select degree mode:

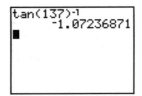

After using $\boxed{\text{CLEAR}}$ to return to the main window, use the keystrokes $\boxed{\text{TAN}}$ $\boxed{1}$ $\boxed{3}$ $\boxed{7}$ $\boxed{)}$ $\boxed{x^{-1}}$ to see that $\cot 137° = -1.07236871$. The screen output is shown here.

Alternatively, you could use $\boxed{1}$ $\boxed{\div}$ $\boxed{\text{TAN}}$ $\boxed{1}$ $\boxed{3}$ $\boxed{7}$ $\boxed{)}$ to obtain the same result.

To compute the arccotangent, arcsecant, and arccosecant functions, use the identities

$$\text{arccot } x = \frac{\pi}{2} - \arctan x, \quad \text{arcsec } x = \arccos\left(\frac{1}{x}\right), \quad \text{and} \quad \text{arccsc } x = \arcsin\left(\frac{1}{x}\right).$$

Functions

1.1 Functions and Graphs

1. $f(-5) = (-5)^2 - 1 = 25 - 1 = 24$

$f(-\sqrt{3}) = (-\sqrt{3})^2 - 1 = 3 - 1 = 2$

$f(3) = (3)^2 - 1 = 9 - 1 = 8$

$f(6) = (6)^2 - 1 = 36 - 1 = 35$

3. $f(-1) = \sqrt{-1+1} = \sqrt{0} = 0$

$f(0) = \sqrt{0+1} = \sqrt{1} = 1$

$f(3) = \sqrt{3+1} = \sqrt{4} = 2$

$f(5) = \sqrt{5+1} = \sqrt{6}$

5. $f(-1) = \dfrac{3(-1)}{(-1)^2 + 1} = \dfrac{-3}{1+1} = -\dfrac{3}{2}$

$f(0) = \dfrac{3(0)}{(0)^2 + 1} = 0$

$f(1) = \dfrac{3(1)}{(1)^2 + 1} = \dfrac{3}{2}$

$f(\sqrt{2}) = \dfrac{3(\sqrt{2})}{(\sqrt{2})^2 + 1} = \dfrac{3\sqrt{2}}{2+1} = \sqrt{2}$

7. $f(x) = -2x^2 + 3x$

$f(2a) = -2(2a)^2 + 3(2a) = -2(4a^2) + 6a = -8a^2 + 6a$

$f(a^2) = -2(a^2)^2 = 3(a^2) = -2a^4 + 3a^2$

$f(-5x) = -2(-5x)^2 + 3(-5x) = -2(25x^2) - 15x = -50x^2 - 15x$

$f(2a+1) = -2(2a+1)^2 + 3(2a+1) = -2(4a^2 + 4a + 1) + 6a + 3 = -8a^2 - 8a - 2 + 6a + 3$

$\qquad = -8a^2 - 2a + 1$

$f(x+h) = -2(x+h)^2 + 3(x+h) = -2(x^2 + 2xh + h^2) + 3x + 3h$

$\qquad = -2x^2 - 4xh - 2h^2 + 3x + 3h$

1

9. Setting $f(x) = 23$ and solving for x, we find

$$6x^2 - 1 = 23$$
$$6x^2 = 24$$
$$x^2 = 4$$
$$x = \pm 2.$$

When we compute $f(-2)$ and $f(2)$ we obtain 23 in both cases, so $x = \pm 2$ is the answer.

11. We need $4x - 2 \geq 0$:

$$4x \geq 2$$
$$x \geq \frac{1}{2}.$$

The domain is $[\frac{1}{2}, \infty)$.

13. We need $1 - x > 0$. This implies $x < 1$, so the domain is $(-\infty, 1)$.

15. The domain of $f(x) = (2x - 5)/x(x - 3)$ is the set of all x for which $x(x - 3) \neq 0$. Since $x(x - 3) = 0$ when $x = 0$ or $x = 3$, the domain of $f(x)$ is $\{x \mid x \neq 0, x \neq 3\}$.

17. We need $x^2 - 10x + 25 \neq 0$ or $(x - 5)^2 \neq 0$. Thus, $x \neq 5$ and the domain is $(-\infty, 5) \cup (5, \infty)$.

19. We need $x^2 - x + 1 \neq 0$. Applying the quadratic formula, we have

$$x = \frac{-(-1) \pm \sqrt{(-1)^2 - 4(1)(1)}}{2(1)} = \frac{1 \pm \sqrt{-3}}{2}.$$

Neither value of x is real, so $x^2 - x + 1 \neq 0$ for all x, and the domain is $(-\infty, \infty)$.

21. The domain of $f(x) = \sqrt{25 - x^2}$ is the set of all x for which $25 - x^2 \geq 0$. Since $25 - x^2 = (5 + x)(5 - x)$, we have $25 - x^2 \geq 0$ when

$$5 + x \geq 0 \text{ and } 5 - x \geq 0 \qquad \text{or} \qquad 5 + x \leq 0 \text{ and } 5 - x \leq 0.$$

Rewriting these inequalities for x, we get $x \geq -5$ and $x \leq 5$ for the first set of conditions or $x \leq -5$ and $x \geq 5$ for the second set of conditions. $x \leq -5$ and $x \geq 5$ at the same time can never happen, so the domain is determined by $-5 \leq x \leq 5$, or $[-5, 5]$.

23. We need $x^2 - 5x = x(x - 5) \geq 0$, which is true when

$$x \geq 0 \text{ and } x - 5 \geq 0 \qquad \text{or} \qquad x \leq 0 \text{ and } x - 5 \leq 0.$$

Rewriting these inequalities for x, we get $x \geq 0$ and $x \geq 5$ for the first set of conditions or $x \leq 0$ and $x \leq 5$ for the second set of conditions. The domain therefore requires that $x \geq 5$ or $x \leq 0$, so it is $(-\infty, 0] \cup [5, \infty)$.

25. We need $(3 - x)/(x + 2) \geq 0$ with $x \neq -2$, which is true when

$$3 - x \geq 0 \text{ and } x + 2 \geq 0 \qquad \text{or} \qquad 3 - x \leq 0 \text{ and } x + 2 \leq 0.$$

Rewriting these inequalities for x, we get $x \leq 3$ and $x \geq -2$ for the first set of conditions or $x \geq 3$ and $x \leq -2$ for the second set of conditions. $x \geq 3$ and $x \leq -2$ at the same time can never happen, so the domain is determined by $-2 \leq x \leq 3$ and $x \neq -2$, or $(-2, 3]$.

27. Since the y-axis (a vertical line) intersects the graph in more than one point (three points in this case), the graph is not that of a function.

29. This is the graph of a function by the vertical line test.

31. Projecting the graph onto the x-axis, we see that the domain is $[-4, 4]$. Projecting the graph onto the y-axis, we see that the range is $[0, 5]$.

33. Horizontally, the graph extends between $x = 1$ and $x = 9$ and terminates at both ends, as indicated by the solid dots. Thus, the domain is $[1, 9]$. Vertically, the graph extends between $y = 1$ and $y = 6$, so the range is $[1, 6]$.

35. ***x*-intercept:** Solving $\frac{1}{2}x - 4 = 0$ we get $x = 8$. The x-intercept is $(8, 0)$.

   ***y*-intercept:** Since $f(0) = \frac{1}{2}(0) - 4 = -4$, the y-intercept is $(0, -4)$.

37. ***x*-intercepts:** We solve $f(x) = 4(x-2)^2 - 1 = 0$:

$$4(x-2)^2 - 1 = 0$$
$$4(x-2)^2 = 1$$
$$(x-2)^2 = \frac{1}{4}$$
$$x - 2 = \pm\sqrt{\frac{1}{4}} = \pm\frac{1}{2}$$
$$x = 2 \pm \frac{1}{2}.$$

   Both of these check with the original equation, so the x-intercepts are $\left(\frac{3}{2}, 0\right)$ and $\left(\frac{5}{2}, 0\right)$.

   ***y*-intercept:** Since $f(0) = 4(0-2)^2 - 1 = 4(4) - 1 = 15$, the y-intercept is $(0, 15)$.

39. ***x*-intercepts:** We solve $f(x) = x^3 - x^2 - 2x = 0$:

$$x^3 - x^2 - 2x = 0$$
$$x(x^2 - x - 2) = 0$$
$$x(x+1)(x-2) = 0$$
$$x = 0, -1, 2.$$

   The x-intercepts are $(0, 0)$, $(-1, 0)$, and $(2, 0)$.

   ***y*-intercept:** Since $f(0) = 0$, the y-intercept is $(0, 0)$.

41. ***x*-intercepts:** We solve $x^2 + 4 = 0$. Since $x^2 + 4$ is never 0, there are no x-intercepts.

   ***y*-intercept:** Since $f(0) = [(0)^2 + 4]/[(0)^2 - 16] = 4/(-16) = -\frac{1}{4}$, the y-intercept is $\left(0, -\frac{1}{4}\right)$.

43. ***x*-intercepts:** We solve $f(x) = \frac{3}{2}\sqrt{4 - x^2} = 0$:

$$\frac{3}{2}\sqrt{4 - x^2} = 0$$
$$4 - x^2 = 0$$
$$(2 - x)(2 + x) = 0$$
$$x = \pm 2.$$

   The x-intercepts are $(-2, 0)$ and $(2, 0)$.

   ***y*-intercept:** Since $f(0) = \frac{3}{2}\sqrt{4 - (0)^2} = \frac{3}{2}(2) = 3$, the y-intercept is $(0, 3)$.

45. To find $f(a)$ for any number a, first locate a on the x-axis and then approximate the signed vertical distance to the graph from $(a, 0)$: $f(-3) \approx 0.5$ (because the graph is so steep at $x = -3$, $f(-3)$ could be reasonably approximated by any number from 0 to 1); $f(-2) \approx -3.4$; $f(-1) \approx 0.3$; $f(1) \approx 2$; $f(2) \approx 3.8$; $f(3) \approx 2.9$. The y-intercept is $(0, 2)$.

47. To find $f(a)$ for any number a, first locate a on the x-axis and then approximate the signed vertical distance to the graph from $(a, 0)$:

$$f(-2) \approx 3.6; \ f(-1.5) \approx 2; \ f(0.5) \approx 3.3; \ f(1) \approx 4.1; \ f(2) \approx 2; \ f(3.2) \approx -4.1.$$

   The x-intercepts are approximately $(-3.2, 0)$, $(2.3, 0)$, and $(3.8, 0)$.

49. Solving $x = y^2 - 5$ for y, we obtain

$$x = y^2 - 5$$
$$x + 5 = y^2$$
$$y^2 = x + 5$$
$$y = \pm\sqrt{x+5}.$$

The two functions are $f_1(x) = -\sqrt{x+5}$ and $f_2(x) = \sqrt{x+5}$. The domains are both $[-5, \infty)$.

51. (a) $f(2) = 2! = 2 \cdot 1 = 2$

$f(3) = 3! = 3 \cdot 2 \cdot 1 = 6$

$f(5) = 5! = 5 \cdot 4 \cdot 3 \cdot 2 \cdot 1 = 120$

$f(7) = 7! = 7 \cdot 6 \cdot 5 \cdot 4 \cdot 3 \cdot 2 \cdot 1 = 5040$

Note that we could have simplified the computation of 7! in this case by writing

$$7! = 7 \cdot 6 \cdot 5! = 7 \cdot 6 \cdot 120 = 5040.$$

(b) $f(n+1) = (n+1)! = (n+1)n! = n!(n+1) = f(n)(n+1)$

(c) Using the result from (b), we can simplify as follows:

$$\frac{f(5)}{f(4)} = \frac{f(4) \cdot 5}{f(4)} = 5$$
$$\frac{f(7)}{f(5)} = \frac{f(6) \cdot 7}{f(5)} = \frac{f(5) \cdot 6 \cdot 7}{f(5)} = 42.$$

(d) $\dfrac{f(n+3)}{f(n)} = \dfrac{(n+3)!}{n!} = \dfrac{(n+3)(n+2)(n+1)n!}{n!} = (n+1)(n+2)(n+3)$

53. Generally, when the domain is a semi-infinite interval, the function can be one of $\sqrt{x-a}$, or $\sqrt{a-x}$. One of these should work when the number a is included in the interval. To exclude a number a, simply use the reciprocal of $\sqrt{x-a}$, or $\sqrt{a-x}$.

(a) We try $f(x) = \sqrt{3-x}$. Since $x = 0$ is in the domain of this function, but not in the interval $[3, \infty)$, this is not the correct choice for $f(x)$. We then try $f(x) = \sqrt{x-3}$ and see that it does work.

(b) Since 3 is not part of the interval, we let $f(x) = 1/\sqrt{x-3}$. This function has domain $(3, \infty)$.

55. The graph indicates that $f(x) < 0$ when $x < -1$ or $x > 3$, so the domain of $g(x) = \sqrt{f(x)}$ must be $[-1, 3]$. Within this domain, $0 \leq f(x) \leq 4$, so the range of $g(x)$ must therefore be $[0, 2]$.

57. $g(x) = \lceil x \rceil = \begin{cases} \vdots \\ -2, & -3 < x \leq -2 \\ -1, & -2 < x \leq -1 \\ 0, & -1 < x \leq 0 \\ 1, & 0 < x \leq 1 \\ 2, & 1 < x \leq 2 \\ 3, & 2 < x \leq 3 \\ \vdots \end{cases}$

59.

61. Since $\dfrac{x^2-9}{x-3} = \dfrac{(x+3)(x-3)}{x-3} = x+3, \quad x \neq 3,$

    the graph of $f(x)$ is a line with a hole at $x = 3$. The graph of $g(x)$ is the same line with a hole at $x = 3$ and a dot at $(3,4)$. Since $x+3 = 6$ when $x = 3$, the graph of $h(x)$ is the same line with no holes.

1.2 Combining Functions

1. $(f+g)(x) = -2x+13 \qquad (f-g)(x) = 6x-3$

   $(fg)(x) = -8x^2-4x+40 \quad (f/g)(x) = \dfrac{2x+5}{-4x+8}, \, x \neq 2$

3. $(f+g)(x) = \dfrac{x^2+x+1}{x^2+x} \qquad (f-g)(x) = \dfrac{x^2-x-1}{x^2+x}$

   $(fg)(x) = \dfrac{1}{x+1}, x \neq 0 \quad (f/g)(x) = \dfrac{x^2}{x+1}, x \neq -1, 0$

5. $(f+g)(x) = 2x^2+5x-7 \qquad\qquad (f-g)(x) = -x+1$

   $(fg)(x) = x^4+5x^3-x^2-17x+12 \quad (f/g)(x) = \dfrac{x+3}{x+4}, x \neq -4, 1$

7. $(f+g)(x) = \sqrt{x-1}+\sqrt{2-x}$; the domain is $[1,2]$.

9. $(f/g)(x) = \sqrt{x-1}/\sqrt{2-x}$; the domain is $[1,2)$.

11. $(f \circ g)(x) = 3x+16; \quad (g \circ f)(x) = 3x+4$

13. $(f \circ g)(x) = x^6+2x^5+x^4; \quad (g \circ f)(x) = x^6+x^4$

15. $(f \circ g)(x) = \dfrac{3x+3}{x}; \quad (g \circ f)(x) = \dfrac{3}{3+x}$

17. $(f \circ g)(x) = f(x^2+2) = \sqrt{(x^2+2)-3} = \sqrt{x^2-1}$

    The domain of f, determined by $x \geq 3$, is $[3,\infty)$. Since the domain of g is all real numbers and $g(x) = x^2+2 \geq 3$ when either $x \leq -1$ or $x \geq 1$, the domain of $f \circ g$ is $(-\infty, -1] \cup [1,\infty)$.

19. $(g \circ f)(x) = g(5-x^2) = 2-\sqrt{5-x^2}$

    The domain of g, determined by $x \geq 0$, is $[0,\infty)$. Since the domain of f is all real numbers and $f(x) = 5-x^2 \geq 0$ when $-\sqrt{5} \leq x \leq \sqrt{5}$, the domain of $g \circ f$ is $[-\sqrt{5}, \sqrt{5}]$.

21. $(f \circ (2f))(x) = 2(4x^3)^3 = 128x^9; \quad (f \circ (1/f))(x) = 2\left(\dfrac{1}{2x^3}\right)^3 = \dfrac{1}{4x^9}$

23. $(f \circ g \circ h)(x) = 36x^2-36x+15$

25. $2g(x)-5 = -4x+13; \quad g(x) = -2x+9$

27. $f(x) = 2x^2-x; \quad g(x) = x^2$

29. The point (x,y) on the graph of f corresponds to the point $(x, y+2)$ on the shifted graph. Thus, $(-2,1)$ corresponds to $(-2,3)$ and $(3,-4)$ corresponds to $(3,-2)$.

31. The point (x,y) on the graph of f corresponds to the point $(x-6, y)$ on the shifted graph. Thus, $(-2,1)$ corresponds to $(-8,1)$ and $(3,-4)$ corresponds to $(-3,-4)$.

33. The point (x,y) on the graph of f corresponds to the point $(x-4, y+1)$ on the shifted graph. Thus, $(-2,1)$ corresponds to $(-6,2)$ and $(3,-4)$ corresponds to $(-1,-3)$.

35. The point (x,y) on the graph of f corresponds to the point $(-x, y)$ on the shifted graph. Thus, $(-2,1)$ corresponds to $(2,1)$ and $(3,-4)$ corresponds to $(-3,-4)$.

37. In (a) the graph is shifted up 2 units; in (b) it is shifted down 2 units; in (c) it is shifted left 2 units; in (d) it is shifted right 5 units; in (e) it is reflected in the x-axis; and in (f) it is reflected in the y-axis.

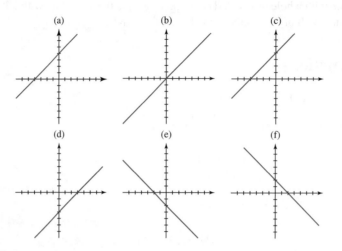

39. In (a) the graph is shifted up 2 units; in (b) it is shifted down 2 units; in (c) it is shifted left 2 units; in (d) it is shifted right 5 units; in (e) it is reflected in the x-axis; and in (f) it is reflected in the y-axis.

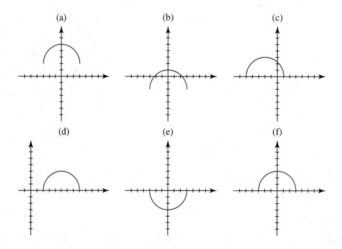

41. In (a) the graph is shifted up 1 unit; in (b) it is shifted down 1 unit; in (c) it is shifted left π units; in (d) it is shifted right $\pi/2$ units; in (e) it is reflected in the x-axis; in (f) it is reflected in the y-axis; in (g) it is stretched vertically by a factor of 3; and in (h) it is compressed vertically by a factor of $1/2$ and then reflected in the x-axis.

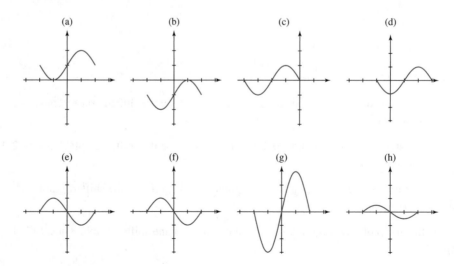

43. If $f(x)$ is shifted up 5 units and right 1 unit, the new function is $y = f(x-1)+5$. Since $f(x) = x^3$, this becomes $y = (x-1)^3 + 5$.

45. The function is first multiplied by -1 and then x is replaced by $x+7$. The equation of the new graph is thus $y = -(x+7)^4$.

47. (a) An even function is symmetric with respect to the y-axis.

(b) An odd function is symmetric with respect to the origin.

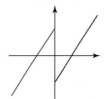

49. To fill in the bottom row, we use the fact that since f is an even function then $f(-x) = f(x)$, and so $(f \circ g)(-x) = (f \circ g)(x)$.

x	0	1	2	3	4
$f(x)$	-1	2	10	8	0
$g(x)$	2	-3	0	1	-4
$(f \circ g)(x)$	10	8	-1	2	0

51.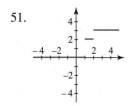

53. $2U(x-0) - 2U(x-2) + (-1)U(x-2) - (-1)U(x-3) = 2 - 3U(x-2) + U(x-3)$

55.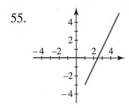

57. False; let $f(x) = x^2$ and $g(x) = h(x) = 1$.

59. If a function is symmetric with respect to the x-axis, then, for any x, both (x, y) and $(x, -y)$ are on the graph. If $y \neq 0$ (as is the case for a function that is nonzero for at least one value of x), then the vertical line test fails and the graph is not the graph of a function.

61. Since $|x| = x$ when $x \geq 0$ and $|x| = -x$ when $x < 0$, the graph of $f(|x|)$ is the same as the graph of $f(x)$ when $x \geq 0$. When $x < 0$, $f(|x|) = f(-x)$, so the graph of $f(|x|)$ in this case is the reflection of the graph of $f(x), x > 0$, in the y-axis. To summarize: to obtain the graph of $f(|x|)$ from the graph of $f(x)$, simply ignore the portion of the graph of $f(x)$ to the left of the y-axis, and then reflect the portion of the graph of $f(x)$ to the right of the y-axis through the y-axis.

63. frac(x) is so named because its value is the noninteger part of x; i.e., the value that follows the decimal point. Its graph is shown on the right.

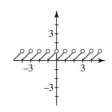

1.3 Polynomial and Rational Functions

1. The form of the equation of the line is $y = \frac{2}{3}x + b$. Letting $x = 1$ and $y = 2$, we have $2 = \frac{2}{3} + b$, so $b = 2 - \frac{2}{3} = \frac{4}{3}$. The equation of the line is $y = \frac{2}{3}x + \frac{4}{3}$.

3. A line with slope 0 is horizontal and has the form $y = b$. In this case, $b = 2$, so the equation is $y = 2$.

5. Letting $m = -1$, $x_1 = 1$, and $y_1 = 2$, we obtain from the point-slope form of the equation of a line that

$$y - 2 = -1(x - 1) = -x + 1$$
$$y = -x + 3.$$

7. To find the x-intercept we set $y = 0$. This gives $3x + 12 = 0$ or $x = -4$. The x-intercept is $(-4, 0)$. Now, write the equation in slope-intercept form by solving for y:

$$3x - 4y + 12 = 0$$
$$-4y = -3x - 12$$
$$y = \frac{3}{4}x + 3.$$

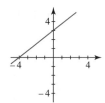

The slope of the line is $m = \frac{3}{4}$ and the y-intercept is $(0, 3)$.

9. Solving for y, we have $3y = 2x - 9$ or $y = \frac{2}{3}x - 3$, so the slope is $\frac{2}{3}$ and the y-intercept is $(0, -3)$. Setting $y = 0$, we have $0 = \frac{2}{3}x - 3$ or $x = \frac{9}{2}$, so the x-intercept is $\left(\frac{9}{2}, 0\right)$.

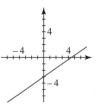

11. The slope of the line is $m = \dfrac{-5 - 3}{6 - 2} = \dfrac{-8}{4} = -2$. The form of the equation of the line is $y = -2x + b$. Letting $x = 2$ and $y = 3$, we have $3 = -2(2) + b$, so $b = 3 + 4 = 7$. The equation of the line is $y = -2x + 7$.

13. Solving $3x + y - 5 = 0$ for y, we obtain $y = -3x + 5$. The slope of this line is $m = -3$, so the form of the line through $(-2, 4)$ is $y = -3x + b$. Letting $x = -2$ and $y = 4$, we have $4 = -3(-2) + b$, so $b = 4 - 6 = -2$ and the equation of the line is $y = -3x - 2$.

15. Solving for $x - 4y + 1 = 0$ for y, we obtain $-4y = -x - 1$ and $y = \frac{1}{4}x + \frac{1}{4}$. Thus, the slope of a line perpendicular to this one is $m = -1/(1/4) = -4$. We identify $x_1 = 2$ and $y_1 = 3$ and use the point-slope form of the equation of a line:

$$y - 3 = -4(x - 2) = -4x + 8$$
$$y = -4x + 11.$$

17. Letting $x = -1$, $y = 5$, and $x = 1$, $y = 6$ in $f(x) = ax + b$, we have

$$
\begin{aligned}
5 &= a(-1) + b \\
6 &= a(1) + b
\end{aligned}
\qquad \text{or} \qquad
\begin{aligned}
-a + b &= 5 \\
a + b &= 6.
\end{aligned}
$$

Adding these equations, we find $2b = 11$, so $b = \frac{11}{2}$. Then $a = 6 - b = 6 - \frac{11}{2} = \frac{1}{2}$ and the function is $f(x) = \frac{1}{2}x + \frac{11}{2}$.

19. When $x = -1$, the corresponding point on the blue curve has y-coordinate $y = (-1)^2 + 1 = 2$. When $x = 2$, the corresponding point on the blue curve has y-coordinate $y = 2^2 + 1 = 5$. The slope of the line through $(-1, 2)$ and $(2, 5)$ is $m = (5 - 2)/[2 - (-1)] = 3/3 = 1$. The form of the equation of the line is then $y = 1x + b = x + b$. Using $x = -1$ and $y = 2$ we have $2 = -1 + b$ or $b = 3$. Thus, the equation of the line is $y = x + 3$.

21. (a) **x-intercepts:** Solving $x(x + 5) = 0$ we get $x = 0, -5$, so the x-intercepts are $(0, 0)$ and $(-5, 0)$.
    y-intercept: Since $f(0) = 0$, the y-intercept is $(0, 0)$.

    (b) $f(x) = x(x + 5) = x^2 + 5x = \left[x^2 + 5x + \left(\frac{5}{2}\right)^2\right] - \left(\frac{5}{2}\right)^2 = (x + 5/2)^2 - (25/4)$

(c) Identifying $h = -\dfrac{5}{2}$ and $k = -\dfrac{25}{4}$ in part (b), we see that the vertex is $\left(-\dfrac{5}{2}, -\dfrac{25}{4}\right)$ and the axis of symmetry is $x = -\dfrac{5}{2}$.

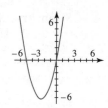

(d) Since $a = 1 > 0$ in part (b), the parabola opens up.

(e) The range of $f(x)$ is $[-25/4, \infty)$.

(f) $f(x)$ is increasing on $[-5/2, \infty)$ and decreasing on $(-\infty, -5/2]$.

23. (a) **x-intercepts:** Solving $(3-x)(x+1) = 0$ we get $x = 3, -1$, so the x-intercepts are $(-1, 0)$ and $(3, 0)$.
    y-intercept: Since $f(0) = (3-0)(0+1) = 3$, the y-intercept is $(0, 3)$.

    (b) $f(x) = (3-x)(x+1) = -x^2 + 2x + 3 = -(x^2 - 2x) + 3$
    $$= -(x^2 - 2x + 1) + 3 + 1 = -(x-1)^2 + 4$$

    (c) Identifying $h = 1$ and $k = 4$ in part (b), we see that the vertex is $(1, 4)$ and the axis of symmetry is $x = 1$.

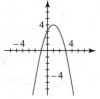

    (d) Since $a = -1 < 0$ in part (b), the parabola opens down.

    (e) The range of $f(x)$ is $(-\infty, 4]$.

    (f) $f(x)$ is increasing on $(-\infty, 1]$ and decreasing on $[1, \infty)$.

25. (a) **x-intercepts:** Solving $x^2 - 3x + 2 = (x-1)(x-2) = 0$ we get $x = 1, 2$, so the x-intercepts are $(1, 0)$ and $(2, 0)$.
    y-intercept: Since $f(0) = 2$, the y-intercept is $(0, 2)$.

    (b) $f(x) = x^2 - 3x + 2 = [x^2 - 3x + (-3/2)^2] - (-3/2)^2 + 2 = (x - 3/2)^2 - (1/4)$

    (c) Identifying $h = 3/2$ and $k = -1/4$ in part (b), we see that the vertex is $(3/2, -1/4)$ and the axis of symmetry is $x = 3/2$.

    (d) Since $a = 1 > 0$ in part (b), the parabola opens up.

    (e) The range of $f(x)$ is $[-1/4, \infty)$.

    (f) $f(x)$ is increasing on $[3/2, \infty)$ and decreasing on $(-\infty, 3/2]$.

27. The vertex of the graph is $(-10, 0)$, so the graph of $f(x)$ is the graph of $y = x^2$ shifted to the right by 10 units.

29. The vertex of the graph is $(-4, 9)$, so the graph of $f(x)$ is the graph of $y = x^2$ reflected in the x-axis, compressed by a factor of $1/3$, shifted to the left by 4 units, and shifted up by 9 units.

31. Since $f(x) = (-x - 6)^2 - 4 = (x + 6)^2 - 4$, the vertex of the graph is $(-6, -4)$, so the graph of $f(x)$ is the graph of $y = x^2$ shifted to the left by 6 units and shifted down by 4 units.

33. **End behavior:** For large x, the graph is like that of $y = x^3$.

    Symmetry: Since the powers of x are all odd, the graph is symmetric with respect to the origin.

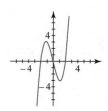

    Intercepts: Since $f(0) = 0$, the y-intercept is $(0, 0)$. Solving $f(x) = x^3 - 4x = x(x-2)(x+2) = 0$, we see that the x-intercepts are $(-2, 0)$, $(0, 0)$, and $(2, 0)$.

    Graph: From $f(x) = x^1(x-2)^1(x+2)^1$ we see that 0, 2, and -2 are all simple zeros.

35. **End behavior:** For large x, the graph is like that of $y = -x^3$.

    Symmetry: Since the powers of x are both even and odd, the graph has no symmetry with respect to the origin or y-axis.

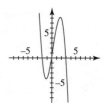

    Intercepts: Since $f(0) = 0$, the y-intercept is $(0, 0)$. Solving $f(x) = -x^3 + x^2 + 6x = -x(x+2)(x-3) = 0$, we see that the x-intercepts are $(0, 0)$, $(-2, 0)$, and $(3, 0)$.

    Graph: From $f(x) = -x^1(x+2)^1(x-3)^1$ we see that 0, -2, and 3 are all simple zeros.

37. **End behavior:** For large x, the graph is like that of $y = x^3$.

Symmetry: Since the powers of x in $f(x) = (x+1)(x-2)(x-4) = x^3 - 5x^2 + 2x + 8$ are both even and odd, the graph has no symmetry with respect to the origin or y-axis.

Intercepts: Since $f(0) = 8$, the y-intercept is $(0,8)$. Solving $f(x) = (x+1)(x-2)(x-4) = 0$, we see that the x-intercepts are $(-1,0)$, $(2,0)$, and $(4,0)$.

Graph: From $f(x) = (x+1)^1(x-2)^1(x-4)^1$ we see that -1, 2, and 4 are all simple zeros.

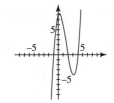

39. **End behavior:** For large x, the graph is like that of $y = x^4$.

    Symmetry: Since the powers of x are both even and odd, the graph has no symmetry with respect to the origin or y-axis.

    Intercepts: Since $f(0) = 0$, the y-intercept is $(0,0)$. Solving $f(x) = x^4 - 4x^3 + 3x^2 = x^2(x-1)(x-3) = 0$, we see that the x-intercepts are $(0,0)$, $(1,0)$, and $(3,0)$.

    Graph: From $f(x) = x^2(x-1)^1(x-3)^1$ we see that 1 and 3 are simple zeros and the graph is tangent to the x-axis at $x = 0$.

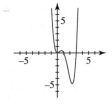

41. **End behavior:** For large x, the graph is like that of $y = -x^4$.

    Symmetry: Since the powers of x are all even, the graph is symmetric with respect to the y-axis.

    Intercepts: Since $f(0) = -1$, the y-intercept is $(0,-1)$. Solving $f(x) = -(x^4 - 2x^2 + 1) = -(x+1)^2(x-1)^2 = 0$, we see that the x-intercepts are $(-1,0)$ and $(1,0)$.

    Graph: From $f(x) = -(x+1)^2(x-1)^2$ we see that the graph is tangent to the x-axis at $x = -1$ and $x = 1$.

To solve Problems 43–48, first note that all of the functions have zeros at $x = 0$ and $x = 1$. Next, note whether the exponents of x and $x-1$ are 1, even, or odd and greater than 1. Finally, use the facts that when the exponent of $x - a$ is

1: the graph passes directly through the x-axis at $x = a$;

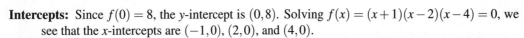

even: the graph is tangent to, but does not pass through the x-axis at $x = a$;

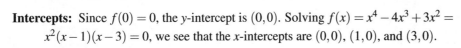

odd and greater than 1: the graph is tangent to, and passes through the x-axis.

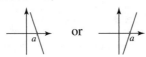

The forms of the functions in (a)–(f): are

    (a) $x^{\text{even}}(x-1)^{\text{even}}$ (b) $x^{\text{odd}}(x-1)^1$
    (c) $x^{\text{odd}}(x-1)^{\text{odd}}$ (d) $x^1(x-1)^{\text{odd}}$
    (e) $x^{\text{even}}(x-1)^1$ (f) $x^{\text{odd}}(x-1)^{\text{even}}$

Because each of these is distinct, it is not necessary for this set of problems to consider whether the lead coefficient is positive or negative.

43. The form of the function must be $x^{\text{odd}}(x-1)^{\text{even}}$, so this graph corresponds to (f).

45. The form of the function must be $x^{\text{even}}(x-1)^1$, so this graph corresponds to (e).

47. The form of the function must be $x^{\text{odd}}(x-1)^1$, so this graph corresponds to (b).

49. **Vertical asymptotes:** Setting $2x + 3 = 0$, we see that $x = -3/2$ is a vertical asymptote.

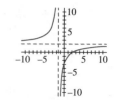

    Horizontal asymptote: The degree of the numerator equals the degree of the denominator, so $y = 4/2 = 2$ is the horizontal asymptote.

    Intercepts: Since $f(0) = -3$, the y-intercept is $(0, -3)$. Setting $4x - 9 = 0$, we see that $x = 9/4$, so $(9/4, 0)$ is the x-intercept.

51. **Vertical asymptote:** Setting $(x - 1)^2 = 0$, we see that $x = 1$ is the vertical asymptote.

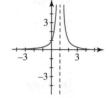

    Horizontal asymptote: The degree of the numerator is less than the degree of the denominator, so $y = 0$ is the horizontal asymptote.

    Intercepts: Since $f(0) = 1$, the y-intercept is $(0, 1)$. The numerator is never zero, so there are no x-intercepts.

53. **Vertical asymptotes:** Setting $x^2 - 1 = (x + 1)(x - 1) = 0$, we see that $x = -1$ and $x = 1$ are vertical asymptotes.

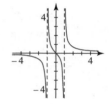

    Horizontal asymptote: The degree of the numerator is less than the degree of the denominator, so $y = 0$ is the horizontal asymptote.

    Intercepts: Since $f(0) = 0$, the y-intercept is $(0, 0)$. Since the numerator is simply x, $(0, 0)$ is also the only x-intercept.

    Graph: We use the facts that the only x-intercept is $(0, 0)$ and the x-axis is a horizontal asymptote. For $x < -1$, $f(x) < 0$, so the left branch is below the x-axis. For $-1 < x < 0$, $f(x) > 0$, and for $0 < x < 1$, $f(x) < 0$, so the middle branch passes through the origin (as opposed to being tangent to the origin and lying strictly above or below the x-axis). For $x > 1$, $f(x) > 0$, so the right branch is above the x-axis.

55. **Vertical asymptotes:** Setting $x^2 = 0$, we see that $x = 0$, or the y-axis is a vertical asymptote.

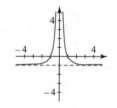

    Horizontal asymptote: The degree of the numerator equals the degree of the denominator, so $y = -1/1 = -1$ is the horizontal asymptote.

    Intercepts: Since the y-axis is a vertical asymptote, there is no y-intercept. Setting $1 - x^2 = (1 + x)(1 - x) = 0$, we see that $x = -1$ and $x = 1$, so $(-1, 0)$ and $(1, 0)$ are x-intercepts.

57. **Vertical asymptotes:** Setting the denominator equal to zero, we see that $x = 0$, or the y-axis is a vertical asymptote.

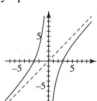

    Slant asymptote: Since the degree of the numerator is one greater than the degree of the denominator, the graph of $f(x)$ possesses a slant asymptote. From $f(x) = (x^2 - 9)/x = x - 9/x$, we see that $y = x$ is a slant asymptote.

    Intercepts: Since the y-axis is a vertical asymptote, the graph has no y-intercept. Setting $x^2 - 9 = (x + 3)(x - 3) = 0$, we see that $x = -3$ and $x = 3$, so $(-3, 0)$ and $(3, 0)$ are the x-intercepts.

    Graph: We need to determine if the graph crosses the slant asymptote. To do this, we solve

$$\frac{x^2 - 9}{x} = x; \quad x^2 - 9 = x^2; \quad -9 = 0.$$

    Since there is no solution, the graph does not cross its slant asymptote.

59. **Vertical asymptotes:** Setting $x + 2 = 0$, we see that $x = -2$ is a vertical asymptote.

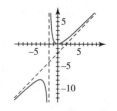

    Slant asymptote: Since the degree of the numerator is one greater than the degree of the denominator, the graph of $f(x)$ possesses a slant asymptote. Using synthetic division, we see that $f(x) = x^2/(x + 2) = x - 2 + 4/(x + 2)$, and the slant asymptote is $y = x - 2$.

    Intercepts: Since $f(0) = 0$, the y-intercept is $(0, 0)$. Setting $x^2 = 0$, we see that $x = 0$, so $(0, 0)$ is also the only x-intercept.

Graph: We need to determine if the graph crosses the slant asymptote. To do this, we solve

$$\frac{x^2}{x+2} = x-2; \quad x^2 = x^2 - 4; \quad 0 = -4.$$

Since there is no solution, the graph does not cross its slant asymptote.

61. **Vertical asymptotes:** Setting $x - 1 = 0$, we see that $x = 1$ is a vertical asymptote.

    Slant asymptote: Since the degree of the numerator is one greater than the degree of the denominator, the graph of $f(x)$ possesses a slant asymptote. Using synthetic division, we see that $f(x) = (x^2 - 2x - 3)/(x-1) = x - 1 + 4/(x-1)$, and the slant asymptote is $y = x - 1$.

    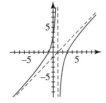

    Intercepts: Since $f(0) = -3/(-1) = 3$, the y-intercept is $(0,3)$. Setting $(x+1)(x-3) = 0$, we see that $x = -1$ and $x = 3$ or $(-1,0)$ and $(3,0)$ are the x-intercepts.

    Graph: We can just about find the graph from the asymptotes and intercepts, but we need to determine if the graph crosses the slant asymptote. To do this, we solve

    $$\frac{x^2 - 2x - 3}{x-1} = x-1; \quad x^2 - 2x - 3 = x^2 - 2x + 1; \quad -3 = 1.$$

    Since there is no solution, the graph does not cross its slant asymptote.

63. Set $f(x) = -1$:   Set $f(x) = 2$:

    $$\frac{2x-1}{x+4} = -1 \qquad \frac{2x-1}{x+4} = 2$$

    $$2x - 1 = -x - 4 \qquad 2x - 1 = 2x + 8$$

    $$x = -1 \qquad\qquad -1 = 8$$

    Thus, -1 is in the range and 2 is not.

65. Begin by calculating $\dfrac{\Delta T_F}{\Delta T_C} = \dfrac{140 - 32}{60 - 0} = \dfrac{9}{5}$. Then:

    $$T_F - 32 = \frac{9}{5}(T_C - 0)$$
    $$T_C = \frac{9}{5}T_C + 32$$

    Try it out: When $T_C = 100$, $T_F = \dfrac{9}{5}(100) + 32 = 212$.

67. Identifying $t = 20$, $P = 1000$, and $r = 0.034$, we have

    $$A = P + Prt = 1000 + 1000(0.034)(20) = 1000 + 680 = \$1680.$$

    Assuming that P and r remain the same, we solve $220 = 1000 + 1000(0.034)t$ for t. This gives $t = 12/0.34 \approx 35.29$ years.

69. The ball is on the ground when $s(t) = 0$. Solving $-16t^2 + 96t = 0$, we find $t = 0$ seconds and 6 seconds.

71. The slope of a line is its rate of change, which means the change in output for each (positive) unit change in input. In this case, the slope is $5/2$, so when x is changed by one unit, y will change by $5/2 = 2.5$ units. When x is changed by 2 units, y will change by $2(5/2) = 5$ units, and when x is changed by n units, y will change by $(5/2)n$ units.

73. First, find the slope of the line through $(\frac{1}{2}, 10)$ and $(\frac{3}{2}, 4)$; this is $m = -6$. The slope of a perpendicular line is then $-1/m$; in this case, $\frac{1}{6}$. To find the point the line passes through, find the midpoint of $(\frac{1}{2}, 10)$ and $(\frac{3}{2}, 4)$; this is $(1, 7)$. Now, find the equation of the line through the midpoint with slope $-1/m$; this is $y = \frac{1}{6}x + \frac{41}{6}$.

1.4 Transcendental Functions

1.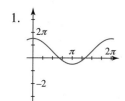

3.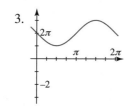

5.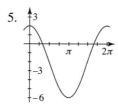

7. The amplitude of $y = 4\sin\pi x$ is $A = 4$ and the period is $\dfrac{2\pi}{\pi} = 2$.

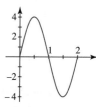

9. The amplitude of $y = -3\cos 2\pi x$ is $A = |-3| = 3$ and the period is $\dfrac{2\pi}{2\pi} = 1$.

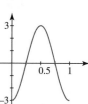

11. The amplitude of $y = 2 - 4\sin x$ is $A = |-4| = 4$ and the period is $\dfrac{2\pi}{1} = 2\pi$.

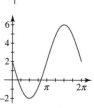

13. The amplitude of $y = 1 + \cos\dfrac{2x}{3}$ is $A = 1$ and the period is $\dfrac{2\pi}{\left(\frac{2}{3}\right)} = 3\pi$.

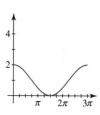

15. We begin with $y = A\sin x$ since the graph flattens out at $\pi/2$ and $3\pi/2$. The amplitude is $\frac{1}{2}[3-(-3)] = 3$ and the graph has been reflected through the line $y = 0$, so $A = -3$. Thus, $y = -3\sin x$.

17. Since the graph flattens out at $x = 0$, we use $y = A\cos x + D$. The amplitude is $\frac{1}{2}[4-(-2)] = 3$ and the graph has been reflected through the line $y = 1$, so $A = -3$ and $D = 1$. Thus $y = -3\cos x + 1$.

19. Since the y-intercept is $(0,0)$, the equation has the form $y = A\sin Bx$. The amplitude of the graph is $A = 3$ and the period is $\pi = 2\pi/B$, so $B = 2$ and $y = 3\sin 2x$.

21. Since the y-intercept is $(0,\frac{1}{2})$ and not $(0,0)$, the equation has the form $y = A\cos Bx$. The amplitude of the graph is $A = \frac{1}{2}$ and the period is $2 = 2\pi/B$, so $B = \pi$ and $y = \frac{1}{2}\cos\pi x$.

23. Since the y-intercept is $(0,0)$, the equation has the form $y = A\sin Bx$. The amplitude of the graph is 1, and the graph has been reflected through the line $y = 0$, so $A = 1$. The period is $2 = 2\pi/B$, so $B = \pi$ and $y = -\sin\pi x$.

25. The amplitude of $y = \sin(x - \pi/6)$ is $A = 1$ and the period is $2\pi/1 = 2\pi$. The phase shift is $|-\pi/6|/1 = \pi/6$. Since $C = -\pi/6 < 0$, the shift is to the right.

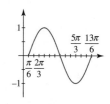

27. The amplitude of $y = \cos(x + \pi/4)$ is $A = 1$ and the period is $2\pi/1 = 2\pi$. The phase shift is $|\pi/4|/1 = \pi/4$. Since $C = \pi/4 > 0$, the shift is to the left.

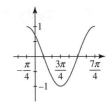

29. The amplitude of $y = 4\cos(2x - 3\pi/2)$ is $A = 4$ and the period is $2\pi/2 = \pi$. The phase shift is $|-3\pi/2|/2 = 3\pi/4$. Since $C = -3\pi/2 < 0$, the shift is to the right.

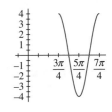

31. The amplitude of $y = 3\sin(x/2 - \pi/3)$ is $A = 3$ and the period is $2\pi/(1/2) = 4\pi$. The phase shift is $|-\pi/3|/(1/2) = 2\pi/3$. Since $C = -\pi/3 < 0$, the shift is to the right.

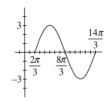

33. The amplitude of $y = -4\sin(\pi x/3 - \pi/3)$ is $A = |-4| = 4$ and the period is $2\pi/(\pi/3) = 6$. The phase shift is $|-\pi/3|/(\pi/3) = 1$. Since $C = -\pi/3 < 0$, the shift is to the right.

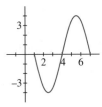

35. $y = 5\sin\left[\pi\left(x - \dfrac{1}{2}\right)\right] = 5\sin\left(\pi x - \dfrac{\pi}{2}\right)$

37. Setting $-1 + \sin x = 0$, we have $\sin x = 1$, which is true for $x = \pi/2$ in $[0, 2\pi]$. By periodicity, the x-intercepts of $-1 + \sin x$ are $(\pi/2 + 2n\pi, 0)$, where n is an integer.

39. Setting $\sin \pi x = 0$, we have by (3) in this section of the text that the x-intercepts are determined by $\pi x = n\pi$, n an integer. Thus, $x = n$, and the x-intercepts are $(n, 0)$, where n is an integer.

41. Setting $10\cos(x/2) = 0$, we have by (4) in this section of the text that the x-intercepts are determined by $x/2 = (2n+1)\pi/2$, n an integer. Thus, $x = (2n+1)\pi$, so the intercepts are $(\pi + 2n\pi, 0)$, where n is an integer.

43. Setting $\sin(x - \pi/4) = 0$, we have by (3) in this section of the text that the x-intercepts are determined by $x - \pi/4 = n\pi$, n an integer. Thus, $x = n\pi + \pi/4 = (n + 1/4)\pi$, and the x-intercepts are $(\pi/4 + n\pi, 0)$, where n is an integer.

45. The period of $y = \tan \pi x$ is $\pi/\pi = 1$. Since

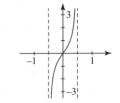

$$\tan \pi x = \frac{\sin \pi x}{\cos \pi x},$$

the x-intercepts of $\tan \pi x$ occur at the zeros of $\sin \pi x$; namely, at $\pi x = n\pi$ or $x = n$ for n an integer. The vertical asymptotes occur at the zeros of $\cos \pi x$; namely, at

$$\pi x = \frac{(2n+1)\pi}{2} \qquad \text{or} \qquad x = \frac{1}{2}(2n+1) \qquad \text{for } n \text{ an integer.}$$

47. The period of $y = \cot 2x$ is $\pi/2$. Since

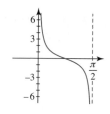

$$\cot 2x = \frac{\cos 2x}{\sin 2x},$$

the x-intercepts of $\cot 2x$ occur at the zeros of $\cos 2x$; namely, at

$$2x = \frac{(2n+1)\pi}{2} \qquad \text{or} \qquad x = \frac{(2n+1)\pi}{4} \qquad \text{for } n \text{ an integer.}$$

The vertical asymptotes occur at the zeros of $\sin 2x$; namely, at

$$2x = n\pi \qquad \text{or} \qquad x = \frac{n\pi}{2} \qquad \text{for } n \text{ an integer.}$$

49. The period of $y = \tan(x/2 - \pi/4)$ is $\pi/(1/2) = 2\pi$. Since

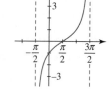

$$\tan\left(\frac{x}{2} - \frac{\pi}{4}\right) = \frac{\sin(x/2 - \pi/4)}{\cos(x/2 - \pi/4)},$$

the x-intercepts of $\tan(x/2 - \pi/4)$ occur at the zeros of $\sin(x/2 - \pi/4)$; namely, at

$$\frac{x}{2} - \frac{\pi}{4} = n\pi \qquad \text{or} \qquad x = 2n\pi + \frac{\pi}{2} \qquad \text{for } n \text{ an integer.}$$

The vertical asymptotes occur at the zeros of $\cos(x/2 - \pi/4)$; namely, at

$$\frac{x}{2} - \frac{\pi}{4} = \frac{(2n+1)\pi}{2} \qquad \text{or} \qquad x = (2n+1)\pi + \frac{\pi}{2} = \frac{3\pi}{2} + 2n\pi \qquad \text{for } n \text{ an integer.}$$

Since the graph has vertical asymptotes at $-\pi/2$ and $3\pi/2$ (using $n = -1$ and $n = 0$), we graph one cycle on the interval $(-\pi/2, 3\pi/2)$.

51. The period of $y = -1 + \cot \pi x$ is $\pi/\pi = 1$. To find the x-intercepts, we solve $-1 + \cot \pi x = 0$, or $\cot \pi x = \cos \pi x / \sin \pi x = 1$, which is equivalent to solving for $\cos \pi x = \sin \pi x$. This occurs when

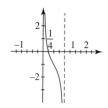

$$\pi x = (4n+1)\frac{\pi}{4} \qquad \text{or} \qquad x = n + \frac{1}{4} \qquad \text{for } n \text{ an integer.}$$

The vertical asymptotes occur at the zeros of $\sin \pi x$; namely, $\pi x = n\pi$ or $x = n$, for n an integer.

53. The period of $y = 3\csc \pi x$ is $2\pi/\pi = 2$. Since $3\csc \pi x = 3/\sin \pi x$, the vertical asymptotes occur at the zeros of $\sin \pi x$; namely, at $\pi x = n\pi$ or $x = n$, for n an integer. We plot one cycle of the graph on $(-1, 1)$, since the period of the function is $2 = 1 - (-1)$ and vertical asymptotes occur at $x = -1$ and $x = 1$ (taking $n = -1$ and 1).

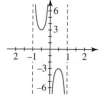

55. The period of $y = \sec(3x - \pi/2)$ is $2\pi/3$. Since

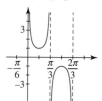

$$\sec\left(3x - \frac{\pi}{2}\right) = \frac{1}{\cos(3x - \pi/2)},$$

the vertical asymptotes occur at the zeros of $\cos(3x - \pi/2)$; namely, at

$$3x - \frac{\pi}{2} = (2n+1)\frac{\pi}{2} \qquad \text{or} \qquad x = (n+1)\frac{\pi}{3} \qquad \text{for } n \text{ an integer.}$$

57. The amplitude is $A = \frac{1}{2}(18-6) = 6$. The tidal period is $2\pi/B = 12$, so $B = \pi/6$, and the average depth $D = (18+6)/2 = 12$. Thus, the function is

$$d(t) = 12 + 6\sin\frac{\pi}{6}\left(t - \frac{\pi}{2}\right).$$

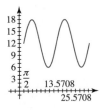

59. (a) When $\theta = 0°$, $\sin\theta = 0$ and $\sin 2\theta = 0$, so $g = 978.0309$ cm/s^2.

   (b) At the north pole, $\theta = 90°$, so $\sin\theta = 1$ and $\sin 2\theta = \sin 180° = 0$, so $g = 978.0309 + 5.18552 = 983.2164$ cm/s^2.

   (c) When $\theta = 45°$, $\sin\theta = \sqrt{2}/2$ and $\sin 2\theta = \sin 90° = 1$, so

$$g = 978.0309 + 5.18552(\sqrt{2}/2)^2 - 0.00570(1)^2$$
$$= 978.0309 + 2.59276 - 0.00570 = 980.618 \text{ cm/s}^2.$$

61. The period of $\sin\frac{1}{2}x$ is $2\pi/(1/2) = 4\pi$. The period of $\sin 2x$ is $2\pi/2 = \pi$. The period of a sum of periodic functions is equal to the least common multiple of the individual periods, so the period of $f(x)$ is 4π.

63. The graph of the absolute value of a function is the graph of the function with any portions of the graph that lie below the x-axis reflected through the x-axis.

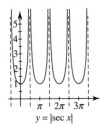

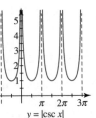

65. The graph of $y = \cot x$ is shown below as a red curve, while the graphs of $y = A\tan(x+C)$, for various choices of A and C, are shown as blue curves.

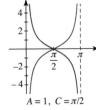 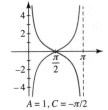 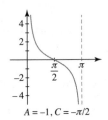

We see from the third graph that $\cot x = -\tan(x - \pi/2)$.

1.5 Inverse Functions

1. For $f(x) = 1 + x(x-5)$, the value $y = 1$ in the range of f occurs at either $x = 0$ or $x = 5$ in the domain of f. Thus, $f(x)$ is not one-to-one.

3.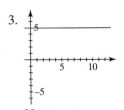

   Not one-to-one.

5.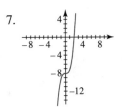

   One-to-one.

7.

   One-to-one.

9. $y = 3x^3 + 7$, $x = \left(\dfrac{y-7}{3}\right)^{1/3}$; $f^{-1}(x) = \left(\dfrac{x-7}{3}\right)^{1/3}$

11. $y = \dfrac{2-x}{1-x}$, $x = \dfrac{y-2}{y-1}$; $f^{-1}(x) = \dfrac{x-2}{x-1}$

13. $f(f^{-1}(x)) = f\left(\dfrac{1}{5}x + 2\right) = 5\left(\dfrac{1}{5}x + 2\right) - 10 = x + 10 - 10 = x$

$f^{-1}(f(x)) = f^{-1}(5x - 10) = \dfrac{1}{5}(5x - 10) + 2 = x - 2 + 2 = x$

15. Domain: $[0, \infty)$; Range: $[-2, \infty)$

17. Domain: $(-\infty, 0) \cup (0, \infty)$; Range: $(-\infty, -3) \cup (-3, \infty)$

19. When $x = 2$, $y = f(2) = 2(2)^3 + 2(2) = 2(8) + 4 = 20$, so the point on the graph of f is $(2, 20)$. The corresponding point on the graph of f^{-1} is then $(20, 2)$.

21. Since $f(9) = 12$, $(9, 12)$ is a point on the graph of f. The corresponding point on the graph of f^{-1} is then $(12, 9)$, and so $f^{-1}(12) = 9$.

23. The graph of f^{-1}, shown in red, is obtained from the graph of f, shown in blue, by reflection through the line $y = x$.

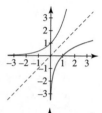

25. To graph f, shown in blue, we use the fact that f is the inverse of f^{-1}, shown in red, and that the graph of the inverse of a function is the reflection of the graph of the original function reflected through the line $y = x$.

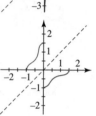

27. By restricting f's domain to $x \geq 5/2$, we get $f^{-1}(x) = \dfrac{5 - \sqrt{x}}{2}, x \geq 0$.

29. $f(x)$ can be rewritten as $(x+1)^2 + 3$. Thus, by restricting the domain of f to $x \geq -1$, we get $f^{-1}(x) = \sqrt{x-3} - 1$.

31. For $f(x) = x^3$ and $g(x) = 4x + 5$, we get $(f \circ g)(x) = (4x+5)^3$ and thus $(f \circ g)^{-1}(x) = \frac{1}{4}(x^{1/3} - 5)$. $f^{-1}(x) = x^{1/3}$ and $g^{-1}(x) = \frac{1}{4}(x - 5)$, so $(g^{-1} \circ f^{-1})(x) = \frac{1}{4}(x^{1/3} - 5)$, which is the same as $(f \circ g)^{-1}$.

33. $3\pi/4$

35. $\pi/4$

37. $3\pi/4$

39. $-\pi/3$

41. $\sin\left(\arctan\dfrac{4}{3}\right) = \dfrac{4}{5}$

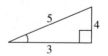

43. $\tan\left(\cot^{-1}\dfrac{1}{2}\right) = 2$

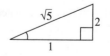

45. $\sin\left(2\sin^{-1}\dfrac{1}{3}\right) = 2\sin\left(\sin^{-1}\dfrac{1}{3}\right)\cos\left(\sin^{-1}\dfrac{1}{3}\right) = 2\left(\dfrac{1}{3}\right)\left(\dfrac{\sqrt{8}}{3}\right) = \dfrac{4\sqrt{2}}{9}$

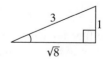

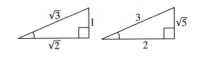

47. $\sin\left(\arcsin\dfrac{\sqrt{3}}{3}+\arccos\dfrac{2}{3}\right)$

$$=\sin\left(\arcsin\dfrac{\sqrt{3}}{3}\right)\cos\left(\arccos\dfrac{2}{3}\right)+\cos\left(\arcsin\dfrac{\sqrt{3}}{3}\right)\sin\left(\arccos\dfrac{2}{3}\right)$$

$$=\left(\dfrac{\sqrt{3}}{3}\right)\left(\dfrac{2}{3}\right)+\left(\dfrac{\sqrt{6}}{3}\right)\left(\dfrac{\sqrt{5}}{3}\right)=\dfrac{2\sqrt{3}}{9}+\dfrac{\sqrt{30}}{9}=\dfrac{(2+\sqrt{10})\sqrt{3}}{9}$$

49. $\sqrt{1-x^2}$

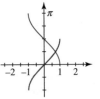

51. $\sqrt{1+x^2}$

53. From the graph on the right, it can be seen that $\sin^{-1}x$, shown in red, when reflected through the x-axis (thus multiplying by -1) then shifted up by $\pi/2$ units (thus adding $\pi/2$), is the graph of $\cos^{-1}x$, shown in blue. Thus,

$$\cos^{-1}x=-\sin^{-1}x+\dfrac{\pi}{2}\qquad\text{or}\qquad\sin^{-1}x+\cos^{-1}x=\dfrac{\pi}{2}$$

55. Let $y=\sec^{-1}x$. Then $x=\sec y=1/\cos y$ and $\cos y=1/x$. This implies that $y=\cos^{-1}(1/x)$. Thus, $\sec^{-1}x=\cos^{-1}(1/x)$. The domain of both $\sec^{-1}x$ and $\cos^{-1}(1/x)$ is $|x|\geq 1$.

57. Since $t-\sin^{-1}(-2/\sqrt{5})$ and $-\pi/2<t<0$, t is in the fourth quadrant. $\cos t=1/\sqrt{5}$, $\tan t=-2$, $\cot t=-1/2$, $\sec t=\sqrt{5}$, $\csc t=-\sqrt{5}/2$.

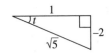

59. (a) $\sec^{-1}(-\sqrt{2})=\cos^{-1}(-\sqrt{2}/2)\approx 2.3562$

    (b) $\csc^{-1}2=\sin^{-1}(1/2)\approx 0.5236$

61. (b) The range of the arctangent function is $(-\pi/2,\pi/2)$, and 5 is not in this interval.

63. $\dfrac{x(1+c^2)}{Lc}-c=\tan\beta;\quad \beta=\tan^{-1}\left[\dfrac{(1+c^2)x}{cL}-c\right]$

    (a) $\beta=\tan^{-1}\left[\dfrac{2L}{3L}-1\right]=\tan^{-1}(1)=\dfrac{\pi}{4}$

    (b) $\beta=\tan^{-1}\left[\dfrac{(1.25)(\frac{3}{4}L)}{0.5L}-0.5\right]=\tan^{-1}(1.375)\approx 0.942\text{ radian}\approx 53.97°$

65. Theorem 1.5.2(i) is not violated, because the range of the arcsine function is $[-\pi/2,\pi/2]$, while the range of $f(x)=x$ is $(-\infty,\infty)$.

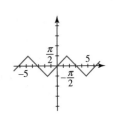

67. A periodic function cannot be one-to-one, since periodic functions have repeated y-values over regular intervals of x-values. For a function to be one-to-one, every y in its range must correspond to a single x in its domain.

1.6 Exponential and Logarithmic Functions

1. Since $f(0) = (3/4)^0 = 1$, the y-intercept is $(0,1)$. The x-axis is a horizontal asymptote.

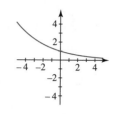

3. Since $f(0) = -2^0 = -1$, the y-intercept is $(0,-1)$. The x-axis is a horizontal asymptote and the graph of $f(x) = -2^x$ is the graph of $y = 2^x$ reflected in the x-axis.

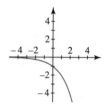

5. Since $f(0) = -5 + e^0 = -5 + 1 = -4$, the y-intercept is $(0,-4)$. The line $y = -5$ is a horizontal asymptote.

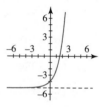

7. Letting $x = 3$ and $f(3) = 216$, we have $f(3) = 216 = b^3$, so $b = 6$ and $f(x) = 6^x$.

9. Letting $x = -1$ and $f(-1) = e^2$, we have $f(-1) = e^2 = b^{-1}$, so $b = e^{-2}$ and $f(x) = (e^{-2})^x = e^{-2x}$.

11. Graphing $y = 2^x$ and $y = 16$, we see that $2^x > 16$ for $x > 4$.

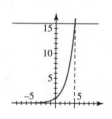

13. Graphing $y = e^{x-2}$ and $y = 1$, we see that $e^{x-2} < 1$ for $x < 2$.

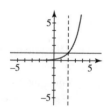

15. Since $f(-x) = e^{(-x)^2} = e^{x^2} = f(x)$, we see that $f(x)$ is even.

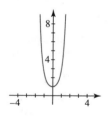

17. The graph of $f(x) = 1 - e^{x^2}$ is the graph of $y = e^{x^2}$ reflected through the x-axis and shifted up 1 unit.

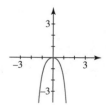

19. Since $f(-x) = 2^{-x} + 2^{-(-x)} = 2^{-x} + 2^x = f(x)$, we see that $f(x)$ is even.

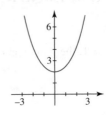

21.

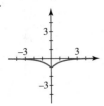

23. $4^{-1/2} = \frac{1}{2}$ is equivalent to $\log_4 \frac{1}{2} = -\frac{1}{2}$.

25. $10^4 = 10,000$ is equivalent to $\log_{10} 10,000 = 4$.

27. $\log_2 128 = 7$ is equivalent to $2^7 = 128$.

29. $\log_{\sqrt{3}} 81 = 8$ is equivalent to $(\sqrt{3})^8 = 81$.

31. We solve $2 = \log_b 49$ or $b^2 = 49$, so $b = 7$ and $f(x) = \log_7 x$.

33. $\ln e^e = e \ln e = e$

35. $10^{\log_{10} 6^2} = 6^2 = 36$

37. $e^{-\ln 7} = e^{\ln 7^{-1}} = 7^{-1} = \dfrac{1}{7}$

39. The domain of $\ln x$ is determined by $x > 0$, so the domain is $(0, \infty)$. The x-intercept is the solution of $-\ln x = 0$. This is equivalent to $e^0 = x$, so $x = 1$ and the x-intercept is $(1, 0)$. The vertical asymptote is $x = 0$ or the y-axis.

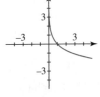

41. The domain of $\ln x$ is determined by $x > 0$, so the domain of $-\ln(x+1)$ is determined by $x + 1 > 0$ or $x > -1$. Thus, the domain is $(-1, \infty)$. The x-intercept is the solution of $-\ln(x+1) = 0$. This is equivalent to $e^0 = x + 1$, or $1 = x + 1$, so $x = 0$ and the x-intercept is $(0, 0)$. The vertical asymptote is $x + 1 = 0$ or $x = -1$.

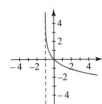

43. The domain of $\ln(9 - x^2)$ is determined by $9 - x^2 > 0$. This is equivalent to $(3 + x)(3 - x) > 0$. Thus, $3 + x$ and $3 - x$ must be both > 0 or both < 0, so the domain is $(-3, 3)$.

45. Since $f(-x) = \ln|-x| = \ln|x| = f(x)$, we see that $f(x)$ is even. The x-intercepts are the solutions of $\ln|x| = 0$, so $x = 1$ or $x = -1$, and the x-intercepts are $(1, 0)$ and $(-1, 0)$. The vertical asymptote is $x = 0$ or the y-axis.

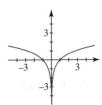

47. Using (*ii*) of the laws of logarithms (Theorem 1.6.1) in the text, we have

$$\ln(x^4 - 4) - \ln(x^2 + 2) = \ln \frac{x^4 - 4}{x^2 + 2} = \ln \frac{(x^2 - 2)(x^2 + 2)}{x^2 + 2} = \ln(x^2 - 2).$$

49. Using (i) of the laws of logarithms (Theorem 1.6.1) in the text, we have

$$\ln 5 + \ln 5^2 + \ln 5^3 - \ln 5^6 = \ln(5 \cdot 5^2 \cdot 5^3) - \ln 5^6 = \ln 5^6 - \ln 5^6 = 0 = \ln 1.$$

51. Using the laws of logarithms (Theorem 1.6.1) in the text, we have

$$\ln y = \ln \frac{x^{10}\sqrt{x^2+5}}{\sqrt[3]{8x^3+2}} = \ln x^{10} + \ln(x^2+5)^{1/2} - \ln(8x^3+2)^{1/3}$$

$$= 10\ln x + \frac{1}{2}\ln(x^2+5) - \frac{1}{3}\ln(8x^3+2).$$

53. Using the laws of logarithms (Theorem 1.6.1) in the text, we have

$$\ln y = \ln \frac{(x^3-3)^5(x^4+3x^2+1)^8}{\sqrt{x}(7x+5)^9} = \ln(x^3-3)^5 + \ln(x^4+3x^2+1)^8 - \ln x^{1/2} - \ln(7x+5)^9$$

$$= 5\ln(x^3-3) + 8\ln(x^4+3x^2+1) - \frac{1}{2}\ln x - 9\ln(7x+5).$$

55. We want to solve $6^x = 51$. This is equivalent to

$$\ln 6^x = \ln 51, \quad x\ln 6 = \ln 51$$

$$x = \frac{\ln 51}{\ln 6} \approx 2.1944.$$

57. Taking the natural logarithm of both sides, we have

$$\ln 2^{x+5} = (x+5)\ln 2 = \ln 9$$

$$x+5 = \frac{\ln 9}{\ln 2}, \quad x = -5 + \frac{\ln 9}{\ln 2} \approx -1.8301.$$

59. Taking the natural logarithm of both sides, we have

$$\ln 5^x = \ln(2e^{x+1}) = \ln 2 + \ln e^{x+1} = \ln 2 + x + 1$$

$$x\ln 5 = \ln 2 + x + 1$$

$$x\ln 5 - x = x(\ln 5 - 1) = 1 + \ln 2$$

$$x = \frac{1 + \ln 2}{\ln 5 - 1} \approx 2.7782.$$

In Problems 61–62, it is necessary to check that the solutions obtained actually satisfy the original equation. This is because equations involving logarithms may lead to extraneous solutions. An extraneous solution, in fact, occurs in Problem 61.

61. We use the laws of logarithms and the fact that $\ln x$ is one-to-one:

$$\ln x + \ln(x-2) = \ln[x(x-2)] = \ln 3$$

$$x^2 - 2x = 3$$

$$x^2 - 2x - 3 = 0$$

$$(x-3)(x+1) = 0.$$

Thus, $x = 3$ and $x = -1$. We disregard $x = -1$ since it is outside the domains of both $\ln x$ and $\ln(x-2)$. We see that $x = 3$ checks.

63. (a) Since the population doubles after 2 hours, we write $P(2) = P_0 e^{2k} = 2P_0$. Solving for k gives

$$e^{2k} = 2, \quad 2k = \ln 2, \quad k = \frac{\ln 2}{2} \approx 0.3466.$$

Thus, $P(t) = P_0 e^{0.3466t}$.

(b) In 5 hours, $P(5) = P_0 e^{(0.3446)(5)} \approx 5.66 P_0$.

(c) Solving $P(t) = P_0 e^{0.3466t} = 20 P_0$ for t, we have

$$e^{0.3466t} = 20, \ \ 0.3466t = \ln 20, \ \ t = \frac{\ln 20}{0.3466} \approx 8.64 \text{ hours.}$$

65. (a) $P(5) = \dfrac{2000}{1 + 1999 e^{-0.8905(5)}} \approx 82$ students

(b) Solving $P(t) = \dfrac{2000}{1 + 1999 e^{-0.8905t}} = 1000$ for t, we have

$$1 + 1999 e^{-0.8905t} = 2, \ \ 1999 e^{-0.8905t} = 1$$

$$e^{-0.8905t} = \frac{1}{1999}$$

$$-0.8905t = \ln\left(\frac{1}{1999}\right) = -\ln 1999$$

$$t = \frac{-\ln 1999}{-0.8905} \approx 8.53 \text{ days.}$$

(c) As $t \to \infty$, $e^{-0.8905t} \to 0$, so $P(t) \to \dfrac{2000}{1+0} = 2000$.

(d) We note that as $t \to \infty$, $e^{-0.8905t} \to 0$, so the graph has a horizontal asymptote at $P = 2000$.

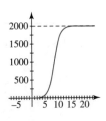

67. (a) Since $y = \ln 5x = \ln 5 + \ln x$, we can obtain the graph of $y = \ln 5x$ by shifting the graph of $y = \ln x$ up $\ln 5$ units.

(b) Since $y = \ln(x/4) = \ln x - \ln 4$, we can obtain the graph of $y = \ln(x/4)$ by shifting the graph of $y = \ln x$ down $\ln 4$ units.

(c) Since $y = \ln x^{-1} = -\ln x$, we can obtain the graph of $y = \ln x^{-1}$ by reflecting the graph of $y = \ln x$ in the x-axis.

(d) The graph of $y = \ln(-x)$ is the reflection of $y = \ln x$ in the y-axis.

1.7   From Words to Functions

1. Let x and y be the positive numbers. Then $xy = 50$ and their sum is $S = x + y$. From $xy = 50$ we have $y = 50/x$, so

$$S(x) = x + \frac{50}{x} = \frac{x^2 + 50}{x}.$$

Since x is positive, the domain of S is $(0, \infty)$.

3. Let x and y be the nonnegative numbers. Then $x + y = 1$. Now, the sum of the square of x and twice the square of y is $x^2 + 2y^2$. From $x + y = 1$ we have $y = 1 - x$, so the function in this case is

$$s(x) = x^2 + 2(1-x)^2 = x^2 + 2(1 - 2x + x^2) = 3x^2 - 4x + 2.$$

Since x and y are both nonnegative, we must have $0 \le x \le 1$ and $0 \le y \le 1$. (If, say, $y > 1$, then we would have $x < 0$.) Thus

$$s(x) = 3x^2 - 4x + 2, \ \ 0 \le x \le 1.$$

Alternatively, if we choose the independent variable of s to be y, we have $x = 1 - y$ and

$$s(y) = (1 - y)^2 + 2y^2 = 1 - 2y + y^2 + 2y^2 = 3y^2 - 2y + 1, \ \ 0 \le y \le 1.$$

5. Let x and y be the sides of the rectangle. Then the perimeter is $2x + 2y = 200$ and the area is $A = xy$. Solving $2x + 2y = 200$ for y, we have $y = 100 - x$, so $A(x) = x(100 - x) = 100x - x^2$. The domain of A is $[0, 100]$.

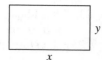

7. The lengths of the sides of the rectangle are x and y, so its area is $A = xy$. Since the lengths are related by $x + 2y = 4$, we have $x = 4 - 2y$ and $A(y) = (4 - 2y)y$. The y-intercept of the line is $(0, 2)$, so the domain of A is $[0, 2]$.

9. The distance between (x, y) and $(2, 3)$ is given by

$$d = \sqrt{(x-2)^2 + (y-3)^2}.$$

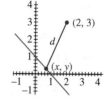

Since $x + y = 1$, we have $y = 1 - x$, so

$$d = \sqrt{(x-2)^2 + [(1-x) - 3]^2} = \sqrt{(x-2)^2 + (-x-2)^2}$$
$$= \sqrt{x^2 - 4x + 4 + x^2 + 4x + 4} = \sqrt{2x^2 + 8}.$$

The domain of d is $(-\infty, \infty)$.

11. If the side of a square is x, then its area is $A = x^2$ and its perimeter is $P = 4x$. Solving $A = x^2$ for x, we have $x = \sqrt{A}$, so $P = 4x = 4\sqrt{A}$. The domain of P is $(0, \infty)$.

13. If the diameter of a circle is d, then its circumference is $C = \pi d$. Solving for d, we have $d = C/\pi$. The domain of $d(C)$ is $(0, \infty)$.

15. Let the sides of the equilateral triangle each be of length s. Then, referring to the figure and using the Pythagorean Theorem, we have

$$\left(\frac{s}{2}\right)^2 + h^2 = s^2$$

$$h^2 = s^2 - \frac{s^2}{4} = \frac{3}{4}s^2$$

$$s^2 = \frac{4}{3}h^2, \quad s = \frac{2}{\sqrt{3}}h.$$

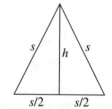

The area of the triangle is $A = \frac{1}{2}sh = \frac{1}{2}\left(\frac{2}{\sqrt{3}}h\right)h = \frac{1}{\sqrt{3}}h^2 = \frac{\sqrt{3}}{3}h^2$. The domain of $A(h)$ is $(0, \infty)$.

17. If r is the radius of a circle, then its circumference, x in this case, is $x = 2\pi r$, and its area is $A = \pi r^2$. Solving $x = 2\pi r$ for r, we have $r = x/2\pi$. Then $A = \pi r^2 = \pi\left(\frac{x}{2\pi}\right)^2 = \frac{x^2}{4\pi}$. The domain of $A(x)$ is $(0, \infty)$.

19. Let x be the length of one \$4-per-foot side of the fence. The length of one \$1.60-per-foot side is therefore $1000/x$. Thus, the total cost $C(x)$ to enclose the corral is

$$C(x) = 2(4x) + 2(1.6)\left(\frac{1000}{x}\right) = 8x + \frac{3200}{x}.$$

The domain of $C(x)$ is $(0, \infty)$ — though increasingly higher values of x imply quite a narrow corral!

21. Let w be the width of the box, and h the height of the box. Then the length of the box is $3w$ and the volume of the box is $V = (3w)(w)(h) = 3w^2 h = 450$.

Since the box is open, its surface area is $S = 2wh + 2(3w)h + 3w(w) = 8wh + 3w^2$. From $3w^2 h = 450$ we have $h = 150/w^2$, so

$$S = 8w\left(\frac{150}{w^2}\right) + 3w^2 = \frac{1200}{w} + 3w^2 = \frac{1200 + 3w^2}{w}.$$

In this problem, $w > 0$.

23. After 1 hour, car A is 40 miles from point O. Let $t = 0$ correspond to this point in time. After t hours, car B has travelled $y = 60t$ miles from point O and car A has travelled $40 + x = 40 + 40t$ miles from point O. By the Pythagorean theorem,

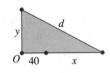

$$d = \sqrt{(40+x)^2 + y^2} = \sqrt{(40+40t)^2 + (60t)^2} = \sqrt{20^2(2+2t)^2 + 20^2(3t)^2}$$
$$= 20\sqrt{4 + 8t + 4t^2 + 9t^2} = 20\sqrt{13t^2 + 8t + 4}.$$

The domain of $d(t)$ is $(0, \infty)$.

25. A cross section of the pool as viewed from the side is shown. When the level of the water is h feet above the bottom of the deep end of the pool, where $0 \leq h \leq 5$, the volume of the water in the pool is the area A of the darker triangle times the width of the pool, 30 feet.

To find A, we let x be the distance shown in the figure and use similar triangles:

$$\frac{x}{h} = \frac{40}{5} = 8 \quad \text{so} \quad x = 8h.$$

Then $A = \frac{1}{2}hx = \frac{1}{2}h(8h) = 4h^2$. Thus, $V(h) = 30A = 30(4h^2) = 120h^2$, $0 \leq h \leq 5$.

When $h = 5$, the volume of water in the pool is $120(5^2) = 3000$ ft³. For each foot y of water above this level, there is an additional $40 \times 30 \times y = 1200y$ ft³ of water in the pool. Thus, using the fact that $y = h - 5$ for $h > 5$,

$$V(h) = \begin{cases} 120h^2, & 0 \leq h < 5 \\ 3000 + 1200(h-5), & 5 \leq h \leq 8 \end{cases} = \begin{cases} 120h^2, & 0 \leq h < 5 \\ 1200h - 3000, & 5 \leq h \leq 8. \end{cases}$$

The domain of $V(h)$ is $[0, 8]$.

27. If θ is the angle of elevation, then $h(\theta) = 300\tan\theta$, $\theta \geq 0$.

29. Start by labeling Figure 1.7.17 such that the distance between the end of the plank that rests on the ground to the center of the sawhorse is c. Thus, we see that $\cot\theta = c/3$ and $\sec\theta = L/(4+c)$. Thus

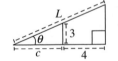

$$L(\theta) = (4+c)\sec\theta = (4 + 3\cot\theta)\sec\theta = 4\sec\theta + 3\csc\theta.$$

The domain of $L(\theta)$ is theoretically $(0, 90)$ degrees, but as stated in the problem, θ cannot get too close to 0 or 90 degrees because the sawhorse would no longer fit beneath the plank.

31. See Figure 1.7.19 in the text. Let θ_1 denote the angle of elevation from eye level to the top of the pedestal, just below the base of the statue. Then

$$\tan\theta_1 = \frac{1/2}{x} \quad \text{so} \quad \theta_1 = \arctan\left(\frac{1}{2x}\right).$$

Let θ_2 denote the angle of elevation from eye level to the top of the statue. Then

$$\tan\theta_2 = \frac{1}{x} \quad \text{so} \quad \theta_2 = \arctan\left(\frac{1}{x}\right).$$

Thus, we see that $\theta = \theta_2 - \theta_1 = \arctan\left(\frac{1}{x}\right) - \arctan\left(\frac{1}{2x}\right)$, where x is measured in meters. The domain of $\theta(x)$ is $(0, \infty)$.

33. If the building is 60 ft high, then from $y = \frac{10(x+5)}{x}$, we get $y = 60$ when $x = 1$. Since we want the ladder to be no higher than the building, then

$$\frac{10(x+5)}{x} \leq 60$$
$$x + 5 \leq 6x$$
$$5x \geq 5, \quad x \geq 1.$$

Thus, for $0 < x < 1$, we get $y > 60$, which is higher than the building (for example, if $x = 1/2$, then $y = 110$). The domain of the function

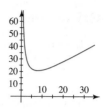

$$L(x) = \frac{x+5}{x}\sqrt{x^2 + 100}$$

must therefore be $[1, \infty)$. L decreases until it attains its absolute minimum at $x = \sqrt[3]{500} \approx 7.94$ ft.

Chapter 1 in Review

A. True/False

1. False; consider $f(x) = x^2$ with $x = -1$ and $x = 1$.

3. True

5. False; $f(3/2) = 0$.

7. True; one for $x \to -\infty$ and another for $x \to +\infty$.

9. False; amplitude is not defined for $\sec x$.

11. True

13. True; if f is even, then $f(-x) = f(x)$ for all x, so f cannot be one-to-one (unless the domain of f is $\{0\}$, which is precluded by the fact that $a > 0$).

15. True; the range of $\sec x$ is $(-\infty, -1] \cup [1, \infty)$.

17. True, since $y = 10^{-x} = (10^{-1})^x = (1/10)^x = (0.1)^x$.

19. True, since $ln\dfrac{e^b}{e^a} = ln e^b - \ln e^a = b \ln e - a \ln e = b - a$.

B. Fill In the Blanks

1. $[-2, 0) \cup (0, \infty)$

3. $(-8, 6)$

5. The graph is tangent to the x-axis at $(1, 0)$, since it is a zero of multiplicity 2. The graph passes through the x-axis at $(0, 0)$ (multiplicity 3) and $(5, 0)$ (simple zero).

7. $-4/5$

9. The period is $\dfrac{2\pi}{B} = \dfrac{2\pi}{\pi/3} = 6$.

11. Since $\sin \pi = 0$, $\sin^{-1}(\sin \pi) = \sin^{-1} 0$. Letting $y = \sin^{-1} 0$, we must find y such that $\sin y = 0$ and $-\pi/2 \le y \le \pi/2$. Within this interval, we have $\sin 0 = 0$, so $\sin^{-1}(\sin \pi) = 0$.

13. $(3, 5)$, since the graph of $y = 4 + e^{x-3}$ is obtained by shifting the graph of $y = e^x$ up four units and to the right three units.

15. $\log_3 5$

17. If $\log_3 x = -2$, then $x = 3^{-2} = 1/9$.

19. $y = \ln x$

C. Exercises

1. (a) $f(-4) = 3$ (b) $f(-3) = 0$ (c) $f(-2) = -2$ (d) $f(-1) = 0$
    (e) $f(0) = 2.5$ (f) $f(1) = 2$ (g) $f(1.5) = 1$ (h) $f(2) = 0$
    (i) $f(3.5) = 3$ (j) $f(4) = 4$

3. 1: in range ($f(1/2) = 1$); 5: not in range; 8: in range ($f(4) = 8$)

5. $$\frac{f(x+h) - f(x)}{h} = \frac{[-(x+h)^3 + 2(x+h)^2 - (x+h) + 5] - (-x^3 + 2x^2 - x + 5)}{h}$$

$$= \frac{(-x^3 - 3x^2h - 3xh^2 - h^3 + 2x^2 + 4xh + 2h^2 - x - h + 5) - (-x^3 + 2x^2 - x + 5)}{h}$$

$$= \frac{-3x^2h - 3xh^2 - h^3 + 4xh + 2h^2 - h}{h} = \frac{h(-3x^2 - 3xh - h^2 + 4x + 2h - 1)}{h}$$

$$= -3x^2 - 3xh - h^2 + 4x + 2h - 1$$

7. Since

$$f(x) = \frac{2x}{x^2 + 1}$$

has no vertical asymptotes and its horizontal asymptote is $y = 0$, its graph must be either (a) or (f). Because $f(x)$ is negative for $x < 0$, it must be (f).

9. Since

$$f(x) = \frac{2x}{x - 2}$$

has vertical asymptote $x = 2$ and its horizontal asymptote is $y = 2$, its graph must be (d).

11. Since

$$f(x) = \frac{x}{(x - 2)^2}$$

has vertical asymptote $x = 2$ and its horizontal asymptote is $y = 0$, its graph must be (h).

13. Since

$$f(x) = \frac{x^2 - 10}{2x - 4} = \frac{1}{2} \cdot \frac{x^2 - 10}{x - 2} = \frac{1}{2}\left(x + 2 - \frac{6}{x - 2}\right)$$

using synthetic division, we see that the graph of $f(x)$ has vertical asymptote $x = 2$ and slant asymptote $y = \frac{1}{2}x + 1$. Thus, the graph of f must be (c).

15. Since

$$f(x) = \frac{2x}{x^3 + 1} = \frac{2x}{(x + 1)(x^2 - x + 1)}$$

has vertical asymptote $x = -1$ and horizontal asymptote $y = 0$, its graph must be (b).

17. Since $f(-2 + h) = 3^{-(-2+h+1)} = 3^{1-h}$, the line passes through $(-2 + h, 3^{1-h})$ and $(-2, 3)$. Its slope is

$$m = \frac{3^{1-h} - 3}{-2 + h - (-2)} = \frac{3^{1-h} - 3}{h}.$$

19. (a) $12^t = (2 \cdot 6)^t = 2^t \cdot 6^t = 5 \cdot 2 = 10$

    (b) $3^t = \left(\frac{6}{2}\right)^t = \frac{6^t}{2^t} = \frac{2}{5}$

(c) $6^{-t} = \dfrac{1}{6^t} = \dfrac{1}{2}$

21. Since $(0,5)$ is on the graph, $5 = Ae^{k \cdot 0} = A$ and the function is $f(x) = 5e^{kx}$. Since the graph passes through $(6,1)$, $1 = 5e^{6k}$ and $k = \dfrac{1}{6}\ln\dfrac{1}{5} = -\dfrac{1}{6}\ln 5$. Thus,

$$f(x) = 5e^{(-\frac{1}{6}\ln 5)x} = 5e^{-0.2682x}.$$

23. The graph of $f(x) = b^x$, where $0 < b < 1$, has a horizontal asymptote at $y = 0$, so the graph of $f(x) = 5 + b^x$ has horizontal asymptote $y = 5$. The graph passes through $(1,5.5)$ so $5.5 = 5 + b$, and $b = 0.5$. Thus, $f(x) = 5 + (1/2)^x$.

25. This looks like the graph of $y = \ln x$ revolved around the x-axis (giving $-\ln x$) and shifted up 2 units. Thus, this is the graph of (b) $y = 2 - \ln x$.

27. This looks like the graph of $y = \ln x$ revolved around the x-axis (giving $-\ln x$) and shifted left 2 units (giving $-\ln(x+2)$). But then the graph should pass through $(-1,0)$. Instead, it appears to pass through $(-1,-2)$. Thus, this is the graph of (d), $y = -2 - \ln(x+2)$.

29. This looks like the graph of $y = \ln x$ shifted to the left 2 units (giving $\ln(x+2)$). But then the graph should pass through $(-1,0)$. Instead, it appears to pass through $(-1,2)$. Thus, this is the graph of (c), $y = 2 + \ln(x+2)$.

31. (a) $V(l) = l(2l)(3l) = 6l^3$

   (b) $V(w) = w\left(\dfrac{1}{3}w\right)\left(\dfrac{2}{3}w\right) = \dfrac{2}{9}w^3$

   (c) $V(h) = h\left(\dfrac{1}{2}h\right)\left(\dfrac{3}{2}h\right) = \dfrac{3}{4}h^3$

33. To find the volume of the box, we begin by finding the area of the side of the box shown in the figure at the right. We see that $\tan\theta = 5/x$, so $x = 5\cot\theta$. Then the area of the figure is given by

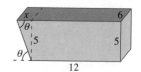

$$A = (12)(5) + \dfrac{1}{2}(5\cot\theta)(5) = 60 + \dfrac{25}{2}\cot\theta.$$

The volume of the box is $V = 6\left(60 + \dfrac{25}{2}\cot\theta\right) = 360 + 75\cot\theta.$

35. Let x and y be as shown in the figure to the right. The cross section is a trapezoid with parallel sides having lengths 10 and $10 + 2x$. The area of the trapezoid is

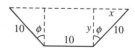

$$A = \dfrac{1}{2}[10 + (2x+10)]y = \dfrac{1}{2}(20+2x)y = (10+x)y,$$

so we need to express x and y in terms of ϕ. Using $\cos\phi = y/10$ and $\sin\phi = x/10$, we have $y = 10\cos\phi$ and $x = 10\sin\phi$, so

$$A = (10+x)y = (10 + 10\sin\phi)(10\cos\phi) = 100(1 + \sin\phi)(\cos\phi)$$
$$= 100\cos\phi + 50\sin 2\phi.$$

37. Let A be the area of a triangular side and t be the thickness of the prism (i.e., the distance between the two parallel faces). The prism's triangular sides are equilateral and its rectangular base is inscribed within the circle $x^2 + y^2 = 1$, so $A = \frac{1}{2}(2y)(\sqrt{3}\cdot y)$ and $t = 2x$, resulting in the volume

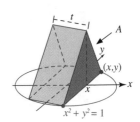

$$V = At = \left(\dfrac{1}{2}(2y)(\sqrt{3}\cdot y)\right)(2x).$$

Since $x^2 + y^2 = 1$, then $y = \sqrt{1 - x^2}$, and we can substitute y above:

$$V(x) = \left[\dfrac{1}{2}(2\sqrt{1-x^2})(\sqrt{3}\cdot\sqrt{1-x^2})\right](2x)$$
$$= 2\sqrt{3}(1 - x^2).$$

Limit of a Function

2.1 Limits—An Informal Approach

1. $\lim\limits_{x \to 2}(3x+2) = 8$

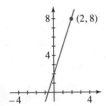

3. No limit as $x \to 0$.

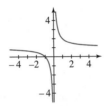

5. $\lim\limits_{x \to 1}\dfrac{x^2-1}{x-1} = \lim\limits_{x \to 1}(x+1) = 2$

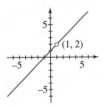

7. No limit as $x \to 3$.

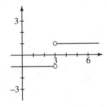

9. $\lim\limits_{x \to 0}\dfrac{x^3}{x} = 0$

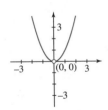

11. $\lim\limits_{x \to 0} f(x) = 3$

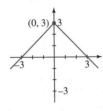

13. $\lim\limits_{x \to 2} f(x) = 0$

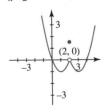

15. (a) 1   (b) -1 (c) 2   (d) doesn't exist

17. (a) 2 (b) −1 (c) −1 (d) −1

19. Correct

21. Incorrect; $\lim\limits_{x\to 1^-} \sqrt{1-x} = 0$

23. Incorrect; $\lim\limits_{x\to 0^+} \lfloor x \rfloor = 0$

25. Correct

27. Incorrect; $\lim\limits_{x\to 3^-} \sqrt{9-x^2} = 0$

29. (a) Does not exist (b) 0 (c) 3 (d) −2 (e) 0 (f) 1

31.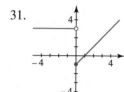

33.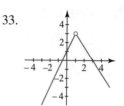

35.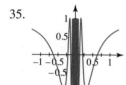

The limit does not exist.

37.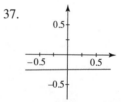

The limit is −0.25.

39.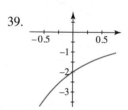

The limit is −2.

41.

$x \to 1^-$	0.9	0.99	0.999	0.9999
$f(x)$	−3.25536642	−3.02276607	−3.00225263	−3.00022503
$x \to 1^+$	1.1	1.01	1.001	1.0001
$f(x)$	−2.79817601	−2.97775903	−2.99775260	−2.99977503

$\lim\limits_{x\to 1} f(x) = -3$

43.

$x \to 0^-$	−0.1	−0.01	−0.001	−0.0001
$f(x)$	−0.04995835	−0.00499996	−0.00050000	−0.00005000
$x \to 0^+$	0.1	0.01	0.001	0.0001
$f(x)$	0.04995835	0.00499996	0.00050000	0.00005000

$\lim\limits_{x\to 0} f(x) = 0$

45. Since $\dfrac{x}{\sin 3x}$ is an even function, it suffices to consider only $x \to 0^+$.

$x \to 0^+$	0.1	0.01	0.001	0.0001
$f(x)$	0.33838634	0.33338334	0.33333383	0.33333334

$\lim\limits_{x\to 0} f(x) = 0.33333333$

47.

$x \to 4^-$	3.9	3.99	3.999	3.9999
$f(x)$	0.25158234	0.25015645	0.25001563	0.25000156
$x \to 4^+$	4.1	4.01	4.001	4.0001
$f(x)$	0.24845673	0.24984395	0.24998438	0.24999844

$\lim\limits_{x \to 4} f(x) = 0.25$

49.

$x \to 1^-$	0.9	0.99	0.999	0.9999
$f(x)$	4.43900000	4.94039900	4.99400400	4.99940004
$x \to 1^+$	1.1	1.01	1.001	1.0001
$f(x)$	5.64100000	5.06040010	5.00600400	5.00060004

$\lim\limits_{x \to 1} f(x) = 5$

2.2   Limit Theorems

1. 15

3. -12

5. 4

7. 4

9. $-8/5$

11. 14

13. 28/9

15. -1

17. $\sqrt{7}$

19. does not exist

21. $\lim\limits_{y \to -5} \dfrac{y^2 - 25}{y + 5} = \lim\limits_{y \to -5} (y - 5) = -10$

23. $\lim\limits_{x \to 1} \dfrac{x^3 - 1}{x - 1} = \lim\limits_{x \to 1} \dfrac{(x - 1)(x^2 + x + 1)}{x - 1} = \lim\limits_{x \to 1} (x^2 + x + 1) = 3$

25. $\lim\limits_{x \to 10} \dfrac{(x - 2)(x + 5)}{x - 8} = \dfrac{8(15)}{2} = 60$

27. $\lim\limits_{x \to 2} \dfrac{x^3 + 3x^2 - 10x}{x - 2} = \lim\limits_{x \to 2} \dfrac{x(x + 5)(x - 2)}{x - 2} = \lim\limits_{x \to 2} x(x + 5) = 14$

29. $\lim\limits_{t \to 1} \dfrac{t^3 - 2t + 1}{t^3 + t^2 - 2} = \lim\limits_{t \to 1} \dfrac{(t - 1)(t^2 + t - 1)}{(t - 1)(t^2 + 2t + 2)} = \lim\limits_{t \to 1} \dfrac{t^2 + t - 1}{t^2 + 2t + 2} = \dfrac{1}{5}$

31. $\lim\limits_{x \to 0^+} \dfrac{(x + 2)(x^5 - 1)^3}{(\sqrt{x} + 4)^2} = \dfrac{2(-1)}{16} = -\dfrac{1}{8}$

33. $\lim\limits_{x \to 0} \left[\dfrac{x^2 + 3x - 1}{x} + \dfrac{1}{x} \right] = \lim\limits_{x \to 0} \dfrac{x^2 + 3x}{x} = \lim\limits_{x \to 0} (x + 3) = 3$

35. does not exist

37. 2

39. $\lim\limits_{h \to 4} \sqrt{\dfrac{h}{h + 5}} \left[\dfrac{h^2 - 16}{h - 4} \right]^2 = \lim\limits_{h \to 4} \sqrt{\dfrac{h}{h + 5}} (h^2 + 8h + 16) = \dfrac{128}{3}$

41. $\displaystyle\lim_{x\to 0^-}\sqrt[5]{\frac{x^3-64x}{x^2+2x}}=\lim_{x\to 0^-}\sqrt[5]{\frac{x^2-64}{x+2}}=-2$

43. $a^2-2ab+b^2$

45. $\displaystyle\lim_{h\to 0}\frac{(8+h)^2-64}{h}=\lim_{h\to 0}\frac{16h+h^2}{h}=\lim_{h\to 0}(16+h)=16$

47. $\displaystyle\lim_{h\to 0}\frac{1}{h}\left(\frac{1}{x+h}-\frac{1}{x}\right)=\lim_{h\to 0}\frac{1}{h}\left(\frac{x-(x+h)}{(x+h)x}\right)=\lim_{h\to 0}\frac{-h}{hx(x+h)}$

$\displaystyle\qquad\qquad =\lim_{h\to 0}-\frac{1}{x^2+hx}=-\frac{1}{x^2}$

49. $\displaystyle\lim_{t\to 1}\frac{\sqrt{t}-1}{t-1}=\lim_{t\to 1}\frac{\sqrt{t}-1}{t-1}\frac{\sqrt{t}+1}{\sqrt{t}+1}=\lim_{t\to 1}\frac{1}{\sqrt{t}+1}=\frac{1}{2}$

51. $\displaystyle\lim_{v\to 0}\frac{\sqrt{25+v}-5}{\sqrt{1+v}-1}=\lim_{v\to 0}\left[\frac{\sqrt{25+v}-5}{\sqrt{1+v}-1}\frac{\sqrt{25+v}+5}{\sqrt{1+v}+1}\right]\frac{\sqrt{1+v}+1}{\sqrt{25+v}+5}$

$\displaystyle\qquad\qquad =\lim_{v\to 0}\frac{v}{v}\frac{\sqrt{1+v}+1}{\sqrt{25+v}+5}=\frac{1}{5}$

53. 32

55. $\dfrac{1}{2}$

57. does not exist

59. $8a$

61. (a) $\displaystyle\lim_{x\to 1}\frac{x^{100}-1}{x^2-1}=\lim_{x\to 1}\frac{x^{100}-1}{(x+1)(x-1)}$

$\displaystyle\qquad\qquad =\lim_{x\to 1}\frac{1}{x+1}\cdot\frac{x^{100}-1}{x-1}=\frac{1}{2}\cdot 100=50$

(b) $\displaystyle\lim_{x\to 1}\frac{x^{50}-1}{x-1}=\lim_{x\to 1}\frac{x^{50}-1}{x-1}\cdot\frac{x^{50}+1}{x^{50}+1}$

$\displaystyle\qquad\qquad =\lim_{x\to 1}\frac{x^{100}-1}{x-1}\cdot\frac{1}{x^{50}+1}=100\cdot\frac{1}{2}=50$

(c) $\displaystyle\lim_{x\to 1}\frac{(x^{100}-1)^2}{(x-1)^2}=\lim_{x\to 1}\frac{x^{100}-1}{x-1}\cdot\frac{x^{100}-1}{x-1}=100\cdot 100=10{,}000$

63. $\displaystyle\lim_{x\to 0}\sin x=\lim_{x\to 0}\left(x\cdot\frac{\sin x}{x}\right)=\lim_{x\to 0}x\cdot\lim_{x\to 0}\frac{\sin x}{x}=0\cdot 1=0$

2.3 Continuity

1. Continuous everywhere

3. Discontinuous at 3 and 6

5. Discontinuous at $\dfrac{n\pi}{2}$, for $n=0,1,2,\dots$

7. Discontinuous at 2

9. Continuous everywhere

11. Discontinuous at e^{-2}

13. (a) yes (b) yes

15. (a) yes (b) yes

17. (a) no (b) no

19. (a) yes (b) no

21. Solving $2 + \sec x = 0$, we obtain $\cos x = -\frac{1}{2}$, so $x = \frac{2\pi}{3} + 2n\pi$ or $x = \frac{4\pi}{3} + 2n\pi$. Thus, $f(x)$ is discontinuous on $(-\infty, \infty)$ and on $[\frac{\pi}{2}, \frac{3\pi}{2}]$.

23. Since $f(x)$ is discontinuous only at $x = 2$, it is discontinuous on $[-1, 3]$ and continuous on $(2, 4]$.

25. Since $\lim\limits_{x \to 4^-} f(x) = 4m$ and $\lim\limits_{x \to 4^+} f(x) = 16$, we have $4m = 16$ and $m = 4$.

27. Since $\lim\limits_{x \to 3^-} f(x) = 3m$, $\lim\limits_{x \to 3^+} f(x) = 3$, and $f(3) = n$, we have $3m = 3 = n$, so $m = 1$ and $n = 3$.

29. Discontinuous at $\dfrac{n}{2}$, n an integer

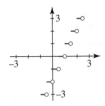

31. Since $\lim\limits_{x \to 9} \dfrac{x - 9}{\sqrt{x} - 3} = \lim\limits_{x \to 9} \dfrac{(\sqrt{x} + 3)(\sqrt{x} - 3)}{\sqrt{x} - 3} = \lim\limits_{x \to 9}(\sqrt{x} + 3) = 6$, define $f(9) = 6$.

33. $\lim\limits_{x \to \pi/6} \sin(2x + \dfrac{\pi}{3}) = \sin\left[\lim\limits_{x \to \pi/6}(2x + \dfrac{\pi}{3})\right] = \sin\dfrac{2\pi}{3} = \dfrac{\sqrt{3}}{2}$

35. $\lim\limits_{x \to \pi/2} \sin(\cos x) = \sin\left(\lim\limits_{x \to \pi/2} \cos x\right) = \sin(\cos\dfrac{\pi}{2}) = \sin 0 = 0$

37. $\lim\limits_{t \to \pi} \cos\left(\dfrac{t^2 - \pi^2}{t - \pi}\right) = \cos\left[\lim\limits_{t \to \pi} \dfrac{(t - \pi)(t + \pi)}{t - \pi}\right] = \cos 2\pi = 1$

39. $\lim\limits_{t \to \pi} \sqrt{t - \pi + \cos^2 t} = \sqrt{\cos^2 \pi} = \sqrt{1} = 1$

41. $\lim\limits_{x \to -3} \sin^{-1}\left(\dfrac{x + 3}{x^2 + 4x + 3}\right) = \sin^{-1}\left(\lim\limits_{x \to -3} \dfrac{x + 3}{x^2 + 4x + 3}\right)$

$= \sin^{-1}\left[\lim\limits_{x \to -3} \dfrac{x + 3}{(x + 3)(x + 1)}\right]$

$= \sin^{-1}\left(\lim\limits_{x \to -3} \dfrac{1}{x + 1}\right) = \sin^{-1}\left(-\dfrac{1}{2}\right) = -\dfrac{\pi}{6}$

43. Since $(f \circ g)(x) = \dfrac{1}{\sqrt{x + 3}}$, $f \circ g$ is continuous for $x + 3 > 0$ or on $(-3, \infty)$.

45. $f(1) = -1$, $f(5) = 15$. By the Intermediate Value Theorem, since $-1 \leq 8 \leq 15$, there exists $c \in [1, 5]$ such that $c^2 - 2c = 8$. Setting $c^2 - 2c - 8 = 0$ gives us $(c - 4)(c + 2) = 0$ or $c = -2, 4$. On $[1, 5]$, $c = 4$.

47. $f(-2) = -3$, $f(2) = 5$. By the Intermediate Value Theorem, since $-3 \le 1 \le 5$, there exists $c \in [-2, 2]$ such that $c^3 - 2x + 1 = 1$. Setting $c^3 - 2c = 0$ gives us $c(c^2 - 2) = 0$. On $[-2, 2]$, $c = 0, \pm\sqrt{2}$.

49. Since $f(0) = -7$, $f(3) = 242$, and $-7 \le 50 \le 242$, then by the Intermediate Value Theorem there exists $c \in [0, 3]$ such that $f(c) = 50$.

51. The equation will have a solution on $(0, 1)$ if $f(x) = 2x^7 + x - 1$ is 0 on $(0, 1)$. Since $f(0) = -1$ and $f(1) = 2$, then by the Intermediate Value Theorem $f(c) = 0$ for some $c \in (0, 1)$.

53. Let $f(x) = e^{-x} - \ln x$. Then $f(1) = e^{-1} - \ln 1 = e^{-1} > 0$ and $f(2) = e^{-2} - \ln 2 < 0$. Thus, by the corollary to the Intermediate Value Theorem, $f(c) = 0$ for some $c \in (1, 2)$.

55. In $[-2, -1]$ the zero is approximately -1.21. In $[-1, 0]$ the zero is approximately -0.64. In $[1, 2]$ the zero is approximately 1.34.

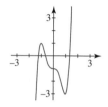

57. We want to solve $f(x) = x^5 + 2x - 7 = 50$ or $x^5 + 2x - 57 = 0$. It is easily seen that the expression on the left side of this equation is negative when $x = 2$ and positive when $x = 3$. Applying the bisection method on $[2, 3]$, we find $c \approx 2.21$.

59. In the solution of Problem 52 we saw that there is a zero in $[0, 1]$. Applying the bisection method on this interval, we find $c \approx 0.78$.

61. Since f and g are continuous at a, then $\lim_{x \to a} f(x) = f(a)$ and $\lim_{x \to a} g(x) = g(a)$. From this, we get

$$\lim_{x \to a}(f+g)(x) = \lim_{x \to a}[f(x) + g(x)] = \lim_{x \to a} f(x) + \lim_{x \to a} g(x)$$
$$= f(a) + g(a) = (f+g)(a).$$

Thus, $f + g$ is continuous at a.

63. $f \circ g$ will be discontinuous whenever $\cos x$ is an integer. In the interval $[0, 2\pi)$, this will be the case whenever $x = 0$, $\pi/2$, π, or $3\pi/2$. Thus, $f \circ g$ will be discontinuous for $x = n\pi/2$, n an integer.

65. (a) For any real a, $\lim_{x \to a} f(x)$ does not exist since f takes on the values 0 and 1 arbitrarily close to any real number. Therefore, the Dirichlet function is discontinuous at every real number.

    (b) The graph consists of infinitely many points on each of the lines $y = 0$ and $y = 1$. In fact, between any two real numbers, there are infinitely many points of the graph on the line $y = 1$ and infinitely many points of the graph on the line $y = 0$.

    (c) Let r be a positive rational number. If x is rational, then $x + r$ is rational so that $f(x + r) = 1 = f(x)$. If x is irrational, then $x + r$ is irrational so that $f(x + r) = 0 = f(x)$.

2.4 Trigonometric Limits

1. $\lim\limits_{t \to 0} \dfrac{\sin 3t}{2t} = \dfrac{1}{2} \lim\limits_{t \to 0} \dfrac{\sin 3t}{t} = \dfrac{3}{2}$

3. $\lim\limits_{x \to 0} \dfrac{\sin x}{4 + \cos x} = \dfrac{0}{4 + 1} = 0$

5. $\lim\limits_{x \to 0} \dfrac{\cos 2x}{\cos 3x} = 1$

7. $\lim\limits_{t \to 0} \dfrac{1}{t \sec t \csc 4t} = \lim\limits_{t \to 0} \left(\dfrac{\sin 4t}{t} \cdot \cos t \right) = 4 \cdot 1 = 4$

9. $\lim\limits_{t \to 0} \dfrac{2 \sin^2 t}{t \cos^2 t} = 2 \lim\limits_{t \to 0} \left(\dfrac{\sin t}{t} \cdot \dfrac{\sin t}{\cos^2 t} \right) = 2 \cdot 1 \cdot 0 = 0$

11. $\lim\limits_{t \to 0} \dfrac{\sin^2 6t}{t^2} = \lim\limits_{t \to 0} \left(\dfrac{\sin 6t}{t} \right)^2 = 6^2 = 36$

13. $\lim\limits_{x \to 1} \dfrac{\sin(x-1)}{2x-2} = \dfrac{1}{2} \lim\limits_{x \to 1} \dfrac{\sin(x-1)}{x-1} = \dfrac{1}{2}$

15. $\lim\limits_{x \to 0} \dfrac{\cos x}{x}$ does not exist.

17. $\lim\limits_{x \to 0} \dfrac{\cos(3x - \pi/2)}{x} = \lim\limits_{x \to 0} \dfrac{\sin 3x}{x} = 3$

19. $\lim\limits_{t \to 0} \dfrac{\sin 3t}{\sin 7t} = \lim\limits_{t \to 0} \left(\dfrac{\sin 3t}{t} \cdot \dfrac{t}{\sin 7t} \right) = \left(\lim\limits_{t \to 0} \dfrac{\sin 3t}{t} \right) \left[\lim\limits_{t \to 0} \dfrac{1}{(\sin 7t)/t} \right] = 3 \cdot \dfrac{1}{7} = \dfrac{3}{7}$

21. $\lim\limits_{t \to 0^+} \dfrac{\sin t}{\sqrt{t}} = \lim\limits_{t \to 0^+} \left(\sqrt{t} \cdot \dfrac{\sin t}{t} \right) = \left(\lim\limits_{t \to 0^+} \sqrt{t} \right) \left(\lim\limits_{t \to 0^+} \dfrac{\sin t}{t} \right) = 0 \cdot 1 = 0$

23. $\lim\limits_{t \to 0} \dfrac{t^2 - 5t \sin t}{t^2} = \lim\limits_{t \to 0} \left[1 - 5 \left(\dfrac{\sin t}{t} \right) \right] = 1 - 5 = -4$

25. $\lim\limits_{x \to 0^+} \dfrac{(x + 2\sqrt{\sin x})^2}{x} = \lim\limits_{x \to 0^+} \dfrac{x^2 + 4x\sqrt{\sin x} + 4 \sin x}{x}$

   $= \lim\limits_{x \to 0^+} \left(x + 4\sqrt{\sin x} + \dfrac{4 \sin x}{x} \right) = 0 + 0 + 4 = 4$

27. $\lim\limits_{x \to 0} \dfrac{\cos x - 1}{\cos^2 x - 1} = \left(\lim\limits_{x \to 0} \dfrac{\cos x - 1}{\cos x - 1} \right) \left(\lim\limits_{x \to 0} \dfrac{1}{\cos x + 1} \right) = 1 \cdot \dfrac{1}{2} = \dfrac{1}{2}$

29. Letting $u = x^2$, we have $\lim\limits_{x \to 0} \dfrac{\sin 5x^2}{x^2} = \lim\limits_{u \to 0} \dfrac{\sin 5u}{u} = 5$.

31. First, rewrite $\lim\limits_{x \to 2} \dfrac{\sin(x-2)}{x^2 + 2x - 8}$ as $\lim\limits_{x \to 2} \dfrac{\sin(x-2)}{(x-2)(x+4)}$.

   Letting $u = x - 2$, we get $\lim\limits_{u \to 0} \left(\dfrac{\sin u}{u} \cdot \dfrac{1}{u+6} \right) = 1 \cdot \dfrac{1}{6} = \dfrac{1}{6}$.

33. $\lim\limits_{x \to 0} \dfrac{2 \sin 4x + 1 - \cos x}{x} = \lim\limits_{x \to 0} \left(\dfrac{2 \sin 4x}{x} + \dfrac{1 - \cos x}{x} \right) = 8 + 0 = 8$

35. Start by multiplying the function by $\dfrac{1 + \tan x}{1 + \tan x}$, producing:

$$\lim\limits_{x \to \pi/4} \dfrac{1 - \tan x}{\cos x - \sin x} = \lim\limits_{x \to \pi/4} \left(\dfrac{1 - \tan x}{\cos x - \sin x} \cdot \dfrac{1 + \tan x}{1 + \tan x} \right)$$

$$= \lim\limits_{x \to \pi/4} \dfrac{1 - \tan^2 x}{(\cos x - \sin x)(1 + \tan x)}$$

Focusing first on the denominator, we multiply out and simplify:

$$(\cos x - \sin x)(1 + \tan x) = \cos x + \cos x \tan x - \sin x - \sin x \tan x$$

$$= \cos x + \cos x \left(\dfrac{\sin x}{\cos x} \right) - \sin x \left(\dfrac{\cos x}{\cos x} \right) - \sin x \left(\dfrac{\sin x}{\cos x} \right)$$

$$= \cos x - \dfrac{\sin^2 x}{\cos x} = \dfrac{\cos^2 x - \sin^2 x}{\cos x}$$

Substituting this result back into the function, we get:

$$\frac{1-\tan^2 x}{(\cos x - \sin x)(1+\tan x)} = (1-\tan^2 x)\left(\frac{\cos x}{\cos^2 x - \sin^2 x}\right)$$

$$= \frac{\cos x - \cos x\left(\dfrac{\sin^2 x}{\cos^2 x}\right)}{\cos^2 x - \sin^2 x} = \frac{\cos x - \left(\dfrac{\sin^2 x}{\cos x}\right)}{\cos^2 x - \sin^2 x}$$

$$= \left(\frac{\cos^2 x - \sin^2 x}{\cos x}\right)\left(\frac{1}{\cos^2 x - \sin^2 x}\right) = \frac{1}{\cos x}$$

Finally, returning to the limit, we have:

$$\lim_{x\to\pi/4}\frac{1-\tan x}{\cos x - \sin x} = \lim_{x\to\pi/4}\frac{1}{\cos x} = \frac{1}{\sqrt{2}/2} = \sqrt{2}$$

37. $\displaystyle\lim_{h\to 0}\frac{f\left(\dfrac{\pi}{4}+h\right)-f\left(\dfrac{\pi}{4}\right)}{h} = \lim_{h\to 0}\frac{\sin\left(\dfrac{\pi}{4}+h\right)-\sin\left(\dfrac{\pi}{4}\right)}{h}$

$\displaystyle = \lim_{h\to 0}\frac{\sin(\pi/4)\cos h + \cos(\pi/4)\sin h - \sin(\pi/4)}{h}$

$\displaystyle = \lim_{h\to 0}\frac{(\sqrt{2}/2)\cos h + (\sqrt{2}/2)\sin h - (\sqrt{2}/2)}{h}$

$\displaystyle = \frac{\sqrt{2}}{2}\lim_{h\to 0}\frac{\cos h + \sin h - 1}{h} = \frac{\sqrt{2}}{2}\lim_{h\to 0}\left(\frac{\cos h - 1}{h} + \frac{\sin h}{h}\right) = \frac{\sqrt{2}}{2}$

39. Since $-1 \le \sin\dfrac{1}{x} \le 1$, then $-|x| \le x\sin\dfrac{1}{x} \le |x|$. Since $\lim_{x\to 0}(-|x|) = 0$ and $\lim_{x\to 0}|x| = 0$, then by the Squeeze Theorem, $\lim_{x\to 0}x\sin\dfrac{1}{x} = 0$.

41. For both limits, we use the result from Problem 39, $\displaystyle\lim_{x\to 0}x\sin\frac{1}{x} = 0$:

(a) $\displaystyle\lim_{x\to 0}x^3\sin\frac{1}{x} = \lim_{x\to 0}\left(x^2\cdot x\sin\frac{1}{x}\right) = \lim_{x\to 0}x^2\cdot\lim_{x\to 0}x\sin\frac{1}{x} = 0\cdot 0 = 0$

(b) $\displaystyle\lim_{x\to 0}x^2\sin^2\frac{1}{x} = \left(\lim_{x\to 0}x\sin\frac{1}{x}\right)\left(\lim_{x\to 0}x\sin\frac{1}{x}\right) = 0\cdot 0 = 0$

43. Since $\lim_{x\to 2}(2x-1) = 3$ and $\lim_{x\to 2}(x^2 - 2x + 3) = 3$, then by the Squeeze Theorem, $\lim_{x\to 2}f(x) = 3$.

45. Let $t = x - \dfrac{\pi}{4}$. Thus, $x = t + \dfrac{\pi}{4}$ and we have the following substitutions:

$$\sin x = \sin(t + \frac{\pi}{4}) = \sin t\cos\frac{\pi}{4} + \cos t\sin\frac{\pi}{4} = \frac{\sqrt{2}}{2}\sin t + \frac{\sqrt{2}}{2}\cos t$$

$$\cos x = \cos(t + \frac{\pi}{4}) = \cos t\cos\frac{\pi}{4} - \sin t\sin\frac{\pi}{4} = \frac{\sqrt{2}}{2}\cos t - \frac{\sqrt{2}}{2}\sin t$$

$$\sin x - \cos x = \left(\frac{\sqrt{2}}{2}\sin t + \frac{\sqrt{2}}{2}\cos t\right) - \left(\frac{\sqrt{2}}{2}\cos t - \frac{\sqrt{2}}{2}\sin t\right) = \sqrt{2}\sin t$$

With these substitutions, $\displaystyle\lim_{x\to\pi/4}\frac{\sin x - \cos x}{x - (\pi/4)} = \lim_{t\to 0}\frac{\sqrt{2}\sin t}{t} = \sqrt{2}$.

47. Let $t = \pi - (\pi/x)$. Therefore $\pi/x = \pi - t$ and $\sin(\pi/x) = \sin(\pi - t) = \sin t$. In addition, we can derive $x - 1 = \dfrac{t}{\pi - t}$, giving us:

$$\lim_{x\to 1}\frac{\sin(\pi/x)}{x-1} = \lim_{t\to 0}\frac{(\sin t)(\pi - t)}{t} = \lim_{t\to 0}\frac{\sin t}{t}\cdot\lim_{t\to 0}(\pi - t) = 1\cdot\pi = \pi$$

49. f is continuous at $x = 0$ because $\displaystyle\lim_{x\to 0}\frac{\sin x}{x} = 1 = f(0)$.

2.5   Limits That Involve Infinity

1. $-\infty$

3. ∞

5. ∞

7. ∞

9. $\displaystyle\lim_{x\to\infty}\frac{x^2-3x}{4x^2+5}=\lim_{x\to\infty}\frac{1-3/x}{4+5/x^2}=\frac{1}{4}$

11. 5

13. $\displaystyle\lim_{x\to\infty}\frac{8-\sqrt{x}}{1+4\sqrt{x}}=\lim_{x\to\infty}\frac{(8/\sqrt{x})-1}{(1/\sqrt{x})+4}=-\frac{1}{4}$

15. $\displaystyle\lim_{x\to\infty}\left(\frac{3x}{x+2}-\frac{x-1}{2x+6}\right)=\lim_{x\to\infty}\left(\frac{3}{1+2/x}-\frac{1-1/x}{2+6/x}\right)=3-\frac{1}{2}=\frac{5}{2}$

17. $\displaystyle\lim_{x\to\infty}\sqrt{\frac{3x+2}{6x-8}}=\lim_{x\to\infty}\sqrt{\frac{3+2/x}{6-8/x}}=\sqrt{\frac{1}{2}}=\frac{\sqrt{2}}{2}$

19. $\displaystyle\lim_{x\to\infty}\left(x-\sqrt{x^2+1}\right)=\lim_{x\to\infty}\left(x-\sqrt{x^2+1}\right)\cdot\frac{x+\sqrt{x^2+1}}{x+\sqrt{x^2+1}}=\lim_{x\to\infty}\frac{-1}{x+\sqrt{x^2+1}}=0$

21. $\displaystyle\lim_{x\to\infty}\cos\left(\frac{5}{x}\right)=\cos\left[\lim_{x\to\infty}\left(\frac{5}{x}\right)\right]=1$

23. $\displaystyle\lim_{x\to-\infty}\sin^{-1}\left(\frac{x}{\sqrt{4x^2+1}}\right)=\lim_{x\to-\infty}\sin^{-1}\left[\frac{\left(\frac{x}{|x|}\right)}{\sqrt{4+1/x^2}}\right]=\lim_{x\to-\infty}\sin^{-1}\left[\frac{\left(\frac{x}{-x}\right)}{\sqrt{4+1/x^2}}\right]$

$\displaystyle\qquad\qquad=\sin^{-1}\left[\lim_{x\to-\infty}\left(\frac{-1}{\sqrt{4+1/x^2}}\right)\right]=\sin^{-1}\left(-\frac{1}{2}\right)=-\frac{\pi}{6}$

25. Start with $\displaystyle\frac{4x+1}{\sqrt{x^2+1}}=\frac{\left(\frac{4x}{|x|}+\frac{1}{|x|}\right)}{\sqrt{1+1/x^2}}$. From this, $\displaystyle\lim_{x\to-\infty}f(x)=\lim_{x\to-\infty}\frac{-4-1/x}{\sqrt{1+1/x^2}}=-4$ and $\displaystyle\lim_{x\to\infty}f(x)=\lim_{x\to\infty}\frac{4+1/x}{\sqrt{1+1/x^2}}=4$.

27. Start with $\displaystyle\frac{2x+1}{\sqrt{3x^2+1}}=\frac{\left(\frac{2x}{|x|}+\frac{1}{|x|}\right)}{\sqrt{3+1/x^2}}$. From this, $\displaystyle\lim_{x\to-\infty}f(x)=\lim_{x\to-\infty}\frac{-2-1/x}{\sqrt{3+1/x^2}}=-\frac{2}{\sqrt{3}}=-\frac{2\sqrt{3}}{3}$ and $\displaystyle\lim_{x\to\infty}f(x)=\lim_{x\to\infty}\frac{2+1/x}{\sqrt{3+1/x^2}}=$ $\displaystyle\frac{2}{\sqrt{3}}=\frac{2\sqrt{3}}{3}$.

29. $\displaystyle\lim_{x\to-\infty}\frac{e^x-e^{-x}}{e^x+e^{-x}}=\frac{\left(\lim_{x\to-\infty}e^x\right)-\left(\lim_{x\to-\infty}e^{-x}\right)}{\left(\lim_{x\to-\infty}e^x\right)+\left(\lim_{x\to-\infty}e^{-x}\right)}=\frac{0-\left(\lim_{x\to-\infty}e^{-x}\right)}{0+\left(\lim_{x\to-\infty}e^{-x}\right)}$

$\displaystyle\qquad\qquad=\lim_{x\to-\infty}\frac{-e^{-x}}{e^{-x}}=\lim_{x\to-\infty}-1=-1$

$\displaystyle\lim_{x\to\infty}\frac{e^x-e^{-x}}{e^x+e^{-x}}=\frac{\left(\lim_{x\to\infty}e^x\right)-\left(\lim_{x\to\infty}e^{-x}\right)}{\left(\lim_{x\to\infty}e^x\right)+\left(\lim_{x\to\infty}e^{-x}\right)}=\frac{\left(\lim_{x\to\infty}e^x\right)-0}{\left(\lim_{x\to\infty}e^x\right)+0}$

$\displaystyle\qquad\qquad=\lim_{x\to\infty}\frac{e^x}{e^x}=\lim_{x\to\infty}1=1$

31. $\displaystyle\lim_{x\to-\infty}\frac{|x-5|}{x-5}=\lim_{x\to-\infty}\frac{-x+5}{x-5}=\lim_{x\to-\infty}\frac{-1+5/x}{1-5/x}=-1$

$\displaystyle\lim_{x\to\infty}\frac{|x-5|}{x-5}=\lim_{x\to\infty}\frac{x-5}{x-5}=1$

33.
Vertical asymptote: none
Horizontal asymptote: $y=0$

35.
Vertical asymptote: $x=-1$
Horizontal asymptote: none

37.
Vertical asymptote: $x=0,\,x=2$
Horizontal asymptote: $y=0$

39.
Vertical asymptote: $x=1$
Horizontal asymptote: $y=1$

41.
Vertical asymptote: none
Horizontal asymptote: $y=-1,\,y=1$

43.   (a) 2   (b) $-\infty$ (c) 0   (d) 2

45.   (a) $-\infty$ (b) $-3/2$ (c) ∞ (d) 0

47.

49.

51. $\lim\limits_{x\to\infty} x\sin\dfrac{3}{x} = \lim\limits_{x\to\infty}\left(x\sin\dfrac{3}{x}\right)\left(\dfrac{3/x}{3/x}\right) = \lim\limits_{x\to\infty} x(3/x)\left(\dfrac{\sin 3/x}{3/x}\right)$

$\qquad = \left(\lim\limits_{x\to\infty} x\cdot\dfrac{3}{x}\right)\left(\lim\limits_{x\to\infty}\dfrac{\sin 3/x}{3/x}\right) = \left(\lim\limits_{x\to\infty} 3\right)\left(\lim\limits_{x\to\infty}\dfrac{\sin 3/x}{3/x}\right)$

At this point, we substitute $t = 3/x$, resulting in:

$$\left(\lim\limits_{x\to\infty} 3\right)\left(\lim\limits_{x\to\infty}\dfrac{\sin 3/x}{3/x}\right) = 3\lim\limits_{t\to 0}\dfrac{\sin t}{t} = 3$$

53.

$x\to\infty$	10	100	1000	10000
$f(x)$	1.99986667	1.99999999	2.00000000	2.00000000

$\lim\limits_{x\to\infty} x^2\sin\dfrac{2}{x^2} = 2$

55.

   (a) $\lim\limits_{x\to -1^+} f(x) = \infty$ (b) $\lim\limits_{x\to 0} f(x) \approx 2.7$ (c) $\lim\limits_{x\to\infty} f(x) = 1$

57.   (a) $\lim\limits_{x\to\pm\infty} [f(x) - g(x)] = \lim\limits_{x\to\pm\infty}\left[\dfrac{x^2}{x+1} - (x-1)\right]$

$\qquad = \lim\limits_{x\to\pm\infty}\left[\dfrac{x^2}{x+1} - \dfrac{(x-1)(x+1)}{x+1}\right] = \lim\limits_{x\to\pm\infty}\dfrac{x^2-(x^2-1)}{x+1}$

$\qquad = \lim\limits_{x\to\pm\infty}\dfrac{1}{x+1} = 0$

  (b) The graphs of f and g get closer and closer to each other when $|x|$ is large.

  (c) g is a *slant asymptote* to f.

2.6 Limits—A Formal Approach

1. $|10-10| = 0 < \varepsilon$ for any choice of δ.

3. $|x-3| < \varepsilon$ whenever $0 < |x-3| < \varepsilon$. Choose $\delta = \varepsilon$.

5. $|x+6-5| = |x+1| < \varepsilon$ whenever $0 < |x-(-1)| < \varepsilon$. Choose $\delta = \varepsilon$.

7. $|3x+7-7| = 3|x-0| < \varepsilon$ whenever $0 < |x-0| < \varepsilon/3$. Choose $\delta = \varepsilon/3$.

9. $\left|\dfrac{2x-3}{4} - \dfrac{1}{4}\right| = \dfrac{1}{4}|2x-4| = \dfrac{1}{2}|x-2| < \varepsilon$ whenever $0 < |x-2| < 2\varepsilon$. Choose $\delta = 2\varepsilon$.

11. $\left|\dfrac{x^2-25}{x+5} - (-10)\right| = |x-5+10| = |x-(-5)| < \varepsilon$ whenever $0 < |x-(-5)| < \varepsilon$. Choose $\delta = \varepsilon$.

13. $\left|\dfrac{8x^5+12x^4}{x^4} - 12\right| = |8x+12-12| = 8|x-0| < \varepsilon$ whenever $0 < |x-0| < \varepsilon/8$. Choose $\delta = \varepsilon/8$.

15. $|x^2-0| = |x-0|^2 < \varepsilon$ whenever $0 < |x-0| < \sqrt{\varepsilon}$. Choose $\delta = \sqrt{\varepsilon}$.

17. $|\sqrt{5x}-0| = \sqrt{5}|x-0|^{1/2} < \varepsilon$ whenever $0 < x < \varepsilon^2/5$. Choose $\delta = \varepsilon^2/5$.

19. $|2x-1-(-1)| = |2x| = 2|x-0| < \varepsilon$ whenever $0 - \varepsilon/2 < x < 0$. Choose $\delta = \varepsilon/2$.

21. Note that $|x^2 - 9| = |x-3||x+3|$ and consider only values of x for which $|x-3| < 1$. Then $2 < x < 4$ and $5 < x+3 < 7$, so $|x+3| < 7$. Thus, $|x^2 - 9| = |x-3||x+3| < 7|x-3| < \varepsilon$ whenever $|x-3| < \varepsilon/7$. Choose $\delta = \min\{1, \varepsilon/7\}$.

23. Note that $|x^2 - 2x + 4 - 3| = |x-1|^2 < \varepsilon$ whenever $|x-1| < \sqrt{\varepsilon}$. Choose $\delta = \sqrt{\varepsilon}$.

25. We need to show $|\sqrt{x} - \sqrt{a}| < \varepsilon$ whenever $0 < |x-a| < \delta$ for an appropriate choice of δ. For $\delta = \sqrt{a}\varepsilon$, we have

$$|\sqrt{x} - \sqrt{a}| = |\sqrt{x} - \sqrt{a}| \cdot \frac{\sqrt{x} + \sqrt{a}}{\sqrt{x} + \sqrt{a}} = \frac{|x-a|}{\sqrt{x} + \sqrt{a}} < \frac{|x-a|}{\sqrt{a}} < \frac{\sqrt{a}\varepsilon}{\sqrt{a}} = \varepsilon$$

whenever $0 < |x-a| < \delta$. Thus, $\lim\limits_{x \to a} \sqrt{x} = \sqrt{a}$.

27. Assume $\lim\limits_{x \to 1} f(x) = L$. Take $\varepsilon = 1$. Then there exists $\delta > 0$ such that $|f(x) - L| < 1$ whenever $0 < |x-1| < \delta$. To the right of 1, choose $x = 1 + \delta/2$.

    Since $0 < |1 + \delta/2 - 1| = |\delta/2| < \delta$,

    we must have $|f(1 + \delta/2) - L| = |0 - L| = |L| < 1$,

    or $-1 < L < 1$.

    To the left of 1, choose $x = 1 - \delta/2$.

    Since $0 < |1 - \delta/2 - 1| = |-\delta/2| < \delta$,

    we must have $|f(1 - \delta/2) - L| = |2 - L| < 1$,

    or $1 < L < 3$.

    Since no L can satisfy the conditions that $-1 < L < 1$ *and* $1 < L < 3$, we conclude that $\lim\limits_{x \to 1} f(x)$ does not exist.

29. Assume $\lim\limits_{x \to 0} f(x) = L$. Take $\varepsilon = 1$. Then there exists $\delta > 0$ such that $|f(x) - L| < 1$ whenever $0 < |x-0| < \delta$. To the right of 0, choose $x = \delta/2$.

    Since $0 < |\delta/2 - 0| = |\delta/2| < \delta$,

    we must have $|f(\delta/2) - L| = |2 - \delta/2 - L| < 1$,

    or $1 - \delta/2 < L < 3 - \delta/2$.

    To the left of 0, choose $x = -\delta/2$.

    Since $0 < |-\delta/2 - 0| = |-\delta/2| < \delta$,

    we must have $|f(-\delta/2) - L| = |-\delta/2 - L| < 1$,

    or $-1 - \delta/2 < L < 1 - \delta/2$.

    Since no L can satisfy the conditions that $1 - \delta/2 < L < 3 - \delta/2$ *and* $-1 - \delta/2 < L < 1 - \delta/2$, we conclude that $\lim\limits_{x \to 0} f(x)$ does not exist.

31. By Definition 2.6.5(i), for any $\varepsilon > 0$ we must find an $N > 0$ such that

$$\left| \frac{5x-1}{2x+1} - \frac{5}{2} \right| < \varepsilon \text{ whenever } x > N.$$

Now by considering $x > 0$,

$$\left| \frac{5x-1}{2x+1} - \frac{5}{2} \right| = \left| \frac{-7}{4x+2} \right| = \frac{7}{4x+2} < \frac{7}{4x} < \varepsilon$$

whenever $x > 7/4\varepsilon$. Hence, choose $N = 7/4\varepsilon$.

33. By Definition 2.6.5(ii), for any $\varepsilon > 0$ we must find an $N < 0$ such that

$$\left| \frac{10x}{x-3} - 10 \right| < \varepsilon \text{ whenever } x < N.$$

Now by considering $x < 0$,

$$\left| \frac{10x}{x-3} - 10 \right| = \left| \frac{30}{x-3} \right| = \left| \frac{30}{-(-x+3)} \right| = \frac{30}{-x+3} < \frac{30}{x} < \varepsilon$$

whenever $x < -30/\varepsilon$. Hence, choose $N = -30/\varepsilon$.

35. We need to show $|f(x) - 0| = |f(x)| < \varepsilon$ whenever $0 < |x - 0| = |x| < \delta$ for an appropriate choice of δ. For $\delta = \varepsilon$,

$$|f(x)| = \begin{cases} |x|, & x \text{ rational} \\ 0, & x \text{ irrational} \end{cases} < \varepsilon \text{ whenever } 0 < |x| < \delta.$$

Thus, $\lim_{x \to 0} f(x) = 0$.

2.7 The Tangent Line Problem

1. change in $x = h = 2.5 - 2 = 0.5$
   change in $y = f(2 + 0.5) - f(2) = 2.75 - 5 = -2.25$
   $$m_{\text{sec}} = \frac{\text{change in } y}{\text{change in } x} = \frac{-2.25}{0.5} = -4.5$$

3. change in $x = h = -1 - (-2) = 1$
   change in $y = f(-2 + 1) - f(-2) = -1 - (-8) = 7$
   $$m_{\text{sec}} = \frac{\text{change in } y}{\text{change in } x} = \frac{7}{1} = 7$$

5. 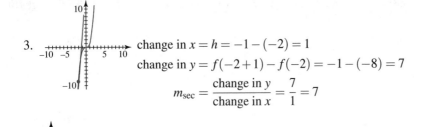 change in $x = h = \frac{2\pi}{3} - \frac{\pi}{2} = \frac{\pi}{6}$
   change in $y = f\left(\frac{\pi}{2} + \frac{\pi}{6}\right) - f\left(\frac{\pi}{2}\right) = \sin\frac{2}{3}\pi - 1 = \sqrt{3}/2 - 1$
   $$m_{\text{sec}} = \frac{\text{change in } y}{\text{change in } x} = \frac{\sqrt{3}/2 - 1}{\pi/6} = \frac{3\sqrt{3} - 6}{\pi}$$

7. $f(a) = f(3) = 3;\ f(a + h) = f(3 + h) = (h + 3)^2 - 6$
   $f(a + h) - f(a) = [(h + 3)^2 - 6] - 3 = [(h^2 + 6h + 9) - 6] - 3 = h^2 + 6h = h(h + 6)$
   $$m_{\text{tan}} = \lim_{h \to 0} \frac{f(a + h) - f(a)}{h} = \lim_{h \to 0} \frac{h(h + 6)}{h} = \lim_{h \to 0} (h + 6) = 6$$
   With point of tangency $(3, 3)$, we have $y - 3 = 6(x - 3)$ or $y = 6x - 15$.

9. $f(a) = f(1) = -2;\ f(a + h) = f(1 + h) = (h + 1)^2 - 3(h + 1)$
   $f(a + h) - f(a) = [(h + 1)^2 - 3(h + 1)] - (-2) = (h^2 - h - 2) - (-2) = h^2 - h = h(h - 1)$
   $$m_{\text{tan}} = \lim_{h \to 0} \frac{f(a + h) - f(a)}{h} = \lim_{h \to 0} \frac{h(h - 1)}{h} = \lim_{h \to 0} (h - 1) = -1$$
   With point of tangency $(1, -2)$, we have $y + 2 = -(x - 1)$ or $y = -x - 1$.

11. $\quad f(a) = f(2) = -14;\ f(a+h) = f(2+h) = -2(h+2)^3 + (h+2)$

$$f(a+h) - f(a) = [-2(h+2)^3 + (h+2)] - (-14)$$
$$= (-2h^3 - 12h^2 - 23h - 14) - (-14) = h(-2h^2 - 12h - 23)$$
$$m_{\tan} = \lim_{h \to 0} \frac{f(a+h) - f(a)}{h} = \lim_{h \to 0} \frac{h(-2h^2 - 12h - 23)}{h}$$
$$= \lim_{h \to 0}(-2h^2 - 12h - 23) = -23$$

With point of tangency $(2, -14)$, we have $y + 14 = -23(x - 2)$ or $y = -23x + 32$.

13. $\quad f(a) = f(-1) = -1/2;\ f(a+h) = f(-1+h) = \dfrac{1}{2(h-1)}$

$$f(a+h) - f(a) = \frac{1}{2(h-1)} - \left(-\frac{1}{2}\right) = \frac{1+h-1}{2(h-1)} = \frac{h}{2(h-1)}$$
$$m_{\tan} = \lim_{h \to 0} \frac{f(a+h) - f(a)}{h} = \lim_{h \to 0}\left[\frac{h}{2(h-1)} \cdot \frac{1}{h}\right]$$
$$= \lim_{h \to 0} \frac{1}{2(h-1)} = -\frac{1}{2}$$

With point of tangency $(-1, -1/2)$, we have $y + \dfrac{1}{2} = -\dfrac{1}{2}(x+1)$ or $y = -\dfrac{x}{2}$.

15. $\quad f(a) = f(0) = 1;\ f(a+h) = f(h) = \dfrac{1}{(h-1)^2}$

$$f(a+h) - f(a) = \frac{1}{(h-1)^2} - 1 = \frac{-h^2 + 2h}{(h-1)^2} = \frac{h(2-h)}{(h-1)^2}$$
$$m_{\tan} = \lim_{h \to 0} \frac{f(a+h) - f(a)}{h} = \lim_{h \to 0}\left[\frac{h(2-h)}{(h-1)^2} \cdot \frac{1}{h}\right] = \lim_{h \to 0} \frac{2-h}{(h-1)^2} = 2$$

With point of tangency $(0, 1)$, we have $y - 1 = 2(x - 0)$ or $y = 2x + 1$.

17. $\quad f(a) = f(4) = 2;\ f(a+h) = f(4+h) = \sqrt{4+h}$

$$f(a+h) - f(a) = \sqrt{4+h} - 2 = (\sqrt{4+h} - 2)\frac{\sqrt{4+h}+2}{\sqrt{4+h}+2} = \frac{4+h-4}{\sqrt{4+h}+2} = \frac{h}{\sqrt{4+h}+2}$$
$$m_{\tan} = \lim_{h \to 0} \frac{f(a+h) - f(a)}{h} = \lim_{h \to 0}\left(\frac{h}{\sqrt{4+h}+2} \cdot \frac{1}{h}\right)$$
$$= \lim_{h \to 0} \frac{1}{\sqrt{4+h}+2} = \frac{1}{4}$$

With point of tangency $(4, 2)$, we have $y - 2 = \dfrac{1}{4}(x - 4)$ or $y = \dfrac{1}{4}x + 1$.

19. $\quad f(a) = f(\pi/6) = 1/2;\ f(a+h) = f(\pi/6 + h) = \sin(\pi/6 + h)$

$$f(a+h) - f(a) = \sin\left(\frac{\pi}{6} + h\right) - \frac{1}{2} = \sin\frac{\pi}{6}\cos h + \cos\frac{\pi}{6}\sin h - \frac{1}{2}$$
$$= \frac{1}{2}\cos h + \frac{\sqrt{3}}{2}\sin h - \frac{1}{2} = \frac{1}{2}(\cos h - 1) + \frac{\sqrt{3}}{2}\sin h$$
$$m_{\tan} = \lim_{h \to 0} \frac{f(a+h) - f(a)}{h} = \lim_{h \to 0}\left(\frac{1}{2} \cdot \frac{\cos h - 1}{h} + \frac{\sqrt{3}}{2} \cdot \frac{\sin h}{h}\right)$$
$$= (1/2)(0) + (\sqrt{3}/2)(1) = \sqrt{3}/2$$

With point of tangency $(\pi/6, 1/2)$, we have $y - \dfrac{1}{2} = \dfrac{\sqrt{3}}{2}\left(x - \dfrac{\pi}{6}\right)$ or $y = \dfrac{\sqrt{3}}{2}x - \dfrac{\sqrt{3}\pi}{12} + \dfrac{1}{2}$.

21. $f(a) = f(1) = 1$; $f(a+h) = f(1+h) = (h+1)^2$

$f(a+h) - f(a) = [(h+1)^2] - 1 = (h^2 + 2h + 1) - 1 = h(h+2)$

$$m_{\tan} = \lim_{h \to 0} \frac{f(a+h) - f(a)}{h} = \lim_{h \to 0} \frac{h(h+2)}{h} = \lim_{h \to 0} (h+2) = 2$$

The slope of the tangent at the blue point $(1,1)$ is 2. The slope of the line through $(1,1)$ and $(4,6)$ is $m = (6-1)/(4-1) = 5/3$. Since the slopes are not equal, then this line is not tangent to the graph.

23. We know that the points $(2,0)$ and $(6,4)$ are on the tangent line, so its equation is

$$y - 0 = \frac{0-4}{2-6}(x-2) \qquad \text{or} \qquad y = x - 2$$

The line's y-intercept is $(0,-2)$.

25. $f(a) = -a^2 + 6a + 1$; $f(a+h) = -(h+a)^2 + 6(h+a) + 1$

$f(a+h) - f(a) = [-(h+a)^2 + 6(h+a) + 1] - (-a^2 + 6a + 1)$

$= -h^2 - 2ha - a^2 + 6h + 6a + 1 - (-a^2) - 6a - 1$

$= -h^2 - 2ha + 6h = h(-h - 2a + 6)$

$$m_{\tan} = \lim_{h \to 0} \frac{f(a+h) - f(a)}{h} = \lim_{h \to 0} \frac{h(-h - 2a + 6)}{h}$$

$$= \lim_{h \to 0} (-h - 2a + 6) = -2a + 6$$

The tangent line is horizontal when $m_{\tan} = 0$, so we substitute and solve $m_{\tan} = 0 = -2a + 6$, yielding $2a = 6$ and $a = 3$. Thus, the tangent line is horizontal at $(3, f(3)) = (3, 10)$.

27. $f(a) = a^3 - 3a$; $f(a+h) = (h+a)^3 - 3(h+a)$

$f(a+h) - f(a) = [(h+a)^3 - 3(h+a)] - (a^3 - 3a)$

$= h^3 + 3h^2 a + 3ha^2 + a^3 - 3h - 3a - a^3 - (-3a)$

$= h^3 + 3h^2 a + 3ha^2 - 3h = h(h^2 + 3ah + 3a^2 - 3)$

$$m_{\tan} = \lim_{h \to 0} \frac{f(a+h) - f(a)}{h} = \lim_{h \to 0} \frac{h(h^2 + 3ah + 3a^2 - 3)}{h}$$

$$= \lim_{h \to 0} (h^2 + 3ah + 3a^2 - 3) = 3a^2 - 3$$

The tangent line is horizontal when $m_{\tan} = 0$, so we substitute and solve $m_{\tan} = 0 = 3a^2 - 3$, yielding $3a^2 = 3$ and $a = \pm 1$. Thus, the tangent line is horizontal at $(-1, f(-1)) = (-1, 2)$ and $(1, f(1)) = (1, -2)$.

29. $v_{\text{ave}} = \dfrac{\text{change of distance}}{\text{change in time}} = \dfrac{290 \text{ mi}}{5 \text{ h}} = 58 \text{ mi/h}$

31. $v_{\text{ave}} = \dfrac{\text{change of distance}}{\text{change in time}}$; $920 \text{ km/h} = \dfrac{3500 \text{ km}}{t}$; $t \approx 3.8 \text{ h} = 3 \text{ h } 48 \text{ min}$

33. $\Delta s = s(t_0 + \Delta t) - s(t_0) = f(3 + \Delta t) - f(3) = [-4(3 + \Delta t)^2 + 10(3 + \Delta t) + 6] - 0 = -14\Delta t - 4\Delta t^2$

The instantaneous velocity at $t = 3$ is

$$v(3) = \lim_{\Delta t \to 0} \frac{\Delta s}{\Delta t} = \lim_{\Delta t \to 0} \frac{-14\Delta t - 4\Delta t^2}{\Delta t} = \lim_{\Delta t \to 0} (-14 - 4\Delta t) = -14.$$

35. (a) $\Delta s = s(t_0 + \Delta t) - s(t_0) = f(1/2 + \Delta t) - f(1/2) = -4.9(1/2 + \Delta t)^2 + 122.5 - 121.275$

$= -4.9\Delta t^2 - 4.9\Delta t$

The instantaneous velocity at $t = 1/2$ is

$$v(1/2) = \lim_{\Delta t \to 0} \frac{\Delta s}{\Delta t} = \lim_{\Delta t \to 0} \frac{-4.9\Delta t^2 - 4.9\Delta t}{\Delta t} = \lim_{\Delta t \to 0} (-4.9\Delta t - 4.9) = -4.9 \text{ m/s}.$$

(b) The ball hits the ground when $s(t) = 0$:

$$-4.9t^2 + 122.5 = 0; \ t^2 = 122.5/4.9; \ t = 5 \text{ s}.$$

(c) Since the ball impacts at $t = 5$,

$$\Delta s = s(t_0 + \Delta t) - s(t_0) = f(5 + \Delta t) - f(5) = [-4.9(5 + \Delta t)^2 + 122.5] - [-4.9(5)^2 + 122.5]$$
$$= -49\Delta t^2 - 49\Delta t.$$

The impact velocity at $t = 5$ is

$$v(5) = \lim_{\Delta t \to 0} \frac{\Delta s}{\Delta t} = \lim_{\Delta t \to 0} \frac{-49\Delta t^2 - 49\Delta t}{\Delta t} = \lim_{\Delta t \to 0} (-49\Delta t - 49) = -49 \text{ m/s}.$$

37. (a) $s(t) = -16t^2 + 256t$

$s(2) = -16(2^2) + 256(2) = 448$ ft

$s(6) = -16(6^2) + 256(6) = 960$ ft

$s(9) = -16(9^2) + 256(9) = 1008$ ft

$s(10) = -16(10^2) + 256(10) = 960$ ft

(b) $s(5) = -16(5^2) + 256(5) = 880$ ft

$s(2) = 448$ ft [from (a)]

$$v_{ave} = \frac{\text{change of distance}}{\text{change in time}} = \frac{880 \text{ ft} - 448 \text{ ft}}{5 \text{ s} - 2 \text{ s}} = 144 \text{ ft/s}$$

(c) $s(7) = -16(7^2) + 256(7) = 1008$ ft

$s(9) = 1008$ ft [from (a)]

$$v_{ave} = \frac{\text{change of distance}}{\text{change in time}} = \frac{1008 \text{ ft} - 1008 \text{ ft}}{9 \text{ s} - 7 \text{ s}} = \frac{0}{2} = 0 \text{ ft/s}$$

At $t = 7$ s, the projectile is at a height of 1008 ft on its way upward. After it reaches a maximum height, it begins to fall downward and, at $t = 9$ s, the height is once again 1008 ft. Since distance upward is positive and distance downward is negative, the net distance is zero.

(d) The projectile hits the ground when $s(t) = 0$:

$$-16t^2 + 256t = 0; \ 16t^2 = 256t; \ t = 256/16 = 16 \text{ s}$$

(e) For some general time t:

$\Delta s = s(t + \Delta t) - s(t) = [-16(t + \Delta t)^2 + 256(t + \Delta t)] - (-16t^2 + 256t)$

$= -16\Delta t^2 + 256\Delta t - 32t\Delta t = \Delta t(-16\Delta t + 256 - 32t)$

The instantaneous velocity at a general time t is

$$v(t) = \lim_{\Delta t \to 0} \frac{\Delta s}{\Delta t} = \lim_{\Delta t \to 0} \frac{\Delta t(-16\Delta t + 256 - 32t)}{\Delta t}$$
$$= \lim_{\Delta t \to 0} (-16\Delta t + 256 - 32t) = (256 - 32t) \text{ ft/s}.$$

(f) From (d), the projectile impacts at $t = 16$ s. From (e), $v(t) = 256 - 32t$, so $v(16) = 256 - 32(16) = -256$ ft/s.

(g) The maximum height is reached when $v(t) = 0$: $256 - 32t = 0$ gives us $t = 8$ s. Since $s(t) = -16t^2 + 256t$, we have $s(8) = -16(8^2) + 256(8) = 1024$ ft.

39. The slopes m of a tangent line at $(a, f(a))$ and m' of a tangent line at $(-a, f(-a))$ are:

$$m = \lim_{h \to 0} \frac{f(a + h) - f(a)}{h}; \ m' = \lim_{h' \to 0} \frac{f(-a + h') - f(-a)}{h'}$$

As defined in Section 1.2, an even function is a function that is symmetric with respect to the y-axis: $f(-x) = f(x)$ for all x. Since f is even, then $f(-a) = f(a)$ and $f(-a+h') = f(-[-a+h']) = f(a-h')$, resulting in:

$$m' = \lim_{h' \to 0} \frac{f(a-h') - f(a)}{h'} = \lim_{h' \to 0} \frac{f(a+[-h']) - f(a)}{h'}$$

Without loss of generality, we apply the substitution $h' = -h$ to obtain:

$$m' = \lim_{h' \to 0} \frac{f(a+[-h']) - f(a)}{h'} = \lim_{h \to 0} \frac{f(a+h) - f(a)}{-h} = -m$$

41. To show that the graph of $f(x) = x^2 + |x|$ does not possess a tangent line at $(0,0)$, we examine

$$\lim_{h \to 0} \frac{f(0+h) - f(0)}{h} = \lim_{h \to 0} \frac{[(0+h)^2 + |0+h|] - 0}{h} = \lim_{h \to 0} \frac{h^2 + |h|}{h}$$

From the definition of absolute value, we see that

$$\lim_{h \to 0^+} \frac{h^2 + |h|}{h} = \frac{h^2 + h}{h} = h + 1 = 1$$

whereas

$$\lim_{h \to 0^-} \frac{h^2 + |h|}{h} = \frac{h^2 - h}{h} = h - 1 = -1$$

Since the right-hand and left-hand limits are not equal, we conclude that $\lim_{h \to 0} \frac{f(0+h) - f(0)}{h} = \lim_{h \to 0} \frac{h^2 + |h|}{h}$ does not exist, and therefore f has no tangent line at $(0,0)$.

Chapter 2 in Review

A. True/False

1. True

3. False; $\lim_{x \to 0^-} \frac{|x|}{x} = -1$.

5. False; $\lim_{x \to 0^+} \left(\tan^{-1} \frac{1}{x} \right) = \frac{\pi}{2}$.

7. True

9. False; consider $f(x) = \frac{1}{x^2}$, $g(x) = \frac{1}{x^4}$, and $a = 0$.

11. False; consider $f(x) = -x$.

13. True, since $f(-1) < 0$ and $f(1) > 0$.

15. True

17. False; consider $f(x) = \begin{cases} 1, & x \le 3 \\ 2, & x > 3 \end{cases}$.

19. True

21. False, since $\frac{\sqrt{x}}{x+1}$ is undefined for $x < 0$.

B. Fill In the Blanks

1. 4

3. $-1/5$

5. 0

7. ∞

9. 1

11. 3^-

13. $-\infty$

15. -2

17. 10

19. continuous

21. 9

C. Exercises

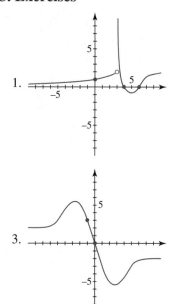

1.

3.

5. (a), (e), (f), (h)

7. (c), (h)

9. (b), (c), (d), (e), (f)

11. 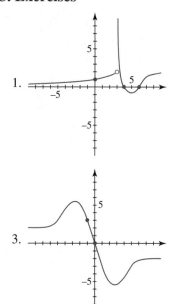 The function is continuous everywhere.

13. $(-\infty,-1)$, $(-1,0)$, $(0,1)$, and $(1,\infty)$

15. $(-\infty,-\sqrt{5})$ and $(\sqrt{5},\infty)$

17. For $f(x)$ to be continuous at the number 3, we must have $f(3) = 3k + 1 = \lim\limits_{x \to 3^+} (2 - kx)$. Thus, we must solve for k in the equation $3k + 1 = 2 - 3k$, resulting in $k = 1/6$. Therefore:

$$f(x) = \begin{cases} \dfrac{x}{6} + 1, & x \le 3 \\ 2 - \dfrac{x}{6}, & x > 3 \end{cases}$$

19. $f(a) = f(2) = 32$; $f(a + h) = f(2 + h) = -3(h + 2)^2 + 16(h + 2) + 12$

$f(a + h) - f(a) = [-3(h + 2)^2 + 16(h + 2) + 12] - 32$

$\qquad\qquad = -3h^2 - 12h - 12 + 16h + 32 + 12 - 32 = -3h^2 + 4h = h(-3h + 4)$

$m_{\tan} = \lim\limits_{h \to 0} \dfrac{f(a + h) - f(a)}{h} = \lim\limits_{h \to 0} \dfrac{h(-3h + 4)}{h} = 4$

With point of tangency $(2, 32)$, we have $y - 32 = 4(x - 2)$ or $y = 4x + 24$.

21. $f(a) = f(1/2) = -2$; $f(a + h) = f(1/2 + h) = \dfrac{-1}{2(h + 1/2)^2}$

$f(a + h) - f(a) = \dfrac{-1}{2(h + 1/2)^2} - (-2) = \dfrac{-1}{2h^2 + 2h + 1/2} - (-2)$

$\qquad = \dfrac{-1 + 4h^2 + 4h + 1}{2h^2 + 2h + 1/2} = \dfrac{4h(h + 1)}{2h^2 + 2h + 1/2}$

$m_{\tan} = \lim\limits_{h \to 0} \dfrac{f(a + h) - f(a)}{h} = \lim\limits_{h \to 0} \dfrac{4h(h + 1)}{(2h^2 + 2h + 1/2)h} = 8$

With point of tangency $(1/2, -2)$, we have $y + 2 = 8(x - 1/2)$ or $y = 8x - 6$.

23. $f(a) = f(1) = 2$; $f(a + h) = f(1 + h) = -4(h + 1)^2 + 6(h + 1)$

$f(a + h) - f(a) = [-4(h + 1)^2 + 6(h + 1)] - 2 = -4h^2 - 8h - 4 + 6h + 6 - 2$

$\qquad\qquad = -4h^2 - 2h = h(-4h - 2)$

$m_{\tan} = \lim\limits_{h \to 0} \dfrac{f(a + h) - f(a)}{h} = \lim\limits_{h \to 0} \dfrac{h(-4h - 2)}{h} = -2$

With point of tangency $(1, 2)$, we have $y - 2 = -2(x - 1)$ or $y = -2x + 4$. Thus, the line that is perpendicular to this line would have a slope of $1/2$ and also pass through $(1, 2)$, resulting in the equation $y - 2 = (x - 1)/2$ or $y = (x + 3)/2$.

The Derivative

3.1 The Derivative

1. $f'(x) = \lim\limits_{h \to 0} \dfrac{f(x+h) - f(x)}{h} = \lim\limits_{h \to 0} \dfrac{10 - 10}{h} = 0$

3. $f'(x) = \lim\limits_{h \to 0} \dfrac{f(x+h) - f(x)}{h} = \lim\limits_{h \to 0} \dfrac{[-3(x+h)+5] - (-3x+5)}{h} = \lim\limits_{h \to 0} \dfrac{-3h}{h} = \lim\limits_{h \to 0} -3 = -3$

5. $f'(x) = \lim\limits_{h \to 0} \dfrac{f(x+h) - f(x)}{h} = \lim\limits_{h \to 0} \dfrac{[3(x+h)^2] - 3x^2}{h} = \lim\limits_{h \to 0} \dfrac{6xh + 3h^2}{h} = \lim\limits_{h \to 0}(6x+3h) = 6x$

7. $f'(x) = \lim\limits_{h \to 0} \dfrac{f(x+h) - f(x)}{h} = \lim\limits_{h \to 0} \dfrac{[-(x+h)^2 + 4(x+h) + 1] - (-x^2 + 4x + 1)}{h}$

   $= \lim\limits_{h \to 0} \dfrac{-2xh - h^2 + 4h}{h} = \lim\limits_{h \to 0}(-2x - h + 4) = -2x + 4$

9. $f'(x) = \lim\limits_{h \to 0} \dfrac{f(x+h) - f(x)}{h} = \lim\limits_{h \to 0} \dfrac{[(x+h)+1]^2 - (x+1)^2}{h}$

   $= \lim\limits_{h \to 0} \dfrac{2xh + h^2 + 2h}{h} = \lim\limits_{h \to 0}(2x + h + 2) = 2x + 2 = 2(x+1)$

11. $f'(x) = \lim\limits_{h \to 0} \dfrac{f(x+h) - f(x)}{h} = \lim\limits_{h \to 0} \dfrac{[(x+h)^3 + (x+h)] - (x^3 + x)}{h}$

    $= \lim\limits_{h \to 0} \dfrac{3x^2 h + 3xh^2 + h^3 + h}{h} = \lim\limits_{h \to 0}(3x^2 + 3xh + h^2 + 1) = 3x^2 + 1$

13. $f'(x) = \lim\limits_{h \to 0} \dfrac{f(x+h) - f(x)}{h} = \lim\limits_{h \to 0} \dfrac{[-(x+h)^3 + 15(x+h)^2 - (x+h)] - (-x^3 + 15x^2 - x)}{h}$

    $= \lim\limits_{h \to 0} \dfrac{-3x^2 h - 3xh^2 - h^3 + 30xh + 15h^2 - h}{h}$

    $= \lim\limits_{h \to 0}(-3x^2 - 3xh - h^2 + 30x + 15h - 1) = -3x^2 + 30x - 1$

15. $f'(x) = \lim\limits_{h \to 0} \dfrac{f(x+h) - f(x)}{h} = \lim\limits_{h \to 0} \dfrac{\dfrac{2}{(x+h)+1} - \dfrac{2}{x+1}}{h}$

    $= \lim\limits_{h \to 0} \dfrac{-2h}{h(x+1)(x+h+1)} = \lim\limits_{h \to 0} \dfrac{-2}{(x+1)(x+h+1)} = -\dfrac{2}{(x+1)^2}$

17. $f'(x) = \lim\limits_{h \to 0} \dfrac{f(x+h) - f(x)}{h} = \lim\limits_{h \to 0} \dfrac{\dfrac{2(x+h)+3}{(x+h)+4} - \dfrac{2x+3}{x+4}}{h}$

$= \lim\limits_{h \to 0} \dfrac{5h}{h(x+4)(x+h+4)} = \lim\limits_{h \to 0} \dfrac{5}{(x+4)(x+h+4)} = \dfrac{5}{(x+4)^2}$

19. $f'(x) = \lim\limits_{h \to 0} \dfrac{f(x+h) - f(x)}{h} = \lim\limits_{h \to 0} \dfrac{\dfrac{1}{\sqrt{x+h}} - \dfrac{1}{\sqrt{x}}}{h} = \lim\limits_{h \to 0} \dfrac{\sqrt{x} - \sqrt{x+h}}{h\sqrt{x^2 + xh}}$

$= \lim\limits_{h \to 0} \left(\dfrac{\sqrt{x} - \sqrt{x+h}}{h\sqrt{x^2 + xh}} \cdot \dfrac{\sqrt{x} + \sqrt{x+h}}{\sqrt{x} + \sqrt{x+h}} \right)$

$= \lim\limits_{h \to 0} \dfrac{-h}{h \left(\sqrt{x^3 + x^2 h} + \sqrt{x^3 + 2x^2 h + xh^2} \right)}$

$= \lim\limits_{h \to 0} \dfrac{-1}{\sqrt{x^3 + x^2 h} + \sqrt{x^3 + 2x^2 h + xh^2}} = -\dfrac{1}{2\sqrt{x^3}} = -\dfrac{1}{2x\sqrt{x}}$

21. $f'(x) = \lim\limits_{h \to 0} \dfrac{f(x+h) - f(x)}{h} = \lim\limits_{h \to 0} \dfrac{[4(x+h)^2 + 7(x+h)] - (4x^2 + 7x)}{h}$

$= \lim\limits_{h \to 0} \dfrac{8xh + 4h^2 + 7h}{h} = \lim\limits_{h \to 0} (8x + 4h + 7) = 8x + 7$

$m_{\tan} = f'(-1) = 8(-1) + 7 = -1$

With point of tangency $(-1, f(-1))$ or $(-1, -3)$, we have $y + 3 = -1(x+1)$ or $y = -x - 4$.

23. $f'(x) = \lim\limits_{h \to 0} \dfrac{f(x+h) - f(x)}{h} = \lim\limits_{h \to 0} \dfrac{\left[(x+h) - \dfrac{1}{x+h} \right] - \left(x - \dfrac{1}{x} \right)}{h}$

$= \lim\limits_{h \to 0} \dfrac{x^2 h + xh^2 + h}{h(x^2 + xh)} = \lim\limits_{h \to 0} \dfrac{x^2 + xh + 1}{x^2 + xh} = 1 + \dfrac{1}{x^2}$

$m_{\tan} = f'(1) = 1 + \dfrac{1}{1^2} = 2$

With point of tangency $(1, f(1))$ or $(1, 0)$, we have $y - 0 = 2(x-1)$ or $y = 2(x-1)$.

25. $f'(x) = \lim\limits_{h \to 0} \dfrac{f(x+h) - f(x)}{h} = \lim\limits_{h \to 0} \dfrac{[(x+h)^2 + 8(x+h) + 10] - (x^2 + 8x + 10)}{h}$

$= \lim\limits_{h \to 0} \dfrac{2xh + h^2 + 8h}{h} = \lim\limits_{h \to 0} (2x + h + 8) = 2x + 8$

The tangent is horizontal when $2x + 8 = 0$ or $x = -4$. Since $f(-4) = -6$, the tangent is horizontal at $(-4, -6)$.

27. $f'(x) = \lim\limits_{h \to 0} \dfrac{f(x+h) - f(x)}{h} = \lim\limits_{h \to 0} \dfrac{[(x+h)^3 - 3(x+h)] - (x^3 - 3x)}{h}$

$= \lim\limits_{h \to 0} \dfrac{3x^2 h + 3xh^2 + h^3 - 3h}{h} = \lim\limits_{h \to 0} (3x^2 + 3xh + h^2 - 3) = 3x^2 - 3$

The tangent is horizontal when $3x^2 - 3 = 3(x+1)(x-1) = 0$ or $x = \pm 1$. Since $f(-1) = 2$ and $f(1) = -2$, the tangent is horizontal at $(-1, 2)$ and $(1, -2)$.

29. $f'(x) = \lim\limits_{h \to 0} \dfrac{f(x+h) - f(x)}{h} = \lim\limits_{h \to 0} \dfrac{\left[\frac{1}{2}(x+h)^2 - 1 \right] - \left(\frac{1}{2}x^2 - 1 \right)}{h}$

$= \lim\limits_{h \to 0} \dfrac{xh + \frac{1}{2}h^2}{h} = \lim\limits_{h \to 0} (x + \frac{1}{2}h) = x$

The given line $3x - y = 1$ has a slope of 3, and so the tangent line is parallel to it when $x = 3$, or at $(3, f(3)) = (3, 7/2)$.

31. $f'(x) = \lim_{h \to 0} \dfrac{f(x+h) - f(x)}{h} = \lim_{h \to 0} \dfrac{[-(x+h)^3 + 4] - (-x^3 + 4)}{h}$

$= \lim_{h \to 0} \dfrac{-3x^2 h - 3xh^2 - h^3}{h} = \lim_{h \to 0} (-3x^2 - 3xh - h^2) = -3x^2$

The given line $12x + y = 4$ has a slope of -12, and so the tangent line is parallel to it when $-3x^2 = -12$, $x = \pm 2$, or at $(-2, f(-2)) = (-2, 12)$ and $(2, f(2)) = (2, -4)$.

33. $f'_+(2) = \lim_{h \to 0^+} \dfrac{f(2+h) - f(2)}{h} = \lim_{h \to 0^+} \dfrac{[2(2+h) - 4] - [2(2) - 4]}{h} = \lim_{h \to 0^+} \dfrac{2h}{h} = 2$

$f'_-(2) = \lim_{h \to 0^-} \dfrac{f(2+h) - f(2)}{h} = \lim_{h \to 0^-} \dfrac{[-(2+h) + 2] - (-2 + 2)}{h} = \lim_{h \to 0^-} \dfrac{-h}{h} = -1$

Since $f'_+(2) \neq f'_-(2)$, f is not differentiable at 2.

35. $f'(a) = \lim_{x \to a} \dfrac{f(x) - f(a)}{x - a} = \lim_{x \to a} \dfrac{(10x^2 - 3) - (10a^2 - 3)}{x - a}$

$= \lim_{x \to a} \dfrac{10(x+a)(x-a)}{x - a} = \lim_{x \to a} 10(x+a) = 20a$

37. $f'(a) = \lim_{x \to a} \dfrac{f(x) - f(a)}{x - a} = \lim_{x \to a} \dfrac{(x^3 - 4x^2) - (a^3 - 4a^2)}{x - a}$

$= \lim_{x \to a} \dfrac{(x^2 + ax + a^2)(x - a) - 4(x+a)(x-a)}{x - a}$

$= \lim_{x \to a} (x^2 + ax + a^2 - 4x - 4a) = 3a^2 - 8a$

39. $f'(a) = \lim_{x \to a} \dfrac{f(x) - f(a)}{x - a} = \lim_{x \to a} \dfrac{\dfrac{4}{3 - x} - \dfrac{4}{3 - a}}{x - a}$

$= \lim_{x \to a} \dfrac{12 - 4a - 12 + 4x}{(x-a)(3-x)(3-a)} = \lim_{x \to a} \dfrac{4(x-a)}{(x-a)(3-x)(3-a)}$

$= \lim_{x \to a} \dfrac{4}{(3-x)(3-a)} = \dfrac{4}{(3-a)^2}$

41. Since $(0, 3)$ and $(-6, 0)$ are points on the line, and the line is tangent to f at $x = -3$, then its slope is $f'(-3) = m = \dfrac{3 - 0}{0 - (-6)} = \dfrac{1}{2}$. Using $(0, 3)$, an equation of the tangent line is $y - 3 = \dfrac{1}{2}x$ or $y = \dfrac{1}{2}x + 3$. Thus, $f(-3) = \dfrac{3}{2}$.

43.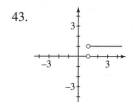

45.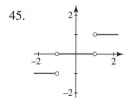

47.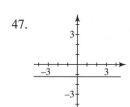

49. (e); consider $f(x) = \tan^{-1} x$.

51. (b); consider $f(x) = \dfrac{1}{3}x^3 - 2x$.

53. (a); consider $f(x) = 3\sqrt{|x|}$.

55. $f'(a) = \lim_{x \to a} \dfrac{f(x) - f(a)}{x - a} = \lim_{x \to a} \dfrac{x^{1/3} - a^{1/3}}{x - a} = \lim_{x \to a} \dfrac{x^{1/3} - a^{1/3}}{(x^{1/3})^3 - (a^{1/3})^3}$

$= \lim_{x \to a} \dfrac{x^{1/3} - a^{1/3}}{(x^{1/3} - a^{1/3})(x^{2/3} + x^{1/3}a^{1/3} + a^{2/3})} = \lim_{x \to a} \dfrac{1}{x^{2/3} + x^{1/3}a^{1/3} + a^{2/3}} = \dfrac{1}{3a^{2/3}}$

57. $f'(x) = \lim\limits_{h \to 0} \dfrac{f(x+h) - f(x)}{h} = \lim\limits_{h \to 0} \dfrac{f(x)f(h) - f(x)}{h} = \lim\limits_{h \to 0} \dfrac{f(x)[f(h) - 1]}{h}$

Since $f(0) = 1$ and $f'(0) = 1$, we can substitute $x = 0$ above to obtain:

$$f'(0) = \lim\limits_{h \to 0} \dfrac{f(0)[f(h) - 1]}{h} = \lim\limits_{h \to 0} \dfrac{1[f(h) - 1]}{h} = \lim\limits_{h \to 0} \dfrac{f(h) - 1}{h} = 1$$

Going back to the general expression of the limit for all x, we get:

$$f'(x) = \lim\limits_{h \to 0} \dfrac{f(x)[f(h) - 1]}{h} = \left[\lim\limits_{h \to 0} f(x)\right] \cdot \left[\lim\limits_{h \to 0} \dfrac{f(h) - 1}{h}\right] = \left[\lim\limits_{h \to 0} f(x)\right] \cdot [1]$$

$$= \lim\limits_{h \to 0} f(x) = f(x)$$

Thus, $f'(x) = f(x)$ for all x.

59.   The statement is true because, given $f(a) = f(b) = 0$, every "upward" or "downward" movement from $x = a$ must eventually have an equal but opposite amount of movement before reaching $x = b$. Since the function is differentiable on $[a, b]$, there must be a point c where the function "transitions" from going up or down to the opposite direction, and so the slope at that point must be horizontal, or $f'(c) = 0$.

61.   For $n = 1$, $f(x) = \begin{cases} x + x = 2x, & x > 0 \\ x + (-x) = 0, & x < 0 \end{cases}$

$$f'_+(x) = \lim\limits_{h \to 0^+} \dfrac{f(x+h) - f(x)}{h} = \lim\limits_{h \to 0^+} \dfrac{2(x+h) - 2x)}{h} = \lim\limits_{h \to 0^+} \dfrac{2h}{h} = 2; \; f'_+(0) = 2$$

$$f'_-(x) = \lim\limits_{h \to 0^-} \dfrac{f(x+h) - f(x)}{h} = \lim\limits_{h \to 0^-} \dfrac{(0 - 0)}{h} = 0; \; f'_-(0) = 0$$

Since $f'_+(0) \neq f'_-(0)$, then f is not differentiable at 0 for $n = 1$.

For $n = 2$, $f(x) = \begin{cases} x^2 + x, & x > 0 \\ x^2 - x, & x < 0 \end{cases}$

$$f'_+(x) = \lim\limits_{h \to 0^+} \dfrac{f(x+h) - f(x)}{h} = \lim\limits_{h \to 0^+} \dfrac{[(x+h)^2 + (x+h)] - (x^2 + x)}{h}$$

$$= \lim\limits_{h \to 0^+} \dfrac{2xh + h^2 + h}{h} = 2x + 1; \; f'_+(0) = 1$$

$$f'_-(x) = \lim\limits_{h \to 0^-} \dfrac{f(x+h) - f(x)}{h} = \lim\limits_{h \to 0^-} \dfrac{[(x+h)^2 - (x+h)] - (x^2 - x)}{h}$$

$$= \lim\limits_{h \to 0^+} \dfrac{2xh + h^2 - h}{h} = 2x - 1; \; f'_-(0) = -1$$

Since $f'_+(0) \neq f'_-(0)$, then f is not differentiable at 0 for $n = 2$. Proceeding similarly for $n = 3, 4$, and 5, we get:

$n = 3$:	$f'_+(x) = 3x^2 + 1$	$f'_-(x) = 3x^2 - 1$
$n = 4$:	$f'_+(x) = 4x^3 + 1$	$f'_-(x) = 4x^3 - 1$
$n = 5$:	$f'_+(x) = 5x^4 + 1$	$f'_-(x) = 5x^4 - 1$

The general case n yields $f'_+(x) = nx^{n-1} + 1$ and $f'_-(x) = nx^{n-1} - 1$. In all cases for $n > 1$, $f' + -(0) = 1$ and $f'_-(0) = -1$, and so f is not differentiable at 0 for any positive integer n.

3.2   Power and Sum Rules

1. $\dfrac{dy}{dx} = 0$

3. $\dfrac{dy}{dx} = 9x^8$

5. $\dfrac{dy}{dx} = 14x - 4$

7. $\dfrac{dy}{dx} = 2x^{-1/2} + 4x^{-5/3} = \dfrac{2}{\sqrt{x}} + \dfrac{4\sqrt[3]{x}}{x^2}$

9. $f'(x) = x^4 - 12x^3 + 18x$

11. $f(x) = 4x^5 - 5x^4 - 6x^3$
    $f'(x) = 20x^4 - 20x^3 - 18x^2$

13. $f(x) = x^2(x^2 + 5)^2 = x^6 + 10x^4 + 25x^2; \ f'(x) = 6x^5 + 40x^3 + 50x$

15. $f(x) = 16x + 8\sqrt{x} + 1; \ f'(x) = 16 + \dfrac{4}{\sqrt{x}}$

17. $h(u) = 64u^3; \ h'(u) = 192u^2$

19. $g(r) = r^{-1} + r^{-2} + r^{-3} + r^{-4}; \ g'(r) = -r^{-2} - 2r^{-3} - 3r^{-4} - 4r^{-5} = -\dfrac{1}{r^2} - \dfrac{2}{r^3} - \dfrac{3}{r^4} - \dfrac{4}{r^5}$

21. $y' = 6x^2$. When $x = -1$, the slope of the tangent line is 6 and the point of tangency is $(-1, y(-1))$ or $(-1, -3)$. Hence, an equation of the tangent line is $y + 3 = 6(x + 1)$ or $y = 6x + 3$.

23. $y' = -2x^{-3/2} + x^{-1/2}$. When $x = 4$, the slope of the tangent line is $1/4$ and the point of tangency is $(4, y(4))$ or $(4, 6)$. Hence, an equation of the tangent line is $y - 6 = (x - 4)/4$ or $y = x/4 + 5$.

25. $y' = 2x - 8$. The tangent is horizontal when $2x - 8 = 0$ or $x = 4$. Since $y(4) = -11$, the tangent is horizontal at $(4, -11)$.

27. $y' = 3x^2 - 6x - 9$. The tangent is horizontal when $3x^2 - 6x - 9 = 0$ or $x = 3, -1$. Since $y(3) = -25$ and $y(-1) = 7$, the tangent is horizontal at $(3, -25)$ and $(-1, 7)$.

29. $y' = -2x$, so $m_{\tan} = -4$ at $(2, -3)$. Thus, the slope of the normal line is $m = 1/4$ and its equation is $y + 3 = \dfrac{1}{4}(x - 2)$ or $y = \dfrac{1}{4}x - \dfrac{7}{2}$.

31. $y' = x^2 - 4x$, so $m_{\tan} = 0$ at $(4, -32/3)$. Thus, tangent line is horizontal and the normal line is vertical. Its equation is $x = 4$.

33. $\dfrac{dy}{dx} = -2x + 3; \ \dfrac{d^2y}{dx^2} = -2$

35. $\dfrac{dy}{dx} = 32x - 72; \ \dfrac{d^2y}{dx^2} = 32$

37. $\dfrac{dy}{dx} = -20x^{-3}; \ \dfrac{d^2y}{dx^2} = 60x^{-4}$

39. $f'(x) = 24x^5 + 5x^4 - 3x^2; \quad f''(x) = 120x^4 + 20x^3 - 6x;$
    $f'''(x) = 480x^3 + 60x^2 - 6; \quad f^{(4)}(x) = 1440x^2 + 120x$

41. $f'(x) = 2x + 8$. Solving $2x + 8 > 0$ we obtain $x > -4$. Thus $f'(x) > 0$ or $(-4, \infty)$. Solving $2x + 8 < 0$ we obtain $x < -4$. Thus $f'(x) < 0$ on $(-\infty, -4)$.

43. $f'(x) = 3x^2 + 24x + 20; \ f''(x) = 6x + 24$. Solving $6x + 24 = 0$, we obtain $x = -4$. Thus the point on the graph is $(-4, f(-4))$ or $(-4, 48)$.

45. $f'(x) = 3(x-1)^2$; $f''(x) = 6(x-1)$. $f''(x) > 0$ for $x > 1$ and $f''(x) < 0$ for $x < 1$. Thus, $f''(x) > 0$ on $(1, \infty)$ and $f''(x) < 0$ on $(-\infty, 1)$.

47. $y' = -x^{-2} + 4x^3$; $y'' = 2x^{-3} + 12x^2$. Substituting into the differential equation, we get:

$$x^2 y'' - 2xy' - 4y = x^2(2x^{-3} + 12x^2) - 2x(-x^{-2} + 4x^3) - 4(x^{-1} + x^4)$$
$$= 2x^{-1} + 12x^4 + 2x^{-1} - 8x^4 - 4x^{-1} - 4x^4 = 0$$

Thus, the function satisfies $x^2 y'' - 2xy' - 4y = 0$.

49. $f'(x) = 4x - 3$. Since the slope of the tangent line is 5, $4x - 3 = 5$ and $x = 2$. Since $f(2) = 8$, the point on the graph is $(2, 8)$.

51. Since the slope of the normal line is 2, the slope of the tangent line is $-1/2$. Thus $f'(x) = 2x - 1 = -1/2$ and $x = 1/4$. Since $f(1/4) = -3/16$, the point on the graph is $(1/4, -3/16)$.

53. $y' = 3x^2 + 6x - 4$; $y'' = 6x + 6$. The second derivative is zero when $x = -1$. Since $y(-1) = 7$ and $y'(-1) = -7$, the point on the graph is $(-1, 7)$ and the slope of the tangent line is -7. Hence, an equation of the tangent line is $y - 7 = -7(x+1)$ or $y = -7x$.

55. $V(r) = \dfrac{4\pi}{3} r^3$; $S = V'(r) = 4\pi r^2$

57. $U(x) = \dfrac{k}{2}x^2$; $F = -\dfrac{dU}{dx} = -\dfrac{k}{2} 2x = -kx$

Given $k = 30$ N/m and $x = \dfrac{m}{2}$, $F = -(30 \text{ N/m})\dfrac{m}{2} = -15$ N

59. When $n = 1$, $\dfrac{d}{dx}x = 1$; when $n = 2$, $\dfrac{d^2}{dx^2}x^2 = 2$; when $n = 3$, $\dfrac{d^3}{dx^3} = 3 \cdot 2 = 3!$; when $n = 4$, $\dfrac{d^4}{dx^4} = 4 \cdot 3 \cdot 2 = 4!$. We can verify by induction that $\dfrac{d^n}{dx^n}x^n = n!$:

$$\frac{d^n}{dx^n}x^n = \frac{d^{n-1}}{dx^{n-1}}\left(\frac{d}{dx}x^n\right) = \frac{d^{n-1}}{dx^{n-1}}(nx^{n-1}) = n\left(\frac{d^{n-1}}{dx^{n-1}}x^{n-1}\right) = n(n-1)! = n!.$$

61. $f(x) = 0$ at 3 points in the figure, and the tangents to $g(x)$ are horizontal at the same locations. In contrast, $g(x) = 0$ at 4 points in the figure, while the tangents to $f(x)$ are horizontal at only two. This implies that $f(x)$ is the derivative of $g(x)$. Similarly, $f(x) < 0$ exactly at the intervals where $g(x)$ is moving downward and $f(x) > 0$ where $g(x)$ is moving upward, but not vice versa; this also implies that $f(x)$ is the derivative of $g(x)$.

63. Let $f(x) = ax^2 + bx + c$. Then $f'(x) = 2ax + b$ and $f''(x) = 2a$. From $f''(-1) = 2a = -4$ we see that $a = -2$ and $f'(x) = -4x + b$. From $f'(-1) = -4(-1) + b = 7$ we see that $b = 3$ and $f(x) = -2x^2 + 3x + c$. From $f(-1) = -2 - 3 + c = -11$ we see that $c = -6$ and $f(x) = -2x^2 + 3x - 6$.

65. $f(-3) = 9 - 3b$, $f'(x) = 2x + b$, and $f'(-3) = -6 + b$. The equation of the tangent line at $(-3, 9 - 3b)$ is $y - (9 - 3b) = (-6 + b)(x + 3)$ or $y = (-6 + b)x - 9$. We are given that the tangent line is $y = 2x + c$. Thus, $-6 + b = 2$ or $b = 8$ and $c = -9$.

67. Let $(a, f(a))$ or $(a, a^2 - 5)$ be the point on the graph. The slope of the line through $(a, a^2 - 5)$ and $(-3, 0)$ is $\dfrac{(a^2 - 5) - 0}{a - (-3)}$ or $\dfrac{a^2 - 5}{a + 3}$. Now we find the slope of the tangent line by finding $f'(a)$ and setting it equal to $\dfrac{a^2 - 5}{a + 3}$. Since $f'(x) = 2x$, $m_{\tan} = f'(a) = 2a$. Then $2a = \dfrac{a^2 - 5}{a + 3}$ or $a^2 + 6a + 5 = 0$, and $a = -1, -5$. Since $f(-1) = -4$ and $f(-5) = 20$, the points on the graph are $(-1, -4)$ and $(-5, 20)$.

69. $f'(x) = x^4 + x^2$. Thus, $f'(x)$ is never negative, and therefore $f(x)$ cannot possibly have a tangent line with slope -1.

71. $f(1) = a + b = 4$, $f'(x) = 2ax + b$, and $f'(1) = 2a + b = -5$. Solving for the two equations in two unknowns, we see that $a = -9$ and $b = 13$.

73. $f'(x) = 2x + 1$ and $g'(x) = 4x + 4$. At $(a, f(a))$ on the graph of f, the slope of the tangent line is $2a + 1$. Solving $g'(x) = 4x + 4 = 2a + 1$ for x we obtain $x = a/2 - 3/4$. Thus, points on the graphs of f and g where the tangent lines are parallel are $(a, f(a))$ and $(a/2 - 3/4, g(a/2 - 3/4))$.

75. $f'(x) = 3ax^2 + 2bx + c$. Thus, $f(x)$ has exactly one, two, or no horizontal tangents if and only if the equation $3ax^2 + 2bx + c = 0$ has exactly one, two, or no solutions, respectively. By the quadratic formula, the solutions to this equation are

$$x = \frac{-(2b) \pm \sqrt{(2b)^2 - 4(3a)(c)}}{2(3a)} = \frac{-2b \pm \sqrt{4b^2 - 12ac}}{6a} = \frac{-b \pm \sqrt{b^2 - 3ac}}{3a}.$$

Thus, $f(x)$ will have exactly one horizontal tangent when $b^2 - 3ac = 0$, and it will have no horizontal tangents when either $a = 0$ or $b^2 - 3ac < 0$. $f(x)$ will have two horizontal tangents when $a \neq 0$ and $b^2 - 3ac > 0$.

77. Since $f'(x) - f(x) = 0$, then $f'(x) = f(x)$. That means $f''(x) = (f')'(x) = f'(x) = f(x)$, $f'''(x) = (f'')'(x) = f'(x) = f(x)$, and so on. Thus, $f^{(100)}(x) = f(x)$.

79. (a)

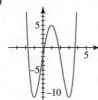

(b) $f'(x) = 4x^3 - 12x^2 - 4x + 12$; $f''(x) = 12x^2 - 24x - 4$

x	-2	-1	0	1	2	3	4
$f''(x)$	92	32	-4	-16	-4	32	92

(c) Where the graph is concave up, the second derivative is positive; where it is concave down, the second derivative is negative.

3.3 Product and Quotient Rules

1. $\dfrac{dy}{dx} = (x^2 - 7)(3x^2 + 4) + (x^3 + 4x + 2)(2x) = 5x^4 - 9x^2 + 4x - 28$

3. $y = (4x^{1/2} + x^{-1})(2x - 6x^{-1/3})$

$\dfrac{dy}{dx} = (4x^{1/2} + x^{-1})(2 + 2x^{-4/3}) + (2x - 6x^{-1/3})(2x^{-1/2} - x^{-2})$

$\quad = 8x^{1/2} + 8x^{-5/6} + 2x^{-1} + 2x^{-7/3} + 4x^{1/2} - 2x^{-1} - 12x^{-5/6} + 6x^{-7/3}$

$\quad = 12x^{1/2} - 4x^{-5/6} + 8x^{-7/3} = 12\sqrt{x} - \dfrac{4}{\sqrt[6]{x^5}} + \dfrac{8}{\sqrt[3]{x^7}}$

5. $\dfrac{dy}{dx} = \dfrac{(x^2 + 1)(0) - (10)(2x)}{(x^2 + 1)^2} = -\dfrac{20x}{(x^2 + 1)^2}$

7. $\dfrac{dy}{dx} = \dfrac{(2x - 5)(3) - (3x + 1)(2)}{(2x - 5)^2} = -\dfrac{17}{(2x - 5)^2}$

9. $y = (6x - 1)(6x - 1);\quad \dfrac{dy}{dx} = (6x - 1)(6) + (6x - 1)(6) = 12(6x - 1) = 72x - 12$

11. $f(x) = (x^{-1} - 4x^{-3})(x^3 - 5x - 1)$

$\quad f'(x) = (x^{-1} - 4x^{-3})(3x^2 - 5) + (x^3 - 5x - 1)(-x^{-2} + 12x^{-4})$

$\quad\quad = 2x - 40x^{-3} + x^{-2} - 12x^{-4} = 2x + \dfrac{1}{x^2} - \dfrac{40}{x^3} - \dfrac{12}{x^4}$

13. $f'(x) = \dfrac{(2x^2 + x + 1)(2x) - (x^2)(4x + 1)}{(2x^2 + x + 1)^2} = \dfrac{x^2 + 2x}{(2x^2 + x + 1)^2}$

15. $f'(x) = [(x+1)(2x+1)](3) + (3x+1)[(x+1)(2) + (2x+1)(1)] = 18x^2 + 22x + 6$

17. $f'(x) = \dfrac{(3x+2)[(2x+1)(1) + (x-5)(2)] - [(2x+1)(x-5)](3)}{(3x+2)^2} = \dfrac{6x^2 + 8x - 3}{(3x+2)^2}$

19. $f'(x) = (x^2 - 2x - 1)\left[\dfrac{(x+3)(1) - (x+1)(1)}{(x+3)^2}\right] + \left(\dfrac{x+1}{x+3}\right)(2x-2)$

$= \dfrac{2x^2 - 4x - 2}{(x+3)^2} + \dfrac{2x^2 - 2}{x+3} = \dfrac{2x^3 + 8x^2 - 6x - 8}{(x+3)^2}$

21. $y' = \dfrac{(x-1)(1) - (x)(1)}{(x-1)^2} = -\dfrac{1}{(x-1)^2}$

When $x = 1/2$, the slope of the tangent line is $-1/(1/2 - 1)^2$ or -4. The point of tangency is $(1/2, y(1/2))$ or $(1/2, -1)$. Hence, an equation of the tangent line is $y - (-1) = -4(x - 1/2)$ or $y = -4x + 1$.

23. $y = (2x^{1/2} + x)(-2x^2 + 5x - 1)$; $y' = (2x^{1/2} + x)(-4x + 5) + (-2x^2 + 5x - 1)(x^{-1/2} + 1)$

When $x = 1$, the slope of the tangent line is $(2+1)(-4+5) + (-2+5-1)(1+1)$ or 7. The point of tangency is $(1, y(1))$ or $(1, 6)$. Hence, an equation of the tangent line is $y - 6 = 7(x - 1)$ or $y = 7x - 1$.

25. $y' = (x^2 - 4)(2x) + (x^2 - 6)(2x) = 4x^3 - 20x$

The tangent line is horizontal when $4x^3 - 20x = 4x(x^2 - 5) = 0$, or $x = 0, \pm\sqrt{5}$. Given $y = (x^2 - 4)(x^2 - 6)$ we see that for $x = 0$, $y = 24$, and for $x = \pm\sqrt{5}$, $y = -1$. Thus, the points on the graph are $(0, 24)$, $(\sqrt{5}, -1)$, and $(-\sqrt{5}, -1)$.

27. $y' = \dfrac{(x^4 + 1)(2x) - (x^2)(4x^3)}{(x^4 + 1)^2} = \dfrac{2x - 2x^5}{(x^4 + 1)^2}$

The tangent is horizontal when $y' = 0$ or $2x - 2x^5 = 0$. Then $2x(1 - x^4) = 0$ and $x = 0, \pm 1$. Given $y = \dfrac{x^2}{x^4 + 1}$, we see that for $x = 0$, $y = 0$, and for $x = \pm 1$, $y = 1/2$. Thus, the points on the graph are $(0, 0)$, $(1, 1/2)$, and $(-1, 1/2)$.

29. $y' = \dfrac{(x+1)(1) - (x+3)(1)}{(x+1)^2} = \dfrac{-2}{(x+1)^2}$

Since the slope of the tangent line is $-1/8$, $-2/(x+1)^2 = -1/8$, $x^2 + 2x - 15 = 0$ and $x = 3, -5$. Given $y = \dfrac{x+3}{x+1}$, we see that for $x = 3$, $y = 3/2$, and for $x = -5$, $y = 1/2$. Thus, the points on the graph are $(3, 3/2)$ and $(-5, 1/2)$.

31. $y' = \dfrac{(x+5)(1) - (x+4)(1)}{(x+5)^2} = \dfrac{1}{(x+5)^2}$. Since the tangent line is supposed to be perpendicular to $y = -x$, its slope is 1, so

$\dfrac{1}{(x+5)^2} = 1$ and $x = -4, -6$. Given $y = \dfrac{x+4}{x+5}$, we see that for $x = -4$, $y = 0$, and for $x = -6$, $y = 2$. Thus, the points on the graph are $(-4, 0)$ and $(-6, 2)$.

33. $f'(x) = \dfrac{(x^2)(1) - (k+x)(2x)}{(x^2)^2} = \dfrac{-x - 2k}{x^3}$. Solving $\dfrac{-2 - 2k}{2^3} = 5$, we obtain $k = -21$.

35. $F'(x) = 2[f(x)g'(x) + g(x)f'(x)]$; $F'(1) = 2[f(1)g'(1) + g(1)f'(1)] = 2[2(2) + 6(-3)] = -28$

37. $F'(x) = \dfrac{2[f(x)g'(x) - g(x)f'(x)]}{3[f(x)]^2}$; $F'(1) = \dfrac{2[f(1)g'(1) - g(1)f'(1)]}{3[f(1)]^2} = \dfrac{2[2(2) - 6(-3)]}{3(2^2)} = \dfrac{11}{3}$

39. $F'(x) = [4x^{-1} + f(x)]g'(x) + g(x)[-4x^{-2} + f'(x)]$

$F'(1) = [4(1^{-1}) + f(1)]g'(1) + g(1)[-4(1^{-2}) + f'(1)] = (4 + 2)(2) + (6)(-4 - 3) = -30$

41. $F'(x) = (x^{1/2})f'(x) + f(x)\left(\frac{1}{2}x^{-1/2}\right)$

$F''(x) = (x^{1/2})f''(x) + f'(x)\left(\frac{1}{2}x^{-1/2}\right) + f(x)\left(-\frac{1}{4}x^{-3/2}\right) + \left(\frac{1}{2}x^{-1/2}\right)f'(x)$

$F''(4) = (4^{1/2})f''(4) + f'(4)\left(\frac{1}{2}\cdot 4^{-1/2}\right) + f(4)\left(-\frac{1}{4}\cdot 4^{-3/2}\right) + \left(\frac{1}{2}\cdot 4^{-1/2}\right)f'(4)$

$= 2(3) + 2\left(\frac{1}{4}\right) - 16\left(-\frac{1}{32}\right) + \frac{1}{4}(2) = 6 + \frac{1}{2} + \frac{1}{2} + \frac{1}{2} = \frac{15}{2}$

43. $F'(x) = \dfrac{xf'(x) - f(x)}{x^2} = \dfrac{f'(x)}{x} - \dfrac{f(x)}{x^2}$

$F''(x) = \dfrac{xf''(x) - f'(x)}{x^2} - \dfrac{x^2 f'(x) - f(x)(2x)}{x^4} = \dfrac{f''(x)}{x} - \dfrac{2f'(x)}{x^2} + \dfrac{2f(x)}{x^3}$

45. $f'(x) = \dfrac{(x^2 - 2x)(0) - 5(2x - 2)}{(x^2 - 2x)^2} = \dfrac{10 - 10x}{x^2(x-2)^2}$

   For $f'(x)$ to be positive, we need $10 - 10x > 0$ or $x < 1$, and $x \neq 0, 2$. So $f'(x) > 0$ on $(-\infty, 0) \cup (0, 1)$.

   For $f'(x)$ to be negative, we need $10 - 10x < 0$ or $x > 1$, and $x \neq 0, 2$. So $f'(x) < 0$ on $(1, 2) \cup (2, \infty)$.

47. $f'(x) = (-2x + 6)(4) + (4x + 7)(-2) = 10 - 16x$

   For $f'(x)$ to be positive, we need $10 - 16x > 0$ or $x < 5/8$, so $f'(x) > 0$ on $(-\infty, 5/8)$.

   For $f'(x)$ to be negative, we need $10 - 16x < 0$ or $x > 5/8$, so $f'(x) < 0$ on $(5/8, \infty)$.

49. $F(r) = km_1 m_2 r^{-2}$; $F'(r) = -2km_1 m_2 r^{-3}$; $F'(1/2) = -2km_1 m_2 (1/2)^{-3} = -16km_1 m_2$

51. Solving for P and differentiating, we obtain

$$P = \frac{RT}{V - b} - \frac{a}{V^2} = \frac{RT}{V - b} - aV^{-2}$$

$$\frac{dP}{dV} = \frac{-RT}{(V - b)^2} - (-2aV^{-3}) = -\frac{RT}{(V - b)^2} + \frac{2a}{V^3}.$$

53. (a)

   (b) $f'(x) = \dfrac{-2(2x)}{(x^2 + 1)^2} = -\dfrac{4x}{(x^2 + 1)^2}$. At $x = a$ the slope of the normal line is $\dfrac{(a^2 + 1)^2}{4a}$ and the equation of the normal line

   through $\left(a, \dfrac{2}{a^2 + 1}\right)$ is $y - \dfrac{2}{a^2 + 1} = \dfrac{(a^2 + 1)^2}{4a}(x - a)$. When the line passes through the origin,

$$-\frac{2}{a^2 + 1} = \frac{(a^2 + 1)^2}{4a}(-a)$$
$$8 = (a^2 + 1)^3, \quad 2 - a^2 + 1$$

   and $a = \pm 1$. Thus, the points on the graph where the normal line passes through the origin are $(-1, 1)$ and $(1, 1)$. From the graph we see that the y-axis is also a normal line, so that another point is $(0, 2)$.

55. For $y = u(x)y_1(x)$, $y' = u(x)y_1'(x) + y_1(x)u'(x)$. Substituting these into the differential equation, we get $u(x)y_1'(x) + u'(x)y_1(x) + P(x)u(x)y_1(x) = f(x)$.

Since $du/dx = f(x)/y_1(x)$, we can substitute $u'(x)$ above to obtain

$$u(x)y_1'(x) + \left[\frac{f(x)}{y_1(x)}\right]y_1(x) + P(x)u(x)y_1(x) = f(x)$$
$$u(x)y_1'(x) + f(x) + P(x)u(x)y_1(x) = f(x)$$
$$u(x)y_1'(x) + P(x)u(x)y_1(x) = u(x)[y_1'(x) + P(x)y_1(x)] = 0.$$

Since $y_1(x)$ satisfies $y' + P(x)y = 0$, we have $u(x) \cdot 0 = 0$, and so $y = u(x)y_1(x)$ satisfies $y' + P(x)y = f(x)$.

3.4 Trigonometric Functions

1. $\dfrac{dy}{dx} = 2x + \sin x$

3. $\dfrac{dy}{dx} = 7\cos x - \sec^2 x$

5. $\dfrac{dy}{dx} = (x)(\cos x) + (\sin x)(1) = x\cos x + \sin x$

7. $\dfrac{dy}{dx} = (x^3 - 2)(\sec^2 x) + (\tan x)(3x^2) = (x^3 - 2)\sec^2 x + 3x^2 \tan x$

9. $y = x^2 \sec x + \sin x \sec x = x^2 \sec x + \tan x;\quad \dfrac{dy}{dx} = x^2 \sec x \tan x + 2x \sec x + \sec^2 x$

11. $y = \cos^2 x + \sin^2 x = 1;\quad \dfrac{dy}{dx} = 0$

13. $f(x) = \sin x;\quad f'(x) = \cos x$

15. $f'(x) = \dfrac{(x+1)(-\csc^2 x) - (\cot x)(1)}{(x+1)^2} = \dfrac{-(x+1)\csc^2 x - \cot x}{(x+1)^2} = -\dfrac{x\csc^2 x + \csc^2 x + \cot x}{(x+1)^2}$

17. $f'(x) = \dfrac{(1+2\tan x)(2x) - (x^2)(2\sec^2 x)}{(1+2\tan x)^2} = \dfrac{2x + 4x\tan x - 2x^2 \sec^2 x}{(1+2\tan x)^2}$

19. $f'(x) = \dfrac{(1+\cos x)(\cos x) - (\sin x)(-\sin x)}{(1+\cos x)^2} = \dfrac{\cos x + 1}{(1+\cos x)^2} = \dfrac{1}{1+\cos x}$

21. $f'(x) = (x^4 \sin x)(\sec^2 x) + (\tan x)[(x^4)(\cos x) + (\sin x)(4x^3)][t]$
$= x^4 \sin x \sec^2 x + x^4 \sin x + 4x^3 \sin x \tan x$

23. $f'(x) = -\sin x$. When $x = \pi/3$, the slope of the tangent line is $f'(\pi/3) = -\sin \pi/3 = -\sqrt{3}/2$. The point of tangency is $(\pi/3, f(\pi/3))$ or $(\pi/3, 1/2)$. Hence, an equation of the tangent line is $y - \dfrac{1}{2} = -\dfrac{\sqrt{3}}{2}\left(x - \dfrac{\pi}{3}\right)$ or $y = -\dfrac{\sqrt{3}}{2}x + \dfrac{\pi\sqrt{3}+3}{6}$.

25. $f'(x) = \sec x \tan x$. When $x = \pi/6$, the slope of the tangent line is $f'(\pi/6) = \sec(\pi/6)\tan(\pi/6) = 2/3$. The point of tangency is $(\pi/6, f(\pi/6))$ or $(\pi/6, 2\sqrt{3}/3)$. Hence, an equation of the tangent line is $y - \dfrac{2\sqrt{3}}{3} = \dfrac{2}{3}(x - \dfrac{\pi}{6})$ or $y = \dfrac{2}{3}x - \dfrac{6\sqrt{3}-\pi}{9}$.

27. The tangent is horizontal when $f'(x) = 0$. $f'(x) = 1 - 2\sin x = 0$; $\sin x = 1/2$; therefore $x = \pi/6, 5\pi/6$ in $[0, 2\pi]$.

29. The tangent is horizontal when $f'(x) = 0$:

$$f'(x) = \frac{(x+\cos x)(0) - (1)(1-\sin x)}{(x+\cos x)^2} = \frac{\sin x - 1}{(x+\cos x)^2} = 0$$

Therefore $\sin x = 1$, and $x = \pi/2$ in $[0, 2\pi]$.

31. $f'(x) = \cos x$. When $x = 4\pi/3$, the slope of the tangent line is $f'(4\pi/3) = \cos 4\pi/3 = -1/2$. Thus, the slope of the normal line at $(4\pi/3, f(4\pi/3))$ or $(4\pi/3, -\sqrt{3}/2)$ is 2. The equation of the normal line is $y + \dfrac{\sqrt{3}}{2} = 2\left(x - \dfrac{4\pi}{3}\right)$ or $y = 2x - \dfrac{16\pi + 3\sqrt{3}}{6}$.

33. $f'(x) = \cos x - x\sin x$. When $x = \pi$, the slope of the tangent line is $f'(\pi) = \cos \pi - \pi \sin \pi = -1$. Thus, the slope of the normal line at $(\pi, f(\pi))$ or $(\pi, -\pi)$ is 1. The equation of the normal line is $y + \pi = 1(x - \pi)$ or $y = x - 2\pi$.

35. $f(x) = \sin 2x = 2\sin x \cos x$

$f'(x) = 2[(\sin x)(-\sin x) + (\cos x)(\cos x)] = 2(\cos^2 x - \sin^2 x) = 2\cos 2x$

37. $f'(x) = x\cos x + \sin x$; $f''(x) = [x(-\sin x) + \cos x] + \cos x = -x\sin x + 2\cos x$

39. $f'(x) = \dfrac{(x)(\cos x) - (\sin x)(1)}{x^2} = \dfrac{\cos x}{x} - \dfrac{\sin x}{x^2}$

$f''(x) = \dfrac{(x)(-\sin x) - (\cos x)(1)}{x^2} - \dfrac{(x^2)(\cos x) - (\sin x)(2x)}{x^4} = -\dfrac{\sin x}{x} - \dfrac{2\cos x}{x^2} + \dfrac{2\sin x}{x^3}$

41. $f'(x) = -\csc x \cot x$

$f''(x) = (-\csc x)(-\csc^2 x) + (\cot x)(\csc x \cot x) = \csc^3 x + \csc x \cot^2 x$

43. $y' = C_1(-\sin x) + C_2 \cos x - \dfrac{1}{2}[(x)(-\sin x) + (\cos x)(1)]$

$\quad = -C_1 \sin x + C_2 \cos x + \dfrac{1}{2}x\sin x - \dfrac{1}{2}\cos x$

$y'' = -C_1 \cos x + C_2(-\sin x) + \dfrac{1}{2}[(x)(\cos x) + (\sin x)(1)] - \dfrac{1}{2}(-\sin x)$

$\quad = -C_1 \cos x - C_2 \sin x + \dfrac{1}{2}x\cos x + \sin x$

Substituting into the differential equation,

$$\left(-C_1 \cos x - C_2 \sin x + \dfrac{1}{2}x\cos x + \sin x\right) + \left(C_1 \cos x + C_2 \sin x - \dfrac{1}{2}x\cos x\right) = \sin x.$$

45. From $s = 40\cot\theta$ we obtain $\dfrac{ds}{d\theta} = -40\csc^2\theta$. When $\theta = \pi/3$ radians $\left.\dfrac{ds}{d\theta}\right|_{\theta = \pi/3} = -40\csc^2\dfrac{\pi}{3} = -40\left(\dfrac{2}{\sqrt{3}}\right)^2 = -\dfrac{160}{3}$ ft.
The rate of change is negative because the length of the shadow decreases as θ increases.

47. (a) We observe by successive differentiation that since

$$\sin x \to \cos x \to -\sin x \to -\cos x \to \sin x \to \cos x \to \ldots,$$

both $\dfrac{d^n}{dx^n}\sin x = \sin x$ and $\dfrac{d^n}{dx^n}\cos x = \cos x$ when n is a multiple of 4;

$\dfrac{d^n}{dx^n}\cos x = \sin x$ for $n = 4k + 3$, k some integer ≥ 0;

$\dfrac{d^n}{dx^n}\sin x = \cos x$ for $n = 4k + 1$, k some integer ≥ 0.

(b) Based on part (a), $\dfrac{d^{21}}{dx^{21}}\sin x = \dfrac{d}{dx}\left(\dfrac{d^{20}}{dx^{20}}\sin x\right) = \dfrac{d}{dx}\sin x = \cos x$,

$\dfrac{d^{30}}{dx^{30}}\sin x = \dfrac{d^2}{dx^2}\left(\dfrac{d^{28}}{dx^{28}}\sin x\right) = \dfrac{d^2}{dx^2}\sin x = -\sin x$,

$\dfrac{d^{40}}{dx^{40}}\cos x = \cos x$, and

$\dfrac{d^{67}}{dx^{67}}\cos x = \dfrac{d^3}{dx^3}\left(\dfrac{d^{64}}{dx^{64}}\cos x\right) = \dfrac{d^3}{dx^3}\cos x = \sin x$.

49. Since $y' = \cos x$, the slope of the tangent line at $P_1 = (x_1, \sin x_1)$ is $m_1 = \cos x_1$ while the slope of the tangent line at $P_2 = (x_2, \sin x_2)$ is $m_2 = \cos x_2$. Since we are looking for points such that the tangents at those points are parallel, we are looking for x_1, x_2 such that $\cos x_1 = \cos x_2$. This is true for any $x_1 = x_2 + 2\pi n$, n an integer, as well as any $x_1 = -x_2$. For example, the tangent lines at $(0,0)$ and $(2\pi, 0)$ are parallel, as are $(\pi/2, 1)$ and $(-\pi/2, -1)$.

51. $\dfrac{d}{dx}\cot x = \dfrac{d}{dx}\dfrac{\cos x}{\sin x} = \dfrac{(\sin x)\left(\dfrac{d}{dx}\cos x\right) - (\cos x)\left(\dfrac{d}{dx}\sin x\right)}{(\sin x)^2}$

$$= \frac{(\sin x)(-\sin x) - (\cos x)(\cos x)}{\sin^2 x} = \frac{-\sin^2 x - \cos^2 x}{\sin^2 x}$$

$$= \frac{-(\sin^2 x + \cos^2 x)}{\sin^2 x} = -\frac{1}{\sin^2 x} = -\csc^2 x$$

53.

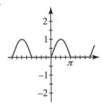

Not differentiable at $x = k\pi$ for k an integer.

55. (a)

(b) $\dfrac{dF}{d\theta} = \dfrac{-70(0.2)(0.2\cos\theta - \sin\theta)}{(0.2\sin\theta + \cos\theta)^2} = \dfrac{-2.8\cos\theta + 14\sin\theta}{(0.2\sin\theta + \cos\theta)^2}$

(c) Solving $-2.8\cos\theta + 14\sin\theta = 0$, we obtain $\tan\theta = 0.2$ and $\theta \approx 0.1974$ radians.

(d) $F(0.1974) \approx 13.7281$

(e) The minimum force required to pull the sled is about 13.73 pounds when θ is about 0.1974 radians or 11.3°.

3.5   Chain Rule

1. $\dfrac{dy}{dx} = 30(-5x)^{29}(-5) = 150(5x)^{29}$

3. $\dfrac{dy}{dx} = 200(2x^2 + x)^{199}(4x + 1)$

5. $y = (x^3 - 2x^2 + 7)^{-4}; \quad \dfrac{dy}{dx} = -4(x^3 - 2x^2 + 7)^{-5}(3x^2 - 4x)$

7. $\dfrac{dy}{dx} = (3x - 1)^4[5(-2x + 9)^4(-2)] + (-2x + 9)^5[4(3x - 1)^3(3)]$

$\quad = -10(3x - 1)^4(-2x + 9)^4 + 12(-2x + 9)^5(3x - 1)^3$

9. $y = \sin(\sqrt{2}x^{1/2}); \quad \dfrac{dy}{dx} = (\cos\sqrt{2x})[(\sqrt{2}/2)x^{-1/2}] = \dfrac{\sqrt{2}}{2\sqrt{x}}\cos\sqrt{2x} = \dfrac{\cos\sqrt{2x}}{\sqrt{2x}}$

11. $y = \left(\dfrac{x^2 - 1}{x^2 + 1}\right)^{1/2}$

$\dfrac{dy}{dx} = \dfrac{1}{2}\left(\dfrac{x^2 - 1}{x^2 + 1}\right)^{-1/2}\left[\dfrac{(x^2 + 1)(2x) - (x^2 - 1)(2x)}{(x^2 + 1)^2}\right] = \dfrac{2x}{(x^2 + 1)^2}\sqrt{\dfrac{x^2 + 1}{x^2 - 1}}$

13. $\dfrac{dy}{dx} = 10[x + (x^2 - 4)^3]^9[1 + 3(x^2 - 4)^2(2x)] = 10[x + (x^2 - 4)^3]^9[1 + 6x(x^2 - 4)^2]$

15. $\dfrac{dy}{dx} = x[-4(x^{-1} + x^{-2} + x^{-3})^{-5}(-x^{-2} - 2x^{-3} - 4x^{-4})] + (x^{-1} + x^{-2} + x^{-3})^{-4}(1)$

$= (x^{-1} + x^{-2} + x^{-3})^{-4} + 4x(x^{-1} + x^{-2} + x^{-3})^{-5}(x^{-2} + 2x^{-3} + 4x^{-4})$

$= \dfrac{(x^{-1} + x^{-2} + x^{-3}) + 4x^{-1} + 8x^{-2} + 16x^{-3}}{(x^{-1} + x^{-2} + x^{-3})^5} = \dfrac{5x^{-1} + 9x^{-2} + 17x^{-3}}{(x^{-1} + x^{-2} + x^{-3})^5}$

17. $\dfrac{dy}{dx} = [\cos(\pi x + 1)](\pi) = \pi \cos(\pi x + 1)$

19. $\dfrac{dy}{dx} = (3\sin^2 5x)(\cos 5x)(5) = 15\sin^2 5x \cos 5x$

21. $f'(x) = (x^3)(-3x^2 \sin x^3) + (\cos x^3)(3x^2) = -3x^5 \sin x^3 + 3x^2 \cos x^3$

23. $f'(x) = 10(2 + x\sin 3x)^9[(x)(\cos 3x)(3) + (\sin 3x)(1)] = 10(2 + x\sin 3x)^9(3x\cos 3x + \sin 3x)$

25. $f(x) = \tan x^{-1}; \quad f'(x) = (\sec^2 x^{-1})(-x^{-2}) = -\dfrac{1}{x^2}\sec^2\dfrac{1}{x}$

27. $f'(x) = (\sin 2x)(-\sin 3x)(3) + (\cos 3x)(\cos 2x)(2) = 2\cos 2x \cos 3x - 3\sin 2x \sin 3x$

29. $f'(x) = 5(\sec 4x + \tan 2x)^4(4\sec 4x \tan 4x + 2\sec^2 2x)$

31. $f'(x) = [\cos(\sin 2x)](\cos 2x)(2) = 2\cos 2x \cos(\sin 2x)$

33. $f(x) = \cos[\sin(2x + 5)^{1/2}]; \quad f'(x) = [-\sin(\sin\sqrt{2x+5})](\cos\sqrt{2x+5})\left[\dfrac{1}{2}(2x+5)^{-1/2}\right](2)$

$= -\dfrac{\sin(\sin\sqrt{2x+5})\cos\sqrt{2x+5}}{\sqrt{2x+5}}$

35. $f'(x) = [3\sin^2(4x^2 - 1)][\cos(4x^2 - 1)](8x) = 24x\sin^2(4x^2 - 1)\cos(4x^2 - 1)$

37. $f'(x) = 6\left(1 + \{1 + [1 + (1 + x^3)^4]^5\}\right)^5 \{5[1 + (1 + x^3)^4]^4\}[4(1 + x^3)^3](3x^2)$

$= 360\left\{1 + [1 + (1 + x^3)^4]^5\right\}^5[1 + (1 + x^3)^4]^4(1 + x^3)^3 x^2$

39. $y' = [3(x^2 + 2)^2](2x) = 6x(x^2 + 2)^2$. When $x = -1$, the slope of the tangent line is $y'(-1) = -6(1 + 2)^2 = -54$.

41. $y' = 3\cos 3x - 20x\sin 5x + 4\cos 5x$. When $x = \pi$, the slope of the tangent line is $y'(\pi) = 3\cos 3\pi - 20\pi \sin 5\pi + 4\cos 5\pi = -7$.

43. $y' = 2\left(\dfrac{x}{x+1}\right)\left[\dfrac{(x+1)(1) - (x)(1)}{(x+1)^2}\right] = \dfrac{2x}{(x+1)^3}$. When $x = -1/2$, $y = \left(\dfrac{-1/2}{-1/2+1}\right)^2 = 1$ and $y' = \dfrac{2(-1/2)}{(-1/2+1)^3} = -8$.
Thus, an equation of the tangent line is $y - 1 = -8(x + 1/2)$ or $y = -8x - 3$.

45. $y' = 3\sec^2 3x$. When $x = \pi/4$, $y = \tan 3\pi/4 = -1$ and $y' = 3\sec^2 3\pi/4 = 6$. Thus, an equation of the tangent line is $y + 1 = 6\left(x - \dfrac{\pi}{4}\right)$ or $y = 6x - \dfrac{3\pi + 2}{2}$.

47. $y' = \sin\dfrac{\pi}{6x}(-2\pi x\sin \pi x^2) + \cos \pi x^2\left(-\dfrac{\pi}{6x^2}\cos\dfrac{\pi}{6x}\right) = -2\pi x\sin\dfrac{\pi}{6x}\sin \pi x^2 - \dfrac{\pi}{6x^2}\cos \pi x^2\cos\dfrac{\pi}{6x}$. When $x = \dfrac{1}{2}$, $y = \dfrac{\sqrt{6}}{4}$ and $y' = -\dfrac{3\pi\sqrt{6} + 2\pi\sqrt{2}}{12}$. Thus, an equation of the normal line is $y - \dfrac{\sqrt{6}}{4} = \dfrac{12}{3\pi\sqrt{6} + 2\pi\sqrt{2}}\left(x - \dfrac{1}{2}\right)$ or $y = \dfrac{12}{3\pi\sqrt{6} + 2\pi\sqrt{2}}x - \dfrac{6}{3\pi\sqrt{6} + 2\pi\sqrt{2}} + \dfrac{\sqrt{6}}{4}$.

49. $f'(x) = \pi \cos \pi x; \quad f''(x) = -\pi^2 \sin \pi x; \quad f'''(x) = -\pi^3 \cos \pi x$

51. $\dfrac{dy}{dx} = (x)(\cos 5x)(5) + (\sin 5x)(1) = 5x\cos 5x + \sin 5x$

$\dfrac{d^2y}{dx^2} = (5x)(-\sin 5x)(5) + (\cos 5x)(5) + (\cos 5x)(5) = -25x\sin 5x + 10\cos 5x$

$\dfrac{d^3y}{dx^3} = (-25x)(\cos 5x)(5) + (\sin 5x)(-25) + 10(-\sin 5x)(5) = -125x\cos 5x - 75\sin 5x$

53. $f'(x) = \dfrac{(x^2+1)^2(1) - (x)[2(x^2+1)(2x)]}{(x^2+1)^4} = \dfrac{1-3x^2}{(x^2+1)^3}$. To find where the tangent line is horizontal, solve $f'(x) = 0$. This gives

$x = \pm\sqrt{\dfrac{1}{3}} = \pm\dfrac{\sqrt{3}}{3}$. The tangent line is horizontal at $\left(-\dfrac{\sqrt{3}}{3}, -\dfrac{3\sqrt{3}}{16}\right)$ and $\left(\dfrac{\sqrt{3}}{3}, \dfrac{3\sqrt{3}}{16}\right)$. Since x^2+1 is never 0, the graph has no vertical tangents.

55. $f'(x) = -\dfrac{1}{3}\sin\dfrac{x}{3}$; $f''(x) = -\dfrac{1}{9}\cos\dfrac{x}{3}$. The slope of the tangent line to the graph of f' at $x = 2\pi$ is $f''(2\pi) = -\dfrac{1}{9}\cos\dfrac{2\pi}{3} = -\dfrac{1}{9}\left(-\dfrac{1}{2}\right) = \dfrac{1}{18}$.

57. $\dfrac{dR}{d\theta} = 2\left(\dfrac{v_0^2}{g}\right)(\cos 2\theta)(2) = 4\left(\dfrac{v_0^2}{g}\right)\cos 2\theta$.

Setting $\dfrac{dR}{d\theta} = 0$, and assuming that $0 \le \theta \le \pi$, we obtain $\theta = \pi/4, 3\pi/4$.

59. The volume of a sphere is $V = (4\pi/3)r^3$, so that $dV/dt = 4\pi r^2(dr/dt)$. Using $dV/dt = 10$ and $r = 2$, we obtain $dr/dt = 10/4\pi(2)^2 = 5/8\pi$ in/min.

61. $\dfrac{d}{dx}F(3x) = F'(3x)\dfrac{d}{dx}(3x) = 3F'(3x)$

63. $\dfrac{d}{dx}f(-10x+7) = [f'(-10x+7)](-10) = \dfrac{1}{-10x+7}(-10) = \dfrac{10}{10x-7}$

65. The derivatives for $n = 1$ to 4 are:

$\dfrac{d}{dx}(1+2x)^{-1} = -2(1+2x)^{-2}$; $\dfrac{d^2}{dx^2}(1+2x)^{-1} = 2!(2^2)(1+2x)^{-3}$;

$\dfrac{d^3}{dx^3}(1+2x)^{-1} = (-3!)(2^3)(1+2x)^{-4}$; $\dfrac{d^4}{dx^4}(1+2x)^{-1} = (4!)(2^4)(1+2x)^{-5}$.

We can verify by induction that $\dfrac{d^n}{dx^n}(1+2x)^{-1} = (-1)^n\, n!\, 2^n\, (1+2x)^{-n-1}$:

$$\dfrac{d^n}{dx^n}(1+2x)^{-1} = \dfrac{d}{dx}\left[\dfrac{d^{n-1}}{dx^{n-1}}(1+2x)^{-1}\right] = \dfrac{d}{dx}(-1)^{n-1}(n-1)!\, 2^{n-1}\, (1+2x)^{-(n-1)-1}$$

$$= \dfrac{d}{dx}(-1)^{n-1}(n-1)!\, 2^{n-1}\, (1+2x)^{-n}$$

$$= (-1)^{n-1}(-1)(n)(n-1)!\, 2^{n-1}\, (1+2x)^{-n-1}(2)$$

$$= (-1)^n\, n!\, 2^n\, (1+2x)^{-n-1}$$

67. $g'(1) = h'(f(1))f'(1) = h'(3)\cdot 6 = -2\cdot 6 = -12$

69. Since f is odd, $f(-x) = -f(x)$. Then $f'(x) = -\dfrac{d}{dx}f(-x) = -[f'(-x)](-1) = f'(-x)$.

Since $f'(-x) = f'(x)$, f' is an even function.

3.6 Implicit Differentiation

1. $\dfrac{d}{dx}x^2y^4 = (x^2)(4y^3\dfrac{dy}{dx}) + (y^4)(2x) = 4x^2y^3\dfrac{dy}{dx} + 2xy^4$

3. $\dfrac{d}{dx}\cos y^2 = (-\sin y^2)(2y)\dfrac{dy}{dx} = -2y(\sin y^2)\dfrac{dy}{dx}$

5. $2y\dfrac{dy}{dx} - 2\dfrac{dy}{dx} = 1 \quad \dfrac{dy}{dx} = \dfrac{1}{2y-2}$

7. $x\left(2y\dfrac{dy}{dx}\right) + y^2 - 2x = 0; \quad \dfrac{dy}{dx} = \dfrac{2x - y^2}{2xy}$

9. $3\dfrac{dy}{dx} + (-\sin y)\dfrac{dy}{dx} = 2x; \quad \dfrac{dy}{dx} = \dfrac{2x}{3 - \sin y}$

11. $x^3\left(2y\dfrac{dy}{dx}\right) + y^2(3x^2) = 4x + 2y\dfrac{dy}{dx}; \quad \dfrac{dy}{dx} = \dfrac{4x - 3x^2y^2}{2x^3y - 2y}$

13. $6(x^2+y^2)^5\left(2x + 2y\dfrac{dy}{dx}\right) = 3x^2 - 3y^2\dfrac{dy}{dx}$

$12y(x^2+y^2)^5\dfrac{dy}{dx} + 3y^2\dfrac{dy}{dx} = 3x^2 - 12x(x^2+y^2)^5; \quad \dfrac{dy}{dx} = \dfrac{x^2 - 4x(x^2+y^2)^5}{y^2 + 4y(x^2+y^2)^5}$

15. $(y^{-3})(6x^5) + (x^6)\left(-3y^{-4}\dfrac{dy}{dx}\right) + (y^6)(-3x^{-4}) + (x^{-3})\left(6y^5\dfrac{dy}{dx}\right) = 2$

$(-3x^6y^{-4} + 6x^{-3}y^5)\dfrac{dy}{dx} = 2 - 6x^5y^{-3} + 3x^{-4}y^6; \quad \dfrac{dy}{dx} = \dfrac{2 - 6x^5y^{-3} + 3x^{-4}y^6}{6x^{-3}y^5 - 3x^6y^{-4}}$

17. $2(x-1) + 2(y+4)\dfrac{dy}{dx} = 0; \quad \dfrac{dy}{dx} = -\dfrac{x-1}{y+4} = \dfrac{1-x}{y+4}$

19. $2y\dfrac{dy}{dx} = \dfrac{x+2-(x-1)}{(x+2)^2} = \dfrac{3}{(x+2)^2}; \quad \dfrac{dy}{dx} = \dfrac{3}{2y(x+2)^2}$

21. $x\dfrac{dy}{dx} + y = [\cos(x+y)]\left(1 + \dfrac{dy}{dx}\right) = \cos(x+y) + \dfrac{dy}{dx}\cos(x+y)$

$x\dfrac{dy}{dx} - \dfrac{dy}{dx}\cos(x+y) = \cos(x+y) - y; \quad \dfrac{dy}{dx} = \dfrac{\cos(x+y) - y}{x - \cos(x+y)}$

23. $1 = (\tan y \sec y)\dfrac{dy}{dx}; \quad \dfrac{dy}{dx} = \dfrac{1}{\tan y \sec y} = \cos y \cot y$

25. $2r\dfrac{dr}{d\theta} = 2\cos 2\theta; \quad \dfrac{dr}{d\theta} = \dfrac{\cos 2\theta}{r}$

27. $x\left(2y\dfrac{dy}{dx}\right) + y^2 + 12y^2\dfrac{dy}{dx} + 3 = 0; \quad 2xy\dfrac{dy}{dx} + 12y^2\dfrac{dy}{dx} = -3 - y^2; \quad \dfrac{dy}{dx} = \dfrac{-3 - y^2}{2xy + 12y^2}$

$\left.\dfrac{dy}{dx}\right|_{(1,-1)} = \dfrac{-3 - (-1)^2}{2(1)(-1) + 12(-1)^2} = -\dfrac{2}{5}$

29. Letting $x = 1/2$ in $2y^2 + 2xy - 1 = 0$, we obtain $2y^2 + y - 1 = 0$. Factoring, we have $(2y-1)(y+1) = 0$. Thus, $y = -1, 1/2$. Differentiating gives

$$4y\dfrac{dy}{dx} + 2x\dfrac{dy}{dx} + 2y = 0; \quad (2y+x)\dfrac{dy}{dx} = -y; \quad \dfrac{dy}{dx} = -\dfrac{y}{2y+x}$$

$$\left.\dfrac{dy}{dx}\right|_{(1/2,-1)} = -\dfrac{-1}{-2 + 1/2} = -\dfrac{2}{3}; \quad \left.\dfrac{dy}{dx}\right|_{(1/2,1/2)} = -\dfrac{1/2}{1 + 1/2} = -\dfrac{1}{3}.$$

31. $4x^3 + 3y^2 \dfrac{dy}{dx} = 0$; $\dfrac{dy}{dx} = -\dfrac{4x^3}{3y^2}$. The slope of the tangent line is $\dfrac{dy}{dx}\bigg|_{(-2,2)} = -\dfrac{4(-2)^3}{3(2)^2} = \dfrac{8}{3}$. Hence an equation of the tangent

line is $y - 2 = \dfrac{8}{3}(x + 2)$ or $y = \dfrac{8}{3}x + \dfrac{22}{3}$.

33. $(\sec^2 y)\dfrac{dy}{dx} = 1$; $\dfrac{dy}{dx} = \dfrac{1}{\sec^2 y} = \cos^2 y$. The slope of the tangent line is $\dfrac{dy}{dx}\bigg|_{y=\pi/4} = \cos^2 \pi/4 = 1/2$. When $y = \pi/4$, $\tan \pi/4 = x$.

Thus, $x = 1$ and the equation of the tangent line is $y - \dfrac{\pi}{4} = \dfrac{1}{2}(x - 1)$ or $y = \dfrac{1}{2}x + \dfrac{\pi - 2}{4}$.

35. $2x - x\dfrac{dy}{dx} - y + 2y\dfrac{dy}{dx} = 0$; $(2y - x)\dfrac{dy}{dx} = y - 2x$; $\dfrac{dy}{dx} = \dfrac{y - 2x}{2y - x}$. Setting $\dfrac{dy}{dx} = 0$, we obtain $y = 2x$. Substituting $y = 2x$ into the

equation of the curve, we have $x^2 - x(2x) + (2x)^2 = 3$, $3x^2 = 3$, or $x = \pm 1$. Given $y = 2x$, we see that for $x = 1$, $y = 2$, and for $x = -1$, $y = -2$. Thus, the curve has horizontal tangent lines at $(1, 2)$ and $(-1, -2)$.

37. $2x + 2y\dfrac{dy}{dx} = 0$; $\dfrac{dy}{dx} = -\dfrac{x}{y}$. Setting $\dfrac{dy}{dx} = \dfrac{1}{2}$ gives us $-\dfrac{x}{y} = \dfrac{1}{2}$ or $y = -2x$. Substituting this into the equation of the curve, we

have $x^2 + (-2x)^2 = 25$, $5x^2 = 25$, or $x = \pm\sqrt{5}$. Since $y = -2x$, the points on the graph are $(\sqrt{5}, -2\sqrt{5})$ and $(-\sqrt{5}, 2\sqrt{5})$.

39. $3y^2 \dfrac{dy}{dx} = 2x$; $\dfrac{dy}{dx} = \dfrac{2x}{3y^2}$. The slope of the tangent line perpendicular to $y + 3x - 5 = 0$, or $y = -3x + 5$, is $1/3$. Setting $\dfrac{dy}{dx} = \dfrac{1}{3}$

gives us $\dfrac{2x}{3y^2} = \dfrac{1}{3}$, $y^2 = 2x$, or $y = \pm\sqrt{2x}$. Substituting this into the equation of the curve, we have $(\pm\sqrt{2x})^3 = x^2$, so we can

eliminate $y = -\sqrt{2x}$. Thus, $2^{3/2}x^{3/2} = x^2$, $\sqrt{x} = \sqrt{8}$, or $x = 8$. Since $y = \sqrt{2x}$, the point on the graph is $(8, 4)$.

41. $12y^2 \dfrac{dy}{dx} = 12x$; $\dfrac{dy}{dx} = \dfrac{x}{y^2}$

$\dfrac{d^2y}{dx^2} = \dfrac{y^2(1) - x(2y \, dy/dx)}{y^4} = \dfrac{y - 2x \, dy/dx}{y^3} = \dfrac{y - 2x(x/y^2)}{y^3} = \dfrac{y^3 - 2x^2}{y^5}$

43. $2x - 2y\dfrac{dy}{dx} = 0$; $\dfrac{dy}{dx} = \dfrac{x}{y}$; $\dfrac{d^2y}{dx^2} = \dfrac{y - x \, dy/dx}{y^2} = \dfrac{y - x(x/y)}{y^2} = \dfrac{y^2 - x^2}{y^3} = -\dfrac{25}{y^3}$

45. $1 + \dfrac{dy}{dx} = (\cos y)\dfrac{dy}{dx}$; $\dfrac{dy}{dx} = \dfrac{1}{\cos y - 1}$

$\dfrac{d^2y}{dx^2} = \dfrac{-(-\sin y) \, dy/dx}{(\cos y - 1)^2} = \dfrac{(\sin y)[1/(\cos y - 1)]}{(\cos y - 1)^2} = \dfrac{\sin y}{(\cos y - 1)^3}$

47. $2x + 2x\dfrac{dy}{dx} + 2y - 2y\dfrac{dy}{dx} = 0$; $\dfrac{dy}{dx} = -\dfrac{x + y}{x - y} = \dfrac{y + x}{y - x}$

$\dfrac{d^2y}{dx^2} = \dfrac{(y - x)(dy/dx + 1) - (y + x)(dy/dx - 1)}{(y - x)^2}$

$= \dfrac{(y - x)\left(\dfrac{y + x}{y - x} + 1\right) - (y + x)\left(\dfrac{y + x}{y - x} - 1\right)}{(y - x)^2}$

$= \dfrac{(y - x)(y + x + y - x) - (y + x)[y + x - (y - x)]}{(y - x)^3} = \dfrac{2y^2 - 4xy - 2x^2}{(y - x)^3}$

$= \dfrac{-2(x^2 + 2xy - y^2)}{(y - x)^3} = \dfrac{-2}{(y - x)^3}$

49. Using implicit differentiation, we see

$$2x - 2y\dfrac{dy}{dx} = 1 \qquad \text{and} \qquad \dfrac{dy}{dx} = \dfrac{2x - 1}{2y}.$$

Solving $x^2 - y^2 = x$ for y and differentiating, we have

$$y = \sqrt{x^2 - x} = (x^2 - x)^{1/2} \qquad \text{and} \qquad \dfrac{dy}{dx} = \dfrac{1}{2}(x^2 - x)^{-1/2}(2x - 1).$$

Substituting the expression for y into $\dfrac{2x-1}{2y}$, we obtain

$$\frac{dy}{dx} = \frac{2x-1}{2(x^2-x)^{1/2}} = \frac{1}{2}(x^2-x)^{-1/2}(2x-1).$$

51. Using implicit differentiation, we see

$$x^3\frac{dy}{dx} + y(3x^2) = 1 \qquad \text{and} \qquad \frac{dy}{dx} = \frac{1-3x^2y}{x^3}.$$

Solving $x^3y = x+1$ for y and differentiating, we have

$$y = \frac{x+1}{x^3} \qquad \text{and} \qquad \frac{dy}{dx} = \frac{x^3-(x+1)(3x^2)}{x^6} = \frac{x-3(x+1)}{x^4} = \frac{-2x-3}{x^4}.$$

Substituting the expression for y into $\dfrac{1-3x^2y}{x^3}$, we obtain

$$\frac{dy}{dx} = \frac{1-3x^2\left(\dfrac{x+1}{x^3}\right)}{x^3} = \frac{1-\dfrac{3(x+1)}{x}}{x^3} = \frac{-2x-3}{x^4}.$$

In Problems 53–56, we solve the given equations for the appropriate values of y.

53. Since the graph lies below the line $y = 1$, we have

$$y - 1 = -\sqrt{x-2} \qquad \text{or} \qquad y = 1 - \sqrt{x-2}.$$

55. $y^2 = 4 - x^2$; $y = \pm\sqrt{4-x^2}$. Since the graph is in the second and fourth quadrants,

$$y = \begin{cases} \sqrt{4-x^2}, & -2 \le x < 0 \\ -\sqrt{4-x^2}, & 0 \le x \le 2 \end{cases}.$$

57. $2x\dfrac{dx}{dt} + 2y\dfrac{dy}{dt} = 0$; $\quad \dfrac{dy}{dt} = -\left(\dfrac{x}{y}\right)\dfrac{dx}{dt}$

59. (a) $3x^2 + 3y^2\dfrac{dy}{dx} = 3x\dfrac{dy}{dx} + 3y$; $\dfrac{dy}{dx} = \dfrac{x^2-y}{x-y^2}$. Setting $y = x$ in $x^3 + y^3 = 3xy$, we obtain

$$x^3 + x^3 = 3x^2; \qquad 2x^3 = 3x^2; \qquad x^2(2x-3) = 0; \qquad \text{so} \quad x = 0, \frac{3}{2}.$$

Since $\dfrac{dy}{dx}\bigg|_{(3/2,3/2)} = \dfrac{9/4-3/2}{3/2-9/4} = -1$, the slope of the tangent line at $(3/2, 3/2)$ is -1 and the equation of the tangent line is

$$y - \frac{3}{2} = -\left(x-\frac{3}{2}\right) \qquad \text{or} \qquad y = -x+3.$$

(b) Setting $\dfrac{dy}{dx} = \dfrac{x^2-y}{x-y^2} = 0$, we obtain $x^2 - y = 0$ or $y = x^2$. Substituting $y = x^2$ in $x^3 + y^3 = 3xy$, we obtain

$$x^3 + x^6 = 3x^3; \qquad x^3(x^3-2) = 0; \qquad \text{so} \quad x = 0, \sqrt[3]{2}.$$

In the first quadrant the tangent line is horizontal at $(\sqrt[3]{2}, \sqrt[3]{4})$.

61. $2y\dfrac{dy}{dx} = 2x^2$; $\qquad \dfrac{dy}{dx} = \dfrac{3x^2}{2y}$; $\qquad \dfrac{dy}{dx}\bigg|_{(1,1)} = \dfrac{3}{2}$

$4x + 6y\dfrac{dy}{dx} = 0$; $\qquad \dfrac{dy}{dx} = -\dfrac{2x}{3y}$; $\qquad \dfrac{dy}{dx}\bigg|_{(1,1)} = -\dfrac{2}{3}$

Since $\dfrac{3}{2}\left(-\dfrac{2}{3}\right) = -1$, the graphs are orthogonal at $(1,1)$.

63. Solving for $\dfrac{dy}{dx}$ for each family of curves, we get

$$2x - 2y\frac{dy}{dx} = 0; \quad \frac{dy}{dx} = \frac{x}{y} \quad \text{and} \quad x\frac{dy}{dx} + y = 0; \quad \frac{dy}{dx} = -\frac{y}{x}.$$

Since $\dfrac{x}{y}\left(-\dfrac{y}{x}\right) = -1$, the families are orthogonal trajectories of each other. The graph on the right
shows $x^2 - y^2 = c_1$ in red, and $xy = c_2$ in blue, for selected values of c_1 and c_2.

65. (a) We use the formula for the tangent of the sum of two angles:

$$\tan(\alpha + \theta) = \frac{\tan\alpha + \tan\theta}{1 - \tan\alpha\tan\theta}.$$

From the figure we see that $\tan\alpha = 14/x$ and

$$\frac{14 + 4}{x} = \tan(\alpha + \theta) = \frac{14/x + \tan\theta}{1 - (14/x)\tan\theta} = \frac{14 + x\tan\theta}{x - 14\tan\theta}$$

$$18(x - 14\tan\theta) = x(14 + x\tan\theta)$$

$$18x - 252\tan\theta = 14x + x^2\tan\theta$$

$$4x = (x^2 + 252)\tan\theta$$

$$\tan\theta = \frac{4x}{x^2 + 252}.$$

(b) $\sec^2\theta\dfrac{d\theta}{dx} = \dfrac{(x^2 + 252)4 - 4x(2x)}{(x^2 + 252)^2} = \dfrac{1008 - 4x^2}{(x^2 + 252)^2}$

From $\sec^2\theta = 1 + \tan^2\theta = 1 + \dfrac{16x^2}{(x^2 + 252)^2} = \dfrac{(x^2 + 252)^2 + 16x^2}{(x^2 + 252)^2}$, we obtain

$$\frac{d\theta}{dx} = \frac{1008 - 4x^2}{(x^2 + 252)^2\sec^2\theta} = \frac{1008 - 4x^2}{(x^2 + 252)^2\{[(x^2 + 252)^2 + 16x^2]/(x^2 + 252)^2\}}$$

$$= \frac{4(252 - x^2)}{(x^2 + 252)^2 + 16x^2}.$$

(c) Setting $\dfrac{d\theta}{dx} = 0$, we obtain $x^2 = 252$ or $x = 6\sqrt{7} \approx 15.87$ ft.

67. We use implicit differentiation to find the slopes of the tangent lines at $(1, 1)$.

$$x^2 + y^2 + 4y = 6 \qquad\qquad\qquad x^2 + 2x + y^2 = 4$$

$$2x + 2y\frac{dy}{dx} + 4\frac{dy}{dx} = 0 \qquad\qquad 2x + 2 + 2y\frac{dy}{dx} = 0$$

$$\frac{dy}{dx} = -\frac{x}{y + 2} \qquad\qquad\qquad \frac{dy}{dx} = -\frac{x + 1}{y}$$

$$\left.\frac{dy}{dx}\right|_{(1, 1)} = -\frac{1}{3} \qquad\qquad\qquad \left.\frac{dy}{dx}\right|_{(1, 1)} = -2$$

Letting $m_1 = -1/3$ and $m_2 = -2$, we have $\tan\theta = \dfrac{-1/3 + 2}{1 + 2/3} = 1$. The angle between the graphs is $\theta = \pi/4$.

69. One more function defined by $x^2 + y^2 = 4$ includes the parts of the circle in the *first* and *third*
quadrants (see graph at right):

$$y = \begin{cases} -\sqrt{4 - x^2}, & -2 \le x < 0 \\ \sqrt{4 - x^2}, & 0 \le x \le 2 \end{cases}.$$

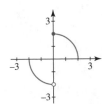

3.7 Derivatives of Inverse Functions

1. Since $f'(x) = 30x^2 + 8 > 0$ for all x, f is increasing on $(-\infty, \infty)$. It follows from Theorem 3.7.2 that f^{-1} exists.

3. Writing $f(x) = x(x^2 + x - 2) = x(x+2)(x-1)$, we see that f has three distinct zeros. It follows that f is not one-to-one and that f^{-1} does not exist.

5. Since $f(x) = 2x^3 + 8$, $f(1/2) = 33/4$, $f'(x) = 6x^2$ and $f'(1/2) = 3/2$. Now $f(1/2) = 33/4$ implies $f^{-1}(33/4) = 1/2$. Thus, $(f^{-1})'(33/4) = 1/f'(f^{-1}(33/4)) = 1/f'(1/2) = 1/(3/2) = 2/3$.

7. $f(x) = 2 + \dfrac{1}{x}$; $f'(x) = -\dfrac{1}{x^2}$; $f^{-1}(x) = \dfrac{1}{x-2}$. Using Theorem 3.7.4, we have

$$(f^{-1})'(x) = \frac{1}{f'(f^{-1}(x))} = \frac{1}{-1/\left(\dfrac{1}{x-2}\right)^2} = -\frac{1}{(x-2)^2}.$$

By direct differentiation, $(f^{-1})(x) = \dfrac{0-1}{(x-2)^2} = -\dfrac{1}{(x-2)^2}$.

9. At $x = 3$, $y = 5$, so $(5, 3)$ is the corresponding point on the graph of f^{-1}. Since $\dfrac{dy}{dx} = x^2 + 1$ and $\left.\dfrac{dy}{dx}\right|_{x=3} = 10$, then by (4) $\left.\dfrac{dx}{dy}\right|_{y=5} = \dfrac{1}{10}$. Thus, an equation of the tangent line is $y - 3 = \dfrac{1}{10}(x - 5)$ or $y = \dfrac{1}{10}x + \dfrac{5}{2}$.

11. At $x = 1$, $y = 8$, so $(8, 1)$ is the corresponding point on the graph of f^{-1}. Since $\dfrac{dy}{dx} = 15x^4(x^5 + 1)^2$ and $\left.\dfrac{dy}{dx}\right|_{x=1} = 60$, then by (4) $\left.\dfrac{dx}{dy}\right|_{y=8} = \dfrac{1}{60}$. Thus, an equation of the tangent line is $y - 1 = \dfrac{1}{60}(x - 8)$ or $y = \dfrac{1}{60}x + \dfrac{13}{15}$.

13. $y' = \dfrac{5}{\sqrt{1 - (5x-1)^2}} = \dfrac{5}{\sqrt{10x - 25x^2}}$

15. $y' = 4\left[\dfrac{-1/2}{1 + (x/2)^2}\right] = \dfrac{-2}{1 + x^2/4} = -\dfrac{8}{4 + x^2}$

17. $y' = 2\sqrt{x}\left[\dfrac{1}{1 + (\sqrt{x})^2}\left(\dfrac{1}{2\sqrt{x}}\right)\right] + (\tan^{-1}\sqrt{x})\left(\dfrac{1}{\sqrt{x}}\right) = \dfrac{1}{1+x} + \dfrac{\tan^{-1}\sqrt{x}}{\sqrt{x}}$

19. $y' = \dfrac{(\cos^{-1} 2x)\left[\dfrac{2}{\sqrt{1 - (2x)^2}}\right] - (\sin^{-1} 2x)\left[\dfrac{-2}{\sqrt{1 - (2x)^2}}\right]}{(\cos^{-1} 2x)^2} = \dfrac{2(\cos^{-1} 2x + \sin^{-1} 2x)}{\sqrt{1 - 4x^2}(\cos^{-1} 2x)^2}$

21. $y = (\tan^{-1} x^2)^{-1}$; $y' = -(\tan^{-1} x^2)^{-2}\left[\dfrac{2x}{1 + (x^2)^2}\right] = -\dfrac{2x}{(\tan^{-1} x^2)^2(1 + x^4)}$

23. $y' = \dfrac{2}{\sqrt{1 - x^2}} + x\left(\dfrac{-1}{\sqrt{1 - x^2}}\right) + (\cos^{-1} x)(1) = \dfrac{2 - x}{\sqrt{1 - x^2}} + \cos^{-1} x$

25. $y' = 3\left(x^2 - 9\tan^{-1}\dfrac{x}{3}\right)^2\left[2x - \dfrac{9}{1 + (x/3)^2}\left(\dfrac{1}{3}\right)\right] = 3\left(x^2 - 9\tan^{-1}\dfrac{x}{3}\right)^2\left(2x - \dfrac{27}{9 + x^2}\right)$

27. $F'(t) = \dfrac{1}{1 + \left(\dfrac{t-1}{t+1}\right)^2}\left[\dfrac{(t+1) - (t-1)}{(t+1)^2}\right] = \dfrac{(t+1)^2}{2(t^2 + 1)} \cdot \dfrac{2}{(t+1)^2} = \dfrac{1}{t^2 + 1}$

29. $f'(x) = \dfrac{1}{\sqrt{1 - (\cos 4x)^2}}(-4\sin 4x) = \dfrac{-4\sin 4x}{\sqrt{1 - \cos^2 4x}} = -\dfrac{4\sin 4x}{|\sin 4x|}$

31. $f'(x) = \sec^2(\sin^{-1}x^2)\left(\dfrac{2x}{\sqrt{1-(x^2)^2}}\right) = \dfrac{2x\sec^2(\sin^{-1}x^2)}{\sqrt{1-x^4}}$

33. $\dfrac{1}{1+y^2}\dfrac{dy}{dx} = 2x + 2y\dfrac{dy}{dx};\quad \dfrac{dy}{dx} = \dfrac{2x}{1/(1+y^2)-2y} = \dfrac{2x(1+y^2)}{1-2y-2y^3}$

35. $f'(x) = \dfrac{1}{\sqrt{1-x^2}} - \dfrac{1}{\sqrt{1-x^2}} = 0$. Since $f'(x) = 0$, $f(x)$ is constant.

37. $y' = \dfrac{1/2}{\sqrt{1-x^2/4}} = \dfrac{1}{\sqrt{4-x^2}};\quad m = y'(1) = \dfrac{1}{\sqrt{3}}$

39. $f'(x) = \dfrac{x}{1+x^2} + \tan^{-1}x;\quad f'(1) = \dfrac{1}{2} + \dfrac{\pi}{4} = \dfrac{2+\pi}{4};\quad f(1) = \dfrac{\pi}{4}.$

    The point on the graph is $(1,\pi/4)$ and the slope is $\dfrac{2+\pi}{4}$. Thus, an equation of the tangent line is $y - \dfrac{\pi}{4} = \dfrac{2+\pi}{4}(x-1)$ or
    $y = \dfrac{2+\pi}{4}x - \dfrac{1}{2}.$

41. Lines with slope $\sqrt{3}$ are parallel to $y = \sqrt{3}x + 1$. Solving $f'(x) = -2\cos x = \sqrt{3}$, we get $\cos x = -\dfrac{\sqrt{3}}{2}$ and $x = \dfrac{5\pi}{6}$ or $\dfrac{7\pi}{6}$.
    Thus, the tangent line to f is parallel to $y = \sqrt{3}x + 1$ at $(5\pi/6, 4)$ and $(7\pi/6, 6)$.

43. Multiple applications of (3) yields $(f^{-1})''(x) = \dfrac{d}{dx}(f^{-1})'(x) = \dfrac{d}{dx}\dfrac{1}{f'(f^{-1}(x))}$, resulting in

$$\frac{d}{dx}\frac{1}{f'(f^{-1}(x))} = \frac{-f''(f^{-1}(x))(f^{-1})'(x)}{[f'(f^{-1}(x))]^2} = \frac{-f''(f^{-1}(x)) \cdot \dfrac{1}{f'(f^{-1}(x))}}{[f'(f^{-1}(x))]^2} = -\frac{f''(f^{-1}(x))}{[f'(f^{-1}(x))]^3}.$$

3.8 Exponential Functions

1. $y' = -e^{-x}$

3. $y' = \dfrac{e^{\sqrt{x}}}{2\sqrt{x}}$

5. $y' = 5^{2x}(\ln 5)(2) = 2(\ln 5)5^{2x}$

7. $y' = (x^3)(e^{4x})(4) + (e^{4x})(3x^2) = 4x^3e^{4x} + 3x^2e^{4x}$

9. $f'(x) = \dfrac{(x)(-2e^{-2x}) - (e^{-2x})(1)}{x^2} = -\dfrac{(2x+1)e^{-2x}}{x^2}$

11. $y' = \dfrac{1}{2}(1+e^{-5x})^{-1/2}(-5e^{-5x}) = -\dfrac{5e^{-5x}}{2\sqrt{1+e^{-5x}}}$

13. $y = 2(e^{x/2}+e^{-x/2})^{-1};\ y' = -2(e^{x/2}+e^{-x/2})^{-2}\left(\dfrac{1}{2}e^{x/2} - \dfrac{1}{2}e^{-x/2}\right) = -\dfrac{e^{x/2}-e^{-x/2}}{(e^{x/2}+e^{-x/2})^2}$

15. $y = e^{8x};\ y' = 8e^{8x}$

17. $y = e^{3x-3};\ y' = 3e^{3x-3}$

19. $f(x) = e^{x^{1/3}} + e^{x/3};\ f'(x) = \dfrac{1}{3}x^{-2/3}e^{x^{1/3}} + \dfrac{1}{3}e^{x/3}$

21. $f'(x) = e^{-x}(e^x\sec^2 e^x) - e^{-x}\tan e^x = \sec^2 e^x - e^{-x}\tan e^x$

23. $f'(x) = e^{x\sqrt{x^2+1}}\left[\dfrac{x}{2}2x(x^2+1)^{-1/2} + \sqrt{x^2+1}\right] = \dfrac{2x^2+1}{\sqrt{x^2+1}}e^{x\sqrt{x^2+1}}$

25. $y' = 2xe^{x^2}e^{e^{x^2}}$

27. $y' = 2(e^x + 1)(e^x) = 2e^x(e^x + 1)$. The point on the graph is $(0,4)$ and the slope is $y'(0) = 4$. An equation of the tangent line is $y - 4 = 4(x - 0)$ or $y = 4x + 4$.

29. $y' = e^x$. The slope of the line $3x - y = 7$ is 3. Solving $e^x = 3$, we obtain $x = \ln 3$. The point on the graph of $y = e^x$ at which the tangent line is parallel to $3x - y = 7$ is $(\ln 3, e^{\ln 3})$ or $(\ln 3, 3)$.

31. $f'(x) = e^{-x}(\cos x) + -e^{-x}\sin x = e^{-x}(\cos x - \sin x)$. The tangent line is horizontal when $f'(x) = e^{-x}(\cos x - \sin x) = 0$, or when $\cos x = \sin x$; $\dfrac{\sin x}{\cos x} = 1$; $\tan x = 1$. Thus, $x = \pi/4 + k\pi$, where k is an integer.

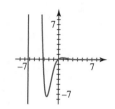

33. $\dfrac{dy}{dx} = 2xe^{x^2}$; $\dfrac{d^2y}{dx^2} = 2x(2xe^{x^2}) + 2e^{x^2} = e^{x^2}(4x^2 + 2)$

$\dfrac{d^3y}{dx^3} = e^{x^2}(8x) + 2xe^{x^2}(4x^2 + 2) = 8x^3e^{x^2} + 12xe^{x^2}$

35. $\dfrac{dy}{dx} = 2e^{2x}\cos e^{2x}$; $\dfrac{d^2y}{dx^2} = 2e^{2x}(-2e^{2x}\sin e^{2x}) + 4e^{2x}\cos e^{2x} = -4e^{4x}\sin e^{2x} + 4e^{2x}\cos e^{2x}$

37. $y' = -3C_1e^{-3x} + 2C_2e^{2x}$; $y'' = 9C_1e^{-3x} + 4C_2e^{2x}$

$y'' + y' - 6y = 9C_1e^{-3x} + 4C_2e^{2x} - 3C_1e^{-3x} + 2C_2e^{2x} - 6C_1e^{-3x} - 6C_2e^{2x} = 0$

39. For $y = Ce^{kx}$, $y' = kCe^{kx} = k(Ce^{kx}) = ky$.

41. $\dfrac{dy}{dx} = \left(1 + \dfrac{dy}{dx}\right)e^{x+y}$; $\dfrac{dy}{dx} - \dfrac{dy}{dx}e^{x+y} = e^{x+y}$; $\dfrac{dy}{dx} = \dfrac{e^{x+y}}{1 - e^{x+y}}$

43. $\dfrac{dy}{dx} = -\left(x\dfrac{dy}{dx} + y\right)e^{xy}\sin e^{xy}$; $\dfrac{dy}{dx} + x\dfrac{dy}{dx}e^{xy}\sin e^{xy} = -ye^{xy}\sin e^{xy}$; $\dfrac{dy}{dx} = -\dfrac{ye^{xy}\sin e^{xy}}{1 + xe^{xy}\sin e^{xy}}$

45. $1 + 2y\dfrac{dy}{dx} = e^{x/y}\left[\dfrac{y - x(dy/dx)}{y^2}\right]$; $2y\dfrac{dy}{dx} + \dfrac{x}{y^2}\left(\dfrac{dy}{dx}\right)e^{x/y} = \dfrac{1}{y}e^{x/y} - 1$;

$2y^3\dfrac{dy}{dx} + x\dfrac{dy}{dx}e^{x/y} = ye^{x/y} - y^2$; $\dfrac{dy}{dx} = \dfrac{ye^{x/y} - y^2}{2y^3 + xe^{x/y}}$

47. (a)

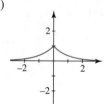

(b) Since $f(x) = e^{-|x|} = \begin{cases} e^x, & x < 0 \\ e^{-x}, & x \geq 0 \end{cases}$, we get $f'(x) = \begin{cases} e^x, & x < 0 \\ -e^{-x}, & x \geq 0 \end{cases}$.

(c)

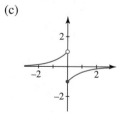

(d) Because $\lim\limits_{x \to 0^-} f'(x) = \lim\limits_{x \to 0^-} e^x = 1$ and $\lim\limits_{x \to 0^+} f'(x) = \lim\limits_{x \to 0^+} (-e^{-x}) = -1$, the function is not differentiable at $x = 0$.

49. (a)
$$\frac{dP}{dt} = \frac{0[bP_0 + (a - bP_0)e^{-at}] - [-a(a - bP_0)e^{-at}]aP_0}{[bP_0 + (a - bP_0)e^{-at}]^2} = \frac{a^2 P_0(a - bP_0)e^{-at}}{[bP_0 + (a - bP_0)e^{-at}]^2}$$

$$= \left(\frac{aP_0}{bP_0 + (a - bP_0)e^{-at}}\right)\left(\frac{a^2 e^{-at} - abP_0 e^{-at}}{bP_0 + (a - bP_0)e^{-at}}\right)$$

$$= \left(\frac{aP_0}{bP_0 + (a - bP_0)e^{-at}}\right)\left[\frac{(abP_0 - abP_0) + a^2 e^{-at} - abP_0 e^{-at}}{bP_0 + (a - bP_0)e^{-at}}\right]$$

$$= \left(\frac{aP_0}{bP_0 + (a - bP_0)e^{-at}}\right)\left(\frac{abP_0 + a^2 e^{-at} - abP_0 e^{-at} - abP_0}{bP_0 + (a - bP_0)e^{-at}}\right)$$

$$= \left(\frac{aP_0}{bP_0 + (a - bP_0)e^{-at}}\right)\left[\frac{a(bP_0 + ae^{-at} - bP_0 e^{-at})}{bP_0 + (a - bP_0)e^{-at}} - \frac{b(aP_0)}{bP_0 + (a - bP_0)e^{-at}}\right]$$

$$= \left(\frac{aP_0}{bP_0 + (a - bP_0)e^{-at}}\right)\left\{a\left[\frac{bP_0 + (a - bP_0)e^{-at}}{bP_0 + (a - bP_0)e^{-at}}\right] - b\left[\frac{aP_0}{bP_0 + (a - bP_0)e^{-at}}\right]\right\}$$

$$= P(a - bP)$$

(b) $\lim\limits_{t \to -\infty} P(t) = 0$

$\lim\limits_{t \to \infty} P(t) = a/b = 2$

(c)

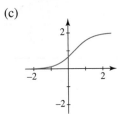

(d) For $a = 2$, $b = 1$, and $P_0 = 1$, $P(t) = \dfrac{2}{1 + e^{-2t}}$, $P'(t) = \dfrac{4e^{-2t}}{(1 + e^{-2t})^2}$, and

$$P''(t) = \frac{-8e^{-2t}(1 + e^{-2t})^2 - 2(-2e^{-2t})(1 + e^{-2t})(4e^{-2t})}{(1 + e^{-2t})^4}$$

$$= \frac{(1 + e^{-2t})(-8e^{-2t} - 8e^{-4t} + 16e^{-4t})}{(1 + e^{-2t})^4} = \frac{8e^{-4t} - 8e^{-2t}}{(1 + e^{-2t})^3}.$$

Thus, $P''(t) = 0$ only when $e^{-4t} = e^{-2t}$ or $t = 0$.

51. The slope of the tangent line at $x = x_0$ is $-e^{-x_0}$. The equation of the line is $y - e^{-x_0} = -e^{-x_0}(x - x_0)$. To find the x-intercept we set $y = 0$ and solve for x: $0 - e^{-x_0} = -e^{-x_0}(x - x_0)$, so $1 = x - x_0$, and $x = x_0 + 1$. Thus, the x-intercept is $x_0 + 1$.

53. Rewriting $2x + y = 1$ as $y = 1 - 2x$, we see that the slope of this line is -2. Since the derivative of $y = e^x$ is e^x, which is never less than zero, then there is no point on the graph of $y = e^x$ at which the tangent line is parallel to $2x + y = 1$.

55. The derivatives for $n = 1$ to 6 are:

$$\frac{d}{dx}\sqrt{e^x} = \frac{1}{2}(e^x)^{-1/2}e^x = \frac{1}{2}\sqrt{e^x}; \qquad \frac{d^2}{dx^2}\sqrt{e^x} = \frac{1}{4}(e^x)^{-1/2}e^x = \frac{1}{4}\sqrt{e^x};$$

$$\frac{d^3}{dx^3}\sqrt{e^x} = \frac{1}{8}(e^x)^{-1/2}e^x = \frac{1}{8}\sqrt{e^x}; \qquad \frac{d^4}{dx^4}\sqrt{e^x} = \frac{1}{16}(e^x)^{-1/2}e^x = \frac{1}{16}\sqrt{e^x};$$

$$\frac{d^5}{dx^5}\sqrt{e^x} = \frac{1}{32}(e^x)^{-1/2}e^x = \frac{1}{32}\sqrt{e^x}; \qquad \frac{d^6}{dx^6}\sqrt{e^x} = \frac{1}{64}(e^x)^{-1/2}e^x = \frac{1}{64}\sqrt{e^x}.$$

We can verify by induction that $\dfrac{d^n}{dx^n}\sqrt{e^x} = \dfrac{1}{2^n}\sqrt{e^x}$:

$$\frac{d^n}{dx^n}\sqrt{e^x} = \frac{d}{dx}\left(\frac{d^{n-1}}{dx^{n-1}}\sqrt{e^x}\right) = \frac{d}{dx}\frac{1}{2^{n-1}}\sqrt{e^x} = \frac{1}{2} \cdot \frac{1}{2^{n-1}}(e^x)^{-1/2} \cdot e^x = \frac{1}{2^n}\sqrt{e^x}.$$

57.

$h \to 0$	0.1	0.01	0.001	0.0001	0.00001	0.000001
$\dfrac{(1.5)^h - 1}{h}$	0.413797	0.406288	0.405547	0.405473	0.405466	0.405465

$$m(b) = \lim_{h \to 0} \frac{b^h - 1}{h} \approx 0.405465$$

59.

$h \to 0$	0.1	0.01	0.001	0.0001	0.00001	0.000001
$\dfrac{3^h - 1}{h}$	1.161232	1.104669	1.099216	1.098673	1.098618	1.098613

$$m(b) = \lim_{h \to 0} \frac{b^h - 1}{h} \approx 1.098613$$

61. $f'(x) = 2x^{-3}e^{-1/x^2}$ exists for all $x \neq 0$, and so is differentiable on $(-\infty, 0)$ and $(0, \infty)$. For $x = 0$, we use the definition of the derivative:

$$f'(0) = \lim_{h \to 0} \frac{f(0+h) - f(0)}{h} = \lim_{h \to 0} \frac{e^{-1/(0+h)^2} - 0}{h}$$
$$= \lim_{h \to 0} \frac{e^{-1/h^2}}{h} = 0.$$

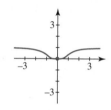

The limit exists, and thus f is differentiable for all x.

3.9 Logarithmic Functions

1. $y' = \dfrac{10}{x}$

3. $y = \dfrac{1}{2} \ln x; \quad y' = \dfrac{1}{2x}$

5. $y' = \dfrac{4x^3 + 6x}{x^4 + 3x^2 + 1}$

7. $y = 3x^2 \ln x; \quad y' = 3x^2 \left(\dfrac{1}{x}\right) + 6x \ln x = 3x + 6x \ln x$

9. $y' = \dfrac{x\left(\dfrac{1}{x}\right) - \ln x}{x^2} = \dfrac{1 - \ln x}{x^2}$

11. $y = \ln x - \ln(x+1); \quad y' = \dfrac{1}{x} - \dfrac{1}{x+1} = \dfrac{1}{x(x+1)}$

13. $y' = -\dfrac{-\sin x}{\cos x} = \tan x$

15. $y' = \dfrac{-1/x}{(\ln x)^2} = -\dfrac{1}{x(\ln x)^2}$

17. $f'(x) = \dfrac{x\left(\dfrac{1}{x}\right) + \ln x}{x \ln x} = \dfrac{1 + \ln x}{x \ln x}$

19. $g(x) = \left(\dfrac{1}{2} \ln x\right)^{1/2} = \dfrac{1}{\sqrt{2}} (\ln x)^{1/2}; \quad g'(x) = \dfrac{1}{\sqrt{2}} \cdot \dfrac{1}{2} (\ln x)^{-1/2} \cdot \dfrac{1}{x} = \dfrac{1}{2x\sqrt{2 \ln x}}$

21. $H(t) = 2\ln t + \ln(3t^2 + 6);$ $\quad H'(t) = \dfrac{2}{t} + \dfrac{6t}{3t^2 + 6} = \dfrac{2}{t} + \dfrac{2t}{t^2 + 2}$

23. $f(x) = \ln(x+1) + \ln(x+2) - \ln(x+3);$ $\quad f'(x) = \dfrac{1}{x+1} + \dfrac{1}{x+2} - \dfrac{1}{x+3}$

25. $y' = 1/x$. The point on the graph is $(1,0)$ and the slope of the line is $y'(1) = 1$. Thus, an equation of the tangent line is $y - 0 = 1(x-1)$ or $y = x - 1$.

27. $y' = \dfrac{3e^{3x} + 1}{e^{3x} + x}$. The slope of the tangent to the graph of y at $x = 0$ is $\dfrac{3e^{3 \cdot 0} + 1}{e^{3 \cdot 0} + 0} = 4$.

29. $f'(x) = 2/x$. The slope of the tangent to the graph of f is 4 when $x = 1/2$. Since $f''(x) = -2/x^2$, the slope of the tangent to the graph of f' at $x = 1/2$ is $f''(1/2) = -8$.

31. The tangent line to f is horizontal when $f'(x) = \dfrac{\left(\frac{1}{x}\right)x - \ln x}{x^2} = \dfrac{1 - \ln x}{x^2}$ is 0. Solving for x, we get $\dfrac{1 - \ln x}{x^2} = 0$; $\ln x = 1$; $x = e$. Thus, the tangent is horizontal at $\left(e, \dfrac{\ln e}{e}\right) = (e, 1/e)$.

33. $\dfrac{d}{dx}\ln\left(x + \sqrt{x^2 - 1}\right) = \dfrac{1 + 2x(x^2-1)^{-1/2}\left(\frac{1}{2}\right)}{x + \sqrt{x^2 - 1}} = \dfrac{1 + \dfrac{x}{\sqrt{x^2 - 1}}}{x + \sqrt{x^2 - 1}} = \dfrac{\dfrac{\sqrt{x^2-1} + x}{\sqrt{x^2-1}}}{x + \sqrt{x^2 - 1}}$

$\qquad = \dfrac{\sqrt{x^2 - 1} + x}{\sqrt{x^2 - 1}} \cdot \dfrac{1}{x + \sqrt{x^2 - 1}} = \dfrac{1}{\sqrt{x^2 - 1}}$

35. $\dfrac{d}{dx}\ln(\sec x + \tan x) = \dfrac{\sec x \tan x + \sec^2 x}{\sec x + \tan x} = \dfrac{(\sec x)(\tan x + \sec x)}{\sec x + \tan x} = \sec x$

37. $\dfrac{dy}{dx} = \dfrac{1}{x};$ $\quad \dfrac{d^2 y}{dx^2} = -\dfrac{1}{x^2};$ $\quad \dfrac{d^3 y}{dx^3} = \dfrac{2}{x^3}$

39. $\dfrac{dy}{dx} = \dfrac{2\ln|x|}{x};$ $\quad \dfrac{d^2 y}{dx^2} = \dfrac{\left(\frac{2}{x}\right)x - 2\ln|x|}{x^2} = \dfrac{2(1 - \ln|x|)}{x^2}$

41. $y' = -\dfrac{1}{2}C_1 x^{-3/2} + C_2\left(x^{-3/2} - \dfrac{1}{2}x^{-3/2}\ln x\right)$

$\quad y'' = \dfrac{3}{4}C_1 x^{-5/2} + C_2\left(-\dfrac{3}{2}x^{-5/2} - \dfrac{1}{2}x^{-5/2} + \dfrac{3}{4}x^{-5/2}\ln x\right)$

$\qquad = \dfrac{3}{4}C_1 x^{-5/2} + C_2\left(-2x^{-5/2} + \dfrac{3}{4}x^{-5/2}\ln x\right)$

$\quad 4x^2 y'' + 8xy' + y = 4x^2\left[\dfrac{3}{4}C_1 x^{-5/2} + C_2\left(-2x^{-5/2} + \dfrac{3}{4}x^{-5/2}\ln x\right)\right]$

$\qquad + 8x\left[-\dfrac{1}{2}C_1 x^{-3/2} + C_2\left(x^{-3/2} - \dfrac{1}{2}x^{-3/2}\ln x\right)\right] + (C_1 x^{-1/2} + C_2 x^{-1/2}\ln x)$

$\qquad = 3C_1 x^{-1/2} - 8C_2 x^{-1/2} + 3C_2 x^{-1/2}\ln x - 4C_1 x^{-1/2} + 8C_2 x^{-1/2} - 4C_2 x^{-1/2}\ln x$

$\qquad\quad + C_1 x^{-1/2} + C_2 x^{-1/2}\ln x$

$\qquad = (3C_1 x^{-1/2} - 4C_1 x^{-1/2} + C_1 x^{-1/2} - 8C_2 x^{-1/2} + 8C_2 x^{-1/2})$

$\qquad\quad + (3C_2 x^{-1/2}\ln x - 4C_2 x^{-1/2}\ln x + C_2 x^{-1/2}\ln x) = 0$

43. $y^2 = \ln x + \ln y;$ $\quad 2yy' = \dfrac{1}{x} + \dfrac{1}{y}\cdot y';$ $\quad 2xy^2 y' = y + xy';$ $\quad (2xy^2 - x)y' = y;$ $\quad y' = \dfrac{y}{2xy^2 - x}$

45. $x + y^2 = \ln x - \ln y;$ $\quad 1 + 2yy' = \dfrac{1}{x} - \dfrac{1}{y}\cdot y';$ $\quad xy + 2xy^2 y' = y - xy';$ $\quad y' = \dfrac{y - xy}{2xy^2 + x}$

47. $xy' + y = \dfrac{2x + 2yy'}{x^2 + y^2}$; $(x^3 + xy^2)y' + x^2y + y^3 = 2x + 2yy'$; $y' = \dfrac{2x - x^2y - y^3}{x^3 + xy^2 - 2y}$

49. $\ln y = (\sin x)\ln x$; $\dfrac{1}{y}\left(\dfrac{dy}{dx}\right) = (\sin x)\left(\dfrac{1}{x}\right) + (\cos x)(\ln x)$; $\dfrac{dy}{dx} = x^{\sin x}\left[\dfrac{\sin x}{x} + (\cos x)\ln x\right]$

51. $\ln y = \ln x + x\ln(x-1)$; $\dfrac{1}{y}\left(\dfrac{dy}{dx}\right) = \dfrac{1}{x} + x\left(\dfrac{1}{x-1}\right) + \ln(x-1)$

$$\dfrac{dy}{dx} = x(x-1)^x\left[\dfrac{1}{x} + \dfrac{x}{x-1} + \ln(x-1)\right]$$

53. $\ln|y| = \dfrac{1}{2}\ln|2x+1| + \dfrac{1}{2}\ln|3x+2| - \ln|4x+3|$

$\dfrac{1}{y}\left(\dfrac{dy}{dx}\right) = \dfrac{2}{2(2x+1)} + \dfrac{3}{2(3x+2)} - \dfrac{4}{4x+3}$

$\dfrac{dy}{dx} = \dfrac{\sqrt{(2x+1)(3x+2)}}{4x+3}\left(\dfrac{1}{2x+1} + \dfrac{3}{6x+4} - \dfrac{4}{4x+3}\right)$

55. $\ln|y| = 5\ln|x^3-1| + 4\ln|x^4+3x^3| - 9\ln|7x+5|$

$\dfrac{1}{y}\left(\dfrac{dy}{dx}\right) = \dfrac{5(3x^2)}{x^3-1} + \dfrac{4(4x^3+9x^2)}{x^4+3x^3} - \dfrac{9(7)}{7x+5}$

$\dfrac{dy}{dx} = \dfrac{(x^3-1)^5(x^4+3x^3)^4}{(7x+5)^9}\left(\dfrac{15x^2}{x^3-1} + \dfrac{16x+36}{x^2+3x} - \dfrac{63}{7x+5}\right)$

57. $\ln y = (x+2)\ln x$; $\dfrac{1}{y}\left(\dfrac{dy}{dx}\right) = (x+2)\dfrac{1}{x} + \ln x$; $\dfrac{dy}{dx} = x^{x+2}\left(\dfrac{x+2}{x} + \ln x\right)$

The point on the graph is $(1,1)$ and the slope of the tangent line is $y'(1) = 3$. An equation of the tangent line is $y - 1 = 3(x - 1)$ or $y = 3x - 2$.

59. $\ln y = x\ln x$; $\dfrac{1}{y}\left(\dfrac{dy}{dx}\right) = x\left(\dfrac{1}{x}\right) + \ln x$; $\dfrac{dy}{dx} = x^x(1 + \ln x)$

The tangent line is horizontal when $dy/dx = x^x(1 + \ln x) = 0$. x^x is never 0, so we need only solve $1 + \ln x = 0$, which yields $\ln x = -1$ and $x = 1/e$. Thus, the tangent line is horizontal at $(1/e, 1/e^{1/e})$.

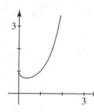

61. From Problem 59, $\dfrac{d}{dx}x^x = x^x(1 + \ln x)$:

(a) $\dfrac{dy}{dx} = (\sec^2 x^x)\dfrac{d}{dx}x^x = x^x(1 + \ln x)\sec^2 x^x$

(b) $\dfrac{dy}{dx} = x^x e^{x^x}[x^x(1 + \ln x)] + e^{x^x}[x^x(1 + \ln x)] = (1 + x^x)e^{x^x}x^x(1 + \ln x)$

(c) $\ln y = x^x \ln x$; $\dfrac{1}{y}\left(\dfrac{dy}{dx}\right) = x^x\left(\dfrac{1}{x}\right) + x^x(1 + \ln x)\ln x$; $\dfrac{dy}{dx} = x^{x^x}x^x\left(\dfrac{1}{x} + \ln x + \ln^2 x\right)$

63. $g(x) = |\ln x|$ is also not differentiable at $x = 1$. When $x > 1$, $\ln x > 0$, so $g(x) = \ln x$, $g'(x) = 1/x$, and $g'_+(1) = 1$. When $x < 1$, $\ln x < 0$, so $g(x) = -\ln x$, $g'(x) = -1/x$, and $g'_-(1) = -1$. Since $g'_+(1) \neq g'_-(1)$, g is not differentiable at 1.

65. (a)

(b) The function is not defined at intervals $[k\pi, (k+1)\pi]$, where k is an odd positive integer, because $\sin x < 0$ at those intervals.

67. $f(x)$ is smallest at the sole point where its tangent is horizontal; $f'(x) = 3x^2 - 12/x$, so solving $3x^2 - 12/x = 0$ we get $x^3 = 4$ or $x = \sqrt[3]{4}$.

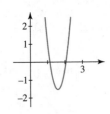

3.10   Hyperbolic Functions

1. $\cosh x = \sqrt{1 + \sinh^2 x} = \sqrt{1 + (-1/2)^2} = \dfrac{\sqrt{5}}{2};$ $\tanh x = \dfrac{\sinh x}{\cosh x} = \dfrac{-1/2}{\sqrt{5}/2} = -\dfrac{\sqrt{5}}{5}$

  $\coth x = \dfrac{\cosh x}{\sinh x} = -\sqrt{5};$ $\operatorname{sech} x = \dfrac{1}{\cosh x} = \dfrac{1}{\sqrt{5}/2} = \dfrac{2\sqrt{5}}{5}$

  $\operatorname{csch} x = \dfrac{1}{\sinh x} = \dfrac{1}{-1/2} = -2$

3. $y' = 10\sinh 10x$

5. $y' = \dfrac{1}{2\sqrt{x}}\operatorname{sech}^2\sqrt{x}$

7. $y' = -6(3x-1)\operatorname{sech}(3x-1)^2\tanh(3x-1)^2$

9. $y' = 3\sinh 3x[-\operatorname{csch}^2(\cosh 3x)] = -3\sinh 3x\operatorname{csch}^2(\cosh 3x)$

11. $y' = (\sinh 2x)(3\sinh 3x) + (\cosh 3x)(2\cosh 2x) = 3\sinh 2x\sinh 3x + 2\cosh 2x\cosh 3x$

13. $y' = x(2x\sinh x^2) + \cosh x^2 = 2x^2\sinh x^2 + \cosh x^2$

15. $y' = 3\sinh^2 x\cosh x$

17. $f'(x) = \dfrac{2}{3}(x-\cosh x)^{-1/3}(1-\sinh x)$

19. $f'(x) = \dfrac{4\sinh 4x}{\cosh 4x} = 4\tanh 4x$

21. $f'(x) = \dfrac{(1+\cosh x)e^x - e^x\sinh x}{(1+\cosh x)^2} = \dfrac{e^x(1+\cosh x - \sinh x)}{(1+\cosh x)^2} = \dfrac{1+e^x}{(1+\cosh x)^2}$

23. $F'(t) = e^{\sinh t}\cosh t$

25. $g'(t) = \dfrac{(1+\sinh 2t)\cos t - 2\sin t\cosh 2t}{(1+\sinh 2t)^2}$

27. $y' = 3\cosh 3x$. The point on the graph is $(0,0)$ and the slope of the tangent line is $y'(0) = 3$. An equation of the tangent line is $y = 3x$.

29. The tangent is horizontal when $f'(x) = (x^2 - 2)\sinh x + 2x\cosh x - 2x\cosh x - 2\sinh x = (x^2 - 4)\sinh x = 0$, or when $x = 0$ or ± 2, yielding the points $(0,-2)$, $(2,-e^2 + 3e^{-2})$, and $(-2,-e^2 + 3e^{-2})$.

31. $\dfrac{dy}{dx} = \operatorname{sech}^2 x;$ $\dfrac{d^2 y}{dx^2} = 2(\operatorname{sech} x)(-\operatorname{sech} x\tanh x) = -2\operatorname{sech}^2 x\tanh x$

33. $y' = kC_1\sinh kx + kC_2\cosh kx;$ $y'' = k^2 C_1\cosh kx + k^2 C_2\sinh kx$

  $y'' - k^2 y = k^2 C_1\cosh kx + k^2 C_2\sinh kx - k^2(C_1\cosh kx + C_2\sinh kx) = 0$

35. $y' = \dfrac{3}{\sqrt{9x^2 + 1}}$

37. $y' = \dfrac{-2x}{1-(1-x^2)^2}$

39. $y' = \dfrac{-\csc x \cot x}{1 - \csc^2 x} = \dfrac{-\csc x \cot x}{-\cot^2 x} = \sec x$

41. $y' = x \dfrac{3x^2}{\sqrt{x^6 + 1}} + \sinh^{-1} x^3 = \dfrac{3x^3}{\sqrt{x^6 + 1}} + \sinh^{-1} x^3$

43. $y' = \dfrac{x\left(\dfrac{-1}{x\sqrt{1-x^2}}\right) - \operatorname{sech}^{-1} x}{x^2} = \dfrac{1 + \sqrt{1-x^2}\,\operatorname{sech}^{-1} x}{x^2\sqrt{1-x^2}}$

45. $y' = \dfrac{1}{\operatorname{sech}^{-1} x} \cdot \dfrac{-1}{x\sqrt{1-x^2}} = \dfrac{-1}{x\sqrt{1-x^2}\,\operatorname{sech}^{-1} x}$

47. $y' = \dfrac{1}{2}(\cosh^{-1} 6x)^{-1/2} \dfrac{6}{\sqrt{36x^2 - 1}} = \dfrac{3}{\sqrt{\cosh^{-1} 6x}\sqrt{36x^2 - 1}}$

49. (a) $\dfrac{dv}{dt} = \sqrt{\dfrac{mg}{k}}\sqrt{\dfrac{kg}{m}}\operatorname{sech}^2\left(\sqrt{\dfrac{kg}{m}}t\right) = g\operatorname{sech}^2\left(\sqrt{\dfrac{kg}{m}}t\right)$

$m\dfrac{dv}{dt} - mg + kv^2 = mg\operatorname{sech}^2\left(\sqrt{\dfrac{kg}{m}}t\right) - mg + k\left[\dfrac{mg}{k}\tanh^2\left(\sqrt{\dfrac{kg}{m}}t\right)\right]$

$= mg\left[\operatorname{sech}^2\left(\sqrt{\dfrac{kg}{m}}t\right) - 1 + \tanh^2\left(\sqrt{\dfrac{kg}{m}}t\right)\right]$

$= mg\left[\operatorname{sech}^2\left(\sqrt{\dfrac{kg}{m}}t\right) - \operatorname{sech}^2\left(\sqrt{\dfrac{kg}{m}}t\right)\right] = 0$

(b) From Figure 3.10.2(a) in the text, we see that $\lim\limits_{t\to\infty}\tanh t = 1$, so $\lim\limits_{t\to\infty} v(t) = \sqrt{mg/k}$.

(c) Using $v_{\text{ter}} = \sqrt{mg/k}$, $m = 80$, and $k = 0.25$, we find $v_{\text{ter}} = \sqrt{80(9.8)/0.25} = 56$ m/s.

51. $\cosh(\ln 4) = \dfrac{e^{\ln 4} + e^{-\ln 4}}{2} = \dfrac{4 + 1/4}{2} = 2.125$

53. $\sinh(\ln x) = \dfrac{e^{\ln x} - e^{-\ln x}}{2} = \dfrac{x - 1/x}{2} = \dfrac{x^2 - 1}{2x}, \quad x > 0$

55. First, we evaluate $\cosh x + \sinh x = \dfrac{e^x + e^{-x}}{2} + \dfrac{e^x - e^{-x}}{2} = e^x$. Applying this result twice, we have $(\cosh x + \sinh x)^n = (e^x)^n = e^{nx} = \cosh nx + \sinh nx$.

Chapter 3 in Review

A. True/False

1. False; consider $y = |x|$ at $x = 0$.

3. False. $y = x^{1/3}$ has a vertical tangent at $x = 0$, but is not differentiable there.

5. True

7. True

9. True

11. True. $f'(x) = \cos x$, whose range is $[-1, 1]$.

13. False. $\dfrac{d}{dx}\cos^{-1} x = \dfrac{-1}{\sqrt{1-x^2}}$.

15. True

17. False. Trivially, $f(x) = 0$ is the same as its derivative, but so is $f(x) = ce^x$ for any non-zero constant c.

19. True. $\dfrac{d}{dx}\cosh^2 x = 2\cosh x \sinh x = \dfrac{d}{dx}\sinh^2 x.$

B. Fill In the Blanks

1. 0

3. $-1/4$

5. $y = -\dfrac{5}{4}x - \dfrac{3}{2}$

7. -3

9. 5

11. $16(\cos^2 4x)F''(\sin 4x) - 16(\sin 4x)F'(\sin 4x)$

13. $a = 6, b = -9$

15. $(1,5)$

17. $\dfrac{1}{x\ln 10}$

19. catenary

C. Exercises

1. $f(x) = \dfrac{4}{5}x^{0.1}; \quad f'(x) = \dfrac{0.4}{5}x^{-0.9} = 0.08x^{-0.9}$

3. $F'(t) = 10[t + (t^2+1)^{1/2}]^9[1 + (1/2)(t^2+1)^{-1/2}(2t)] = 10[t + (t^2+1)^{1/2}]^9[1 + t(t^2+1)^{-1/2}]$

5. $y = (x^4+16)^{1/4}(x^3+8)^{1/3}$

$y' = (x^4+16)^{1/4}\left[\dfrac{1}{3}(x^3+8)^{-2/3}(3x^2)\right] + (x^3+8)^{1/3}\left[\dfrac{1}{4}(x^4+16)^{-3/4}(4x^3)\right]$

$= x^2(x^4+16)^{1/4}(x^3+8)^{-2/3} + x^3(x^4+16)^{-3/4}(x^3+8)^{1/3}$

7. $y' = \dfrac{(4x+1)(-4\sin 4x) - (\cos 4x)(4)}{(4x+1)^2} = -\dfrac{16x\sin 4x + 4\sin 4x + 4\cos 4x}{(4x+1)^2}$

9. $f'(x) = x^3(2\sin 5x)(\cos 5x)(5) + (\sin^2 5x)(3x^2) = 10x^3\sin 5x\cos 5x + 3x^2\sin^2 5x$

11. $y' = \dfrac{-3/x^2}{\sqrt{1 - (3/x)^2}} = -\dfrac{3}{\dfrac{x^2}{|x|}\sqrt{x^2-9}} = -\dfrac{3}{|x|\sqrt{x^2-9}}$

13. $y' = -(\cot^{-1}x)^{-2}\left(-\dfrac{1}{1+x^2}\right) = \dfrac{1}{(1+x^2)(\cot^{-1}x)^2}$

15. $y' = -\dfrac{2}{\sqrt{1-x^2}} + 2x\dfrac{-2x}{2\sqrt{1-x^2}} + 2\sqrt{1-x^2} = \dfrac{-2(1+x^2)}{\sqrt{1-x^2}} + 2\sqrt{1-x^2} = \dfrac{-4x^2}{\sqrt{1-x^2}}$

17. $y = \dfrac{x+1}{e^x}; \quad y' = \dfrac{e^x - (x+1)e^x}{e^{2x}} = \dfrac{-xe^x}{e^{2x}} = -xe^{-x}$

19. $y' = 7x^6 + 7^x\ln 7 + 7e^{7x}$

21. $y = \ln x + \dfrac{1}{2}\ln(4x-1); \quad y' = \dfrac{1}{x} + \dfrac{4}{2(4x-1)} = \dfrac{1}{x} + \dfrac{2}{4x-1}$

23. $y' = \dfrac{1/\sqrt{1-x^2}}{\sqrt{(\sin^{-1}x)^2+1}} = \dfrac{1}{\sqrt{1-x^2}\sqrt{(\sin^{-1}x)^2+1}}$

25. $y' = xe^{x\cosh^{-1}x}\left(x\dfrac{1}{\sqrt{x^2-1}}+\cosh^{-1}x\right)+e^{x\cosh^{-1}x}$

$\quad = \left(\dfrac{x^2}{\sqrt{x^2-1}}+x\cosh^{-1}x+1\right)e^{x\cosh^{-1}x}$

27. $y' = 3x^2e^{x^3}\sinh e^{x^3}$

29. $\dfrac{dy}{dx} = \dfrac{5}{2}(3x+1)^{3/2}(3) = \dfrac{15}{2}(3x+1)^{3/2}; \quad \dfrac{d^2y}{dx^2} = \dfrac{45}{4}(3x+1)^{1/2}(3) = \dfrac{135}{4}(3x+1)^{1/2};$

$\quad \dfrac{d^3y}{dx^3} = \dfrac{135}{8}(3x+1)^{-1/2}(3) = \dfrac{405}{8}(3x+1)^{-1/2}$

31. $\dfrac{ds}{dt} = 2t-2t^{-3}; \quad \dfrac{d^2s}{dt^2} = 2+6t^{-4}; \quad \dfrac{d^3s}{dt^3} = -24t^{-5}; \quad \dfrac{d^4s}{dt^4} = 120t^{-6}$

33. $y' = e^{\sin 2x}(\cos 2x)(2) = 2e^{\sin 2x}\cos 2x$

$\quad y'' = 2e^{\sin 2x}(-\sin 2x)(2)+2e^{\sin 2x}(\cos 2x)(2)(\cos 2x) = 4e^{\sin 2x}(\cos^2 2x - \sin 2x)$

35. $y = \ln|(x+5)^4(2-x)^3| - \ln|(x+8)^{10}\sqrt[3]{6x+4}|$

$\quad = \ln(x+5)^4 + \ln|(2-x)^3| - \ln(x+8)^{10} - \ln|\sqrt[3]{6x+4}|$

$\quad = 4\ln|x+5| + 3\ln|2-x| - 10\ln|x+8| - \dfrac{1}{3}\ln|6x+4|$

$\quad y' = \dfrac{4}{x+5} - \dfrac{3}{2-x} - \dfrac{10}{x+8} - \dfrac{1}{3x+2}$

37. $y' = 3x^2+1$. The slope of the tangent at $x = 1$ is 4, so the slope of the tangent to the inverse of y is $1/4$.

39. $x(2yy')+y^2 = e^x - e^y y'; \quad (2xy+e^y)y' = e^x - y^2; \quad y' = \dfrac{e^x - y^2}{2xy+e^y}$

41. A line that is perpendicular to $y = -3x$ will have slope $\dfrac{1}{3}$. Thus, we need x such that $f'(x) = 3x^2 = \dfrac{1}{3}$, so $x = \pm\dfrac{1}{3}$. Equations of tangent lines that are perpendicular to $y = -3x$ are $y\pm\dfrac{1}{27} = \dfrac{1}{3}\left(x\pm\dfrac{1}{3}\right)$ or $y = \dfrac{1}{3}x\pm\dfrac{2}{27}$.

43. $y' = 2x$. If (a, a^2) is the point of tangency, then the slope of the tangent line through $(0, -9)$ is $\dfrac{a^2+9}{a-0}$. The slope is also $y'(a) = 2a$. Thus, $\dfrac{a^2+9}{a} = 2a$, $9 = a^2$, and $a = \pm 3$. The tangent line through $(-3, 9)$ is $y-9 = -6(x+3)$ or $y = -6x-9$. The tangent line through $(3, 9)$ is $y-9 = 6(x-3)$ or $y = 6x-9$.

45. $f'(x) = \dfrac{1}{2\sqrt{x}}$. The slope of the line through $(1, 1)$ and $(9, 3)$ is $1/4$. Solving $\dfrac{1}{2\sqrt{x}} = \dfrac{1}{4}$, we obtain $x = 4$. The point on the graph is $(4, 2)$.

47. $f'(x) = -2\sin x - 2\sin 2x$. The slope of a horizontal line is 0, so we solve $-2\sin x - 2\sin 2x = 0$. Using $\sin 2x = 2\sin x\cos x$, we have $\sin x + 2\sin x\cos x = 0$, $(\sin x)(1+2\cos x) = 0$, and x must therefore satisfy $\sin x = 0$ or $\cos x = -1/2$. For $0 \le x \le 2\pi$, this gives $x = 0, \pi, 2\pi, 2\pi/3$, and $4\pi/3$.

49. Evaluating $q(t)$ when $t = 0$, we get

$$q(0) = E_0C + (q_0 - E_0C)\left(\dfrac{k_1}{k_1 + k_2 \cdot 0}\right)^{1/Ck_2}$$

$$= E_0C + (q_0 - E_0C)(1) = q_0.$$

So $q(t)$ satisfies the initial condition $q(0) = q_0$. Now, rewriting

$$q(t) = E_0 C + (q_0 - E_0 C) k_1^{1/Ck_2} (k_1 + k_2 t)^{-1/Ck_2},$$

we get

$$\frac{dq}{dt} = -\frac{(q_0 - E_0 C) k_1^{1/Ck_2}}{Ck_2} (k_1 + k_2 t)^{-1/Ck_2 - 1} (k_2) = -\frac{q_0 - E_0 C}{C(k_1 + k_2 t)} \left(\frac{k_1}{k_1 + k_2 t} \right)^{1/Ck_2}.$$

Evaluating $(k_1 + k_2 t) \dfrac{dq}{dt}$ yields:

$$(k_1 + k_2 t) \left[-\frac{q_0 - E_0 C}{C(k_1 + k_2 t)} \left(\frac{k_1}{k_1 + k_2 t} \right)^{1/Ck_2} \right] = -\left(\frac{q_0 - E_0 C}{C} \right) \left(\frac{k_1}{k_1 + k_2 t} \right)^{1/Ck_2}$$

Then, evaluating $\dfrac{1}{C} q$ results in:

$$\frac{1}{C} \left[E_0 C + (q_0 - E_0 C) \left(\frac{k_1}{k_1 + k_2 t} \right)^{1/Ck_2} \right] = E_0 + \left(\frac{q_0 - E_0 C}{C} \right) \left(\frac{k_1}{k_1 + k_2 t} \right)^{1/Ck_2}$$

Substituting into the left side of the differential equation, we find

$$-\left(\frac{q_0 - E_0 C}{C} \right) \left(\frac{k_1}{k_1 + k_2 t} \right)^{1/Ck_2} + E_0 + \left(\frac{q_0 - E_0 C}{C} \right) \left(\frac{k_1}{k_1 + k_2 t} \right)^{1/Ck_2} = E_0.$$

51. $y' = -C_1 e^{-x} + C_2 e^x + C_3 [x(-e^{-x}) + e^{-x}] + C_4 (xe^x + e^x)$

$= (C_3 - C_1) e^{-x} + (C_2 + C_4) e^x - C_3 xe^{-x} + C_4 xe^x$

$y'' = -(C_3 - C_1) e^{-x} + (C_2 + C_4) e^x - C_3 [x(-e^{-x}) + e^{-x}] + C_4 (xe^x + e^x)$

$= (C_1 - 2C_3) e^{-x} + (C_2 + 2C_4) e^x + C_3 xe^{-x} + C_4 xe^x$

$y''' = -(C_1 - 2C_3) e^{-x} + (C_2 + 2C_4) e^x + C_3 [x(-e^{-x}) + e^{-x}] + C_4 (xe^x + e^x)$

$= (C_3 - C_1) e^{-x} + (C_2 + 3C_4) e^x - C_3 xe^{-x} + C_4 xe^x$

$y^{(4)} = -(C_3 - C_1) e^{-x} + (C_2 + 3C_4) e^x - C_3 [x(-e^{-x}) + e^{-x}] + C_4 (xe^x + e^x)$

$= (C_1 - 2C_3) e^{-x} + (C_2 + 4C_4) e^x + C_3 xe^{-x} + C_4 xe^x$

$y^{(4)} - 2y'' + y = [(C_1 - 2C_3) e^{-x} + (C_2 + 4C_4) e^x + C_3 xe^{-x} + C_4 xe^x] - 2[(C_1 - 2C_3) e^{-x}$

$+ (C_2 + 2C_4) e^x + C_3 xe^{-x} + C_4 xe^x] + C_1 e^{-x} + C_2 e^x + C_3 xe^{-x} + C_4 xe^x$

$= (C_1 - 2C_3 - 2C_1 + 2C_3 + C_1) e^{-x} + (C_2 + 4C_4 - 2C_2 - 4C_4 + C_2) e^x$

$+ (C_3 - 2C_3 + C_3) xe^{-x} + (C_4 - 2C_4 + C_4) xe^x = 0$

53. (a) Setting $x = 2$, we have $y^3 - y - 2^2 - 4 = y(y+1)(y-1) = 0$. Thus, we find that $(2,0)$, $(2,1)$, and $(2,-1)$ lie on the graph.

(b) Using implicit differentiation, we obtain

$$3y^2 y' - y' + 2x = 0; \qquad y'(3y^2 - 1) = -2x; \qquad y' = \frac{2x}{1 - 3y^2}.$$

Thus $y'|_{x=2,\, y=0} = 4$, $y'|_{x=2,\, y=1} = -2$, and $y'|_{x=2,\, y=-1} = -2$.

55. Setting $x = \dfrac{1}{8}$, we have $\left(\dfrac{1}{8} \right)^{2/3} + y^{2/3} = 1$; $y^{2/3} = \dfrac{3}{4}$. Thus, the points corresponding to $x = 1/8$ are $(1/8, \pm 3\sqrt{3}/8)$.

Using implicit differentiation, we obtain $\dfrac{2}{3} x^{-1/3} + \dfrac{2}{3} y^{-1/3} y' = 0$; $y' = -x^{-1/3} y^{1/3}$. Thus $y'|_{x=1/8,\, y=3\sqrt{3}/8} = -\sqrt{3}$ and

$y'|_{x=1/8,\, y=-3\sqrt{3}/8} = -\sqrt{3}$. Equations of the tangent lines to the graph at these points are $y \pm \dfrac{3\sqrt{3}}{8} = \pm \sqrt{3}(x - 1/8)$ or

$y = \pm \sqrt{3} x \mp \dfrac{\sqrt{3}}{2}$.

57. For $x \neq 0$, $f'(x) = \begin{cases} 2x, & x < 0 \\ \dfrac{1}{2}x^{-1/2}, & x > 0. \end{cases}$ Using (2) of Section 3.1:

$$f'_-(0) = \lim_{h \to 0^-} \frac{(0+h)^2 - 0^2}{h} = \lim_{h \to 0^-} \frac{h^2}{h} = \lim_{h \to 0^-} h = 0$$

$$f'_+(0) = \lim_{h \to 0^+} \frac{\sqrt{0+h} - \sqrt{0}}{h} = \lim_{h \to 0^+} \frac{\sqrt{h}}{h} = \lim_{h \to 0^+} \frac{1}{\sqrt{h}} = \infty$$

$f'_+(0)$ does not exist, so $f'(0)$ does not exist.

Applications of the Derivative

4.1 Rectilinear Motion

1. $s(1/2) = -1$, $s(3) = 19$; $v(t) = 8t - 6$, $v(1/2) = -2$, $v(3) = 18$, $|v(1/2)| = 2$, $|v(3)| = 18$; $a(t) = 8$, $a(1/2) = 8$, $a(3) = 8$

3. $s(-2) = 18$, $s(2) = 6$; $v(t) = -3t^2 + 6t + 1$, $v(-2) = -23$, $v(2) = 1$, $|v(-2)| = 23$, $|v(2)| = 1$; $a(t) = -6t + 6$, $a(-2) = 18$, $a(2) = -6$

5. $s(1/4) = -15/4$, $s(1) = 0$; $v(t) = 1 + 1/t^2$, $v(1/4) = 17$, $v(1) = 2$, $|v(1/4)| = 17$, $|v(1)| = 2$; $a(t) = -2/t^3$, $a(1/4) = -128$, $a(1) = -2$

7. $s(1) = 1$, $s(3/2) = 1/2$; $v(t) = 1 + \pi \cos \pi t$, $v(1) = 1 - \pi$, $v(3/2) = 1$, $|v(1)| = \pi - 1$, $|v(3/2)| = 1$; $a(t) = -\pi^2 \sin \pi t$, $a(1) = 0$, $a(3/2) = \pi^2$

9. $v(t) = 2t - 4$

   (a) Solving $t^2 - 4t - 5 = 0$ gives $t = -1, 5$. The velocity when $s(t) = 0$ is $v(-1) = -6$, $v(5) = 6$.

   (b) Solving $t^2 - 4t - 5 = 7$ gives $t = -2, 6$. The velocity when $s(t) = 7$ is $v(-2) = -8$, $v(6) = 8$.

11. $v(t) = 3t^2 - 4$; $\quad a(t) = 6t$

   (a) Solving $3t^2 - 4 = 2$ gives $t = \pm\sqrt{2}$. When $v(t) = 2$, $a(-\sqrt{2}) = -6\sqrt{2}$, $a(\sqrt{2}) = 6\sqrt{2}$.

   (b) Solving $6t = 18$ gives $t = 3$. Then $s(3) = 15$.

   (c) Solving $t^3 - 4t = t(t+2)(t-2) = 0$ gives $t = 0, \pm 2$. Then $v(0) = -4$, $v(-2) = 8$, $v(2) = 8$.

13. $v(t) = 3t^2 - 27 = 3(t-3)(t+3)$; $\quad a(t) = 6t$

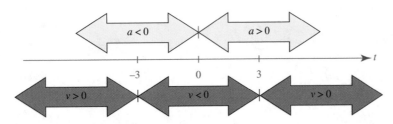

The particle is slowing down on $(-\infty, -3)$ and on $(0, 3)$; it is speeding up on $(-3, 0)$ and on $(3, \infty)$.

In order to draw the graphs in Problems 15–28 we need to determine when the particle changes direction. For a continuous position function, this will occur when the velocity is 0. This is a *necessary* condition; it is not *sufficient*. That is, the velocity may be 0 without the particle changing direction (see, for example, Problem 16). The arrows $\rightarrow$ and $\leftarrow$ in the charts indicate the direction of motion on the specified interval, as determined by the sign of the velocity on that interval.

15. $v(t) = 2t$; $a(t) = 2$. Solving $v = 0$ we obtain $t = 0$.

t	-1		0		3
s	1	$\leftarrow$	0	$\rightarrow$	9
v	$-$		0	$+$	
a			$+$		

The particle is slowing down on $(-\infty, 0)$ and speeding up on $(0, \infty)$.

17. $v(t) = 2t - 4$; $a(t) = 2$. Solving $v = 0$ we obtain $t = 2$.

t	-1		2		5
s	3	$\leftarrow$	-6	$\rightarrow$	3
v	$-$		0	$+$	
a			$+$		

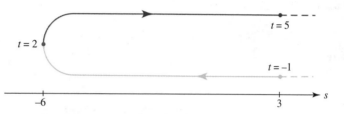

The particle is slowing down on $(-\infty, 2)$ and speeding up on $(2, \infty)$.

19. $v(t) = 6t^2 - 12t = 6t(t - 2)$; $a(t) = 12t - 12$. Solving $v = 0$ we obtain $t = 0, 2$; solving $a = 0$ we obtain $t = 1$.

t	-2		0		1		2		3
s	-40	$\rightarrow$	0	$\leftarrow$	-4	$\leftarrow$	-8	$\rightarrow$	0
v	$+$		0		$-$		0	$+$	
a		$-$			0		$+$		

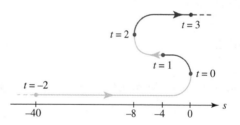

The particle is slowing down on $(-\infty, 0)$ and $(1, 2)$; it is speeding up on $(0, 1)$ and $(2, \infty)$.

21. $v(t) = 12t^3 - 24t^2 = 12t^2(t - 2)$; $a(t) = 36t^2 - 48t$. Solving $v = 0$ we obtain $t = 0, 2$.

t	-1		0		2		3
s	11	$\leftarrow$	0	$\leftarrow$	-16	$\rightarrow$	27
v	$-$		0	$-$	0	$+$	

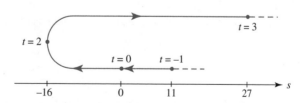

23. $v(t) = 1 - 2t^{-1/2}$; $a(t) = t^{-3/2}$. Setting $v = 0$ we obtain $2/\sqrt{t} = 1$, so $\sqrt{t} = 2$ and $t = 4$.

t	1		4		9
s	-3	$\leftarrow$	-4	$\rightarrow$	-3
v	$-$		0	$+$	

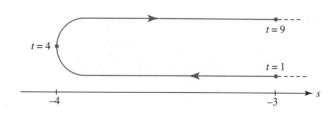

25. $v(t) = \dfrac{\pi}{2}\cos\dfrac{\pi}{2}t$; $a(t) = -\dfrac{\pi^2}{4}\sin\dfrac{\pi}{2}t$. Setting $v = 0$ we obtain $\cos\dfrac{\pi}{2}t = 0$. Thus, for $0 \le t \le 4$, we have $t = 1, 3$.

t	0		1		3		4
s	0	$\rightarrow$	1	$\leftarrow$	-1	$\rightarrow$	0
v		$+$	0	$-$	0	$+$	

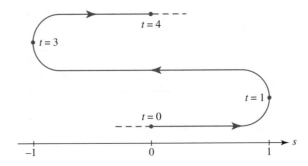

27. $v(t) = t^3(-e^{-t}) + 3t^2(e^{-t}) = e^{-t}(3t^2 - t^3) = t^2 e^{-t}(3 - t)$; $a(t) = e^{-t}(6t - 3t^2) + (-e^{-t})(3t^2 - t^3) = te^{-t}(t^2 - 6t + 6)$. Setting $v = 0$ we obtain $t = 0, 3$. In addition, $\lim\limits_{t \to \infty} v(t) = 0$.

t	0		3		∞
s	0	$\rightarrow$	$27e^{-3}$	$\leftarrow$	0
v	0	$+$	0		$-$

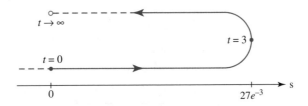

29.

Interval	$v(t)$	$a(t)$
(a, b)	$+$	$-$
(b, c)	0	0
(c, d)	$+$	$+$
(d, e)	$+$	$-$
(e, f)	$-$	$-$
(f, g)	$-$	$+$

The particle is slowing down on (a, b), (d, e), and (f, g); it is speeding up on (c, d) and (e, f).

31.   (a) $v(t) = -32t + 48$. Solving $v = 0$ we obtain $t = 3/2$. The velocity is positive on $(-\infty, 3/2)$ and negative on $(3/2, \infty)$.

   (b) The maximum is attained when the velocity is 0. This height is then $s(3/2) = 42$ ft.

33. $s(t) = 16t^2 \sin 30° = 8t^2$; $v(t) = 16t$; $a(t) = 16$. At the bottom of the hill, $s = 8t^2 = 256$ and $t = \sqrt{32} = 4\sqrt{2}$. The velocity is $v(4\sqrt{2}) = 64\sqrt{2}$ ft/s and the acceleration is $a(4\sqrt{2}) = 16$ ft/s^2.

35. We are given $\theta = 16t^2$. Since the circle has radius 1, $y = \sin\theta = \sin 16t^2$ and $dy/dt = 32t\cos 16t^2$. For $t = \sqrt{\pi}/4$, $dy/dt = 8\sqrt{\pi}\cos\pi = -8\sqrt{\pi}$ ft/s. Since dy/dt is negative, the y-coordinate is decreasing.

4.2   Related Rates

1. Let V be the volume and x the length of a side. Then $V = x^3$ and $\dfrac{dV}{dt} = 3x^2 \dfrac{dx}{dt}$.

3. Let A be the area and let x be the length of a side. Then $A = \dfrac{\sqrt{3}x^2}{4}$ and $\dfrac{dA}{dt} = \dfrac{\sqrt{3}}{2}x\dfrac{dx}{dt}$. When $\dfrac{dx}{dt} = 2$ and $x = 8$ we have $\dfrac{dA}{dt} = \dfrac{\sqrt{3}}{2}(8)(2) = 8\sqrt{3}$ cm^2/h.

5. Let x be the length, y the width, and s the diagonal of the rectangle. Then $s^2 = x^2 + y^2$ or $y^2 = s^2 - x^2$, and $2y\dfrac{dy}{dt} = 2s\dfrac{ds}{dt} - 2x\dfrac{dx}{dt}$ or $\dfrac{dy}{dt} = \dfrac{s}{y}\left(\dfrac{ds}{dt}\right) - \dfrac{x}{y}\left(\dfrac{dx}{dt}\right)$. When $x = 8$ in and $y = 6$ in, $s = 10$ in. Then $\dfrac{dy}{dt} = \dfrac{10}{6}(1) - \dfrac{8}{6}\left(\dfrac{1}{4}\right) = \dfrac{4}{3}$ in/h.

7. $\sin\theta = x/s$ or $x = s\sin\theta$. Differentiating with respect to t gives $\dfrac{dx}{dt} = s\dfrac{d}{dt}\sin\theta + (\sin\theta)\dfrac{ds}{dt} = s\cos\theta\dfrac{d\theta}{dt} + \sin\theta\dfrac{ds}{dt}$.

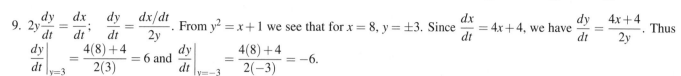

9. $2y\dfrac{dy}{dt} = \dfrac{dx}{dt}$; $\dfrac{dy}{dt} = \dfrac{dx/dt}{2y}$. From $y^2 = x+1$ we see that for $x = 8$, $y = \pm 3$. Since $\dfrac{dx}{dt} = 4x+4$, we have $\dfrac{dy}{dt} = \dfrac{4x+4}{2y}$. Thus $\left.\dfrac{dy}{dt}\right|_{y=3} = \dfrac{4(8)+4}{2(3)} = 6$ and $\left.\dfrac{dy}{dt}\right|_{y=-3} = \dfrac{4(8)+4}{2(-3)} = -6$.

11. If T is the area of the triangle then $T = \dfrac{1}{2}xy = \dfrac{1}{2}x^{4/3}$ and $\dfrac{dT}{dt} = \dfrac{2}{3}x^{1/3}\dfrac{dx}{dt}$. When $x = 8$, then $\dfrac{dT}{dt} = \dfrac{2}{3}(8)^{1/3}\left(\dfrac{1}{3}\right) = \dfrac{4}{9}$ cm^2/h.

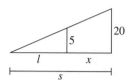

13. (a) Since the lengths of corresponding sides in similar triangles are proportional, $\dfrac{5}{20} = \dfrac{l}{l+x}$ or $l = x/3$. When $\dfrac{dx}{dt} = 3$ ft/s, differentiating gives $\dfrac{dl}{dt} = \dfrac{1}{3}\cdot\dfrac{dx}{dt} = \dfrac{1}{3}(3) = 1$ ft/s.

(b) Differentiating $s = l+x$ gives $\dfrac{ds}{dt} = \dfrac{dl}{dt} + \dfrac{dx}{dt}$. Since $\dfrac{dx}{dt} = 3$ ft/s and from (a) $\dfrac{dl}{dt} = 1$ ft/s, the tip of the shadow is moving at a rate of $\dfrac{ds}{dt} = 1+3 = 4$ ft/s.

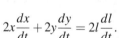

15. From the Pythagorean Theorem, $15^2 = h^2 + x^2$. Differentiating gives

$$0 = 2h\dfrac{dh}{dt} + 2x\dfrac{dx}{dt} \qquad \text{or} \qquad \dfrac{dh}{dt} = -\dfrac{x}{h}\cdot\dfrac{dx}{dt}.$$

When $x = 5$ ft, $h = 10\sqrt{2}$ ft, and $\dfrac{dx}{dt} = 2$ ft/min, we have $\dfrac{dh}{dt} = -\dfrac{5}{10\sqrt{2}}(2) = -\dfrac{1}{\sqrt{2}}$ ft/min.

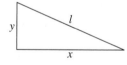

17. Since $\theta_1 = \dfrac{\pi}{2} - \theta_2$, $\dfrac{d\theta_1}{dt} = -\dfrac{d\theta_2}{dt}$ and θ_1 is increasing at the same rate θ_2 is decreasing.

19. From the Pythagorean Theorem, $x^2 + y^2 = l^2$. Differentiating gives

$$2x\dfrac{dx}{dt} + 2y\dfrac{dy}{dt} = 2l\dfrac{dl}{dt}.$$

We are given $\dfrac{dx}{dt} = 10$ knots and $\dfrac{dy}{dt} = 15$ knots, so $\dfrac{dl}{dt} = \dfrac{10x + 15y}{l}$. At 2:00 PM, $x = 20$ nautical miles, $y = 15$ nautical miles, and $l = \sqrt{20^2 + 15^2} = 25$ nautical miles. Thus

$$\dfrac{dl}{dt} = \dfrac{10(20) + 15(15)}{25} = 17 \text{ knots.}$$

21. Let x be the distance from the boat to the base of the dock and y be the distance from the boat to the pulley. From the Pythagorean Theorem, $x^2 + 144 = y^2$. Thus, $2x\dfrac{dx}{dt} = 2y\dfrac{dy}{dt}$ and $\dfrac{dx}{dt} = \left(\dfrac{y}{x}\right)\dfrac{dy}{dt}$. We are given $\dfrac{dy}{dt} = -1$ and $x = 16$. At this time, $y = 20$ and $\dfrac{dx}{dt} = \dfrac{20}{16}(-1) = -\dfrac{5}{4}$ ft/s. Hence, the boat is approaching the dock at a rate of $\dfrac{5}{4}$ ft/s.

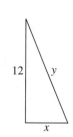

23. We are given $\dfrac{dx}{dt} = 15$ and we want to find $\dfrac{d\theta}{dt}$ when $x = 1/2$. Then $\theta = \tan^{-1}\dfrac{x}{1/2} = \tan^{-1}2x$, $\dfrac{d\theta}{dt} = \dfrac{2}{1+4x^2}\cdot\dfrac{dx}{dt}$, and $\left.\dfrac{d\theta}{dt}\right|_{x=1/2} = 1(15) = 15$ rad/h.

25. From the Pythagorean Theorem, $l^2 = x^2 + 4$. Differentiating gives

$$2l\dfrac{dl}{dt} = 2x\dfrac{dx}{dt} \qquad \text{or} \qquad \dfrac{dl}{dt} = \dfrac{x}{l}\cdot\dfrac{dx}{dt}.$$

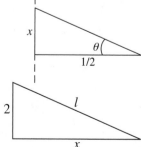

We are given $\dfrac{dx}{dt} = -600$ mi/h, and when $x = 1.5$ mi, $l = 2.5$ mi. Then

$$\frac{dl}{dt} = \frac{1.5}{2.5}(-600) = -360 \text{ mi/h}.$$

Thus the distance is decreasing at a rate of 360 mi/h.

27. Differentiating $x = 4\cot\theta$, we obtain $\dfrac{dx}{dt} = -4\csc^2\theta\,\dfrac{d\theta}{dt}$. Converting $30°$ to $\pi/6$ radians, we are given $\dfrac{d\theta}{dt} = -\dfrac{\pi}{6}$. Thus, when $\theta = 60°$,

$$\frac{dx}{dt} = -4(\csc^2 60°)\left(-\frac{\pi}{6}\right) = \frac{8\pi}{9} \approx 2.79 \text{ km/min.}$$

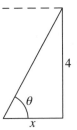

29. Let y be the altitude of the rocket, x the distance along the ground from the point of launch, and s the distance the rocket has travelled.

(a) $y = s\sin 60° = \dfrac{\sqrt{3}}{2}s;\ \dfrac{dy}{dt} = \dfrac{\sqrt{3}}{2}\cdot\dfrac{ds}{dt}$. When $\dfrac{ds}{dt} = 1000$, $\dfrac{dy}{dt} = \dfrac{\sqrt{3}}{2}(1000) = 500\sqrt{3}$ mi/h.

(b) $x = s\cos 60° = \dfrac{1}{2}s;\ \dfrac{dx}{dt} = \dfrac{1}{2}\cdot\dfrac{ds}{dt}$. When $\dfrac{ds}{dt} = 1000$, $\dfrac{dx}{dt} = \dfrac{1}{2}(1000) = 500$ mi/h.

31. $V = \pi r^2 h$. Since r is a constant, differentiating with respect to t gives $\dfrac{dV}{dt} = \pi r^2 h\dfrac{dh}{dt}$ or $\dfrac{dh}{dt} = \dfrac{1}{\pi r^2}\cdot\dfrac{dV}{dt}$. When $r = 8$ m and $\dfrac{dV}{dt} = 10$ m^3/min, the oil level rises at a rate of $\dfrac{dh}{dt} = \dfrac{10}{\pi(8)^2} = \dfrac{5}{32\pi}$ m/min.

33. (a) Since the lengths of corresponding sides in similar triangles are proportional, $\dfrac{h}{9} = \dfrac{r}{3}$ or $h = 3r$.

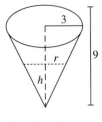

The volume of water is $V = \dfrac{1}{3}\pi r^2 h = \dfrac{1}{3}\pi\left(\dfrac{h}{3}\right)^2 h = \dfrac{1}{27}\pi h^3$. Differentiating gives

$$\frac{dV}{dt} = \frac{1}{9}\pi h^2\frac{dh}{dt} \qquad \text{or} \qquad \frac{dh}{dt} = \frac{9}{\pi h^2}\cdot\frac{dV}{dt}.$$

We are given $\dfrac{dV}{dt} = -1$. When $h = 6$, the water level is changing at a rate of $\dfrac{dh}{dt} = \dfrac{9}{\pi(36)}(-1) = -\dfrac{1}{4\pi}$ ft/min.

(b) From $h = 3r$ we have $\dfrac{dh}{dt} = 3\dfrac{dr}{dt}$, so when $h = 6$, the radius of water is changing at a rate of $\dfrac{dr}{dt} = \dfrac{1}{3}\cdot\dfrac{dh}{dt} = \dfrac{1}{3}\left(-\dfrac{1}{4\pi}\right) = -\dfrac{1}{12\pi}$ ft/min.

(c) The initial volume of water is $V_0 = \dfrac{1}{3}\pi(3)^2(9) = 27\pi$ ft^3. At time t the volume of water is $V(t) = V_0 - t = 27\pi - t$. We also have from part (a) that $V = \dfrac{1}{3}\pi r^2 h = \dfrac{1}{3}\pi r^2(3r) = \pi r^3$. Thus, $\pi r^3 = 27\pi - t$ or $r = \left(27 - \dfrac{t}{\pi}\right)^{1/3}$. Then $\dfrac{dr}{dt} = -\dfrac{1}{3\pi}\left(27 - \dfrac{t}{\pi}\right)^{-2/3}$ and $\dfrac{dr}{dt}\bigg|_{t=6} = -\dfrac{1}{3\pi(27 - 6/\pi)^{2/3}} \approx -0.0124$ ft/min.

35. (a) From the Pythagorean Theorem, $s^2 = h^2 + (s/2)^2$ or $s = \dfrac{2h}{\sqrt{3}}$. The volume of the water is

$V = \dfrac{1}{2}sh(20) = 10\dfrac{2h}{\sqrt{3}}h = \dfrac{20h^2}{\sqrt{3}}$. Differentiating with respect to t gives $\dfrac{dV}{dt} = \dfrac{40h}{\sqrt{3}}\cdot\dfrac{dh}{dt}$ so $\dfrac{dh}{dt} = \dfrac{\sqrt{3}}{40h}\cdot\dfrac{dV}{dt}$. When $h = 1$ ft and $\dfrac{dV}{dt} = 4$ ft^3/min, the rate at which the water level rises is $\dfrac{dh}{dt} = \dfrac{\sqrt{3}}{10}$ ft/min.

(b) From part (a) we see that the initial volume of water is $V_0 = \dfrac{20h_0^2}{\sqrt{3}}$. At time t, the volume of water is $V = 4t + V_0 = 4t + \dfrac{20h_0^2}{\sqrt{3}}$. In terms of h, we saw in part (a) that $V = \dfrac{20h^2}{\sqrt{3}}$. Thus, $\dfrac{20h^2}{\sqrt{3}} = 4t + \dfrac{20h_0^2}{\sqrt{3}}$. Solving for h and differentiating, we find

$$h = \sqrt{\frac{\sqrt{3}}{5}t + h_0^2} \quad \text{and} \quad \frac{dh}{dt} = \frac{1}{2}\left(\frac{\sqrt{3}}{5}t + h_0^2\right)^{-1/2}\frac{\sqrt{3}}{5} = \frac{\sqrt{3}}{10}\left(\frac{\sqrt{3}}{5}t + h_0^2\right)^{-1/2}.$$

(c) Setting $h = 5$ and $h_0 = \dfrac{1}{2}$ in part (b), we have $5 = \sqrt{\dfrac{\sqrt{3}}{5}t + \dfrac{1}{4}}$ or $t = \dfrac{165\sqrt{3}}{4} \approx 71.45$ min. The rate at which the water is rising when $t = \dfrac{165\sqrt{3}}{4}$ is

$$\left.\frac{dh}{dt}\right|_{t=165\sqrt{3}/4} = \frac{\sqrt{3}}{10}\left(\frac{\sqrt{3}}{5}\cdot\frac{165\sqrt{3}}{4} + \frac{1}{4}\right)^{-1/2} = \frac{\sqrt{3}}{50} \approx 0.035 \text{ ft/min}.$$

37. The volume of a sphere is $V = \dfrac{4}{3}\pi r^3$. Differentiating gives $\dfrac{dV}{dt} = 4\pi r^2\dfrac{dr}{dt}$. The surface area of a sphere is $S = 4\pi r^2$, so $\dfrac{dV}{dt} = S\dfrac{dr}{dt}$. Since we are given that $\dfrac{dV}{dt} = kS$, we have $\dfrac{dr}{dt} = k$. Thus, the radius changes at a constant rate.

39. $V = x^3$, $\dfrac{dV}{dt} = 3x^2\dfrac{dx}{dt}$. The surface area of the cube is $S = 6x^2$, so when $S = 54$, $x = 3$. Now when $\dfrac{dV}{dt} = -\dfrac{1}{4}$ we have $-\dfrac{1}{4} = 3(3)^2\dfrac{dx}{dt}$ and $\dfrac{dx}{dt} = -\dfrac{1}{108}$. From $\dfrac{dS}{dt} = 12x\dfrac{dx}{dt}$ we use $x = 3$ and $\dfrac{dx}{dt} = -\dfrac{1}{108}$ to compute $\dfrac{dS}{dt} = 12(3)\left(-\dfrac{1}{108}\right) = -\dfrac{1}{3}$ in^2/min.

41. Place the origin at the center of the ferris wheel with the x-axis parallel to the ground. The coordinates of the point P are $x = 60\cos\theta$ and $y = 60\sin\theta$. If the coordinates of the point Q are $(q, -64)$ then the slope of the line through PQ is $\dfrac{60\sin\theta + 64}{60\cos\theta - q}$. Since this line is perpendicular to the line through the center of the wheel and P, its slope is also $-\dfrac{1}{\tan\theta}$.

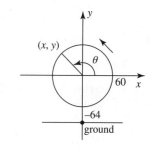

Solving $\dfrac{60\sin\theta + 64}{60\cos\theta - q} = -\dfrac{1}{\tan\theta}$ for q we obtain $q = 60\cos\theta + 64\tan\theta + 60\sin\theta\tan\theta$ and $\dfrac{dq}{dt} = \left(-60\sin\theta + 64\sec^2\theta + 60\sin\theta\sec^2\theta + 60\sin\theta\right)\dfrac{d\theta}{dt}$. When $\theta = \pi/4$ and $\dfrac{d\theta}{dt} = \pi$ we have $\left.\dfrac{dq}{dt}\right|_{\theta=\pi/4} = (60\sqrt{2} + 128)\pi \approx 668.7$ radians/min.

43. $R^{-1} = R_1^{-1} + R_2^{-1}$. Differentiating with respect to t gives

$$-R^{-2}\frac{dR}{dt} = -R_1^{-2}\frac{dR_1}{dt} - R_2^{-2}\frac{dR_2}{dt} \quad \text{and} \quad \frac{1}{R^2}\cdot\frac{dR}{dt} = \frac{1}{R_1^2}\cdot\frac{dR_1}{dt} + \frac{1}{R_2^2}\cdot\frac{dR_2}{dt}$$

so $\dfrac{dR}{dt} = \dfrac{R^2}{R_1^2}\cdot\dfrac{dR_1}{dt} + \dfrac{R^2}{R_2^2}\cdot\dfrac{dR_2}{dt}$.

45. (a) From $R = \dfrac{C}{T} = \dfrac{0.493T - 0.913}{T} = 0.493 - \dfrac{0.913}{T}$ we find $\dfrac{dR}{dt} = \dfrac{0.913}{T^2}\cdot\dfrac{dT}{dt} > 0$. Thus, the ratio increases.

(b) To find the value of T when $C = \dfrac{T}{3}$, we solve $\dfrac{T}{3} = 0.493T - 0.913$, obtaining $T \approx 5.718$. Then $\left.\dfrac{dR}{dt}\right|_{T=5.718} \approx \dfrac{0.913}{(5.718)^2}(1) \approx 0.028 = 2.8\%$/day.

47. (a) From $\dfrac{dP}{dt} = 800\dfrac{dm}{dt}$ and $\dfrac{dm}{dt} = 30$ kg/h, we see that the momentum is changing at a rate of $800(400) = 24{,}000$ kg km/h.

(b) In this case, both m and v are variables so $\dfrac{dP}{dt} = m\dfrac{dv}{dt} + v\dfrac{dm}{dt}$. At $t = 1$ hour the mass of the airplane is $10^5 + 30 = 100{,}030$ kg and the velocity is 750 km/h. Thus,

$$\left.\frac{dP}{dt}\right|_{t=1} = 100{,}030(20) + 750(30) = 2{,}023{,}100 \text{ kg km/h}^2.$$

4.3 Extrema of Functions

1. (a) Absolute maximum: $f(2) = -2$; absolute minimum: $f(-1) = -5$

   (b) Absolute maximum: $f(7) = 3$; absolute minimum: $f(3) = -1$

   (c) No extrema

   (d) Absolute maximum: $f(4) = 0$; absolute minimum: $f(1) = -3$

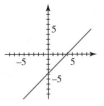

3. (a) Absolute maximum: $f(4) = 0$; absolute minimum: $f(2) = -4$

   (b) Absolute maximum: $f(1) = f(3) = -3$;
   absolute minimum: $f(2) = -4$

   (c) Absolute minimum: $f(2) = -4$

   (d) Absolute maximum: $f(5) = 5$

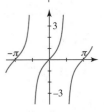

5. (a) No extrema

   (b) Absolute maximum: $f(\pi/4) = 1$; absolute minimum: $f(-\pi/4) = -1$

   (c) Absolute maximum: $f(\pi/3) = \sqrt{3}$; absolute minimum: $f(0) = 0$

   (d) No extrema

7. Solving $f'(x) = 4x - 6 = 0$ we obtain critical number $3/2$.

9. Solving $f'(x) = 6x^2 - 30x - 36 = 6(x - 6)(x + 1) = 0$ we obtain the critical numbers 6 and -1.

11. Solving $f'(x) = (x-2)^2(1) + (x-1)[2(x-2)] = (x-2)[(x-2) + 2(x-1)] = (x-2)(3x-4) = 0$ we obtain the critical numbers 2 and $4/3$.

13. Solving $f'(x) = \dfrac{x^{1/2} - (1+x)\left(\dfrac{1}{2}x^{-1/2}\right)}{x} = \dfrac{x-1}{2x^{3/2}}$ we obtain the critical number 1. $f'(x)$ does not exist when $x = 0$, but 0 is not in the domain of $f(x)$, so the only critical number is 1.

15. We note that $f'(x) = \dfrac{4}{3(4x-3)^{2/3}} \neq 0$ for all x and $f'(3/4)$ does not exist. Since $3/4$ is in the domain of $f(x)$, it is a critical number.

17. Solving $f'(x) = \dfrac{1}{3}(x-1)^2(x+2)^{-2/3} + 2(x+2)^{1/3}(x-1) = 0$ we observe $(x-1)^2(x+1)^{-2/3} + 6(x+2)^{1/3}(x-1) = 0$ or $(x+2)^{1/3}(x-1)[(x-1)(x+2)^{-1} + 6] = 0$. Thus, -2 and 1 are critical numbers. Since we also have $\dfrac{x-1}{x+2} + 6 = 0$ or $x - 1 = -6(x+2)$, then $x = -11/7$ and $-11/7$ is also a critical number.

19. Solving $f'(x) = -1 + \cos x = 0$ we obtain the critical numbers $2n\pi$ where $n = 0, \pm 1, \pm 2, \ldots$.

21. Solving $f'(x) = 2x - 8/x = 0$ we obtain the critical number 2. $f'(x) = 0$ when $x = -2$, but -2 is not in the domain of $f(x)$, so it is not a critical number. $f'(x)$ does not exist when $x = 0$, but 0 is not in the domain of $f(x)$, so the only critical number is 2.

23. Solving $f'(x) = -2x + 6 = 0$ we obtain the critical number 3. The absolute maximum is $f(3) = 9$ and the absolute minimum is $f(1) = 5$.

x	1	3	4
$f(x)$	5	9	8

25. We note that $f'(x) = \frac{2}{3}x^{-1/3}$ does not exist at $x = 0$. Since 0 is in the domain, it is a critical number. The absolute maximum is $f(8) = 4$ and the absolute minimum is $f(0) = 0$.

x	-1	0	8
$f(x)$	1	0	4

27. Solving $f'(x) = 3x^2 - 12x = 0$ we obtain the critical numbers 0 and 4. However, only 0 is in $[-3, 2]$. The absolute maximum is $f(0) = 2$ and the absolute minimum is $f(-3) = -79$.

x	-3	0	2
$f(x)$	-79	2	-14

29. Solving $f'(x) = 3x^2 - 6x + 3 = 0$ we obtain the critical number 1. The absolute maximum is $f(3) = 8$ and the absolute minimum is $f(-4) = -125$.

x	-4	1	3
$f(x)$	-125	0	8

31. Write $f(x) = x^6 - 2x^5 + x^4$. Then solving $f'(x) = 6x^5 - 10x^4 + 4x^3 = 2x^3(3x - 2)(x - 1) = 0$ we obtain 0, 2/3, and 1. The absolute maximum is $f(2) = 16$ and the absolute minimum is 0 and occurs at $x = 0$ and $x = 1$.

x	-1	0	2/3	1	2
$f(x)$	4	0	16/729	0	16

33. Solving $f'(x) = -4\sin 2x + 4\sin 4x = -4\sin 2x + 8\sin 2x\cos 2x = 4\sin 2x(2\cos 2x - 1) = 0$ on $[0, 2\pi]$ we obtain the critical numbers 0, $\pi/2$, π, $3\pi/2$, 2π, $\pi/6$, $5\pi/6$, $7\pi/6$, and $11\pi/6$. The absolute maximum is 3/2 and occurs at $x = \pi/6$, $5\pi/6$, $7\pi/6$, and $11\pi/6$. The absolute minimum is -3 and occurs at $x = \pi/2$ and $3\pi/2$.

x	0	$\pi/6$	$\pi/2$	$5\pi/6$	π	$7\pi/6$	$3\pi/2$	$11\pi/6$	2π
$f(x)$	1	3/2	-3	3/2	1	3/2	-3	3/2	1

35. Solving $f'(x) = 96\sin 24x\cos 24x = 48\sin 48x = 0$ on $[0, \pi]$ we obtain the critical numbers $k\pi/48$, where k is an integer from 0 to 48. The absolute maximum is 5 and occurs when k is odd. The absolute minimum is 3 and occurs when k is even.

x	0	$\pi/48$	$\pi/24$	...	$23\pi/24$	$47\pi/48$	π
$f(x)$	3	5	3	...	3	5	3

37. Solving $f'(x) = \begin{cases} 2x+2, & x < 0 \\ 2x-2, & x > 0 \end{cases} = 0$ we obtain the critical numbers 1 and -1. We note that $f'(x)$ does not exist when $x = 0$. Since 0 is in the domain of $f(x)$, it is also a critical number. The absolute minimum is $f(-1) = f(1) = -1$, the endpoint absolute maximum is $f(3) = 3$, and the relative maximum is $f(0) = 0$.

x	-2	-1	0	1	3
$f(x)$	0	-1	0	-1	3

39. (a) c_1, c_3, c_4, c_{10}

    (b) $c_2, c_5, c_6, c_7, c_8, c_9$

    (c) Endpoint absolute maximum: $f(b)$;   absolute minimum: $f(c_7)$

    (d) Relative maxima: $f(c_3)$, $f(c_5)$, $f(c_9)$;   relative minima: $f(c_2)$, $f(c_4)$, $f(c_7)$, $f(c_{10})$

41. (a) $-16t^2 + 320t$ is negative outside the interval $[0, 20]$.

    (b) Solving $s'(t) = -32t + 320 = 0$ we obtain the critical number $t = 10$. From the data in the accompanying table, we see that the maximum height attained by the projectile on $[0, 20]$ is 1600 ft.

t	0	10	20
$s(t)$	0	1600	0

43.

45. For every x that is not an integer, $f'(x) = 0$, and for every integer value of x, $f'(x)$ does not exist. Therefore, since $f(x)$ is defined for all real x, every value of x is a critical number.

47. Solving $f'(x) = nx^{n-1} = 0$, we see that 0 is a critical number and $f(0) = 0$ is the only possible relative extremum. When n is even, $f(x)$ is positive for all non-zero x, and $f(0) = 0$ is a relative minimum. When n is odd, $f(x) < 0$ for $x < 0$ and $f(x) > 0$ for $x > 0$, so $f(0) = 0$ is not a relative extremum in this case.

49. Since $f(a)$ is a relative minimum, there is an interval (c_1, c_2) around a in which $f(x) \geq f(a)$. Consider the interval $(-c_2, -c_1)$ around $-a$. Since $f(x)$ is even, $f(-x) = f(x)$, and for $-x$ in $(-c_2, -c_1)$, $f(-x) = f(x) \geq f(a) = f(-a)$. Therefore $f(-a)$ is a relative minimum.

51. Since $f(x)$ is even and everywhere differentiable, we have $f(-x) = f(x)$ and $-f'(-x) = f'(x)$ through implicit differentiation. When $x = 0$, we have $-f'(0) = f'(0)$ and so $f'(0) = 0$. Thus, 0 is a critical number of f.

53. (a)

(b) Solving $f'(x) = 2\sin x - 2\sin 2x = 2\sin x - 4\sin x \cos x = 2\sin x(1 - 2\cos x) = 0$ on $[0, 2\pi]$, we obtain the critical numbers $0, \pi/3, \pi, 5\pi/3$, and 2π.

(c) Computing $f(0) = f(2\pi) = -1$, $f(\pi/3) = f(-5\pi/3) = -3/2$, and $f(\pi) = 3$, we see that the absolute maximum is $f(\pi) = 3$ and the absolute minimum is $f(\pi/3) = f(-5\pi/3) = -3/2$.

4.4   Mean Value Theorem

1. $f(x)$ is continuous and differentiable on $[-2, 2]$ and $f(-2) = f(2) = 0$, so Rolle's Theorem applies. Solving $f'(c) = 2c = 0$ we obtain $c = 0$.

3. Since $f(-2) = 19 \neq 0$, Rolle's Theorem does not apply.

5. $f(x)$ is continuous and differentiable on $[-1, 0]$ and $f(-1) = f(0) = 0$, so Rolle's Theorem applies. Solving $f'(c) = 3c^2 + 2c = 0$ we obtain $c = 0, -2/3$. Only $c = -2/3$ is in the interval $(-1, 0)$.

7. $f(x)$ is continuous and differentiable on $[-\pi, 2\pi]$ and $f(-\pi) = f(2\pi) = 0$, so Rolle's Theorem applies. Solving $f'(c) = \cos c = 0$ on $(-\pi, 2\pi)$, we obtain $c = -\pi/2, \pi/2, 3\pi/2$.

9. Since $f'(x) = x^{-1/3}$, $f(x)$ is not differentiable on $(-1, 1)$ and Rolle's Theorem does not apply.

11. $f(a) \neq 0$

13. $f(x)$ is continuous and differentiable on $[-1, 7]$, so the Mean Value Theorem applies. Setting $f'(c) = 2c = \dfrac{f(7) - f(-1)}{7 + 1} = 6$, we obtain $c = 3$.

15. $f(x)$ is continuous and differentiable on $[2, 5]$, so the Mean Value Theorem applies. Setting $f'(c) = 3c^2 + 1 = \dfrac{f(5) - f(2)}{5 - 2} = 40$, we obtain $3c^2 = 39$. Then on $(2, 5)$, $c = \sqrt{13}$.

17. $f(x)$ is not continuous at 0, so the Mean Value Theorem does not apply.

19. $f(x)$ is continuous and differentiable on $[0, 9]$, so the Mean Value Theorem applies. Setting $f'(c) = 1/2\sqrt{c} = \dfrac{f(9) - f(0)}{9 - 0} = 1/3$, we obtain $c = 9/4$.

21. $f(x)$ is continuous and differentiable on $[-2, -1]$, so the Mean Value Theorem applies. Setting $f'(c) = -2/(c - 1)^2 = \dfrac{f(-1) - f(-2)}{-1 + 2} = -1/3$, we obtain $(c - 1)^2 = 6$. Then on $[-2, -1]$, $c = 1 - \sqrt{6}$.

23. f is not continuous at b.

25. $f'(x) = 2x$. Solving $f'(x) = 0$, we obtain the critical number 0. The function is decreasing on $(-\infty, 0]$ and increasing on $[0, \infty)$.

x		0	
f	$\searrow$		$\nearrow$
f'	$-$	0	$+$

27. $f'(x) = 2x + 6$. Solving $f'(x) = 0$, we obtain the critical number -3. The function is decreasing on $(-\infty, -3]$ and increasing on $[-3, \infty)$.

x		-3	
f	$\searrow$		$\nearrow$
f'	$-$	0	$+$

29. $f'(x) = 3x^2 - 6x = 3x(x - 2)$. Solving $f'(x) = 0$, we obtain the critical numbers 0 and 2. The function is decreasing on $[0, 2]$ and increasing on $(-\infty, 0]$ and $[2, \infty)$.

x		0		2	
f	$\nearrow$		$\searrow$		$\nearrow$
f'	$+$	0	$-$	0	$+$

31. $f'(x) = 4x^3 - 12x^2 = 4x^2(x - 3)$. Solving $f'(x) = 0$, we obtain the critical numbers 0 and 3. The function is decreasing on $(-\infty, 0]$ and $[0, 3]$, and increasing on $[3, \infty)$.

x		0		3	
f	$\searrow$		$\searrow$		$\nearrow$
f'	$-$	0	$-$	0	$+$

33. $f'(x) = -\dfrac{1}{3}x^{-2/3}$. The only critical number is at 0 where $f'(x)$ does not exist. The function is decreasing on $(-\infty, 0]$ and $[0, \infty)$.

x		0	
f	$\searrow$		$\searrow$
f'	$-$	undefined	$-$

35. $f'(x) = 1 - 1/x^2 = (x^2 - 1)/x^2$. Solving $f'(x) = 0$, we obtain the critical numbers -1 and 1. At $x = 0$, $f(x)$ is undefined. The function is decreasing on $[-1, 0)$ and $(0, 1]$, and increasing on $(-\infty, 1]$ and $[1, \infty)$.

x		-1		0		1	
f	$\nearrow$		$\searrow$	undefined	$\searrow$		$\nearrow$
f'	$+$	0	$-$	undefined	$-$	0	$+$

37. $f'(x) = x\left(\dfrac{-2x}{2\sqrt{8 - x^2}}\right) + \sqrt{8 - x^2} = \dfrac{-x^2 + 8 - x^2}{\sqrt{8 - x^2}} = \dfrac{8 - 2x^2}{\sqrt{8 - x^2}}$. Solving $f'(x) = 0$, we obtain the critical numbers -2 and 2. Also, $f(x)$ is only defined for $-\sqrt{8} \le x \le \sqrt{8}$ and $f'(x)$ is only defined for $-\sqrt{8} < x < \sqrt{8}$. The function is decreasing on $[-\sqrt{8}, -2]$ and $[2, \sqrt{8}]$, and increasing on $[-2, 2]$.

x	$-\sqrt{8}$		-2		2		$\sqrt{8}$
f		$\searrow$		$\nearrow$		$\searrow$	
f'	undefined	$-$	0	$+$	0	$-$	undefined

39. $f'(x) = -10x/(x^2 + 1)^2$. Solving $f'(x) = 0$, we obtain the critical number 0. The function is decreasing on $[0, \infty)$ and increasing on $(-\infty, 0]$.

x		0	
f	$\nearrow$		$\searrow$
f'	$+$	0	$-$

41. $f'(x) = 3x^2 - 12x + 9 = 3(x - 1)(x - 3)$. Solving $f'(x) = 0$, we obtain the critical numbers 1 and 3. The function is decreasing on $[1, 3]$ and increasing on $(-\infty, 1]$ and $[3, \infty)$.

x		1		3	
f	$\nearrow$		$\searrow$		$\nearrow$
f'	$+$	0	$-$	0	$+$

43. $f'(x) = \cos x$. Solving $f'(x) = 0$, we obtain the critical numbers $\pi/2 + k\pi$ for $k = 0, \pm1, \pm2, \ldots$. The sign of $f'(x) = \cos x$ is positive on $(-\pi/2 + 2k\pi, \pi/2 + 2k\pi)$ for $k = 0, \pm1, \pm2, \ldots$, and negative on the other intervals. Thus, $f(x) = \sin x$ is increasing on $[-\pi/2 + 2k\pi, \pi/2 + 2k\pi]$ and decreasing on $[\pi/2 + 2k\pi, 3\pi/2 + 2k\pi]$ for $k = 0, \pm1, \pm2, \ldots$.

45. $f'(x) = 1 - e^{-x}$. Solving $f'(x) = 0$, we obtain the critical number 0. The function is decreasing on $(-\infty, 0]$ and increasing on $[0, \infty)$.

x		0	
f	$\searrow$		$\nearrow$
f'	$-$	0	$+$

47. Since $f'(x) = 12x^2 + 1 > 0$ for all x, the function is increasing and has relative extrema.

49. Let $s(t)$ denote the distance travelled since 1:15 P.M. At 2:15 P.M., we have $t = 1$. By the Mean Value Theorem applied to the interval $[0, 1]$, we have $v(c) = s'(c) = \dfrac{s(1) - s(0)}{1 - 0} = 70$, for some c between 0 and 1. That is, at some time between 1:15 and 2:15 P.M. the motorist was speeding at 70 mi/h.

51. $f'(x) = 4x^3 + 3x^2 - 1$. Since f and f' are continuous and differentiable on $[-1, 1]$ and $f(-1) = f(1) = 0$, Rolle's Theorem applies. Thus, there exists c in $(-1, 1)$ such that $f'(c) = 4c^3 + 3c^2 - 1 = 0$.

53. We want $(fg)'(x) = f(x)g'(x) + f'(x)g(x) > 0$ for all x in (a, b). Since $f'(x) > 0$ and $g'(x) > 0$, the condition will hold if $f(x) > 0$ and $g(x) > 0$ for all x in (a, b).

55. We have $f'(x) = 2ax + b$. If there exist numbers $r_1 < r_2 < r_3$ such that $f(r_1) = f(r_2) = f(r_3) = 0$, then by Rolle's Theorem there exist c_1 in (r_1, r_2) and c_2 in (r_2, r_3) such that $f'(c_1) = f'(c_2) = 0$. But this is impossible since $f'(x)$ has only the single real root $-b/2a$. Therefore $f(x)$ can have at most two real roots.

57. The polynomial function f is continuous and differentiable everywhere; if it has four distinct x-intercepts, then there exist values $x_1 < x_2 < x_3 < x_4$ such that $f(x_1) = f(x_2) = f(x_3) = f(x_4) = 0$. Since the values are distinct, there are three distinct intervals (x_1, x_2), (x_2, x_3), and (x_3, x_4) for which f satisfies the hypotheses of Rolle's Theorem, and so there exist c_1 in (x_1, x_2), c_2 in (x_2, x_3), and c_3 in (x_3, x_4) such that $f'(c_1) = f'(c_2) = f'(c_3) = 0$. Because $x_1 < x_2 < x_3 < x_4$, then $c_1 < c_2 < c_3$, and as such there are at least three points at which a tangent line to the graph of f is horizontal.

59. $f'(x) = x\cos x + \sin x$. The hypotheses of Rolle's Theorem apply on $[0, \pi]$, so for some x in $(0, \pi)$, $x\cos x + \sin x = 0$ or $\cot x = -1/x$.

61. We have $f'(x) = -2\sin 2x$. Setting $f'(c) = -2\sin 2c = \dfrac{f(\pi/4) - f(0)}{\pi/4 - 0} = -4/\pi$, we obtain $\sin 2c = 2/\pi$. Then $2c \approx 0.6901$ and $c \approx 0.3451$.

4.5 Limits Revisited—L'Hôpital's Rule

In this exercise set, the symbol "$\overset{h}{=}$" is used to denote the fact that L'Hôpital's Rule was applied to obtain the equality.

1. $\lim\limits_{x \to 0} \dfrac{\cos x - 1}{x} \overset{h}{=} \dfrac{-\sin x}{1} = 0$

3. $\lim\limits_{x \to 1} \dfrac{2x - 2}{\ln x} \overset{h}{=} \lim\limits_{x \to 1} \dfrac{2}{1/x} = 2$

5. $\lim\limits_{x \to 0} \dfrac{e^{2x} - 1}{3x + x^2} \overset{h}{=} \lim\limits_{x \to 0} \dfrac{2e^{2x}}{3 + 2x} = \dfrac{2}{3}$

7. $\lim\limits_{t \to \pi} \dfrac{5\sin^2 t}{1 + \cos t} \overset{h}{=} \lim\limits_{t \to \pi} \dfrac{10\sin t \cos t}{-\sin t} = \lim\limits_{t \to \pi} \dfrac{10\cos t}{-1} = 10$

9. $\lim\limits_{x \to 0} \dfrac{6 + 6x + 6x^2 - 6e^x}{x - \sin x} \overset{h}{=} \lim\limits_{x \to 0} \dfrac{6 + 6x - 6e^x}{1 - \cos x} \overset{h}{=} \lim\limits_{x \to 0} \dfrac{6 - 6e^x}{\sin x} \overset{h}{=} \lim\limits_{x \to 0} \dfrac{-6e^x}{\cos x} = -6$

11. It is not necessary to use L'Hôpital's Rule here.
$$\lim\limits_{x \to 0^+} \dfrac{\cot 2x}{\cot x} = \lim\limits_{x \to 0^+} \dfrac{\cos 2x \sin x}{\sin 2x \cos x} = \lim\limits_{x \to 0^+} \dfrac{(\cos^2 x - \sin^2 x)\sin x}{2\sin x \cos x \cos x} = \lim\limits_{x \to 0^+} \dfrac{\cos^2 x - \sin^2 x}{2\cos^2 x} = \dfrac{1}{2}$$

13. $\lim\limits_{t \to 2} \dfrac{t^2 + 3t - 10}{t^3 - 2t^2 + t - 2} \overset{h}{=} \lim\limits_{t \to 2} \dfrac{2t + 3}{3t^2 - 4t + 1} = \dfrac{7}{5}$

15. $\lim\limits_{x \to 0} \dfrac{x - \sin x}{x^3} \overset{h}{=} \lim\limits_{x \to 0} \dfrac{1 - \cos x}{3x^2} \overset{h}{=} \lim\limits_{x \to 0} \dfrac{\sin x}{6x} \overset{h}{=} \lim\limits_{x \to 0} \dfrac{\cos x}{6} = \dfrac{1}{6}$

17. $\lim\limits_{x \to 0} \dfrac{\cos 2x}{x^2}$ has the form $1/0$ and does not exist.

19. $\displaystyle\lim_{x\to 1^+} \frac{\ln\sqrt{x}}{x-1} = \lim_{x\to 1^+} \frac{\frac{1}{2}\ln x}{x-1} \overset{h}{=} \lim_{x\to 1^+} \frac{1/2x}{1} = \frac{1}{2}$

21. $\displaystyle\lim_{x\to 2} \frac{e^{x^2}-e^{2x}}{x-2} \overset{h}{=} \lim_{x\to 2} \frac{2xe^{x^2}-2e^{2x}}{1} = 2e^4$

23. $\displaystyle\lim_{x\to\infty} \frac{x\ln x}{x^2+1} \overset{h}{=} \frac{1+\ln x}{2x} \overset{h}{=} \lim_{x\to\infty} \frac{1/x}{2} = 0$

25. $\displaystyle\lim_{x\to 0} \frac{1-\tan^{-1}x}{x^3} \overset{h}{=} \lim_{x\to 0} \frac{1-\dfrac{1}{1+x^2}}{3x^2} = \lim_{x\to 0} \frac{x^2}{3(x^2+x^4)} = \lim_{x\to 0} \frac{1}{3(1+x^2)} = \frac{1}{3}$

27. $\displaystyle\lim_{x\to\infty} \frac{e^x}{x^4} \overset{h}{=} \lim_{x\to\infty} \frac{e^x}{4x^3} \overset{h}{=} \lim_{x\to\infty} \frac{e^x}{12x^2} \overset{h}{=} \lim_{x\to\infty} \frac{e^x}{24x} \overset{h}{=} \lim_{x\to\infty} \frac{e^x}{24}$

The limit has the form $\infty/24$ and does not exist.

29. $\displaystyle\lim_{x\to 0} \frac{x-\tan^{-1}x}{x-\sin^{-1}x} \overset{h}{=} \lim_{x\to 0} \frac{1-\dfrac{1}{1+x^2}}{1-\dfrac{1}{\sqrt{1-x^2}}} = \lim_{x\to 0} \frac{1-(1+x^2)^{-1}}{1-(1-x^2)^{-1/2}}$

$\overset{h}{=} \displaystyle\lim_{x\to 0} \frac{2x(1+x^2)^{-2}}{-x(1-x^2)^{-3/2}} = \lim_{x\to 0} \frac{2(1-x^2)^{3/2}}{-(1+x^2)^2} = -2$

31. $\displaystyle\lim_{u\to\pi/2} \frac{\ln(\sin u)}{(2u-\pi)^2} \overset{h}{=} \lim_{u\to\pi/2} \frac{\cot u}{4(2u-\pi)} \overset{h}{=} \lim_{u\to\pi/2} \frac{-\csc^2 u}{8} = -\frac{1}{8}$

33. $\displaystyle\lim_{x\to-\infty} \frac{1+e^{-2x}}{1-e^{-2x}} \overset{h}{=} \lim_{x\to-\infty} \frac{-2e^{-2x}}{2e^{-2x}} = -1$

35. $\displaystyle\lim_{r\to 0} \frac{r-\cos r}{r-\sin r}$ has the form $-1/0$ and does not exist.

37. $\displaystyle\lim_{x\to 0^+} \frac{x^2}{\ln^2(1+3x)} \overset{h}{=} \lim_{x\to 0^+} \frac{2x}{\dfrac{6\ln(1+3x)}{1+3x}} = \lim_{x\to 0^+} \frac{x(1+3x)}{3\ln(1+3x)} \overset{h}{=} \lim_{x\to 0^+} \frac{1+6x}{\dfrac{9}{1+3x}}$

$= \displaystyle\lim_{x\to 0^+} \frac{(1+6x)(1+3x)}{9} = \frac{1}{9}$

39. $\displaystyle\lim_{x\to 0} \frac{3x^2+e^x-e^{-x}-2\sin x}{x\sin x} \overset{h}{=} \lim_{x\to 0} \frac{6x+e^x+e^{-x}-2\cos x}{x\cos x+\sin x} \overset{h}{=} \lim_{x\to 0} \frac{6+e^x-e^{-x}+2\sin x}{-x\sin x+2\cos x} = 3$

41. Indeterminate form: $\infty-\infty$

$$\lim_{x\to 0}\left(\frac{1}{e^x-1}-\frac{1}{x}\right) = \lim_{x\to 0} \frac{x-e^x+1}{xe^x-x} \overset{h}{=} \lim_{x\to 0} \frac{1-e^x}{xe^x+e^x-1} \overset{h}{=} \lim_{x\to 0} \frac{-e^x}{xe^x+2e^x} = -\frac{1}{2}$$

43. Indeterminate form: $\infty-\infty$

$$\lim_{x\to\infty} x(e^{1/x}-1) = \lim_{x\to\infty} \frac{e^{1/x}-1}{1/x} \overset{h}{=} \lim_{x\to\infty} \frac{-e^{1/x}/x^2}{-1/x^2} = \lim_{x\to\infty} e^{1/x} = 1$$

45. Indeterminate form: 0^0. Set $y=x^x$. Then $\ln y = x\ln x$ and $\displaystyle\lim_{x\to 0^+} x\ln x = \lim_{x\to 0^+} \frac{\ln x}{1/x} \overset{h}{=} \lim_{x\to 0^+} \frac{1/x}{-1/x^2} = \lim_{x\to 0^+}(-x) = 0$. Thus,

$\displaystyle\lim_{x\to 0^+} x^x = e^0 = 1$.

47. Indeterminate form: $\infty-\infty$

$$\lim_{x\to 0}\left(\frac{1}{x}-\frac{1}{\sin x}\right) = \lim_{x\to 0} \frac{\sin x-x}{x\sin x} \overset{h}{=} \lim_{x\to 0} \frac{\cos x-1}{x\cos x+\sin x} \overset{h}{=} \lim_{x\to 0} \frac{-\sin x}{-x\sin x+2\cos x} = 0$$

49. Indeterminate form: $\infty - \infty$

$$\lim_{t \to 3} \left(\frac{\sqrt{t+1}}{t^2-9} - \frac{2}{t^2-9} \right) = \lim_{t \to 3} \frac{\sqrt{t+1}-2}{t^2-9} \overset{h}{=} \lim_{t \to 3} \frac{1/2\sqrt{t+1}}{2t} = \lim_{t \to 3} \frac{1}{4t\sqrt{t+1}} = \frac{1}{24}$$

51. Indeterminate form: $0 \cdot \infty$

$$\lim_{\theta \to 0} \theta \csc 4\theta = \lim_{\theta \to 0} \frac{\theta}{\sin 4\theta} \overset{h}{=} \lim_{\theta \to 0} \frac{1}{4\cos 4\theta} = \frac{1}{4}$$

53. Indeterminate form: ∞^0. Set $y = (2+e^x)^{e^{-x}}$. Then $\ln y = e^{-x} \ln(2+e^x)$ and

$$\lim_{x \to \infty} e^{-x} \ln(2+e^x) = \lim_{x \to \infty} \frac{\ln(2+e^x)}{e^x} \overset{h}{=} \lim_{x \to \infty} \frac{e^x/(2+e^x)}{e^x} = \lim_{x \to \infty} \frac{1}{2+e^x} = 0.$$

Thus, $\lim_{x \to \infty} (2+e^x)^{e^{-x}} = e^0 = 1.$

55. Indeterminate form: 1^∞. Set $y = (1+3/t)^t$. Then $\ln y = t \ln(1+3/t)$ and

$$\lim_{t \to \infty} t \ln(1+3/t) = \lim_{t \to \infty} \frac{\ln(1+3/t)}{1/t} \overset{h}{=} \lim_{t \to \infty} \frac{\dfrac{-3/t^2}{1+3/t}}{-1/t^2} = \lim_{t \to \infty} \frac{3}{1+3/t} = 3.$$

Thus, $\lim_{t \to \infty} (1+3/t)^t = e^3.$

57. Indeterminate form: 0^0. Set $y = x^{(1-\cos x)}$. Then $\ln y = (1-\cos x) \ln x$ and

$$\lim_{x \to 0} (1-\cos x) \ln x = \lim_{x \to 0} \frac{\ln x}{1/(1-\cos x)} \overset{h}{=} \lim_{x \to 0} \frac{1/x}{-\sin x/(1-\cos x)^2} = \lim_{x \to 0} \frac{-(1-\cos x)^2}{x\sin x}$$

$$\overset{h}{=} \lim_{x \to 0} \frac{-2(1-\cos x)\sin x}{x\cos x + \sin x} = \lim_{x \to 0} \frac{-2\sin x + \sin 2x}{x\cos x + \sin x}$$

$$\overset{h}{=} \lim_{x \to 0} \frac{-2\cos x + 2\cos 2x}{-x\sin x + 2\cos x} = 0.$$

Thus, $\lim_{x \to 0} x^{(1-\cos x)} = e^0 = 1.$

59. Indeterminate form: $0 \cdot \infty$

$$\lim_{x \to \infty} \frac{1}{x^2 \sin^2(2/x)} = \lim_{x \to \infty} \frac{1/x^2}{\sin^2(2/x)} \overset{h}{=} \lim_{x \to \infty} \frac{-2/x^3}{[2\sin(2/x)\cos(2/x)](-2/x^2)}$$

$$= \lim_{x \to \infty} \frac{1}{2x\sin(2/x)\cos(2/x)} = \lim_{x \to \infty} \frac{1/x}{\sin(4/x)} \overset{h}{=} \lim_{x \to \infty} \frac{-1/x^2}{\cos(4/x)(-4/x^2)}$$

$$= \lim_{x \to \infty} \frac{1}{4\cos(4/x)} = \frac{1}{4}$$

61. Indeterminate form: $\infty - \infty$

$$\lim_{x \to 1} \left(\frac{1}{x-1} - \frac{5}{x^2+3x-4} \right) = \lim_{x \to 1} \left[\frac{1}{x-1} - \frac{5}{(x-1)(x+4)} \right] = \lim_{x \to 1} \frac{x+4-5}{(x-1)(x+4)}$$

$$= \lim_{x \to 1} \frac{x-1}{(x-1)(x+4)} = \lim_{x \to 1} \frac{1}{x+4} = \frac{1}{5}$$

63. Indeterminate form: $0 \cdot \infty$

$$\lim_{x \to \infty} x^5 e^{-x} = \lim_{x \to \infty} \frac{x^5}{e^x} \overset{h}{=} \lim_{x \to \infty} \frac{5x^4}{e^x} \overset{h}{=} \cdots \overset{h}{=} \lim_{x \to \infty} \frac{5!}{e^x} = 0$$

65. Indeterminate form: $0 \cdot \infty$

$$\lim_{x \to \infty} x \left(\frac{\pi}{2} - \arctan x \right) = \lim_{x \to \infty} \frac{\pi/2 - \tan^{-1} x}{1/x} \overset{h}{=} \lim_{x \to \infty} \frac{-1/(1+x^2)}{-1/x^2} = \lim_{x \to \infty} \frac{x^2}{1+x^2} \overset{h}{=} \lim_{x \to \infty} \frac{2x}{2x} = 1$$

67. Indeterminate form: $0 \cdot \infty$

$$\lim_{x \to \infty} x \tan \left(\frac{5}{x} \right) = \lim_{x \to \infty} \frac{\tan(5/x)}{1/x} \overset{h}{=} \lim_{x \to \infty} \frac{\sec^2(5/x)(-5/x^2)}{-1/x^2} = 5 \lim_{x \to \infty} \sec^2(5/x) = 5$$

69. Indeterminate form: $\infty - \infty$. $\lim_{x \to -\infty} \left(\frac{1}{e^x} - x^2 \right) = \lim_{x \to -\infty} \frac{1 - x^2 e^x}{e^x}$. Now

$$\lim_{x \to -\infty} x^2 e^x = \lim_{x \to -\infty} \frac{x^2}{e^{-x}} \overset{h}{=} \lim_{x \to -\infty} \frac{2x}{-e^{-x}} \overset{h}{=} \lim_{x \to -\infty} \frac{2}{e^{-x}} = 0,$$

so $\lim_{x \to -\infty} \frac{1 - x^2 e^x}{e^x}$ has the form $1/0$ and $\lim_{x \to -\infty} \left(\frac{1}{e^x} - x^2 \right)$ does not exist.

71. Indeterminate form: 1^∞. Set $y = \left(\frac{3x}{3x+1} \right)^x$. Then $\ln y = x \ln \left(\frac{3x}{3x+1} \right)$ and

$$\lim_{x \to \infty} x \ln \left(\frac{3x}{3x+1} \right) = \lim_{x \to \infty} \frac{\ln \left(\frac{3x}{3x+1} \right)}{1/x} \overset{h}{=} \lim_{x \to \infty} \frac{\left(\frac{3x+1}{3x} \right) \left[\frac{3(3x+1) - 3(3x)}{(3x+1)^2} \right]}{-1/x^2}$$

$$= \lim_{x \to \infty} \left(\frac{-x}{3x+1} \right) \overset{h}{=} \lim_{x \to \infty} \frac{-1}{3} = -\frac{1}{3}.$$

Thus, $\lim_{x \to \infty} \left(\frac{3x}{3x+1} \right)^x = e^{-1/3}$.

73. Indeterminate form: 0^0. Set $y = (\sinh x)^{\tan x}$. Then $\ln y = \tan x \ln(\sinh x)$ and

$$\lim_{x \to 0} \tan x \ln(\sinh x) = \lim_{x \to 0} \frac{\ln(\sinh x)}{\cot x} \overset{h}{=} \lim_{x \to 0} \frac{\cosh x / \sinh x}{-\csc^2 x} = \lim_{x \to 0} \frac{-\sin^2 x}{\tanh x}$$

$$\overset{h}{=} \lim_{x \to 0} \frac{-2 \sin x \cos x}{\operatorname{sech}^2 x} = 0.$$

Thus, $\lim_{x \to 0} (\sinh x)^{\tan x} = e^0 = 1$.

75. Since $\lim_{x \to 0^+} \frac{e^x - 1}{x} \overset{h}{=} \lim_{x \to 0^+} \frac{e^x}{1} = 1$, $\lim_{x \to 0^+} \ln \frac{e^x - 1}{x} = 0$, and $\frac{1}{x} \ln \left(\frac{e^x - 1}{x} \right)$ has the form $0/0$ as $x \to 0$. Now

$$\lim_{x \to 0^+} \frac{\ln[(e^x - 1)/x]}{x} = \lim_{x \to 0^+} \frac{\ln(e^x - 1) - \ln x}{x} \overset{h}{=} \lim_{x \to 0^+} \frac{\frac{e^x}{e^x - 1} - \frac{1}{x}}{1} = \lim_{x \to 0^+} \frac{xe^x - e^x + 1}{xe^x - x}$$

$$\overset{h}{=} \lim_{x \to 0^+} \frac{xe^x}{xe^x + e^x - 1} \overset{h}{=} \lim_{x \to 0^+} \frac{xe^x + e^x}{xe^x + 2e^x} = \frac{1}{2}.$$

77.

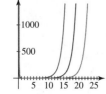

In all three cases, it appears that $\lim_{x \to \infty} f(x) = \infty$.

79. Since n is a positive integer, then by repeated applications of L'Hôpital's Rule,

$$\lim_{x \to \infty} \frac{x^n}{e^x} \overset{h}{=} \lim_{x \to \infty} \frac{nx^{n-1}}{e^x} \overset{h}{=} \lim_{x \to \infty} \frac{n(n-1)x^{n-2}}{e^x} \overset{h}{=} \cdots \overset{h}{=} \lim_{x \to \infty} \frac{n!}{e^x} = 0.$$

81. (a) Letting r be the radius of the circle, we see that the area of the circular sector is $A_C = \frac{1}{2}r^2(2\theta) = r^2\theta$ and the area of the isosceles triangle is $A_T = 2\left(\frac{1}{2}r\sin\theta \cdot r\cos\theta\right) = \frac{1}{2}r^2\sin 2\theta$. Then the area of the shaded region is $A = A_C - A_T = r^2\left(\theta - \frac{1}{2}\sin 2\theta\right)$. Now the length of the arc is 5, so $r\theta = 5$ and $r = 5/\theta$. Thus, $A = 25\left(\theta - \frac{1}{2}\sin 2\theta\right)/\theta^2$.

(b) $\displaystyle\lim_{\theta \to 0} A \overset{h}{=} \lim_{\theta \to 0} \frac{25(1 - \cos 2\theta)}{2\theta} \overset{h}{=} \lim_{\theta \to 0} \frac{50\sin 2\theta}{2} = 0$

(c) $\displaystyle\frac{dA}{d\theta} = 25\left[\frac{\theta^2(1 - \cos 2\theta) - 2\theta\left(\theta - \frac{1}{2}\sin 2\theta\right)}{\theta^4}\right] = 25\left(\frac{-\theta - \theta\cos 2\theta + \sin 2\theta}{\theta^3}\right)$

$$\lim_{\theta \to 0} \frac{dA}{dh} \overset{h}{=} 25\lim_{\theta \to 0}\left[\frac{-1 - (-2\theta\sin 2\theta + \cos 2\theta) + 2\cos 2\theta}{3\theta^2}\right]$$

$$= 25\lim_{\theta \to 0}\left(\frac{-1 + 2\theta\sin 2\theta + \cos 2\theta}{3\theta^2}\right) \overset{h}{=} 25\lim_{\theta \to 0}\left(\frac{4\theta\cos 2\theta + 2\sin 2\theta - 2\sin 2\theta}{6\theta}\right)$$

$$= 25\lim_{\theta \to 0}\left(\frac{2\cos 2\theta}{3}\right) = \frac{50}{3}$$

83. (a) From $p_1 v_1^{\gamma} = k = p_2 v_2^{\gamma}$ we have $p_2 = p_1(v_1/v_2)^{\gamma}$. Then

$$W = \frac{p_2 v_2 - p_1 v_1}{1 - \gamma} = \frac{p_1(v_1/v_2)^{\gamma}v_2 - p_1 v_1}{1 - \gamma} = p_1 v_1\left[\frac{(v_1/v_2)^{\gamma-1} - 1}{1 - \gamma}\right]$$

$$= p_1 v_1\left[\frac{(v_2/v_1)^{1-\gamma} - 1}{1 - \gamma}\right].$$

(b) $\displaystyle\lim_{\gamma \to 1}\left\{p_1 v_1\left[\frac{(v_2/v_1)^{1-\gamma} - 1}{1 - \gamma}\right]\right\} \overset{h}{=} p_1 v_1 \lim_{\gamma \to 1}\frac{(v_2/v_1)^{1-\gamma}\ln(v_2/v_1)(-1)}{-1} = p_1 v_1 \ln\left(\frac{v_2}{v_1}\right).$

85. $\displaystyle\lim_{h \to 0}\frac{f(x+h) - 2f(x) + f(x-h)}{h^2} \overset{h}{=} \lim_{h \to 0}\frac{f'(x+h) - f'(x-h)}{2h}$

$$\overset{h}{=} \lim_{h \to 0}\frac{f''(x+h) + f''(x-h)}{2} = f''(x)$$

4.6 Graphing and the First Derivative

1. The x-intercepts are $1 \pm \sqrt{2}$. The y-intercept is 1. Solving $f'(x) = -2x + 2 = 0$ we obtain the critical number 1. The relative maximum is $f(1) = 2$.

x		1	
f	↗	2	↘
f'	+	0	−

3. The x-intercepts are 0 and $\pm\sqrt{3}$. The y-intercept is 0. Solving $f'(x) = 3x^2 - 3 = 0$ we obtain the critical numbers -1 and 1. The relative maximum is $f(-1) = 2$ and the relative minimum is $f(1) = -2$.

x		-1		1	
f	↗	2	↘	-2	↗
f'	+	0	−	0	+

5. The x-intercepts are 0 and 2. The y-intercept is 0. Writing $f(x) = x^3 - 4x^2 + 4x$ and solving $f'(x) = 3x^2 - 8x + 4 = (3x-2)(x-2) = 0$ we obtain the critical numbers $2/3$ and 2. The relative maximum is $f(2/3) = 32/27$ and the relative minimum is $f(2) = 0$.

x		$2/3$		2	
f	↗	$32/27$	↘	0	↗
f'	$+$	0	$-$	0	$+$

7. There is no easily determined x-intercept. The y-intercept is -3. Since $f'(x) = 3x^2 + 1 > 0$ for all x, there are no critical numbers and no relative extrema.

9. The x-intercepts are 0 and $-4^{1/3}$. The y-intercept is 0. Solving $f'(x) = 4x^3 + 4 = 0$ we obtain the critical number -1. The relative minimum is $f(-1) = -3$.

x		-1	
f	↘	-3	↗
f'	$-$	0	$+$

11. The x- and y-intercepts are 0. Solving $f'(x) = x^3 + 4x^2 + 4x = x(x+2)^2 = 0$ we obtain the critical numbers -2 and 0. The relative minimum is $f(0) = 0$.

x		-2		0	
f	↘	$4/3$	↘	0	↗
f'	$-$	0	$-$	0	$+$

13. The x-intercepts are 0 and 3. The y-intercept is 0. Solving $f'(x) = -2x^2(x-3) + (x-3)(-2x) = -2x(x-3)(2x-3) = 0$ we obtain the critical numbers 0, $3/2$, and 3. The relative maxima are $f(0) = f(3) = 0$ and the relative minimum is $f(3/2) = -81/16$.

x		0		$3/2$		3	
f	↗	0	↘	$-81/16$	↗	0	↘
f'	$+$	0	$-$	0	$+$	0	$-$

15. The x-intercepts are 0 and $5/4$. The y-intercept is 0. Solving $f'(x) = 20x^4 - 20x^3 = 20x^3(x-1) = 0$ we obtain the critical numbers 0 and 1. The relative maximum is $f(0) = 0$ and the relative minimum is $f(1) = -1$.

x		0		1	
f	↗	0	↘	-1	↗
f'	$+$	0	$-$	0	$+$

17. The y-intercept is 3. The function is undefined for $x = -1$. Solving $f'(x) = (x^2 + 2x - 3)/(x+1)^2 = (x+3)(x-1)/(x+1)^2 = 0$ we obtain the critical numbers -3 and 1. The relative maximum is $f(-3) = -6$ and the relative minimum is $f(1) = 2$.

x		-3		-1		1	
f	↗	-6	↘	undefined	↘	2	↗
f'	$+$	0	$-$	undefined	$-$	0	$+$

19. We write $f(x) = (x^2 - 1)/x^3$. The x-intercepts are ± 1. The function is undefined for $x = 0$. Solving $f'(x) = -1/x^2 + 3/x^4 = (3 - x^2)/x^4 = 0$ we obtain the critical numbers $-\sqrt{3}$ and $\sqrt{3}$. The relative maximum is $f(\sqrt{3}) = 2/3\sqrt{3}$ and the relative minimum is $f(-\sqrt{3}) = -2/3\sqrt{3}$.

x		$-\sqrt{3}$		0		$\sqrt{3}$	
f	↘	$-2/3\sqrt{3}$	↗	undefined	↗	$2/3\sqrt{3}$	↘
f'	−	0	+	undefined	+	0	−

21. The y-intercept is 10. Solving $f'(x) = -20x/(x^2+1) = 0$ we obtain the critical number 0. The relative maximum is $f(0) = 10$.

x		0	
f	↗	10	↘
f'	+	0	−

23. The x-intercepts are 2 and -2. The y-intercept is $2\sqrt[3]{2}$. Solving $f'(x) = (4/3)x(x^2-4)^{-1/3} = 0$ we obtain the critical number 0. Also, $f'(x)$ is undefined for $x = \pm 2$, which are critical numbers. The relative maximum is $f(0) \approx 2.5$ and the relative minima are $f(\pm 2) = 0$.

x		-2		0		2	
f	↘	0	↗	2.5	↘	0	↗
f'	−	undefined	+	0	−	undefined	+

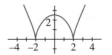

25. The x-intercepts are -1, 0, and 1. The y-intercept is 0. The function is undefined for $x < -1$ and $x > 1$. Solving $f'(x) = (1 - 2x^2)/\sqrt{1-x^2} = 0$ we obtain the critical numbers $-1/\sqrt{2}$ and $1/\sqrt{2}$. The relative maximum is $f(1/\sqrt{2}) = 1/2$ and the relative minimum is $f(-1/\sqrt{2}) = -1/2$.

x	-1		$-1/\sqrt{2}$		$1/\sqrt{2}$		1
f	0	↘	$-1/2$	↗	$1/2$	↘	0
f'	undefined	−	0	+	0	−	undefined

27. The x-intercepts are $-24\sqrt{3}$, 0, and $24\sqrt{3}$. The y-intercept is 0. Solving $f'(x) = 1 - 4x^{-2/3} = (x^{2/3} - 4)/x^{2/3} = 0$ we obtain the critical numbers -8 and 8. Also, $f'(x)$ is undefined for $x = 0$, which is a critical number. The relative maximum is $f(-8) = 16$ and the relative minimum is $f(8) = -16$.

x		-8		0		8	
f	↗	16	↘	0	↘	-16	↗
f'	+	0	−	undefined	−	0	+

29. We have $f(x) = \begin{cases} x^3 - 24\ln x, & x > 0 \\ x^3 - 24\ln(-x), & x < 0 \end{cases}$. There are no easily-determined x-intercepts. The function is undefined for $x = 0$. Solving $f'(x) = 3x^2 - 24/x = 0$ we obtain the critical number 2. Also, $f'(x)$ is undefined for $x = 0$, which is a critical number. The relative minimum is $f(2) = 8 - 24\ln 2 \approx -8.6355$.

x		0		2	
f	↗	undefined	↘	-8.6355	↗
f'	+	undefined	−	0	+

31. The x-intercept is 3. The y-intercept is 9. Solving

$$f'(x) = -(x+3)^2 e^{-x} + 2e^{-x}(x+3)$$
$$= e^{-x}(-x^2 - 4x - 3)$$
$$= -e^{-x}(x+1)(x+3) = 0$$

x		-3		-1	
f	$\searrow$	0	$\nearrow$	$4e$	$\searrow$
f'	$-$	0	$+$	0	$-$

we obtain the critical numbers -3 and -1. The relative maximum is $f(-1) = 4e$ and the relative minimum is $f(-3) = 0$.

33. 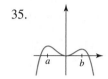 35. 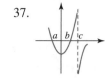 37. 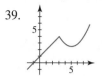 39. 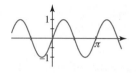 41.

43. The slopes of the tangent lines are given by $f'(x) = 3x^2 + 12x - 1$. Solving $f''(x) = 6x + 12 = 0$ we obtain the point -2. The relative minimum is $f'(-2) = -13$.

x		-2	
f'	$\searrow$	-13	$\nearrow$
f''	$-$	0	$+$

45. (a) $g(x) > 0$ for x in $(k\pi, k\pi + \pi/2)$, where k is an integer. $g(x) < 0$ for x in $(k\pi - \pi/2, k\pi)$, where k is an integer.

(b) Solving $f'(x) = 2\sin x \cos x = \sin 2x = g(x) = 0$ we obtain the critical numbers $k\pi/2$, where k is an integer. The relative maxima of $f(x)$ occur at $k\pi + \pi/2$, where k is an integer. The relative minima of $f(x)$ occur at $k\pi$, where k is an integer.

(c)

47. (a) Setting $f'(x) = 2(x - a_1) + 2(x - a_2) + \cdots + 2(x - a_n) = 0$ we obtain $nx = a_1 + a_2 + \cdots + a_n$ or $x = (a_1 + a_2 + \cdots + a_n)///n = \bar{x}$. Thus, $\bar{x}$ is a critical number of $f(x)$.

(b) To see that $f(\bar{x})$ is a relative minimum write $f(x) = nx^2 - 2(a_1 + a_2 + \cdots + a_n)x + (a_1^2 + a_2^2 + \cdots + a_n^2)$. Then $f'(x) = 2nx - 2(a_1 + a_2 + \cdots + a_n) = 2n(x - \bar{x})$. When $x < \bar{x}$, $f'(x) < 0$ and when $x > \bar{x}$, $f'(x) > 0$. Thus, $f(\bar{x})$ is a relative minimum.

49. Solving $f'(x) = 2ax + b = 0$ we obtain the critical number $-b/2a$. We want $-b/2a = 2$ or $b = -4a$. We are given $f(2) = 4a + 2b + c = 6$ and $f(0) = c = 4$. Solving these three equations, we obtain $a = -1/2$, $b = 2$, and $c = 4$. Thus $f(x) = -\dfrac{1}{2}x^2 + 2x + 4$.

51. If $f'(0) > 0$, then $f'(x) > 0$ on some interval $(-a, a)$. But then $f(x)$ is increasing on $(-a, a)$ and cannot be symmetric about the y-axis. Similarly, if $f'(0) < 0$, then $f(x)$ is decreasing on some open interval containing 0 and cannot be symmetric about the y-axis. Since $f'(0)$ exists, we must have $f'(0) = 0$. Since $f(x)$ is neither increasing nor decreasing in an open interval around 0, it must have a relative extremum at 0.

53. (a) Since f and g are differentiable and have a critical number at c, $f'(c) = g'(c) = 0$. Then $(f + g)'(c) = f'(c) + g'(c) = 0$, $(f - g)'(c) = f'(c) - g'(c) = 0$, and $(fg)'(c) = f(c)g'(c) + f'(c)g(c) = 0$. Thus, $f + g$, $f - g$, and fg have critical numbers c.

(b) Suppose $f'(x) > 0$ and $g'(x) > 0$ for $a < x < c$, and $f'(x) < 0$ and $g'(x) < 0$ for $c < x < b$. Then $(f + g)'(x) = f'(x) + g'(x) > 0$ for $a < x < c$ and $(f + g)'(x) < 0$ for $c < x < b$. Thus, $f + g$ has a relative maximum at c.

To see that neither $f - g$ nor fg necessarily have relative maxima at c let $f(x) = -x^4$ and $g(x) = -x^2$. Both have relative maxima at $c = 0$. However $(f - g)(x) = -x^4 + x^2$ and $(fg)(x) = x^6$ both have relative minima at $c = 0$.

4.7 Graphing and the Second Derivative

1. $f'(x) = -2x + 7$; $f''(x) = -2$. Since $f''(x) < 0$ for all x, the graph is concave downward on $(-\infty, \infty)$.

3. $f'(x) = -3x^2 + 12x + 1$; $f''(x) = -6x + 12$. Solving $f''(x) = 0$ we obtain $x = 2$. The graph is concave upward on $(-\infty, 2)$ and concave downward on $(2, \infty)$.

x		2	
f''	+	0	−

5. $f'(x) = (x-4)^2(4x-4)$; $f''(x) = 12(x-4)(x-2)$. Solving $f''(x) = 0$ we obtain $x = 2$ and 4. The graph is concave upward on $(-\infty, 2)$, $(4, \infty)$, and concave downward on $(2, 4)$.

x		2		4	
f''	+	0	−	0	+

7. $f'(x) = \frac{1}{3}x^{-2/3} + 2$; $f''(x) = -\frac{2}{9}x^{-5/3}$. $f''(x)$ is undefined for $x = 0$. The graph is concave upward on $(-\infty, 0)$ and concave downward on $(0, \infty)$.

x		0	
f''	+	undefined	−

9. $f'(x) = 1 - 9/x^2$; $f''(x) = 18/x^3$. $f''(x)$ is undefined for $x = 0$. The graph is concave upward on $(0, \infty)$ and concave downward on $(-\infty, 0)$.

x		0	
f''	−	undefined	+

11. $f'(x) = \frac{-2x}{(x^2+3)^2}$; $f''(x) = \frac{(x^2+3)^2(-2) + 2x[4x(x^2+3)]}{(x^2+3)^4} = \frac{6(x^2-1)}{(x^2+3)^3}$.

Solving $f''(x) = 0$ we obtain $x = \pm 1$. The graph is concave upward on $(-\infty, -1)$ and $(1, \infty)$, and concave downward on $(-1, 1)$.

x		−1		1	
f''	+	0	−	0	+

13. f' is increasing on $(-2, 2)$.
    f' is decreasing on $(-\infty, -2)$ and $(2, \infty)$.

x		−2		2	
f''	−	0	+	0	−

15. f' is increasing on $(-\infty, -1)$ and $(4, \infty)$.
    f' is decreasing on $(-1, 4)$.

x		−1		4	
f''	+	0	−	0	+

17. $f'(x) = \sec x \tan x$;

    $f''(x) = (\sec x)(\sec^2 x) + (\tan x)(\sec x \tan x) = (\sec x)(\sec^2 x + \tan^2 x) = (1 + \sin^2 x)/\cos^3 x$. $f''(x)$ is positive when $\cos x > 0$ and negative when $\cos x < 0$. Thus, the graph of $f(x) = \sec x$ is concave upward when $\cos x > 0$ and concave downward when $\cos x < 0$.

19. $f'(x) = 4x^3 - 24x + 1$; $f''(x) = 12x^2 - 24$. Solving $f''(x) = 0$ we obtain $x = -\sqrt{2}$ and $\sqrt{2}$. The inflection points are $(-\sqrt{2}, -21 - \sqrt{2})$ and $(\sqrt{2}, -21 + \sqrt{2})$.

x		$-\sqrt{2}$		$\sqrt{2}$	
f''	+	0	−	0	+

21. $f'(x) = \cos x$; $f''(x) = -\sin x$. Solving $f''(x) = 0$ we obtain $x = k\pi$, where k is an integer. Since $f(x) = \sin x$ is 2π-periodic, the graph has inflection points at $(k\pi, 0)$, where k is an integer.

x	0		π		2π
f''	0	−	0	+	0

23. $f'(x) = 1 - \cos x$; $f''(x) = \sin x$. Solving $f''(x) = 0$ we obtain $x = k\pi$, where k is an integer. Since the sign of $f''(x) = \sin x$ changes around each $k\pi$, the graph of $f(x) = x - \sin x$ has inflection points at $(k\pi, k\pi)$, where k is an integer.

25. $f'(x) = 1 - xe^{-x} + e^{-x} = 1 + (1-x)e^{-x}$; $f''(x) = -(1-x)e^{-x} - e^{-x} = (x-2)e^{-x}$. Solving $f''(x) = 0$ we obtain $x = 2$. The inflection point is $(2, 2 + 2/e^2)$.

x		2	
f''	−	0	+

27. The x-intercept is $5/2$. The y-intercept is -25. $f'(x) = -4(2x - 5)$; $f''(x) = -8$. Solving $f'(x) = 0$ we obtain the critical number $5/2$. The relative maximum is $(5/2, 0)$.

x		5/2	
f		0	
f'		0	
f''	−	−	−

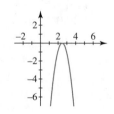

29. $f(x) = (x+1)^3$. The x-intercept is -1. The y-intercept is 1. $f'(x) = 3(x+1)^2$; $f''(x) = 6(x+1)$. Solving $f'(x) = 0$ we obtain the critical number -1. Solving $f''(x) = 0$ we obtain $x = -1$. Since $f''(-1) = 0$ the second derivative test does not apply. There are no relative extrema. The inflection point is $(-1,0)$.

x		-1	
f	↗	0	↗
f'	$+$	0	$+$
f''	$-$	0	$+$

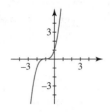

31. $f(x) = 2x^3(3x^2 - 5)$. The x-intercepts are $-\sqrt{5/3}$, 0, and $\sqrt{5/3}$. The y-intercept is 0. $f'(x) = 30x^4 - 30x^2 = 30x^2(x^2 - 1)$; $f''(x) = 120x^3 - 60x = 60x(2x^2 - 1)$. Solving $f'(x) = 0$ we obtain the critical numbers -1, 0, and 1. Solving $f''(x) = 0$ we obtain $x = -1/\sqrt{2}$, 0, and $1/\sqrt{2}$. The second derivative test does not apply at 0. The relative maximum is $(-1,4)$ and the relative minimum is $(1,-4)$. The inflection points are $(-1/\sqrt{2}, 7/\sqrt{8})$, $(0,0)$, and $(1/\sqrt{2}, -7/\sqrt{8})$.

x		-1		$-1/\sqrt{2}$		0		$1/\sqrt{2}$		1	
f		4		$7/\sqrt{8}$	↘	0	↘	$-7/\sqrt{8}$		-4	
f'		0			$-$	0	$-$			0	
f''	$-$	$-$	$-$	0	$+$	0	$-$	0	$+$	$+$	$+$

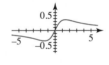

33. The x- and y-intercepts are 0. $f'(x) = \dfrac{2 - 2x^2}{(x^2 + 2)^2}$; $f''(x) = \dfrac{2x(x^2 - 6)}{(x^2 + 2)^3}$. Solving $f'(x) = 0$ we obtain the critical numbers $-\sqrt{2}$ and $\sqrt{2}$. Solving $f''(x) = 0$ we obtain $-\sqrt{6}$, 0, and $\sqrt{6}$. The relative maximum is $(\sqrt{2}, \sqrt{2}/4)$ and the relative minimum is $(-\sqrt{2}, -\sqrt{2}/4)$. The inflection points are $(-\sqrt{6}, -\sqrt{6}/8)$, $(0,0)$, and $(\sqrt{6}, \sqrt{6}/8)$.

x		$-\sqrt{6}$		$-\sqrt{2}$		0		$\sqrt{2}$		$\sqrt{6}$	
f		$-\sqrt{6}/8$		$-\sqrt{2}/4$		0		$\sqrt{2}/4$		$\sqrt{6}/8$	
f'				0				0			
f''	$-$	0	$+$	$+$	$+$	0	$-$	$-$	$-$	0	$+$

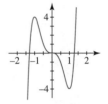

35. The domain of the function is $[-3,3]$. The x-intercepts are -3 and 3. The y-intercept is 3. $f'(x) = -x/\sqrt{9 - x^2}$; $f''(x) = -9/(9 - x^2)^{3/2}$. Solving $f'(x) = 0$ we obtain the critical number 0. $f''(x) = 0$ has no real solution. The relative maximum is $(0,3)$.

x		-3		0		3	
f		0		3		0	
f'		undefined		0		undefined	
f''		undefined	$-$	$-$	$-$	undefined	

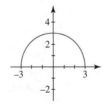

37. The x-intercepts are -1 and 0. The y-intercept is 0. $f'(x) = \dfrac{4x + 1}{3x^{2/3}}$; $f''(x) = \dfrac{4x - 2}{9x^{5/3}}$. Solving $f'(x) = 0$ we obtain the critical point $-1/4$. Also, $f'(x)$ is undefined for $x = 0$. Solving $f''(x) = 0$ we obtain $x = 1/2$. The relative minimum is $(-1/4, -3/4^{4/3})$. The inflection points are $(1/2, 3/2^{4/3})$ and $(0,0)$.

x		$-1/4$		0		$1/2$	
f		$-3/4^{4/3}$		0		$3/2^{4/3}$	
f'		0		undefined			
f''	$+$	$+$	$+$	undefined	$-$	0	$+$

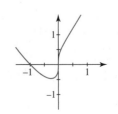

39. The x-intercepts are $(k\pi/6, 0)$ for $k = 1, 3, 5, 7, 9$, and 11. The y-intercept is 1. $f'(x) = -3\sin 3x$; $f''(x) = -9\cos 3x$. Solving $f'(x) = 0$ we obtain the critical numbers $k\pi/3$ for $k = 0$, 1, 2, 3, 4, 5, and 6. Computing $f''(x)$ at these values we see that $f(x)$ has relative maxima at $(2\pi/3, 1)$ and $(4\pi/3, 1)$, and relative minima at $(\pi/3, -1)$, $(\pi, -1)$, and $(5\pi/3, -1)$.

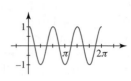

Solving $f''(x) = 0$ we obtain $(k\pi/6, 0)$ for $k = 1, 3, 5, 7, 9$, and 11. These are all points of inflection since the sign of $f''(x)$ changes around each one.

41. Solving $\sin x = -\cos x$ or $\tan x = -1$ we obtain the x-intercepts $3\pi/4$ and $7\pi/4$. The y-intercept is 1. $f'(x) = -\sin x + \cos x$; $f''(x) = -\cos x - \sin x$. Solving $f'(x) = 0$ we obtain the critical numbers $\pi/4$ and $5\pi/4$. Solving $f''(x) = 0$ we obtain $x = 3\pi/4$ and $7\pi/4$. The relative maximum is $(\pi/4, \sqrt{2})$ and the relative minimum is $(5\pi/4, -\sqrt{2})$. The inflection points are $(3\pi/4, 0)$ and $(7\pi/4, 0)$.

x	0		$\pi/4$		$3\pi/4$		$5\pi/4$		$7\pi/4$		2π
f	1		$\sqrt{2}$		0		$-\sqrt{2}$		0		1
f'			0				0				
f''		$-$	$-$	$-$	0	$+$	$+$	$+$	0	$-$	

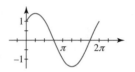

43. The domain of the function is $(0, \infty)$. Solving $2x = x\ln x$ or $\ln x = 2$ we obtain the x-intercept e^2. There is no y-intercept. $f'(x) = 1 - \ln x$; $f''(x) = -1/x$. Solving $f'(x) = 0$ we obtain the critical number e. $f''(x) = 0$ has no solution. The relative maximum is (e, e).

x	0		e		e^2
f	undefined		e		0
f'	undefined		0		
f''	undefined	$-$	$-$	$-$	$-$

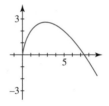

45. $f(x) = \dfrac{1}{2}\sin 2x$; $f'(x) = \cos 2x$; $f''(x) = -2\sin 2x$. Since $f''(\pi/4) = -2 < 0$, the function has a relative maximum at $(\pi/4, 1/2)$.

47. $f'(x) = 2\tan x \sec^2 x$; $f''(x) = 2\sec^4 x + 4\tan^2 x \sec^2 x$. Since $f''(\pi) = 2 > 0$, the function has a relative minimum at $(\pi, 0)$.

49.

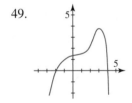

51.

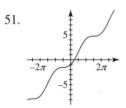

53. $f'(x) = 3ax^2 + 2bx + c$; $f''(x) = 6ax + 2b$. Since the graph has an inflection point at $(1, 1)$, $f''(1) = 6a + 2b = 0$. Using the fact that $(-1, 0)$ and $(1, 1)$ lie on the graph, we have $-a + b - c = 0$ and $a + b + c = 1$. Solving these three equations, we obtain $a = -1/6$, $b = 1/2$, and $c = 2/3$.

55. Since $f(x)$ is an odd function, the graph is symmetric with respect to the origin. $f'(x) = -\cos(1/x)/x^2$; $f''(x) = [2x\cos(1/x) - \sin(1/x)]/x^4$. Solving $f'(x) = 0$ we obtain the positive critical numbers $2/\pi, 2/3\pi, 2/5\pi, 2/7\pi, \ldots$. Since the sign of $f''(x)$ alternates at these points, they are alternately relative maxima and relative minima. For $x > 3\pi$, $f''(x) > 0$ and the graph is concave upwards.

57. $f'(x) = n(x - x_0)^{n-1}$; $f''(x) = n(n-1)(x - x_0)^{n-2}$.

   (a) If n is odd, then the sign of $f''(x)$ changes around x_0 and $(x_0, 0)$ is a point of inflection.

(b) If n is even, then $f''(x) > 0$ for $x \neq x_0$ and $(x_0, 0)$ is not a point of inflection. Using the first derivative test, we see that $f'(x) < 0$ for $x < x_0$ and $f'(x) > 0$ for $x > x_0$. Thus $(x_0, 0)$ is a relative minimum.

59. Since $f'''(c) \neq 0$, the graph of $f''(x)$ is either increasing (when $f'''(c) > 0$) or decreasing (when $f'''(c) < 0$) through $(c, 0)$. Thus, the sign of $f''(x)$ changes around $x = c$, and $(c, f(c))$ is a point of inflection.

61. The statement is false: Consider $f(x) = x^{1/3}$. Since $f'(x) = 1/3x^{2/3}$ is not defined at $x = 0$, f' cannot have a critical number at 0. (Recall that a critical number must be in the domain of the function.) From $f''(x) = -2/9x^{5/3}$ we see that $f''(x) > 0$ for $x < 0$, and $f''(x) < 0$ for $x > 0$. Thus $(0,0)$ is a point of inflection.

63. The graph of $f(x) = \begin{cases} 4x^2 - x, & x \leq 0 \\ -x^3, & x > 0 \end{cases}$ is shown on the right. The function has no tangent line

at $(0,0)$ since $f'_+(0) = 0$ and $f'_-(0) = -1$. Although concavity changes at $(0,0)$, according to Definition 4.7.2 there is no point of inflection because there is no tangent line at the point.

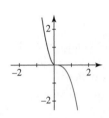

65. As stated in *Notes from the Classroom*, textbooks disagree on the precise definition of a point of inflection. With sufficient research, the student should find definitions for which some function f will have different points of inflection.

4.8 Optimization

1. Let x and $60 - x$ be the two numbers. We want to maximize $P(x) = x(60 - x) = 60x - x^2$ on $[0, 60]$. Solving $P'(x) = 60 - 2x = 0$ we obtain the critical number 30. Since $P''(x) = -2 < 0$, the product is maximized by the numbers 30, 30.

3. Let x be the number. We want to maximize $f(x) = x - x^2$. Solving $f'(x) = 1 - 2x = 0$ we obtain the critical number $1/2$. Since $f''(x) = -2 < 0$, the product is maximized by the number $1/2$.

5. Let x and $1 - x$ be the two numbers. We want to maximize $S(x) = x^2 + 2(1-x)^2 = 2 - 4x + 3x^2$. Solving $S'(x) = -4 + 6x = 0$ we obtain the critical number $2/3$. Since $S''(x) = 6 > 0$, the sum is minimized by $x = 2/3$ and $1 - x = 1/3$.

7. Let $(x, \sqrt{6x})$ be on the graph. We will minimize the square of the distance.

For $(5, 0)$, $D(x) = (x - 5)^2 + (\sqrt{6x} - 0)^2 = x^2 - 4x + 25$. Solving $D'(x) = 2x - 4 = 0$ we obtain the critical number 2. Since $D''(x) = 2 > 0$, the distance is minimized by the points $(2, \pm 2\sqrt{3})$ on the graph.

For $(3, 0)$, $D(x) = (x - 3)^2 + (\sqrt{6x} - 0)^2 = x^2 + 9$. Solving $D'(x) = 2x = 0$ we obtain the critical number 0. Since $D''(x) = 2 > 0$, the distance is minimized by the point $(0, 0)$ on the graph.

9. The slope of the tangent line at x is $s(x) = 3x^2 - 8x$. To minimize $s(x)$, we solve $s'(x) = 6x - 8 = 0$. This gives $x = 4/3$. Since $s''(x) = 6 > 0$, the slope is minimized at the point $(4/3, -128/27)$.

11. Let (x, y) be the corner of the rectangle lying on the line. Then $y = 2 - \frac{2}{3}x$, and we want to maximize $A(x) = xy = 2x - \frac{2}{3}x^2$ on $[0, 3]$. Solving $A'(x) = 2 - \frac{4}{3}x = 0$ we obtain the critical number $3/2$. Since $A''(x) = -4/3 < 0$, the area is maximized when the base of the rectangle is $3/2$ and the height is 1.

13. Triangle OXY is similar to triangle AXP, so

$$\frac{y}{x} = \frac{4}{x-2} \qquad \text{and} \qquad y = \frac{4x}{x-2}.$$

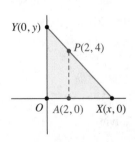

We want to minimize the area of the triangle $A(x) = \frac{1}{2}xy = \frac{1}{2}x\left(\frac{4x}{x-2}\right) = \frac{2x^2}{x-2}$ on $(2, \infty)$.

Solving $A'(x) = \dfrac{4x(x-2) - 2x^2}{(x-2)^2} = \dfrac{2x(x-4)}{(x-2)^2} = 0$ we obtain the critical number 4. Since $A''(4) > 0$, the area is minimized when the vertices are $(4, 0)$ and $(0, 8)$.

15. Let x and $1500 - x$ be the two sides of the corral. We want to maximize $A(x) = x(1500 - x)$ on $[0, 1500]$. Solving $A'(x) = 1500 - 2x = 0$ we obtain the critical number 750. Since $A''(x) = -2 < 0$, the area is maximized when the corral is 750 ft × 750 ft.

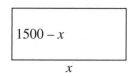

17. Let x and y be as shown in the figure. Since $4x + 2y = 8000$, $y = 4000 - 2x$, and we want to maximize $A(x) = xy = 4000x - 2x^2$. Solving $A'(x) = 4000 - 4x = 0$ we obtain the critical number 1000. Since $A''(x) = -4 < 0$, the area is maximized when $x = 1000$ m and $y = 2000$ m.

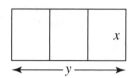

19. Let x and y be as shown in the figure. Then $2x + 40 + 2y = 80$ and $y = 20 - x$. We want to maximize $A(x) = (x + 40)y = (x + 40)(20 - x) = 800 - 20x - x^2$ for x in $[0, 20]$. Solving $A'(x) = -20 - 2x = 0$ we obtain the critical number -10. Comparing $A(0) = 800$ and $A(20) = 0$ we see that the maximum area is obtained when $x = 0$ and $y = 20$. Thus, the yard will be a rectangle 40 feet long by 20 feet wide.

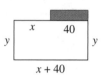

21. Let x be a side of the base and y the height of the box. Then $x^2 y = 32,000$ and $y = 32,000/x^2$. We want to minimize $A(x) = x^2 + 4xy = x^2 + 128,000/x$. Solving $A'(x) = 2x - 128,000/x^2 = 0$ we obtain the critical number 40. Since $A''(40) > 0$, the amount of material is minimized when $x = 40$ cm and $y = 20$ cm.

23. Let x be the side of the square cut-out. We want to maximize $V(x) = x(40 - 2x)^2$ on $[0, 20]$. Solving $V'(x) = 12x^2 - 320x + 1600 = 4(3x - 20)(x - 20) = 0$ we obtain the critical numbers $20/3$ and 20. Since $V(20) = 0$ and $V''(20) < 0$, the volume is maximized when the height is $20/3$ cm and the base is $80/3$ cm × $80/3$ cm. The maximum volume is $V(20/3) = 128,000/27$ cm^3.

25. Let x be the height of the gutter. We want to maximize $A(x) = x(30 - 2x)$ on $[0, 15]$. Solving $A'(x) = 30 - 4x = 0$ we obtain the critical number $15/2$. Since $A''(x) = -4 < 0$, the cross-sectional area and hence the volume, is maximized when the gutter is 7.5 cm high and 15 cm wide.

27. Let x be the distance from the 10-foot flagpole. We want to minimize $L(x) = \sqrt{400 + (30 - x)^2} + \sqrt{100 + x^2}$ on $[0, 30]$. Setting

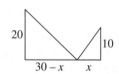

$$L'(x) = \frac{-(30 - x)}{\sqrt{400 + (30 - x)^2}} + \frac{x}{\sqrt{100 + x^2}} = 0$$

we obtain $x^2 + 20x - 300 = (x + 30)(x - 10) = 0$. The positive critical number is 10. Comparing $L(0) = 10 + 10\sqrt{13}$, $L(10) = 30\sqrt{2}$, and $L(30) = 20 + 10\sqrt{10}$, we see that the length of wire is minimized when it is attached 10 feet from the 10-foot flagpole.

29. Let the radius of the semicircle be r and the height of the rectangle h. Then the perimeter is $2r + 2h + \pi r = 10$ and $h = \frac{1}{2}[10 - (2 + \pi)r]$. We want to maximize

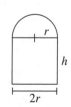

$$A(r) = 2rh + \frac{\pi r^2}{2} = 10r - \left(2 + \frac{\pi}{2}\right)r^2.$$

Solving $A'(r) = 10 - (4 + \pi)r = 0$ we obtain the critical number $\dfrac{10}{4 + \pi}$. Since $A''(r) = -4 - \pi < 0$, the area is maximized when the base of the window is $\dfrac{20}{4 + \pi}$ m, the height of the rectangular portion is $\dfrac{10}{4 + \pi}$ m, and the radius of the circular portion is $\dfrac{10}{4 + \pi}$ m.

31. By the Pythagorean Theorem, $L^2 = (x+5)^2 + y^2$. Using similar triangles, we have $\dfrac{y}{10} = \dfrac{x+5}{x}$ and $y = \dfrac{10(x+5)}{x}$. We want to minimize

$$L^2 = (x+5)^2 + \frac{100(x+5)^2}{x^2} = (x+5)^2\left(1+\frac{100}{x^2}\right)$$

for $x > 0$. Setting $\dfrac{dL^2}{dx} = 0$ we obtain

$$\frac{dL^2}{dx} = (x+5)^2\left(-\frac{200}{x^3}\right) + \left(1+\frac{100}{x^2}\right)[2(x+5)] = 2(x+5)\left[1+\frac{100}{x^2}-(x+5)\frac{100}{x^3}\right]$$

$$= 2(x+5)\left(1-\frac{500}{x^3}\right) = 2(x+5)\left(\frac{x^3-500}{x^3}\right) = 0$$

so $x = \sqrt[3]{500}$. Using the first derivative test, $\dfrac{dL^2}{dx} < 0$ for $0 < x < \sqrt[3]{500}$ and $\dfrac{dL^2}{dx} > 0$ for $x > \sqrt[3]{500}$, we see that L^2 and hence L is minimized when $x = \sqrt[3]{500}$. In this case $L = (5+\sqrt[3]{500})\sqrt{1+100/500^{2/3}} \approx 20.81$ ft.

33. As seen in Figure 4.8.24, let r be the radius and h the height of the cylinder. Then, using similar triangles, we have $\dfrac{h}{12} = \dfrac{8-r}{8}$ or $h = 12 - \dfrac{12r}{8} = 12 - \dfrac{3r}{2}$. We want to maximize $V(r) = \pi r^2 h = 12\pi r^2 - \dfrac{3\pi r^3}{2}$ on $[0,8]$. Solving $V'(r) = 24\pi r - \dfrac{9\pi r^2}{2} = 0$ we obtain the critical numbers 0 and 16/3. Comparing $V(0)$, $V(16/3)$, and $V(8)$ we see that the volume is maximized when $r = 16/3$ and $h = 4$.

35. As seen in Figure 4.8.26, let r be the radius and h the height of the can. Then $\pi r^2 h = 32$ and $h = 32/\pi r^2$. We want to minimize $A(r) = 2\pi r^2 + 2\pi rh = 2\pi r^2 + 64/r$ on $(0,\infty)$. Solving $A'(r) = 4\pi r - 64/r^2 = 0$ we obtain the critical number $\sqrt[3]{16/\pi}$. Since $A''(r) > 0$ for $r > 0$, the surface area of the can is minimized when $r = \sqrt[3]{16/\pi}$ in and $h = 2\sqrt[3]{16/\pi}$ in.

37. Let x be the distance on land across from the island to the point where the bird first intersects the shore. We want to minimize $T(x) = \dfrac{1}{6}\sqrt{x^2+9} + \dfrac{1}{10}(20-x)$ on $[0,20]$. Solving $T'(x) = \dfrac{1}{6}x/\sqrt{x^2+9} - \dfrac{1}{10} = 0$ we obtain the critical number 9/4. Since $T'(0) = -1/10 < 0$ and $T'(4) = 1/30 > 0$, the time is minimized when $x = 9/4$ km. Therefore, the bird should fly over water to a point on land $20 - 9/4 = 17.75$ km from the nest.

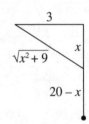

39. Let x be the distance along the opposite bank to which the pipeline is run. Let h be the length of pipe across the swamp. Then $h^2 = x^2 + 16$. We want to minimize $C(x) = 2h + 4 - x = 2\sqrt{x^2+16} + 4 - x$ on $[0,4]$, where the cost is measured in units of $1000. Solving $C'(x) = 2x/\sqrt{x^2+16} - 1 = 0$ we obtain the critical number $4/\sqrt{3}$. Comparing $C(0) = 12$, $C(4/\sqrt{3}) = 4\sqrt{3}+4$, and $C(4) = 8\sqrt{2}$, we see that the cost is minimized when $x = 4\sqrt{3}+4$ mi.

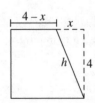

41. As seen in Figure 4.8.30, let r be the radius of the cylinder and h the length of the cylinder without the hemispherical ends. The volume of the container is $\pi r^2 h + \dfrac{4}{3}\pi r^3 = 30\pi$. Solving for h we obtain $h = \dfrac{30}{r^2} - \dfrac{4r}{3}$. We want to minimize

$$C(r) = 2\pi rh + \frac{3}{2}(4\pi r^2) = 2\pi r\left(\frac{30}{r^2} - \frac{4r}{3}\right) + 6\pi r^2 = \frac{60\pi}{r} - \frac{10\pi r^2}{3}.$$

Setting $C'(r) = 0$ we have

$$C'(r) = -\frac{60\pi}{r^2} + \frac{20\pi}{3}r = 0; \qquad \frac{-3}{r^2} + \frac{r}{3} = 0; \qquad \frac{r^3-9}{3r^2} = 0;$$

so $r = 9^{1/3}$. Since $C''(r) = 120\pi/r^3 + 20\pi/3$ and $C''(9^{1/3}) = 20\pi\left(\dfrac{6}{9} + \dfrac{1}{3}\right) > 0$, the cost is minimized when $r = 9^{1/3}$ and $h = 30/9^{2/3} - 4(9^{1/3})/3 = 2(9^{1/3})$.

43. Label the figure as shown. Note that $\triangle ADE$ and $\triangle DBC$ are similar and that $\triangle ABD$ is a right triangle. Using similarity, we have

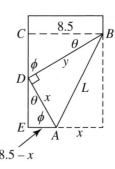

$$\frac{y}{8.5} = \frac{x}{\sqrt{x^2 - (8.5 - x)^2}} \quad \text{or} \quad y = \frac{\sqrt{8.5}x}{\sqrt{2x - 8.5}} = \frac{\sqrt{4.25}x}{\sqrt{x - 4.25}}.$$

We want to minimize $L^2 = x^2 + y^2 = x^2 + \dfrac{4.25x^2}{x - 4.25}$ on $(4, 8]$.

Solving $\dfrac{dL^2}{dx} = 2x + \dfrac{4.25x^2 - 36.125x}{(x - 4.25)^2} = \dfrac{2x^2(x - 6.375)}{(x - 4.25)^2} = 0$, we obtain the critical number 6.375. Since $\dfrac{dL^2}{dx}\Big|_{x=6} = -8.8163 <$ 0 and $\dfrac{dL^2}{dx}\Big|_{x=7} = 8.0992 > 0$, we see that the length is minimized when the width of the fold is 6.375 inches.

45. We want to maximize $A(\theta) = (a \sin\theta + b \cos\theta)(a\cos\theta + b\sin\theta)$

$$= (a^2 + b^2)\sin\theta\cos\theta + ab = \frac{1}{2}(a^2 + b^2)\sin 2\theta + ab$$

on $(0, \pi/2)$. Solving $A'(\theta) = (a^2 + b^2)\cos 2\theta = 0$ we obtain the critical number $\pi/4$. Since $A''(\pi/4) < 0$, the area is maximized when $\theta = \pi/4$. The circumscribed rectangle is a square whose side is $a\sin\pi/4 + b\cos\pi/4 = (a+b)/\sqrt{2}$.

47. As seen in Figure 4.8.36, let x be the width of the beam, y the height, and d the diameter of the log and diagonal of the wooden beam (shown by the dotted line). From the Pythagorean Theorem, we have $d^2 = x^2 + y^2$. We want to maximize $S(x) = xy^2 = x(d^2 - x^2)$. Solving $S'(x) = d^2 - 3x^2 = 0$ we obtain the critical number $d/\sqrt{3}$. Since $S''(d/\sqrt{3}) = -6d/\sqrt{3} < 0$, the strength is maximized when the length is $d/\sqrt{3}$ and the width is $d\sqrt{2/3}$.

49. Let x be the distance from P to the bulb with intensity I_1. We want to minimize $E(x) = \dfrac{125}{x^2} + \dfrac{216}{(10-x)^2}$ on $(0, 10)$. Setting $E'(x) = -\dfrac{250}{x^3} + \dfrac{432}{(10-x)^3} = 0$, we have $\dfrac{(10-x)^3}{x^3} = \dfrac{432}{250} = \dfrac{216}{125}$ or $\dfrac{10-x}{x} = \dfrac{6}{5}$.

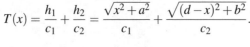

Thus, 50/11 is a critical number, and by the first derivative test, we see that the total illuminance will be a minimum at 50/11 m from the bulb with intensity $I_1 = 125$.

51. Let x be the point on the x-axis where the light crosses from one medium to the other. We want to minimize

$$T(x) = \frac{h_1}{c_1} + \frac{h_2}{c_2} = \frac{\sqrt{x^2 + a^2}}{c_1} + \frac{\sqrt{(d-x)^2 + b^2}}{c_2}.$$

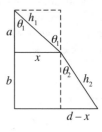

To find the critical numbers, we set

$$T'(x) = \frac{x}{c_1\sqrt{x^2 + a^2}} - \frac{d-x}{c_2\sqrt{(d-x)^2 + b^2}} = 0.$$

Then $\dfrac{x/\sqrt{x^2 + a^2}}{c_1} = \dfrac{(d-x)/\sqrt{(d-x)^2 + b^2}}{c_2}$, and since $\sin\theta_1 = \dfrac{x}{h_1}$ and $\sin\theta_2 = \dfrac{d-x}{h_2}$, we have $\sin\dfrac{\theta_1}{c_1} = \sin\dfrac{\theta_2}{c_2}$ at the critical number. To see that the time is actually minimized at this point, we compute the second derivative $T''(x) = \dfrac{a^2}{c_1(x^2 + a^2)^{3/2}} + \dfrac{b^2}{c_2[(d-x)^2 + b^2]^{3/2}}$. Since $T''(x) > 0$ for all x, we do have a minimum at the critical number.

53. $U'(x) = -24/x^{13} + 6/x^7 = 6(x^6 - 4)/x^{13}$. Solving $U'(x) = 0$ we obtain the critical numbers $\pm\sqrt[3]{2}$. From the first derivative test, we see that relative minima occur at both of these points. Thus, the minimum potential energy is $U(\sqrt[3]{2}) = -1/8$.

x		$-\sqrt[3]{2}$		0		$\sqrt[3]{2}$	
f	↘		↗		↘		↗
f'	$-$	0	$+$	undefined	$-$	0	$+$

55. (a) We want to maximize

$$y(x) = \frac{w_0}{24EI}(L^2x^2 - 2Lx^3 + x^4) = \frac{w_0}{24EI}x^2(x-L)^2.$$

Now

$$y'(x) = \frac{w_0}{24EI}\left[2x^2(x-L) + 2x(x-L)^2\right] = \frac{w_0}{12EI}x(x-L)(2x-L).$$

Solving $y'(x) = 0$ we obtain the critical numbers 0, $L/2$, and L. Using the first derivative test we see that $y(L/2) = w_0L^4/384EI$.

(b)

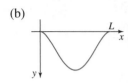

57. We want to minimize $m(x) = \dfrac{\pi\rho M^{2/3}}{2K^{2/3}}\left[\dfrac{2-x^2}{(1-x^4)^{2/3}}\right]$. Solving

$$m'(x) = -\frac{\pi\rho M^{2/3}x}{3K^{2/3}}\left[\frac{x^4 - 8x^2 + 3}{(1-x^4)^{5/3}}\right] = 0$$

we obtain the critical numbers 0, $\sqrt{4-\sqrt{13}}$, and $\sqrt{4+\sqrt{13}}$. Since $x = r/R$ where $0 < r < R$, we must have x in the interval $(0,1)$. Thus, the only appropriate critical number is $x = \sqrt{4-\sqrt{13}} \approx 0.63$. Use the first derivative test with $x = 0.6$ and 0.7 to show that the mass m is minimized for $x = r/R \approx 0.63$ or $r \approx 0.63R$.

59. Let x be the length of wire formed into a circle and $1-x$ the length formed into a square. Since x is the circumference of the circle, its radius is $x/2\pi$. We want to maximize $A(x) = \pi\left(\dfrac{x}{2\pi}\right)^2 + \left(\dfrac{1-x}{4}\right)^2 = \dfrac{x^2}{4\pi} + \dfrac{(1-x)^2}{16}$ on $[0,1]$.

Solving $A'(x) = \dfrac{x}{2\pi} - \dfrac{1-x}{8} = 0$ we obtain the critical number $\dfrac{\pi}{4+\pi}$. Comparing $A(0) = 1/6$, $A\left(\dfrac{\pi}{4+\pi}\right) = \dfrac{1}{4(4+\pi)}$, and $A(1) = 1/4\pi$, we see that the maximum area is obtained when the entire wire is formed into a circle of radius $1/2\pi$ m.

61. (a) By the Pythagorean Theorem, $R^2 = r^2 + h^2$ or $r^2 = R^2 - h^2$. We want to maximize $V(h) = \dfrac{1}{3}\pi r^2 h = \dfrac{1}{3}\pi(R^2 - h^2)h$ on $[0,R]$. Solving $V'(h) = \dfrac{1}{3}\pi\left(R^2 - \dfrac{1}{3}h^2\right) = 0$ we obtain the critical numbers $h = \pm R/\sqrt{3}$. Since $V(0) = V(R) = 0$ and $V''(R/\sqrt{3}) = -\dfrac{2}{9}R/\sqrt{3} < 0$, the volume is maximized when $h = r/\sqrt{3}$ or

$$r = \sqrt{R^2 - h^2} = \sqrt{R^2 - R^2/3} = \sqrt{2/3}R.$$

(b) The maximum volume is $V = \dfrac{1}{3}\pi\left(\dfrac{2}{3}R^2\right)R/\sqrt{3} = \dfrac{2}{9}\pi R^3/\sqrt{3}$.

(c) The circumference of the circular piece of paper is $2\pi R$. The circumference of the base of the cone is $2\pi r = 2\pi\sqrt{2/3}R$. Thus, $s = 2\pi R - 2\pi r = 2\pi R(1 - \sqrt{2/3})$ and $\theta = s/R = 2\pi(1 - \sqrt{2/3}) \approx 1.15$ radians.

63. Problem 27 showed that the optimal amount of wire (the least amount) is used when it is attached 10 feet from the 10-foot flagpole. From the figure, this means that the right triangles formed by the flagpoles, the wire, and the ground are isosceles right triangles, and therefore similar. Thus, the non-right angles θ_1 and θ_2 of each triangle are the same.

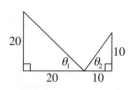

65. The total time it takes the swimmer to reach C from A is $T = \dfrac{\sqrt{x^2+1}}{3} + \dfrac{\sqrt{x^2-8x+17}}{2}$ on $[0,4]$. Differentiating then solving for 0 gives

$$\frac{dT}{dx} = \frac{x}{3\sqrt{x^2+1}} + \frac{x-4}{2\sqrt{x^2-8x+17}} = 0$$

$$2x\sqrt{x^2-8x+17} = -3(x-4)\sqrt{x^2+1}$$

With CAS help, the foregoing equation has only one real root in the interval $[0,4]$, namely $x \approx 3.176$. Now, $T(0) \approx 2.395$, $T(3.176) \approx 1.758$, and $T(4) \approx 1.874$. Therefore, to minimize her time in the race, she should swim from point A to point B about 3.18 miles down the beach from the point of the beach closest to A, then proceed directly to C.

67. (a) Let x be the length of the cable AB. Then $L(x) = x + 2\sqrt{(4-x)^2+4}$.

   (b)

   (c) Solving $L'(x) = 1 + -2(4-x)/\sqrt{(4-x)^2+4} = 0$ we obtain the critical numbers $4 \pm \dfrac{2}{3}\sqrt{3}$. Since $4 + \dfrac{2}{3}\sqrt{3} > 4$, we use $x = 4 - \dfrac{2}{3}\sqrt{3}$. We know from the graph that this gives a minimum.

   (d) Let x be the length of the cable AB. Then $L(x) = x + \sqrt{1+(4-x)^2} + \sqrt{4+(4-x)^2}$.

   (e)

   (f) From the graph in part (e) we estimate that $L(x)$ is minimized when $x \approx 3.2$. (Using a numerical procedure to solve $L'(x) = 0$ gives $x \approx 3.1955$.)

4.9 Linearization and Differentials

1. Using $f(9) = 3$, $f'(x) = 1/2\sqrt{x}$, and $f'(9) = 1/6$, the tangent line to the graph is $y - 3 = (1/6)(x-9)$. The linearization of $f(x)$ at $a = 9$ is $L(x) = \dfrac{1}{6}(x-9) + 3$.

3. Using $f(\pi/4) = 1$, $f'(x) = \sec^2 x$, and $f'(\pi/4) = 2$, the tangent line to the graph is $y - 1 = 2(x - \pi/4)$. The linearization of $f(x)$ at $a = \pi/4$ is $L(x) = 2(x - \pi/4) + 1$.

5. Using $f(1) = 0$, $f'(x) = 1/x$, and $f'(1) = 1$, the tangent line to the graph is $y - 0 = 1(x-1)$. The linearization of $f(x)$ at $a = 1$ is $L(x) = x - 1$.

7. Using $f(3) = 2$, $f'(x) = 1/2\sqrt{1+x}$, and $f'(3) = 1/4$, the tangent line to the graph is $y - 2 = (1/4)(x-3)$. The linearization of $f(x)$ at $a = 3$ is $L(x) = \dfrac{1}{4}(x-3) + 2$.

9. Using $f(0) = 1$, $f'(x) = e^x$, and $f'(0) = 1$, the tangent line to the graph is $y - 1 = 1(x-0)$, yielding $L(x) = x + 1$ or $e^x \approx 1 + x$ whenever x is close to 0.

11. Using $f(0) = 1$, $f'(x) = 10(1+x)^9$, and $f'(0) = 10$, the tangent line to the graph is $y - 1 = 10(x - 0)$, yielding $L(x) = 10x + 1$ or $(1+x)^{10} \approx 1 + 10x$ whenever x is close to 0.

13. Using $f(0) = 1$, $f'(x) = -1/2\sqrt{1-x}$, and $f'(0) = -1/2$, the tangent line to the graph is $y - 1 = -\dfrac{1}{2}(x - 0)$, yielding $L(x) = 1 - \dfrac{1}{2}x$ or $\sqrt{1-x} \approx 1 - \dfrac{1}{2}x$ whenever x is close to 0.

15. Using $f(0) = 1/3$, $f'(x) = -\dfrac{1}{(3+x)^2}$, and $f'(0) = -1/9$, the tangent line to the graph is $y - 1/3 = -\dfrac{1}{9}(x - 0)$, yielding $L(x) = -\dfrac{1}{9}x + \dfrac{1}{3}$ or $\dfrac{1}{3+x} \approx \dfrac{1}{3} - \dfrac{1}{9}x$ whenever x is close to 0.

17. From Problem 2 we have $\dfrac{1}{x^2} \approx -2x + 3$ whenever x is close to 1. Thus,

$$(1.01)^{-2} = f(1.01) \approx -2(1.01) + 3 = 0.98.$$

19. From Problem 6 we have $5x + e^{x-2} \approx 6x - 1$ whenever x is close to 2. Thus,

$$10.5 + e^{0.1} = f(2.1) \approx 6(2.1) - 1 = 11.6.$$

21. From Problem 12 we have $(1 + 2x)^{-3} \approx 1 - 6x$ whenever x is close to 0. Thus,

$$(1.1)^{-3} = f(0.05) \approx 1 - 6(0.05) = 0.7.$$

23. From Problem 16 we have $\sqrt[3]{1 - 4x} \approx 1 - \dfrac{4}{3}x$ whenever x is close to 0. Thus,

$$(0.88)^{1/3} = f(0.03) \approx 1 - \dfrac{4}{3}(0.03) = 0.96.$$

25. To find an approximation for $(1.8)^5$ we choose $f(x) = x^5$; $a = 2$. Using $f(2) = 32$, $f'(x) = 5x^4$, and $f'(2) = 80$, the tangent line to the graph is $y - 32 = 80(x - 2)$. The linearization of $f(x)$ at $a = 2$ is $L(x) = 80x - 128$. Thus,

$$(1.8)^5 = f(1.8) \approx 80(1.8) - 128 = 16.$$

27. To find an approximation for $\dfrac{(0.9)^4}{(0.9) + 1}$ we choose $f(x) = \dfrac{x^4}{x+1}$; $a = 1$. Using $f(1) = 1/2$, $f'(x) = \dfrac{4x^3(x+1) - x^4}{(x+1)^2} = \dfrac{x^3(3x+4)}{(x+1)^2}$, and $f'(1) = 7/4$, the tangent line to the graph is $y - \dfrac{1}{2} = \dfrac{7}{4}(x - 1)$. The linearization of $f(x)$ at $a = 1$ is $L(x) = \dfrac{7}{4}(x - 1) + \dfrac{1}{2}$. Thus,

$$\dfrac{(0.9)^4}{(0.9) + 1} = f(0.9) \approx \dfrac{7}{4}(0.9 - 1) + \dfrac{1}{2} = \dfrac{13}{40} = 0.325.$$

29. To find an approximation for $\cos(\pi/2 - 0.4)$ we choose $f(x) = \cos x$; $a = \pi/2$. Using $f(\pi/2) = 0$, $f'(x) = -\sin x$, and $f'(\pi/2) = -1$, the tangent line to the graph is $y - 0 = -1(x - \pi/2)$. The linearization of $f(x)$ at $a = \pi/2$ is $L(x) = -x + \pi/2$. Thus,

$$\cos(\pi/2 - 0.4) = f(\pi/2 - 0.4) \approx -(\pi/2 - 0.4) + \pi/2 = 0.4.$$

31. To find an approximation for $\sin 33°$ we choose $f(x) = \sin(x + \pi/6)$; $a = 0$. Using $f(0) = 1/2$, $f'(x) = \cos(x + \pi/6)$, and $f'(0) = \dfrac{\sqrt{3}}{2}$, the tangent line to the graph is $y - \dfrac{1}{2} = \dfrac{\sqrt{3}}{2}(x - 0)$. The linearization of $f(x)$ at $a = 0$ is $L(x) = \dfrac{\sqrt{3}}{2}x + \dfrac{1}{2}$. Thus,

$$\sin 33° = \sin(\pi/6 + 3\pi/180) = f(\pi/60) \approx \dfrac{\sqrt{3}}{2}\left(\dfrac{\pi}{60}\right) + \dfrac{1}{2} = \dfrac{\sqrt{3}\pi + 60}{120}.$$

33. According to the graph, $f(1) = 4$ and $f'(1) = 2$, so the tangent line to the graph is $y - 4 = 2(x - 1)$. The linearization of $f(x)$ at $a = 1$ is $L(x) = 2x + 2$. Thus, $f(1.04) \approx 2(1.04) + 2 = 4.08$.

35. $\Delta y = (x + \Delta x)^2 + 1 - (x^2 + 1) = 2x\Delta x + (\Delta x)^2$; $dy = 2x\, dx$

37. $\Delta y = (x + \Delta x + 1)^2 - (x + 1)^2 = (x + \Delta x)^2 + 2(x + \Delta x) + 1 - (x + 1)^2$

$\quad = x^2 + 2x\Delta x + (\Delta x)^2 + 2x + 2\Delta x + 1 - x^2 - 2x - 1 = 2x\Delta x + 2\Delta x + (\Delta x)^2$

$dy = 2(x + 1)dx$

39. $\Delta y = \dfrac{3(x + \Delta x) + 1}{x + \Delta x} - \dfrac{3x + 1}{x} = \dfrac{3x(x + \Delta x) + x - [3x(x + \Delta x) + (x + \Delta x)]}{x(x + \Delta x)} = -\dfrac{\Delta x}{x(x + \Delta x)}$

$dy = \dfrac{x(3) - (3x + 1)}{x^2}\, dx = -\dfrac{dx}{x^2}$

41. $\Delta y = \sin(x + \Delta x) - \sin x = \sin x \cos \Delta x + \cos x \sin \Delta x - \sin x$; $dy = \cos x\, dx$

43. $\Delta y = 5(x + \Delta x)^2 - 5x^2 = 10x\Delta x + 5(\Delta x)^2$

$\Delta y|_{x=2} = 20\Delta x + 5(\Delta x)^2$;

$\quad dy = 10x\, dx$; $dy|_{x=2} = 20\, dx$

x	Δx	Δy	dy	$\Delta y - dy$
2	1.00	25.0000	20.0	5.0000
2	0.50	11.2500	10.0	1.2500
2	0.10	2.0500	2.0	0.0500
2	0.01	0.2005	0.2	0.0005

(Recall that $dx = \Delta x$.)

45. (a) $df = (8x + 5)\, dx$. When $x = 4$ and $dx = 0.03$, $df = [8(4) + 5](0.03) = 1.11$.

(b) When $x = 3$ and $dx = -0.1$, $df = [8(3) + 5](-0.1) = -2.9$.

47. (a) $A(4) = 16\pi$ cm; $A(5) = 25\pi$ cm; $A(5) - A(4) = 9\pi$ cm.

(b) $dA = 2\pi r\, dr$. When $r = 4$ cm and $dr = 1$ cm, $dA = 8\pi$ cm.

49. The exact volume of the cover is

$$\Delta V = \frac{4}{3}\pi(r + t)^3 - \frac{4}{3}\pi r^3 = \frac{4}{3}\pi(3r^2 t + 3rt^2 + t^3).$$

$dV = 4\pi r^2\, dr$. When $dr = t$, the approximate volume of the cover is $4\pi r^2 t$. If $r = 0.8$ in and $t = 0.04$ in, then the approximate volume of the cover is $4\pi(0.8)^2(0.04) = 0.1024\pi$ in^3.

51. $A = s^2$; $dA = 2s\, ds$. Letting $s = 10$ cm and $ds = 0.3$ cm, we obtain $dA = 2(10)(0.3) = 6$ cm^2. The maximum error in the area is ± 6 cm^2. The approximate relative error is $\pm 6/10^2 = \pm 0.06$ cm^2 and the approximate percentage error is $\pm 6\%$.

53. $P = cV^{-\gamma}$; $dP = -\gamma cV^{-\gamma-1}dV$. The approximate relative error in P is

$$\frac{dP}{P} = \frac{-\gamma cV^{-\gamma-1}dV}{cV^{-\gamma}} = -\gamma\frac{dV}{V},$$

so the approximate relative error in P is proportional to the approximate relative error in V.

55. For $v_0 = 256$ ft/s, $\theta = 45°$, and $g = 32$ ft/s^2, the range is $R = \dfrac{256^2}{32}\sin 90° = 2048$ ft. The approximate change in R with respect to v_0 is $dV = \dfrac{2v_0}{g}\sin 2\theta dv_0$. For $v_0 = 256$ ft/s, $g = 32$ ft/s^2, $\theta = 45°$, and $dv_0 = 10$ ft/s, the approximate change in the range is

$dv = \dfrac{2(256)}{32}(1)(10) = 160$ ft.

57. (a) Setting $g' = 978.0318(53.024 \times 10^{-4} \times 2\sin\theta\cos\theta - 5.9 \times 10^{-6} \times 4\sin 2\theta\cos 2\theta) = 0$ we obtain $\sin 2\theta(53.024 \times 10^{-4} - 5.9 \times 10^{-6} \times 4\cos 2\theta) = 0$. From $\sin 2\theta = 0$ we find $\theta = 0°$ or $90°$, and since $4\cos 2\theta = 53.024 \times 10^{-4}/5.9 \times 10^{-6} \times 4 > 1$, we see that these are the only critical numbers. By inspection of g we find that g is minimum on the equator ($\theta = 0°$) and maximum at the poles ($\theta = 90°$).

(b) $g(60°) \approx 981.9169$ cm/s^2

(c) $dg = 978.0318 \sin 2\theta (53.024 \times 10^{-4} - 5.9 \times 10^{-6} \times 4 \cos 2\theta) d\theta$.

Using $\theta = \pi/3$ and $d\theta = \pi/180$ we find $dg \approx 0.07856$ cm/s^2.

59. Writing $T = 2\pi \sqrt{L} g^{-1/2}$, we have $dT = -\pi \sqrt{L} g^{-3/2} dg$. For $L = 4$ m, $g = 9.8$ m/s^2, and $dg = -0.05$ m/s^2, we obtain the approximate change $dT = -\pi \sqrt{4}(9.8)^{-3/2}(-0.05) \approx 0.0102$ s.

61. By the definition in (2), $L(x) = f(a) + f'(a)(x - a)$. Since $p(a) = f(a)$ and $p'(a) = f'(a)$, we can rewrite $L(x) = p(a) + p'(a)(x - a) = (c_1 a + c_0) + (c_1)(x - a)$. Simplifying, $c_1 a$ cancels out, and we get $L(x) = c_1 x + c_0$, which is $p(x)$.

63. If $f''(x) > 0$ for all x in some open interval containing a, then that interval is concave up. Thus, the tangent at a will lie below $f(x)$ for all x within that interval, and so $L(x)$ will underestimate $f(x)$ for x near a.

65. $\Delta A = (x + \Delta x)^2 - x^2 = 2x\Delta x + (\Delta x)^2$

   $dA = 2x\, dx = 2x\Delta x$ (Recall that $dx = \Delta x$.)

$\Delta A - dA = [2x\Delta x + (\Delta x)^2] - 2x\Delta x = (\Delta x)^2$

Thus, ΔA is the combined area of the beige and green regions, dA is the combined area of the beige regions only, and $\Delta A - dA$ is the area of the green region.

4.10   Newton's Method

1.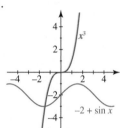

The equation has one real root.

3.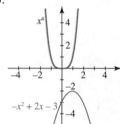

The equation has no real roots.

5.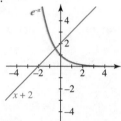

The equation has one real root.

7. Let $f(x) = x^2 - 10$. Then $f'(x) = 2x$ and

$$x_{n+1} = x_n - \frac{x_n^2 - 10}{2x_n} = \frac{x_n^2 + 10}{2x_n}.$$

Choosing $x_0 = 3$ we obtain $x_1 \approx 3.1667$, $x_2 \approx 3.1623$, $x_3 \approx 3.1623$. Thus, $\sqrt{10} \approx 3.1623$.

9. Let $f(x) = x^3 - 4$. Then $f'(x) = 3x^2$ and

$$x_{n+1} = x_n - \frac{x_n^3 - 4}{3x_n^2} = \frac{2x_n^3 + 4}{3x_n^2}.$$

Choosing $x_0 = 1$ we obtain $x_1 \approx 2.0000$, $x_2 \approx 1.6667$, $x_3 \approx 1.5911$, $x_4 \approx 1.5874$, $x_5 \approx 1.5874$. Thus, $\sqrt[3]{4} \approx 1.5874$.

11. Let $f(x) = x^3 + x - 1$. Then $f'(x) = 3x^2 + 1$ and

$$x_{n+1} = x_n - \frac{x_n^3 + x_n - 1}{3x_n^2 + 1} = \frac{2x_n^3 + 1}{3x_n^2 + 1}.$$

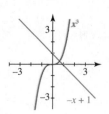

From the graph we see that $f(x)$ has a single root near $x_0 = 1$. Then $x_1 \approx 0.7500$, $x_2 \approx 0.6860$, $x_3 \approx 0.6823$, $x_4 \approx 0.6823$, and the only real root is approximately 0.6823.

13. From the quadratic formula, $x^2 = \dfrac{-1 \pm \sqrt{1 + 12}}{2} = \dfrac{-1 \pm \sqrt{13}}{2}$. Since x^2 must be positive for x real, we have $x = \pm\sqrt{\dfrac{-1 \pm \sqrt{13}}{2}} \approx \pm 1.1414$. Newton's Method is not necessary.

15. Let $f(x) = x^2 - \sin x$. Then $f'(x) = 2x - \cos x$ and

$$x_{n+1} = x_n - \frac{x_n^2 + \sin x_n}{2x_n - \cos x_n} = \frac{x_n^2 - x_n \cos x_n + \sin x_n}{2x_n - \cos x_n}.$$

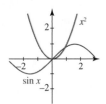

From the graph we see that $f(x)$ has one root at $x = 0$ and another one near $x_0 = 1$. Then $x_1 \approx 0.8914$, $x_2 \approx 0.8770$, $x_3 \approx 0.8767$, $x_4 \approx 0.8767$. Thus, the two real roots are 0 and approximately 0.8767.

17. $f'(x) = -3\sin x + 4\cos x$ and

$$x_{n+1} = x_n - \frac{3\cos x_n + 4\sin x_n}{-3\sin x_n + 4\cos x_n} = \frac{(4x_n - 3)\cos x_n - (3x_n + 4)\sin x_n}{4\cos x_n - 3\sin x_n}.$$

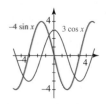

From the graphs of $3\cos x$ and $-4\sin x$ we see that the first positive root of $f(x)$ is near $x_0 = 2$. Then $x_1 \approx 2.5438$, $x_2 \approx 2.4981$, $x_3 \approx 2.4981$. Thus, the smallest positive x-intercept is approximately 2.4981.

19. We want to solve $\dfrac{60x^2 - x^3}{16000} = 0.01$ or $x^3 - 60x^2 + 160 = 0$. Let $f(x) = x^3 - 60x^2 + 160$. Then $f'(x) = 3x^2 - 120x$ and

$$x_{n+1} = x_n - \frac{x_n^3 - 60x_n^2 + 160}{3x_n^2 - 120x_n} = \frac{2x_n^3 - 60x_n^2 - 160}{3x_n^2 - 120x_n}.$$

Choosing $x_0 = 1$ we obtain $x_1 \approx 1.8632$, $x_2 \approx 1.6670$, $x_3 \approx 1.6560$, $x_4 \approx 1.6560$. Thus, $x \approx 1.6560$ ft.

21. We want to solve $\sin\theta = \dfrac{1}{1.5} = \dfrac{2}{3}$ or $3\sin\theta = 2$. Let $f(\theta) = 3\sin\theta - 2$. Then $f'(\theta) = 3\cos\theta$ and

$$\theta_{n+1} = \theta_n - \frac{3\sin\theta_n - 2}{3\cos\theta_n} = \frac{3\theta_n\cos\theta_n - 3\sin\theta_n + 2}{3\cos\theta_n}.$$

Choosing $\theta_0 = 0$ we obtain $\theta_1 \approx 0.6667$, $\theta_2 \approx 0.7281$, $\theta_3 \approx 0.7297$, $\theta_4 \approx 0.7297$. Thus, $\theta \approx 0.7297$ radian.

23. (a) The volume of water displaced is $V_w = 3(7)(2) = 42$ ft^3. The volume of steel in the tub is $V_s = 42 - (3 - 2t)(7 - 2t)(2 - t) = 4t^3 - 28t^2 + 61t$. Since the weight of water displaced is equal to the weight of the tub,

$$62.4(42) = 490(4t^3 - 28t^2 + 61t) \quad \text{or} \quad f(t) = t^3 - 7t^2 + \frac{61}{4}t - \frac{1638}{1225} = 0.$$

(b) Then $f'(t) = 3t^2 - 14t + 61/4$ and

$$t_{n+1} = t_n - \frac{f(t)}{f'(t)}.$$

From the graph we see that $f(t)$ has its only root near $t_0 = 0$. Then $t_1 \approx 0.0877$, $t_2 \approx 0.0915$, $t_3 \approx 0.0915$, and $t \approx 0.915$ ft.

25. (a) Since θ subtends an arc of length $L/2$ on a circle of radius R, we have $L/2 = R\theta$ and $R = L/2\theta$. From Figure 4.10.11 we see that

$$\sin\theta = \frac{(L-l)/2}{R} = \frac{(L-l)/2}{L/2\theta} = \left(1 - \frac{l}{L}\right)\theta \quad \text{and} \quad \cos\theta = \frac{R-h}{R}.$$

Then

$$h = R(1 - \cos\theta) = \frac{L}{2\theta}\left(1 - \sqrt{1 - \sin^2\theta}\right) = \frac{L}{2\theta}\left[1 - \sqrt{1 - (1 - l/L)^2\theta^2}\right]$$
$$= \frac{L}{2\theta}\frac{1 - \left[1 - (1 - l/L)^2\theta^2\right]}{1 + \sqrt{1 - (1 - l/L)^2\theta^2}} = \frac{L(1 - l/L)^2\theta}{2\left[1 + \sqrt{1 - (1 - l/L)^2\theta^2}\right]}.$$

(b) Setting $L = 5280$ and $l = 1$ we have $\sin\theta = \left(1 - \dfrac{1}{5280}\right)\theta$ or $f(\theta) = \sin\theta - \dfrac{5279}{5280}\theta = 0$. The formula for Newton's Method is

$$\theta_{n+1} = \theta_n - \frac{f(\theta_n)}{f'(\theta_n)} = \theta_n - \frac{\sin\theta_n - \dfrac{5279}{5280}\theta_n}{\cos\theta_n - \dfrac{5279}{5280}} = \frac{\theta_n\cos\theta_n - \sin\theta_n}{\cos\theta_n - 5279/5280}.$$

Taking $\theta_0 = 0.1$ we obtain $\theta_1 \approx 0.069282$, $\theta_2 \approx 0.050143$, $\theta_3 \approx 0.039358$, $\theta_4 \approx 0.034732$, $\theta_5 \approx 0.033754$, $\theta_6 \approx 0.033711$, $\theta_7 \approx 0.033711$, so $\theta \approx 0.033711$ and $h \approx 44.494$ ft.

(c) From $\sin\theta = (1 - l/L)\theta \approx \theta - \theta^3/6$ we obtain $l/L \approx \theta^2/6$ and $\theta \approx \sqrt{6l/L}$. Then

$$h \approx \frac{L\theta}{4} \approx \frac{L\sqrt{6l/L}}{4} \approx \sqrt{\frac{3lL}{8}}.$$

Setting $l = 1$ and $L = 5280$ we find $h \approx 44.4972$, which is very close to the result obtained in (b).

27. Setting $M = 4m$ we have $4mg\dfrac{r}{2}\sin\theta - mgr\theta = 0$ or $2\sin\theta - \theta = 0$. Setting $f(\theta) = 2\sin\theta - \theta$ we obtain $f'(\theta) = 2\cos\theta - 1$ and

$$\theta_{n+1} = \theta_n - \frac{2\sin\theta_n - \theta_n}{2\cos\theta_n - 1} = \frac{2\theta_n\cos\theta_n - 2\sin\theta_n}{2\cos\theta_n - 1}.$$

Taking $\theta_0 = 2$ we find $\theta_1 \approx 1.9010$, $\theta_2 \approx 1.8955$, $\theta_3 \approx 1.8955$, and $\theta \approx 1.8955$ radians.

29. $f(x) = 2x^5 + 3x^4 - 7x^3 + 2x^2 + 8x - 8$; $f'(x) = 10x^4 + 12x^3 - 21x^2 + 4x + 8$

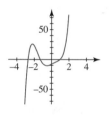

$x_{n+1} = x_n - \dfrac{f(x_n)}{f'(x_n)}$

$x_0 = -1$, $x_1 \approx -1.3158$, $x_2 \approx -1.2517$, $x_3 \approx -1.2494$, $x_4 \approx -1.2494$

$x_0 = -3$, $x_1 \approx -2.7679$, $x_2 \approx -2.6776$, $x_3 \approx -2.6641$, $x_4 \approx -2.6638$,

$x_5 \approx -2.6638$

$x_0 = 1$, $x_1 = 1$

31. (a)

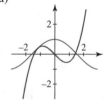

(b)

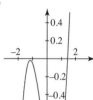

(c) The number of roots appears to be two in part (a), but the "zoomed in" graph in part (b) shows there is only one root.

(d) $y = 0.5x^3 - x - \cos x$

$y' = 1.5x^2 - 1 + \sin x$

$x_{n+1} = x_n - \dfrac{0.5x_n^3 - x_n - \cos x_n}{1.5x_n^2 - 1 + \sin x_n}$

$x_0 = 1.5$, $x_1 \approx 1.4654$, $x_2 \approx 1.4645$,

$x_3 \approx 1.4645$

33. $f'(x) = \begin{cases} \dfrac{1}{2\sqrt{4-x}}, & x < 4 \\[2mm] \dfrac{1}{2\sqrt{x-4}}, & x > 4 \end{cases}$

Since $f'(4)$ does not exist, Newton's Method will fail for $x_0 = 4$. For any choice of $x_0 < 4$,

$$x_1 = x_0 - \frac{f(x_0)}{f'(x_0)} = x_0 - (2x_0 - 8) = 8 - x_0.$$

Since $x_1 = 8 - x_0 > 4$, $f(x_1) = \sqrt{x_1 - 4} = \sqrt{8 - x_0 - 4} = \sqrt{4 - x_0} = -f(x_0)$, and $f'(x_1) = \dfrac{1}{2\sqrt{x_1 - 4}} = \dfrac{1}{2\sqrt{8 - x_0 - 4}} = \dfrac{1}{2\sqrt{4 - x_0}} = f'(x_0)$. By Problem 32, then, $x_2 = x_0$. The same result is obtained for any choice of $x_0 > 4$. Newton's Method will therefore yield the sequence of iterates $x_0, x_1, x_0, x_1, \ldots$, and thus fail to converge.

Chapter 4 in Review

A. True/False

1. False; the function may not be differentiable, or $f'(x)$ may be 0 for some x on the interval.

3. False; this is only true when the velocity is positive.

5. True

7. False; $f(x)$ need not be differentiable at c.

9. True

11. True

13. True

15. False; this is an indeterminate form.

17. True

19. False; let a be ∞, $f(x) = x^2$ and $g(x) = e^x$.

B. Fill In the Blanks

1. Velocity

3. $f(x) = x^{1/3}$

5. 0

7. 2

9. $\Delta y = (x + \Delta x)^2 - (x + \Delta x) - (x^2 - x) = (2x - 1)\Delta x + (\Delta x)^2$

C. Exercises

1. Solving $f'(x) = 3x^2 - 75 = 0$, we obtain the critical numbers -5 and 5, neither of which is in the interval $[-3, 4]$. Comparing $f(-3) = 348$ and $f(4) = -86$, we see that the absolute maximum is 348 and the absolute minimum is -86.

3. Solving $f'(x) = x(x + 8)/(x + 4)^2 = 0$, we obtain the critical numbers -8 and 0. Comparing $f(-1) = 1/3$, $f(0) = 0$, and $f(3) = 9/7$, we see that the absolute maximum is $9/7$ and the absolute minimum is 0.

5.

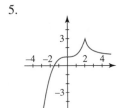

7. $v(t) = -3t^2 + 12t$. Solving $v(t) = 0$ we obtain $t = 0, 4$. To find the maximum velocity, we solve $v'(t) = -6t + 12 = 0$ and obtain $t = 2$. Comparing $v(-1) = -15$, $v(2) = 12$, and $v(5) = -15$, we see that the maximum velocity is 12. Since speed is the absolute value of velocity, the maximum speed on the interval is 15 when $t = -1$ and $t = 5$.

t	-1		0		4		5
s	7	$\leftarrow$	0	$\rightarrow$	32	$\leftarrow$	25
v		$-$	0	$+$	0	$-$	

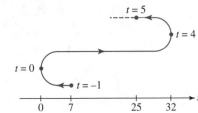

9. (a) $f'(x) = (x-a)^2[(x-b)^2g'(x) + 2(x-b)g(x)] + 2(x-a)(x-b)^2g(x)$

$= (x-a)(x-b)[(x-a)(x-b)g'(x) + 2(x-a)g(x) + 2(x-b)g(x)]$

We see immediately that $f(a) = f(b) = 0$, so by Rolle's Theorem there exists c in (a,b) such that $f'(c) = 0$.

(b) If $g(x) = C$ then $g'(x) = 0$ and $f'(x) = (x-a)(x-b)[2C(x-a+x-b)] =$

$2C(x-a)(x-b)(2x-a-b)$. Solving $f'(x) = 0$ we obtain the critical numbers a, b, and $\dfrac{a+b}{2}$.

11. Solving $f'(x) = 6x^2 + 6x - 36 = 6(x+3)(x-2) = 0$ we obtain the critical numbers -3 and 2. The relative maximum is $(-3, 81)$ and the relative minimum is $(2, -44)$.

x		-3		2	
f	↗	81	↘	-44	↗
f'	$+$	0	$-$	0	$+$

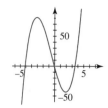

13. Solving $f'(x) = 4 - 4x^{-1/3} = 4(x^{1/3} - 1)/x^{1/3} = 0$ we obtain the critical number 1. Since $f'(0)$ does not exist, 0 is also a critical number. The relative maximum is $(0, 2)$ and the relative minimum is $(1, 0)$.

x		0		1	
f	↗	2	↘	0	↗
f'	$+$	undefined	$-$	0	$+$

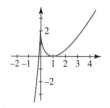

15. Solving $f'(x) = 4x^3 + 24x^2 + 36x = 4x(x+3)^2 = 0$ we obtain the critical numbers -3 and 0. Solving $f''(x) = 12x^2 + 48x + 36 = 12(x+1)(x+3) = 0$ we obtain $x = -1$ and -3. The relative minimum is $(0, 0)$. The inflection points are $(-3, 27)$ and $(-1, 11)$.

x		-3		-1		0	
f	↘	27	↘	11	↘	0	↗
f'	$-$	0	$-$	$-$	$-$	0	$+$
f''	$+$	0	$-$	0	$+$	$+$	$+$

17. Since $f'(x) = -\dfrac{1}{3}(x-3)^{-2/3}$, we see that the only critical number is 3. $f''(x) = \dfrac{2}{9}(x-3)^{-5/3}$ is undefined at $x = 3$. There are no relative extrema. The inflection point is $(3, 10)$.

x		3	
f		10	
f'	$-$	undefined	$-$
f''	$-$	undefined	$+$

19. (c), (d)

21. (c), (d), (e)

23. (a), (c)

25. $f'(x) = (x-a)[(x-b)+(x-c)] + (x-b)(x-c) = (x-a)(x-b) + (x-a)(x-c) + (x-b)(x-c)$

$f''(x) = x-a+x-b+x-a+x-c+x-b+x-c = 6x - 2(a+b+c)$

Solving $f'(x) = 0$ we obtain $x = \dfrac{1}{3}(a+b+c)$. Since $f''(\dfrac{1}{6}(a+b+c)) = -(a+b+c)$ and

$f''(a+b+c) = 4(a+b+c)$, we see that $f''(x)$ has opposite signs on either side of $\dfrac{1}{3}(a+b+c)$. Thus, the graph of $f(x)$ has a

point of inflection at $\dfrac{1}{3}(a+b+c)$.

27. Assume that the center of the circle is at the origin so that $x^2 + y^2 = r^2$. In the first quadrant a vertex of the square is $(x,y) = (x,x)$. Therefore $2x^2 = r^2$, $A = (2x)(2x)$, $A(r) = 2r^2$, and $\dfrac{dA}{dt} = 4r\dfrac{dr}{dt}$. At the instant when $r = 2$,

$$\frac{dA}{dt} = 4(2)(4) = 32 \text{ in}^2/\text{min}.$$

29. $B'(x) = \frac{1}{2}\mu_0 r_0^2 I \left\{ -\frac{3}{2}\left[r_0^2 + \left(x + \frac{r_0}{2}\right)^2\right]^{-5/2}\left[2\left(x + \frac{r_0}{2}\right)\right]\right.$

$$\left. -\frac{3}{2}\left[r_0^2 + \left(x - \frac{r_0}{2}\right)^2\right]^{-5/2}\left[2\left(x - \frac{r_0}{2}\right)\right]\right\}$$

$$= -\frac{3}{2}\mu_0 r_0^2 I\left\{\frac{x + r_0/2}{\left[r_0^2 + \left(x + \frac{r_0}{2}\right)^2\right]^{5/2}} + \frac{x - r_0/2}{\left[r_0^2 + \left(x - \frac{r_0}{2}\right)^2\right]^{5/2}}\right\}$$

We see that $B'(0) = 0$. To apply the first derivative test, we compute

$$B'\left(-\frac{r_0}{4}\right) = -\frac{384\mu_0 I}{r_0^2}\left(\frac{1}{17^{5/2}} - \frac{3}{25^{5/2}}\right) \quad \text{and} \quad B'\left(\frac{r_0}{4}\right) = -\frac{384\mu_0 I}{r_0^2}\left(\frac{3}{25^{5/2}} - \frac{1}{17^{5/2}}\right).$$

Since $B'\left(-\frac{r_0}{4}\right) > 0$ and $B'\left(\frac{r_0}{4}\right) < 0$, B has a maximum at $x = 0$.

31. $\frac{dx}{dy} = \frac{h - 2y}{\sqrt{y(h - y)}}$. Solving $\frac{dx}{dy} = 0$ we obtain the critical number $h/2$. Since $0 \le y \le h$, we compare $z(0) = 0$, $x(h/2) = h$, $x(h) = 0$, and observe that the maximum distance of h ft is obtained for $y = h/2$.

33. Since 585 feet of fence is available, $y = 585 - x$. We want to maximize $A(x) = x(y - x) + \frac{1}{2}x^2 = x(585 - 2x) + \frac{1}{2}x^2 = 585x - \frac{3}{2}x^2$. Solving $A'(x) = 585 - 3x$ we obtain the critical number 195. Since $A''(x) = -3 < 0$, the maximum area is obtained when $x = 195$ ft and $y = 390$ ft. The maximum area is $A(195) = 57037.5$ ft^2.

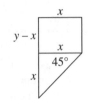

35. We want to minimize $T(x) = \frac{\sqrt{x^2 + h_1^2}}{c} + \frac{\sqrt{(d - x)^2 + h_2^2}}{c}$ on $[0, d]$. Setting

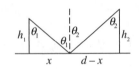

$$T'(x) = \frac{1}{c}\left[\frac{x}{\sqrt{x^2 + h_1^2}} - \frac{d - x}{\sqrt{(d - x)^2 + h_2^2}}\right] = 0,$$

we obtain $x\sqrt{(d - x)^2 + h_2^2} = (d - x)\sqrt{x^2 + h_1^2}$ or $x^2[(d - x)^2 + h_2^2] = (d - x)^2(x^2 + h_1^2)$. Simplifying, we have $\frac{d - x}{h_2} = \frac{x}{h_1}$ or $\tan\theta_2 = \tan\theta_1$. Solving for x, we find $x = \frac{h_1 d}{h_1 + h_2}$. Since

$$T''(x) = \frac{1}{c}\left\{\frac{h_1^2}{(x^2 + h_1^2)^{3/2}} + \frac{h_2^2}{[(d - x)^2 + h_2^2]^{3/2}}\right\} > 0,$$

the time is minimized when $\tan\theta_2 = \tan\theta_1$.

37. Let r be the radius and h the height. Then $\pi r^2 h = 100$ and $h = 100/\pi r^2$. We want to minimize $C(r) = 3\pi r^2 + \pi r^2 + 2\pi rh = 4\pi r^2 + 200/r$. Solving $C'(r) = 8\pi r - 200/r^2 = 0$ we obtain the critical number $\sqrt[3]{25/\pi}$. Since $C''(\sqrt[3]{25/\pi}) > 0$, the cost is minimized when $r = \sqrt[3]{25/\pi}$ and $h = 4\sqrt[3]{25/\pi}$. (In this case $h = 4r$.)

39. $\lim\limits_{x \to \sqrt{3}} \frac{\sqrt{3} - \tan(\pi/x^2)}{x - \sqrt{3}} \overset{h}{=} \lim\limits_{x \to \sqrt{3}} \frac{[\sec^2(\pi/x^2)](2\pi/x^3)}{1} = \frac{8\pi}{3\sqrt{3}}$

41. $\lim\limits_{x \to \infty} x\left(\cos\frac{1}{x} - e^{2/x}\right) = \lim\limits_{x \to \infty} \frac{\cos(1/x) - e^{2/x}}{1/x} \overset{h}{=} \lim\limits_{x \to \infty} \frac{[\sin(1/x)](1/x^2) - e^{2/x}(-2/x^2)}{-1/x^2}$

$$= \lim\limits_{x \to \infty} \frac{\sin(1/x) + 2e^{2/x}}{-1} = -2$$

43. $\lim\limits_{t \to 0} \frac{(\sin t)^2}{\sin t^2} \overset{h}{=} \lim\limits_{t \to 0} \frac{2\sin t\cos t}{2t\cos t^2} = \lim\limits_{t \to 0} \frac{\sin 2t}{2t\cos t^2} \overset{h}{=} \lim\limits_{t \to 0} \frac{2\cos 2t}{-4t^2\sin t^2 + 2\cos t^2} = 1$

45. Set $y = (3x)^{-1/\ln x}$. Then $\ln y = -\dfrac{\ln 3x}{\ln x}$ and $\displaystyle\lim_{x \to 0^+} \left(-\dfrac{\ln 3x}{\ln x} \right) \overset{h}{=} \lim_{x \to 0^+} \dfrac{-1/x}{1/x} = -1$. Thus, $\displaystyle\lim_{x \to 0^+} (3x)^{-1/\ln x} = e^{-1}$.

47. $\displaystyle\lim_{x \to \infty} \ln \left(\dfrac{x + e^{2x}}{1 + e^{4x}} \right) = \ln \left(\lim_{x \to \infty} \dfrac{x + e^{2x}}{1 + e^{4x}} \right) \overset{h}{=} \ln \left(\lim_{x \to \infty} \dfrac{1 + 2e^{2x}}{4e^{4x}} \right) \overset{h}{=} \ln \left(\lim_{x \to \infty} \dfrac{4e^{2x}}{16e^{4x}} \right)$

$\qquad = \ln \left(\displaystyle\lim_{x \to \infty} \dfrac{1}{4e^{2x}} \right) = \ln 0 = -\infty$

The limit does not exist.

49. Let $f(x) = x^3 - 4x + 2$. Then $f'(x) = 3x^2 - 4$ and

$$x_{n+1} = x_n - \dfrac{x_n^3 - 4x_n + 2}{3x_n^2 - 4} = \dfrac{2x_n^3 - 2}{3x_n^2 - 4}.$$

From the graph we see that $f(x)$ has its largest positive root near $x_0 = 2$. Then $x_1 \approx 1.7500$, $x_2 \approx 1.6807$, $x_3 \approx 1.6752$, $x_4 \approx 1.6751$, $x_5 \approx 1.6751$, and the largest positive root is approximately 1.6751.

Integrals

5.1 The Indefinite Integral

1. $\displaystyle\int 3\,dx = 3x + C$

3. $\displaystyle\int x^5\,dx = \frac{1}{6}x^6 + C$

5. $\displaystyle\int \frac{dx}{\sqrt[3]{x}} = \int x^{-1/3}\,dx = \frac{3}{2}x^{2/3} + C$

7. $\displaystyle\int (1 - t^{-0.52})\,dt = t - \frac{1}{0.48}t^{0.48} + C$

9. $\displaystyle\int (3x^2 + 2x - 1)\,dx = x^3 + x^2 - x + C$

11. $\displaystyle\int \sqrt{x}(x^2 - 2)\,dx = \int (x^{5/2} - 2x^{1/2})\,dx = \frac{2}{7}x^{7/2} - \frac{4}{3}x^{3/2} + C$

13. $\displaystyle\int (4x+1)^2\,dx = \int (16x^2 + 8x + 1)\,dx = \frac{16}{3}x^3 + 4x^2 + x + C$

15. $\displaystyle\int (4w - 1)^3\,dw = \int (64w^3 - 48w^2 + 12w - 1)\,dw = 16w^4 - 16w^3 + 6w^2 - w + C$

17. $\displaystyle\int \frac{r^2 - 10r + 4}{r^3}\,dr = \int (r^{-1} - 10r^{-2} + 4r^{-3})\,dr = \ln|r| + 10r^{-1} - 2r^{-2} + C$

19. $\displaystyle\int \frac{x^{-1} - x^{-2} + x^{-3}}{x^2}\,dx = \int (x^{-3} - x^{-4} + x^{-5})\,dx = -\frac{1}{2}x^{-2} + \frac{1}{3}x^{-3} - \frac{1}{4}x^{-4} + C$

21. $\displaystyle\int (4\sin x - 1 + 8x^{-5})\,dx = -4\cos x - x - 2x^{-4} + C$

23. $\displaystyle\int \csc x(\csc x - \cot x)\,dx = \int (\csc^2 x - \csc x \cot x)\,dx = -\cot x + \csc x + C$

25. $\displaystyle\int \frac{2 + 3\sin^2 x}{\sin^2 x}\,dx = \int (2\csc^2 x + 3)\,dx = -2\cot x + 3x + C$

27. $\displaystyle\int (8x + 1 - 9e^x)\,dx = 4x^2 + x - 9e^x + C$

29. $\displaystyle\int \frac{2x^3 - x^2 + 2x + 4}{1 + x^2}\,dx = \int \left(2x - 1 + \frac{5}{x^2 + 1} \right)\,dx = x^2 - x + 5\tan^{-1}x + C$

31. $\int \tan^2 x\,dx = \int (\sec^2 x - 1)\,dx = \tan x - x + C$

33. $\frac{d}{dx}(\sqrt{2x+1} + C) = \frac{2}{2\sqrt{2x+1}} = \frac{1}{\sqrt{2x+1}}$

35. $\frac{d}{dx}\left(\frac{1}{4}\sin 4x + C\right) = \frac{4}{4}\cos 4x = \cos 4x$

37. $\frac{d}{dx}\left(-\frac{1}{2}\cos x^2 + C\right) = \frac{2x}{2}\sin x^2 = x\sin x^2$

39. $\frac{d}{dx}(x\ln x - x + C) = x\left(\frac{1}{x}\right) + \ln x - 1 = \ln x$

41. $\frac{d}{dx}\int (x^2 - 4x + 5)\,dx = \frac{d}{dx}\left(\frac{1}{3}x^3 - 2x^2 + 5x + C\right) = x^2 - 4x + 5$

43. $y = \int (6x^2 + 9)\,dx = 2x^3 + 9x + C$

45. $y = \int x^{-2}\,dx = -x^{-1} + C = -\frac{1}{x} + C$

47. $y = \int (1 - 2x + \sin x)\,dx = x - x^2 - \cos x + C$

49. We have $f(x) = \int (2x - 1)\,dx = x^2 - x + C$. Solving $3 = f(2) = 4 - 2 + C = 2 + C$ we obtain $C = 1$. Thus $f(x) = x^2 - x + 1$.

51. $f'(x) = \int 2x\,dx = x^2 + C_1; \quad f(x) = \int (x^2 + C_1)\,dx = \frac{1}{3}x^3 + C_1 x + C_2$

53. We have $f'(x) = \int (12x^2 + 2)\,dx = 4x^3 + 2x + C$. Solving $3 = f'(1) = 6 + C$ we obtain $C = -3$. Then $f'(x) = 4x^3 + 2x - 3$ and $f(x) = \int (4x^3 + 2x - 3)\,dx = x^4 + x^2 - 3x + C$. Solving $1 = f(1) = -1 + C$ we obtain $C = 2$. Thus $f(x) = x^4 + x^2 - 3x + 2$.

55. G is an antiderivative of f. In other words, since $G'(x) = f(x)$, f is the slope function for G. Observe where G is increasing, and the graph of f is always positive. Also, G appears to have no relative extrema on the interval shown, and correspondingly the graph of f does not cross the x-axis.

57. $y = \int \left(\frac{\omega^2}{g}x\right)dx = \frac{\omega^2}{2g}x^2 + C$. From Figure 5.1.5 we see that $y(0) = 0$. Thus, $0 = y(0) = C$, and $y = \frac{\omega^2 x^2}{2g}$.

59. $\frac{d}{dx}(\ln|\ln x| + C) = \frac{1}{\ln x}\left(\frac{1}{x}\right) = \frac{1}{x\ln x}$

61. Since $f'(x) = x^2$, $f(x) = \int x^2\,dx = \frac{1}{3}x^3 + C$. Since $y = 4x + 7$ is a tangent line to the graph of f, then $4x + 7 = \frac{1}{3}x^3 + C$ at some point on f. In addition, the slope at this point is $4 = f'(x) = x^2$, so $x = \pm 2$. Thus, $4(\pm 2) + 7 = \frac{1}{3}(\pm 2)^3 + C$, so $C = 37/3$ or $5/3$. Thus, $f(x) = \frac{1}{3}x^3 + \frac{37}{3}$ or $f(x) = \frac{1}{3}x^3 + \frac{5}{3}$.

63. $\frac{d}{dx}\left[\frac{1}{4}(x+1)^4 + C\right] = (x+1)^3$

$\frac{d}{dx}\left(\frac{1}{4}x^4 + x^3 + \frac{3}{2}x^2 + x + C\right) = x^3 + 3x^2 + 3x + 1 = (x+1)^3$

Thus, both results are correct.

5.2 Integration by the *u*-Substitution

1. $\displaystyle\int \sqrt{1-4x}\,dx = -\frac{1}{4}\int (1-4x)^{1/2}(-4\,dx)$ $\boxed{u=1-4x,\ du=-4\,dx}$

$$= -\frac{1}{4}\int u^{1/2}\,du = \frac{1}{6}u^{3/2}+C = -\frac{1}{6}(1-4x)^{3/2}+C$$

3. $\displaystyle\int \frac{1}{(5x+1)^3}\,dx = \frac{1}{5}\int (5x+1)^{-3}(5\,dx)$ $\boxed{u=5x+1,\ du=5\,dx}$

$$= \frac{1}{5}\int u^{-3}\,du = -\frac{1}{10}u^{-2}+C = -\frac{1}{10(5x+1)^2}+C$$

5. $\displaystyle\int x\sqrt{x^2+4}\,dx = \frac{1}{2}\int \sqrt{x^2+4}\,(2x\,dx)$ $\boxed{u=x^2+4,\ du=2x\,dx}$

$$= \frac{1}{2}\int u^{1/2}\,du = \frac{1}{3}u^{3/2}+C = \frac{1}{3}(x^2+4)^{3/2}+C$$

7. $\displaystyle\int \sin^5 3x\cos 3x\,dx = \frac{1}{3}\int (\sin^5 3x)(3\cos 3x\,dx)$ $\boxed{u=\sin 3x,\ du=3\cos 3x\,dx}$

$$= \frac{1}{3}\int u^5\,du = \frac{1}{18}u^6+C = \frac{1}{18}\sin^6 3x+C$$

9. $\displaystyle\int \tan^2 2x\sec^2 2x\,dx = \frac{1}{2}\int (\tan^2 2x)(2\sec^2 2x\,dx)$ $\boxed{u=\tan 2x,\ du=2\sec^2 2x\,dx}$

$$= \frac{1}{2}\int u^2\,du = \frac{1}{6}u^3+C = \frac{1}{6}\tan^3 2x+C$$

11. $\displaystyle\int \sin 4x\,dx = \frac{1}{4}\int (\sin 4x)(4\,dx) = -\frac{1}{4}\cos 4x+C$

13. $\displaystyle\int (\sqrt{2t}-\cos 6t)\,dt = \sqrt{2}\int t^{1/2}\,dt - \frac{1}{6}\int (\cos 6t)(6\,dt) = \frac{2\sqrt{2}}{3}t^{3/2} - \frac{1}{6}\sin 6t+C$

$$= \frac{1}{3}(2t)^{3/2} - \frac{1}{6}\sin 6t+C$$

15. $\displaystyle\int x\sin x^2\,dx = \frac{1}{2}\int (\sin x^2)(2x\,dx) = -\frac{1}{2}\cos x^2+C$

17. $\displaystyle\int x^2\sec^2 x^3\,dx = \frac{1}{3}\int (\sec^2 x^3)(3x^2\,dx) = \frac{1}{3}\tan x^3+C$

19. $\displaystyle\int \frac{\csc\sqrt{x}\cot\sqrt{x}}{\sqrt{x}}\,dx = 2\int \frac{\csc\sqrt{x}\cot\sqrt{x}}{2\sqrt{x}}\,dx$ $\boxed{u=\sqrt{x},\ du=dx/2\sqrt{x}}$

$$= 2\int \csc u\cot u\,du = -2\csc u+C = -2\csc\sqrt{x}+C$$

21. $\displaystyle\int \frac{1}{7x+3}\,dx = \frac{1}{7}\int \frac{1}{7x+3}(7\,dx)$ $\boxed{u=7x+3,\ du=7\,dx}$

$$= \frac{1}{7}\int \frac{1}{u}\,du = \frac{1}{7}\ln|u|+C = \frac{1}{7}\ln|7x+3|+C$$

23. $\displaystyle\int \frac{x}{x^2+1}\,dx = \frac{1}{2}\int \frac{2x\,dx}{x^2+1} = \frac{1}{2}\ln(x^2+1)+C$

25. $\displaystyle\int \frac{x}{x+1}\,dx = \int \frac{x+1-1}{x+1}\,dx = \int dx - \int \frac{dx}{x+1} = x - \ln|x+1|+C$

27. $\displaystyle\int \frac{1}{x\ln x}\,dx = \frac{1}{\ln x}\left(\frac{1}{x}\,dx\right)$ $\boxed{u=\ln x,\ du=\frac{1}{x}\,dx}$

$$= \int \frac{1}{u}\,du = \ln|u|+C = \ln|\ln x|+C$$

29. $\int \dfrac{\sin(\ln x)}{x}\,dx = \int \sin(\ln x)\left(\dfrac{1}{x}\,dx\right)$ $\boxed{u = \ln x,\ du = \dfrac{1}{x}\,dx}$

$$= \int \sin u\,du = -\cos u + C = -\cos(\ln x) + C$$

31. $\int e^{10x}\,dx = \dfrac{1}{10}\int e^{10x}(10\,dx) = \dfrac{1}{10}e^{10x} + C$

33. $\int x^2 e^{-2x^3}\,dx = -\dfrac{1}{6}\int e^{-2x^3}(-6x^2\,dx) = -\dfrac{1}{6}e^{-2x^3} + C$

35. $\int \dfrac{e^{-\sqrt{x}}}{\sqrt{x}}\,dx = -2\int e^{-\sqrt{x}}\left(-\dfrac{1}{2\sqrt{x}}\,dx\right) = -2e^{-\sqrt{x}} + C$

37. $\int \dfrac{e^x - e^{-x}}{e^x + e^{-x}}\,dx = \int \dfrac{1}{e^x + e^{-x}}[(e^x - e^{-x})\,dx]$ $\boxed{u = e^x + e^{-x},\ du = (e^x - e^{-x})\,dx}$

$$= \int \dfrac{1}{u}\,du = \ln|u| + C = \ln(e^x + e^{-x}) + C$$

39. $\int \dfrac{1}{\sqrt{5 - x^2}}\,dx = \sin^{-1}\dfrac{x}{\sqrt{5}} + C$

41. $\int \dfrac{1}{1 + 25x^2}\,dx = \dfrac{1}{5}\int \dfrac{1}{1 + (5x)^2}(5\,dx) = \dfrac{1}{5}\tan^{-1}5x + C$

43. $\int \dfrac{e^x}{1 + e^{2x}}\,dx = \int \dfrac{1}{1 + (e^x)^2}(e^x\,dx)$ $\boxed{u = e^x,\ du = e^x\,dx}$

$$= \int \dfrac{1}{1 + u^2}\,du = \tan^{-1}u + C = \tan^{-1}e^x + C$$

45. $\int \dfrac{2x - 3}{\sqrt{1 - x^2}}\,dx = \int \dfrac{1}{\sqrt{1 - x^2}}(2x\,dx) - \int \dfrac{3}{\sqrt{1 - x^2}}\,dx$ $\boxed{u = x^2,\ du = 2x\,dx}$

$$= \int \dfrac{1}{\sqrt{1 - u}}\,du - 3\int \dfrac{1}{\sqrt{1 - x^2}}\,dx = -2(1 - u)^{1/2} - 3\sin^{-1}x + C$$

$$= -2(1 - x^2)^{1/2} - 3\sin^{-1}x + C$$

47. $\int \dfrac{\tan^{-1}x}{1 + x^2}\,dx = \int (\tan^{-1}x)\left(\dfrac{1}{1 + x^2}\,dx\right)$ $\boxed{u = \tan^{-1}x,\ du = \dfrac{1}{1 + x^2}\,dx}$

$$= \int u\,du = \dfrac{1}{2}u^2 + C = \dfrac{1}{2}(\tan^{-1}x)^2 + C$$

49. $\int \tan 5x\,dx = \dfrac{1}{5}\int (\tan 5x)(5\,dx) = -\dfrac{1}{5}\ln|\cos 5x| + C$

51. $\int \sin^2 x\,dx = \int \dfrac{1}{2}(1 - \cos 2x)\,dx = \dfrac{1}{2}\left(x - \dfrac{1}{2}\sin 2x\right) + C$

53. $\int \cos^2 4x\,dx = \int \dfrac{1}{2}(1 + \cos 8x)\,dx = \dfrac{1}{2}\left(x + \dfrac{1}{8}\sin 8x\right) + C$

55. $\int (3 - 2\sin x)^2\,dx = \int (9 - 12\sin x + 4\sin^2 x)\,dx = 9x + 12\cos x + 4\int \dfrac{1}{2}(1 - \cos 2x)\,dx$

$$= 9x + 12\cos x + 2\left(x - \dfrac{1}{2}\sin 2x\right) + C = 11x + 12\cos x - \sin 2x + C$$

57. $y = \int \sqrt[3]{1 - x}\,dx = -\int (1 - x)^{1/3}(-dx) = -\dfrac{3}{4}(1 - x)^{4/3} + C$

59. We have $f(x) = \int (1 - 6\sin 3x)\,dx = x + 2\cos 3x + C$. Solving $-1 = f(\pi) = \pi + 2\cos 3\pi + C = \pi - 2 + C$ we obtain $C = 1 - \pi$. Thus $f(x) = x + 2\cos 3x + 1 - \pi$.

61. (a) $\displaystyle\int \sin x \cos x\,dx = \int \sin x(\cos x\,dx)$ $\boxed{u = \sin x,\ du = \cos x\,dx}$

$$= \int u\,du = \frac{1}{2}u^2 + C_1 = \frac{1}{2}\sin^2 x + C_1$$

(b) $\displaystyle\int \sin x \cos x\,dx = -\int \cos x(-\sin x\,dx)$ $\boxed{u = \cos x,\ du = -\sin x\,dx}$

$$= -\int u\,du = -\frac{1}{2}u^2 + C_2 = -\frac{1}{2}\cos^2 x + C_2$$

(c) $\displaystyle\int \sin x \cos x\,dx = \frac{1}{2}\int \sin 2x\,dx = -\frac{1}{4}\cos 2x + C_3$

63. (a) From the given derivative, we have $t(s) = \sqrt{\dfrac{L}{g}}\sin^{-1}\left(\dfrac{s}{s_C}\right) + C$. Solving $t(0) = 0$, we obtain $C = 0$.

(b) $t(s_C) = \sqrt{\dfrac{L}{g}}\sin^{-1}\left(\dfrac{s_C}{s_C}\right) = \sqrt{\dfrac{L}{g}}\sin^{-1}1 = \dfrac{\pi}{2}\sqrt{\dfrac{L}{g}}$

(c) By symmetry, $T = 4t(s_C) = 4\left(\dfrac{\pi}{2}\sqrt{\dfrac{L}{g}}\right) = 2\pi\sqrt{\dfrac{L}{g}}$.

65. $\displaystyle\int \cos^4 x\,dx = \int (\cos^2 x)^2\,dx = \int \left[\frac{1}{2}(1 + \cos 2x)\right]^2 dx = \frac{1}{4}\int (1 + 2\cos 2x + \cos^2 2x)\,dx$

$$= \frac{1}{4}x + \frac{1}{4}\sin 2x + \frac{1}{4}\int \frac{1}{2}(1 + \cos 4x)\,dx = \frac{1}{4}x + \frac{1}{4}\sin 2x + \frac{1}{8}\int (1 + \cos 4x)\,dx$$

$$= \frac{1}{4}x + \frac{1}{4}\sin 2x + \frac{1}{8}x + \frac{1}{32}\sin 4x + C = \frac{3}{8}x + \frac{1}{4}\sin 2x + \frac{1}{32}\sin 4x + C$$

67. $\displaystyle\int \frac{1}{x\sqrt{x^4 - 16}}\,dx = \int \frac{1}{2x^2\sqrt{x^4 - 16}}(2x\,dx)$ $\boxed{u = x^2,\ du = 2x\,dx}$

$$= \frac{1}{2}\int \frac{1}{u\sqrt{u^2 - 4^2}}\,du = \frac{1}{4}\sec^{-1}\left|\frac{u}{4}\right| + C = \frac{1}{4}\sec^{-1}\frac{x^2}{4} + C$$

69. $\displaystyle\int \frac{1}{1 - \cos x}\,dx = \int \frac{1}{1 - \cos x}\left(\frac{1 + \cos x}{1 + \cos x}\right)dx = \int \frac{1 + \cos x}{1 - \cos^2 x}\,dx = \int \frac{1 + \cos x}{\sin^2 x}\,dx$

$$= \int \left[\frac{1}{\sin^2 x} + \frac{\cos x}{(\sin x)(\sin x)}\right]dx = \int (\csc^2 x + \csc x \cot x)\,dx$$

$$= -\cot x - \csc x + C$$

71. $\displaystyle\int f'(8x)\,dx = \frac{1}{8}\int f'(8x)(8\,dx)$ $\boxed{u = 8x,\ du = 8\,dx}$

$$= \frac{1}{8}\int f'(u)\,du = \frac{1}{8}f(u) + C = \frac{1}{8}f(8x) + C$$

73. $\displaystyle\int \sqrt{f(2x)}\,f'(2x)\,dx = \frac{1}{2}\int [f(2x)]^{1/2}[2f'(2x)\,dx]$ $\boxed{u = f(2x),\ du = 2f'(2x)\,dx}$

$$= \frac{1}{2}\int u^{1/2}\,du = \frac{1}{3}u^{3/2} + C = \frac{1}{3}[f(2x)]^{3/2} + C$$

75. For any f, $\displaystyle\int f''(4x)\,dx = \frac{1}{4}\int f''(4x)(4\,dx)$ $\boxed{u = 4x,\ du = 4\,dx}$

$$= \frac{1}{4}\int f''(u)\,du = \frac{1}{4}f'(u) + C = \frac{1}{4}f'(4x) + C$$

Given $f(x) = \sqrt{x^4 + 1} = (x^4 + 1)^{1/2}$, we have $f'(x) = 2x^3(x^4 + 1)^{-1/2}$. Thus,

$$\int f''(4x)\,dx = \frac{1}{4}f'(4x) + C = \frac{1}{4}\{2(4x)^3[(4x)^4 + 1]^{-1/2}\} + C = \frac{32x^3}{\sqrt{256x^4 + 1}} + C.$$

To check this, take the derivative of the above function, yielding $\dfrac{96x^2}{\sqrt{256x^4+1}} - \dfrac{16384x^6}{\sqrt{(256x^4+1)^3}}$, which should be the same as

$f''(4x)$. Since $f''(x) = \dfrac{6x^2}{\sqrt{x^4+1}} - \dfrac{4x^6}{\sqrt{(x^4+1)^3}}$, we have $f''(4x) = \dfrac{6(4x)^2}{\sqrt{(4x)^4+1}} - \dfrac{4(4x)^6}{\sqrt{[(4x)^4+1]^3}} = \dfrac{96x^2}{\sqrt{256x^4+1}} - \dfrac{16384x^6}{\sqrt{(256x^4+1)^3}}.$

5.3 The Area Problem

1. $3+6+9+12+15$

3. $2+2+8/3+4$

5. $-\dfrac{1}{7}+\dfrac{1}{9}-\dfrac{1}{11}+\dfrac{1}{13}-\dfrac{1}{15}+\dfrac{1}{17}-\dfrac{1}{19}+\dfrac{1}{21}-\dfrac{1}{23}+\dfrac{1}{25}$

7. $0+3+8+15$

9. $-1+1-1+1-1$

11. $\displaystyle\sum_{k=1}^{7}(2k+1)$

13. $\displaystyle\sum_{k=1}^{13}(3k-2)$

15. $\displaystyle\sum_{k=1}^{5}\dfrac{(-1)^{k+1}}{k}$

17. $\displaystyle\sum_{k=1}^{8}6$

19. $\displaystyle\sum_{k=1}^{4}\dfrac{(-1)^{k+1}}{k^2}\cos\dfrac{k\pi}{p}x$

21. $\displaystyle\sum_{k=1}^{20}2k = 2\sum_{k=1}^{20}k = 2\left(\dfrac{20\cdot 21}{2}\right) = 420$

23. $\displaystyle\sum_{k=1}^{10}(k+1) = \sum_{k=1}^{10}k + \sum_{k=1}^{10}1 = \dfrac{10\cdot 11}{2} + 10\cdot 1 = 65$

25. $\displaystyle\sum_{k=1}^{6}(k^2+3) = \sum_{k=1}^{6}k^2 + \sum_{k=1}^{6}3 = \dfrac{6\cdot 7\cdot 13}{6} + 6\cdot 3 = 109$

27. $\displaystyle\sum_{p=0}^{10}(p^3+4) = 0+4+\sum_{p=1}^{10}p^3 + \sum_{p=1}^{10}4 = 4 + \dfrac{10^2\cdot 11^2}{4} + 10\cdot 4 = 3069$

29. Using $\Delta x = \dfrac{6-0}{n} = \dfrac{6}{n}$ and $f\left(a+k\dfrac{b-a}{n}\right) = f\left(\dfrac{6k}{n}\right) = \dfrac{6k}{n}$ we have

$$A = \lim_{n\to\infty}\sum_{k=1}^{n}\left(\dfrac{6k}{n}\right)\dfrac{6}{n} = \lim_{n\to\infty}\dfrac{36}{n^2}\sum_{k=1}^{n}k = \lim_{n\to\infty}\dfrac{36}{n^2}\cdot\dfrac{n(n+1)}{2} = \lim_{n\to\infty}18\left(1+\dfrac{1}{n}\right) = 18.$$

31. Using $\Delta x = \dfrac{4}{n}$ and $f\left(a+k\dfrac{b-a}{n}\right) = 3+\dfrac{8k}{n}$ we have

$$A = \lim_{n\to\infty}\sum_{k=1}^{n}\left(3+\dfrac{8k}{n}\right)\dfrac{4}{n} = \lim_{n\to\infty}\sum_{k=1}^{n}\left(\dfrac{12}{n}+\dfrac{32k}{n^2}\right) = \lim_{n\to\infty}\left(\dfrac{12}{n}\sum_{k=1}^{n}1 + \dfrac{32k}{n^2}\sum_{k=1}^{n}k\right)$$

$$= \lim_{n\to\infty}\left[\dfrac{12}{n}\cdot n + \dfrac{32k}{n^2}\cdot\dfrac{n(n+1)}{2}\right] = \lim_{n\to\infty}\left[12 + 16\left(1+\dfrac{1}{n}\right)\right] = 28.$$

33. Using $\Delta x = \dfrac{2}{n}$ and $f\left(a + k\dfrac{b-a}{n}\right) = \dfrac{4k^2}{n^2}$ we have

$$A = \lim_{n\to\infty} \sum_{k=1}^{n} \left(\frac{4k^2}{n^2} \cdot \frac{2}{n}\right) = \lim_{n\to\infty} \frac{8}{n^3} \sum_{k=1}^{n} k^2 = \lim_{n\to\infty} \left[\frac{8}{n^3} \cdot \frac{n(n+1)(2n+1)}{6}\right]$$

$$= \lim_{n\to\infty} \frac{4}{3}\left(2 + \frac{3}{n} + \frac{1}{n^2}\right) = \frac{8}{3}.$$

35. Using $\Delta x = \dfrac{2}{n}$ and $f\left(a + k\dfrac{b-a}{n}\right) = \dfrac{4k}{n} - \dfrac{4k^2}{n^2}$ we have

$$A = \lim_{n\to\infty} \sum_{k=1}^{n} \left(\frac{4k}{n} - \frac{4k^2}{n^2}\right)\frac{2}{n} = \lim_{n\to\infty}\left(\frac{8}{n^2}\sum_{k=1}^{n}k - \frac{8}{n^3}\sum_{k=1}^{n}k^2\right)$$

$$= \lim_{n\to\infty}\left[4\left(1 + \frac{1}{n}\right) - \frac{4}{3}\left(2 + \frac{3}{n} + \frac{1}{n^2}\right)\right] = 4 - \frac{8}{3} = \frac{4}{3}.$$

37. Using $\Delta x = \dfrac{1}{n}$ and $f\left(a + k\dfrac{b-a}{n}\right) = 3 + \dfrac{4k}{n} + \dfrac{k^2}{n^2}$ we have

$$A = \lim_{n\to\infty} \sum_{k=1}^{n} \left(3 + \frac{4k}{n} + \frac{k^2}{n^2}\right)\frac{1}{n} = \lim_{n\to\infty}\left(\frac{3}{n}\sum_{k=1}^{n}1 + \frac{4}{n^2}\sum_{k=1}^{n}k + \frac{1}{n^3}\sum_{k=1}^{n}k^2\right)$$

$$= \lim_{n\to\infty}\left[3 + 2\left(1 + \frac{1}{n}\right) + \frac{1}{6}\left(2 + \frac{3}{n} + \frac{1}{n^2}\right)\right] = 3 + 2 + \frac{1}{3} = \frac{16}{3}.$$

39. Using $\Delta x = \dfrac{1}{n}$ and $f\left(a + k\dfrac{b-a}{n}\right) = \dfrac{k^3}{n^3}$ we have

$$A = \lim_{n\to\infty} \sum_{k=1}^{n} \left(\frac{k^3}{n^3} \cdot \frac{1}{n}\right) = \lim_{n\to\infty} \frac{1}{n^4}\sum_{k=1}^{n}k^3 = \lim_{n\to\infty} \frac{1}{4}\left(1 + \frac{2}{n} + \frac{1}{n^2}\right) = \frac{1}{4}.$$

41. Let $A = A_1 + A_2$ where A_1 is the area under $f(x) = 2$ on $[0,1)$ and A_2 is the area under $f(x) = x + 1$ on $[1,4]$. For A_1, we have $\Delta x = \dfrac{1}{n}$, $f\left(a + k\dfrac{b-a}{n}\right) = 2$, and

$$A_1 = \lim_{n\to\infty} \sum_{k=1}^{n} \left(2 \cdot \frac{1}{n}\right) = \lim_{n\to\infty} \frac{2}{n}\sum_{k=1}^{n}1 = \lim_{n\to\infty} \frac{2n}{n} = 2.$$

For A_2, we have $\Delta x = \dfrac{3}{n}$, $f\left(a + k\dfrac{b-a}{n}\right) = 2 + \dfrac{3k}{n}$, and

$$A_2 = \lim_{n\to\infty} \sum_{k=1}^{n} \left(2 + \frac{3k}{n}\right)\frac{3}{n} = \lim_{n\to\infty}\left(\frac{6}{n}\sum_{k=1}^{n}1 + \frac{9}{n^2}\sum_{k=1}^{n}k\right)$$

$$= \lim_{n\to\infty}\left[\frac{6n}{n} + \frac{9}{2}\left(1 + \frac{1}{n}\right)\right] = 6 + \frac{9}{2} = \frac{21}{2}.$$

Then $A = 2 + \dfrac{21}{2} = \dfrac{25}{2}$.

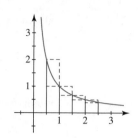

43. $A_R = 1 \cdot \dfrac{1}{2} + \dfrac{2}{3} \cdot \dfrac{1}{2} + \dfrac{1}{2} \cdot \dfrac{1}{2} + \dfrac{2}{5} \cdot \dfrac{1}{2} = \dfrac{77}{60}$

   $A_L = 2 \cdot \dfrac{1}{2} + 1 \cdot \dfrac{1}{2} + \dfrac{2}{3} \cdot \dfrac{1}{2} + \dfrac{1}{2} \cdot \dfrac{1}{2} = \dfrac{25}{12}$

45. Using $\Delta x = \dfrac{2 - (-1)}{n} = \dfrac{3}{n}$ and $x_k^* = -1 + (k-1)\dfrac{3}{n}$ we obtain

$$A = \lim_{n \to \infty} \sum_{k=1}^{n} f(x_k^*)\Delta x = \lim_{n \to \infty} \sum_{k=1}^{n} \left\{ 4 - \left[-1 + (k-1)\frac{3}{n} \right]^2 \right\} \frac{3}{n}$$

$$= \lim_{n \to \infty} \frac{3}{n} \sum_{k=1}^{n} \left[3 + 6\frac{k-1}{n} - 9\frac{(k-1)^2}{n^2} \right]$$

$$= \lim_{n \to \infty} \frac{3}{n} \left[3 \sum_{k=1}^{n} 1 + \frac{6}{n} \sum_{k=1}^{n} (k-1) - \frac{9}{n^2} \sum_{k=1}^{n} (k^2 - 2k + 1) \right]$$

$$= \lim_{n \to \infty} \frac{3}{n} \left[3 \sum_{k=1}^{n} 1 + \frac{6}{n} \sum_{k=1}^{n} k - \frac{6}{n} \sum_{k=1}^{n} 1 - \frac{9}{n^2} \sum_{k=1}^{n} k^2 + \frac{18}{n^2} \sum_{k=1}^{n} k - \frac{9}{n^2} \sum_{k=1}^{n} 1 \right]$$

$$= \lim_{n \to \infty} \left[\frac{9}{n}n + \frac{18}{n^2}\frac{n(n+1)}{2} - \frac{18}{n^2}n - \frac{27}{n^3}\frac{n(n+1)(2n+1)}{6} + \frac{54}{n^3}\frac{n(n+1)}{2} - \frac{27}{n^3}n \right]$$

$$= \lim_{n \to \infty} \left[9 + 9\left(1 + \frac{1}{n}\right) - \frac{18}{n} - \frac{9}{2}\left(1 + \frac{1}{n}\right)\left(2 + \frac{1}{n}\right) + \frac{27}{n}\left(1 + \frac{1}{n}\right) - \frac{27}{n^2} \right]$$

$$= 9 + 9 - 0 - 9 + 0 - 0 = 9.$$

47. Identify $b - a = 2$. Taking $a = 0$, we have $f\left(a + k\dfrac{b-a}{n}\right) = f\left(\dfrac{2k}{n}\right) = \sqrt{4 - \left(\dfrac{2k}{n}\right)^2}$. Then
   A is the area under $f(x) = \sqrt{4 - x^2}$ from $x = 0$ to $x = 2$.

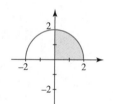

49. $0.11111111 = \dfrac{1}{10} + \dfrac{1}{10^2} + \cdots + \dfrac{1}{10^8} = \displaystyle\sum_{k=1}^{8} \dfrac{1}{10^k}$

51. $\displaystyle\sum_{k=21}^{60} k^2 = \sum_{k=1}^{60} k^2 - \sum_{k=1}^{20} k^2 = \dfrac{60 \cdot 61 \cdot 121}{6} - \dfrac{20 \cdot 21 \cdot 41}{6} = 70{,}940$

53. $0 = \displaystyle\sum_{k=1}^{n} x_k - \sum_{k=1}^{n} \bar{x} = \sum_{k=1}^{n} x_k - n\bar{x}; \quad \bar{x} = \dfrac{1}{n}\sum_{k=1}^{n} x_k$

55. (a) Identifying $f(k) = (k+1)^2$ in part (a) of Problem 54, we have $\displaystyle\sum_{k=1}^{n} [(k+1)^2 - k^2] = (n+1)^2 - 1^2 = n^2 + 2n$.

   (b) $\displaystyle\sum_{k=1}^{n} [(k+1)^2 - k^2] = \sum_{k=1}^{n} (2k+1) = \sum_{k=1}^{n} 2k + \sum_{k=1}^{n} 1 = 2\sum_{k=1}^{n} k + n$

   (c) Comparing the results of (a) and (b), we find that equating them leads to summation formula (ii):

$$2\sum_{k=1}^{n} k + n = n^2 + 2n; \qquad \sum_{k=1}^{n} k = \frac{n^2 + 2n - n}{2} = \frac{n^2 + n}{2} = \frac{n(n+1)}{2}$$

   Using $f(k) = (k+1)^3$ similarly to (a), we obtain

$$\sum_{k=1}^{n} [(k+1)^3 - k^3] = (n+1)^3 - 1^3 = n^3 + 3n^2 + 3n.$$

Analogously for (b), we also have

$$\sum_{k=1}^{n} [(k+1)^3 - k^3] = \sum_{k=1}^{n} (3k^2 + 3k + 1) = 3\sum_{k=1}^{n} k^2 + 3\sum_{k=1}^{n} k + n.$$

Combining these, we obtain

$$3\sum_{k=1}^{n} k^2 + 3\sum_{k=1}^{n} k + n = n^3 + 3n^2 + 3n$$

$$3\sum_{k=1}^{n} k^2 + \frac{3n(n+1)}{2} + n = n^3 + 3n^2 + 3n$$

$$3\sum_{k=1}^{n} k^2 = n^3 + 3n^2 + 2n - \frac{3n^2 + 3n}{2}$$

$$\sum_{k=1}^{n} k^2 = \frac{2n^3 + 6n^2 + 4n - 3n^2 - 3n}{6} = \frac{2n^3 + 3n^2 + n}{6}$$

$$= \frac{n(2n^2 + 3n + 1)}{6} = \frac{n(n+1)(2n+1)}{6}.$$

57. The equation of the line through $(0, h_1)$ and (b, h_2) is $f(x) = \dfrac{h_2 - h_1}{b}x + h_1$. Using $\Delta x = \dfrac{b}{n}$ and

$$f\left(a + k\frac{b-a}{n}\right) = f\left(\frac{kb}{n}\right) = \frac{k(h_2 - h_1)}{n} + h_1 \text{ we find}$$

$$A = \lim_{n\to\infty} \sum_{k=1}^{n} \left[\frac{k(h_1 - h_2)}{n} + h_1\right] \frac{b}{n} = \lim_{n\to\infty} \left[\frac{b}{n^2}(h_2 - h_1)\sum_{k=1}^{n} k + \frac{bh_1}{n}\sum_{k=1}^{n} 1\right]$$

$$= \lim_{n\to\infty} \left[\frac{b(h_2 - h_1)}{2}\left(1 + \frac{1}{n}\right) + \frac{bh_1 n}{n}\right] = \frac{b(h_2 - h_1)}{2} + bh_1$$

$$= \frac{bh_2 - bh_2 + 2bh_1}{2} = \left(\frac{h_1 + h_2}{2}\right)b.$$

59. Using $\Delta x = \dfrac{4}{n}$ and $f\left(a + k\dfrac{b-a}{n}\right) = \dfrac{128k}{n} - \dfrac{384k^2}{n^2} + \dfrac{512k^3}{n^3} - \dfrac{256k^4}{n^4}$ we have

$$A = \lim_{n\to\infty} \sum_{k=1}^{n} \left(\frac{128k}{n} - \frac{384k^2}{n^2} + \frac{512k^3}{n^3} - \frac{256k^4}{n^4}\right)\frac{4}{n}$$

$$= \lim_{n\to\infty} \left(\frac{512}{n^2}\sum_{k=1}^{n} k - \frac{1536}{n^3}\sum_{k=1}^{n} k^2 + \frac{2048}{n^4}\sum_{k=1}^{n} k^3 - \frac{1024}{n^5}\sum_{k=1}^{n} k^4\right)$$

$$= \lim_{n\to\infty} \left[256\left(1 + \frac{1}{n}\right) - 256\left(2 + \frac{3}{n} + \frac{1}{n^2}\right) + 512\left(1 + \frac{2}{n} + \frac{1}{n^2}\right) - \frac{512}{15}\left(6 + \frac{15}{n} + \frac{10}{n^2} - \frac{1}{n^4}\right)\right]$$

$$= 256 - 512 + 512 - \frac{1024}{5} = \frac{256}{5}.$$

61. We note that $A_2 = 16 - A_1$ where A_1 is the area under $y = x^3$ from 0 to 2. Using $\Delta x = \dfrac{2}{n}$ and

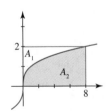

$$f\left(a + k\frac{b-a}{n}\right) = \frac{8k^3}{n^3} \text{ we find}$$

$$A_1 = \lim_{n\to\infty} \sum_{k=1}^{n} \left(\frac{8k^3}{n^3}\frac{2}{n}\right) = \lim_{n\to\infty} \frac{16}{n^4}\sum_{k=1}^{n} k^3 = \lim_{n\to\infty} 4\left(1 + \frac{2}{n} + \frac{1}{n^2}\right) = 4.$$

Thus, $A_2 = 16 - 4 = 12$.

63. By (8) of this section,

$$A = \lim_{n \to \infty} \sum_{k=1}^{n} f\left(0 + [k-1]\frac{1}{n}\right) \frac{1}{n} = \lim_{n \to \infty} \sum_{k=1}^{n} e^{(k-1)/n} \cdot \frac{1}{n}$$

$$= \lim_{n \to \infty} \frac{1}{n}\left[1 + e^{1/n} + e^{2/n} + \cdots + e^{(n-1)/n}\right]$$

$$= \lim_{n \to \infty} \frac{1}{n}\left[1 + e^{1/n} + (e^{1/n})^2 + \cdots + (e^{1/n})^{n-1}\right].$$

Using $a = 1$, $r = e^{1/n}$, we obtain

$$A = \lim_{n \to \infty} \frac{1}{n} \cdot 1 \cdot \left[\frac{1 - (e^{1/n})^n}{1 - e^{1/n}}\right] = (1 - e)\lim_{n \to \infty} \frac{1}{n(1 - e^{1/n})}$$

Now, $\displaystyle \lim_{n \to \infty} n(1 - e^{1/n}) = \lim_{n \to \infty} \frac{1 - e^{1/n}}{1/n}$ (form $\infty \cdot 0$)

$$\overset{h}{=} \lim_{n \to \infty} \frac{-e^{1/n}(-1/n^2)}{-1/n^2} = -\lim_{n \to \infty} e^{1/n} = -1,$$

so $A = (1 - e)\left(\dfrac{1}{-1}\right) = e - 1.$

5.4 The Definite Integral

1. From $\Delta x_1 = 1$, $\Delta x_2 = 2/3$, $\Delta x_3 = 2/3$, and $\Delta x_4 = 2/3$ we see that the norm of the partition is $\|P\| = 1$. Using $f(x_1^*) = 5/2$, $f(x_2^*) = 5$, $f(x_3^*) = 7$, and $f(x_4^*) = 9$ we compute the Riemann sum

$$\sum_{k=1}^{4} f(x_k^*)\Delta x_k = \frac{5}{2}(1) + 5\left(\frac{2}{3}\right) + 7\left(\frac{2}{3}\right) + 9\left(\frac{2}{3}\right) = \frac{33}{2}.$$

3. From $\Delta x_1 = 3/4$, $\Delta x_2 = 1/2$, $\Delta x_3 = 1/2$, and $\Delta x_4 = 1/4$ we see that the norm of the partition is $\|P\| = 3/4$. Using $f(x_1^*) = 9/16$, $f(x_2^*) = 0$, $f(x_3^*) = 1/4$, and $f(x_4^*) = 49/64$ we compute the Riemann sum

$$\sum_{k=1}^{4} f(x_k^*)\Delta x_k = \frac{9}{15}\left(\frac{3}{4}\right) + 0\left(\frac{1}{2}\right) + \frac{1}{4}\left(\frac{1}{2}\right) + \frac{49}{64}\left(\frac{1}{4}\right) = \frac{189}{256}.$$

5. From $\Delta x_1 = \pi$, $\Delta x_2 = \pi/2$, and $\Delta x_3 = \pi/2$ we see that the norm of the partition is $\|P\| = \pi$. Using $f(x_1^*) = 1$, $f(x_2^*) = -1/2$, and $f(x_3^*) = -\sqrt{2}/2$ we compute the Riemann sum

$$\sum_{k=1}^{3} f(x_k^*)\Delta x_k = 1(\pi) + \left(-\frac{1}{2}\right)\left(\frac{\pi}{2}\right) + \left(-\frac{\sqrt{2}}{2}\right)\left(\frac{\pi}{2}\right) = \frac{(3 - \sqrt{2})\pi}{4}.$$

7. We have $\Delta x_k = 1$ and $x_k^* = k$ for $k = 1, 2, 3, 4, 5$. Using $f(x_1^*) = -1$, $f(x_2^*) = 0$, $f(x_3^*) = 1$, $f(x_4^*) = 2$, and $f(x_5^*) = 3$ we compute the Riemann sum

$$\sum_{k=1}^{5} f(x_k^*)\Delta x_n = -1(1) + 0(1) + 1(1) + 2(1) + 3(1) = 5.$$

9. $\displaystyle \int_{-2}^{4} \sqrt{9 + x^2}\,dx$

11. Identify $a = 0$ and $b = 2$. Then

$$\left(1 + \frac{2k}{n}\right)\frac{2}{n} = \left[f\left(a + k\frac{b-a}{n}\right)\right]\frac{b-a}{n} \quad \text{and} \quad f\left(a + k\frac{b-a}{n}\right) = f\left(\frac{2k}{n}\right) = 1 + \frac{2k}{n}.$$

Taking $f(x) = x + 1$ we have $\displaystyle \lim_{n \to \infty} \sum_{k=1}^{n} \left(1 + \frac{2k}{n}\right)\frac{2}{n} = \int_{0}^{2} (x + 1)\,dx.$

13. Using $\dfrac{b-a}{n} = \dfrac{4}{n}$ and $f\left(a+k\dfrac{b-a}{n}\right) = -3 + \dfrac{4k}{n}$ we have

$$\int_{-3}^{1} x\,dx = \lim_{n\to\infty} \sum_{k=1}^{n} \left(-3 + \frac{4k}{n}\right)\frac{4}{n} = \lim_{n\to\infty}\left(-\frac{12}{n}\sum_{k=1}^{n} 1 + \frac{16}{n^2}\sum_{k=1}^{n} k\right)$$

$$= \lim_{n\to\infty}\left[-\frac{12n}{n} + 8\left(1 + \frac{1}{n}\right)\right] = -12 + 8 = -4.$$

15. Using $\dfrac{b-a}{n} = \dfrac{1}{n}$ and $f\left(a+k\dfrac{b-a}{n}\right) = \left(1 + \dfrac{k}{n}\right)^2 - \left(1 + \dfrac{k}{n}\right) = \dfrac{k}{n} + \dfrac{k^2}{n^2}$ we have

$$\int_{1}^{2} (x^2 - x)\,dx = \lim_{n\to\infty}\sum_{k=1}^{n}\left(\frac{k}{n} + \frac{k^2}{n^2}\right)\frac{1}{n} = \lim_{n\to\infty}\left(\frac{1}{n^2}\sum_{k=1}^{n} k + \frac{1}{n^3}\sum_{k=1}^{n} k^2\right)$$

$$= \lim_{n\to\infty}\left[\frac{1}{2}\left(1 + \frac{1}{n}\right) + \frac{1}{6}\left(2 + \frac{3}{n} + \frac{1}{n^2}\right)\right] = \frac{1}{2} + \frac{1}{3} = \frac{5}{6}.$$

17. Using $\dfrac{b-a}{n} = \dfrac{1}{n}$ and $f\left(a+k\dfrac{b-a}{n}\right) = \dfrac{k^3}{n^3} - 1$ we have

$$\int_{0}^{1} (x^3 - 1)\,dx = \lim_{n\to\infty}\sum_{k=1}^{n}\left(\frac{k^3}{n^3} - 1\right)\frac{1}{n} = \lim_{n\to\infty}\left(\frac{1}{n^4}\sum_{k=1}^{n} k^3 - \frac{1}{n}\sum_{k=1}^{n} 1\right)$$

$$= \lim_{n\to\infty}\left[\frac{1}{4}\left(1 + \frac{2}{n} + \frac{1}{n^2}\right) - \frac{n}{n}\right] = \frac{1}{4} - 1 = -\frac{3}{4}.$$

19. Using $f\left(a+k\dfrac{b-a}{n}\right) = a + \dfrac{k(b-a)}{n}$ we have

$$\int_{a}^{b} x\,dx = \lim_{n\to\infty}\sum_{k=1}^{n}\left[a + \frac{k(b-a)}{n}\right]\frac{b-a}{n} = \lim_{n\to\infty}\left[\frac{a(b-a)}{n}\sum_{k=1}^{n} 1 + \frac{(b-a)^2}{n^2}\sum_{k=1}^{n} k\right]$$

$$= \lim_{n\to\infty}\left[\frac{a(b-a)n}{n} + \frac{(b-a)^2}{2}\left(1 + \frac{1}{n}\right)\right] = a(b-a) + \frac{(b-a)^2}{2}$$

$$= \frac{b-a}{2}(2a+b-a) = \frac{b-a}{2}(b+a) = \frac{b^2 - a^2}{2}.$$

21. $\displaystyle\int_{-1}^{3} x\,dx = \frac{1}{2}[3^2 - (-1)^2] = 4$

23. $\displaystyle\int_{3}^{6} 4\,dx = 4(6-3) = 12$

25. $\displaystyle\int_{4}^{-2} \frac{1}{2}\,dx = \frac{1}{2}(-2-4) = -3$

27. $\displaystyle-\int_{3}^{-1} 10x\,dx = 10\int_{-1}^{3} x\,dx = 10(4) = 40$

29. $\displaystyle\int_{3}^{-1} t^2\,dt = -\int_{-1}^{3} t^2\,dt = -\frac{28}{3}$

31. $\displaystyle\int_{-1}^{3} (-3x^2 + 4x - 5)\,dx = -3\int_{-1}^{3} x^2\,dx + 4\int_{-1}^{3} x\,dx - \int_{-1}^{3} 5\,dx$

$$= -3\frac{28}{3} + 4(4) - 5[3 - (-1)] = -32$$

33. $\displaystyle\int_{-1}^{0} x^2\,dx + \int_{0}^{3} x^2\,dx = \int_{-1}^{3} x^2\,dx = \frac{28}{3}$

35. $\displaystyle\int_{0}^{4} x\,dx + \int_{0}^{4} (9-x)\,dx = \int_{0}^{4} [x+(9-x)]\,dx = \int_{0}^{4} 9\,dx = 9(4-0) = 36$

37. $\displaystyle\int_{0}^{3} x^3\,dx + \int_{3}^{0} t^3\,dt = \int_{0}^{3} x^3\,dx - \int_{0}^{3} x^3\,dx = 0$

39. $\displaystyle\int_{2}^{5} f(x)\,dx = \int_{0}^{5} f(x)\,dx - \int_{0}^{2} f(x)\,dx = 8.5 - 6 = 2.5$

41. $\displaystyle\int_{-1}^{2} [2f(x)+g(x)]\,dx = \int_{-1}^{2} 2f(x)\,dx + \int_{-1}^{2} g(x)\,dx = 2\int_{-1}^{2} f(x)\,dx + \frac{1}{3}(3)\int_{-1}^{2} g(x)\,dx$

$$= 2(3.4) + \frac{1}{3}\int_{-1}^{2} 3g(x)\,dx = 6.8 + \frac{1}{3}(12.6) = 6.8 + 4.2 = 11$$

43.   (a) $\displaystyle\int_{a}^{b} f(x)\,dx = -2.5$

     (b) $\displaystyle\int_{b}^{c} f(x)\,dx = 3.9$

     (c) $\displaystyle\int_{c}^{d} f(x)\,dx = -1.2$

     (d) $\displaystyle\int_{a}^{c} f(x)\,dx = \int_{a}^{b} f(x)\,dx + \int_{b}^{c} f(x)\,dx = -2.5 + 3.9 = 1.4$

     (e) $\displaystyle\int_{b}^{d} f(x)\,dx = \int_{b}^{c} f(x)\,dx + \int_{c}^{d} f(x)\,dx = 3.9 - 1.2 = 2.7$

     (f) $\displaystyle\int_{a}^{d} f(x)\,dx = \int_{a}^{b} f(x)\,dx + \int_{b}^{c} f(x)\,dx + \int_{c}^{d} f(x)\,dx = -2.5 + 3.9 - 1.2 = 0.2$

45.

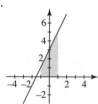

47.

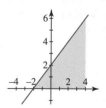

49. From the figure, we see that the area under the graph is a triangle with a base and height of 6. Thus, the area from geometry is

$$A = \frac{bh}{2} = \frac{6(6)}{2} = 18.$$

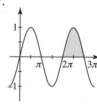

51. From the figure, we see that the area under the graph consists of one-fourth of a circle of radius 3. Thus, the area from geometry is

$$A = \frac{\pi r^2}{4} = \frac{\pi(3)^2}{4} = \frac{9\pi}{4}.$$

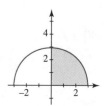

53.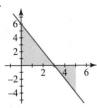

55.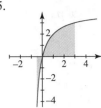

57. From the figure, we see that the net signed area under the graph is the area of a triangle with a base of 1 and a height of 2 subtracted from the area of a triangle with a base of 4 and a height of 8. Thus, the net signed area from geometry is

$$A = \frac{b_1 h_1}{2} - \frac{b_2 h_2}{2} = \frac{4(8)}{2} - \frac{1(2)}{2} = 15.$$

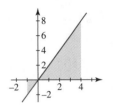

59. $\int_{-1}^{1}(x - \sqrt{1 - x^2})\,dx$ can be rewritten as $\int_{-1}^{1} x\,dx - \int_{-1}^{1}\sqrt{1 - x^2}\,dx$, so the net signed area of the graph below left is the same as the difference between the net signed areas of the graphs below right. This difference, in turn, is the area of a semicircle of radius 1 subtracted from the net signed area of two triangles with bases and heights of 1. From geometry, this is

$$A = \left(\frac{b_1 h_1}{2} - \frac{b_2 h_2}{2}\right) - \frac{\pi r^2}{2} = \left[\frac{1(1)}{2} - \frac{1(1)}{2}\right] - \frac{\pi(1)^2}{2} = -\frac{\pi}{2}.$$

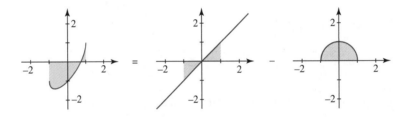

61. From the figure, we see that the net signed area under the graph is the negative of the area of a triangle with a base of 2 and a height of 2. Thus, the net signed area from geometry is

$$A = -\frac{bh}{2} = -\frac{2(2)}{2} = -2.$$

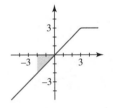

63. From the figure, we see that the net signed area under the graph is the area of a triangle with a base of 4 and a height of 4 subtracted from the sum of the areas of a triangle with a base of 3 and a height of 3, and a rectangle of width 2 and height 3. Thus, the net signed area from geometry is

$$A = \left(wh + \frac{b_1 h_1}{2}\right) - \frac{b_2 h_2}{2} = [2(3) + \frac{3(3)}{2}] - \frac{4(4)}{2} = \frac{5}{2} = 2.5.$$

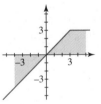

65. For $-1 \le x \le 0$, $e^x \le 1$ and $e^{-x} \ge 1$. Then $e^{-x} \ge e^x$ on $[-1, 0]$ and by Theorem 5.4.7(i) we have $\int_{-1}^{0} e^x\,dx \le \int_{-1}^{0} e^{-x}\,dx$.

67. Letting $f(x) = (x^3 + 1)^{1/2}$ we have $f'(x) = \frac{3}{2}x^2(x^3 + 1)^{-1/2}$. For $0 \le x \le 1$, $f'(x) \ge 0$ and $f(0) \le f(x) \le f(1)$. Since $f(0) = 1$ and $f(1) = \sqrt{2} < 1.42$, we identify $m = 1$ and $M = 1.42$. Then by Theorem 5.4.7(ii)

$$1(1 - 0) \le \int_{0}^{1} (x^3 + 1)^{1/2}\,dx \le 1.42(1 - 0) \qquad \text{and} \qquad 1 \le \int_{0}^{1} (x^3 + 1)^{1/2}\,dx \le 1.42.$$

69. On $[0,1]$, $x^2 - x^3 = x^2(1-x) \geq 0$, so $x^2 \geq x^3$. Thus by Theorem 5.4.7(i), $\int_0^1 x^2\,dx \geq \int_0^1 x^3\,dx$.

71. Since $f^2(x) \geq 0$ on $[a,b]$, by (12), $\int_a^b f^2(x)\,dx \geq 0$.

73. Using $\Delta x = \dfrac{k^2}{n^2} - \dfrac{(k-1)^2}{n^2} = \dfrac{2k-1}{n^2}$ we have

$$\int_0^1 \sqrt{x}\,dx = \lim_{n\to\infty} \sum_{k=1}^n \sqrt{\frac{k^2}{n^2}}\left(\frac{2k-1}{n^2}\right) = \lim_{n\to\infty} \frac{1}{n^3}\sum_{k=1}^n (2k^2 - k)$$

$$= \lim_{n\to\infty}\left(\frac{2}{n^3}\sum_{k=1}^n k^2 - \frac{1}{n^3}\sum_{k=1}^n k\right) = \lim_{n\to\infty}\left[\frac{2}{n^3}\cdot\frac{n(n+1)(2n+1)}{6} - \frac{1}{n^3}\cdot\frac{n(n+1)}{2}\right]$$

$$= \lim_{n\to\infty}\left[\frac{1}{3}\left(1+\frac{1}{n}\right)\left(2+\frac{1}{n}\right) - \frac{1}{2n}\left(1+\frac{1}{n}\right)\right] = \frac{2}{3} - 0 = \frac{2}{3}.$$

5.5 Fundamental Theorem of Calculus

1. $\int_3^7 dx = x\big]_3^7 = 7 - 3 = 4$

3. $\int_{-1}^2 (2x+3)\,dx = (x^2 + 3x)\big]_{-1}^2 = 10 - (-2) = 12$

5. $\int_1^3 (6x^2 - 4x + 5)\,dx = (2x^3 - 2x^2 + 5x)\big]_1^3 = 51 - 5 = 46$

7. $\int_0^{\pi/2} \sin x\,dx = -\cos x\big]_0^{\pi/2} = 0 - (-1) = 1$

9. $\int_{\pi/4}^{\pi/2} \cos 3t\,dt = \frac{1}{3}\sin 3t\big]_{\pi/4}^{\pi/2} = \frac{1}{3}\sin\frac{3\pi}{2} - \frac{1}{3}\sin\frac{3\pi}{4} = -\frac{1}{3} - \frac{\sqrt{2}}{6} = -\frac{2+\sqrt{2}}{6}$

11. $\int_{1/2}^{3/4} \frac{1}{u^2}\,du = -\frac{1}{u}\Big]_{1/2}^{3/4} = -\frac{4}{3} - (-2) = \frac{2}{3}$

13. $\int_{-1}^1 e^x\,dx = e^x\big]_{-1}^1 = e - \frac{1}{e}$

15. $\int_0^2 x(1-x)\,dx = \int_0^2 (x - x^2)\,dx = \left(\frac{1}{2}x^2 - \frac{1}{3}x^3\right)\Big]_0^2 = 2 - \frac{8}{3} - 0 = -\frac{2}{3}$

17. $\int_{-1}^1 (7x^3 - 2x^2 + 5x - 4)\,dx = \left(\frac{7}{4}x^4 - \frac{2}{3}x^3 + \frac{5}{2}x^2 - 4x\right)\Big]_{-1}^1$

$$= \left(\frac{7}{4} - \frac{2}{3} + \frac{5}{2} - 4\right) - \left(\frac{7}{4} + \frac{2}{3} + \frac{5}{2} + 4\right) = -\frac{28}{3}$$

19. $\int_1^4 \frac{x-1}{\sqrt{x}}\,dx = \int_1^4 (x^{1/2} - x^{-1/2})\,dx = \left(\frac{2}{3}x^{3/2} - 2x^{1/2}\right)\Big]_1^4 = \left(\frac{16}{3} - 4\right) - \left(\frac{2}{3} - 2\right) = \frac{8}{3}$

21. $\int_1^{\sqrt{3}} \frac{1}{1+x^2}\,dx = \tan^{-1} x\big]_1^{\sqrt{3}} = \frac{\pi}{3} - \frac{\pi}{4} = \frac{\pi}{12}$

23. $\int_{-4}^{12} \sqrt{z+4}\,dz$ $\boxed{u = z+4,\ du = dz}$

$$= \int_0^{16} u^{1/2}\,du = \frac{2}{3}u^{3/2}\Big]_0^{16} = \frac{128}{3}$$

25. $\displaystyle\int_0^3 \frac{x}{\sqrt{x^2+16}}\,dx$ $\quad\boxed{u = x^2+16,\ du = 2x\,dx}$

$$= \frac{1}{2}\int_{16}^{25}\frac{1}{\sqrt{u}}\,du = \frac{1}{2}\int_{16}^{25}u^{-1/2}\,du = \sqrt{u}\Big]_{16}^{25} = 5-4 = 1$$

27. $\displaystyle\int_{1/2}^1\left(1+\frac{1}{x}\right)^2\frac{1}{x^2}\,dx$ $\quad\boxed{u = 1+\frac{1}{x},\ du = -\frac{1}{x^2}\,dx}$

$$= -\int_3^2 u^3\,du = \int_2^3 u^3\,du = \frac{1}{4}u^4\Big]_2^3 = \frac{81}{4}-4 = \frac{65}{4}$$

29. $\displaystyle\int_0^1 \frac{x+1}{\sqrt{x^2+2x+3}}\,dx$ $\quad\boxed{u = x^2+2x+3,\ du = 2(x+1)\,dx}$

$$= \frac{1}{2}\int_3^6\frac{1}{\sqrt{u}}\,du = \frac{1}{2}\int_3^6 u^{-1/2}\,du = \sqrt{u}\Big]_3^6 = \sqrt{6}-\sqrt{3}$$

31. $\displaystyle\int_0^{\pi/8}\sec^2 2x\,dx$ $\quad\boxed{u = 2x,\ du = 2\,dx}$

$$= \frac{1}{2}\int_0^{\pi/4}\sec^2 u\,du = \frac{1}{2}\tan u\Big]_0^{\pi/4} = \frac{1}{2}$$

33. $\displaystyle\int_{-1/2}^{3/2}(x-\cos\pi x)\,dx = \left(\frac{1}{2}x^2 - \frac{1}{\pi}\sin\pi x\right)\Big]_{-1/2}^{3/2} = \left[\frac{9}{8}-\left(-\frac{1}{\pi}\right)\right] - \left[\frac{1}{8}-\left(-\frac{1}{\pi}\right)\right] = 1$

35. $\displaystyle\int_0^{\pi/2}\sqrt{\cos x}\sin x\,dx$ $\quad\boxed{u = \cos x,\ du = -\sin x\,dx}$

$$= -\int_1^0\sqrt{u}\,du = \int_0^1 u^{1/2}\,du = \frac{2}{3}u^{3/2}\Big]_0^1 = \frac{2}{3}$$

37. $\displaystyle\int_{\pi/6}^{\pi/2}\frac{1+\cos\theta}{(\theta+\sin\theta)^2}\,d\theta$ $\quad\boxed{u = \theta+\sin\theta,\ du = (1+\cos\theta)\,d\theta}$

$$= \int_{(\pi+3)/6}^{(\pi+2)/2} u^{-2}\,du = -\frac{1}{u}\Big]_{(\pi+3)/6}^{(\pi+2)/2} = -\frac{2}{\pi+2}+\frac{6}{\pi+3} = \frac{4\pi+6}{(\pi+3)(\pi+2)}$$

39. $\displaystyle\int_0^{3/4}\sin^2\pi x\,dx = \int_0^{3/4}\frac{1}{2}(1-\cos 2\pi x)\,dx = \left(\frac{1}{2}x - \frac{1}{4\pi}\sin 2\pi x\right)\Big]_0^{3/4}$

$$= \frac{3}{8}-\frac{1}{4\pi}\sin\frac{3\pi}{2} = \frac{3}{8}+\frac{1}{4\pi}$$

41. $\displaystyle\int_1^5\frac{1}{1+2x}\,dx$ $\quad\boxed{u = 1+2x,\ du = 2\,dx}$

$$= \frac{1}{2}\int_3^{11}\frac{1}{u}\,du = \frac{1}{2}\ln|u|\,\Big]_3^{11} = \frac{1}{2}(\ln 11 - \ln 3)$$

43. $\displaystyle\frac{d}{dx}\int_0^x te^t\,dt = xe^x$

45. $\displaystyle\frac{d}{dt}\int_2^t (3x^2-2x)^6\,dx = (3t^2-2t)^6$

47. $\displaystyle\frac{d}{dx}\int_3^{6x-1}\sqrt{4t+9}\,dt$ $\quad\boxed{u = 6x-1,\ du = 6\,dx}$

$$= \frac{d}{du}\left(\int_3^u\sqrt{4t+9}\,dt\right)\frac{du}{dx} = \sqrt{4u+9}\,\frac{du}{dx}$$

$$= \sqrt{4(6x-1)+9}\cdot 6 = 6\sqrt{24x+5}$$

49. $F'(x) = \dfrac{d}{dx}\left(\displaystyle\int_{3x}^{0}\dfrac{1}{t^3+1}\,dt + \int_{0}^{x^2}\dfrac{1}{t^3+1}\,dt\right) = \dfrac{d}{dx}\left(-\displaystyle\int_{0}^{3x}\dfrac{1}{t^3+1}\,dt + \int_{0}^{x^2}\dfrac{1}{t^3+1}\,dt\right)$

$$\boxed{u=3x,\ du=3\,dx;\ z=x^2,\ dz=2x\,dx}$$

$= \dfrac{d}{du}\left(-\displaystyle\int_{0}^{u}\dfrac{1}{t^3+1}\,dt\right)\dfrac{du}{dx} + \dfrac{d}{dz}\left(\displaystyle\int_{0}^{z}\dfrac{1}{t^3+1}\,dt\right)\dfrac{dz}{dx}$

$= -\dfrac{1}{(3x)^3+1}(3) + \dfrac{1}{(x^2)^3+1}(2x) = \dfrac{2x}{x^6+1} - \dfrac{3}{27x^3+1}$

51. $\dfrac{d}{dx}\displaystyle\int_{1}^{x}(6t^2-8t+5)\,dt = \dfrac{d}{dx}\left.(2t^3-4t^2+5t)\right]_1^x$

$\qquad\qquad = \dfrac{d}{dx}[(2x^3-4x^2+5x)-(2-4+5)] = 6x^2-8x+5$

53. (a) $f(1) = \displaystyle\int_{1}^{1}\ln(2t+1)\,dt = 0$

(b) $f'(x) = \ln(2x+1)$, so $f'(1) = \ln[2(1)+1] = \ln 3$.

(c) $f''(x) = \dfrac{2}{2x+1}$, so $f''(1) = \dfrac{2}{2(1)+1} = \dfrac{2}{3}$.

(d) $f'''(x) = -\dfrac{4}{(2x+1)^2}$, so $f'''(1) = -\dfrac{4}{[2(1)+1]^2} = -\dfrac{4}{9}$.

55. $\displaystyle\int_{-1}^{2}f(x)\,dx = \int_{-1}^{0}f(x)\,dx + \int_{0}^{2}f(x)\,dx = \int_{-1}^{0}-x\,dx + \int_{0}^{2}x^2\,dx = \left.-\dfrac{1}{2}x^2\right]_{-1}^{0} + \left.\dfrac{1}{3}x^3\right]_0^2$

$= -\left(0-\dfrac{1}{2}\right) + \dfrac{1}{3}(8-0) = \dfrac{19}{6}$

57. $\displaystyle\int_{0}^{3}f(x)\,dx = \int_{0}^{2}f(x)\,dx + \int_{2}^{3}f(x)\,dx = \int_{0}^{2}4\,dx + \int_{2}^{3}dx = \left.4x\right]_0^2 + \left.x\right]_2^3$

$= (8-0)+(3-2) = 9$

59. Using the fact that $f(x)$ is an even function on $[-2,2]$, we have

$$\int_{-2}^{2}f(x)\,dx = 2\int_{0}^{2}f(x)\,dx = 2\left[\int_{0}^{1}f(x)\,dx + \int_{1}^{2}f(x)\,dx\right] = 2\left(\int_{0}^{1}4\,dx + \int_{1}^{2}x^2\,dx\right)$$

$$= 2\left(\left.4x\right]_0^1 + \left.\dfrac{1}{3}x^3\right]_1^2\right) = 2\left[(4-0)+\dfrac{1}{3}(8-1)\right] = \dfrac{38}{3}.$$

61. $\displaystyle\int_{-3}^{1}|x|\,dx = \int_{-3}^{0}-x\,dx + \int_{0}^{1}x\,dx = \left.-\dfrac{1}{2}x^2\right]_{-3}^{0} + \left.\dfrac{1}{2}x^2\right]_0^1 = \dfrac{9}{2}+\dfrac{1}{2} = 5$

63. $\displaystyle\int_{-8}^{3}\sqrt{|x|+1}\,dx = \int_{-8}^{0}\sqrt{-x+1}\,dx + \int_{0}^{3}\sqrt{x+1}\,dx = \left.-\dfrac{2}{3}(1-x)^{3/2}\right]_{-8}^{0} + \left.\dfrac{2}{3}(x+1)^{3/2}\right]_0^3$

$= -\dfrac{2}{3}(1-27) + \dfrac{2}{3}(8-1) = 22$

65. Using the fact that $f(x) = |\sin x|$ is an even function on $[-\pi,\pi]$ and $\sin x > 0$ for $0 \le x \le \pi$,

$$\int_{-\pi}^{\pi}|\sin x|\,dx = 2\int_{0}^{\pi}|\sin x|\,dx = 2\int_{0}^{\pi}\sin x\,dx = \left.-2\cos x\right]_0^\pi = -2(-1-1) = 4.$$

67. $\displaystyle\int_{1/2}^{e}\dfrac{(\ln 2t)^5}{t}\,dt$ $\boxed{u=\ln 2t,\ du=\dfrac{1}{t}\,dt;\ u(1/2)=0,\ u(e)=1+\ln 2}$

$= \displaystyle\int_{0}^{1+\ln 2}u^5\,du = \left.\dfrac{1}{6}u^6\right]_0^{1+\ln 2} = \dfrac{1}{6}[(1+\ln 2)^6 - 0] = \dfrac{(1+\ln 2)^6}{6} \approx 3.9266$

69. $\int_0^1 \dfrac{e^{-2x}}{e^{-2x}+1}\,dx$ $\boxed{u = e^{-2x}+1,\ du = -2e^{-2x};\ u(0) = 2,\ u(1) = 1+e^{-2}}$

$$= -\frac{1}{2}\int_2^{1+e^{-2}} \frac{1}{u}\,du = -\frac{1}{2}\ln|u|\Big]_2^{1+e^{-2}} = -\frac{1}{2}[\ln(1+e^{-2}) - \ln 2] \approx 0.2831$$

71. (a) Since $\operatorname{erf}'(x) = \dfrac{2}{\sqrt{\pi}}e^{-x^2} > 0$, $\operatorname{erf}(x)$ is increasing for all x.

(b) The derivative of $y = e^{x^2}[1+\sqrt{\pi}\,\operatorname{erf}(x)]$ is $\dfrac{dy}{dx} = 2 + 2e^{x^2}x[1+\sqrt{\pi}\,\operatorname{erf}(x)]$, so

$$\frac{dy}{dx} - 2xy = 2 + 2e^{x^2}x[1+\sqrt{\pi}\,\operatorname{erf}(x)] - 2xe^{x^2}[1+\sqrt{\pi}\,\operatorname{erf}(x)] = 2$$

Also, $y(0) = e^0[1+\sqrt{\pi}\,\operatorname{erf}(0)] = 1+\sqrt{\pi}\cdot 0 = 1$.

73. $\displaystyle\lim_{\|P\|\to 0}\sum_{k=1}^n (2x_k^* + 5)\Delta x_k = \int_{-1}^3 (2x+5)\,dx = (x^2+5x)\Big]_{-1}^3 = 24 - (-4) = 28$

75. Letting $\Delta x_k = \pi/n$ we have

$$\lim_{n\to\infty}\frac{\pi}{n}\sum_{k=1}^n \sin x_k^* = \lim_{n\to\infty}\sum_{k=1}^n (\sin x_k^*)\Delta x_k = \int_0^\pi \sin x\,dx = -\cos x\Big]_0^\pi = -(-1-1) = 2.$$

77. $\displaystyle\int_{-1}^2 \left\{\int_1^x 12t^2\,dt\right\}dx = \int_{-1}^2 \left(4t^3\Big]_1^x\right)dx = \int_{-1}^2 (4x^3 - 4)\,dx = (x^4 - 4x)\Big]_{-1}^2 = 8 - 5 = 3$

79. Since $f(x)$ is even, $f(-x) = f(x)$. Then

$$\int_{-a}^a f(x)\,dx = \int_{-a}^0 f(x)\,dx + \int_0^a f(x)\,dx \qquad \boxed{t = -x,\ dt = -dx}$$

$$= \int_a^0 f(-t)(-dt) + \int_0^a f(x)\,dx = -\int_a^0 f(t)\,dt + \int_0^a f(x)\,dx$$

$$= \int_0^a f(t)\,dt + \int_0^a f(x)\,dx = 2\int_0^a f(x)\,dx.$$

81. The reasoning is flawed at the point that $\sin t$ is substituted with $\sqrt{1-\cos^2 t}$. The use of the square root loses $\sin t$'s sign changes.

83. (a)

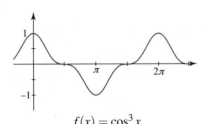

$f(x) = \cos^3 x$

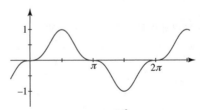

$f(x) = \sin^3 x$

(b) $\displaystyle\int_0^{2\pi} \cos^3 x\,dx = 0;\qquad \int_0^{2\pi} \sin^3 x\,dx = 0$

85. (a) At time n the radius of the circle is $r_0 + cu$ and the area is $A(u) = \pi(r_0 + cu)^2$. Then

$$\frac{RT}{Pv} = \int_0^t \frac{k\pi(r_0 + cu)^2}{V_0}\, du = \frac{K\pi}{cV_0}\int_0^t (r_0 + cu)^2\,(c\,du)$$

$$= \frac{K\pi}{cV_0}\frac{1}{3}(r_0 + cu)^3 \Big]_0^t = \frac{K\pi}{3cV_0}\left[(r_0 + ct)^3 - r_0^3\right]$$

$$\frac{3cV_0 RT}{PKv\pi} = (r_0 + ct)^3 - r_0^3$$

$$(r_0 + ct)^3 = \frac{3cV_0 RT}{PKv\pi} + r_0^3$$

$$r_0 + ct = \sqrt[3]{\frac{3cV_0 RT}{PKv\pi} + r_0^3}$$

$$t = \frac{1}{c}\sqrt[3]{\frac{3cV_0 RT}{PKv\pi} + r_0^3} - \frac{r_0}{c}.$$

(b) Substituting $RT/Pv = 1.9 \times 10^6$, $K = 0.01 \times 10^{-3}$, $c = 0.01$, $r_0 = 100$, and $V_0 = 10,000$, we find $t \approx 2,617,695$ seconds, or $t \approx 30$ days and 7 hours.

(c) The final area is $A(2,617,695) = \pi[100 + 0.01(2,617,695)]^2 \approx 2.169 \times 10^9$ m^2 = 2169 km^2.

Chapter 5 in Review

A. True/False

1. False. Consider $f(x) = x^3 + x^2 + 1$.

3. True

5. True

7. True, since no portion of the graph of $y = x - x^3$ lies below the x-axis on $[0, 1]$.

9. False. Consider the partition $\left\{0, \dfrac{1}{n}, \dfrac{1}{n-1}, \ldots, \dfrac{1}{2}, 1\right\}$ of $\{0, 1\}$.

11. True

13. False. $\int \sin x\, dx = -\cos x + C$.

15. True

B. Fill In the Blanks

1. $f(x)$

3. $\dfrac{\ln x}{x}$

5. $-f(g(x))g'(x)\, dx$

7. $\displaystyle\sum_{k=1}^{5} \frac{k}{2k+1}$

9. $\displaystyle\int_5^{17}$

11. 5/2

13. $\displaystyle\int_0^4 \sqrt{x}\, dx$; $\dfrac{2}{3}x^{3/2}\Big|_0^4 = \dfrac{16}{3}$

15. $\int_{-1}^{1} \left\{ \int_{0}^{x} e^{-t} dt \right\} dx = \int_{-1}^{1} \left(-e^{-t} \right]_{0}^{x} \right) dx = -\int_{-1}^{1} \left(e^{-x} - 1 \right) dx$

$$= -(-e^{-x} - x)\Big]_{-1}^{1} = -[(-e^{-1} - 1) - (-e^{1} + 1)] = \frac{1}{e} - e + 2$$

$\int_{-1}^{1} \frac{d}{dx} \left\{ \int_{0}^{x} e^{-t} dt \right\} dx = \int_{-1}^{1} e^{-x} dx = -e^{-x}\Big]_{-1}^{1} = e - \frac{1}{e}$

C. Exercises

1. $\int_{-1}^{1} (4x^3 - 6x^2 + 2x - 1) dx = (x^4 - 2x^3 + x^2 - x)\Big]_{-1}^{1} = -1 - 5 = -6$

3. $\int (5t + 1)^{100} dt = \frac{1}{505} (5t + 1)^{101} + C$

5. $\int_{0}^{\pi/4} (\sin 2x - 5\cos 4x) dx = \left(-\frac{1}{2}\cos 2x - \frac{5}{4}\sin 4x \right)\Big]_{0}^{\pi/4} = 0 - \left(-\frac{1}{2} \right) = \frac{1}{2}$

7. $\int_{4}^{4} (-2x^2 + x^{1/2}) dx = 0$

9. $\int \cot^6 8x \csc^2 8x\, dx$ $\boxed{u = \cot 8x,\ du = -8\csc^2 x\, dx}$

$$= -\frac{1}{8} \int u^6\, du = -\frac{1}{56} u^7 + C = -\frac{1}{56} \cot^7 8x + C$$

11. $\int (4x^2 - 16x + 7)^4 (x - 2) dx = \frac{1}{8} \int (4x^2 - 16x + 7)^4 [8(x - 2) dx]$

$$\boxed{u = 4x^2 - 16x + 7,\ du = 8(x - 2) dx}$$

$$= \frac{1}{8} \int u^4\, du = \frac{1}{40} u^5 + C = \frac{1}{40} (4x^2 - 16x + 7)^5 + C$$

13. $\int \frac{x^2 + 1}{\sqrt[3]{x^3 + 3x - 16}} dx = \frac{1}{3} \int (x^3 + 3x - 16)^{-1/3} [3(x^2 + 1) dx]$

$$\boxed{u = x^3 + 3x - 16,\ du = 3(x^2 + 1) dx}$$

$$= \frac{1}{3} \int u^{-1/3}\, du = \frac{1}{2} u^{2/3} + C = \frac{1}{2} (x^3 + 3x - 16)^{2/3} + C$$

15. $\int_{0}^{4} \frac{x}{16 + x^2} dx$ $\boxed{u = x^2 + 16,\ du = 2x\, dx}$

$$= \frac{1}{2} \int_{16}^{32} \frac{1}{u}\, du = \frac{1}{2} \ln|u|\Big]_{16}^{32} = \frac{1}{2} (\ln 32 - \ln 16) = \frac{1}{2} \ln \frac{32}{16} = \frac{1}{2} \ln 2$

17. $\int_{0}^{2} \frac{1}{\sqrt{16 - x^2}} dx = \sin^{-1} \frac{x}{4}\Big]_{0}^{2} = \sin^{-1} \frac{1}{2} - \sin^{-1} 0 = \frac{\pi}{6}$

19. $\int \tan 10x\, dx = -\frac{1}{10} \ln|\cos 10x| + C$

21. $\int_{0}^{7} f(x) dx = \int_{0}^{5} f(x) dx + \int_{5}^{7} f(x) dx;\quad 2 = -3 + \int_{5}^{7} f(x) dx;\quad \int_{5}^{7} f(x) dx = 5$

23. Since $|x-1| = \begin{cases} -x+1, & 0 \le x < 1 \\ x-1, & 1 \le x \le 3 \end{cases}$, we have

$$\int_0^3 (1+|x-1|)\,dx = \int_0^1 (1+|x-1|)\,dx + \int_1^3 (1+|x-1|)\,dx$$

$$= \int_0^1 (1-x+1)\,dx + \int_1^3 (1+x-1)\,dx$$

$$= \left(2x - \frac{1}{2}x^2\right)\Big]_0^1 + \frac{1}{2}x^2\Big]_1^3 = \left(\frac{3}{2}-0\right)+\left(\frac{9}{2}-\frac{1}{2}\right) = \frac{11}{2}.$$

25. $\displaystyle\int_{\pi/2}^{\pi/2} \frac{\sin^{10}t}{16t^7+1}\,dt = 0$

27. Since $f(x) = \dfrac{1}{1+3x^2}$ is an even function, $\displaystyle\int_{-1}^1 \frac{1}{1+3x^2}\,dx = 2\int_0^1 \frac{1}{1+3x^2}\,dx$. Therefore

$$2\int_0^1 \frac{1}{1+3x^2}\,dx \qquad \boxed{u = \sqrt{3}x,\ du = \sqrt{3}\,dx}$$

$$= \frac{2}{\sqrt{3}}\int_0^{\sqrt{3}} \frac{1}{1+u^2}\,du = \frac{2}{\sqrt{3}}\tan^{-1}u\Big]_0^{\sqrt{3}}$$

$$= \frac{2}{\sqrt{3}}(\tan^{-1}\sqrt{3} - \tan^{-1}0) = \frac{2}{\sqrt{3}}\left(\frac{\pi}{3}-0\right) = \frac{2\pi}{3\sqrt{3}}.$$

29. $\displaystyle\lim_{n\to\infty} \frac{1+2+3+\cdots+n}{n^2} = \lim_{n\to\infty}\left(\sum_{k=1}^n k\cdot\frac{1}{n^2}\right) = \lim_{n\to\infty}\left[\frac{n(n+1)}{2}\cdot\frac{1}{n^2}\right]$

$$= \lim_{n\to\infty}\left(\frac{n^2}{2n^2}+\frac{n}{2n^2}\right) = \lim_{n\to\infty}\left(\frac{1}{2}+\frac{1}{2n}\right) = \frac{1}{2}$$

31. Since $\dfrac{dV}{dt} = \dfrac{1}{4}$, $V = \dfrac{1}{4}t+C$. When $t = 0$, $V = \dfrac{1}{2}$ ft^3 and $C = \dfrac{1}{2}$. Thus, $V(t) = \dfrac{1}{4}t+\dfrac{1}{2}$ and $V(8) = \dfrac{5}{2}$ ft^3. The scale at this time

will read $\dfrac{5}{2}(62.4) = 156$ lbs. The volume of the bucket is $V = \dfrac{\pi}{3}\cdot 3\left[\left(\dfrac{1}{2}\right)^2+\dfrac{1}{2}(1)+1^2\right] = \dfrac{7\pi}{4}$ ft^3. Solving $\dfrac{7\pi}{4} = \dfrac{1}{4}t+\dfrac{1}{2}$ for

t, we obtain $t \approx 20$ min.

33. From the figure we note that $\displaystyle\int_{f(1)}^{f(2)} f^{-1}(x)\,dx = 20 - 2 - \int_1^2 f(x)\,dx$. Since

$$\int_1^2 f(x)\,dx = \int_1^2 (x^3+x)\,dx = \left(\frac{1}{4}x^4+\frac{1}{2}x^2\right)\Big]_1^2$$

$$= \left(\frac{16}{4}+\frac{4}{2}\right)-\left(\frac{1}{4}+\frac{1}{2}\right) = \frac{21}{4},$$

we have $\displaystyle\int_{f(1)}^{f(2)} f^{-1}(x)\,dx = 20 - 2 - \frac{21}{4} = \frac{51}{4}.$

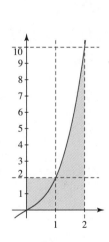

Applications of the Integral

6.1   Rectilinear Motion Revisited

1. $s(t) = \displaystyle\int 6\,dt = 6t + c; \quad 5 = s(2) = 6(2) + c; \quad c = -7; \quad s(t) = 6t - 7$

3. $s(t) = \displaystyle\int (t^2 - 4t)\,dt = \frac{1}{3}t^3 - 2t^2 + c; \quad 6 = s(3) = -9 + c; \quad c = 15; \quad s(t) = \frac{1}{3}t^3 - 2t^2 + 15$

5. $s(t) = \displaystyle\int -10\cos\left(4t + \frac{\pi}{6}\right)dt = \frac{5}{2}\sin\left(4t + \frac{\pi}{6}\right) + c; \quad \frac{5}{4} = s(0) = -\frac{5}{2}\left(\frac{1}{2}\right) + c = -\frac{5}{4} + c;$

   $c = \frac{5}{2}; \quad s(t) = -\frac{5}{2}\sin\left(4t + \frac{\pi}{6}\right) + \frac{5}{2}$

7. $v(t) = \displaystyle\int -5\,dt = -5t + c; \quad 4 = v(1) = -5 + c; \quad c = 9; \quad v(t) = -5t + 9;$

   $s(t) = \displaystyle\int (-5t + 9)\,dt = -\frac{5}{2}t^2 + 9t + c; \quad 2 = s(1) = \frac{13}{2} + c; \quad c = -\frac{9}{2};$

   $s(t) = -\frac{5}{2}t^2 + 9t - \frac{9}{2}$

9. $v(t) = \displaystyle\int (3t^2 - 4t + 5)\,dt = t^3 - 2t^2 + 5t + c; \quad -3 = v(0) = c; \quad v(t) = t^3 - 2t^2 + 5t - 3;$

   $s(t) = \displaystyle\int (t^3 - 2t^2 + 5t - 3)\,dt = \frac{1}{4}t^3 - \frac{2}{3}t^3 + \frac{5}{2}t^2 - 3t + c; \quad 10 = s(0) = c;$

   $s(t) = \frac{1}{4}t^4 - \frac{2}{3}t^3 + \frac{5}{2}t^2 - 3t + 10$

11. $v(t) = \displaystyle\int (7t^{1/3} - 1)\,dt = \frac{21}{4}t^{4/3} - t + c; \quad 50 = v(8) = 76 + c; \quad c = -26;$

   $v(t) = \frac{21}{4}t^{4/3} - t + 26;$

   $s(t) = \displaystyle\int \left(\frac{21}{4}t^{4/3} - t - 26\right)dt = \frac{9}{4}t^{7/3} - \frac{1}{2}t^2 - 26t + c; \quad 0 = s(8) = 48 + c; \quad c = -48;$

   $s(t) = \frac{9}{4}t^{7/3} - \frac{1}{2}t^2 - 26t - 48$

13. $v(t) = 2t - 2 = 2(t - 1)$

   $\text{dist.} = \displaystyle\int_0^5 |2(t-1)|\,dt = 2\int_0^1 -(t-1)\,dt + 2\int_1^5 (t-1)\,dt$

   $= 2\left(-\frac{1}{2}t^2 + t\right)\Big]_0^1 + 2\left(\frac{1}{2}t^2 - t\right)\Big]_1^5 = 2\left(\frac{1}{2} - 0\right) + 2\left[\frac{15}{2} - \left(-\frac{1}{2}\right)\right] = 17 \text{ cm}$

15. $v(t) = 3t^2 - 6t - 9 = 3(t+1)(t-3)$

dist. $= \int_0^4 |13t^2 - 6t - 9| \, dt = 3 \int_0^3 -(t^2 - 2t - 3) \, dt + 3 \int_3^4 (t^2 - 2t - 3) \, dt$

$= 3\left(-\dfrac{1}{3}t^3 + t^2 + 3t\right)\Big]_0^3 + 3\left(\dfrac{1}{3}t^3 - t^2 - 3t\right)\Big]_3^4 = 3(9-0) + 3\left[-\dfrac{20}{3} - (-9)\right] = 34 \text{ cm}$

17. $v(t) = 6\pi \cos \pi t$

dist. $= \int_1^3 |6\pi \cos \pi t| \, dt = 6\int_1^{3/2} -\pi \cos \pi t \, dt + 6\int_{3/2}^{5/2} \pi \cos \pi t \, dt + 6\int_{5/2}^3 -\pi \cos \pi t \, dt$

$= 6(-\sin \pi t)\big]_1^{3/2} + 6(\sin \pi t)\big]_{3/2}^{5/2} + 6(-\sin \pi t)\big]_{5/2}^3$

$= 6[-(-1) - 0] + 6[1 - (-1)] + 6[0 - (-1)] = 24 \text{ cm}$

19. We first convert mi/h to mi/s: 60 mi/h = 60/3600 mi/s. Then the distance traveled is

$$\int_0^2 \frac{60}{3600} \, dt = \frac{60}{3600}t\Big]_0^2 = \frac{60}{1800} \text{ mi} = \frac{1}{30} \text{ mi} \times 5280 \text{ ft/mi} = 176 \text{ ft}.$$

21. $a(t) = -32$; $v(0) = 0$; $s(4) = 0$; $v(t) = \int -32 \, dt = -32t + c$; $0 = v(0) = c$; $v(t) = -32t$;

$s(t) = \int -32t \, dt = -16t^2 + c$; $0 = s(4) = -256 + c$; $c = 256$; $s(t) = -16t^2 + 256$

The height of the building is $s(0) = 256$ ft.

23. $a(t) = -9.8$; $v(0) = 24.5$; $s(0) = 0$; $v(t) = \int -9.8 \, dt = -9.8t + c$; $24.5 = v(0) = c$;

$v(t) = -9.8t + 24.5$; $s(t) = \int (-9.8t + 24.5) \, dt = -4.9t^2 + 24.5t + c$; $0 = s(0) = c$;

$s(t) = -4.9t^2 + 24.5t$

Solving $v(t) = -9.8t + 24.5 = 0$, we see that the maximum height is attained when $t = 2.5$ seconds. The maximum height is $s(2.5) = 30.625$ m.

25. $a(t) = -32$; $v(0) = 32$; $s(0) = 384$; $v(t) = \int -32 \, dt = -32t + c$; $32 = v(0) = c$;

$v(t) = -32t + 32$; $s(t) = \int (-32t + 32) \, dt = -16t^2 + 32t + c$; $384 = s(0) = c$;

$s(t) = -16t^2 + 32t + 384$

Solving $v(t) = -32t + 32 = 0$ we see that the maximum height is attained when $t = 1$ second. The maximum height is $s(1) = 400$ ft. Setting $s(t) = -16t^2 + 32t + 384 = 0$, we have $t^2 - 2t - 24 = (t-6)(t+4) = 0$. Thus, the ball hits the ground at 6 seconds.

27. $a(t) = -32$; $v(0) = -16$; $s(0) = 102$; $v(t) = \int -32 \, dt = -32t + c$; $-16 = v(0) = c$;

$v(t) = -32t - 16$; $s(t) = \int (-32t - 16) \, dt = -16t^2 - 16t + c$; $102 = s(0) = c$;

$s(t) = -16t^2 - 16t + 102$

Solving $s(t) = -16t^2 - 16t + 102 = 6$, we see that the marshmallow hits the person at $t = 2$ seconds. The impact velocity is $v(2) = -80$ ft/s.

29. We measure upward from the top of the volcano, so that $s(0) = 0$. From $a(t) = g = -1.8$ we obtain $v(t) = -1.8t + v_0$ and $s(t) = -0.9t^2 + v_0 t$. If the rock attains its maximum height at time t_1, then $v(t_1) = 0 = -1.8t + v_0$ and $t_1 = v_0/1.8$. Solving

$$200,000 = -0.9t_1^2 + v_0 t_1 = -0.9\left(\frac{v_0}{1.8}\right)^2 + v_0\left(\frac{v_0}{1.8}\right) = 0.9\left(\frac{v_0}{1.8}\right)^2 = \frac{v_0^2}{3.6}$$

gives $v_0 \sqrt{3.6(200,000)} \approx 848.5$ m/s.

31. From the hint, $a = \dfrac{dv}{dt} = \dfrac{dv}{ds}v$, and integrating with respect to s gives $\displaystyle\int a\,ds = \int\left(v\dfrac{dv}{ds}\right)ds$. Then $as = \dfrac{1}{2}v^2 + c$, and solving for v we have $v^2 = 2as - 2c$. Since $v = v_0$ when $s = 0$, $v_0^2 = -2c$ and $v^2 = 2as + v_0^2$.

33. Let a be the acceleration of gravity on the earth, $v(0) = v_0$, and $s(0) = 0$.

$$v(t) = \int a\,dt = at + c; \quad v_0 = v(0) = c; \quad v(t) = at + v_0;$$

$$s(t) = \int (at + v_0)\,dt = \frac{1}{2}at^2 + v_0 t + c \quad 0 = s(0) = c; \quad s(t) = \frac{1}{2}at^2 + v_0 t$$

To find the maximum height reached on earth, we solve $v(t) = at + v_0 = 0$. The maximum height is reached when $t = -v_0/a$ and is $s(-v_0/a) = v_0^2/2a - v_0^2/a = -v_0^2/2a$. On the planet, the acceleration of gravity is $a/2$. Proceeding as on the earth, we obtain $v(t) = \frac{1}{2}at + v_0$, and $s(t) = \frac{1}{4}at^2 + v_0 t$. To find the maximum height reached on the planet, we solve $v(t) = \frac{1}{2}at + v_0 = 0$. The maximum height is reached when $t = -2v_0/a$ and is $s(-2v_0/a) = v_0^2/a - 2v_0^2/a = -v_0^2/a$. Thus, the maximum height reached on the planet is twice that reached on earth.

6.2 Area Revisited

1. $A = \displaystyle\int_{-1}^{1} -(x^2 - 1)\,dx = \left(-\frac{1}{3}x^3 + x\right)\Big]_{-1}^{1} = \frac{2}{3} - \left(-\frac{2}{3}\right) = \frac{4}{3}$

3. $A = \displaystyle\int_{-3}^{0} -x^3\,dx = -\frac{1}{4}x^4\Big]_{-3}^{0} = 0 - \left(-\frac{81}{4}\right) = \frac{81}{4}$

5. $A = \displaystyle\int_{0}^{3} -(x^2 - 3x)\,dx = \left(-\frac{1}{3}x^3 - \frac{3}{2}x^2\right)\Big]_{0}^{3} = \frac{9}{2} - 0 = \frac{9}{2}$

7. $A = \displaystyle\int_{-1}^{0} (x^3 - 6x)\,dx + \int_{0}^{1} -(x^3 - 6x)\,dx$

$= \left(\frac{1}{4}x^4 - 3x^2\right)\Big]_{-1}^{0} + \left(-\frac{1}{4}x^4 - 3x^2\right)\Big]_{0}^{1}$

$= \left[0 - \left(-\frac{11}{4}\right)\right] + \left(\frac{11}{4} - 0\right) = \frac{11}{2}$

9. $A = \displaystyle\int_{0}^{1} -(x^3 - 6x^2 + 11x - 6)\,dx + \int_{1}^{2} (x^3 - 6x^2 + 11x - 6)\,dx$

$\qquad + \displaystyle\int_{2}^{3} -(x^3 - 6x^2 + 11x - 6)\,dx$

$= \left(-\frac{1}{4}x^4 + 2x^3 - \frac{11}{2}x^2 + 6x\right)\Big]_{0}^{1} + \left(\frac{1}{4}x^4 - 2x^3 + \frac{11}{2}x^2 - 6x\right)\Big]_{1}^{2}$

$\qquad + \left(-\frac{1}{4}x^4 + 2x^3 - \frac{11}{2}x^2 + 6x\right)\Big]_{2}^{3}$

$= \left(\frac{9}{4} - 0\right) + \left[(-2) - \left(-\frac{9}{4}\right)\right] + \left(\frac{9}{4} - 2\right) = \frac{11}{4}$

11. $A = \int_{1/2}^{1} -(1-x^{-2})\,dx + \int_{1}^{3}(1-x^{-2})\,dx = \left(-x-\frac{1}{x}\right)\Big]_{1/2}^{1} + \left(x+\frac{1}{x}\right)\Big]_{1}^{3}$

$\qquad = \left[-2 - \left(-\frac{5}{2}\right)\right] + \left(\frac{10}{3} - 2\right) = \frac{11}{6}$

13. $A = \int_{0}^{1} -(x^{1/2}-1)\,dx + \int_{1}^{4}(x^{1/2}-1)\,dx$

$\qquad = \left(-\frac{2}{3}x^{3/2}+x\right)\Big]_{0}^{1} + \left(\frac{2}{3}x^{3/2}-x\right)\Big]_{1}^{4} = \left(\frac{1}{3}-0\right) + \left[\frac{4}{3} - \left(-\frac{1}{3}\right)\right] = 2$

15. $A = \int_{-2}^{0} -x^{1/3}\,dx + \int_{0}^{3} x^{1/3}\,dx = -\frac{3}{4}x^{4/3}\Big]_{-2}^{0} + \frac{3}{4}x^{4/3}\Big]_{0}^{3}$

$\qquad = \left[0 + \frac{3}{4}(2^{4/3})\right] + \left[\frac{3}{4}(3^{4/3}) - 0\right] = \frac{3}{4}(2^{4/3} + 3^{4/3})$

17. $A = \int_{-\pi}^{0} -\sin x\,dx + \int_{0}^{\pi}\sin x\,dx = \cos x]_{-\pi}^{0} - \cos x]_{0}^{\pi}$

$\qquad = [1-(-1)] - (-1-1) = 4$

19. $A = \int_{-3\pi/2}^{\pi/2} -(-1+\sin x)\,dx = (x+\cos x)]_{-3\pi/2}^{\pi/2} = \frac{\pi}{2} - \left(-\frac{3\pi}{2}\right) = 2\pi$

21. $A = \int_{-2}^{0} -x\,dx + \int_{0}^{1} x^{2}\,dx = -\frac{1}{2}x^{2}\Big]_{-2}^{0} + \frac{1}{3}x^{3}\Big]_{0}^{1} = -(0-2) + \left(\frac{1}{3}-0\right) = \frac{7}{3}$

23. $A = \int_{0}^{3}[x-(-2x)]\,dx = \int_{0}^{3} 3x\,dx = \frac{3}{2}x^{2}\Big]_{0}^{3} = \frac{27}{2}$

25. $A = \int_{-2}^{2}(4-x^{2})\,dx = \left(4x-\frac{1}{3}x^{3}\right)\Big]_{-2}^{2} = \frac{16}{3} - \left(-\frac{16}{3}\right) = \frac{32}{3}$

27. $A = \int_{-1}^{2}(8-x^{3})\,dx = \left(8x-\frac{1}{4}x^{4}\right)\Big]_{-1}^{2} = 12 - \left(-\frac{33}{4}\right) = \frac{81}{4}$

29. $A = \int_{-1}^{1}[4(1-x^{2})-(1-x^{2})]\,dx = \int_{-1}^{1}(3-3x^{2})\,dx = (3x-x^{3})]_{-1}^{1}$

$\qquad = 2-(-2) = 4$

31. $A = \int_{1}^{3}(x-x^{-2})\,dx = \left(\frac{1}{2}x^{2}+\frac{1}{x}\right)\Big]_{1}^{3} = \frac{29}{6} - \frac{3}{2} = \frac{10}{3}$

33. $A = \int_{-3}^{1} [(-x^2 + 6) - (x^2 + 4x)]\, dx = \int_{-3}^{1} (6 - 4x - 2x^2)\, dx$

$= \left(6x - 2x^2 - \dfrac{2}{3}x^3\right)\bigg]_{-3}^{1} = \dfrac{10}{3} - (-18) = \dfrac{64}{3}$

35. $A = \int_{-8}^{8} (4 - x^{2/3})\, dx = \left(4x - \dfrac{3}{5}x^{5/3}\right)\bigg]_{-8}^{8} = \dfrac{64}{5} - \left(-\dfrac{64}{5}\right) = \dfrac{128}{5}$

37. $A = \int_{-1}^{5} [(2x + 2) - (x^2 - 2x - 3)]\, dx + \int_{5}^{6} [(x^2 - 2x - 3) - (2x + 2)]\, dx$

$= \int_{-1}^{5} (5 + 4x - x^2)\, dx + \int_{5}^{6} (x^2 - 4x - 5)\, dx$

$= \left(5x + 2x^2 + \dfrac{1}{3}x^3\right)\bigg]_{-1}^{5} + \left(\dfrac{1}{3}x^3 - 2x^2 - 5x\right)\bigg]_{5}^{6}$

$= \dfrac{100}{3} - \left(-\dfrac{8}{3}\right) + (-30) - \left(-\dfrac{100}{3}\right) = \dfrac{118}{3}$

39. $A = \int_{-4}^{0} \left(x + 6 + \dfrac{1}{2}x\right)\, dx + \int_{0}^{2} (x + 6 - x^3)\, dx$

$= \left(\dfrac{3}{4}x^2 + 6x\right)\bigg]_{-4}^{0} + \left(\dfrac{1}{2}x^2 + 6x - \dfrac{1}{4}x^4\right)\bigg]_{0}^{2} = (0 + 12) + (10 - 0) = 22$

41. $A = \int_{-1}^{2} [(2 - y^2) - (-y)]\, dy = \int_{-1}^{2} (2 + y - y^2)\, dy$

$= \left(2y + \dfrac{1}{2}y^2 - \dfrac{1}{3}y^3\right)\bigg]_{-1}^{2} = \dfrac{10}{3} - \left(-\dfrac{7}{6}\right) = \dfrac{9}{2}$

43. $A = \int_{-2}^{0} [(-y^2 - 2y + 2) - (y^2 + 2y + 2)]\, dy = \int_{-2}^{0} (-2y^2 - 4y)\, dy$

$= \left(-\dfrac{2}{3}y^3 - 2y^2\right)\bigg]_{-2}^{0} = 0 - \left(-\dfrac{8}{3}\right) = \dfrac{8}{3}$

45. $A = \int_{-1}^{1} [(x + 4) - (x^3 - x)]\, dx = \int_{-1}^{1} (4 + 2x - x^3)\, dx$

$= \left(4x - x^2 - \dfrac{1}{4}x^4\right)\bigg]_{-1}^{1} = \dfrac{19}{4} - \left(-\dfrac{13}{4}\right) = 8$

47. $A = \int_{0}^{\pi/4} (\cos x - \sin x)\, dx + \int_{\pi/4}^{\pi/2} (\sin x - \cos x)\, dx$

$= (\sin x + \cos x)]_{0}^{\pi/4} + (-\cos x - \sin x)]_{\pi/4}^{\pi/2}$

$= \sqrt{2} - 1 + (-1) - (-\sqrt{2}) = 2\sqrt{2} - 2$

49. $A = \int_{\pi/6}^{5\pi/6} (4\sin x - 2)\, dx = (-4\cos x - 2x)]_{\pi/6}^{5\pi/6}$

$= 2\sqrt{3} - \dfrac{5\pi}{3} - \left(-2\sqrt{3} - \dfrac{\pi}{3}\right) = \dfrac{12\sqrt{3} - 4\pi}{3}$

51. Region 1: $y = \sqrt{x},\ y = -x,\ x = 0,\ x = 4$

Region 2: $y = -\sqrt{x}$, $y = x$, $x = 0$, $x = 4$

53. $\displaystyle\int_0^2 \left| \frac{3}{x+1} - 4x \right| dx = \int_0^{1/2} \left(\frac{3}{x+1} - 4x \right) dx + \int_{1/2}^2 \left(4x - \frac{3}{x+1} \right) dx$

$\displaystyle = \left(3\ln|x+1| - 2x^2 \right) \Big]_0^{1/2} + \left(2x^2 - 3\ln|x+1| \right) \Big]_{1/2}^2$

$\displaystyle = 3\ln\frac{3}{2} - \frac{1}{2} + 8 - 3\ln 3 - \frac{1}{2} + 3\ln\frac{3}{2}$

$\displaystyle = 7 + 3\ln\frac{3}{4} \approx 6.1370$

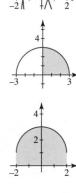

55. $\displaystyle\int_0^3 \sqrt{9-x^2}\,dx = \frac{1}{4}\pi(3)^2 = \frac{9\pi}{4}$

57. $\displaystyle\int_{-2}^2 \left(1 + \sqrt{4-x^2} \right) dx = \int_{-2}^2 1\,dx + \int_{-2}^2 \sqrt{4-x^2}\,dx = 4 + \frac{1}{2}\pi(2)^2 = 4 + 2\pi$

59. The area of the ellipse is four times the area in the first quadrant portion of the ellipse. Thus,

$$A = 4\int_0^a \sqrt{b^2 - b^2 x^2/a^2}\,dx = \frac{4b}{a}\int_0^a \sqrt{a^2 - x^2}\,dx = \frac{4b}{a}\left(\frac{1}{4}\pi a^2 \right) = \pi ab.$$

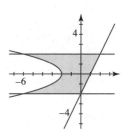

61. $\displaystyle A = \int_{-6}^{-2} \left(2 - \sqrt{-x-2} \right) dx + \int_{-6}^{-2} \left[-\sqrt{-x-2} - (-2) \right] dx$

$\displaystyle \qquad + \int_{-2}^0 \left[2 - (-2) \right] dx + \int_0^2 \left[2 - (2x-2) \right] dx$

$\displaystyle = 2\int_{-6}^{-2} \left(2 - \sqrt{-x-2} \right) dx + \int_{-2}^0 4\,dx + \int_0^2 (4 - 2x)\,dx$

$\displaystyle = 2\left[2x + \frac{2}{3}(-x-2)^{3/2} \right]\Big|_{-6}^{-2} + 4x\,]_{-2}^0 + (4x - x^2)\,]_0^2$

$\displaystyle = 2\left[-4 - \left(-\frac{20}{3} \right) \right] + 8 + 4 - 0 = \frac{52}{3}$

63. The area with respect to x is $\displaystyle A_x = \int_0^{\ln 3/2} (e^x - 1)\,dx + \int_{\ln 3/2}^{\ln 2} (2 - e^x)\,dx.$

The area with respect to y is $\displaystyle A_y = \int_1^2 \left(\ln y - \ln\frac{y+1}{2} \right) dy.$

If integration with respect to x is chosen, we get

$$A_x = \int_0^{\ln 3/2} (e^x - 1)\,dx + \int_{\ln 3/2}^{\ln 2} (2 - e^x)\,dx = (e^x - x)\big]_0^{\ln 3/2} + (2x - e^x)\big]_{\ln 3/2}^{\ln 2}$$

$$= \frac{3}{2} - \ln\frac{3}{2} - 1 + 2\ln 2 - 2 - 2\ln\frac{3}{2} + \frac{3}{2} = -3\ln 3 + 5\ln 2 \approx 0.1699.$$

If integration with respect to y is chosen, we get

$$A_y = \int_1^2 \left(\ln y - \ln\frac{y+1}{2} \right) dy = \left[y\ln y - y - (y+1)\ln\frac{y+1}{2} + (y+1) \right]\Big|_1^2$$

$$= \left[y\ln y - (y+1)\ln\frac{y+1}{2} + 1 \right]\Big|_1^2 = 2\ln 2 - 3\ln\frac{3}{2} + 1 - \ln 1 + 2\ln 1 - 1$$

$$= -3\ln 3 + 5\ln 2 \approx 0.1699$$

(see Problem 5.1.39 for the antiderivative of $\ln x$)

65. At $P(x_0, 1/x_0)$ the slope of the line segment is $-1/x_0^2$. The equation of the line through Q and R is then $y = -x/x_0^2 + 2/x_0$. Setting $y = 0$ we see that the x-intercept is $2x_0$. The area is

$$A = \int_0^{2x_0} \left(-\frac{1}{x_0^2}x + \frac{2}{x_0} \right) dx = \left(-\frac{1}{2x_0^2}x^2 + \frac{2}{x_0}x \right) \Bigg]_0^{2x_0} = -2 + 4 = 2,$$

which does not depend on x_0.

67. By symmetry with respect to the line $y = x$,

$$A = 2\int_0^a (\cos x - x)\, dx = 2\left(\sin x - \frac{1}{2}x^2 \right) \Bigg]_0^a = 2\sin a - a^2$$

(Using *Mathematica* it is easily shown that $a \approx 0.739085$.)

69. The areas are the same. Let w be the length of the line segments $\overline{AB}$ and $\overline{CD}$, and without loss of generality, let $\overline{AB}$ reside on $y = 0$, with $\overline{CD}$ residing on $y = h$. Thus, in Figure 6.2.17(a), the area of the rectangle is wh. Since Figure 6.2.17(b) describes a parallelogram, the line defined by $\overline{AD'}$ can be written as $x = f(y)$. Thus, the line defined by $\overline{BC'}$ is $x = f(y) + w$. The area of the parallelogram is therefore

$$\int_0^h \{[f(y) + w] - f(y)\}\, dy = \int_0^h w\, dy = wh.$$

6.3 Volumes of Solids: Slicing Method

1. $x^2 + y^2 = 16;\quad y = \sqrt{16 - x^2};\quad A(x) = \sqrt{3}y^2 = \sqrt{3}(16 - x^2)$

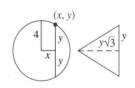

$$V = \int_{-4}^4 \sqrt{3}(16 - x^2)\, dx = \sqrt{3}\left(16x - \frac{1}{3}x^3 \right) \Bigg]_{-4}^4$$

$$= \sqrt{3}\left[\frac{128}{3} - \left(-\frac{128}{3} \right) \right] = \frac{256\sqrt{3}}{3}\ \text{ft}^3$$

3. $x = y^2;\quad A(x) = 2y(8y) = 16y^2 = 16x;\quad V = \int_0^4 16x\, dx = 8x^2 \Big]_0^4 = 128$

5. $y = -\frac{2}{5}x + 2;\quad A(x) = \frac{1}{2}\pi y^2 = \frac{2\pi}{25}(x - 5)^2$

$$V = \int_0^5 \frac{2\pi}{25}(x - 5)^2\, dx = \frac{2\pi}{75}(x - 5)^2 \Bigg]_0^5 = \frac{10\pi}{3}\ \text{ft}^3$$

7. $x = -y + 3;\quad A(y) = x^2 = (y - 3)^2$

$$V = \int_0^3 (y - 3)^2\, dy = \frac{1}{3}(y - 3)^3 \Bigg]_0^3 = 9$$

9. $x = \sqrt{y}$

$$V = \pi \int_0^1 [1^2 - (\sqrt{y})^2]\, dy = \pi \int_0^1 (1 - y)\, dy = \pi \left(y - \frac{1}{2}y^2 \right) \Bigg]_0^1 = \frac{\pi}{2}$$

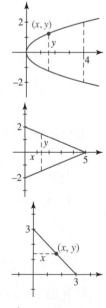

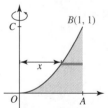

11. $y = x^2$

$$V = \pi \int_0^1 (1^2 - x^4)\, dx = \pi \left(x - \frac{1}{5}x^5 \right) \Big]_0^1 = \frac{4\pi}{5}$$

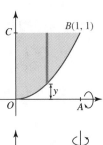

13. $x = \sqrt{y}$

$$V = \pi \int_0^1 (1 - \sqrt{y})^2\, dy = \pi \int_0^1 (1 - 2\sqrt{y} + y)\, dy$$

$$= \pi \left(y - \frac{4}{3}y^{3/2} + \frac{1}{2}y^2 \right) \Big]_0^1 = \frac{\pi}{6}$$

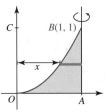

15. $y = 9 - x^2$

$$V = \pi \int_{-3}^3 (9 - x^2)^2\, dx = 2\pi \int_0^3 (81 - 18x^2 + x^4)\, dx$$

$$= 2\pi \left(81x - 6x^3 + \frac{1}{5}x^5 \right) \Big]_0^3 = \frac{1296\pi}{5}$$

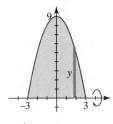

17. $x = \dfrac{1}{y}$

$$V = \pi \int_{1/2}^1 \left[\left(\frac{1}{y} \right)^2 - 1^2 \right] dy = \pi \int_{1/2}^1 (y^{-2} - 1)\, dy$$

$$= \pi \left(-\frac{1}{y} - y \right) \Big]_{1/2}^1 = \pi \left[-2 - \left(-\frac{5}{2} \right) \right] = \frac{\pi}{2}$$

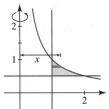

19. $y = (x - 2)^2$

$$V = \pi \int_0^2 (x - 2)^4\, dx = \frac{\pi}{5}(x - 2)^5 \Big]_0^2 = \frac{32\pi}{5}$$

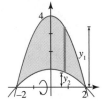

21. $y_1 = 4 - x^2; \quad y_2 = 1 - \dfrac{1}{4}x^2$

$$V = 2\pi \int_0^2 \left[(4 - x^2)^2 - \left(1 - \frac{1}{4}x^2 \right)^2 \right] dx = 2\pi \int_0^2 \left(15 - \frac{15}{2}x^2 + \frac{15}{16}x^4 \right) dx$$

$$= 2\pi \left(15x - \frac{5}{2}x^3 + \frac{3}{16}x^5 \right) \Big]_0^2 = 32\pi$$

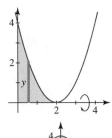

23. $x_1 = y; \quad x_2 = y - 1$

$$V = \pi \int_0^1 y^2\, dy + \pi \int_1^2 [y^2 - (y-1)^2]\, dy$$

$$= \pi \left(\frac{1}{3}y^3 \right) \Big]_0^1 + \pi(y^2 - y) \Big]_1^2 = \pi \left(\frac{1}{3} + 2 \right) = \frac{7\pi}{3}$$

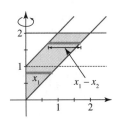

25. $x = y^2 + 1$

$$V = \pi \int_0^2 [5 - (y^2 + 1)]^2 \, dy = \pi \int_0^2 (4 - y^2)^2 \, dy$$

$$= \pi \int_0^2 (16 - 8y^2 + y^4) \, dy = \pi \left(16y - \frac{8}{3} y^3 + \frac{1}{5} y^5 \right) \Big]_0^2 = \frac{256\pi}{15}$$

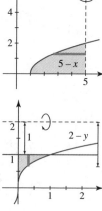

27. $y = x^{1/3}$

$$V = \pi \int_0^1 [(2 - x^{1/3})^2 - 1^2] \, dx = \pi \int_0^1 (3 - 4x^{1/3} + x^{2/3}) \, dx$$

$$= \pi \left(3x - 3x^{4/3} + \frac{3}{5} x^{5/3} \right) \Big]_0^1 = \frac{3\pi}{5}$$

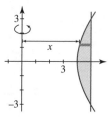

29. $x = \sqrt{y^2 + 16}$

$$V = \pi \int_{-3}^3 \left[5^2 - \left(\sqrt{y^2 + 16} \right)^2 \right] \, dy = \pi \int_{-3}^3 (9 - y^2) \, dy$$

$$= \pi \left(9y - \frac{1}{3} y^3 \right) \Big]_{-3}^3 = \pi [18 - (-18)] = 36\pi$$

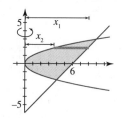

31. $x_1 = y + 6; \quad x_2 = y^2$

$$V = \pi \int_{-2}^3 [(y + 6)^2 - y^4] \, dy = \pi \int_{-2}^3 (36 + 12y + y^2 - y^4) \, dy$$

$$= \pi \left(36y + 6y^2 + \frac{1}{3} y^3 - \frac{1}{5} y^5 \right) \Big]_{-2}^3 = \pi \left[\frac{612}{5} - \left(-\frac{664}{15} \right) \right] = \frac{500\pi}{3}$$

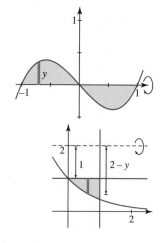

33. $y = x^3 - x$

$$V = \pi \int_{-1}^1 (x^3 - x)^2 \, dx = \pi \int_{-1}^1 (x^6 - 2x^4 + x^2) \, dx$$

$$= \pi \left(\frac{1}{7} x^7 - \frac{2}{5} x^5 + \frac{1}{3} x^3 \right) \Big]_{-1}^1 = \pi \left[\frac{8}{105} - \left(-\frac{8}{105} \right) \right] = \frac{16\pi}{105}$$

35. $y = e^{-x}$

$$V = \pi \int_0^1 [(2 - e^{-x})^2 - 1^2] \, dx = \pi \int_0^1 (3 - 4e^{-x} + e^{-2x}) \, dx$$

$$= \pi \left(3x + 4e^{-x} - \frac{1}{2} e^{-2x} \right) \Big]_0^1 = \pi \left(4e^{-1} - \frac{1}{2} e^{-2} - \frac{1}{2} \right)$$

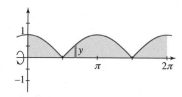

37. $y = |\cos x|$

$$V = \pi \int_0^{2\pi} |\cos x|^2 \, dx = \pi \int_0^{2\pi} \frac{1 + \cos 2x}{2} \, dx$$

$$= \frac{\pi}{2} \int_0^{2\pi} (1 + \cos 2x) \, dx = \frac{\pi}{2} \left(x + \frac{1}{2} \sin 2x \right) \Big]_0^{2\pi} = \pi^2$$

39. $y = \tan x$

$$V = \pi \int_0^{\pi/4} \tan^2 x \, dx = \pi \int_0^{\pi/4} (\sec^2 x - 1) \, dx = \pi(\tan x - x)]_0^{\pi/4}$$

$$= \pi \left(1 - \frac{\pi}{4} - 0\right) = \frac{4\pi - \pi^2}{4}$$

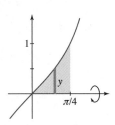

41. The volume of the right circular cylinder is $\pi r^2 h$. Placing the center of the red circular cylinder's base in Figure 6.3.19 on the origin, we see that $A = \pi r^2$ for every slice from $y = 0$ to h. Thus, the volume V of the cylinder is

$$V = \int_0^h \pi r^2 \, dy = \pi r^2 y \Big]_0^h = \pi r^2 h.$$

43. (a) Using *Mathematica*, we obtain with the disk method

$$V = \pi \int_{-1}^1 [P(x)]^2 (1 - x^2) \, dx = \frac{4\pi}{315} (5a^2 + 9b^2 + 21c^2 + 105d^2 + 18ac + 42bd).$$

(b) Setting $a = -0.07$, $b = -0.02$, $c = 0.2$, and $d = 0.56$ we obtain $V \approx 1.32$ cubic units.

(c)

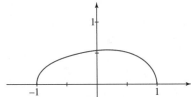

(d) Setting $a = -0.06$, $b = 0.04$, $c = 0.1$, and $d = 0.54$ we obtain $V \approx 1.26$ cubic units.

45. (a) Each eighth of the bicylinder can be sliced into squares whose sides follow the perimeter of a quadrant of the cylinders' base; that is, $x^2 + y^2 = r^2$, one side of the square is $y = \sqrt{r^2 - x^2}$, and its area is $y^2 = r^2 - x^2$. Using symmetry, the volume common to the cylinders is thus

$$V = 8 \int_0^r (r^2 - x^2) \, dx = 8 \left(r^2 x - \frac{x^3}{3}\right)\Big]_0^r = \frac{16r^3}{3}.$$

(b) This item involves a research report, and thus a preset solution is not applicable.

6.4 Volumes of Solids: Shell Method

1. $y = \sqrt{x}$

$$V = 2\pi \int_0^1 x\sqrt{x} \, dx = \frac{4\pi}{5} x^{5/2} \Big]_0^1 = \frac{4\pi}{5}$$

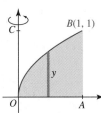

3. $x = y^2$

$$V = 2\pi \int_0^1 (1 - y) y^2 \, dy = 2\pi \left(\frac{1}{3} y^3 - \frac{1}{4} y^4\right)\Big]_0^1 = \frac{\pi}{6}$$

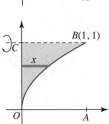

5. $y = \sqrt{x}$

$$V = 2\pi \int_0^1 (1-x)\sqrt{x}\,dx = 2\pi \left(\frac{2}{3}x^{3/2} - \frac{2}{5}x^{5/2} \right) \Big]_0^1 = \frac{8\pi}{15}$$

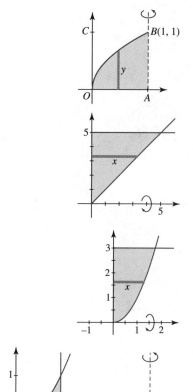

7. $x = y$

$$V = 2\pi \int_0^5 y \cdot y\,dy = \frac{2\pi}{3}y^3 \Big]_0^5 = \frac{250\pi}{3}$$

9. $x = \sqrt{y}$

$$V = 2\pi \int_0^3 y\sqrt{y}\,dy = \frac{4\pi}{5}y^{5/2} \Big]_0^3 = \frac{36\pi\sqrt{3}}{5}$$

11. $y = x^2$

$$V = 2\pi \int_0^1 (3-x)x^2\,dx = 2\pi \left(x^3 - \frac{1}{4}x^4 \right) \Big]_0^1 = \frac{3\pi}{2}$$

13. $y = x^2 + 4$

$$V = 2\pi \int_0^2 x(x^2 + 4 - 2)\,dx = 2\pi \int_0^2 (x^3 + 2x)\,dx$$

$$= 2\pi \left(\frac{1}{4}x^4 + x^2 \right) \Big]_0^2 = 16\pi$$

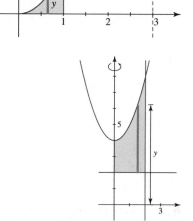

15. $x_1 = 1 + \sqrt{y}; \quad x_2 = 1 - \sqrt{y}$

$$V = 2\pi \int_0^1 y[1 + \sqrt{y} - (1 - \sqrt{y})]\,dy = 2\pi \int_0^1 2y^{3/2}\,dy$$

$$= \frac{8\pi}{5}y^{5/2} \Big]_0^1 = \frac{8\pi}{5}$$

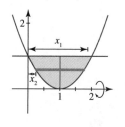

17. $x = y^3$

$$V = 2\pi \int_0^1 (y+1)(1-y^3)\,dy = 2\pi \int_0^1 (1 + y - y^3 - y^4)\,dy$$

$$= 2\pi \left(y + \frac{1}{2}y^2 - \frac{1}{4}y^4 - \frac{1}{5}y^5 \right) \Big]_0^1 = \frac{21\pi}{10}$$

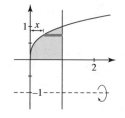

19. $y_1 = x; \quad y_2 = x^2$

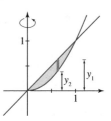

$$V = 2\pi \int_0^1 x(x - x^2)\,dx = 2\pi \left(\frac{1}{3}x^3 - \frac{1}{4}x^4\right)\Big]_0^1 = \frac{\pi}{6}$$

21. $y = -x^3 + 3x^2$

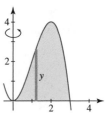

$$V = 2\pi \int_0^3 x(-x^3 + 3x^2)\,dx = 2\pi \int_0^3 (-x^4 + 3x^3)\,dx$$

$$= 2\pi \left(-\frac{1}{5}x^5 + \frac{3}{4}x^4\right)\Big]_0^3 = \frac{243\pi}{10}$$

23. $y_1 = 2 - x^2; \quad y_2 = x^2 - 2$

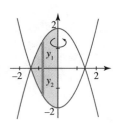

$$V = 2\pi \int_{-\sqrt{2}}^0 -x[2 - x^2 - (x^2 - 2)]\,dx = 2\pi \int_{-\sqrt{2}}^0 (2x^3 - 4x)\,dx$$

$$= 2\pi \left(\frac{1}{2}x^4 - 2x^2\right)\Big]_{-\sqrt{2}}^0 = 4\pi$$

25. $x = y^2 - 5y$

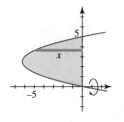

$$V = 2\pi \int_0^5 y(-y^2 + 5y)\,dy = 2\pi \left(-\frac{1}{4}y^4 + \frac{5}{3}y^3\right)\Big]_0^5 = \frac{625\pi}{6}$$

27. $y_1 = x + 6; \quad y_2 = x^3$

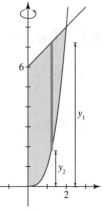

$$V = 2\pi \int_0^2 x(x + 6 - x^3)\,dx = 2\pi \left(\frac{1}{3}x^3 + 3x^2 - \frac{1}{5}x^5\right)\Big]_0^2 = \frac{248\pi}{15}$$

29. $y = \sin x^2$

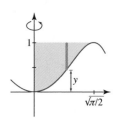

$$V = 2\pi \int_0^{\sqrt{\pi/2}} x(1 - \sin x^2)\,dx = 2\pi \int_0^{\sqrt{\pi/2}} (x - x\sin x^2)\,dx$$

$$= 2\pi \left(\frac{1}{2}x^2 + \frac{1}{2}\cos x^2\right)\Big]_0^{\sqrt{\pi/2}} = 2\pi \left(\frac{\pi}{4} - \frac{1}{2}\right) = \frac{\pi^2 - 2\pi}{2}$$

31. We use the shell method.

$$V = 2\pi \int_0^r (r - x)\left(\frac{h}{r}x\right) dx = 2\pi \int_0^r \left(hx - \frac{h}{r}x^2\right) dx = 2\pi \left(\frac{1}{2}hx^2 - \frac{h}{3r}x^3\right)\Big]_0^r = \frac{1}{3}\pi r^2 h$$

33. We use the disk method.

$$V = \pi \int_{-r}^{r} (\sqrt{r^2 - y^2})^2 \, dy = \pi \int_{-r}^{r} (r^2 - y^2) \, dy$$

$$= \pi \left(r^2 y - \frac{1}{3} y^3 \right) \Big]_{-r}^{r} = \pi \left[\frac{2}{3} r^3 - \left(-\frac{2}{3} r^3 \right) \right] = \frac{4}{3} \pi r^3$$

35. The equation of the ellipse is $y = b \sqrt{1 - \dfrac{x^2}{a^2}}$. We use the disk method.

$$V = \pi \int_{-a}^{a} \left(b \sqrt{1 - \frac{x^2}{a^2}} \right)^2 dx = \pi b^2 \int_{-a}^{a} \left(1 - \frac{x^2}{a^2} \right) dx = \pi b^2 \left(x - \frac{1}{3a^2} x^3 \right) \Big]_{-a}^{a}$$

$$= \pi b^2 \left[\frac{2a}{3} - \left(-\frac{2a}{3} \right) \right] = \frac{4\pi a b^2}{3}$$

37. $y_1 = \dfrac{\omega^2 x^2}{2g}$. The depth of the liquid below the x-axis is $y_2 = h - \dfrac{\omega^2 r^2}{2g}$. So the volume is

$$V = 2\pi \int_0^r x \left(\frac{\omega^2 x^2}{2g} + h - \frac{\omega^2 r^2}{2g} \right) dx$$

$$= 2\pi \int_0^r \left(\frac{\omega^2}{2g} x^3 + \frac{2hg - \omega^2 r^2}{2g} x \right) dx$$

$$= 2\pi \left(\frac{\omega^2}{8g} x^4 + \frac{2hg - \omega^2 r^2}{4g} x^2 \right) \Big]_0^r$$

$$= \frac{\pi \omega^2 r^4}{4g} + \frac{4\pi h g r^2 - 2\pi \omega^2 r^4}{4g} = \pi r^2 h - \frac{\pi \omega^2 r^4}{4g}.$$

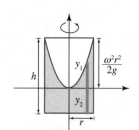

6.5 Length of a Graph

1. $y' = 1$; $\quad s = \displaystyle\int_{-1}^{1} \sqrt{1 + 1^2} \, dx = 2\sqrt{2}$

3. $y' = \dfrac{3}{2} x^{1/2}$

$$s = \int_0^1 \sqrt{1 + \frac{9}{4} x} \, dx = \frac{8}{27} \left(1 + \frac{9}{4} x \right)^{3/2} \Big]_0^1 = \frac{8}{27} \left[\left(\frac{13}{4} \right)^{3/2} - 1 \right] = \frac{13^{3/2} - 8}{27} \approx 1.4397$$

5. $y' = 2x(x^2 + 1)^{1/2}$

$$s = \int_1^4 \sqrt{1 + 4x^2(x^2 + 1)} \, dx = \int_1^4 \sqrt{(2x^2 + 1)^2} \, dx = \int_1^4 (2x^2 + 1) \, dx$$

$$= \left(\frac{2}{3} x^3 + x \right) \Big]_1^4 = \frac{140}{3} - \frac{5}{3} = 45$$

7. $y' = \dfrac{1}{2} x^{1/2} - \dfrac{1}{2} x^{-1/2} = \dfrac{x - 1}{2x^{1/2}}$

$$s = \int_1^4 \sqrt{1 + \frac{(x-1)^2}{4x}} \, dx = \int_1^4 \sqrt{\frac{4x + x^2 - 2x + 1}{4x}} \, dx = \int_1^4 \sqrt{\frac{(x+1)^2}{4x}} \, dx$$

$$= \frac{1}{2} \int_1^4 \frac{x+1}{x^{1/2}} \, dx = \frac{1}{2} \int_1^4 (x^{1/2} + x^{-1/2}) \, dx = \frac{1}{2} \left(\frac{2}{3} x^{3/2} + 2x^{1/2} \right) \Big]_1^4 = \frac{1}{2} \left(\frac{28}{3} - \frac{8}{3} \right) = \frac{10}{3}$$

9. $y' = x^3 - \dfrac{1}{4x^3} = \dfrac{4x^6 - 1}{4x^3}$

$$S = \int_2^3 \sqrt{1 + \dfrac{(4x^6 - 1)^2}{16x^6}}\, dx = \int_2^3 \sqrt{\dfrac{16x^6 + 16x^{12} - 8x^6 + 1}{16x^6}}\, dx = \int_2^3 \sqrt{\dfrac{(4x^6 + 1)^2}{16x^6}}\, dx$$

$$= \int_2^3 \dfrac{4x^6 + 1}{4x^3}\, dx = \dfrac{1}{4}\int_2^3 (4x^3 + x^{-3})\, dx = \dfrac{1}{4}\left(x^4 - \dfrac{1}{2x^2}\right)\Big]_2^3$$

$$= \dfrac{1}{4}\left(\dfrac{1457}{18} - \dfrac{127}{8}\right) = \dfrac{4685}{288} \approx 16.2674$$

11. $y' = -\dfrac{(4 - x^{2/3})^{1/2}}{x^{1/3}};\quad S = \int_1^8 \sqrt{1 + \dfrac{4 - x^{2/3}}{x^{2/3}}}\, dx = \int_1^8 \dfrac{2}{x^{1/3}}\, dx = 3x^{2/3}\Big]_1^8 = 9$

13. $y' = 2x;\quad S = \displaystyle\int_{-1}^3 \sqrt{1 + 4x^2}\, dx$

15. $y' = \cos x;\quad S = \displaystyle\int_0^\pi \sqrt{1 + \cos^2 x}\, dx$

17. $\dfrac{dx}{dy} = -\dfrac{2}{3}y^{-1/3}$

$$S = \int_0^8 \sqrt{1 + \dfrac{4}{9}y^{-2/3}}\, dy = \int_0^8 \sqrt{\dfrac{9y^{2/3} + 4}{9y^{2/3}}}\, dy = \dfrac{1}{3}\int_0^8 y^{-1/3}\sqrt{9y^{2/3} + 4}\, dy$$

$$\boxed{u = 9y^{2/3} + 4,\ du = 6y^{-1/3}\, dy}$$

$$= \dfrac{1}{18}\int_4^{40} u^{1/2}\, du = \dfrac{1}{27}u^{3/2}\Big]_4^{40} = \dfrac{1}{27}(40^{3/2} - 8) \approx 9.0734$$

19. (a) $y = (1 - x^{2/3})^{3/2};\quad y' = \dfrac{3}{2}(1 - x^{2/3})^{1/2}\left(-\dfrac{2}{3}x^{-1/3}\right) = -\dfrac{(1 - x^{2/3})^{1/2}}{x^{1/3}}$

$$S = \int_0^1 \sqrt{1 + \dfrac{1 - x^{2/3}}{x^{2/3}}}\, dx = \int_0^1 \sqrt{\dfrac{1}{x^{2/3}}}\, dx = \int_0^1 \dfrac{1}{x^{1/3}}\, dx$$

At $x = 0$, $\dfrac{1}{x^{1/3}}$ is discontinuous.

(b) The graph is symmetric with respect to both coordinate axes, so

$$s = 4\int_0^1 \dfrac{1}{x^{1/3}}\, dx = 4\left(\dfrac{3}{2}x^{2/3}\right)\Big]_0^1 = 6.$$

21. $y = \sqrt{r^2 - x^2};\quad y' = -x(r^2 - x^2)^{-1/2}$

$$2\pi r = s = 4\int_0^r \sqrt{1 + x^2(r^2 - x^2)^{-1}}\, dr = 4r\int_0^r \left(\dfrac{1}{\sqrt{r^2 - x^2}}\, dx\right) dr$$

Letting $r = 1$ we have $2\pi = 4\displaystyle\int_0^1 \dfrac{1}{\sqrt{1 - x^2}}\, dx$ or $\displaystyle\int_0^1 \dfrac{1}{\sqrt{1 - x^2}}\, dx = \dfrac{\pi}{2}.$

6.6 Area of a Surface of Revolution

1. $y' = x^{-1/2}$

$$S = 2\pi\int_0^8 2\sqrt{x}\sqrt{1 + x^{-1}}\, dx = 4\pi\int_0^8 \sqrt{x + 1}\, dx = \dfrac{8\pi}{3}(x + 1)^{3/2}\Big]_0^8 = \dfrac{8\pi}{3}(27 - 1) = \dfrac{208\pi}{3}$$

3. $y' = 3x^2$

$$S = 2\pi \int_0^1 x^3 \sqrt{1+9x^4}\, dx \quad \boxed{u = 1+9x^4,\ du = 36x^3\, dx}$$

$$= 2\pi \int_1^{10} \sqrt{u}\left(\frac{1}{36}\, du\right) = \frac{\pi}{27} u^{3/2}\bigg]_1^{10} = \frac{\pi}{27}(10^{3/2}-1) \approx 3.5631$$

5. $y' = 2x;\quad S = 2\pi \int_0^3 x\sqrt{1+4x^2}\, dx = \frac{\pi}{6}(1+4x^2)^{3/2}\bigg]_0^3 = \frac{\pi}{6}(37^{3/2}-1) \approx 117.3187$

7. $y' = 2$

$$S = 2\pi \int_2^7 (2x+1)\sqrt{1+4}\, dx = 2\sqrt{5}\pi \int_2^7 (2x+1)\, dx = 2\sqrt{5}\pi\left(x^2+x\right)\bigg]_2^7$$

$$= 2\sqrt{5}\pi(56-6) = 100\sqrt{5}\pi$$

9. $y' = x^3 - \dfrac{1}{4}x^{-3} = \dfrac{4x^6-1}{4x^3}$

$$S = 2\pi \int_1^2 x\sqrt{1+\frac{16x^{12}-8x^6+1}{16x^6}}\, dx = 2\pi \int_1^2 x\sqrt{\frac{16x^6+16x^{12}-8x^6+1}{16x^6}}\, dx$$

$$= \frac{\pi}{2}\int_1^2 \frac{\sqrt{(4x^6+1)^2}}{x^2}\, dx = \frac{\pi}{2}\int_1^2 (4x^4+x^{-2})\, dx = \frac{\pi}{2}\left(\frac{4}{5}x^5 - \frac{1}{x}\right)\bigg]_1^2$$

$$= \frac{\pi}{2}\left[\frac{251}{10} - \left(-\frac{1}{5}\right)\right] = \frac{253\pi}{20}$$

11. (a) $f'(x) = -\dfrac{r}{2h}\left(1-\dfrac{x}{h}\right)^{-1/2};\quad 1+[f'(x)]^2 = \dfrac{4h^2-4hx+r^2}{4h^2(1-x/h)}$

$$f(x)\sqrt{1+[f'(x)]^2} = r\sqrt{1-x/h}\,\frac{\sqrt{4h^2-2hx+r^2}}{2h\sqrt{1-x^2}} = \frac{r}{2h}\sqrt{4h^2-4hx+r^2}$$

$$S = 2\pi \int_0^h \frac{r}{2h}\sqrt{r^2+4h^2-4hx}\, dx = \frac{\pi r}{h}\left(\frac{2}{3}\right)\left(-\frac{1}{4h}\right)(r^2+4h^2-4hx)^{3/2}\bigg]_0^h$$

$$= \frac{\pi r}{6h^2}[(r^2+4h^2)^{3/2}-r^3]$$

(b) With $h = 0.1r$ we have

$$S = \frac{\pi r}{6(0.01)r^2}[r^3(1.04)^{3/2}-r^3] \approx \pi r^2\left(\frac{1.04^{3/2}-1}{0.06}\right) \approx \pi r^2\frac{0.060596}{0.06} \approx \pi r^2.$$

The approximate percentage error is $\dfrac{(0.060596/0.06)\pi r^2 - \pi r^2}{\pi r^2} = \dfrac{0.060596}{0.06} - 1 \approx 0.0099$, or approximately 1%.

13. For $x < -2$, $y = -x-2$ and $y' = -1$. For $x > 2$, $y = x+2$ and $y' = 1$.

$$S = 2\pi \int_{-4}^{-2} (-x-2)\sqrt{1+1}\, dx + 2\pi \int_{-2}^2 (x+2)\sqrt{1+1}\, dx$$

$$= 2\sqrt{2}\pi \int_{-4}^{-2} (-x-2)\, dx + 2\sqrt{2}\pi \int_{-2}^2 (x+2)\, dx$$

$$= 2\sqrt{2}\pi\left(-\frac{1}{2}x^2-2x\right)\bigg]_{-4}^{-2} + 2\sqrt{2}\pi\left(\frac{1}{2}x^2+2x\right)\bigg]_{-2}^2$$

$$= 2\sqrt{2}\pi[(-2+4)-(-8+8)] + 2\sqrt{2}\pi[(2+4)-(2-4)] = 20\sqrt{2}\pi$$

15. Let θ be the angle formed when the cone is cut and flattened out. The length of the arc of the sector is $2\pi r$, the circumference of the base of the cone. Then $\theta/2\pi r = 2\pi/2\pi L$ (the angle subtended by the sector is to the length of the sector as 2π radians is to the circumference of the circle of radius L), and $\theta = 2\pi r/L$. Using the hint in the text, the lateral surface area is $\dfrac{1}{2}L^2\left(\dfrac{2\pi r}{L}\right) = \pi rL$.

17. By similar triangles, $\dfrac{r_1}{L_1} = \dfrac{r_2}{L_2}$, $r_1 L_2 = r_2 L_1$, and $r_1 L_2 - r_2 L_1 = 0$.

    From Problem 15, the lateral surface area of the frustum is

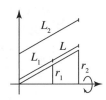

$$S = \pi r_2 L_2 - \pi r_1 L_1 = \pi (r_2 L_2 - r_1 L_1) = \pi[r_2 L_2 + \overbrace{(r_1 L_2 - r_2 L_1)}^{0} - r_1 L_1]$$
$$= \pi[(r_2 + r_1)L_2 - (r_2 + r_1)L_1] = \pi(r_1 + r_2)(L_2 - L_1) = \pi(r_1 + r_2)L.$$

19. We need to extend (3) in Section 6.6 to include functions which are not necessarily non-negative. In this case we have
    $S = 2\pi \displaystyle\int_a^b |f(x)|\sqrt{1 + [f'(x)]^2}\,dx$. Next, we require the fact that the surface area obtained by revolving f around $y = L$ is the same as that obtained by revolving $F(x) = f(x) - L$ around the x-axis. Then $f'(x) = f'(x)$ and

$$S = 2\pi \int_a^b |F(x)|\sqrt{1 + [f'(x)]^2}\,dx = 2\pi \int_a^b |f(x) - L|\sqrt{1 + [f'(x)]^2}\,dx.$$

21. (a) Since $\triangle BCT$ is similar to $\triangle TCS$ we have $\overline{CB}/\overline{TC} = \overline{CT}/\overline{CS}$ or $\dfrac{y_B}{R} = \dfrac{R}{R+h}$, which gives $y_B = \dfrac{R^2}{R+h}$. Now, revolving $x = \sqrt{R^2 - y^2}$ for $y_B \le y \le R$ around the y-axis, we obtain the surface area

$$A_S = 2\pi \int_{y_B}^R \sqrt{R^2 - y^2}\sqrt{1 + \left(\frac{-y}{\sqrt{R^2 - y^2}}\right)^2}\,dy = 2\pi \int_{\frac{R^2}{R+h}}^R R\,dy$$
$$= 2\pi R\left(R - \frac{R^2}{R+h}\right) = \frac{2\pi R^2 h}{R+h}.$$

   Since $A_E = 4\pi R^2$, we have $\dfrac{A_S}{A_E} = \dfrac{h}{2(R+h)}$.

   (b) With $h = 2000$ and $R = 6380$, $\dfrac{A_S}{A_E} = \dfrac{2000}{2(6380 + 2000)} \approx 0.119332 \approx 11.9\%$.

   (c) Setting $\dfrac{h}{2(R+h)} = \dfrac{1}{4}$ we obtain $2h = R + h$ or $h = R = 6380$ km.

   (d) $\displaystyle\lim_{h\to\infty} \frac{A_S}{A_E} = \lim_{h\to\infty} \frac{h}{2R + 2h} = \frac{1}{2}$

   From a large distance we would expect to see half of the earth's surface.

   (e) $\dfrac{A_S}{A_E} = \dfrac{3.76 \times 10^5}{2(6380 + 3.76 \times 10^5)} \approx 0.4917 = 49.17\%$

6.7 Average Value of a Function

1. $f_{\text{ave}} = \dfrac{1}{1 - (-3)} \displaystyle\int_{-3}^1 4x\,dx = \dfrac{1}{4}(2x^2)\Big]_{-3}^1 = \dfrac{1}{4}(2 - 18) = -4$

3. $f_{\text{ave}} = \dfrac{1}{2 - 0} \displaystyle\int_0^2 (x^2 + 10)\,dx = \dfrac{1}{2}\left(\dfrac{1}{3}x^3 + 10x\right)\Big]_0^2 = \dfrac{1}{2}\left(\dfrac{68}{3} - 0\right) = \dfrac{34}{3}$

5. $f_{\text{ave}} = \dfrac{1}{3 - (-1)} \displaystyle\int_{-1}^3 (3x^2 - 4x)\,dx = \dfrac{1}{4}(x^3 - 2x^2)\Big]_{-1}^3 = \dfrac{1}{4}[9 - (-3)] = 3$

7. $f_{\text{ave}} = \dfrac{1}{2 - (-2)} \displaystyle\int_{-2}^2 x^3\,dx = \dfrac{1}{4}\left(\dfrac{1}{4}x^4\right)\Big]_{-2}^2 = \dfrac{1}{4}(4 - 4) = 0$

9. $f_{\text{ave}} = \dfrac{1}{9 - 0} \displaystyle\int_0^9 x^{1/2}\,dx = \dfrac{1}{9}\left(\dfrac{2}{3}x^{3/2}\right)\Big]_0^9 = 2$

11. $f_{ave} = \dfrac{1}{3-0} \displaystyle\int_0^3 x\sqrt{x^2+16}\,dx = \dfrac{1}{3} \cdot \dfrac{1}{3}(x^2+16)^{3/2} \Big]_0^3 = \dfrac{1}{9}(125-64) = \dfrac{61}{9}$

13. $f_{ave} = \dfrac{1}{1/2-1/4} \displaystyle\int_{1/4}^{1/2} x^{-3}\,dx = 4\left(-\dfrac{1}{2}x^{-2}\right)\Big]_{1/4}^{1/2} = -2\left(\dfrac{1}{x^2}\right)\Big]_{1/4}^{1/2} = -2(4-16) = 24$

15. $f_{ave} = \dfrac{1}{5-3} \displaystyle\int_3^5 2(x+1)^{-2}\,dx = \dfrac{1}{2}[-2(x+1)^{-1}]\Big]_3^5 = -\dfrac{1}{x+1}\Big]_3^5 = -\left(\dfrac{1}{6}-\dfrac{1}{4}\right) = \dfrac{1}{12}$

17. $f_{ave} = \dfrac{1}{\pi-(-\pi)} \displaystyle\int_{-\pi}^{\pi} \sin x\,dx = \dfrac{1}{2\pi}(-\cos x)\Big]_{-\pi}^{\pi} = -\dfrac{1}{2\pi}[-1-(-1)] = 0$

19. $f_{ave} = \dfrac{1}{\pi/2-\pi/6} \displaystyle\int_{\pi/6}^{\pi/2} \csc^2 x\,dx = \dfrac{3}{\pi}(-\cot x)\Big]_{\pi/6}^{\pi/2} = -\dfrac{3}{\pi}(0-\sqrt{3}) = \dfrac{3\sqrt{3}}{\pi}$

21. $f_{ave} = \dfrac{1}{1-(-1)} \displaystyle\int_{-1}^{1} (x^2+2x)\,dx = \dfrac{1}{2}\left(\dfrac{1}{3}x^3+x^2\right)\Big]_{-1}^1 = \dfrac{1}{2}\left(\dfrac{4}{3}-\dfrac{2}{3}\right) = \dfrac{1}{3}$

Setting $f(c) = c^2+c = \dfrac{1}{3}$, we obtain $3c^2+6c-1 = 0$. Then $c = \dfrac{-6\pm\sqrt{36+12}}{6} = -1\pm\dfrac{2}{3}\sqrt{3}$. The only solution on $[-1,1]$ is $-1+\dfrac{2}{3}\sqrt{3}$.

23. We are given $\dfrac{1}{5-1} \displaystyle\int_1^5 f(x)\,dx = 3$. The area under the graph is $\displaystyle\int_1^5 f(x)\,dx = 12$.

25. $T_{ave} = \dfrac{1}{6-0} \displaystyle\int_0^6 \left(100+3t-\dfrac{1}{2}t^2\right)dt = \dfrac{1}{6}\left(100t+\dfrac{3}{2}t^2-\dfrac{1}{6}t^3\right)\Big]_0^6 = 103°$

27. Using $s'(t) = v(t)$ we have

$$v_{ave} = \dfrac{1}{t_2-t_1} \int_{t_1}^{t_2} v(t)\,dt = \dfrac{1}{t_2-t_1}s(t)\Big]_{t_1}^{t_2} = \dfrac{s(t_2)-s(t_1)}{t_2-t_1} = \bar{v}.$$

29. $mv_1 - mv_0 = (t_1-0)\overline{F} = \dfrac{t_1}{t_1-0} \displaystyle\int_0^{t_1} k\left[1-\left(\dfrac{2t}{t_1}-1\right)^2\right]dt = k\displaystyle\int_0^{t_1}\left(1-\dfrac{4}{t_1^2}t^2+\dfrac{4}{t_1}t-1\right)dt$

$= k\left(-\dfrac{4}{3t_1^2}t^3+\dfrac{2}{t_1}t^2\right)\Big]_0^{t_1} = k\left(-\dfrac{4}{3}t_1+2t_1\right) = \dfrac{2kt_1}{3}$

31. 0, since $\displaystyle\int_{-a}^{a} f(x)\,dx = 0$.

33. $f'_{ave} = \dfrac{1}{h} \displaystyle\int_x^{x+h} f'(x)\,dx = \dfrac{1}{h}f(x)\Big]_x^{x+h} = \dfrac{f(x+h)-f(x)}{h}$

35. If $\displaystyle\int_a^b [f(x)-f_{ave}]\,dx = 0$, then $\displaystyle\int_a^b f(x)\,dx = \int_a^b f_{ave}\,dx$. Thus, $\displaystyle\int_a^b f(x)\,dx = f_{ave}(b-a)$ and therefore $f_{ave} = \dfrac{1}{b-a}\displaystyle\int_a^b f(x)\,dx$.

37. There is no unique answer to this question; among several possible approaches, here is probably the simplest. Suppose the circle is centered at the origin and that one of the points on the circle is $(-1,0)$. If (x,y) is any other point on the circle, then the length of the chords between $(-1,0)$ and (x,y) is the distance between the points:

$$\sqrt{(x+1)^2+y^2} = \sqrt{x^2+y^2+2x+1} = \sqrt{2x+2}.$$

The average chord length L_{ave} is then

$$L_{ave} = \dfrac{1}{1-(-1)} \int_{-1}^1 \sqrt{2x+2}\,dx = \dfrac{1}{2}\cdot\dfrac{1}{2}\cdot\dfrac{(2x+2)^{3/2}}{3/2}\Big]_{-1}^1 = \dfrac{1}{6}\cdot 4^{3/2} = \dfrac{8}{6} = \dfrac{4}{3}.$$

6.8 Work

1. $W = 55 \cdot 20 = 1100$ yd-lb $= 3300$ ft-lb

3. Since $10 = k\left(\dfrac{1}{2}\right)$, $k = 20$ and $F = 20x$. Solving $8 = 20x$ we obtain $x = \dfrac{2}{5}$ ft.

5. (a) $W = \displaystyle\int_0^{0.2} 500x\,dx = 250x^2\Big]_0^{0.2} = 10$ joules

   (b) $W = \displaystyle\int_{0.5}^{0.6} 500x\,dx = 250x^2\Big]_{0.5}^{0.6} = 27.5$ joules

7. Since $10 = k\left(\dfrac{2}{3}\right)$, $k = 15$ and $F = 15x$.

   (a) $W = \displaystyle\int_0^1 15x\,dx = \dfrac{15}{2}x^2\Big]_0^1 = \dfrac{15}{2}$ ft-lb

   (b) $W = \displaystyle\int_2^3 15x\,dx = \dfrac{15}{2}x^2\Big]_2^3 = \dfrac{75}{2}$ ft-lb

9. We use 500 km $= 0.5 \times 10^6$ m.

$$W = (6.67 \times 10^{-11})(6.0 \times 10^{24})(10^4)\left(\frac{1}{6.4 \times 10^6} - \frac{1}{6.9 \times 10^6}\right) \approx 4.531 \times 10^{10}$$

$$= 453.1 \times 10^8 \text{ joules}$$

11. $W = \displaystyle\int_0^{12} 62.4\pi(3)^2 x\,dx = 9(62.4\pi)\frac{1}{2}x^2\Big]_0^{12} = 9(62.4\pi)(72) \approx 127{,}030.9$ ft-lb

13. $W = \displaystyle\int_0^{10} 62.4\pi\left(\frac{1}{25}x^2\right)(25-x)\,dx = 62.4\pi\int_0^{10}\left(x^2 - \frac{1}{25}x^3\right)dx$

$$= 62.4\pi\left(\frac{1}{3}x^3 - \frac{1}{100}x^4\right)\Big]_0^{10} = 62.4\pi\left(\frac{700}{3}\right) \approx 45{,}741.6 \text{ ft-lb}$$

15. $y = \dfrac{3}{4}x$

$$W = \int_0^4 62.4(2y\cdot10)(x+5)\,dx = 62.4(20)\int_0^4 \frac{3}{4}x(x+5)\,dx$$

$$= 936\int_0^4 (x^2 + 5x)\,dx = 936\left(\frac{1}{3}x^3 + \frac{5}{2}x^2\right)\Big]_0^4 = 936\left(\frac{184}{3}\right) = 57{,}408 \text{ ft-lb}$$

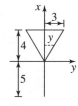

17. The weight of the chain is $F(x) = 20(100-x)$ lb when x feet of chain have been pulled up.

$$W = \int_0^{40} 20(100-x)\,dx = 20\left(100x - \frac{1}{2}x^2\right)\Big]_0^{40} = 64{,}000 \text{ ft-lb}$$

19. (a) $W = 80 \cdot 65 = 5200$ ft-lb

   (b) The weight of the system is $F(x) = 80 + \dfrac{1}{2}(65-x)$ when the bucket has been lifted x feet.

$$W = \int_0^{65}\left[80 + \frac{1}{2}(65-x)\right]dx = \left[80x + \frac{1}{2}\left(65x - \frac{1}{2}x^2\right)\right]\Big|_0^{65} = 6256.25 \text{ ft-lb}$$

21. If x is the distance separating the electron and the nucleus, then the force is $F(x) = k/x^2$.

$$W = \int_1^4 \frac{k}{x^2}\,dx = -\frac{k}{x}\Big]_1^4 = -k\left(\frac{1}{4} - 1\right) = \frac{3k}{4}$$

23. Since $p = kv^{-\gamma}$,

$$W = \int_{v_1}^{v_2} p\,dv = \int_{v_1}^{v_2} kv^{-\gamma}\,dv = \frac{k}{1-\gamma}v^{1-\gamma}\Big]_{v_1}^{v_2} = \frac{k}{1-\gamma}(v_2^{1-\gamma} - v_1^{1-\gamma})$$

$$= \frac{1}{1-\gamma}(kv_2^{-\gamma}v_2 - kv_1^{-\gamma}v_1) = \frac{1}{1-\gamma}(p_2 v_2 - p_1 v_1),$$

where p_1 and p_2 are the pressures corresponding to volumes v_1 and v_2, respectively.

25. Since the distance moved is 0, no work is done.

27. $W = 165 \cdot 1350 = 222{,}750$ ft-lb

29. Using $F = ma = mv'$ and $dx = x'(t)\,dt = v\,dt$, we have

$$W = \int_{x_1}^{x_2} F(x)\,dx = \int_{t_1}^{t_2} mv'v\,dt \qquad \boxed{u = v,\ du = v'\,dt}$$

$$= \int_{v_1}^{v_2} mu\,du = \frac{1}{2}mu^2\Big]_{v_1}^{v_2} = \frac{1}{2}mv_2^2 - \frac{1}{2}mv_1^2.$$

6.9   Fluid Pressure and Force

1. (a) pressure $= 62.4(20) = 1248$ lb/ft^2; $F = (1248)(25\pi) = 31200\pi$ lb
    (b) pressure $= 62.4(20) = 1248$ lb/ft^2; $F = (1248)(4\pi) = 4992\pi$ lb
    (c) pressure $= 62.4(20) = 1248$ lb/ft^2; $F = (1248)(100\pi) = 124800\pi$ lb

3. (a) pressure $= (62.4)(8) = 499.2$ lb/ft^2; $F = (499.2)(30)(15) = 224{,}640$ lb

    (b) sidewall force $= \int_0^8 62.4x(30)\,dx = 62.4(15)x^2\Big]_0^8 = 59{,}904$ lb

    end force $= \int_0^8 62.4x(15)\,dx = 62.4\left(\frac{15}{2}\right)x^2\Big]_0^8 = 29{,}952$ lb

5. The equation of the line through $(5/2, 0)$ and $(1, \sqrt{3}/2)$ is

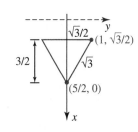

$y = -\dfrac{\sqrt{3}}{3}x + \dfrac{5\sqrt{3}}{6}$. Using symmetry,

$$F = \int_1^{5/2} 62.4x(2y)\,dx = 124.8\int_1^{5/2} x\left(-\frac{\sqrt{3}}{3}x + \frac{5\sqrt{3}}{6}\right)dx$$

$$= 124.8\int_1^{5/2}\left(-\frac{\sqrt{3}}{3}x^2 + \frac{5\sqrt{3}}{6}x\right)dx = 124.8\left(-\frac{\sqrt{3}}{9}x^3 + \frac{5\sqrt{3}}{12}x^2\right)\Big]_1^{5/2}$$

$$\approx 124.8(1.50 - 0.53) = 121.59 \text{ lb}.$$

7. $F = \int_0^4 50x(2\sqrt{x})\,dx = 100\int_0^4 x^{3/2}\,dx = 40x^{5/2}\Big]_0^4 = 1{,}280$ lb

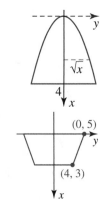

9. The equation of the line through $(0, 5)$ and $(4, 3)$ is $y = -\dfrac{1}{2}x + 5$. Using symmetry,

$$F = \int_0^4 62.4x\left[2\left(-\frac{1}{2}x + 5\right)\right]dx = 124.8\int_0^4\left(-\frac{1}{2}x^2 + 5x\right)dx$$

$$= 124.8\left(-\frac{1}{6}x^3 + \frac{5}{2}x^2\right)\Big]_0^4 = 3660.8 \text{ lb}$$

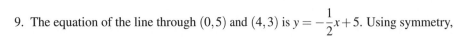

11. The equation of the line through $(8,0)$ and $(4,12)$ is $y=-3x+24$.

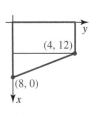

$$F = \int_0^4 62.4x(12)\,dx + \int_4^8 62.4x(-3x+24)\,dx$$

$$= 374.4 \int_0^4 2x\,dx + 62.4 \int_4^8 (-3x^2+24x)\,dx$$

$$= 374.4x^2\big]_0^4 + 62.4(-x^3+12x^2)\big]_4^8 = 5{,}990.4 + 7{,}987.2 = 13{,}977.6 \text{ lb}$$

13. $y = \sqrt{16-x^2}$. Using symmetry,

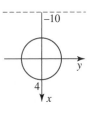

$$F = \int_{-4}^4 62.4(x+10)\left(2\sqrt{16-x^2}\right)dx$$

$$= 124.8 \int_{-4}^4 x\sqrt{16-x^2}\,dx + 624 \int_{-4}^4 2\sqrt{16-x^2}\,dx.$$

The first integral has an odd integrand and is thus 0. The second integral represents the area of the circle and is thus $\pi(4)^2$. Therefore $F = 624(16\pi) = 9984\pi$ lb.

15. $F = F_{\text{top}} + F_{\text{bottom}} + 4F_{\text{side}} = 62.4(3)(4) + 62.4(5)(4) + 4\int_3^5 62.4x(2)\,dx$

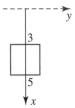

$$= 748.8 + 1248 + 249.6x^2\big]_3^5 = 1996.8 + 3993.6 = 5990.4 \text{ lb}$$

17. The length of the bottom of the pool is $\sqrt{20^2+5^2} = 5\sqrt{17}$. Using similar triangles, $h/x = 5/5\sqrt{17}$ and $h = x/\sqrt{17}$. Then $d = 4 + x/\sqrt{17}$. Now

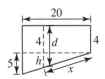

$$F = \lim_{\|P\|\to 0}\sum_{k=1}^n 62.4\left(4+\frac{x_k^*}{\sqrt{17}}\right)(15)\Delta x_k = \int_0^{5\sqrt{17}} 62.4\left(4+\frac{x}{\sqrt{17}}\right)15\,dx$$

$$= 936\int_0^{5\sqrt{17}}\left(4+\frac{1}{\sqrt{17}}x\right)dx = 936\left(4x+\frac{1}{2\sqrt{17}}x^2\right)\Big]_0^{5\sqrt{17}}$$

$$= 936\left(\frac{65\sqrt{17}}{2}\right) = 30{,}420\sqrt{17} \text{ lb}.$$

19. $F = \lim_{\|P\|\to 0}\sum_{k=1}^n 62.4x_k^*(100\sqrt{2}\Delta x_k) = \int_0^{40} 6240\sqrt{2}x\,dx = 3120\sqrt{2}x^2\big]_0^{40}$

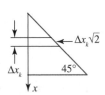

$= 4{,}992{,}000\sqrt{2}$ lb

6.10 Centers of Mass and Centroids

1. $\bar{x} = \dfrac{2(4)+5(-2)}{2+5} = -\dfrac{2}{7}$

3. $\bar{x} = \dfrac{10(-5)+5(2)+8(6)+7(-3)}{10+5+8+7} = -\dfrac{13}{30}$

5. $\bar{x} = \dfrac{10(-5)+15(5)}{10+15} = \dfrac{25}{25} = 1$ m (to the right of the center)

7. $m = \displaystyle\int_0^5 (2x+1)\,dx = (x^2+x)\big]_0^5 = 30;\quad M_0 = \int_0^5 x(2x+1)\,dx = \left(\frac{2}{3}x^3+\frac{1}{2}x^2\right)\Big]_0^5 = \frac{575}{6}$

$\bar{x} = \dfrac{575/6}{30} = \dfrac{115}{36}$

9. $m = \int_0^1 x^{1/3}\,dx = \frac{3}{4}x^{4/3}\Big]_0^1 = \frac{3}{4}$

   $M_0 = \int_0^1 xx^{1/3}\,dx = \int_0^1 x^{4/3}\,dx = \frac{3}{7}x^{7/3}\Big]_0^1 = \frac{3}{7}; \quad \bar{x} = \frac{3/7}{3/4} = \frac{4}{7}$

11. $m = \int_0^4 |x-3|\,dx = \int_0^3 -(x-3)\,dx + \int_3^4 (x-3)\,dx = \left(-\frac{1}{2}x^2 + 3x\right)\Big]_0^3 + \left(\frac{1}{2}x^2 - 3x\right)\Big]_3^4$

    $= \frac{9}{2} + \left(-4 + \frac{9}{2}\right) = 5$

    $M_0 = \int_0^4 x|x-3|\,dx = \int_0^3 -(x^2 - 3x)\,dx + \int_3^4 (x^2 - 3x)\,dx$

    $= \left(-\frac{1}{3}x^3 + \frac{3}{2}x^2\right)\Big]_0^3 + \left(\frac{1}{3}x^3 - \frac{3}{2}x^2\right)\Big]_3^4 = \left(-9 + \frac{27}{2}\right) + \left(\frac{64}{3} - 24\right) - \left(9 - \frac{27}{2}\right) = \frac{19}{3}$

    $\bar{x} = \frac{19/3}{5} = \frac{19}{15}$

13. $m = \int_0^2 \rho(x)\,dx = \int_0^1 x^2\,dx + \int_1^2 (2-x)\,dx = \frac{1}{3}x^3\Big]_0^1 + \left(2x - \frac{1}{2}x^2\right)\Big]_1^2$

    $= \frac{1}{3} + \left(2 - \frac{3}{2}\right) = \frac{5}{6}$

    $M_0 = \int_0^2 x\rho(x)\,dx = \int_0^1 x^3\,dx + \int_1^2 (2x - x^2)\,dx = \frac{1}{4}x^4\Big]_0^1 + \left(x^2 - \frac{1}{3}x^3\right)\Big]_1^2$

    $= \frac{1}{4} + \left(\frac{4}{3} - \frac{2}{3}\right) = \frac{11}{12}$

    $\bar{x} = \frac{11/12}{5/6} = \frac{11}{10}$

15. Since $\rho(x) = kx^2$ and $12.5 = \rho(5) = 25k$, $k = \frac{1}{2}$ and $\rho(x) = \frac{1}{2}x^2$.

    $m = \int_0^{10} \frac{1}{2}x^2\,dx = \frac{1}{6}x^3\Big]_0^{10} = \frac{500}{3}; \quad M_0 = \int_0^{10} \frac{1}{2}x^3\,dx = \frac{1}{8}x^4\Big]_0^{10} = 1250$

    $\bar{x} = \frac{1250}{500/3} = 7.5$ ft (from the left end)

17. $m = 3 + 4 = 7; \quad \bar{x} = \frac{3(-2) + 4(1)}{7} = -\frac{2}{7}; \quad \bar{y} = \frac{3(3) + 4(2)}{7} = \frac{17}{7}$

19. $m = 4 + 8 + 10 = 22; \quad \bar{x} = \frac{4(1) + 8(-5) + 10(7)}{22} = \frac{17}{11}; \quad \bar{y} = \frac{4(1) + 8(2) + 10(-6)}{22} = -\frac{20}{11}$

21. $A = \int_0^2 (2x+4)\,dx = (x^2 + 4x)\Big]_0^2 = 12$

    $M_y = \int_0^2 x(2x+4)\,dx = \int_0^2 (2x^2 + 4x)\,dx = \left(\frac{2}{3}x^3 + 2x^2\right)\Big]_0^2 = \frac{40}{3}$

    $M_x = \frac{1}{2}\int_0^2 (2x+4)^2\,dx = \frac{1}{2}\int_0^2 (4x^2 + 16x + 16)\,dx$

    $= 2\left(\frac{1}{3}x^3 + 2x^2 + 4x\right)\Big]_0^2 = \frac{112}{3}$

    $\bar{x} = \frac{40/3}{12} = \frac{10}{9}; \quad \bar{y} = \frac{112/3}{12} = \frac{28}{9}$

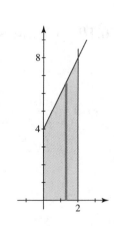

23. $A = \int_0^1 x^2\,dx = \frac{1}{3}x^3\Big]_0^1 = \frac{1}{3}$

$M_y = \int_0^1 x^3\,dx = \int_0^2 (2x^2 + 4x)\,dx = \frac{1}{4}x^4\Big]_0^1 = \frac{1}{4}$

$M_x = \frac{1}{2}\int_0^1 x^4\,dx = \frac{1}{10}x^5\Big]_0^1 = \frac{1}{10};\quad \bar{x} = \frac{1/4}{1/3} = \frac{3}{4};\quad \bar{y} = \frac{1/10}{1/3} = \frac{3}{10}$

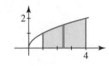

25. $A = \int_0^3 x^3\,dx = \frac{1}{4}x^4\Big]_0^3 = \frac{81}{4}$

$M_y = \int_0^3 x^4\,dx = \frac{1}{5}x^5\Big]_0^3 = \frac{243}{5};\quad M_x = \frac{1}{2}\int_0^3 x^6\,dx = \frac{1}{14}x^7\Big]_0^3 = \frac{2187}{14}$

$\bar{x} = \frac{243/5}{81/4} = \frac{12}{5};\quad \bar{y} = \frac{2187/14}{81/4} = \frac{54}{7}$

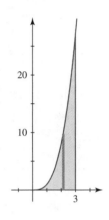

27. $A = \int_1^4 x^{1/2}\,dx = \frac{2}{3}x^{3/2}\Big]_1^4 = \frac{16}{3} - \frac{2}{3} = \frac{14}{3}$

$M_y = \int_1^4 x^{3/2}\,dx = \frac{2}{5}x^{5/2}\Big]_1^4 = \frac{64}{5} - \frac{2}{5} = \frac{62}{5}$

$M_x = \frac{1}{2}\int_1^4 x\,dx = \frac{1}{4}x^2\Big]_1^4 = 4 - \frac{1}{4} = \frac{15}{4};\quad \bar{x} = \frac{62/5}{14/3} = \frac{93}{35};\quad \bar{y} = \frac{15/4}{14/3} = \frac{45}{56}$

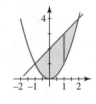

29. $A = \int_{-1}^2 [(x+2) - x^2]\,dx = \left(\frac{1}{2}x^2 + 2x - \frac{1}{3}x^3\right)\Big]_{-1}^2 = \frac{10}{3} - \left(-\frac{7}{6}\right) = \frac{9}{2}$

$M_y = \int_{-1}^2 x(x + 2 - x^2)\,dx = \int_{-1}^2 (x^2 + 2x - x^3)\,dx$

$= \left(\frac{1}{3}x^3 + x^2 - \frac{1}{4}x^4\right)\Big]_{-1}^2 = \frac{8}{3} - \frac{5}{12} = \frac{9}{4}$

$M_x = \frac{1}{2}\int_{-1}^2 [(x+2)^2 - x^4]\,dx = \frac{1}{2}\left[\frac{1}{3}(x+2)^3 - \frac{1}{5}x^5\right]\Big|_{-1}^2 = \frac{1}{2}\left(\frac{224}{15} - \frac{8}{15}\right) = \frac{36}{5}$

$\bar{x} = \frac{9/4}{9/2} = \frac{1}{2};\quad \bar{y} = \frac{36/5}{9/2} = \frac{8}{5}$

31. $A = \int_0^1 (x^{1/3} - x^3)\,dx = \left(\frac{3}{4}x^{4/3} - \frac{1}{4}x^4\right)\Big]_0^1 = \frac{1}{2}$

$M_y = \int_0^1 x(x^{1/3} - x^3)\,dx = \int_0^1 (x^{4/3} - x^4)\,dx = \left(\frac{3}{7}x^{7/3} - \frac{1}{5}x^5\right)\Big]_0^1 = \frac{8}{35}$

$M_x = \frac{1}{2}\int_0^1 [(x^{1/3})^2 - (x^3)^2]\,dx = \frac{1}{2}\int_0^1 (x^{2/3} - x^6)\,dx = \frac{1}{2}\left(\frac{3}{5}x^{5/3} - \frac{1}{7}x^7\right)\Big]_0^1 = \frac{8}{35}$

$\bar{x} = \frac{8/35}{1/2} = \frac{16}{35};\quad \bar{y} = \frac{8/35}{1/2} = \frac{16}{35}$

33. $y = x^{-3}$; $A = \int_1^3 x^{-3}\,dx = -\frac{1}{2x^2}\Big]_1^3 = -\left(\frac{1}{18} - \frac{1}{2}\right) = \frac{4}{9}$

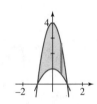

$M_y = \int_1^3 x^{-2}\,dx = -\frac{1}{x}\Big]_1^3 = -\left(\frac{1}{3} - 1\right) = \frac{2}{3}$

$M_x = \frac{1}{2}\int_1^3 x^{-6}\,dx = -\frac{1}{10x^5}\Big]_1^3 = -\left(\frac{1}{2430} - \frac{1}{10}\right) = \frac{121}{1215}$

$\bar{x} = \frac{2/3}{4/9} = \frac{3}{2}$; $\quad \bar{y} = \frac{121/1215}{4/9} = \frac{121}{540}$

35. $A = \int_{-1}^2 (y^2 - 1 + 2)\,dy = \left(\frac{1}{3}y^3 + y\right)\Big]_{-1}^2 = \frac{14}{3} - \left(-\frac{4}{3}\right) = 6$

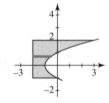

$M_x = \int_{-1}^2 y(y^2 + 1)\,dy = \int_{-1}^2 (y^3 + y)\,dy = \left(\frac{1}{4}y^4 + \frac{1}{2}y^2\right)\Big]_{-1}^2$

$= 6 - \frac{3}{4} = \frac{21}{4}$

$M_y = \frac{1}{2}\int_{-1}^2 [(y^2 - 1)^2 - (-2)^2]\,dy = \frac{1}{2}\int_{-1}^2 (y^4 - 2y^2 - 3)\,dy$

$= \frac{1}{2}\left(\frac{1}{5}y^5 - \frac{2}{3}y^3 - 3y\right)\Big]_{-1}^2 = \frac{1}{2}\left(-\frac{74}{15} - \frac{52}{15}\right) = -\frac{21}{5}$

$\bar{x} = \frac{-21/5}{6} = -\frac{7}{10}$; $\quad \bar{y} = \frac{21/4}{6} = \frac{7}{8}$

37. $A = \int_{-1}^1 [(4 - 4x^2) - (1 - x^2)]\,dx = \int_{-1}^1 (3 - 3x^2)\,dx = (3x - x^3)\Big]_{-1}^1 = 4$

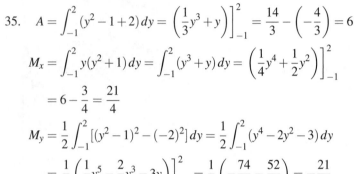

$M_y = \int_{-1}^1 x(3 - 3x^2)\,dx = \int_{-1}^1 (3x - 3x^3)\,dx$

$= \left(\frac{3}{2}x^2 - \frac{3}{4}x^4\right)\Big]_{-1}^1 = \frac{3}{4} - \frac{3}{4} = 0$

$M_x = \frac{1}{2}\int_{-1}^1 [(4 - 4x^2)^2 - (1 - x^2)^2]\,dx = \frac{1}{2}\int_{-1}^1 [16(1 - x^2)^2 - (1 - x^2)^2]\,dx$

$= \frac{15}{2}\int_{-1}^1 (1 - 2x^2 + x^4)\,dx = \frac{15}{2}\left(x - \frac{2}{3}x^3 + \frac{1}{5}x^5\right)\Big]_{-1}^1 = \frac{15}{2}\left[\frac{8}{15} - \left(-\frac{8}{15}\right)\right] = 8$

$\bar{x} = \frac{0}{4} = 0$; $\quad \bar{y} = \frac{8}{4} = 2$

39. By symmetry, $\bar{x} = 0$.

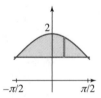

$A = 2\int_0^{\pi/2} (1 + \cos x - 1)\,dx = 2\int_0^{\pi/2} \cos x\,dx = 2\sin x\Big]_0^{\pi/2} = 2$

$M_x = \frac{1}{2}\int_{-\pi/2}^{\pi/2} [(1 + \cos x)^2 - 1^2]\,dx = \frac{1}{2}\int_{-\pi/2}^{\pi/2} (\cos^2 x + 2\cos x)\,dx$

$= \frac{1}{2}\int_{-\pi/2}^{\pi/2} \left(\frac{1}{2}\cos 2x + \frac{1}{2} + 2\cos x\right)\,dx = \frac{1}{2}\left(\frac{1}{4}\sin 2x + \frac{1}{2}x + 2\sin x\right)\Big]_{-\pi/2}^{\pi/2}$

$= \frac{1}{2}\left[\left(\frac{\pi}{4} + 2\right) - \left(-\frac{\pi}{4} - 2\right)\right] = \frac{\pi + 8}{4}$

$\bar{y} = \frac{(\pi + 8)/4}{2} = \frac{\pi + 8}{8}$

41. (a) The circumference of a circle with radius $\bar{y}$ is $2\pi\bar{y}$. Thus, $V = 2\pi\bar{y}A$.

(b) By the same reasoning, $V = 2\pi\bar{x}A$ when the region R is revolved around the y-axis.

43. We identify $A = \pi a^2$. The centroid of the region R is b units from L, so $V = 2\pi b(\pi a^2) = 2\pi^2 a^2 b$.

45. Thinking geometrically, the centroid of a triangle would appear to be the intersection of its three medians (a median is a line segment from one of the triangle's vertices to the midpoint of the opposing side). Doing some research on the centroid of a triangle shows this to be true, and in fact this intersection is the mean of the coordinates of the triangle's vertices.

Chapter 6 in Review

A. True/False

1. False. $\int_a^b f(x)\,dx$ may be positive even though a portion of the graph of f lies below the x-axis.

3. True

5. True

7. True

9. True

11. False. The distance moved is given by $\int_{t_1}^{t_2} |v(t)|\,dt$.

B. Fill In the Blanks

1. Newton-meter of joule

3. $(100\cos 60°)(50) = 2{,}500$ ft-lb

5. 6

7. smooth

C. Exercises

1. $-\int_0^a f(x)\,dx$

3. The line in Figure 6.R.3 is $y = \dfrac{f(a)}{a}x$. Thus, the integral is $\displaystyle\int_0^a \left[f(x) - \dfrac{f(a)}{a}x \right] dx.$

5. $-\displaystyle\int_a^b 2f(x)\,dx + \int_b^c 2f(x)\,dx$

7. $\displaystyle\int_b^c [a - f(y)]\,dy + \int_c^d [f(y) - a]\,dy$

9. $A = -\displaystyle\int_a^0 \dfrac{x}{2}\,dx + \int_0^{2b}\left(b - \dfrac{x}{2}\right) dx = \left. -\dfrac{1}{4}x^2 \right]_a^0 + \left.\left(bx - \dfrac{1}{4}x^2\right)\right]_0^{2b} = \dfrac{a^2}{4} + 2b^2 - b^2 = \dfrac{a^2}{4} + b^2$

11. $\bar{x} = \dfrac{\int_0^2 x[g(x) - f(x)]\,dx}{\int_0^2 [g(x) - f(x)]\,dx}$; $\bar{y} = \dfrac{\frac{1}{2}\int_0^2 \{[g(x)]^2 - [f(x)]^2\}\,dx}{\int_0^2 [g(x) - f(x)]\,dx}$

13. $V = 2\pi \displaystyle\int_0^2 x[g(x) - f(x)]\,dx$

15. $V = 2\pi \int_0^2 (2-x)[g(x) - f(x)]\,dx$

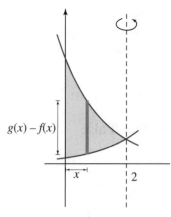

$g(x) - f(x)$

17. Solving $\sin x = \sin 2x$, we get $\sin x = 2\sin x \cos x$, $\cos x = \dfrac{1}{2}$, and $x = \dfrac{\pi}{3}$ on $(0, \pi)$. Thus,

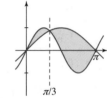

$$A = \int_0^{\pi/3} (\sin 2x - \sin x)\,dx + \int_{\pi/3}^{\pi} (\sin x - \sin 2x)\,dx$$

$$= \left(-\frac{1}{2}\cos 2x + \cos x \right)\Big]_0^{\pi/3} + \left(-\cos x + \frac{1}{2}\cos 2x \right)\Big]_{\pi/3}^{\pi}$$

$$= \frac{1}{4} + \frac{1}{2} - \left(-\frac{1}{2} \right) - 1 + 1 + \frac{1}{2} + \frac{1}{2} + \frac{1}{4} = \frac{5}{2}$$

19. (a) $x = y^2$; $A(x) = (2y)^2 = (2\sqrt{x})^2 = 4x$

$$V = 4 \int_0^{\sqrt{2}} x\,dx = 4\left(\frac{1}{2}x^2 \right)\Big]_0^{\sqrt{2}} = 4$$

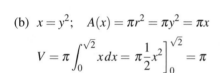

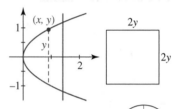

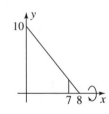

(b) $x = y^2$; $A(x) = \pi r^2 = \pi y^2 = \pi x$

$$V = \pi \int_0^{\sqrt{2}} x\,dx = \pi \frac{1}{2}x^2 \Big]_0^{\sqrt{2}} = \pi$$

21. The equation of the line is $y = -\dfrac{5}{4}x + 10$. Then $y' = -5/4$ and $\sqrt{1 + (y')^2} = \sqrt{1 + 25/16} = \dfrac{1}{4}\sqrt{41}$. The surface area is

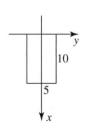

$$S = 2\pi \int_0^7 \left(-\frac{5}{4}x + 10 \right) \frac{\sqrt{41}}{4}\,dx = \frac{\sqrt{41}\pi}{2}\left(-\frac{5}{8}x^2 + 10x \right)\Big]_0^7$$

$$= \frac{\sqrt{41}\pi}{2}\left(-\frac{5}{8}\cdot 49 + 70 \right) = \frac{315\pi\sqrt{41}}{16} \approx 396.03 \text{ ft}^2.$$

23. $f_{\text{ave}} = \dfrac{1}{4-1}\int_1^4 (x^{3/2} + x^{1/2})\,dx = \dfrac{1}{3}\left(\frac{2}{5}x^{5/2} + \frac{2}{3}x^{3/2} \right)\Big]_1^4 = \dfrac{1}{3}\left(\frac{272}{15} - \frac{16}{15} \right) = \dfrac{256}{45}$

25. $50 = k(1/2)$, $k = 100$, $F = 100x$; $W = \displaystyle\int_{1/2}^1 100x\,dx = 50x^2\Big]_{1/2}^1 = 50 - \dfrac{25}{2} = 37.5$ joules

27. $W = \displaystyle\int_0^{10} 62.4(10)^2(x+5)\,dx = 6240\left(\frac{1}{2}x^2 + 5x \right)\Big]_0^{10} = 624,000$ ft-lb

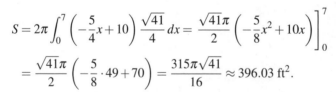

29. At a rate of loss of 1/4 pound per second, it will take 120 seconds to lose the entire 30 pounds. In 120 seconds, the bucket will be raised 120 feet.

$$W = \int_0^{120} \left(32 - \frac{1}{4}x\right) dx = \left(32x - \frac{1}{8}x^2\right)\Big]_0^{120} = 2040 \text{ ft-lb}$$

31. $y = \begin{cases} x+3 & 0 \le x \le 2 \\ 5 & 2 \le x \le 8 \end{cases}$

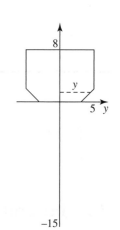

$$W = \int_0^8 62.4(\pi y^2)(x+15)\,dx$$

$$= \int_0^2 62.4\pi(x+3)^2(x+15)\,dx + \int_2^8 62.4\pi(25)(x+15)\,dx$$

$$= 62.4\pi \int_0^2 (x^3 + 21x^2 + 99x + 135)\,dx + 1560\pi \int_2^8 (x+15)\,dx$$

$$= 62.4\pi \left(\frac{1}{4}x^4 + 7x^3 + \frac{99}{2}x^2 + 135x\right)\Big]_0^2 + 1560\pi \left(\frac{1}{2}x^2 + 15x\right)\Big]_2^8$$

$$= 62.4\pi(528) + 1560\pi(152 - 32) = 220,147.2\pi \approx 691,612.83 \text{ ft-lb}$$

33. $y' = \frac{3}{2}(x-1)^{1/2}$

$$s = \int_1^5 \sqrt{1 + \frac{9}{4}(x-1)}\,dx = \frac{1}{2}\int_1^5 \sqrt{9x-5}\,dx = \frac{1}{27}(9x-5)^{3/2}\Big]_1^5 = \frac{1}{27}(40^{3/2} - 4^{3/2})$$

$$= \frac{40^{3/2} - 8}{27} \approx 9.07$$

35. $y = \sqrt{16 - x^2}$

$$F = \int_0^4 800x\sqrt{16-x^2}\,dx = -\frac{800}{3}(16-x^2)^{3/2}\Big]_0^4 = -\frac{800}{3}(0-64) = \frac{51,200}{3} \approx 17,066.67 \text{ N}$$

37. Consider first the system of 1 kg and 3 kg weights. Taking the origin at the left end, we obtain
$$\bar{x}_1 = \frac{1(0) + 3(1)}{1+3} = \frac{3}{4}.$$

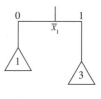

Now consider the system with the left two weights concentrated at the left end of the upper bar. Taking the origin at the left, we obtain $\bar{x}_2 = \dfrac{4(0) + 6(2)}{4+6} = \dfrac{12}{10} = \dfrac{6}{5}.$

Techniques of Integration

7.1 Integration—Three Resources

1. $\displaystyle\int 5^{-5x}\,dx = -\frac{1}{5}\int 5^{-5x}(-5\,dx)$ $\boxed{u=-5x,\; du=-5\,dx}$

$$= -\frac{1}{5}\int 5^{u}\,du = -\frac{1}{5}\left(\frac{1}{\ln 5}5^{u}\right)+C = -\frac{5^{-5x}}{5\ln 5}+C = -\frac{1}{5^{5x+1}\ln 5}+C$$

3. $\displaystyle\int \frac{\sin\sqrt{1+x}}{\sqrt{1+x}} = 2\int \left[\sin(1+x)^{1/2}\right]\left[\frac{1}{2}(1+x)^{-1/2}\right]dx$

$$\boxed{u=(1+x)^{1/2},\; du=\frac{1}{2}(1+x)^{-1/2}\,dx}$$

$$= 2\int \sin u\,du = -2\cos u + C = -2\cos\sqrt{1+x}+C$$

5. $\displaystyle\int \frac{x}{\sqrt{25-4x^2}}\,dx = -\frac{1}{8}\int (25-4x^2)^{-1/2}(-8x\,dx)$ $\boxed{u=25-4x^2,\; du=-8x\,dx}$

$$= -\frac{1}{8}\int u^{-1/2}\,du = -\frac{1}{8}(2u^{1/2})+C = -\frac{1}{4}\sqrt{25-4x^2}+C$$

7. $\displaystyle\int \frac{1}{x\sqrt{4x^2-25}}\,dx = \int \frac{1}{2x\sqrt{(2x)^2-5^2}}(2\,dx)$ $\boxed{u=2x,\; du=2\,dx}$

$$= \int \frac{1}{u\sqrt{u^2-5^2}}\,du = \frac{1}{5}\sec^{-1}\left|\frac{u}{5}\right|+C = \frac{1}{5}\sec^{-1}\left|\frac{2x}{5}\right|+C$$

9. $\displaystyle\int \frac{1}{25+4x^2}\,dx = \frac{1}{2}\int \frac{1}{5^2+(2x)^2}(2\,dx)$ $\boxed{u=2x,\; du=2\,dx}$

$$= \frac{1}{2}\int \frac{1}{5^2+u^2}\,du = \frac{1}{2}\left(\frac{1}{5}\tan^{-1}\frac{u}{5}\right)+C = \frac{1}{10}\tan^{-1}\frac{2x}{5}+C$$

11. $\displaystyle\int \frac{1}{4x^2-25}\,dx = \frac{1}{2}\int \frac{1}{(2x)^2-5^2}(2\,dx)$ $\boxed{u=2x,\; du=2\,dx}$

$$= \frac{1}{2}\int \frac{1}{u^2-5^2}\,du = \frac{1}{2}\left[\frac{1}{2(5)}\ln\left|\frac{u-5}{u+5}\right|\right]+C$$

$$= \frac{1}{20}\ln\left|\frac{2x-5}{2x+5}\right|+C$$

13. $\displaystyle\int \cot 10x\,dx = \frac{1}{10}\int \cot 10x(10\,dx)$ $\boxed{u=10x,\; du=10\,dx}$

$$= \frac{1}{10}\int \cot u\,du = \frac{1}{10}(\ln|\sin u|)+C = \frac{1}{10}\ln|\sin 10x|+C$$

15. $\displaystyle\int \frac{6}{(3-5t)^{2.2}}\,dx = -\frac{6}{5}\int (3-5t)^{-2.2}(-5\,dt)$ $\boxed{u=3-5t,\ du=-5\,dt}$

$$= -\frac{6}{5}\int u^{-2.2}\,du = -\frac{6}{5}\left(-\frac{1}{1.2}u^{-1.2}\right)+C = \frac{1}{(3-5t)^{1.2}}+C$$

17. $\displaystyle\int \sec 3x\,dx = \frac{1}{3}\int (\sec 3x)(3\,dx)$ $\boxed{u=3x,\ du=3\,dx}$

$$= \frac{1}{3}\int \sec u\,du = \frac{1}{3}\left(\ln|\sec u + \tan u|\right)+C = \frac{1}{3}\ln|\sec 3x+\tan 3x|+C$$

19. $\displaystyle\int \frac{\sin^{-1}x}{\sqrt{1-x^2}}\,dx = \int (\sin^{-1}x)\left(\frac{1}{\sqrt{1-x^2}}\,dx\right)$ $\boxed{u=\sin^{-1}x,\ du=\dfrac{1}{\sqrt{1-x^2}}\,dx}$

$$= \int u\,du = \frac{1}{2}u^2+C = \frac{1}{2}(\sin^{-1}x)^2+C$$

21. $\displaystyle\int \frac{\sin x}{1+\cos^2 x}\,dx = -\int \frac{1}{1^2+\cos^2 x}(-\sin x\,dx)$ $\boxed{u=\cos x,\ du=-\sin x\,dx}$

$$= -\int \frac{1}{1^2+u^2}\,du = -\tan^{-1}u+C = -\tan^{-1}(\cos x)+C$$

23. $\displaystyle\int \frac{x^3}{\cosh^2 x^4}\,dx = \frac{1}{4}\int (\mathrm{sech}^2 x^4)(4x^3\,dx)$ $\boxed{u=x^4,\ du=4x^3\,dx}$

$$= \frac{1}{4}\int \mathrm{sech}^2 u\,du = \frac{1}{4}\tanh u+C = \frac{1}{4}\tanh x^4+C$$

25. $\displaystyle\int \tan 2x\sec 2x\,dx = \frac{1}{2}\int (\sec 2x\tan 2x)(2\,dx)$ $\boxed{u=2x,\ du=2\,dx}$

$$= \frac{1}{2}\int \sec u\tan u\,du = \frac{1}{2}\sec u+C = \frac{1}{2}\sec 2x+C$$

27. $\displaystyle\int \sin x\csc(\cos x)\cot(\cos x)\,dx = -\int [\csc(\cos x)\cot(\cos x)](-\sin x\,dx)$

$$\boxed{u=\cos x,\ du=-\sin x}$$

$$= -\int \csc u\cot u\,du = -(-\csc u)+C = \csc(\cos x)+C$$

29. $\displaystyle\int (1+\tan x)^2\sec^2 x\,dx$ $\boxed{u=1+\tan x,\ du=\sec^2 x\,dx}$

$$= \int u^2\,du = \frac{1}{3}u^3+C = \frac{1}{3}(1+\tan x)^3+C$$

31. $\displaystyle\int \frac{e^{2x}}{1+e^{2x}}\,dx = \frac{1}{2}\int (1+e^{2x})^{-1}(2e^{2x}\,dx)$ $\boxed{u=1+e^{2x},\ du=2e^{2x}\,dx}$

$$= \frac{1}{2}\int u^{-1}\,du = \frac{1}{2}\ln|u|+C = \frac{1}{2}\ln(1+e^{2x})+C$$

7.2   Integration by Substitution

1. $\displaystyle\int x(x+1)^3\,dx$ $\boxed{u=x+1,\ x=u-1,\ dx=du}$

$$= \int (u-1)u^3\,du = \int (u^4-u^3)\,du$$

$$= \frac{1}{5}u^5 - \frac{1}{4}u^4+C = \frac{1}{5}(x+1)^5 - \frac{1}{4}(x+1)^4+C$$

3. $\int (2x+1)\sqrt{x-5}\,dx$ $\boxed{u=x-5,\ x=u+5,\ dx=du}$

$$= \int (2u+11)u^{1/2}\,du = \int (2u^{3/2}+11u^{1/2})\,du = \frac{4}{5}u^{5/2} + \frac{22}{3}u^{3/2} + C$$

$$= \frac{4}{5}(x-5)^{5/2} + \frac{22}{3}(x-5)^{3/2} + C$$

5. $\int \frac{x}{\sqrt{x-1}}\,dx$ $\boxed{u=x-1,\ x=u+1,\ dx=du}$

$$= \int \frac{u+1}{u^{1/2}}\,du = \int (u^{1/2}+u^{-1/2})\,du = \frac{2}{3}u^{3/2} + 2u^{1/2} + C$$

$$= \frac{2}{3}(x-1)^{3/2} + 2(x-1)^{1/2} + C$$

7. $\int \frac{x+3}{(3x-4)^{3/2}}\,dx$ $\boxed{u=3x-4,\ x=\frac{1}{3}(u+4),\ dx=\frac{1}{3}\,du}$

$$= \int \frac{u/3+13/3}{u^{3/2}}\left(\frac{1}{3}\,du\right) = \int \frac{1}{9}u^{-1/2} + \frac{13}{9}u^{-3/2} = \frac{2}{9}u^{1/2} - \frac{26}{9}u^{-1/2} + C$$

$$= \frac{2}{9}(3x-4)^{1/2} - \frac{26}{9}(3x-4)^{-1/2} + C$$

9. $\int \frac{\sqrt{x}}{x+1}\,dx$ $\boxed{u=\sqrt{x},\ x=u^2,\ dx=2u\,du}$

$$= \int \frac{u}{u^2+1}(2u\,du) = \int \frac{2u^2}{u^2+1}\,du = \int \left(2 - \frac{2}{u^2+1}\right)du$$

$$= 2u - 2\tan^{-1}u + C = 2\sqrt{x} - 2\tan^{-1}\sqrt{x} + C$$

11. $\int \frac{\sqrt{t}-3}{\sqrt{t}+1}\,dt$ $\boxed{u=\sqrt{t}+1,\ t=(u-1)^2,\ dt=2(u-1)\,du}$

$$= \int \frac{u-4}{u}[2(u-1)\,du] = \int \left(2u - 10 + \frac{8}{u}\right)du = u^2 - 10u + 8\ln|u| + C$$

$$= (\sqrt{t}+1)^2 - 10(\sqrt{t}+1) + 8\ln(\sqrt{t}+1) + C$$

13. $\int \frac{x^3}{\sqrt[3]{x^2+1}}\,dx = \int \frac{x^2}{(x^2+1)^{1/3}}(x\,dx)$ $\boxed{u=x^2+1,\ du=2x\,dx}$

$$= \int \frac{u-1}{u^{1/3}}\left(\frac{1}{2}\,du\right) = \frac{1}{2}\int (u^{2/3}-u^{-1/3})\,du = \frac{3}{10}u^{5/3} - \frac{3}{4}u^{2/3} + C$$

$$= \frac{3}{10}(x^2+1)^{5/3} - \frac{3}{4}(x^2+1)^{2/3} + C$$

15. $\int \frac{x^2}{(x-1)^4}\,dx$ $\boxed{u=x-1,\ du=dx}$

$$= \int \frac{(u+1)^2}{u^4}\,du = \int (u^{-2}+2u^{-3}+u^{-4})\,du = -u^{-1} - u^{-2} - \frac{1}{3}u^{-3} + C$$

$$= -\frac{1}{x-1} - \frac{1}{(x-1)^2} - \frac{1}{3(x-1)^3} + C$$

17. $\int \sqrt{e^x-1}\,dx$ $\boxed{u=\sqrt{e^x-1},\ x=\ln(u^2+1),\ dx=\frac{2u}{u^2+1}\,du}$

$$= \int u\left(\frac{2u}{u^2+1}\right)du = \int \left(2 - \frac{2u}{u^2+1}\right)du = 2u - 2\tan^{-1}u + C$$

$$= 2\sqrt{e^x-1} - 2\tan^{-1}\sqrt{e^x-1} + C$$

19. $\displaystyle\int \sqrt{1-\sqrt{v}}\,dv$ $\boxed{u=1-\sqrt{v},\ v=(1-u)^2,\ dv=-2(1-u)\,du}$

$$=\int u^{1/2}[-2(1-u)\,du]=\int(2u^{3/2}-2u^{1/2})\,du=\frac{4}{5}u^{5/2}-\frac{4}{3}u^{3/2}+C$$

$$=\frac{4}{5}(1-\sqrt{v})^{5/2}-\frac{4}{3}(1-\sqrt{v})^{3/2}+C$$

21. $\displaystyle\int \frac{\sqrt{1+\sqrt{t}}}{\sqrt{t}}\,dt$ $\boxed{u=1+\sqrt{t},\ du=\frac{1}{2\sqrt{t}}\,dt}$

$$=\int\sqrt{u}(2\,du)=\frac{4}{3}u^{3/2}+C=\frac{4}{3}(1+\sqrt{t})^{3/2}+C$$

23. $\displaystyle\int \frac{2x+7}{x^2+2x+5}\,dx=\int\frac{2(x+1)+5}{(x+1)^2+4}\,dx=\int\frac{2(x+1)}{(x+1)^2+4}\,dx+\int\frac{5}{(x+1)^2+4}\,dx$

$$=\ln[(x+1)^2+4]+\frac{5}{2}\tan^{-1}\frac{x+1}{2}+C$$

$$=\ln(x^2+2x+5)+\frac{5}{2}\tan^{-1}\frac{x+1}{2}+C$$

25. $\displaystyle\int\frac{2x+5}{\sqrt{16-16x-x^2}}\,dx=\int\frac{2(x+3)-1}{\sqrt{25-(x+3)^2}}\,dx$

$$=\int\frac{2(x+3)}{\sqrt{25-(x+3)^2}}\,dx-\int\frac{1}{\sqrt{25-(x+3)^2}}\,dx$$

$$\boxed{u=25-(x+3)^2,\ du=-2(x+3)\,dx}$$

$$=\int\frac{-1}{\sqrt{u}}\,du-\sin^{-1}\frac{x+3}{5}+C=-2\sqrt{u}-\sin^{-1}\frac{x+3}{5}+C$$

$$=-2\sqrt{16-6x-x^2}-\sin^{-1}\frac{x+3}{5}+C$$

27. $\displaystyle\int\frac{1}{\sqrt{x}-\sqrt[3]{x}}\,dx$ $\boxed{u=x^{1/6},\ x=u^6,\ dx=6u^5\,du}$

$$=\int\frac{6u^5}{u^3-u^2}\,du=6\int\left(u^2+u+1+\frac{1}{u-1}\right)du$$

$$=2u^3+3u^2+6u+6\ln|u-1|+C$$

$$=2\sqrt{x}+3\sqrt[3]{x}+6\sqrt[6]{x}+6\ln|\sqrt[6]{x}-1|+C$$

29. $\displaystyle\int_0^1 x\sqrt{5x+4}\,dx$ $\boxed{u=5x+4,\ x=\frac{1}{5}(u-4),\ dx=\frac{1}{5}\,du}$

$$=\int_4^9\frac{1}{5}(u-4)u^{1/2}\left(\frac{1}{5}\,du\right)=\frac{1}{25}\int_4^9(u^{3/2}-4u^{1/2})\,du$$

$$=\frac{1}{25}\left(\frac{2}{5}u^{5/2}-\frac{8}{3}u^{3/2}\right)\Big]_4^9=\frac{1}{25}\left[\left(\frac{486}{5}-72\right)-\left(\frac{64}{5}-\frac{64}{3}\right)\right]=\frac{506}{375}$$

31. $\displaystyle\int_1^{16}\frac{1}{10+\sqrt{x}}\,dx$ $\boxed{u=10+\sqrt{x},\ x=(u-10)^2,\ dx=2(u-10)\,du}$

$$=\int_{11}^{14}\frac{2(u-10)}{u}\,du=\int_{11}^{14}\left(2-\frac{20}{u}\right)du=(2u-20\ln|u|)]_{11}^{14}$$

$$=(28-20\ln 14)-(22-20\ln 11)=6-20\ln\frac{14}{11}$$

33. $\int_2^9 \dfrac{5x-6}{\sqrt[3]{x-1}}\,dx$ $\boxed{u=\sqrt[3]{x-1},\ x=u^3+1,\ dx=3u^2\,du}$

$$= \int_1^2 \frac{5(u^3+1)-6}{u}(3u^2\,du) = \int_1^2 (15u^4-3u)\,du = \left(3u^5 - \frac{3}{2}u^2\right)\Big]_1^2$$

$$= 90 - \frac{3}{2} = \frac{177}{2}$$

35. $\int_0^1 (1-\sqrt{x})^{50}\,dx$ $\boxed{u=1-\sqrt{x},\ x=(1-u)^2,\ dx=-2(1-u)\,du}$

$$= \int_1^0 u^{50}[-2(1-u)\,du] = \int_1^0 (2u^{51}-2u^{50})\,du = \left(\frac{1}{26}u^{52}-\frac{2}{51}u^{51}\right)\Big]_1^0$$

$$= 0 - \left(-\frac{1}{1326}\right) = \frac{1}{1326}$$

37. $\int_1^8 \dfrac{1}{x^{1/3}+x^{2/3}}\,dx$ $\boxed{u=x^{1/3},\ x=u^3,\ dx=3u^2\,du}$

$$= \int_1^2 \frac{3u^2}{u+u^2}\,du = \int_1^2 \left(3-\frac{3}{1+u}\right)du = (3u-3\ln|1+u|)]_1^2$$

$$= (6-3\ln 3)-(3-3\ln 2) = 3+3\ln\frac{2}{3}$$

39. $\int_0^1 x^2(1-x)^5\,dx$ $\boxed{u=1-x,\ du=-dx}$

$$= \int_1^0 (1-u)^2 u^5(-du) = \int_0^1 (u^5-2u^6+u^7)\,du$$

$$= \left(\frac{1}{6}u^6-\frac{2}{7}u^7+\frac{1}{8}u^8\right)\Big]_0^1 = \frac{1}{6}-\frac{2}{7}+\frac{1}{8} = \frac{1}{168}$$

41. $\int_1^{x^2} \dfrac{1}{t}\,dt$ $\boxed{u=\sqrt{t},\ t=u^2,\ dt=2u\,du}$

$$= \int_1^x \frac{1}{u^2}(2u\,du) = 2\int_1^x \frac{1}{u}\,du = 2\int_1^x \frac{1}{t}\,dt$$

43. $A = \int_0^1 \dfrac{1}{x^{1/3}+1}\,dx$ $\boxed{u=x^{1/3},\ x=u^3,\ dx=3u^2\,du}$

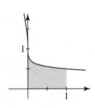

$$= \int_0^1 \frac{3u^2}{u+1}\,du = \int_0^1 \left(3u-3+\frac{3}{u+1}\right)du$$

$$= \left(\frac{3}{2}u^2-3u+3\ln|u+1|\right)\Big]_0^1 = \left(-\frac{3}{2}+3\ln 2\right)-0 \approx 0.5794$$

45. $V = 2\pi \int_0^4 x\dfrac{1}{\sqrt{x}+1}\,dx$ $\boxed{u=\sqrt{x}+1,\ x=(u-1)^2,\ dx=2(u-1)\,du}$

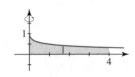

$$= 2\pi \int_1^3 \frac{(u-1)^2}{u}[2(u-1)\,du] = 2\pi \int_1^3 \left(2u^2-6u+6-\frac{2}{u}\right)du$$

$$= 2\pi \left(\frac{2}{3}u^3-3u^2+6u-2\ln|u|\right)\Big]_1^3 = 2\pi \left[(9-2\ln 3)-\frac{11}{3}\right]$$

$$= 2\pi \left(\frac{16}{3}-2\ln 3\right) \approx 19.7048$$

47. $y = x^{1/4}$

$$L = \int_0^9 \sqrt{1 + (x^{1/4})^2}\, dx = \int_0^9 \sqrt{1 + \sqrt{x}}\, dx$$

$$\boxed{u = 1 + \sqrt{x}, \ x = (u-1)^2, \ dx = 2(u-1)\, du}$$

$$= \int_1^4 \sqrt{u}[2(u-1)\, du] = 2\int_1^4 (u^{3/2} - u^{1/2})\, du = 2\left(\frac{2}{5}u^{5/2} - \frac{2}{3}u^{3/2}\right)\Bigg]_1^4$$

$$= 2\left[\left(\frac{64}{5} - \frac{16}{3}\right) - \left(\frac{2}{5} - \frac{2}{3}\right)\right] = \frac{232}{15}$$

7.3   Integration by Parts

1. $\displaystyle\int x\sqrt{x+3}\, dx$ $\quad\boxed{u = x, \ du = dx; \quad dv = (x+3)^{1/2}\, dx, \ v = \frac{2}{3}(x+3)^{3/2}}$

$$= \frac{2}{3}x(x+3)^{3/2} - \int \frac{2}{3}(x+3)^{3/2}\, dx = \frac{2}{3}x(x+3)^{3/2} - \frac{4}{15}(x+3)^{5/2} + C$$

3. $\displaystyle\int \ln 4x\, dx$ $\quad\boxed{u = \ln 4x, \ du = \frac{1}{x}\, dx; \quad dv = dx, \ v = x}$

$$= x\ln 4x - \int dx = x\ln 4x - x + C$$

5. $\displaystyle\int x\ln 2x\, dx$ $\quad\boxed{u = \ln 2x, \ du = \frac{1}{x}\, dx; \quad dv = x\, dx, \ v = \frac{1}{2}x^2}$

$$= \frac{1}{2}x^2 \ln 2x - \int \frac{x^2}{2x}\, dx = \frac{1}{2}x^2 \ln 2x - \frac{1}{2}\int x\, dx = \frac{1}{2}x^2 \ln 2x - \frac{1}{4}x^2 + C$$

7. $\displaystyle\int \frac{\ln x}{x^2}\, dx$ $\quad\boxed{u = \ln x, \ du = \frac{1}{x}\, dx; \quad dv = \frac{1}{x^2}\, dx, \ v = -\frac{1}{x}}$

$$= -\frac{1}{x}\ln x - \int \left(-\frac{1}{x^2}\right) dx = -\frac{1}{x}\ln x - \frac{1}{x} + C$$

9. $\displaystyle\int (\ln t)^2\, dt$ $\quad\boxed{u = (\ln t)^2, \ du = \frac{2\ln t}{t}\, dt; \quad dv = dt, \ v = t}$

$$= t(\ln t)^2 - \int 2\ln t\, dt \quad \boxed{u = 2\ln t, \ du = \frac{2}{t}\, dt; \quad dv = dt, \ v = t}$$

$$= t(\ln t)^2 - \left(2t\ln t - \int 2\, dt\right) = t(\ln t)^2 - 2t\ln t + 2t + C$$

11. $\displaystyle\int \sin^{-1} x\, dx$ $\quad\boxed{u = \sin^{-1} x, \ du = \frac{1}{\sqrt{1-x^2}}\, dx; \quad dv = dx, \ v = x}$

$$= x\sin^{-1} x - \int \frac{x}{\sqrt{1-x^2}}\, dx \quad \boxed{u = 1 - x^2, \ du = -2x\, dx}$$

$$= x\sin^{-1} x - \int \frac{1}{\sqrt{u}}\left(-\frac{1}{2}\, du\right) = x\sin^{-1} x + \sqrt{u} + C$$

$$= x\sin^{-1} x + \sqrt{1-x^2} + C$$

13. $\displaystyle\int xe^{3x}\, dx$ $\quad\boxed{u = x, \ du = dx; \quad dv = e^{3x}\, dx, \ v = \frac{1}{3}e^{3x}}$

$$= \frac{1}{3}xe^{3x} - \int \frac{1}{3}e^{3x}\, dx = \frac{1}{3}xe^{3x} - \frac{1}{9}e^{3x} + C$$

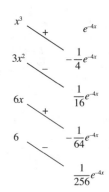

15. $\displaystyle\int x^3 e^{-4x}\,dx = -\frac{1}{4}x^3 e^{-4x} - \frac{3}{16}x^2 e^{-4x} - \frac{3}{32}x e^{-4x} - \frac{3}{128}e^{-4x} + C$

17. $\displaystyle\int x^3 e^{x^2}\,dx$ $\boxed{u = x^2,\ du = 2x\,dx;\quad dv = xe^{x^2}\,dx,\ v = \frac{1}{2}e^{x^2}}$

$\displaystyle = \frac{1}{2}x^2 e^{x^2} - \int xe^{x^2}\,dx = \frac{1}{2}x^2 e^{x^2} - \frac{1}{2}e^{x^2} + C$

19. $\displaystyle\int t\cos 8t\,dt$ $\boxed{u = t,\ du = dt;\quad dv = \cos 8t\,dt,\ v = \frac{1}{8}\sin 8t}$

$\displaystyle = \frac{1}{8}t\sin 8t - \int \frac{1}{8}\sin 8t\,dt = \frac{1}{8}t\sin 8t + \frac{1}{64}\cos 8t + C$

21. $\displaystyle\int x^2 \sin x\,dx$ $\boxed{u = x^2,\ du = 2x\,dx;\quad dv = \sin x\,dx,\ v = -\cos x}$

$\displaystyle = -x^2 \cos x + \int 2x\cos x\,dx$ $\boxed{u = 2x,\ du = 2\,dx;\quad dv = \cos x\,dx,\ v = \sin x}$

$\displaystyle = -x^2 \cos x + 2x\sin x - \int 2\sin x\,dx = -x^2 \cos x + 2x\sin x + 2\cos x + C$

23. $\displaystyle\int x^3 \cos 3x\,dx$ $\boxed{u = x^3,\ du = 3x^2\,dx;\quad dv = \cos 3x\,dx,\ v = \frac{1}{3}\sin 3x}$

$\displaystyle = \frac{1}{3}x^3 \sin 3x - \int x^2 \sin 3x\,dx$

$\boxed{u = x^2,\ du = 2x\,dx;\quad dv = \sin 3x\,dx,\ v = -\frac{1}{3}\cos 3x}$

$\displaystyle = \frac{1}{3}x^3 \sin 3x - \left(-\frac{1}{3}x^2 \cos 3x + \frac{2}{3}\int x\cos 3x\,dx \right)$

$\boxed{u = x,\ du = dx;\quad dv = \cos 3x\,dx,\ v = \frac{1}{3}\sin 3x}$

$\displaystyle = \frac{1}{3}x^3 \sin 3x + \frac{1}{3}x^2 \cos 3x - \frac{2}{3}\left(\frac{1}{3}x\sin 3x - \frac{1}{3}\int \sin 3x\,dx \right)$

$\displaystyle = \frac{1}{3}x^3 \sin 3x + \frac{1}{3}x^2 \cos 3x - \frac{2}{9}x\sin 3x - \frac{2}{27}\cos 3x + C$

25. $\displaystyle\int e^x \sin 4x\,dx$ $\boxed{u = \sin 4x,\ du = 4\cos 4x\,dx;\quad dv = e^x\,dx,\ v = e^x}$

$\displaystyle = e^x \sin 4x - \int 4e^x \cos 4x\,dx$

$\boxed{u = \cos 4x,\ du = -4\sin 4x\,dx;\quad dv = 4e^x\,dx,\ v = 4e^x}$

$\displaystyle = e^x \sin 4x - \left(4e^x \cos 4x + 16\int e^x \sin 4x\,dx \right)$

Solving for the integral, we have $\displaystyle 17\int e^x \sin 4x\,dx = e^x \sin 4x - 4e^x \cos 4x + C$ or

$$\int e^x \sin 4x\,dx = \frac{e^x}{17}(\sin 4x - 4\cos 4x) + C.$$

27. $\displaystyle\int e^{-2\theta}\cos\theta\,d\theta$ $\boxed{u=e^{-2\theta},\ du=-2e^{-2\theta}\,d\theta;\quad dv=\cos\theta\,d\theta,\ v=\sin\theta}$

$$= e^{-2\theta}\sin\theta - \int -2e^{-2\theta}\sin\theta\,d\theta$$

$\boxed{u=e^{-2\theta},\ du=-2e^{-2\theta}\,d\theta;\quad dv=\sin\theta\,d\theta,\ v=-\cos\theta}$

$$= e^{-2\theta}\sin\theta + 2\left(-e^{-2\theta}\cos\theta - \int 2e^{-2\theta}\cos\theta\,d\theta\right)$$

$$= e^{-2\theta}\sin\theta - 2e^{-2\theta}\cos\theta - 4\int e^{-2\theta}\cos\theta\,d\theta$$

Solving for the integral, we have $\displaystyle\int e^{-2\theta}\cos\theta\,d\theta = \frac{\sin\theta - 2\cos\theta}{5e^{2\theta}} + C.$

29. $\displaystyle\int \theta\sec\theta\tan\theta\,d\theta$ $\boxed{u=\theta,\ du=d\theta;\quad dv=\sec\theta\tan\theta\,d\theta,\ v=\sec\theta}$

$$= \theta\sec\theta - \int \sec\theta\,d\theta = \theta\sec\theta - \ln|\sec\theta+\tan\theta| + C$$

31. $\displaystyle\int \sin x\cos 2x\,dx$ $\boxed{u=\cos 2x,\ du=-2\sin 2x\,dx;\quad dv=\sin x\,dx,\ v=-\cos x}$

$$= -\cos x\cos 2x - \int 2\cos x\sin 2x\,dx$$

$\boxed{u=2\sin 2x,\ du=4\cos 2x\,dx;\quad dv=\cos x\,dx,\ v=\sin x}$

$$= -\cos x\cos 2x - \left(2\sin x\sin 2x - 4\int \sin x\cos 2x\,dx\right)$$

Solving for the integral, we have $\displaystyle\int \sin x\cos 2x\,dx = \frac{1}{3}\cos x\cos 2x + \frac{2}{3}\sin x\sin 2x + C.$

33. $\displaystyle\int x^3\sqrt{x^2+4}\,dx$ $\boxed{u=x^2,\ du=2x\,dx;\quad dv=x\sqrt{x^2+4}\,dx,\ v=\frac{1}{3}(x^2+4)^{3/2}}$

$$= \frac{1}{3}x^2(x^2+4)^{3/2} - \frac{2}{3}\int x(x^2+4)^{3/2}\,dx$$

$$= \frac{1}{3}x^2(x^2+4)^{3/2} - \frac{2}{15}(x^2+4)^{5/2} + C$$

35. $\displaystyle\int \sin(\ln x)\,dx$ $\boxed{u=\sin(\ln x),\ du=\frac{1}{x}\cos(\ln x)\,dx;\quad dv=dx,\ v=x}$

$$= x\sin(\ln x) - \int \cos(\ln x)\,dx$$

$\boxed{u=\cos(\ln x),\ du=-\frac{1}{x}\sin(\ln x)\,dx;\quad dv=dx,\ v=x}$

$$= x\sin(\ln x) - \left[x\cos(\ln x) + \int \sin(\ln x)\,dx\right]$$

Solving for the integral, we have $\displaystyle\int \sin(\ln x)\,dx = \frac{1}{2}x\sin(\ln x) - \frac{1}{2}x\cos(\ln x) + C.$

37. $\displaystyle\int \csc^3 x\,dx$ $\boxed{u=\csc x,\ du=-\csc x\cot x\,dx;\quad dv=\csc^2 x\,dx,\ v=-\cot x}$

$$= -\csc x\cot x - \int \csc x\cot^2 x\,dx = -\csc x\cot x - \int \csc x(\csc^2 x - 1)\,dx$$

$$= -\csc x\cot x - \int \csc^3 x\,dx + \int \csc x\,dx$$

$$= -\csc x\cot x - \int \csc^3 x\,dx + \ln|\csc x - \cot x|$$

Solving for the integral, we have $\displaystyle\int \csc^3 x\,dx = -\frac{1}{2}\csc x\cot x + \frac{1}{2}\ln|\csc x - \cot x| + C.$

39. $\displaystyle\int x\sec^2 x\, dx$ $\boxed{u = x,\ du = dx;\quad dv = \sec^2 x\, dx,\ v = \tan x}$

$$= x\tan x - \int \tan x\, dx = x\tan x - \ln|\sec x| + C$$

41. $\displaystyle\int_0^2 x\ln(x+1)\, dx$ $\boxed{u = \ln(x+1),\ du = \dfrac{1}{x+1}dx;\quad dv = x\, dx,\ v = \dfrac{1}{2}x^2}$

$$= \frac{1}{2}x^2\ln(x+1)\Big]_0^2 - \int_0^2 \frac{1}{2}\left(\frac{x^2}{x+1}\right)dx$$

$$= \frac{1}{2}(4\ln 3 - 0) - \frac{1}{2}\int_0^2\left(x - 1 + \frac{1}{x+1}\right)dx$$

$$= 2\ln 3 - \frac{1}{2}\left[\frac{1}{2}x^2 - x + \ln(x+1)\right]\Big|_0^2 = 2\ln 3 - \frac{1}{2}(\ln 3 - 0) = \frac{3}{2}\ln 3$$

43. $\displaystyle\int_2^4 xe^{-x/2}\, dx$ $\boxed{u = x,\ du = dx;\quad dv = e^{-x/2}\, dx,\ v = -2e^{-x/2}}$

$$= -2xe^{-x/2}\Big]_2^4 - \int_2^4 (-2e^{-x/2})\, dx = -2(4e^{-2} - 2e^{-1}) - 4e^{-x/2}\Big]_2^4$$

$$= \frac{4}{e} - \frac{8}{e^2} - 4(e^{-2} - e^{-1}) = \frac{8}{e} - \frac{12}{e^2} = \frac{8e - 12}{e^2}$$

45. $\displaystyle\int_0^1 \tan^{-1}x\, dx$ $\boxed{u = \tan^{-1}x,\ du = \dfrac{1}{1+x^2}dx;\quad dv = dx,\ v = x}$

$$= x\tan^{-1}x\Big]_0^1 - \int_0^1 \frac{x}{1+x^2}\, dx = \left(\frac{\pi}{4} - 0\right) - \frac{1}{2}\ln(1+x^2)\Big]_0^1$$

$$= \frac{\pi}{4} - \frac{1}{2}(\ln 2 - 0) = \frac{\pi}{4} - \frac{1}{2}\ln 2$$

47. $A = \displaystyle\int_{e^{-1}}^3 (1+\ln x)\, dx$ $\boxed{u = 1+\ln x,\ du = \dfrac{1}{x}dx;\quad dv = dx,\ v = x}$

$$= (x + x\ln x)\big]_{e^{-1}}^3 - \int_{e^{-1}}^3 dx = (3 + 3\ln 3) - (e^{-1} - e^{-1}) - x\big]_{e^{-1}}^3$$

$$= 3 + 3\ln 3 - \left(3 - \frac{1}{e}\right) = 3\ln 3 + \frac{1}{e}$$

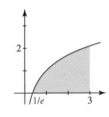

49. $V = \pi\displaystyle\int_1^5 (\ln x)^2\, dx$ $\boxed{u = (\ln x)^2,\ du = \dfrac{2\ln x}{x}dx;\quad dv = dx,\ v = x}$

$$= \pi x(\ln x)^2\big]_1^5 - \pi\int_1^5 2\ln x\, dx$$

$\boxed{u = 2\ln x,\ du = \dfrac{2}{x}dx;\quad dv = dx,\ v = x}$

$$= \pi[5(\ln 5)^2 - 0] - \pi\left(2x\ln x\big]_1^5 - \int_1^5 2\, dx\right)$$

$$= 5\pi(\ln 5)^2 - \pi\left[(10\ln 5 - 0) - 2x\big]_1^5\right] = 5\pi(\ln 5)^2 - 10\pi\ln 5 + 8\pi$$

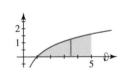

51. $V = 2\pi\displaystyle\int_0^\pi x\sin x\, dx$ $\boxed{u = x,\ du = dx;\quad dv = \sin x\, dx,\ v = -\cos x}$

$$= 2\pi\left(-x\cos x\big]_0^\pi + \int_0^\pi \cos x\, dx\right) = 2\pi\left[-(-\pi - 0) + \sin x\big]_0^\pi\right] = 2\pi^2$$

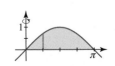

53. $f_{\text{ave}} = \dfrac{1}{2-0} \displaystyle\int_0^2 \tan^{-1}\dfrac{x}{2}\,dx$ $\boxed{u = \tan^{-1}\dfrac{x}{2},\ du = \dfrac{2}{4+x^2}\,dx;\quad dv = dx,\ v = x}$

$= \dfrac{1}{2}\left(x\tan^{-1}\dfrac{x}{2}\Big]_0^2 - \displaystyle\int_0^2 \dfrac{2x}{4+x^2}\,dx\right) = \dfrac{1}{2}\left[2\left(\dfrac{\pi}{4}\right) - \ln(4+x^2)\right]_0^2$

$= \dfrac{1}{2}\left[\dfrac{\pi}{2} - (\ln 8 - \ln 4)\right] = \dfrac{\pi}{4} - \dfrac{1}{2}\ln 2$

55. $v(t) = \displaystyle\int te^{-t}\,dt$ $\boxed{u = t,\ du = dt;\quad dv = e^{-t}\,dt,\ v = -e^{-t}}$

$= -te^{-t} + \displaystyle\int e^{-t}\,dt = -te^{-t} - e^{-t} + C.$

Now $1 = v(0) = -1 + C$, so $C = 2$ and $v(t) = -te^{-t} - e^{-t} + 2$.

$s(t) = \displaystyle\int(-te^{-t} - e^{-t} + 2)\,dt = -\displaystyle\int te^{-t}\,dt + e^{-t} + 2t = -(-te^{-t} - e^{-t}) + e^{-t} + 2t + C$

$= te^{-t} + 2e^{-t} + 2t + C.$

Now $-1 = s(0) = 2 + C$, so $C = -3$ and $s(t) = te^{-t} + 2e^{-t} + 2t - 3$.

57. Using symmetry,

$F = 2\left(62.4\displaystyle\int_0^2 x\cos\dfrac{\pi x}{4}\,dx\right)$ $\boxed{u = x,\ du = dx;\quad dv = \cos\dfrac{\pi x}{4}\,dx,\ v = \dfrac{4}{\pi}\sin\dfrac{\pi x}{4}}$

$= 124.8\left(\dfrac{4x}{\pi}\sin\dfrac{\pi x}{4}\Big]_0^2 - \dfrac{4}{\pi}\displaystyle\int_0^2 \sin\dfrac{\pi x}{4}\,dx\right) = 124.8\left(\dfrac{8}{\pi} + \dfrac{16}{\pi^2}\cos\dfrac{\pi x}{4}\Big]_0^2\right)$

$= 124.8\left(\dfrac{8}{\pi} - \dfrac{16}{\pi^2}\right) \approx 115.4825\text{ lb}$

59. $\displaystyle\int_1^4 \dfrac{\tan^{-1}\sqrt{x}}{\sqrt{x}}\,dx$ $\boxed{t = \sqrt{x},\ dt = \dfrac{1}{2\sqrt{x}}\,dx}$

$= 2\displaystyle\int_1^2 \tan^{-1}t\,dt$ $\boxed{u = \tan^{-1}t,\ du = \dfrac{1}{1+t^2}\,dt;\quad dv = dt,\ v = t}$

$= 2\left(t\tan^{-1}t\Big]_1^2 - \displaystyle\int_1^2 \dfrac{t}{1+t^2}\,dt\right) = 2\left[\left(2\tan^{-1}2 - \dfrac{\pi}{4}\right) - \dfrac{1}{2}\ln(1+t^2)\right]_1^2$

$= 4\tan^{-1}2 - \dfrac{\pi}{2} - (\ln 5 - \ln 2) = 4\tan^{-1}2 - \dfrac{\pi}{2} - \ln\dfrac{5}{2}$

61. $\displaystyle\int \sin\sqrt{x+2}\,dx$ $\boxed{t = \sqrt{x+2},\ x = t^2 - 2,\ dx = 2t\,dt}$

$= \displaystyle\int(\sin t)2t\,dt$ $\boxed{u = 2t,\ du = 2dt;\quad dv = \sin t\,dt,\ v = -\cos t}$

$= -2t\cos t + \displaystyle\int 2\cos t\,dt = -2t\cos t + 2\sin t + C$

$= -2\sqrt{x+2}\cos\sqrt{x+2} + 2\sin\sqrt{x+2} + C$

63. $\displaystyle\int(\ln x)^n\,dx$ $\boxed{u = (\ln x)^n,\ du = \dfrac{n(\ln x)^{n-1}}{x}\,dx;\quad dv = dx,\ v = x}$

$= x(\ln x)^n - n\displaystyle\int(\ln x)^{n-1}\,dx$

65. $\displaystyle\int \cos^n x\,dx$ $\boxed{u = \cos^{n-1} x,\ du = -(n-1)\cos^{n-2} x\sin x\,dx;\quad dv = \cos x\,dx,\ v = \sin x}$

$$= \cos^{n-1} x\sin x + \int (n-1)\sin^2 x\cos^{n-2} x\,dx$$

$$= \cos^{n-1} x\sin x + (n-1)\int (1-\cos^2 x)\cos^{n-2} x\,dx$$

$$= \cos^{n-1} x\sin x + (n-1)\int \cos^{n-2} x\,dx - (n-1)\int \cos^n x\,dx$$

Solving for the integral $\displaystyle\int \cos^n x\,dx$, we have

$$\int \cos^n x\,dx = \frac{\cos^{n-1} x\sin x}{n} + \frac{n-1}{n}\int \cos^{n-2} x\,dx.$$

67. Using $\displaystyle\int \sin^n x\,dx = -\frac{\sin^{n-1} x\cos x}{n} + \frac{n-1}{n}\int \sin^{n-2} x\,dx$ with $n = 3$,

$$\int \sin^3 x\,dx = -\frac{\sin^2 x\cos x}{3} + \frac{2}{3}\int \sin x\,dx = -\frac{\sin^2 x\cos x}{3} - \frac{2}{3}\cos x + C.$$

69. Using $\displaystyle\int \cos^n x\,dx = \frac{\cos^{n-1} x\sin x}{n} + \frac{n-1}{n}\int \cos^{n-2} x\,dx$ with $n = 3$,

$$\int \cos^3 10x\,dx \qquad \boxed{u = 10x,\ du = 10\,dx}$$

$$= \frac{1}{10}\int \cos^3 u\,du = \frac{\cos^2 u\sin u}{10\cdot 3} + \frac{2}{10\cdot 3}\int \cos u\,du$$

$$= \frac{\cos^2 u\sin u}{30} + \frac{1}{15}\sin u + C = \frac{\cos^2 10x\sin 10x}{30} + \frac{1}{15}\sin 10x + C.$$

71. Using $\displaystyle\int \sin^n x\,dx = -\frac{\sin^{n-1} x\cos x}{n} + \frac{n-1}{n}\int \sin^{n-2} x\,dx$, we have

$$\int_0^{\pi/2} \sin^n x\,dx = -\frac{\sin^{n-1} x\cos x}{n}\Big]_0^{\pi/2} + \frac{n-1}{n}\int_0^{\pi/2} \sin^{n-2} x\,dx$$

$$= -\left[\frac{\sin^{n-1}(\pi/2)\cos(\pi/2)}{n} - \frac{\sin^{n-1} 0\cos 0}{n}\right] + \frac{n-1}{n}\int_0^{\pi/2} \sin^{n-2} x\,dx$$

$$= -(0-0) + \frac{n-1}{n}\int_0^{\pi/2} \sin^{n-2} x\,dx = \frac{n-1}{n}\int_0^{\pi/2} \sin^{n-2} x\,dx.$$

73. $\displaystyle\int_0^{\pi/2} \sin^8 x\,dx = \frac{\pi}{2}\cdot\frac{1\cdot 3\cdot 5\cdot 7}{2\cdot 4\cdot 6\cdot 8} = \frac{105\pi}{768} = \frac{35\pi}{256}$

75. $\displaystyle\int e^{2x}\tan^{-1} e^x\,dx$ $\boxed{u = \tan^{-1} e^x,\ du = \dfrac{e^x}{1+e^{2x}}\,dx;\quad dv = e^{2x}\,dx,\ v = \dfrac{1}{2}e^{2x}}$

$$= \frac{1}{2}e^{2x}\tan^{-1} e^x - \int \frac{1}{2}\left(\frac{e^{3x}}{1+e^{2x}}\right)dx \qquad \boxed{t = e^x,\ dt = e^x\,dx}$$

$$= \frac{1}{2}e^{2x}\tan^{-1} e^x - \frac{1}{2}\int \frac{t^2}{1+t^2}\,dt = \frac{1}{2}e^{2x}\tan^{-1} e^x - \frac{1}{2}\int \left(1 - \frac{1}{1+t^2}\right)dt$$

$$= \frac{1}{2}e^{2x}\tan^{-1} e^x - \frac{1}{2}t + \frac{1}{2}\tan^{-1} t + C = \frac{1}{2}(e^{2x}+1)\tan^{-1} e^x - \frac{1}{2}e^x + C$$

77. $\displaystyle\int \frac{xe^x}{(x+1)^2}\,dx$ $\boxed{u = xe^x,\ du = (x+1)e^x\,dx;\quad dv = \dfrac{1}{(x+1)^2}\,dx,\ v = -\dfrac{1}{x+1}}$

$$= -\frac{x}{x+1}e^x + \int e^x\,dx = \left(1 - \frac{x}{x+1}\right)e^x + C = \frac{1}{x+1}e^x + C = \frac{e^x}{x+1} + C$$

79. We first compute $\int e^x \sin x\, dx$:

$$\int e^x \sin x\, dx \qquad \boxed{u = e^x,\ du = e^x\, dx; \quad dv = \sin x\, dx,\ v = -\cos x}$$

$$= -e^x \cos x + \int e^x \cos x\, dx$$

$$\boxed{u = e^x,\ du = e^x\, dx; \quad dv = \cos x\, dx,\ v = \sin x}$$

$$= -e^x \cos x + e^x \sin x - \int e^x \sin x\, dx.$$

Solving for the integral, we have $\int e^x \sin x\, dx = \dfrac{1}{2} e^x(\sin x - \cos x)$. Similarly, $\int e^x \cos x\, dx = \dfrac{1}{2} e^x(\sin x + \cos x)$. Then

$$\int x e^x \sin x\, dx \qquad \boxed{u = x,\ du = dx; \quad dv = e^x \sin x\, dx,\ v = \dfrac{1}{2} e^x(\sin x - \cos x)}$$

$$= \frac{1}{2} x e^x(\sin x - \cos x) - \frac{1}{2}\int (e^x \sin x - e^x \cos x)\, dx$$

$$= \frac{1}{2} x e^x(\sin x - \cos x) - \frac{1}{2}\left[\frac{1}{2} e^x(\sin x - \cos x) - \frac{1}{2} e^x(\sin x + \cos x)\right] + C$$

$$= \frac{1}{2} x e^x \sin x - \frac{1}{2} x e^x \cos x + \frac{1}{2} e^x \cos x + C.$$

81. $\displaystyle\int \ln\left(x + \sqrt{x^2 + 1}\right) dx$

$$\boxed{u = \ln\left(x + \sqrt{x^2+1}\right),\ du = \frac{1 + x/\sqrt{x^2+1}}{x + \sqrt{x^2+1}}\, dx; \quad dv = dx,\ v = x}$$

$$= x\ln\left(x + \sqrt{x^2+1}\right) - \int \frac{1}{x+\sqrt{x^2+1}}\left(1 + \frac{x}{\sqrt{x^2+1}}\right) x\, dx$$

$$= x\ln\left(x + \sqrt{x^2+1}\right) - \int \frac{x - \sqrt{x^2+1}}{x^2 - (x^2+1)}\left(\frac{\sqrt{x^2+1}+x}{\sqrt{x^2+1}}\right) x\, dx$$

$$= x\ln\left(x + \sqrt{x^2+1}\right) - \int \left(\sqrt{x^2+1} - x\right)\left(\sqrt{x^2+1} + x\right)\left(\frac{x}{\sqrt{x^2+1}}\right) dx$$

$$= x\ln\left(x + \sqrt{x^2+1}\right) - \int \frac{x}{\sqrt{x^2+1}}\, dx$$

$$= x\ln\left(x + \sqrt{x^2+1}\right) - \sqrt{x^2+1} + C$$

83. (a) Graph shown at right.

(b) We use the reduction formula

$$\int \sin^n x\, dx = -\frac{\sin^{n-1} x \cos x}{n} + \frac{n-1}{n}\int \sin^{n-2} x\, dx$$

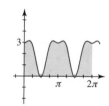

with $n = 4$ and $n = 2$.

$$A = \int_0^{2\pi} (3 + \sin^2 x - 5\sin^4 x)\, dx$$

$$= 3x\Big]_0^{2\pi} + 2\int_0^{2\pi} \sin^2 x\, dx - 5\left(-\frac{\sin^3 x \cos x}{4}\Big]_0^{2\pi} + \frac{3}{4}\int_0^{2\pi} \sin^2 x\, dx\right)$$

$$= 6\pi + 5\left(-\frac{0}{4} + \frac{0}{4}\right) - \frac{7}{4}\int_0^{2\pi}\sin^2 x\, dx = 6\pi - \frac{7}{4}\left(-\frac{\sin x \cos x}{2}\Big]_0^{2\pi} + \frac{1}{2}\int_0^{2\pi} dx\right)$$

$$= 6\pi - \frac{7}{8}x\Big]_0^{2\pi} = \frac{17\pi}{4}.$$

7.4   Powers of Trigonometric Functions

1. $\displaystyle\int \sin^{1/2} x \cos x \, dx$ $\boxed{u = \sin x, \ du = \cos x \, dx}$

$$= \int u^{1/2} \, du = \frac{2}{3} u^{3/2} + C = \frac{2}{3} (\sin x)^{3/2} + C$$

3. $\displaystyle\int \cos^3 x \, dx = \int \cos^2 x \cos x \, dx = \int (1 - \sin^2 x) \cos x \, dx = \int \cos x \, dx - \int \sin^2 x (\cos x \, dx)$

$$= \sin x - \frac{1}{3} \sin^3 x + C$$

5. $\displaystyle\int \sin^5 t \, dt = \int \sin^4 t \sin t \, dt = \int (1 - \cos^2 t)^2 \sin t \, dt = \int (1 - 2\cos^2 t + \cos^4 t) \sin t \, dt$

$$= \int \sin t \, dt + 2 \int \cos^2 t (-\sin t \, dt) - \int \cos^4 t (-\sin t \, dt)$$

$$= -\cos t + \frac{2}{3} \cos^3 t - \frac{1}{5} \cos^5 t + C$$

7. $\displaystyle\int \sin^3 x \cos^3 x \, dx = \int \cos^2 x \sin^3 x \cos x \, dx = \int (1 - \sin^2 x) \sin^3 x \cos x \, dx$

$$= \int \sin^3 x (\cos x \, dx) - \int \sin^5 x (\cos x \, dx) = \frac{1}{4} \sin^4 x - \frac{1}{6} \sin^6 x + C$$

9. $\displaystyle\int \sin^4 t \, dt = \int (\sin^2 t)^2 \, dt = \int \left(\frac{1 - \cos 2t}{2} \right)^2 dt = \frac{1}{4} \int (1 - 2\cos 2t + \cos^2 2t) \, dt$

$$= \frac{1}{4} \int \left(1 - 2\cos 2t + \frac{1 + \cos 4t}{2} \right) dt = \frac{1}{4} \int \left(\frac{3}{2} - 2\cos 2t + \frac{1}{2} \cos 4t \right) dt$$

$$= \frac{3}{8} t - \frac{1}{4} \sin 2t + \frac{1}{32} \sin 4t + C$$

11. $\displaystyle\int \sin^2 x \cos^4 x \, dx = \int (1 - \cos^2 x) \cos^4 x \, dx = \int (\cos^4 x - \cos^6 x) \, dx$

$$= \int \left[\left(\frac{1 + \cos 2x}{2} \right)^2 - \left(\frac{1 + \cos 2x}{2} \right)^3 \right] dx$$

$$= \frac{1}{8} \int [2(1 + 2\cos 2x + \cos^2 2x) - (1 + 3\cos 2x + 3\cos^2 2x + \cos^3 2x)] \, dx$$

$$= \frac{1}{8} \int (1 + \cos 2x + \cos^2 2x - \cos^3 2x) \, dx$$

$$= \frac{1}{8} \int \left[1 + \cos 2x - \frac{1 + \cos 4x}{2} - (1 - \sin^2 2x) \cos 2x \right] dx$$

$$= \frac{1}{8} \int \left(\frac{1}{2} - \frac{1}{2} \cos 4x + \sin^2 2x \cos 2x \right) dx$$

$$= \frac{1}{16} x - \frac{1}{64} \sin 4x + \frac{1}{48} \sin^3 2x + C$$

13. $\displaystyle\int \sin^4 x \cos^4 x \, dx = \frac{1}{16} \int \sin^4 2x \, dx = \frac{1}{16} \int \left(\frac{1 - \cos 4x}{2} \right)^2 dx$

$$= \frac{1}{64} \int (1 - 2\cos 4x + \cos^2 4x) \, dx = \frac{1}{64} \int \left(1 - 2\cos 4x + \frac{1 + \cos 8x}{2} \right) dx$$

$$= \frac{1}{64} \int \left(\frac{3}{2} - 2\cos 4x + \frac{1}{2} \cos 8x \right) dx$$

$$= \frac{3}{128} x - \frac{1}{128} \sin 4x + \frac{1}{1024} \sin 8x + C$$

15. $\int \tan^3 2t \sec^4 2t \, dt = \int \tan^3 2t \sec^2 2t \sec^2 2t \, dt = \int \tan^3 2t (1 + \tan^2 2t) \sec^2 2t \, dt$

$$= \frac{1}{2} \int \tan^3 2t (2 \sec^2 2t \, dt) + \frac{1}{2} \int \tan^5 2t (2 \sec^2 2t \, dt)$$

$$= \frac{1}{8} \tan^4 2t + \frac{1}{12} \tan^6 2t + C$$

17. $\int \tan^2 x \sec^3 x \, dx = \int (\sec^2 x - 1) \sec^3 x \, dx = \sec^5 x \, dx - \int \sec^3 x \, dx$

$$\boxed{u = \sec^3 x, \; du = 3 \sec^2 x \sec x \tan x \, dx; \quad dv = \sec^2 x \, dx, \; v = \tan x}$$

$$= \tan x \sec^3 x - 3 \int \tan^2 x \sec^3 x \, dx - \int \sec^3 x \, dx$$

From Example 5 of Section 7.3, $\int \sec^3 x \, dx = \frac{1}{2} \sec x \tan x + \frac{1}{2} \ln|\sec x + \tan x| + C$, so

$$\int \tan^2 x \sec^3 x \, dx = \tan x \sec^3 x - 3 \int \tan^2 x \sec^3 x \, dx - \frac{1}{2} \sec x \tan x - \frac{1}{2} \ln|\sec x + \tan x|.$$

Solving for the integral, we have

$$\int \tan^2 x \sec^3 x \, dx = \frac{1}{4} \tan x \sec^3 x - \frac{1}{8} \sec x \tan x - \frac{1}{8} \ln|\sec x + \tan x| + C.$$

19. $\int \tan^3 x (\sec x)^{-1/2} \, dx = \int \tan^2 x (\sec x)^{-1/2} \tan x \, dx$

$$= \int (\sec^2 x - 1)(\sec x)^{-3/2} \sec x \tan x \, dx$$

$$= \int (\sec x)^{1/2} (\sec x \tan x \, dx) - \int (\sec x)^{-3/2} (\sec x \tan x \, dx)$$

$$= \frac{2}{3} (\sec x)^{3/2} + 2(\sec x)^{-1/2} + C$$

21. $\int \tan^3 x \sec^5 x \, dx = \int \tan^2 x \sec^4 x \sec x \tan x \, dx = \int (\sec^2 x - 1) \sec^4 x \sec x \tan x \, dx$

$$= \int \sec^6 x (\sec x \tan x \, dx) - \int \sec^4 x (\sec x \tan x \, dx)$$

$$= \frac{1}{7} \sec^7 x - \frac{1}{5} \sec^5 x + C$$

23. $\int \sec^5 x \, dx = \int \sec^3 x \sec^2 x \, dx = \int \sec^3 x (1 + \tan^2 x) \, dx = \int \sec^3 x \, dx + \int \tan^2 x \sec^3 x \, dx$

From Example 5 of Section 7.3, $\int \sec^3 x \, dx = \frac{1}{2} \sec x \tan x + \frac{1}{2} \ln|\sec x + \tan x| + C$.

From Exercise 17, $\int \tan^2 x \sec^3 x \, dx = \frac{1}{4} \tan x \sec^3 x - \frac{1}{8} \sec x \tan x - \frac{1}{8} \ln|\sec x + \tan x| + C_1$.

Thus, $\int \sec^5 x = \left(\frac{1}{2} \sec x \tan x + \frac{1}{2} \ln|\sec x + \tan x| \right)$

$$+ \left(\frac{1}{4} \tan x \sec^3 x - \frac{1}{8} \sec x \tan x - \frac{1}{8} \ln|\sec x + \tan x| \right) + C_2$$

$$= \frac{3}{8} \sec x \tan x + \frac{3}{8} \ln|\sec x + \tan x| + \frac{1}{4} \tan x \sec^3 x + C_2.$$

25. $\int \cos^2 x \cot x \, dx = \int \frac{\cos^3 x}{\sin x} \, dx = \int \frac{(1 - \sin^2 x)}{\sin x} \cos x \, dx$

$$= \int (\sin x)^{-1} (\cos x \, dx) - \int \sin x (\cos x \, dx) = \ln|\sin x| - \frac{1}{2} \sin^2 x + C$$

(Alternatively, $\int \sin x \cos x \, dx = \int \cos x (\sin x \, dx) = -\frac{1}{2} \cos^2 x + C$ or

$$\int \sin x \cos x \, dx = \frac{1}{2} \int \sin 2x \, dx = -\frac{1}{4} \cos 2x + C).$$

27. $\displaystyle\int \cot^{10} x \csc^4 x\,dx = \int \cot^{10} x \csc^2 x \csc^2 x\,dx = \int \cot^{10} x(1+\cot^2 x)\csc^2 x\,dx$

$$= -\int \cot^{10} x(-\csc^2 x)\,dx - \int \cot^{12} x(-\csc^2 x)\,dx$$

$$= -\frac{1}{11}\cot^{11} x - \frac{1}{13}\cot^{13} x + C$$

29. $\displaystyle\int \frac{\sec^4(1-t)}{\tan^8(1-t)}\,dt = \int \frac{\sec^2(1-t)}{\tan^8(1-t)}\sec^2(1-t)\,dt = \int \frac{1+\tan^2(1-t)}{\tan^8(1-t)}\sec^2(1-t)\,dt$

$$= \int [\tan(1-t)]^{-8}\sec^2(1-t)\,dt + \int [\tan(1-t)]^{-6}\sec^2(1-t)\,dt$$

$$= \frac{1}{7}[\tan(1-t)]^{-7} + \frac{1}{5}[\tan(1-t)]^{-5} + C$$

$$= \frac{1}{7\tan^7(1-t)} + \frac{1}{5\tan^5(1-t)} + C$$

31. $\displaystyle\int (1+\tan x)^2 \sec x\,dx = \int (1+2\tan x + \tan^2 x)\sec x\,dx$

$$= \int \sec x\,dx + 2\int \tan x \sec x\,dx + \int \tan^2 x \sec x\,dx$$

The last integral is evaluated in Example 8 of Section 7.4. Thus,

$$\int (1+\tan x)^2 \sec x\,dx = \ln|\sec x + \tan x| + 2\sec x + \frac{1}{2}\sec x \tan x - \frac{1}{2}\ln|\sec x + \tan x| + C$$

$$= \frac{1}{2}\ln|\sec x + \tan x| + 2\sec x + \frac{1}{2}\sec x \tan x + C.$$

33. $\displaystyle\int \tan^4 x\,dx = \int \tan^2 x \tan^2 x\,dx = \int \tan^2 x(\sec^2 x - 1)\,dx$

$$= \int \tan^2 x \sec^2 x\,dx - \int (\sec^2 x - 1)\,dx = \frac{1}{3}\tan^3 x - \tan x + x + C$$

35. $\displaystyle\int \cot^3 t\,dt = \int \cot^2 t \cot t\,dt = \int (\csc^2 t - 1)\cot t\,dt = \int \csc t(\csc t \cot t\,dt) - \int \cot t\,dt$

$$= -\frac{1}{2}\csc^2 t - \ln|\sin t| + C$$

37. $\displaystyle\int (\tan^6 x - \tan^2 x)\,dx = \int (\tan^4 x \tan^2 x - \tan^2 x)\,dx = \int (\tan^4 x - 1)\tan^2 x\,dx$

$$= \int (\tan^4 x - 1)(\sec^2 x - 1)\,dx$$

$$= \int (\tan^4 x \sec^2 x - \tan^4 x - \sec^2 x + 1)\,dx$$

$$= \int \tan^4 x(\sec^2 x\,dx) - \int \tan^4 x\,dx - \int \sec^2 x\,dx + \int dx$$

$$= \frac{1}{5}\tan^5 x - \int \tan^4 x\,dx - \tan x + x$$

From Exercise 33, $\displaystyle\int \tan^4 x = \frac{1}{3}\tan^3 x - \tan x + x + C$, so

$$\int (\tan^6 x - \tan^2 x)\,dx = \frac{1}{5}\tan^5 x - \left(\frac{1}{3}\tan^3 x - \tan x + x\right) - \tan x + x + C_1$$

$$= \frac{1}{5}\tan^5 x - \frac{1}{3}\tan^3 x + C_1.$$

39. $\displaystyle\int x\sin^3 x^2\,dx = \int x\sin^2 x^2\sin x^2\,dx = \int x(1-\cos^2 x^2)\sin x^2\,dx$

$$= \int x\sin x^2\,dx - \int x\cos^2 x^2\sin x^2\,dx$$

$$= \frac{1}{2}\int \sin x^2(2x\,dx) + \frac{1}{2}\int \cos^2 x^2(-2x\sin x^2\,dx)$$

$$\boxed{t = x^2,\ dt = 2x\,dx;\quad u = \cos x^2,\ du = -2x\sin x^2\,dx}$$

$$= \frac{1}{2}\left(\int \sin t\,dt + \int u^2\,du\right) = \frac{1}{2}\left(-\cos t + \frac{1}{3}u^3\right) + C$$

$$= \frac{1}{2}\left(-\cos x^2 + \frac{1}{3}\cos^3 x^2\right) + C = \frac{1}{6}\cos^3 x^2 - \frac{1}{2}\cos x^2 + C$$

41. $\displaystyle\int_{\pi/3}^{\pi/2}\sin^3\theta\sqrt{\cos\theta}\,d\theta = \int_{\pi/3}^{\pi/2}\sin^2\theta(\cos\theta)^{1/2}\sin\theta\,d\theta = \int_{\pi/3}^{\pi/2}(1-\cos^2\theta)(\cos\theta)^{1/2}\sin\theta\,d\theta$

$$= -\int_{\pi/3}^{\pi/2}(\cos\theta)^{1/2}(-\sin\theta\,d\theta) + \int_{\pi/3}^{\pi/2}(\cos\theta)^{5/2}(-\sin\theta\,d\theta)$$

$$= -\frac{2}{3}(\cos\theta)^{3/2}\Big]_{\pi/3}^{\pi/2} + \frac{2}{7}(\cos\theta)^{7/2}\Big]_{\pi/3}^{\pi/2}$$

$$= -\frac{2}{3}\left[0 - \left(\frac{1}{2}\right)^{3/2}\right] + \frac{2}{7}\left[0 - \left(\frac{1}{2}\right)^{7/2}\right] = \frac{\sqrt{2}}{6} - \frac{\sqrt{2}}{56} = \frac{25\sqrt{2}}{168}$$

43. $\displaystyle\int_0^\pi \sin^3 2t\,dt = \int_0^\pi \sin^2 2t\sin 2t\,dt = \int_0^\pi(1-\cos^2 2t)\sin 2t\,dt$

$$= \int_0^\pi \sin 2t\,dt + \frac{1}{2}\int_0^\pi \cos^2 2t(-2\sin 2t\,dt) = -\frac{1}{2}\cos 2t\Big]_0^\pi + \frac{1}{6}\cos^3 2t\Big]_0^\pi$$

$$= -\frac{1}{2}(1-1) + \frac{1}{6}(1-1) = 0$$

45. $\displaystyle\int_0^{\pi/4}\tan y\sec^4 y\,dy = \int_0^{\pi/4}\tan y\sec^2 y\sec^2 y\,dy = \int_0^{\pi/4}\tan y(1+\tan^2 y)\sec^2 y\,dy$

$$= \int_0^{\pi/4}\tan y\sec^2 y\,dy + \int_0^{\pi/4}\tan^3 y\sec^2 y\,dy$$

$$= \frac{1}{2}\tan^2 y\Big]_0^{\pi/4} + \frac{1}{4}\tan^4 y\Big]_0^{\pi/4} = \frac{1}{2}(1-0) + \frac{1}{4}(1-0) = \frac{3}{4}$$

47. $\displaystyle\int \sin x\cos 2x\,dx = \frac{1}{2}\int[\sin 3x + \sin(-x)]\,dx = \frac{1}{2}\int(\sin 3x - \sin x)\,dx$

$$= -\frac{1}{6}\cos 3x + \frac{1}{2}\cos x + C$$

49. $\displaystyle\int \sin 2x\sin 4x\,dx = \frac{1}{2}\int(\cos 2x - \cos 6x)\,dx = \frac{1}{4}\sin 2x - \frac{1}{12}\sin 6x + C$

51. $\displaystyle\int_0^{\pi/6}\cos 2x\cos x\,dx = \frac{1}{2}\int_0^{\pi/6}(\cos x + \cos 3x)\,dx = \frac{1}{2}\left(\sin x + \frac{1}{3}\sin 3x\right)\Big]_0^{\pi/6}$

$$= \frac{1}{2}\left(\frac{1}{2} + \frac{1}{3}\right) = \frac{5}{12}$$

53. If $m \neq n$, then using the fact that $\sin mx\sin nx$ is an even function we have

$$\int_{-\pi}^{\pi}\sin mx\sin nx\,dx = \int_0^\pi[\cos(m-n)x - \cos(m+n)x]\,dx$$

$$= \frac{1}{m-n}\sin(m-n)x\Big]_0^\pi - \frac{1}{m+n}\sin(m+n)x\Big]_0^\pi = 0.$$

If $m = n$, then

$$\int_{-\pi}^{\pi} \sin mx \sin nx\, dx = \int_{-\pi}^{\pi} \sin^2 mx\, dx = 2\int_0^{\pi} \sin^2 mx\, dx = \int_0^{\pi} (1 + \cos 2mx)\, dx$$

$$= \left(x + \frac{1}{2m}\sin 2mx\right)\Bigg]_0^{\pi} = \pi.$$

Thus, $\displaystyle\int_{-\pi}^{\pi} \sin mx \sin nx\, dx = \begin{cases} 0, & m \neq n \\ \pi, & m = n \end{cases}$.

55. $y = \sec^2 \dfrac{x}{2}$

$$V = \pi \int_{-\pi/2}^{\pi/2} \sec^4 \frac{x}{2}\, dx = \pi \int_{-\pi/2}^{\pi/2} \sec^2 \frac{x}{2} \sec^2 \frac{x}{2}\, dx$$

$$= \pi \int_{-\pi/2}^{\pi/2} \left(1 + \tan^2 \frac{x}{2}\right) \sec^2 \frac{x}{2}\, dx$$

$$= 2\pi \left[\int_{-\pi/2}^{\pi/2} \left(\sec^2 \frac{x}{2}\right)\left(\frac{1}{2}\, dx\right) + \int_{-\pi/2}^{\pi/2} \left(\tan^2 \frac{x}{2}\right)\left(\frac{1}{2}\sec^2 \frac{x}{2}\, dx\right)\right]$$

$$= 2\pi \left(\tan \frac{x}{2}\right]_{-\pi/2}^{\pi/2} + \frac{1}{3}\tan^3 \frac{x}{2}\Bigg]_{-\pi/2}^{\pi/2}\right) = 2\pi \left[1 - (-1) + \frac{1}{3} - \frac{1}{3}(-1)\right]$$

$$= 2\pi \left(2 + \frac{2}{3}\right) = \frac{16\pi}{3}$$

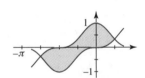

57. $A = \displaystyle\int_{-3\pi/4}^{\pi/4} (\cos^3 x - \sin^3 x)\, dx$

$$= \int_{-3\pi/4}^{\pi/4} (\cos^2 x \cos x - \sin^2 x \sin x)\, dx = \int_{-3\pi/4}^{\pi/4} [(1 - \sin^2 x)\cos x - (1 - \cos^2 x)\sin x]\, dx$$

$$= \int_{-3\pi/4}^{\pi/4} [\cos x - (\sin^2 x)\cos x - \sin x + (\cos^2 x)\sin x]\, dx$$

$$= \left(\sin x - \frac{1}{3}\sin^3 x + \cos x - \frac{1}{3}\cos^3 x\right)\Bigg]_{-3\pi/4}^{\pi/4}$$

$$= \left[\left(\frac{\sqrt{2}}{2} - \frac{\sqrt{2}}{12} + \frac{\sqrt{2}}{2} - \frac{\sqrt{2}}{12}\right) - \left(-\frac{\sqrt{2}}{2} + \frac{\sqrt{2}}{12} - \frac{\sqrt{2}}{2} + \frac{\sqrt{2}}{12}\right)\right] = 2\sqrt{2} - \frac{\sqrt{2}}{3} = \frac{5\sqrt{2}}{3}$$

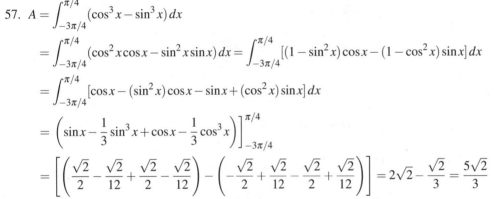

59. Based on the graphs, the values of $\displaystyle\int_0^{\pi} \cos^3 x\, dx$, $\displaystyle\int_0^{\pi} \cos^5 x\, dx$, and $\displaystyle\int_0^{\pi} \cos^7 x\, dx$ all appear to be 0. We note that, for every t such that $0 \leq t \leq \dfrac{\pi}{2}$, $\cos\left(\dfrac{\pi}{2} - t\right) = -\cos\left(\dfrac{\pi}{2} + t\right)$, thus lending credence to this conjecture.

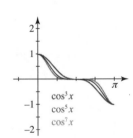

7.5   Trigonometric Substitutions

1. $\displaystyle\int \frac{\sqrt{1 - x^2}}{x^2}\, dx$ $\boxed{x = \sin\theta,\ dx = \cos\theta\, d\theta}$

$$= \int \frac{\sqrt{1 - \sin^2\theta}}{\sin^2\theta}\cos\theta\, d\theta = \int \frac{\cos^2\theta}{\sin^2\theta}\, d\theta = \int \cot^2\theta\, d\theta$$

$$= \int (\csc^2\theta - 1)\, d\theta = -\cot\theta - \theta + C = -\frac{\sqrt{1 - x^2}}{x} - \sin^{-1}x + C$$

3. $\displaystyle\int \frac{1}{\sqrt{x^2 - 36}}\, dx = \cosh^{-1}\frac{x}{6} + C = \ln\left(x + \sqrt{x^2 - 36^2}\right) + C, \quad x > 6$

Alternatively, the substitution $x = 6\sec\theta$ could have been used.

5. $\int x\sqrt{x^2+7}\,dx = \frac{1}{2}\int (x^2+7)^{1/2}(2x\,dx) = \frac{1}{3}(x^2+7)^{3/2}+C$

7. $\int x^3\sqrt{1-x^2}\,dx = -\frac{1}{2}\int x^2\sqrt{1-x^2}(-2x\,dx)$ $\boxed{u=1-x^2,\ x^2=1-u,\ 2x\,dx=-du}$

$\qquad = -\frac{1}{2}\int(1-u)u^{1/2}\,du = -\frac{1}{2}\int(u^{1/2}-u^{3/2})\,du$

$\qquad = -\frac{1}{2}\left(\frac{2}{3}u^{3/2}-\frac{2}{5}u^{5/2}\right)+C = -\frac{1}{3}(1-x^2)^{3/2}+\frac{1}{5}(1-x^2)^{5/2}+C$

9. $\int \frac{1}{(x^2-4)^{3/2}}\,dx$ $\boxed{x=2\sec\theta,\ dx=2\sec\theta\tan\theta\,d\theta}$

$\qquad = \int \frac{2\sec\theta\tan\theta}{(4\sec^2\theta-4)^{3/2}}\,d\theta$

$\qquad = \frac{1}{4}\int \frac{\sec\theta}{\tan^2\theta}\,d\theta = \frac{1}{4}\int(\sin\theta)^{-2}\cos\theta\,d\theta$

$\qquad = -\frac{1}{4}(\sin\theta)^{-1}+C = -\frac{1}{4}\csc\theta+C = -\frac{1}{4}\frac{x}{\sqrt{x^2-4}}+C$

11. $\int \sqrt{x^2+4}\,dx$ $\boxed{x=2\tan\theta,\ dx=2\sec^2\theta\,d\theta}$

$\qquad = \int \sqrt{4\tan^2\theta+4}\,2\sec^2\theta\,d\theta = 4\int \sec^3\theta\,d\theta$

$\qquad \boxed{\text{See Section 7.3, Example 5}}$

$\qquad = 2\sec\theta\tan\theta + 2\ln|\sec\theta+\tan\theta|+C$

$\qquad = 2\frac{\sqrt{x^2+4}}{2}\left(\frac{x}{2}\right) + 2\ln\left|\frac{\sqrt{x^2+4}}{2}+\frac{x}{2}\right|+C$

$\qquad = \frac{x}{2}\sqrt{x^2+4} + 2\ln\left|\sqrt{x^2+4}+x\right|+C_1$

13. $\int \frac{1}{\sqrt{25-x^2}}\,dx = \sin^{-1}\frac{x}{5}+C$

15. $\int \frac{1}{x\sqrt{16-x^2}}\,dx$ $\boxed{x=4\sin\theta,\ dx=4\cos\theta\,d\theta}$

$\qquad = \int \frac{4\cos\theta}{4\sin\theta\sqrt{16-16\sin^2\theta}}\,d\theta = \frac{1}{4}\int \csc\theta\,d\theta$

$\qquad = \frac{1}{4}\ln|\csc\theta-\cot\theta|+C = \frac{1}{4}\ln\left|\frac{4}{x}-\frac{16-x^2}{x}\right|+C$

17. $\int \frac{1}{x\sqrt{1+x^2}}\,dx$ $\boxed{x=\tan\theta,\ dx=\sec^2\theta\,d\theta}$

$\qquad = \int \frac{\sec^2\theta}{\tan\theta\sqrt{1+\tan^2\theta}}\,d\theta = \int \frac{\sec\theta}{\tan\theta}\,d\theta = \int \csc\theta\,d\theta$

$\qquad = \ln|\csc\theta-\cot\theta|+C = \ln\left|\frac{\sqrt{1+x^2}}{x}-\frac{1}{x}\right|+C$

19. $\int \frac{\sqrt{1-x^2}}{x^4}\,dx$ $\boxed{x=\sin\theta,\ dx=\cos\theta\,d\theta}$

$\qquad = \int \frac{\sqrt{1-\sin^2\theta}}{\sin^4\theta}\cos\theta\,d\theta = \int \frac{\cos^2\theta}{\sin^4\theta}\,d\theta = \int \cot^2\theta\csc^2\theta\,d\theta$

$\qquad = -\frac{1}{3}\cot^3\theta+C = -\frac{1}{3}\left(\frac{\sqrt{1-x^2}}{x}\right)^3+C = -\frac{1}{3x^3}(1-x^2)^{3/2}+C$

21. $\int \dfrac{x^2}{(9-x^2)^{3/2}}\,dx$ $\boxed{x=3\sin\theta,\ dx=3\cos\theta\,d\theta}$

$$=\int \frac{9\sin^2\theta}{(9-9\sin^2\theta)^{3/2}}3\cos\theta\,d\theta=\int\frac{\sin^2\theta}{\cos^2\theta}\,d\theta=\int\tan^2\theta\,d\theta$$

$$=\int(\sec^2\theta-1)\,d\theta=\tan\theta-\theta+C=\frac{x}{\sqrt{9-x^2}}-\sin^{-1}\frac{x}{3}+C$$

23. $\int \dfrac{1}{(1+x^2)^2}\,dx$ $\boxed{x=\tan\theta,\ dx=\sec^2\theta\,d\theta}$

$$=\int\frac{\sec^2\theta}{(1+\tan^2\theta)^2}\,d\theta=\int\frac{1}{\sec^2\theta}\,d\theta=\int\cos^2\theta\,d\theta$$

$$=\frac{1}{2}\int(1+\cos2\theta)\,d\theta=\frac{1}{2}\theta+\frac{1}{4}\sin2\theta+C=\frac{1}{2}\theta+\frac{1}{2}\sin\theta\cos\theta+C$$

$$=\frac{1}{2}\tan^{-1}x+\frac{1}{2}\left(\frac{x}{\sqrt{1+x^2}}\right)\frac{1}{\sqrt{1+x^2}}+C=\frac{1}{2}\tan^{-1}x+\frac{1}{2}\left(\frac{x}{1+x^2}\right)+C$$

25. $\int \dfrac{1}{(4+x^2)^{5/2}}\,dx$ $\boxed{x=2\tan\theta,\ dx=2\sec^2\theta\,d\theta}$

$$=\int\frac{2\sec^2\theta}{(4+4\tan^2\theta)^{5/2}}\,d\theta=\frac{1}{16}\int\frac{1}{\sec^3\theta}\,d\theta=\frac{1}{16}\int\cos^3\theta\,d\theta$$

$$=\frac{1}{16}\int\cos^2\theta\cos\theta\,d\theta=\frac{1}{16}\int(1-\sin^2\theta)\cos\theta\,d\theta$$

$$=\frac{1}{16}\int\cos\theta-\frac{1}{16}\int\sin^2\theta\cos\theta\,d\theta=\frac{1}{16}\sin\theta-\frac{1}{48}\sin^3\theta+C$$

$$=\frac{1}{16}\left(\frac{x}{\sqrt{4+x^2}}\right)-\frac{1}{48}\left[\frac{x^3}{(4+x^2)^{3/2}}\right]+C$$

27. $\int \dfrac{1}{\sqrt{x^2+2x+10}}\,dx=\int\dfrac{1}{\sqrt{(x+1)^2+9}}\,dx$

$$\boxed{x+1=3\tan\theta,\ dx=3\sec^2\theta\,d\theta}$$

$$=\int\frac{3\sec^2\theta}{\sqrt{9\tan^2\theta+9}}\,d\theta=\int\sec\theta\,d\theta=\ln|\sec\theta+\tan\theta|+C$$

$$=\ln\left|\frac{\sqrt{x^2+2x+10}}{3}+\frac{x+1}{3}\right|+C=\ln\left|\sqrt{x^2+2x+10}+x+1\right|+C_1$$

29. $\int \dfrac{1}{(x^2+6x+13)^2}\,dx=\int\dfrac{1}{[(x+3)^2+4]^2}\,dx$

$$\boxed{x+3=2\tan\theta,\ dx=2\sec^2\theta\,d\theta}$$

$$=\int\frac{2\sec^2\theta}{(4\tan^2\theta+4)^2}\,d\theta=\int\frac{\sec^2\theta}{8\sec^4\theta}\,d\theta=\frac{1}{8}\int\cos^2\theta\,d\theta$$

$$=\frac{1}{16}\int(1+\cos2\theta)\,d\theta=\frac{1}{16}\left(\theta+\frac{1}{2}\sin2\theta\right)+C$$

$$=\frac{1}{16}\theta+\frac{1}{16}\sin\theta\cos\theta+C$$

$$=\frac{1}{16}\tan^{-1}\frac{x+3}{2}+\frac{1}{16}\left(\frac{x+3}{\sqrt{x^2+6x+13}}\right)\frac{2}{\sqrt{x^2+6x+13}}+C$$

$$=\frac{1}{16}\tan^{-1}\frac{x+3}{2}+\frac{x+3}{8(x^2+6x+13)}+C$$

31.
$$\int \frac{x-3}{(5-4x-x^2)^{3/2}}\,dx = \int \frac{x-3}{[9-(x+2)^2]^{3/2}}\,dx$$

$$\boxed{x+2=3\sin\theta,\ dx=3\cos\theta\,d\theta}$$

$$= \int \frac{3\sin\theta-5}{(9-9\sin^2\theta)^{3/2}}(3\cos\theta\,d\theta)$$

$$= \int \frac{9\sin\theta\cos\theta-15\cos\theta}{27\cos^3\theta}\,d\theta$$

$$= \frac{1}{3}\int \frac{\sin\theta}{\cos^2\theta}\,d\theta - \frac{5}{9}\int \frac{1}{\cos^2\theta}\,d\theta = \frac{1}{3}\int \tan\theta\sec\theta\,d\theta - \frac{5}{9}\int \sec^2\theta\,d\theta$$

$$= \frac{1}{3}\sec\theta - \frac{5}{9}\tan\theta + C$$

$$= \frac{1}{3}\left(\frac{3}{\sqrt{5-4x-x^2}}\right) - \frac{5}{9}\left(\frac{x+2}{\sqrt{5-4x-x^2}}\right) + C$$

$$= \frac{-5x-1}{9\sqrt{5-4x-x^2}} + C$$

33.
$$\int \frac{2x+4}{x^2+4x+13}\,dx = \ln(x^2+4x+13) + C$$

35.
$$\int \frac{x^2}{x^2+16}\,dx = \int \left(1 - \frac{16}{x^2+16}\right)dx = x - 4\tan^{-1}\frac{x}{4} + C$$

37.
$$\int \sqrt{6x-x^2}\,dx = \int \sqrt{9-(x-3)^2}\,dx \quad \boxed{x-3=3\sin\theta,\ dx=3\cos\theta\,d\theta}$$

$$= \int \sqrt{9-9\sin^2\theta}\,3\cos\theta\,d\theta = 9\int \cos^2\theta\,d\theta$$

$$= \frac{9}{2}\int (1+\cos2\theta)\,d\theta = \frac{9}{2}\left(\theta + \frac{1}{2}\sin2\theta\right) + C$$

$$= \frac{9}{2}\theta + \frac{9}{2}\sin\theta\cos\theta + C = \frac{9}{2}\sin^{-1}\frac{x-3}{3} + \frac{9}{2}\left(\frac{x-3}{3}\right)\frac{\sqrt{6x-x^2}}{3} + C$$

$$= \frac{9}{2}\sin^{-1}\frac{x-3}{3} + \frac{1}{2}(x-3)\sqrt{6x-x^2} + C$$

39.
$$\int_{-1}^{1}\sqrt{4-x^2}\,dx \quad \boxed{x=2\sin\theta,\ dx=2\cos\theta\,d\theta}$$

$$= \int_{-\pi/6}^{\pi/6}\sqrt{4-4\sin^2\theta}\,2\cos\theta\,d\theta = 4\int_{-\pi/6}^{\pi/6}\cos^2\theta\,d\theta$$

$$= \int_{-\pi/6}^{\pi/6}(2+2\cos2\theta)\,d\theta = (2\theta+\sin2\theta)\Big]_{-\pi/6}^{\pi/6}$$

$$= \left[\left(\frac{\pi}{3}+\frac{\sqrt{3}}{2}\right) - \left(-\frac{\pi}{3}-\frac{\sqrt{3}}{2}\right)\right] = \frac{2\pi+3\sqrt{3}}{3}$$

41.
$$\int_{0}^{5}\frac{1}{(x^2+25)^{3/2}}\,dx \quad \boxed{x=5\tan\theta,\ dx=5\sec^2\theta\,d\theta}$$

$$= \int_{0}^{\pi/4}\frac{5\sec^2\theta}{(25\tan^2\theta+25)^{3/2}}\,d\theta = \frac{1}{25}\int_{0}^{\pi/4}\frac{1}{\sec\theta}\,d\theta = \frac{1}{25}\int_{0}^{\pi/4}\cos\theta\,d\theta$$

$$= \frac{1}{25}\sin\theta\Big]_{0}^{\pi/4} = \frac{\sqrt{2}}{50}$$

43. $\displaystyle\int_{1}^{6/5} \frac{16}{x^4\sqrt{4-x^2}}\,dx$ $\boxed{x = 2\sin\theta,\ dx = 2\cos\theta\,d\theta}$

$$= \int_{\pi/6}^{\sin^{-1}(3/5)} \frac{32\cos\theta}{16\sin^4\theta\sqrt{4-4\sin^2\theta}}\,d\theta = \int_{\pi/6}^{\sin^{-1}(3/5)} \frac{2\cos\theta}{\sin^4(2\cos\theta)}\,d\theta$$

$$= \int_{\pi/6}^{\sin^{-1}(3/5)} \csc^2\theta\csc^2\theta\,d\theta = \int_{\pi/6}^{\sin^{-1}(3/5)} (1+\cot^2\theta)\csc^2\theta\,d\theta$$

$$= \int_{\pi/6}^{\sin^{-1}(3/5)} \csc^2\theta\,d\theta + \int_{\pi/6}^{\sin^{-1}(3/5)} \cot^2\theta\csc^2\theta\,d\theta$$

$$= -\cot\theta\Big]_{\pi/6}^{\sin^{-1}(3/5)} - \frac{1}{3}\cot^3\theta\Big]_{\pi/6}^{\sin^{-1}(3/5)}$$

$$= -[\cot(\sin^{-1}3/5) - \sqrt{3}] - \frac{1}{3}[\cot^3(\sin^{-1}3/5) - 3\sqrt{3}]$$

$$= -\left(\frac{4}{3} - \sqrt{3}\right) - \frac{1}{3}\left(\frac{64}{27} - 3\sqrt{3}\right) = 2\sqrt{3} - \frac{172}{81}$$

45. $\displaystyle\int x^2\sin^{-1}x\,dx$ $\boxed{u = \sin^{-1}x,\ du = \dfrac{1}{\sqrt{1-x^2}}\,dx;\quad dv = x^2\,dx,\ v = \dfrac{1}{3}x^3}$

$$= \frac{1}{3}x^3\sin^{-1}x - \frac{1}{3}\int \frac{x^3}{\sqrt{1-x^2}}\,dx \quad \boxed{x = \sin\theta,\ dx = \cos\theta\,d\theta}$$

$$= \frac{1}{3}x^3\sin^{-1}x - \frac{1}{3}\int \frac{\sin^3\theta}{\sqrt{1-\sin^2\theta}}\cos\theta\,d\theta = \frac{1}{3}x^3\sin^{-1}x - \frac{1}{3}\int \sin^3\theta\,d\theta$$

$$= \frac{1}{3}x^3\sin^{-1}x - \frac{1}{3}\int (1-\cos^2\theta)\sin\theta\,d\theta$$

$$= \frac{1}{3}x^3\sin^{-1}x - \frac{1}{3}\left(-\cos\theta + \frac{1}{3}\cos^3\theta\right) + C$$

$$= \frac{1}{3}x^3\sin^{-1}x + \frac{1}{3}\sqrt{1-x^2} - \frac{1}{9}(1-x^2)^{3/2} + C$$

47. $\displaystyle A = \int_{1}^{\sqrt{3}} \frac{1}{x\sqrt{3+x^2}}\,dx$ $\boxed{x = \sqrt{3}\tan\theta,\ dx = \sqrt{3}\sec^2\theta\,d\theta}$

$$= \int_{\pi/6}^{\pi/4} \frac{\sqrt{3}\sec^2\theta}{\sqrt{3}\tan\theta\sqrt{3+3\tan^2\theta}}\,d\theta$$

$$= \frac{1}{\sqrt{3}}\int_{\pi/6}^{\pi/4} \frac{\sec\theta}{\tan\theta}\,d\theta = \frac{1}{\sqrt{3}}\int_{\pi/6}^{\pi/4} \csc\theta\,d\theta$$

$$= \frac{1}{\sqrt{3}}\ln|\csc\theta - \cot\theta|\Big]_{\pi/6}^{\pi/4} = \frac{1}{\sqrt{3}}(\ln|\sqrt{2}-1| - \ln|2-\sqrt{3}|) = \frac{1}{\sqrt{3}}\ln\frac{\sqrt{2}-1}{2-\sqrt{3}} \approx 0.2515$$

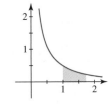

49. We find the area in the first quadrant and use symmetry.

$$A = 4\int_{0}^{a} \sqrt{a^2-x^2}\,dx \quad \boxed{x = a\sin\theta,\ dx = a\cos\theta\,d\theta}$$

$$= 4\int_{0}^{\pi/2} \sqrt{a^2-a^2\sin^2\theta}\,a\cos\theta\,d\theta = 4a^2\int_{0}^{\pi/2} \cos^2\theta\,d\theta$$

$$= 2a^2\int_{0}^{\pi/2} (1+\cos 2\theta)\,d\theta = 2a^2\left(\theta + \frac{1}{2}\sin 2\theta\right)\Big]_{0}^{\pi/2} = 2a^2\left(\frac{\pi}{2}\right) = \pi a^2$$

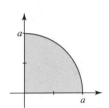

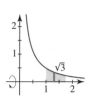

51. $V = \pi \int_0^{\sqrt{3}} \dfrac{1}{x^2(3+x^2)}\,dx$ $\boxed{x = \sqrt{3}\tan\theta,\ dx = \sqrt{3}\sec^2\theta\,d\theta}$

$$= \pi \int_{\pi/6}^{\pi/4} \frac{\sqrt{3}\sec^2\theta}{3\tan^2\theta(3+3\tan^2\theta)}\,d\theta = \frac{\pi\sqrt{3}}{9}\int_{\pi/6}^{\pi/4} \frac{1}{\tan^2\theta}\,d\theta$$

$$= \frac{\pi\sqrt{3}}{9}\int_{\pi/6}^{\pi/4}\cot^2\theta\,d\theta = \frac{\pi\sqrt{3}}{9}\int_{\pi/6}^{\pi/4}(\csc^2\theta - 1)\,d\theta = \frac{\pi\sqrt{3}}{9}(-\cot\theta - \theta)\Big]_{\pi/6}^{\pi/4}$$

$$= -\frac{\pi\sqrt{3}}{9}\left[\left(1+\frac{\pi}{4}\right) - \left(\sqrt{3}+\frac{\pi}{6}\right)\right] = \frac{\pi\sqrt{3}}{9}\left(\sqrt{3}-1-\frac{\pi}{12}\right) \approx 0.2843$$

53. Using the shell method,

$$V = 2\pi \int_0^2 x^2\sqrt{4+x^2}\,dx \quad \boxed{x = 2\tan\theta,\ dx = 2\sec^2\theta\,d\theta}$$

$$= 2\pi \int_0^{\pi/4} 4\tan^2\theta\sqrt{4+4\tan^2\theta}\,2\sec^2\theta\,d\theta = 32\pi\int_0^{\pi/4}\tan^2\theta\sec^3\theta\,d\theta$$

$$= 32\pi\int_0^{\pi/4}(\sec^2\theta - 1)\sec^3\theta\,d\theta = 32\pi\int_0^{\pi/4}\sec^5\theta\,d\theta - 32\pi\int_0^{\pi/4}\sec^3\theta\,d\theta.$$

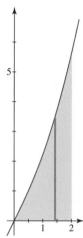

From Section 7.3, Example 5 we obtain

$$\int_0^{\pi/4}\sec^3\theta\,d\theta = \left(\frac{1}{2}\sec\theta\tan\theta + \frac{1}{2}\ln|\sec\theta + \tan\theta|\right)\Big]_0^{\pi/4}$$

$$= \frac{1}{2}(\sqrt{2})(1) + \frac{1}{2}\ln(\sqrt{2}+1)$$

$$= \frac{1}{2}[\sqrt{2} + \ln(\sqrt{2}+1)].$$

To find $\displaystyle\int_0^{\pi/4}\sec^5\theta\,d\theta$ we use integration by parts.

$$\int_0^{\pi/4}\sec^5\theta\,d\theta = \int_0^{\pi/4}\sec^3\theta\sec^2\theta\,d\theta$$

$$\boxed{u = \sec^3\theta,\ du = 3\sec^2\theta\sec\theta\tan\theta\,d\theta;\quad dv = \sec^2\theta\,d\theta,\ v = \tan\theta}$$

$$= \sec^3\theta\tan\theta\Big]_0^{\pi/4} - 3\int_0^{\pi/4}\sec^3\theta\tan^2\theta\,d\theta$$

$$= 2\sqrt{2} - 3\int_0^{\pi/4}\sec^3\theta(\sec^2\theta - 1)\,d\theta$$

$$= 2\sqrt{2} - 3\int_0^{\pi/4}\sec^5\theta\,d\theta + 3\int_0^{\pi/4}\sec^3\theta\,d\theta$$

$$= 2\sqrt{2} - 3\int_0^{\pi/4}\sec^5\theta\,d\theta + \frac{3}{2}[\sqrt{2} + \ln(\sqrt{2}+1)]$$

Solving for $\displaystyle\int_0^{\pi/4}\sec^5\theta\,d\theta$ we obtain $\displaystyle\int_0^{\pi/4}\sec^5\theta\,d\theta = \frac{\sqrt{2}}{2} + \frac{3}{8}[\sqrt{2} + \ln(\sqrt{2}+1)]$. Then

$$V = 32\pi\left\{\frac{\sqrt{2}}{2} + \frac{3}{8}[\sqrt{2} + \ln(\sqrt{2}+1)]\right\} - 32\pi\left[\frac{\sqrt{2}}{2} + \frac{1}{2}\ln(\sqrt{2}+1)\right]$$

$$= 12\pi\sqrt{2} - 4\pi\ln(\sqrt{2}+1).$$

55. $y' = 1/x$

$$L = \int_1^{\sqrt{3}} \sqrt{1 + (1/x)^2}\, dx = \int_1^{\sqrt{3}} \frac{\sqrt{x^2+1}}{x}\, dx \qquad \boxed{x = \tan\theta,\ dx = \sec^2\theta\, d\theta}$$

$$= \int_{\pi/4}^{\pi/3} \frac{\sqrt{\tan^2\theta + 1}}{\tan\theta}\sec^2\theta\, d\theta = \int_{\pi/4}^{\pi/3} \frac{\sec^3\theta}{\tan\theta}\, d\theta = \int_{\pi/4}^{\pi/3} \frac{\sec\theta}{\tan\theta}\sec^2\theta\, d\theta$$

$$= \int_{\pi/4}^{\pi/3} \csc\theta(\tan^2\theta + 1)\, d\theta = \int_{\pi/4}^{\pi/3}\left(\frac{\sin\theta}{\cos^2\theta} + \csc\theta\right) d\theta = \int_{\pi/4}^{\pi/3}(\sec\theta\tan\theta + \csc\theta)\, d\theta$$

$$= (\sec\theta + \ln|\csc\theta - \cot\theta|)\Big]_{\pi/4}^{\pi/3} = \left(2 + \ln\left|\frac{2}{\sqrt{3}} - \frac{1}{\sqrt{3}}\right|\right) - (\sqrt{2} + \ln|\sqrt{2} - 1|)$$

$$= 2 - \sqrt{2} - \ln(\sqrt{6} - \sqrt{3}) \approx 0.9179$$

57. **(a)** The slope at (x,y) is $-\dfrac{\sqrt{a^2 - x^2}}{x}$, which is also $\dfrac{dy}{dx}$.

(b) Separating variables, $\displaystyle\int \frac{\sqrt{a^2-x^2}}{x}\, dx = -\int dy$. Now

$$\int \frac{\sqrt{a^2-x^2}}{x}\, dx \qquad \boxed{x = a\sin\theta,\ dx = a\cos\theta\, d\theta}$$

$$= \int \frac{\sqrt{a^2 - a^2\sin\theta}}{a\sin\theta}\, a\cos\theta\, d\theta = a\int \frac{\cos^2\theta}{\sin\theta}\, d\theta = a\int \frac{1 - \sin^2\theta}{\sin\theta}\, d\theta$$

$$= a\int \csc\theta\, d\theta - a\int \sin\theta\, d\theta = a\ln|\csc\theta - \cot\theta| + a\cos\theta + C$$

$$= a\ln\left|\frac{a}{x} - \frac{\sqrt{a^2-x^2}}{x}\right| + a\left(\frac{\sqrt{a^2-x^2}}{a}\right) + C.$$

Then $a\ln\left|\dfrac{a - \sqrt{a^2-x^2}}{x}\right| + \sqrt{a^2-x^2} = -y + C_1$. Now $y(10) = 0$ and $a = 10$, so $10\ln\left|\dfrac{10 - \sqrt{100 - 100}}{10}\right| + \sqrt{100 - 100} = 0 + C_1$ and $C_1 = 0$. Thus

$$y = -10\ln\left|\frac{10 - \sqrt{100 - x^2}}{x}\right| - \sqrt{100 - x^2}.$$

Note: If the substitution $y = a\cos\theta$ is used, we obtain the equivalent solution

$$y = 10\ln\left|\frac{10 - \sqrt{100 - x^2}}{x}\right| - \sqrt{100 - x^2}.$$

59. $F = 62.4\displaystyle\int_1^2 x\sqrt{\frac{2-x}{x}}\, dx = 62.4\int_1^2 \sqrt{2x - x^2}\, dx = 62.4\int_1^2 \sqrt{1 - (x-1)^2}\, dx$

$$\boxed{x - 1 = \sin\theta,\ dx = \cos\theta\, d\theta}$$

$$= 62.4\int_0^{\pi/2} \sqrt{1 - \sin^2\theta}\,\cos\theta\, d\theta = 62.4\int_0^{\pi/2}\cos^2\theta\, d\theta = 31.2\int_0^{\pi/2}(1 + \cos 2\theta)\, d\theta$$

$$= 31.2\left(\theta + \frac{1}{2}\sin 2\theta\right)\Big]_0^{\pi/2} = 31.2\left(\frac{\pi}{2}\right) = 15.6\pi \approx 49.0088\ \text{lb}$$

61. **(a)** $\displaystyle\int \frac{1}{\sqrt{e^{2x} - 1}}\, dx \qquad \boxed{e^x = \sec\theta,\ e^x\, dx = \tan\theta\sec\theta\, d\theta,\ dx = \tan\theta\, d\theta}$

$$= \int \frac{\tan\theta}{\sqrt{\sec^2\theta - 1}}\, d\theta = \int \frac{\tan\theta}{\tan\theta}\, d\theta = \int d\theta = \theta + C = \sec^{-1} e^x + C$$

(b) $\displaystyle\int \sqrt{e^{2x}-1}\,dx$ $\boxed{e^x = \sec\theta,\ e^x\,dx = \tan\theta\sec\theta\,d\theta,\ dx = \tan\theta\,d\theta}$

$$= \int \sqrt{\sec^2\theta - 1}\,\tan\theta\,d\theta = \int \tan^2\theta\,d\theta$$

$$= \int (\sec^2\theta - 1)\,d\theta = \tan\theta - \theta + C = \sqrt{e^{2x}-1} - \sec^{-1}e^x + C$$

7.6 Partial Fractions

1. Write $\dfrac{x-1}{x^2+x} = \dfrac{x-1}{x(x+1)} = \dfrac{A}{x} + \dfrac{B}{x+1}$.

3. Write $\dfrac{x^3}{(x-1)(x+2)^3} = \dfrac{A}{x-1} + \dfrac{B}{x+2} + \dfrac{C}{(x+2)^2} + \dfrac{D}{(x+2)^3}$.

5. Write $\dfrac{4}{x^3(x^2+3)} = \dfrac{A}{x} + \dfrac{B}{x^2} + \dfrac{C}{x^3} + \dfrac{Dx+E}{x^2+3}$.

7. Write $\dfrac{2x^3-x}{(x^2+9)^2} = \dfrac{Ax+B}{x^2+9} + \dfrac{Cx+D}{(x^2+9)^2}$.

9. Write $\dfrac{1}{x(x-2)} = \dfrac{A}{x} + \dfrac{B}{x-2}$. Then $1 = A(x-2) + Bx$.

Setting $x = 0$ and $x = 2$ gives $A = -1/2$ and $B = 1/2$. Thus

$$\int \frac{1}{x(x-2)}\,dx = -\frac{1}{2}\int \frac{1}{x}\,dx + \frac{1}{2}\int \frac{1}{x-2}\,dx$$

$$= -\frac{1}{2}\ln|x| + \frac{1}{2}\ln|x-2| + C = \frac{1}{2}\ln\left|\frac{x-2}{x}\right| + C.$$

11. Write $\dfrac{x+2}{2x^2-x} = \dfrac{x+2}{x(2x-1)} = \dfrac{A}{x} + \dfrac{B}{2x-1}$. Then $x+2 = A(2x-1) + Bx$.

Setting $x = 0$ and $x = 1/2$ gives $A = -2$ and $B = 5$. Thus

$$\int \frac{x+2}{2x^2-x}\,dx = -2\int \frac{1}{x}\,dx + 5\int \frac{1}{2x-1}\,dx = -2\ln|x| + \frac{5}{2}\ln|2x-1| + C.$$

13. Write $\dfrac{x+1}{x^2-16} = \dfrac{A}{x+4} + \dfrac{B}{x-4}$. Then $x+1 = A(x-4) + B(x+4)$.

Setting $x = -4$ and $x = 4$ gives $A = 3/8$ and $B = 5/8$. Thus

$$\int \frac{x+1}{x^2-16}\,dx = \frac{3}{8}\int \frac{1}{x+4}\,dx + \frac{5}{8}\int \frac{1}{x-4}\,dx = \frac{3}{8}\ln|x+4| + \frac{5}{8}\ln|x-4| + C.$$

15. Write $\dfrac{x}{2x^2+5x+2} = \dfrac{A}{2x+1} + \dfrac{B}{x+2}$.

Then $x = A(x+2) + B(2x+1)$.

Setting $x = -1/2$ and $x = -2$ gives $A = -1/3$ and $B = 2/3$. Thus

$$\int \frac{x}{2x^2+5x+2}\,dx = -\frac{1}{3}\int \frac{1}{2x+1}\,dx + \frac{2}{3}\int \frac{1}{x+2}\,dx = -\frac{1}{6}\ln|2x+1| + \frac{2}{3}\ln|x+2| + C.$$

17. Write $\dfrac{x^2+2x+6}{x^3-x} = \dfrac{A}{x} + \dfrac{B}{x-1} + \dfrac{C}{x+1}$.

Then $x^2 + 2x - 6 = A(x^2-1) + B(x^2+x) + C(x^2-x)$.

Setting $x = 0$, $x = 1$, and $x = -1$ gives $A = 6$, $B = -3/2$, and $C = -7/2$. Thus

$$\int \frac{x^2 + 2x - 6}{x^3 - x} \, dx = 6 \int \frac{1}{x} \, dx - \frac{3}{2} \int \frac{1}{x-1} \, dx - \frac{7}{2} \int \frac{1}{x+1} \, dx$$

$$= 6 \ln |x| - \frac{3}{2} \ln |x-1| - \frac{7}{2} \ln |x+1| + C.$$

19. Write $\dfrac{1}{(x+1)(x+2)(x+3)} = \dfrac{A}{x+1} + \dfrac{B}{x+2} + \dfrac{C}{x+3}$.

Then $1 = A(x+2)(x+3) + B(x+1)(x+3) + C(x+1)(x+2)$.

Setting $x = -1$, $x = -2$, and $x = -3$ gives $A = 1/2$, $B = -1$, and $C = 1/2$. Thus

$$\int \frac{1}{(x+1)(x+2)(x+3)} \, dx = \frac{1}{2} \int \frac{1}{x+1} \, dx - \int \frac{1}{x+2} \, dx + \frac{1}{2} \int \frac{1}{x+3} \, dx$$

$$= \frac{1}{2} \ln |x+1| - \ln |x+2| + \frac{1}{2} \ln |x+3| + C.$$

21. Write $\dfrac{4t^2 + 3t - 1}{t^3 - t^2} = \dfrac{A}{t} + \dfrac{B}{t^2} + \dfrac{C}{t-1}$.

Then $4t^2 + 3t - 1 = A(t^2 - t) + B(t - 1) + Ct^2 = (A+C)t^2 + (-A+B)t - B$.

Solving $\boxed{A + C = 4 \qquad\qquad -A + B = 3 \qquad\qquad -B = -1}$

gives $A = -2$, $B = 1$, and $C = 6$. Thus

$$\int \frac{4t^2 + 3t - 1}{t^3 - t^2} \, dt = -2 \int \frac{1}{t} \, dt + \int \frac{1}{t^2} \, dt + 6 \int \frac{1}{t-1} \, dt = -2 \ln |t| - \frac{1}{t} + 6 \ln |t-1| + C.$$

23. Write $\dfrac{1}{x^3 + 2x^2 + x} = \dfrac{A}{x} + \dfrac{B}{x+1} + \dfrac{C}{(x+1)^2}$.

Then $1 = A(x+1)^2 + B(x^2 + x) + Cx = (A+B)x^2 + (2A + B + C)x + A$.

Solving $\boxed{A + B = 0 \qquad\qquad 2A + B + C = 0 \qquad\qquad A = 1}$

gives $A = 1$, $B = -1$, and $C = -1$. Thus

$$\int \frac{1}{x^3 + 2x^2 + x} \, dx = \int \frac{1}{x} \, dx - \int \frac{1}{x+1} \, dx - \int \frac{1}{(x+1)^2} \, dx$$

$$= \ln |x| - \ln |x+1| + \frac{1}{x+1} + C.$$

25. $\displaystyle \int \frac{2x-1}{(x+1)^3} \, dx = \int \frac{2(x+1)-3}{(x+1)^3} \, dx = \int \frac{2}{(x+1)^2} \, dx - \int \frac{3}{(x+1)^3} \, dx$

$$= -\frac{2}{x+1} + \frac{3}{2} \left[\frac{1}{(x+1)^2} \right] + C = \frac{3}{2(x+1)^2} - \frac{2}{x+1} + C$$

27. Write $\dfrac{1}{(x^2 + 6x + 5)^2} = \dfrac{A}{x+1} + \dfrac{B}{(x+1)^2} + \dfrac{C}{x+5} + \dfrac{D}{(x+5)^2}$.

Then $1 = A(x+1)(x+5)^2 + B(x+5)^2 + C(x+5)(x+1)^2 + D(x+1)^2$

$$= (A+C)x^3 + (11A + B + 7C + D)x^2 + (35A + 10B + 11C + 2D)x$$

$$+ 25A + 25B + 5C + D.$$

Solving $\boxed{\begin{array}{ll} A + C = 0 & 11A + B + 7C + D = 0 \\ 35A + 10B + 11C + 2D = 0 & 25A + 25B + 5C + D = 1 \end{array}}$

gives $A = -1/32$, $B = 1/16$, $C = 1/32$, and $D = 1/16$. Thus

$$\int \frac{1}{(x^2 + 6x + 5)^2}\, dx = -\frac{1}{32}\int \frac{1}{x+1}\, dx + \frac{1}{16}\int \frac{1}{(x+1)^2}\, dx + \frac{1}{32}\int \frac{1}{x+5}\, dx$$

$$+ \frac{1}{16}\int \frac{1}{(x+5)^2}\, dx$$

$$= -\frac{1}{32}\ln|x+1| - \frac{1}{16}\left(\frac{1}{x+1}\right) + \frac{1}{32}\ln|x+5| - \frac{1}{16}\left(\frac{1}{x+5}\right) + C.$$

29. Write $\dfrac{x^4 + 2x^2 - x + 9}{x^5 + 2x^4} = \dfrac{A}{x} + \dfrac{B}{x^2} + \dfrac{C}{x^3} + \dfrac{D}{x^4} + \dfrac{E}{x+2}$.

Then $x^4 + 2x^2 - x + 9 = Ax^3(x+2) + Bx^2(x+2) + Cx(x+2) + D(x+2) + Ex^4$

$$= (A+E)x^4 + (2A+B)x^3 + (2B+C)x^2 + (2C+D)x + 2D.$$

Solving

$A + E = 1$	$2A + B = 0$	$2B + C = 2$
$2C + D = -1$	$2D = 9$	

gives $A = -19/16$, $B = 19/8$, $C = -11/4$, $D = 9/2$, and $E = 35/16$. Thus

$$\int \frac{x^4 + 2x^2 - x + 9}{x^5 + 2x^4}\, dx = -\frac{19}{16}\int \frac{1}{x}\, dx + \frac{19}{8}\int \frac{1}{x^2}\, dx - \frac{11}{4}\int \frac{1}{x^3}\, dx$$

$$+ \frac{9}{2}\int \frac{1}{x^4}\, dx + \frac{35}{16}\int \frac{1}{x+2}\, dx$$

$$= -\frac{19}{16}\ln|x| - \frac{19}{8}\left(\frac{1}{x}\right) + \frac{11}{8}\left(\frac{1}{x^2}\right) - \frac{3}{2}\left(\frac{1}{x^3}\right) + \frac{35}{16}\ln|x+2| + C.$$

31. Write $\dfrac{x-1}{x(x^2+1)} = \dfrac{A}{x} + \dfrac{Bx+C}{x^2+1}$.

Then $x - 1 = A(x^2+1) + (Bx+C)x = (A+B)x^2 + Cx + A$.

Solving

$A + B = 0$	$C = 1$	$A = -1$

gives $A = -1$, $B = 1$, and $C = 1$. Thus

$$\int \frac{x-1}{x(x^2+1)}\, dx = -\int \frac{1}{x}\, dx + \int \frac{x}{x^2+1}\, dx + \int \frac{1}{x^2+1}\, dx$$

$$= -\ln|x| + \frac{1}{2}\ln(x^2+1) + \tan^{-1}x + C.$$

33. Write $\dfrac{x}{(x+1)^2(x^2+1)} = \dfrac{A}{x+1} + \dfrac{B}{(x+1)^2} + \dfrac{Cx+D}{x^2+1}$.

Then $x = A(x+1)(x^2+1) + B(x^2+1) + (Cx+D)(x+1)^2$

$$= (A+C)x^3 + (A+B+2C+D)x^2 + (A+C+2D)x + (A+B+D).$$

Solving

$A + C = 0$	$A + B + 2C + D = 0$
$A + C + 2D = 1$	$A + B + D = 0$

gives $A = 0$, $B = -1/2$, $C = 0$, and $D = 1/2$. Thus

$$\int \frac{x}{(x+1)^2(x^2+1)}\, dx = -\frac{1}{2}\int \frac{1}{(x+1)^2}\, dx + \frac{1}{2}\int \frac{1}{x^2+1}\, dx = \frac{1}{2}\left(\frac{1}{x+1}\right) + \frac{1}{2}\tan^{-1}x + C.$$

35. Write $\dfrac{1}{x^4 + 5x^2 + 4} = \dfrac{Ax+B}{x^2+1} + \dfrac{Cx+D}{x^2+4}$.

Then $1 = (Ax+B)(x^2+4) + (Cx+D)(x^2+1)$

$$= (A+C)x^3 + (B+D)x^2 + (4A+C)x + (4B+D).$$

Solving $\boxed{A+C=0 \qquad B+D=0 \qquad 4A+C=0 \qquad 4B+D=1}$

gives $A = 0$, $B = 1/3$, $C = 0$, and $D = -1/3$. Thus

$$\int \frac{1}{x^4+5x^2+4}\,dx = \frac{1}{3}\int \frac{1}{x^2+1}\,dx - \frac{1}{3}\int \frac{1}{x^2+4}\,dx = \frac{1}{3}\tan^{-1}x - \frac{1}{6}\tan^{-1}\frac{x}{2} + C.$$

37. Write $\dfrac{1}{x^3-1} = \dfrac{A}{x-1} + \dfrac{Bx+C}{x^2+x+1}$.

Then $1 = A(x^2+x+1) + (Bx+C)(x-1) = (A+B)x^2 + (A-B+C)x + (A-C)$.

Solving $\boxed{A+B=0 \qquad A-B+C=0 \qquad A-C=1}$

gives $A = 1/3$, $B = -1/3$, and $C = -2/3$. Thus

$$\begin{aligned}
\int \frac{1}{x^3-1}\,dx &= \frac{1}{3}\int \frac{1}{x-1}\,dx - \frac{1}{6}\int \frac{2x+4}{x^2+x+1}\,dx \\
&= \frac{1}{3}\ln|x-1| - \frac{1}{6}\int \frac{2x+1}{x^2+x+1}\,dx - \frac{1}{2}\int \frac{1}{x^2+x+1}\,dx \\
&= \frac{1}{3}\ln|x-1| - \frac{1}{6}\ln|x^2+x+1| - \frac{1}{2}\int \frac{1}{(x+1/2)^2+3/4}\,dx \\
&= \frac{1}{3}\ln|x-1| - \frac{1}{6}\ln|x^2+x+1| - \frac{1}{\sqrt{3}}\tan^{-1}\frac{2x+1}{\sqrt{3}} + C.
\end{aligned}$$

39. Write $\dfrac{3x^2-x+1}{(x+1)(x^2+2x+2)} = \dfrac{A}{x+1} + \dfrac{Bx+C}{x^2+2x+2}$.

Then $3x^2 - x + 1 = A(x^2+2x+2) + (Bx+C)(x+1)$

$$= (A+B)x^2 + (2A+B+C)x + (2A+C).$$

Solving $\boxed{A+B=3 \qquad 2A+B+C=-1 \qquad 2A+C=1}$

gives $A = 5$, $B = -2$, and $C = -9$. Thus

$$\begin{aligned}
\int \frac{3x^2-x+1}{(x+1)(x^2+2x+2)}\,dx &= 5\int \frac{1}{x+1}\,dx - \int \frac{2x+9}{x^2+2x+2}\,dx \\
&= 5\ln|x+1| - \int \frac{2x+2}{x^2+2x+2}\,dx - \int \frac{7}{x^2+2x+2}\,dx \\
&= 5\ln|x+1| - \ln|x^2+2x+2| - \int \frac{7}{(x+1)^2+1}\,dx \\
&= 5\ln|x+1| - \ln|x^2+2x+2| - 7\tan^{-1}(x+1) + C.
\end{aligned}$$

41. Write $\dfrac{x^2-x+4}{(x^2+4)^2} = \dfrac{Ax+B}{x^2+4} + \dfrac{Cx+D}{(x^2+4)^2}$.

Then $x^2 - x + 4 = (Ax+B)(x^2+4) + Cx + D = Ax^3 + Bx^2 + (4A+C)x + (4B+D)$.

Solving $\boxed{A=0 \qquad B=1 \qquad 4A+C=-1 \qquad 4B+D=4}$

gives $A = 0$, $B = 1$, $C = -1$, and $D = 0$. Thus

$$\int \frac{x^2-x+4}{(x^2+4)^2}\,dx = \int \frac{1}{x^2+4}\,dx - \int \frac{x}{(x^2+4)^2}\,dx = \frac{1}{2}\tan^{-1}\frac{x}{2} + \frac{1}{2}\left(\frac{1}{x^2+4}\right) + C$$

43. For this and possibly later problems, we will encounter $\displaystyle\int \cos^2\theta\,d\theta$. Using Example 12 of Section 5.2 in the text, we have

$$\int \cos^2\theta\,d\theta = \frac{1}{2}\theta + \frac{1}{4}\sin 2\theta + C = \frac{1}{2}\theta + \frac{1}{2}\sin\theta\cos\theta + C.$$

Write $\dfrac{x^3 - 2x^2 + x - 3}{x^4 + 8x^2 + 16} = \dfrac{Ax + B}{x^2 + 4} + \dfrac{Cx + D}{(x^2 + 4)^2}.$

Then $x^3 - 2x^2 + x - 3 = (Ax + B)(x^2 + 4) + Cx + D$

$$= Ax^3 + Bx^2 + (4A + C)x + (4B + D).$$

Solving $\boxed{A = 1 \qquad B = -2 \qquad 4A + C = 1 \qquad 4B + D = -3}$

gives $A = 1$, $B = -2$, $C = -3$, and $D = 5$. Thus

$$\int \frac{x^3 - 2x^2 + x - 3}{x^4 + 8x^2 + 16}\,dx = \int \frac{x}{x^2 + 4}\,dx - 2\int \frac{1}{x^2 + 4}\,dx - 3\int \frac{x}{(x^2 + 4)^2}\,dx$$

$$+ 5\int \frac{1}{(x^2 + 4)^2}\,dx \qquad \boxed{x = 2\tan\theta,\ dx = 2\sec^2\theta\,d\theta}$$

$$= \frac{1}{2}\ln(x^2 + 4) - \tan^{-1}\frac{x}{2} + \frac{3}{2}\left(\frac{1}{x^2 + 4}\right) + 5\int \frac{2\sec^2\theta}{(4\tan^2\theta + 4)^2}\,d\theta$$

$$= \frac{1}{2}\ln(x^2 + 4) - \tan^{-1}\frac{x}{2} + \frac{3}{2(x^2 + 4)} + \frac{5}{8}\int \cos^2\theta\,d\theta$$

$$= \frac{1}{2}\ln(x^2 + 4) - \tan^{-1}\frac{x}{2} + \frac{3}{2(x^2 + 4)} + \frac{5}{16}\theta + \frac{5}{16}\sin\theta\cos\theta + C$$

$$= \frac{1}{2}\ln(x^2 + 4) - \tan^{-1}\frac{x}{2} + \frac{3}{2(x^2 + 4)} + \frac{5}{16}\tan^{-1}\frac{x}{2}$$

$$+ \frac{5}{8}\left(\frac{x}{x^2 + 4}\right) + C$$

$$= \frac{1}{2}\ln(x^2 + 4) - \frac{11}{6}\tan^{-1}\frac{x}{2} + \frac{5x + 12}{8(x^2 + 4)} + C.$$

45. Write $\dfrac{x^4 + 3x^2 + 4}{(x + 1)^2} = x^2 - 2x + 6 - \dfrac{10x + 2}{(x + 1)^2} = x^2 - 2x + 6 - \dfrac{10(x + 1) - 8}{(x + 1)^2}.$

Then $\displaystyle\int \frac{x^4 + 3x^2 + 4}{(x + 1)^2}\,dx = \int (x^2 - 2x + 6)\,dx - 10\int \frac{1}{x + 1}\,dx + 8\int \frac{1}{(x + 1)^2}\,dx$

$$= \frac{1}{3}x^3 - x^2 + 6x - 10\ln|x + 1| - \frac{8}{x + 1} + C.$$

47. Write $\dfrac{1}{x^2 - 6x + 5} = \dfrac{A}{x - 1} + \dfrac{B}{x - 5}.$

Then $1 = A(x - 5) + B(x - 1)$. Setting $x = 1$ and $x = 5$ gives $A = -1/4$ and $B = 1/4$. Thus

$$\int_2^4 \frac{1}{x^2 - 6x + 5}\,dx = -\frac{1}{4}\int_2^4 \frac{1}{x - 1}\,dx + \frac{1}{4}\int_2^4 \frac{1}{x - 5}\,dx = -\frac{1}{4}\ln|x - 1|\Big]_2^4 + \frac{1}{4}\ln|x - 5|\Big]_2^4$$

$$= \frac{1}{4}\ln\left|\frac{x - 5}{x - 1}\right|\Big]_2^4 = \frac{1}{4}\left(\ln\frac{1}{3} - \ln 3\right) = -\frac{1}{2}\ln 3.$$

49. $\displaystyle\int_0^2 \frac{2x - 1}{(x + 3)^2}\,dx = \int_0^2 \frac{2(x + 3) - 7}{(x + 3)^2}\,dx = \int_0^2 \frac{2}{x + 3}\,dx - 7\int_0^2 \frac{1}{(x + 3)^2}\,dx$

$$= 2\ln|x + 3|\Big]_0^2 + \frac{7}{x + 3}\Big]_0^2 = 2(\ln 5 - \ln 3) + \left(\frac{7}{5} - \frac{7}{3}\right) = 2\ln\frac{5}{3} - \frac{14}{15}$$

51. Write $\dfrac{1}{x^3 + x^2 + 2x + 2} = \dfrac{A}{x + 1} + \dfrac{Bx + C}{x^2 + 2}.$

Then $1 = A(x^2 + 2) + (Bx + C)(x + 1) = (A + B)x^2 + (B + C)x + (2A + C).$

Solving $\boxed{A + B = 0 \qquad B + C = 0 \qquad 2A + C = 1}$

gives $A = 1/3$, $B = -1/3$, and $C = 1/3$. Thus

$$\int_0^1 \frac{1}{x^3 + x^2 + 2x + 2}\,dx = \frac{1}{3}\int_0^1 \frac{1}{x+1}\,dx - \frac{1}{6}\int_0^1 \frac{2x}{x^2+2}\,dx + \frac{1}{3}\int_0^1 \frac{1}{x^2+2}\,dx$$

$$= \frac{1}{3}\ln|x+1|\Big]_0^1 + \frac{1}{6}\ln(x^2+2)\Big]_0^1 + \frac{1}{3\sqrt{2}}\tan^{-1}\frac{x}{\sqrt{2}}\Big]_0^1$$

$$= \frac{1}{3}\ln 2 - \frac{1}{6}(\ln 3 - \ln 2) + \frac{1}{3\sqrt{2}}\tan^{-1}\frac{1}{\sqrt{2}}$$

$$= \frac{1}{6}\ln\frac{8}{3} + \frac{1}{3\sqrt{2}}\tan^{-1}\frac{1}{\sqrt{2}}.$$

53. $\displaystyle\int_{-1}^1 \frac{2x^3 + 5x}{x^4 + 5x^2 + 6}\,dx = \frac{1}{2}\int_{-1}^1 \frac{4x^3 + 10x}{x^4 + 5x^2 + 6}\,dx = \frac{1}{2}\ln|x^4 + 5x^2 + 6|\Big]_{-1}^1$

$$= \frac{1}{2}(\ln 12 - \ln 12) = 0$$

55. $\displaystyle\int \frac{\sqrt{1-x^2}}{x^3}\,dx = \int \frac{\sqrt{1-x^2}}{x^4}\,x\,dx \qquad \boxed{u^2 = 1 - x^2,\ 2u\,du = -2x\,dx}$

$$= \int \frac{u}{(1-u^2)^2}(-u\,du) = -\int \frac{u^2}{(1-u^2)^2}\,du$$

Write $\dfrac{u^2}{(1-u^2)^2} = \dfrac{A}{1-u} + \dfrac{B}{(1-u)^2} + \dfrac{C}{1+u} + \dfrac{D}{(1+u)^2}$.

Then $u^2 = A(1-u)(1+u)^2 + B(1+u^2) + C(1+u)(1-u)^2 + D(1-u)^2$

$$= (A+B+C+D) + (A+2B-C-2D)u + (-A+B-C+D)u^2 + (-A+C)u^3.$$

Solving $\boxed{\begin{array}{ll} A+B+C+D=0 & A+2B-C-2D=0 \\ -A+B-C+D=1 & -A+C=0 \end{array}}$

gives $A = -1/4$, $B = 1/4$, $C = -1/4$, and $D = 1/4$. Thus

$$\int \frac{\sqrt{1-x^2}}{x^3}\,dx = -\int \frac{u^2}{(1-u^2)^2}\,du$$

$$= \frac{1}{4}\int \frac{1}{1-u}\,du - \frac{1}{4}\int \frac{1}{(1-u)^2}\,du + \frac{1}{4}\int \frac{1}{1+u}\,du - \frac{1}{4}\int \frac{1}{(1+u)^2}\,du$$

$$= -\frac{1}{4}\ln|1-u| - \frac{1}{4}\left(\frac{1}{1-u}\right) + \frac{1}{4}\ln|1+u| + \frac{1}{4}\left(\frac{1}{1+u}\right) + C$$

$$= \frac{1}{4}\ln\left|\frac{1+u}{1-u}\right| - \frac{1}{2}\left(\frac{u}{1-u^2}\right) + C = \frac{1}{4}\ln\left|\frac{1+\sqrt{1-x^2}}{1-\sqrt{1-x^2}}\right| - \frac{1}{2}\left(\frac{\sqrt{1-x^2}}{x^2}\right) + C.$$

57. $\displaystyle\int \frac{\sqrt[3]{x+1}}{x}\,dx \qquad \boxed{u^3 = x+1,\ 3u^2\,du = dx} \qquad = \int \frac{u}{u^3-1}(3u^2\,du)$

$$= 3\int \frac{u^3-1+1}{u^3-1}\,du = 3\int du + 3\int \frac{1}{u^3-1}\,du = 3u + \int \frac{1}{u^3-1}\,du$$

$$= 3u + \ln|u-1| - \frac{1}{2}\ln|u^2+u+1| - \sqrt{3}\tan^{-1}\frac{2u+1}{\sqrt{3}} + C$$

$$= 3\sqrt[3]{x+1} + \ln|\sqrt[3]{x+1} - 1| - \frac{1}{2}\ln\left|\sqrt[3]{(x+1)^2} + \sqrt[3]{x+1} + 1\right|$$

$$- \sqrt{3}\tan^{-1}\frac{2\sqrt[3]{x+1}+1}{\sqrt{3}} + C.$$

59. Write $\dfrac{1}{x^2+2x-3} = \dfrac{A}{x+3} + \dfrac{B}{x-1}$.

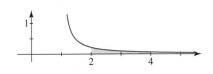

Then $1 = A(x-1) + B(x+3)$. Setting $x = -3$ and $x = 1$ gives $A = -1/4$ and $B = 1/4$. Thus

$$\text{Area} = -\frac{1}{4}\int_2^4 \frac{1}{x+3}\,dx + \frac{1}{4}\int_2^4 \frac{1}{x-1}\,dx = -\frac{1}{4}\ln|x+3|\Big]_2^4 + \frac{1}{4}\ln|x-1|\Big]_2^4$$

$$= \frac{1}{4}\ln\left|\frac{x-1}{x+3}\right|\Big]_2^4 = \frac{1}{4}\left(\ln\frac{3}{7} - \ln\frac{1}{5}\right) = \frac{1}{4}\ln\frac{15}{7} \approx 0.1905.$$

61. Write $\dfrac{x}{(x+2)(x+3)} = \dfrac{A}{x+2} + \dfrac{B}{x+3}$. Then $x = A(x+3) + B(x+2)$.

Setting $x = -2$ and $x = -3$ gives $A = -2$ and $B = 3$. Thus

$$\text{Area} = -\int_{-1}^0 \left(\frac{-2}{x+2} + \frac{3}{x+3}\right)dx + \int_0^1 \left(\frac{-2}{x+2} + \frac{3}{x+3}\right)dx$$

$$= -(-2\ln|x+2| + 3\ln|x+3|)\Big]_{-1}^0 + (-2\ln|x+2| + 3\ln|x+3|)\Big]_0^1$$

$$= -[(-2\ln 2 + 3\ln 3) - (-2\ln 1 + 3\ln 2)] + [(-2\ln 3 + 3\ln 4) - (-2\ln 2 + 3\ln 3)]$$

$$= 7\ln 2 - 8\ln 3 + 3\ln 4 = \ln\frac{8192}{6561} \approx 0.2220.$$

63. $V = \pi \displaystyle\int_1^3 \dfrac{4}{x^2(x+1)^2}\,dx$

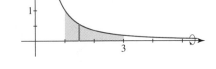

Write $\dfrac{4}{x^2(x+1)^2} = \dfrac{A}{x} + \dfrac{B}{x^2} + \dfrac{C}{x+1} + \dfrac{D}{(x+1)^2}$.

Then $4 = Ax(x+1)^2 + B(x+1)^2 + Cx^2(x+1) + Dx^2$

$\qquad = (A+C)x^3 + (2A+B+C+D)x^2 + (A+2B)x + B.$

Solving

$A+C=0$	$2A+B+C+D=0$	$A+2B=0$	$B=4$

gives $A = -8$, $B = 4$, $C = 8$, and $D = 4$. Thus

$$V = \pi\left[-8\int_1^3 \frac{1}{x}\,dx + 4\int_1^3 \frac{1}{x^2}\,dx + 8\int_1^3 \frac{1}{x+1}\,dx + 4\int_1^3 \frac{1}{(x+1)^2}\,dx\right]$$

$$= \pi\left(-8\ln|x| - \frac{4}{x} + 8\ln|x+1| - \frac{4}{x+1}\right)\Big]_1^3 = \pi\left[8\ln\left|\frac{x+1}{x}\right| - \frac{8x+4}{x(x+1)}\right]\Big]_1^3$$

$$= \pi\left[\left(8\ln\frac{4}{3} - \frac{7}{3}\right) - (8\ln 2 - 6)\right] = 8\pi\ln\frac{2}{3} + \frac{11\pi}{3} \approx 1.3287.$$

65. $V = 2\pi \displaystyle\int_0^1 \dfrac{4x}{(x+1)^2}\,dx = 8\pi\int_0^1 \dfrac{x+1-1}{(x+1)^2}\,dx$

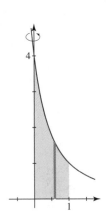

$$= 8\pi\left[\int_0^1 \frac{1}{x+1}\,dx - \int_0^1 \frac{1}{(x+1)^2}\,dx\right]$$

$$= 8\pi\left(\ln|x+1| + \frac{1}{x+1}\right)\Big]_0^1 = 8\pi\left[\left(\ln 2 + \frac{1}{2}\right) - (\ln 1 + 1)\right]$$

$$= 8\pi\ln 2 - 4\pi \approx 4.8543$$

67. $\displaystyle\int \frac{\cos x}{\sin^2 x + 3\sin x + 2}\,dx$ $\boxed{u = \sin x,\ du = \cos x\,dx}$ $= \displaystyle\int \frac{1}{u^2 + 3u + 2}\,du$

Write $\dfrac{1}{(u+1)(u+2)} = \dfrac{A}{u+1} + \dfrac{B}{u+2}$.

Then $1 = A(u+2) + B(u+1)$. Setting $u = -1$ and $u = -2$ gives $A = 1$ and $B = -1$. Thus

$$\int \frac{\cos x}{\sin^2 x + 3\sin x + 2}\,dx = \int \frac{1}{u+1}\,du - \int \frac{1}{u+2}\,du = \ln|u+1| - \ln|u+2| + C$$
$$= \ln\left|\frac{u+1}{u+2}\right| + C = \ln\left|\frac{\sin x + 1}{\sin x + 2}\right| + C.$$

69. $\displaystyle\int \frac{e^t}{(e^t+1)^2(e^t-2)}\,dt$ $\boxed{u = e^t,\ du = e^t\,dt}$ $= \displaystyle\int \frac{1}{(u+1)^2(u-2)}\,du$

Write $\dfrac{1}{(u+1)^2(u-2)} = \dfrac{A}{u+1} + \dfrac{B}{(u+1)^2} + \dfrac{C}{u-2}$.

Then $1 = A(u+1)(u-2) + B(u-2) + C(u+1)^2$
$\qquad = (A+C)u^2 + (-A+B+2C)u + (-2A-2B+C)$.

Solving $\boxed{A+C = 0 \qquad\qquad -A+B+2C = 0 \qquad\qquad -2A-2B+C = 1}$

gives $A = -1/9$, $B = -1/3$, and $C = 1/9$. Thus

$$\int \frac{e^t}{(e^t+1)^2(e^t-2)}\,dt = -\frac{1}{9}\int \frac{1}{u+1}\,du - \frac{1}{3}\int \frac{1}{(u+1)^2}\,du + \frac{1}{9}\int \frac{1}{u-2}\,du$$
$$= -\frac{1}{9}\ln|u+1| + \frac{1}{3}\left(\frac{1}{u+1}\right) + \frac{1}{9}\ln|u-2| + C$$
$$= \frac{1}{9}\ln\left|\frac{e^t-2}{e^t+1}\right| + \frac{1}{3(e^t+1)} + C.$$

71. $y' = e^x$

$$L = \int_0^{\ln 2} \sqrt{1 + e^{2x}}\,dx \qquad \boxed{u^2 = 1 + e^{2x},\ 2u\,du = 2e^{2x}\,dx,\ dx = \frac{u}{e^{2x}}\,du = \frac{u}{u^2-1}\,du}$$

$$= \int_{\sqrt 2}^{\sqrt 5} \frac{u^2}{u^2-1}\,du = \int_{\sqrt 2}^{\sqrt 5}\left(1 + \frac{1}{u^2-1}\right)du$$

Write $\dfrac{1}{u^2-1} = \dfrac{A}{u-1} + \dfrac{B}{u+1}$.

Then $1 = A(u+1) + B(u-1)$. Setting $u = 1$ and $u = -1$ gives $A = 1/2$ and $B = -1/2$.

Thus $L = \displaystyle\int_{\sqrt 2}^{\sqrt 5} du + \frac{1}{2}\int_{\sqrt 2}^{\sqrt 5}\frac{1}{u-1}\,du - \frac{1}{2}\int_{\sqrt 2}^{\sqrt 5}\frac{1}{u+1}\,du$

$$= \left(u + \frac{1}{2}\ln|u-1| - \frac{1}{2}\ln|u+1|\right)\Bigg]_{\sqrt 2}^{\sqrt 5} = \left(u + \frac{1}{2}\ln\left|\frac{u-1}{u+1}\right|\right)\Bigg]_{\sqrt 2}^{\sqrt 5}$$

$$= \left[\left(\sqrt 5 + \frac{1}{2}\ln\frac{\sqrt 5 - 1}{\sqrt 5 + 1}\right) - \left(\sqrt 2 + \frac{1}{2}\ln\frac{\sqrt 2 - 1}{\sqrt 2 + 1}\right)\right] \approx 1.2220.$$

73. Rewrite the integral as

$$\int \frac{x^5}{(x-1)^{10}(x+1)^{10}}\,dx = \int \frac{x^5}{(x^2-1)^{10}}\,dx = \int \frac{x^4}{(x^2-1)^{10}}\,(x\,dx)$$

then integrate using $\boxed{u = x^2 - 1, \ x^2 = u + 1, \ \dfrac{1}{2}u = x\,dx}$:

$$\int \frac{x^4}{(x^2 - 1)^{10}}(x\,dx) = \frac{1}{2}\int \frac{(u+1)^2}{u^{10}}\,du = \frac{1}{2}\int \frac{u^2 + 2u + 1}{u^{10}}\,du$$

$$= \frac{1}{2}\int (u^{-8} + 2u^{-9} + u^{-10})\,du = -\frac{1}{14}u^{-7} - \frac{1}{8}u^{-8} - \frac{1}{18}u^{-9} + C$$

$$= -\frac{1}{14}(x^2 - 1)^{-7} - \frac{1}{8}(x^2 - 1)^{-8} - \frac{1}{18}(x^2 - 1)^{-9} + C$$

7.7 Improper Integrals

In this exercise set, the symbol "$\overset{h}{=}$" is used to denote the fact that L'Hôpital's Rule was applied to obtain the equality.

1. $\displaystyle\int_3^\infty \frac{1}{x^4}\,dx = \lim_{t\to\infty}\int_3^t x^{-4}\,dx = \lim_{t\to\infty}\left(-\frac{1}{3}x^{-3}\right)\Big]_3^t = \lim_{t\to\infty}\left(\frac{1}{81} - \frac{1}{3t^3}\right) = \frac{1}{81}$

3. $\displaystyle\int_1^\infty \frac{1}{x^{0.99}}\,dx = \lim_{t\to\infty}\int_1^t x^{-0.99}\,dx = \lim_{t\to\infty}\frac{x^{0.01}}{0.01}\Big]_1^t = \lim_{t\to\infty}\left(\frac{t^{0.01}}{0.01} - \frac{1}{0.01}\right)$

   The integral diverges.

5. $\displaystyle\int_{-\infty}^3 e^{2x}\,dx = \lim_{s\to-\infty}\int_s^3 e^{2x}\,dx = \lim_{s\to-\infty}\frac{1}{2}e^{2x}\Big]_s^3 = \lim_{s\to-\infty}\left(\frac{1}{2}e^6 - \frac{1}{2}e^{2x}\right) = \frac{1}{2}e^6$

7. $\displaystyle\int_1^\infty \frac{\ln x}{x}\,dx = \lim_{t\to\infty}\int_1^t \frac{\ln x}{x}\,dx = \lim_{t\to\infty}\frac{1}{2}(\ln x)^2\Big]_1^t = \lim_{t\to\infty}\left[\frac{1}{2}(\ln t)^2 - 0\right]$

   The integral diverges.

9. $\displaystyle\int_e^\infty \frac{1}{x(\ln x)^3}\,dx = \lim_{t\to\infty}\int_e^t (\ln x)^{-3}\frac{1}{x}\,dx = \lim_{t\to\infty}\left[-\frac{1}{2}(\ln x)^{-2}\right]\Big|_e^t = \lim_{t\to\infty}\left[\frac{1}{2} - \frac{1}{2(\ln t)^2}\right] = \frac{1}{2}$

11. $\displaystyle\int_{-\infty}^\infty \frac{x}{(x^2 + 1)^{3/2}}\,dx = \int_{-\infty}^0 \frac{x}{(x^2 + 1)^{3/2}}\,dx + \int_0^\infty \frac{x}{(x^2 + 1)^{3/2}}\,dx$

$$= \lim_{s\to-\infty}\int_s^0 \frac{x}{(x^2 + 1)^{3/2}}\,dx + \lim_{t\to\infty}\int_0^t \frac{x}{(x^2 + 1)^{3/2}}\,dx$$

$$= \lim_{s\to-\infty}\left[-(x^2 + 1)^{-1/2}\right]_s^0 + \lim_{t\to\infty}\left[-(x^2 + 1)^{-1/2}\right]_0^t$$

$$= \lim_{s\to-\infty}\left(-1 + \frac{1}{\sqrt{s^2 + 1}}\right) + \lim_{t\to\infty}\left(-\frac{1}{\sqrt{t^2 + 1}} + 1\right) = -1 + 1 = 0$$

13. $\displaystyle\int_{-\infty}^0 \frac{x}{(x^2 + 9)^2}\,dx = \lim_{s\to-\infty}\int_s^0 \frac{x}{(x^2 + 9)^2}\,dx = \lim_{s\to-\infty}\left[-\frac{1}{2}(x^2 + 9)^{-1}\right]\Big|_s^0$

$$= \lim_{s\to-\infty}\left[\frac{1}{2}\left(\frac{1}{s^2 + 9}\right) - \frac{1}{18}\right] = -\frac{1}{18}$$

15. $\displaystyle\int_2^\infty ue^{-u}\,du = \lim_{t\to\infty}\int_2^t ue^{-u}\,du = \lim_{t\to\infty}(-ue^{-u} - e^{-u})\Big]_2^t = \lim_{t\to\infty}(2e^{-2} + e^{-2} - te^{-t} - e^{-t})$

$$= 2e^{-2} + e^{-2} - \lim_{t\to\infty}\frac{t+1}{e^t} \overset{h}{=} 2e^{-2} + e^{-2} - \lim_{t\to\infty}\frac{1}{e^t} = 2e^{-2} + e^{-2} = 3e^{-2}$$

17. $\displaystyle\int_{2/\pi}^\infty \frac{\sin(1/x)}{x^2}\,dx = \lim_{t\to\infty}\int_{2/\pi}^t \frac{\sin(1/x)}{x^2}\,dx = \lim_{t\to\infty}\cos\frac{1}{x}\Big]_{2/\pi}^t = \lim_{t\to\infty}\left(\cos\frac{1}{t} - \cos\frac{\pi}{2}\right) = 1$

19. $\displaystyle\int_{-1}^\infty \frac{1}{x^2 + 2x + 2}\,dx = \lim_{t\to\infty}\int_{-1}^t \frac{1}{(x+1)^2 + 1}\,dx = \lim_{t\to\infty}\tan^{-1}(x+1)\Big]_{-1}^t$

$$= \lim_{t\to\infty}[\tan^{-1}(t+1) - 0] = \frac{\pi}{2}$$

21. $\displaystyle\int_0^\infty e^{-x}\sin x\,dx = \lim_{t\to\infty}\int_0^t e^{-x}\sin x\,dx$

$$\boxed{u = e^{-x},\ du = -e^{-x}\,dx;\quad dv = \sin x\,dx,\ v = -\cos x}$$

$$= \lim_{t\to\infty}\left(-e^{-x}\cos x\Big]_0^t - \int_0^t e^{-x}\cos x\,dx\right)$$

$$\boxed{u = e^{-x},\ du = -e^{-x}\,dx;\quad dv = \cos x\,dx,\ v = \sin x}$$

$$= \lim_{t\to\infty}\left(1 - e^{-t}\cos t - e^{-x}\sin x\Big]_0^t - \int_0^t e^{-x}\sin x\,dx\right)$$

$$= \lim_{t\to\infty}\left(1 - e^{-t}\cos t - e^{-t}\sin t - \int_0^t e^{-x}\sin x\,dx\right) = 1 - \int_0^\infty e^{-x}\sin x\,dx$$

Solving for the integral, $\displaystyle\int_0^\infty e^{-x}\sin x\,dx = \frac{1}{2}$.

23. $\displaystyle\int_{1/2}^\infty \frac{x+1}{x^3}\,dx = \lim_{t\to\infty}\int_{1/2}^t (x^{-2}+x^{-3})\,dx = \lim_{t\to\infty}\left(-x^{-1} - \frac{1}{2}x^{-2}\right)\Big]_{1/2}^t$

$$= \lim_{t\to\infty}\left(2 + 2 - \frac{1}{t} - \frac{1}{2t^2}\right) = 4$$

25. $\displaystyle\int_1^\infty \left(\frac{1}{x} - \frac{1}{x+1}\right)dx = \lim_{t\to\infty}\int_1^t \left(\frac{1}{x} - \frac{1}{x+1}\right)dx = \lim_{t\to\infty}\big(\ln|x| - \ln|x+1|\big)\Big]_1^t$

$$= \lim_{t\to\infty}\big[\ln t - \ln(t+1) + \ln 2\big] = \ln 2 + \lim_{t\to\infty}\ln\frac{t}{t+1}$$

$$= \ln 2 + \ln\left(\lim_{t\to\infty}\frac{t}{t+1}\right) \overset{h}{=} \ln 2 + \ln\left(\lim_{t\to\infty}\frac{1}{1}\right) = \ln 2 + \ln 1 = \ln 2$$

27. $\displaystyle\int_2^\infty \frac{1}{x^2+6x+5}\,dx = \lim_{t\to\infty}\int_2^t \frac{1}{(x+1)(x+5)}\,dx = \lim_{t\to\infty}\int_2^t \left(\frac{1/4}{x+1} - \frac{1/4}{x+5}\right)dx$

$$= \lim_{t\to\infty}\left(\frac{1}{4}\ln|x+1| - \frac{1}{4}\ln|x+5|\right)\Big]_2^t = \lim_{t\to\infty}\frac{1}{4}\ln\left|\frac{x+1}{x+5}\right|\,\Big]_2^t$$

$$= \lim_{t\to\infty}\left(\frac{1}{4}\ln\frac{t+1}{t+5} - \frac{1}{4}\ln\frac{3}{7}\right) = \frac{1}{4}\ln\left(\lim_{t\to\infty}\frac{t+1}{t+5}\right) - \frac{1}{4}\ln\frac{3}{7}$$

$$\overset{h}{=} \frac{1}{4}\ln\left(\lim_{t\to\infty}\frac{1}{1}\right) - \frac{1}{4}\ln\frac{3}{7} = -\frac{1}{4}\ln\frac{3}{7} = \frac{1}{4}\ln\frac{7}{3}$$

29. $\displaystyle\int_{-\infty}^{-2} \frac{x^2}{(x^3+1)^2}\,dx = \lim_{s\to-\infty}\int_s^{-2} \frac{1}{3}\left[\frac{3x^2}{(x^3+1)^2}\right]dx = \lim_{s\to-\infty}\left[-\frac{1}{3}\left(\frac{1}{x^3+1}\right)\right]\Big|_s^{-2}$

$$= -\frac{1}{3}\left(-\frac{1}{7}\right) = \frac{1}{21}$$

31. $\displaystyle\int_0^5 \frac{1}{x}\,dx = \lim_{s\to 0^+}\int_s^5 \frac{1}{x}\,dx = \lim_{s\to 0^+}\ln|x|\,\Big]_s^5 = \lim_{s\to 0^+}(\ln 5 - \ln s)$. The integral diverges.

33. $\displaystyle\int_0^1 \frac{1}{x^{0.99}}\,dx = \lim_{s\to 0^+}\int_s^1 x^{-0.99}\,dx = \lim_{s\to 0^+}100x^{0.01}\,\Big]_s^1 = \lim_{s\to 0^+}(100 - 100s^{0.01}) = 100$

35. $\displaystyle\int_0^2 \frac{1}{\sqrt{2-x}}\,dx = \lim_{t\to 2^-}\int_0^t (2-x)^{-1/2}\,dx = \lim_{t\to 2^-}\left[-2(2-x)^{1/2}\right]\Big]_0^t$

$$= \lim_{t\to 2^-}\left(2\sqrt{2} - 2\sqrt{2-t}\right) = 2\sqrt{2}$$

37. $\displaystyle\int_{-1}^1 \frac{1}{x^{5/3}}\,dx = \lim_{t\to 0^-}\int_{-1}^t \frac{1}{x^{5/3}}\,dx + \lim_{s\to 0^+}\int_s^1 \frac{1}{x^{5/3}}\,dx$

Since $\displaystyle\lim_{t\to 0^-}\int_{-1}^t \frac{1}{x^{5/3}}\,dx = \lim_{t\to 0^-}\left(-\frac{3}{2}x^{-2/3}\right)\Big]_{-1}^t = \lim_{t\to 0^-}\left(\frac{3}{2} - \frac{3}{2t^{2/3}}\right)$, the integral diverges.

39. $\displaystyle\int_0^2 (x-1)^{-2/3}\,dx = \lim_{t\to1^-}\int_0^t (x-1)^{-2/3}\,dx + \lim_{s\to1^+}\int_s^2 (x-1)^{-2/3}\,dx$

$\displaystyle = \lim_{t\to1^-} 3(x-1)^{1/3}\Big]_0^t + \lim_{s\to1^+} 3(x-1)^{1/3}\Big]_s^2 = 3+3 = 6$

41. $\displaystyle\int_0^1 x\ln x\,dx = \lim_{s\to0^+}\int_s^1 x\ln x\,dx$ $\qquad\boxed{u=\ln x,\ du=\frac1x\,dx;\quad dv=x\,dx,\ v=\frac12 x^2}$

$\displaystyle = \lim_{s\to0^+}\left(\frac12 x^2\ln x\Big]_s^1 - \int_s^1 \frac12 x\,dx\right) = \lim_{s\to0^+}\left(-\frac12 s^2\ln x - \frac14 x^2\Big]_s^1\right)$

$\displaystyle = \lim_{s\to0^+}\left(-\frac12 s^2\ln s - \frac14 + \frac{s^2}{4}\right) = \lim_{s\to0^+}\left(-\frac{\ln s}{2/s^2}\right) - \frac14$

$\displaystyle \overset{h}{=} \lim_{s\to0^+} -\frac{1/s}{-4/s^3} - \frac14 = \lim_{s\to0^+} \frac{s^2}{4} - \frac14 = -\frac14$

43. $\displaystyle\int_0^{\pi/2}\tan t\,dt = \lim_{k\to\pi/2^-}\int_0^k \tan t\,dt = \lim_{k\to\pi/2^-}\ln\sec t\Big]_0^k = \lim_{k\to\pi/2^-}\ln\sec k$

The integral diverges.

45. $\displaystyle\int_0^\pi \frac{\sin x}{1+\cos x}\,dx = \lim_{t\to\pi^-}\int_0^t \frac{\sin x}{1+\cos x}\,dx$ $\qquad\boxed{u=1+\cos x,\ du=-\sin x\,dx}$

$\displaystyle = \lim_{t\to\pi^-}\int_2^{1+\cos t} -\frac1u\,du = \lim_{t\to\pi^-}\left(-\ln|u|\right)\Big]_2^{1+\cos t} = \lim_{t\to\pi^-}\left[\ln 2 - \ln(1+\cos t)\right]$

The integral diverges.

47. $\displaystyle\int_{-1}^0 \frac{x}{\sqrt{1+x}}\,dx = \lim_{s\to-1^+}\int_s^0 \frac{x}{\sqrt{1+x}}\,dx$ $\qquad\boxed{u^2=1+x,\ x=u^2-1,\ dx=2u\,du}$

$\displaystyle = \lim_{s\to-1^+}\int_{\sqrt{1+s}}^1 \frac{u^2-1}{u}(2u\,du) = \lim_{s\to-1^+}\int_{\sqrt{1+s}}^1 (2u^2-2)\,du$

$\displaystyle = \lim_{s\to-1^+}\left(\frac23 u^3 - 2u\right)\Big]_{\sqrt{1+s}}^1 = \lim_{s\to-1^+}\left[-\frac43 - \frac23(1+s)^{3/2} + 2\sqrt{1+s}\right] = -\frac43$

49. $\displaystyle\int_0^1 \frac{x^2}{\sqrt{1-x^2}}\,dx = \lim_{t\to1^-}\int_0^t \frac{x^2}{\sqrt{1-x^2}}\,dx$ $\qquad\boxed{x=\sin\theta,\ dx=\cos\theta\,d\theta}$

$\displaystyle = \lim_{t\to1^-}\int_0^{\sin^{-1}t} \frac{\sin^2\theta}{\sqrt{1-\sin^2\theta}}\cos\theta\,d\theta = \lim_{t\to1^-}\int_0^{\sin^{-1}t}\sin^2\theta\,d\theta$

$\displaystyle = \lim_{t\to1^-}\int_0^{\sin^{-1}t}\frac12(1-\cos 2\theta)\,d\theta = \lim_{t\to1^-}\left(\frac12\theta - \frac14\sin 2\theta\right)\Big]_0^{\sin^{-1}t}$

$\displaystyle = \lim_{t\to1^-}\left(\frac12\theta - \frac12\sin\theta\cos\theta\right)\Big]_0^{\sin^{-1}t} = \lim_{t\to1^-}\left(\frac12\sin^{-1}t - \frac12 t\sqrt{1-t^2}\right) = \frac\pi4$

51. $\displaystyle\int_1^3 \frac{1}{\sqrt{3+2x-x^2}}\,dx = \lim_{t\to3^-}\int_1^t \frac{1}{\sqrt{4-(x-1)^2}}\,dx = \lim_{t\to3^-}\sin^{-1}\frac{x-1}{2}\Big]_1^t$

$\displaystyle = \lim_{t\to3^-}\sin^{-1}\frac{t-1}{2} = \sin^{-1}1 = \frac\pi2$

53. $\displaystyle\int_{12}^\infty \frac{1}{\sqrt{x}(x+4)}\,dx = \lim_{t\to\infty}\int_{12}^t \frac{1}{\sqrt{x}(x+4)}\,dx$ $\qquad\boxed{u=\sqrt{x},\ u^2=x,\ dx=2u\,du}$

$\displaystyle = 2\lim_{t\to\infty}\int_{2\sqrt3}^t \frac{u}{u(u^2+4)}\,du = 2\lim_{t\to\infty}\int_{2\sqrt3}^t \frac{1}{u^2+4}\,du$

$\displaystyle = 2\lim_{t\to\infty}\frac12\tan^{-1}\frac{u}{2}\Big]_{2\sqrt3}^t = \lim_{t\to\infty}\left(\tan^{-1}\frac t2 - \tan^{-1}\sqrt3\right) = \frac\pi2 - \frac\pi3 = \frac\pi6$

55. $A = \int_{1}^{\infty} \frac{1}{(2x+1)^2}\,dx = \lim_{t\to\infty} \int_{1}^{t} \frac{1}{(2x+1)^2}\,dx$

$= \lim_{t\to\infty} \left(-\frac{1/2}{2x+1}\right)\Big]_{1}^{t} = \lim_{t\to\infty}\left(\frac{1}{6}-\frac{1}{4t+2}\right) = \frac{1}{6}$

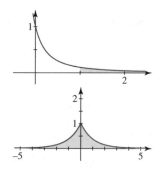

57. $A = \int_{-\infty}^{\infty} e^{-|x|}\,dx = 2\int_{0}^{\infty} e^{-x}\,dx = 2\lim_{t\to\infty}(-e^{-x})\Big]_{0}^{t}$

$= -2\lim_{t\to\infty}(e^{-t}-e^{0}) = -2(0-1) = 2$

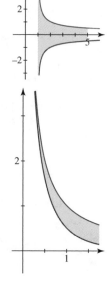

59. $A = \int_{1}^{5}\left[\frac{1}{\sqrt{x-1}} - \left(-\frac{1}{\sqrt{x-1}}\right)\right]dx = \lim_{s\to 1^+}\int_{s}^{5}\frac{2}{\sqrt{x-1}}\,dx$

$= \lim_{s\to 1^+} 4\sqrt{x-1}\Big]_{s}^{5} = \lim_{s\to 1^+}(8-4\sqrt{s-1}) = 8$

61. $A = \int_{0}^{1}\left[\frac{1}{x} - \frac{1}{x(x^2+1)}\right]dx = \lim_{s\to 0^+}\int_{s}^{1}\frac{x}{x^2+1}\,dx = \lim_{s\to 0^+}\frac{1}{2}\ln(x^2+1)\Big]_{s}^{1}$

$= \lim_{s\to 0^+}\left[\frac{1}{2}\ln 2 - \frac{1}{2}\ln(s^2+1)\right] = \frac{1}{2}\ln 2$

63. $W = \int_{1.7\times 10^6}^{\infty}(6.67\times 10^{-11})(7.3\times 10^{22})(10{,}000)\frac{1}{r^2}\,dr \approx 4.87\times 10^{16}\lim_{t\to\infty}\int_{1.7\times 10^6}^{t} r^{-2}\,dr$

$= 4.87\times 10^{16}\lim_{t\to\infty}\left(-\frac{1}{r}\right)\Big]_{1.7\times 10^6}^{t} = 4.87\times 10^{16}\lim_{t\to\infty}\left(\frac{1}{1.7\times 10^6} - \frac{1}{t}\right)$

$= \frac{4.87\times 10^{16}}{1.7\times 10^6} \approx 2.86\times 10^{10}\text{ joules}$

65. $\mathscr{L}\{1\} = \int_{0}^{\infty} e^{-st}\,dt = \lim_{k\to\infty}\int_{0}^{k} e^{-st}\,dt = \lim_{k\to\infty}\left(-\frac{1}{s}e^{-st}\right)\Big]_{0}^{k}$

$= \lim_{k\to\infty}\left(\frac{1}{s}-\frac{1}{s}e^{-sk}\right) = \frac{1}{s},\ s>0$

67. $\mathscr{L}\{e^x\} = \int_{0}^{\infty} e^t e^{-st}\,dt = \lim_{k\to\infty}\int_{0}^{k} e^{(1-s)t}\,dt = \lim_{k\to\infty}\frac{1}{1-s}e^{(1-s)t}\Big]_{0}^{k}$

$= \lim_{k\to\infty}\left[\frac{1}{1-s}e^{(1-s)k} - \frac{1}{1-s}\right] = \frac{1}{1-s},\ s>1$

69. $\mathscr{L}\{\sin x\} = \int_{0}^{\infty} e^{-st}\sin t\,dt = \lim_{k\to\infty}\int_{0}^{k} e^{-st}\sin t\,dt$

$$\boxed{u = e^{-st},\ du = -se^{-st}\,dt;\quad dv = \sin t\,dt,\ v = -\cos t}$$

$= \lim_{k\to\infty}\left(-e^{-st}\cos t\Big]_{0}^{k} - \int_{0}^{k} se^{-st}\cos t\,dt\right) = 1 - \lim_{k\to\infty} s\int_{0}^{k} e^{-st}\cos t\,dt$

$$\boxed{u = e^{-st},\ du = -se^{-st}\,dt;\quad dv = \cos t\,dt,\ v = \sin t}$$

$= 1 - \lim_{k\to\infty} s\left(e^{-st}\sin t\Big]_{0}^{k} + s\int_{0}^{k} e^{-st}\sin t\,dt\right) = 1 - s^2\mathscr{L}\{\sin x\}$

Solving for $\mathscr{L}\{\sin x\}$, we have $\mathscr{L}\{\sin x\} = \dfrac{1}{s^2+1}$, where $s > 0$.

71. $\mathscr{L}\{f(x)\} = \displaystyle\int_1^\infty e^{-st}\,dt = \lim_{k\to\infty}\left(-\frac{1}{s}e^{-st}\right)\Big]_1^k = \frac{1}{s}e^{-s}$, $s > 0$

73. $\displaystyle\int_{-\infty}^\infty f(x)\,dx = \int_{-\infty}^0 f(x)\,dx + \int_0^\infty f(x)\,dx = 0 + \lim_{t\to\infty}\int_0^t ke^{-kx}\,dx$

$\quad = \lim_{t\to\infty} -e^{-kx}\Big]_0^t = \lim_{t\to\infty}(e^0 - e^{-kx}) = 1$

75. $\displaystyle\int_1^\infty \frac{1}{x^k}\,dx = \lim_{t\to\infty}\int_1^t x^{-k}\,dx = \lim_{t\to\infty}\frac{x^{-k+1}}{-k+1}\Big]_1^t = \lim_{t\to\infty}\left(\frac{t^{-k+1}}{-k+1} - \frac{1}{-k+1}\right)$, $k\neq 1$

$\quad = \lim_{t\to\infty}\left[\frac{1}{1-k}\left(\frac{1}{t^{k-1}}\right)\right] + \frac{1}{k-1}$, $k\neq 1$

The integral converges for $k > 1$ and diverges for $k < 1$. If $k = 1$ then

$$\int_1^\infty \frac{1}{x}\,dx = \lim_{t\to\infty}\int_1^t \frac{1}{x}\,dx = \lim_{t\to\infty}\ln x\Big]_1^t = \lim_{t\to\infty}\ln t,$$

and the integral diverges.

77. $\displaystyle\int_0^\infty e^{kx}\,dx = \lim_{t\to\infty}\int_0^t e^{kx}\,dx = \lim_{t\to\infty}\frac{1}{k}e^{kx}\Big]_0^t = \lim_{t\to\infty}\left(\frac{e^{kt}}{k} - \frac{1}{k}\right)$, $k\neq 0$

The integral converges for $k < 0$ and diverges for $k > 0$. If $k = 0$, $\displaystyle\int_0^\infty 1\,dx$ diverges.

79. By Problem 75, $\displaystyle\int_1^\infty \frac{1}{x^2}\,dx$ converges. Since $0 \le \dfrac{\sin^2 x}{x^2} \le \dfrac{1}{x^2}$ for all x in $[1,\infty)$, $\displaystyle\int_1^\infty \frac{\sin^2 x}{x^2}\,dx$ converges.

81. By Problem 77, $\displaystyle\int_0^\infty \frac{1}{e^x}\,dx$ converges. Since $0 < \dfrac{1}{x+e^x} < \dfrac{1}{e^x}$ for all x in $[0,\infty)$, $\displaystyle\int_0^\infty \frac{1}{x+e^x}\,dx$ converges.

83. By Problem 75, $\displaystyle\int_1^\infty \frac{1}{\sqrt{x}}\,dx = \int_1^\infty \frac{1}{x^{1/2}}\,dx$ diverges.

$\quad$ Since $0 < \dfrac{1}{\sqrt{x}} < \dfrac{1+e^{-2x}}{\sqrt{x}}$ for all x in $[1,\infty)$, $\displaystyle\int_1^\infty \frac{1+e^{-2x}}{\sqrt{x}}\,dx$ diverges.

85. $\displaystyle\int_1^\infty \frac{1}{x\sqrt{x^2-1}}\,dx = \lim_{s\to 1^+}\int_s^2 \frac{1}{x\sqrt{x^2-1}}\,dx + \lim_{t\to\infty}\int_2^t \frac{1}{x\sqrt{x^2-1}}\,dx$

$\quad = \lim_{s\to 1^+}\sec^{-1}x\Big]_s^2 + \lim_{t\to\infty}\sec^{-1}x\Big]_2^t$

$\quad = \lim_{s\to 1^+}(\sec^{-1}2 - \sec^{-1}s) + \lim_{t\to\infty}(\sec^{-1}t - \sec^{-1}2)$

$\quad = \sec^{-1}2 - 0 + \dfrac{\pi}{2} - \sec^{-1}2 = \dfrac{\pi}{2}$

87. $\displaystyle\int_{-1}^1 \frac{1}{\sqrt{1-x^2}}\,dx = \lim_{s\to-1^+}\int_s^0 \frac{1}{\sqrt{1-x^2}}\,dx + \lim_{t\to 1^-}\int_0^t \frac{1}{\sqrt{1-x^2}}\,dx$

$\quad = \lim_{s\to-1^+}\sin^{-1}x\Big]_s^0 + \lim_{t\to 1^-}\sin^{-1}x\Big]_0^t$

$\quad = \lim_{s\to-1^+}(0 - \sin^{-1}s) + \lim_{t\to 1^-}(\sin^{-1}t - 0) = -\left(-\dfrac{\pi}{2}\right) + \dfrac{\pi}{2} = \pi$

89. (a) $A = \displaystyle\int_1^\infty \frac{1}{x}\,dx = \lim_{t\to\infty}\int_1^t \frac{1}{x}\,dx = \lim_{t\to\infty}\ln x\Big]_1^t = \lim_{t\to\infty}\ln t$

$\quad$ The integral diverges, so the area is not finite.

(b) $V = \pi\displaystyle\int_1^\infty \frac{1}{x^2}\,dx = \pi\lim_{t\to\infty}\int_1^t \frac{1}{x^2}\,dx = \pi\lim_{t\to\infty}\left(-\frac{1}{x}\right)\Big]_1^t = \pi\lim_{t\to\infty}\left(1 - \frac{1}{t}\right) = \pi$

(c) $y' = -\dfrac{1}{x^2}$; $\quad S = 2\pi \displaystyle\int_1^\infty \dfrac{1}{x}\sqrt{1 + \dfrac{1}{x^4}}\,dx = 2\pi \displaystyle\int_1^\infty \dfrac{\sqrt{x^4 + 1}}{x^3}\,dx$

For all x in $[1, \infty)$, $\dfrac{1}{x} = \dfrac{x^2}{x^3} = \dfrac{\sqrt{x^4}}{x^3} < \dfrac{\sqrt{x^4 + 1}}{x^3}$. By Problem 75, $\displaystyle\int_1^\infty \dfrac{1}{x}\,dx$ diverges. Since $0 \le \dfrac{1}{x} \le \dfrac{\sqrt{x^4 + 1}}{x^3}$ for all x in $[1, \infty)$, $\displaystyle\int_1^\infty \dfrac{\sqrt{x^4 + 1}}{x^3}\,dx$ diverges.

7.8 Approximate Integration

1. Midpoint Rule

k	1	2	3
x_k	3/2	5/2	7/2
$f(x_k)$	39/4	95/4	175/4

$\displaystyle\int_1^4 (3x^2 + 2x)\,dx \approx \dfrac{4 - 1}{3}\left(\dfrac{39}{4} + \dfrac{95}{4} + \dfrac{175}{4}\right) = \dfrac{309}{4} = 77.25$

$\displaystyle\int_1^4 (3x^2 + 2x)\,dx = (x^3 + x^2)\Big]_1^4 = 80 - 2 = 78$

3. Trapezoidal Rule

k	0	1	2	3	4
x_k	1	3/2	2	5/2	3
$f(x_k)$	2	35/8	9	133/8	28

$\displaystyle\int_1^3 (x^3 + 1)\,dx \approx \dfrac{3 - 1}{8}\left[2 + 2\left(\dfrac{35}{8}\right) + 2(9) + 2\left(\dfrac{133}{8}\right) + 28\right] = \dfrac{45}{2} \approx 22.5$

$\displaystyle\int_1^3 (x^3 + 1)\,dx = \left(\dfrac{x^4}{4} + x\right)\Big]_1^3 = \dfrac{93}{4} - \dfrac{5}{4} = 22$

5. Midpoint Rule

k	1	2	3	4	5
x_k	3/2	5/2	7/2	9/2	11/2
$f(x_k)$	2/3	2/5	2/7	2/9	2/11

$\displaystyle\int_1^6 \dfrac{1}{x}\,dx \approx \dfrac{6 - 1}{5}\left(\dfrac{2}{3} + \dfrac{2}{5} + \dfrac{2}{7} + \dfrac{2}{9} + \dfrac{2}{11}\right) = \dfrac{6086}{3465} \approx 1.75642$

Trapezoidal Rule

k	0	1	2	3	4	5
x_k	1	2	3	4	5	6
$f(x_k)$	1	1/2	1/3	1/4	1/5	1/6

$\displaystyle\int_1^6 \dfrac{1}{x}\,dx \approx \dfrac{6 - 1}{12}\left[1 + 2\left(\dfrac{1}{2}\right) + 2\left(\dfrac{1}{3}\right) + 2\left(\dfrac{1}{4}\right) + 2\left(\dfrac{1}{5}\right) + \dfrac{1}{6}\right] = \dfrac{28}{15} \approx 1.86667$

7. Midpoint Rule

k	1	2	3	4	5
x_k	0.05	0.15	0.25	0.35	0.45
$f(x_k)$	1.00125	1.01119	1.03078	1.05948	1.09659

k	6	7	8	9	10
x_k	0.55	0.65	0.75	0.85	0.95
$f(x_k)$	1.14127	1.19269	1.25000	1.31244	1.37931

$$\int_0^1 \sqrt{x^2+1}\,dx \approx \frac{1-0}{10}(1.00125 + \cdots + 1.37931) \approx 1.1475$$

Trapezoidal Rule

k	0	1	2	3	4	5
x_k	0	0.1	0.2	0.3	0.4	0.5
$f(x_k)$	1	1.00499	1.0198	1.04403	1.07703	1.11803

k	6	7	8	9	10
x_k	0.6	0.7	0.8	0.9	1.0
$f(x_k)$	1.16619	1.22066	1.28062	1.34536	1.41421

$$\int_0^1 \sqrt{x^2+1}\,dx \approx \frac{1-0}{20}[1 + 2(1.00499) + \cdots + 2(1.34536) + 1.41421] \approx 1.14838$$

9. Midpoint Rule

k	1	2	3	4	5	6
x_k	$\pi/12$	$\pi/4$	$5\pi/12$	$7\pi/12$	$3\pi/4$	$11\pi/12$
$f(x_k)$	0.0760474	0.180063	0.217033	0.194188	0.128617	0.0429833

$$\int_0^\pi \frac{\sin x}{x+\pi}\,dx \approx \frac{\pi-0}{6}(0.0760474 + \cdots + 0.0429833) \approx 0.439263$$

Trapezoidal Rule

k	0	1	2	3	4	5	6
x_k	0	$\pi/6$	$\pi/3$	$\pi/2$	$2\pi/3$	$5\pi/6$	π
$f(x_k)$	0	0.136419	0.206748	0.212207	0.165399	0.0868118	0

$$\int_0^\pi \frac{\sin x}{x+\pi}\,dx \approx \frac{\pi-0}{12}[1 + 2(0.136419) + \cdots + 2(0.0868118) + 0] \approx 0.42285$$

11. Midpoint Rule

k	1	2	3	4	5	6
x_k	1/6	1/2	5/6	7/6	3/2	11/6
$f(x_k)$	0.999614	0.968912	0.768409	0.208152	−0.628174	−0.976002

$$\int_0^2 \cos x^2\,dx \approx \frac{2-0}{6}(0.999614 + \cdots - 0.976002) \approx 0.446971$$

Trapezoidal Rule

k	0	1	2	3	4	5	6
x_k	0	1/3	2/3	1	4/3	5/3	2
$f(x_k)$	1	0.993834	0.90285	0.540302	−0.205507	−0.934546	−0.653644

$$\int_0^2 \cos x^2\,dx \approx \frac{2-0}{12}[1 + 2(0.993834) + \cdots + 2(0.934546) - 0.653644] \approx 0.490037$$

13. Simpson's Rule

k	0	1	2	3	4
x_k	0	1	2	3	4
$f(x_k)$	1	$\sqrt{3}$	$\sqrt{5}$	$\sqrt{7}$	3

$$\int_0^4 \sqrt{2x+1}\,dx \approx \frac{4-0}{12}(1 + 4\sqrt{3} + 2\sqrt{5} + 4\sqrt{7} + 3) \approx 8.6611$$

$$\int_0^4 \sqrt{2x+1}\,dx \qquad \boxed{u = 2x+1,\ du = 2\,dx}$$

$$= \int_1^9 u^{1/2}\left(\frac{1}{2}\,du\right) = \frac{1}{2}\left(\frac{2}{3}u^{3/2}\right)\Bigg]_1^9 = \frac{26}{3} \approx 0.86667$$

15. Simpson's Rule

k	0	1	2	3	4
x_k	1/2	1	3/2	2	5/2
$f(x_k)$	2	1	2/3	1/2	2/5

$$\int_{1/2}^{5/2} \frac{1}{x}\, dx \approx \frac{5/2 - 1/2}{12}\left[2 + 4(1) + 2\left(\frac{2}{3}\right) + 4\left(\frac{1}{2}\right) + \frac{2}{5}\right] \approx 1.6222$$

17. Simpson's Rule

k	0	1	2	3	4
x_k	0	1/4	1/2	3/4	1
$f(x_k)$	1	16/17	4/5	16/25	1/2

$$\int_0^1 \frac{1}{1+x^2}\, dx \approx \frac{1-0}{12}\left[1 + 4\left(\frac{16}{17}\right) + 2\left(\frac{4}{5}\right) + 4\left(\frac{16}{25}\right) + \frac{1}{2}\right] \approx 0.7854$$

19. Simpson's Rule

k	0	1	2	3	4	5	6
x_k	0	$\pi/6$	$\pi/3$	$\pi/2$	$2\pi/3$	$5\pi/6$	π
$f(x_k)$	0	$3/7\pi$	$3\sqrt{8}/8\pi$	$2/3\pi$	$3\sqrt{3}/10\pi$	$3/11\pi$	0

$$\int_0^\pi \frac{\sin x}{x+\pi}\, dx \approx \frac{\pi - 0}{18}\left[0 + 4\left(\frac{3}{7\pi}\right) + 2\left(\frac{3\sqrt{3}}{8\pi}\right) + 4\left(\frac{2}{3\pi}\right) + 2\left(\frac{3\sqrt{3}}{10\pi}\right) + 4\left(\frac{3}{11\pi}\right) + 0\right]$$
$$\approx 0.4339$$

21. Simpson's Rule

k	0	1	2	3	4
x_k	2	2.5	3	3.5	4
$f(x_k)$	3.1623	4.2573	5.4772	6.8099	8.2462

$$\int_2^4 \sqrt{x^3 + x}\, dx \approx \frac{4-2}{12}[3.1623 + 4(4.2573) + 2(5.4772) + 4(6.8099) + 8.2462] \approx 11.1053$$

23. $f(x) = \dfrac{1}{x+3}$; $f'(x) = -\dfrac{1}{(x+3)^2}$; $f''(x) = \dfrac{2}{(x+3)^3}$. Since $f''(x)$ decreases on $[-1,2]$, $f''(x) \le f''(-1) = 1/4$ on the interval. Taking $M = 1/4$, we want $\dfrac{(1/4)(2+1)^3}{24n^2} < 0.005$ or $n^2 > \dfrac{225}{4} = 56.25$. Take $n = 8$ to obtain the desired accuracy.

25. $f(x) = \dfrac{1}{1+x^2}$; $f'(x) = -\dfrac{2x}{(1+x^2)^2}$; $f''(x) = \dfrac{6x^2 - 2}{(1+x^2)^3}$. To determine an upper bound for $|f''(x)|$ on $[0,2]$, we compute $f'''(x) = \dfrac{24x(1-x^2)}{(1+x^2)^4}$. Setting $f'''(x) = 0$ we obtain the critical numbers 0 and 1 on $[0,2]$. Comparing $f''(0) = -2$, $f''(1) = \dfrac{1}{2}$, and $f''(2) = \dfrac{22}{125}$, we see that $|f''(x)| \le 2$ on the interval. Taking $M = 2$ we want $\dfrac{2(2-0)^3}{12n^2} < 0.005$ or $n^2 > \dfrac{800}{3} \approx 267$. For $n = 17$, the Trapezoidal Rule gives $\displaystyle\int_0^2 \frac{1}{1+x^2}\, dx \approx 1.11$ to two decimal places.

27. $f'(x) = -\dfrac{1}{x^2}$; $f''(x) = \dfrac{2}{x^3}$; $f'''(x) = -\dfrac{6}{x^4}$; $f^{(4)}(x) = \dfrac{24}{x^5}$. Since $f^{(4)}(x)$ is decreasing on $[1,3]$, $|f^{(4)}(x)| \le f^{(4)}(1) = 24$ for all x in $[1,3]$. Taking $M = 24$, we want $\dfrac{24(3-1)^5}{180n^4} < 10^{-5}$ or $n^4 > \dfrac{1,280,000}{3} \approx 426,667$. To have the desired accuracy with Simpson's Rule, we need $n \ge 26$. To obtain the required n for the Trapezoidal Rule, we note that $f''(x) = \dfrac{2}{x^3}$ is decreasing on $[1,3]$. We thus take $M = f''(1) = 2$. We want $\dfrac{2(3-1)^3}{12n^2} < 10^{-5}$ or $n^2 > \dfrac{400,000}{3} \approx 133,333$. For the Trapezoidal Rule to have the desired accuracy, we need $n \ge 366$.

29. Since $n = 5$ is odd we cannot use Simpson's Rule. Because the Midpoint Rule does not readily work with tabular data we use the Trapezoidal Rule:

$$\int_{2.05}^{2.30} f(x)\,dx \approx \frac{2.30 - 2.05}{10}[4.91 + 2(4.80) + 2(4.66) + 2(4.41) + 2(3.93) + 3.58] = 1.10225.$$

31. $\int_0^4 (2x+5)\,dx = (x^2 + 5x)\Big]_0^4 = 36$

$n = 2$: $\int_0^4 (2x+5)\,dx \approx \frac{4-0}{2}[(2\cdot1+5) + (2\cdot3+5)] = 36$

$n = 4$: $\int_0^4 (2x+5)\,dx \approx \frac{4-0}{4}[(2\cdot0.5+5) + (2\cdot1.5+5) + (2\cdot2.5+5) + (2\cdot3.5+5)]$

$$= 36$$

33. (a) $I = \int_{-1}^1 (x^3 + x^2)\,dx = \left(\frac{1}{4}x^4 + \frac{1}{3}x^3\right)\Big]_{-1}^1 = \frac{2}{3}$

(b) $M_8 = \frac{1-(-1)}{8}[f(-7/8) + f(-5/8) + \cdots + f(5/8) + f(7/8)]$

$$= \frac{1}{4}\left(\frac{49}{512} + \frac{75}{512} + \cdots + \frac{325}{512} + \frac{735}{512}\right) = \frac{21}{32}$$

(c) $T_8 = \frac{1-(-1)}{16}[f(-1) + 2f(-3/4) + \cdots + 2f(3/4) + f(1)]$

$$= \frac{1}{8}\left[0 + 2\left(\frac{9}{64}\right) + \cdots + 2\left(\frac{63}{64}\right) + 2\right] = \frac{11}{16}$$

(d) $E_8 = \left|\frac{2}{3} - M_8\right| = \frac{1}{96}$; $E_8 = \left|\frac{2}{3} - T_8\right| = \frac{1}{48}$

The error for the Midpoint Rule is one half the error for the Trapezoidal Rule.

35. The exact value is

$$\int_a^b (c_1x + c_0)\,dx = \left(\frac{c_1}{2}x^2 + c_0x\right)\Big]_a^b = \frac{c_1}{2}(b^2 - a^2) + c_0(b - a)$$

$$= \frac{b-a}{2}[(c_1a + c_0) + (c_1b + c_0)].$$

The Trapezoidal Rule gives

$$\frac{b-a}{2n}\left[(c_1a + c_0) + 2\left[c_1\left(a + \frac{b-a}{n}\right) + c_0\right] + 2\left\{c_1\left[a + \frac{2(b-a)}{n}\right] + c_0\right\} + \cdots \right.$$

$$\left. + 2\left\{c_1\left[a + \frac{(n-1)(b-a)}{n}\right] + c_0\right\} + \left\{c_1\left[a + \frac{n(b-a)}{n}\right] + c_0\right\}\right]$$

$$= \frac{b-a}{2n}\left[2nc_1a + 2nc_0 + 2c_1\frac{b-a}{n}(1 + 2 + \cdots + n - 1) + c_1(b - a)\right]$$

$$= \frac{b-a}{2n}\left\{2nc_1a + 2nc_0 + 2c_1\left(\frac{b-a}{n}\right)\left[\frac{(n-1)n}{2}\right] + c_1(b - a)\right\}$$

$$= \frac{b-a}{2n}[2nc_1a + 2nc_0 + c_1(b-a)(n-1) + c_1(b-a)]$$

$$= \frac{b-a}{2n}[2nc_1a + 2nc_0 + nc_1b - nc_1a - c_1b + c_1a + c_1b - c_1a]$$

$$= \frac{b-a}{2n}(c_1a + c_0 + c_1b + c_0).$$

Since the graph of $f(x) = c_1x + c_0$ is a straight line and the Trapezoidal Rule uses straight line approximations to the curve, it will give the exact value.

37. $\int_1^4 f(x)\,dx \approx \dfrac{4-1}{3(6)}[1.3+4(1.5)+2(3)+4(3.3)+2(2.2)+4(2.4)+1.9] \approx 7.06667$

39. $A \approx \dfrac{18.6-0}{3(10)}[0-4(5.8)+2(7.3)+4(6.9)+2(8.7)+4(8.8)+2(10.3)+4(14.5)$

$$+2(15)+4(10.4)+0]$$

$$= 166.284$$

The volume of the pond is $4(166.284) = 665.136$ ft^3 and the number of gallons of water is $7.48(665.136) = 4975.22$ gallons.

41. Simpson's Rule

k	0	1	2	3	4	5
x_k	-5	-4	-3	-2	-1	0
$f(x_k)$	0	3.55936	4.38712	4.79112	4.96403	5

k	6	7	8	9	10
x_k	1	2	3	4	5
$f(x_k)$	4.96403	4.79112	4.38712	3.55936	0

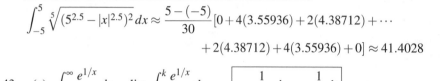

$\int_{-5}^{5} \sqrt[5]{(5^{2.5}-|x|^{2.5})^2}\,dx \approx \dfrac{5-(-5)}{30}[0+4(3.55936)+2(4.38712)+\cdots$

$$+2(4.38712)+4(3.55936)+0] \approx 41.4028$$

43. (a) $\displaystyle\int_1^\infty \frac{e^{1/x}}{x^{5/2}}\,dx = \lim_{k\to\infty}\int_1^k \frac{e^{1/x}}{x^{5/2}}\,dx \qquad \boxed{x = \frac{1}{t},\ dx = -\frac{1}{t^2}\,dt}$

$$= \lim_{k\to\infty}\int_1^{1/k} \frac{e^t}{(1/t)^{5/2}}\left(-\frac{1}{t^2}\,dt\right) = \int_1^0 (-t^{1/2}e^t\,dt) = \int_0^1 t^{1/2}e^t\,dt$$

(b) $\displaystyle\int_1^\infty \frac{e^{1/x}}{x^{5/2}}\,dx = \int_0^1 t^{1/2}e^t\,dt$

$$\approx \frac{1-0}{12}\left[0+4\left(\frac{1}{2}e^{1/4}\right)+2\left(\sqrt{\frac{1}{2}}e^{1/2}\right)+4\left(\frac{\sqrt{3}}{2}e^{3/4}\right)+e\right]$$

$$\approx \frac{1}{12}(2.5681+2.3316+7.3335+2.7183) \approx 1.2460$$

45. $y' = 2x$

$s = \displaystyle\int_0^1 \sqrt{1+4x^2}\,dx \approx \frac{1-0}{20}\Big\{[1+4(0)^2]^{1/2}+2[1+4(0.1)^2]^{1/2}+2[1+4(0.2)^2]^{1/2}+\cdots$

$$+2[1+4(0.9)^2]^{1/2}+[1+4(1)^2]^{1/2}\Big\}$$

$$= \frac{1}{20}(1+2\sqrt{1.04}+2\sqrt{1.16}+\cdots+2\sqrt{4.24}+\sqrt{5}) \approx 1.4804$$

47. $y' = x;\quad S = 2\pi\displaystyle\int_0^2 \frac{1}{2}x^2\sqrt{1+x^2}\,dx = \pi\int_0^2 x^2\sqrt{1+x^2}\,dx.$ Using the Midpoint Rule with $n = 5$, we obtain

$$M_5 = \frac{2}{5}\left(\frac{1}{25}\sqrt{1+\frac{1}{25}}+\frac{9}{25}\sqrt{1+\frac{9}{25}}+\sqrt{2}+\frac{49}{25}\sqrt{1+\frac{49}{25}}+\frac{81}{25}\sqrt{1+\frac{81}{25}}\right) \approx 4.76741.$$

Then $S \approx \pi M_5 \approx 14.9772.$

49. (a) Approximating the graph on each interval between integers with line segments and using the Pythagorean theorem, we have

$$s < \sqrt{1^2+2^2}+\sqrt{1^2+4^2}+\sqrt{1^2+3^2}+\sqrt{1^2+1^2}+\sqrt{1^2+3^2}+\sqrt{1^2+1^2}+\sqrt{1^2+2^2}$$

$$= \sqrt{5}+\sqrt{17}+\sqrt{10}+\sqrt{2}+\sqrt{10}+\sqrt{2}+\sqrt{5} \approx 17.75.$$

(b) Since $y'(x_i) = 0$ for $x_i = 1, 2, \ldots, 8$, the formula for arc length gives $\displaystyle\int_1^8 \sqrt{1+(y')^2}\,dx = \int_1^8 dx = 7$ which is simply the length of the interval.

Chapter 7 in Review

A. True/False

1. True

3. True

5. True

7. False; $\dfrac{1}{(x^2-1)^2} = \dfrac{A}{x-1} + \dfrac{B}{(x-1)^2} + \dfrac{C}{x+1} + \dfrac{D}{(x+1)^2}.$

9. False; substituting $u = 9 - x^2$ will work.

11. True

13. True

15. False; the infinite discontinuity occurs at $x = 1/e \approx 0.3679$, which is outside of $[1/2, 1]$.

17. True

19. False; since

$$\int_2^\infty \left(\frac{e^x}{e^x+1} - \frac{e^x}{e^x-1} \right) dx = \lim_{t\to\infty} \int_2^t \left(\frac{e^x}{e^x+1} - \frac{e^x}{e^x-1} \right) dx$$

$$= \lim_{t\to\infty} \left(\ln|e^x+1| - \ln|e^x-1| \right) \Big]_2^t$$

$$= \lim_{t\to\infty} \ln\left| \frac{e^t+1}{e^t-1} \right| - \ln\frac{e^2+1}{e^2-1} = \ln\left(\lim_{t\to\infty} \frac{e^t+1}{e^t-1} \right) - \ln\frac{e^2+1}{e^2-1}$$

$$= \ln\left(\lim_{t\to\infty} \frac{1+e^{-t}}{1-e^{-t}} \right) - \ln\frac{e^2+1}{e^2-1} = -\ln\frac{e^2+1}{e^2-1},$$

the integral converges.

B. Fill In the Blanks

1. $1/5$ (see Problem 77 in Exercises 7.7)

3. $\displaystyle\int_0^\infty \frac{e^{-x}}{\sqrt{x}}\,dx$ $\boxed{u = \sqrt{x},\ x = u^2,\ dx = 2u\,du}$

$$= \int_0^\infty \frac{e^{-u^2}}{u}(2u\,du) = 2\int_0^\infty e^{-u^2}\,du = 2\left(\frac{\sqrt{\pi}}{2} \right) = \sqrt{\pi}$$

5. Integrating, we obtain the equation $-\dfrac{1}{2}e^{-2t}\Big]_0^x = \lim_{k\to\infty}\left(-\dfrac{1}{2}e^{-2t} \right)\Big]_x^k = -\dfrac{1}{2}\lim_{k\to\infty} 2e^{-2t}\Big]_x^k.$

Then, solving for x, $\quad e^{-2x} - e^{-2(0)} = 0 - e^{-2x}; \quad 2e^{-2x} = 1; \quad -2x = \ln\left(\dfrac{1}{2} \right); \quad$ and

$x = -\dfrac{1}{2}(0 - \ln 2) = \ln\sqrt{2}.$

C. Exercises

1. $\displaystyle\int \frac{1}{\sqrt{x}+9}\,dx$ $\boxed{u = \sqrt{x}+9,\ x = (u-9)^2,\ dx = 2(u-9)\,du}$

$$= \int \frac{2(u-9)}{u}\,du = 2\int du - 18\int \frac{1}{u}\,du = 2u = 18\ln|u| + C$$

$$= 2\sqrt{x} + 18 - 18\ln(\sqrt{x}+9) + C = 2\sqrt{x} - 18\ln(\sqrt{x}+9) + C_1$$

3. $\displaystyle\int \frac{x}{\sqrt{x^2+4}}\,dx = \frac{1}{2}\int (x^2+4)^{-1/2}(2x\,dx) = \frac{1}{2}\left[\frac{(x^2+4)^{1/2}}{1/2}\right]+C = \sqrt{x^2+4}+C$

5. $\displaystyle\int \frac{1}{(x^2+4)^3}\,dx$ $\boxed{x = 2\tan\theta,\ dx = 2\sec^2\theta\,d\theta}$ $= \displaystyle\int \frac{2\sec^2\theta}{(4\tan^2\theta+4)^3}\,d\theta$

$\qquad\qquad = \displaystyle\frac{1}{32}\int \frac{\sec^2\theta}{\sec^6\theta}\,d\theta = \frac{1}{32}\int \cos^4\theta\,d\theta$

$\qquad\qquad \boxed{\text{See Section 7.4, Example 5}}$

$\qquad\qquad = \displaystyle\frac{3}{256}\theta + \frac{1}{128}\sin 2\theta + \frac{1}{1024}\sin 4\theta + C$

$\qquad\qquad = \displaystyle\frac{3}{256}\tan^{-1}\frac{x}{2} + \frac{1}{64}\sin\theta\cos\theta + \frac{1}{256}\sin\theta\cos\theta(\cos^2\theta - \sin^2\theta) + C$

$\qquad\qquad = \displaystyle\frac{3}{256}\tan^{-1}\frac{x}{2} + \frac{x}{32(x^2+4)} + \frac{x}{128(x^2+4)}\left(\frac{4-x^2}{x^2+4}\right) + C$

$\qquad\qquad = \displaystyle\frac{3}{256}\tan^{-1}\frac{x}{2} + \frac{x}{32(x^2+4)} + \frac{x}{32(x^2+4)^2} - \frac{x^3}{128(x^2+4)^2} + C$

7. $\displaystyle\int \frac{x^2+4}{x^2}\,dx = \int dx + 4\int \frac{1}{x^2}\,dx = x - \frac{4}{x} + C$

9. $\displaystyle\int \frac{x-5}{x^2+4}\,dx = \frac{1}{2}\int \frac{2x}{x^2+4}\,dx - 5\int \frac{1}{x^2+4}\,dx = \frac{1}{2}\ln(x^2+4) - \frac{5}{2}\tan^{-1}\frac{x}{2} + C$

11. $\displaystyle\int \frac{(\ln x)^9}{x}\,dx$ $\boxed{u = \ln x,\ du = \frac{1}{x}\,dx}$ $= \displaystyle\int u^9\,du = \frac{1}{10}u^{10}+C = \frac{1}{10}(\ln x)^{10}+C$

13. $\displaystyle\int t\sin^{-1}t\,dt$ $\boxed{u = \sin^{-1}t,\ du = \frac{1}{\sqrt{1-t^2}}\,dt;\quad dv = t\,dt,\ v = \frac{1}{2}t^2}$

$\qquad\qquad = \displaystyle\frac{1}{2}t^2\sin^{-1}t - \frac{1}{2}\int \frac{t^2}{\sqrt{1-t^2}}\,dt$ $\boxed{t = \sin\theta,\ dt = \cos\theta\,d\theta}$

$\qquad\qquad = \displaystyle\frac{1}{2}t^2\sin^{-1}t - \frac{1}{2}\int \frac{\sin^2\theta}{\cos\theta}\cos\theta\,d\theta = \frac{1}{2}t^2\sin^{-1}t - \frac{1}{2}\int \sin^2\theta\,d\theta$

$\qquad\qquad = \displaystyle\frac{1}{2}t^2\sin^{-1}t - \frac{1}{4}\int (1-\cos 2\theta)\,d\theta = \frac{1}{2}t^2\sin^{-1}t - \frac{1}{4}\left(\theta - \frac{1}{2}\sin 2\theta\right) + C$

$\qquad\qquad = \displaystyle\frac{1}{2}t^2\sin^{-1}t - \frac{1}{4}\sin^{-1}t + \frac{1}{4}\sin\theta\cos\theta + C$

$\qquad\qquad = \displaystyle\frac{1}{2}t^2\sin^{-1}t - \frac{1}{4}\sin^{-1}t + \frac{1}{4}t\sqrt{1-t^2} + C$

15. $\displaystyle\int (x+1)^3(x-2)\,dx$ $\boxed{u = x+1,\ x = u-1,\ dx = du}$

$\qquad\qquad = \displaystyle\int u^3(u-3)\,du = \int (u^4-3u^3)\,du$

$\qquad\qquad = \displaystyle\frac{1}{5}u^5 - \frac{3}{4}u^4 + C = \frac{1}{5}(x+1)^5 - \frac{3}{4}(x+1)^4 + C$

17. $\displaystyle\int \ln(x^2+4)\,dx$ $\boxed{u = \ln(x^2+4),\ du = \frac{2x}{x^2+4}\,dx;\quad dv = dx,\ v = x}$

$\qquad\qquad = \displaystyle x\ln(x^2+4) - \int \frac{2x^2}{x^2+4}\,dx = x\ln(x^2+4) - 2\int \frac{x^2+4-4}{x^2+4}\,dx$

$\qquad\qquad = \displaystyle x\ln(x^2+4) - 2\int dx + 8\int \frac{1}{x^2+4}\,dx = x\ln(x^2+4) - 2x + 4\tan^{-1}\frac{x}{2} + C$

19. Write $\displaystyle\frac{1}{x^4+10x^3+25x^2} = \frac{A}{x} + \frac{B}{x^2} + \frac{C}{x+5} + \frac{D}{(x+5)^2}$.

Then $1 = Ax(x+5)^2 + B(x+5)^2 + Cx^2(x+5) + Dx^2$

$= (A+C)x^3 + (10A+B+5C+D)x^2 + (25A+10B)x + 25B.$

Solving

$A+C = 0$	$10A+B+5C+D = 0$
$25A+10B = 0$	$25B = 1$

gives $A = -2/125$, $B = 1/25$, $C = 2/125$, and $D = 1/25$. Thus

$$\int \frac{1}{x^4 + 10x^3 + 25x^2}\,dx = -\frac{2}{125}\int \frac{1}{x}\,dx + \frac{1}{25}\int \frac{1}{x^2}\,dx + \frac{2}{125}\int \frac{1}{x+5}\,dx + \frac{1}{25}\int \frac{1}{(x+5)^2}\,dx$$

$$= -\frac{2}{125}\ln|x| - \frac{1}{25}\left(\frac{1}{x}\right) + \frac{2}{125}\ln|x+5| - \frac{1}{25}\left(\frac{1}{x+5}\right) + C$$

21. Write $\dfrac{x}{x^3 + 3x^2 - 9x - 27} = \dfrac{A}{x+3} + \dfrac{B}{(x+3)^2} + \dfrac{C}{x-3}$.

Then $x = A(x^2 - 9) + B(x-3) + C(x+3)^2 = (A+C)x^2 + (B+6C)x + (-9A - 3B + 9C).$

Solving

$A+C = 0$	$B+6C = 1$	$-9A - 3B + 9C = 0$

gives $A = -1/12$, $B = 1/2$, and $C = 1/12$. Thus

$$\int \frac{x}{x^3 + 3x^2 - 9x - 27}\,dx = -\frac{1}{12}\int \frac{1}{x+3}\,dx + \frac{1}{2}\int \frac{1}{(x+3)^2}\,dx + \frac{1}{12}\int \frac{1}{x-3}\,dx$$

$$= -\frac{1}{12}\ln|x+3| - \frac{1}{2}\left(\frac{1}{x+3}\right) + \frac{1}{12}\ln|x-3| + C$$

$$= \frac{1}{12}\ln\left|\frac{x-3}{x+3}\right| - \frac{1}{2(x+3)} + C.$$

23. $\displaystyle\int \frac{\sin^2 t}{\cos^2 t}\,dt = \int \tan^2 t\,dt = \int (\sec^2 t - 1)\,dt = \tan t - t + C$

25. $\displaystyle\int \tan^{10} x \sec^4 x\,dx = \int \tan^{10} x (\tan^2 x + 1)\sec^2 x\,dx$ $\boxed{u = \tan x,\ du = \sec^2 x\,dx}$

$$= \int u^{10}(u^2 + 1)\,du = \int (u^{12} + u^{10})\,du = \frac{1}{13}u^{13} + \frac{1}{11}u^{11} + C$$

$$= \frac{1}{13}\tan^{13} x + \frac{1}{11}\tan^{11} x + C$$

27. $\displaystyle\int y\cos y\,dy$ $\boxed{u = y,\ du = dy;\quad dv = \cos y\,dy,\ v = \sin y}$

$$= y\sin y - \int \sin y\,dy = y\sin y + \cos y + C$$

29. $\displaystyle\int (1 + \sin^2 t)\cos^3 t\,dt = \int (1 + \sin^2 t)(1 - \sin^2 t)\cos t\,dt = \int (1 - \sin^4 t)\cos t\,dt$

$$\boxed{u = \sin t,\ du = \cos t\,dt}$$

$$= \int (1 - u^4)\,du = u - \frac{1}{5}u^5 + C = \sin t - \frac{1}{5}\sin^5 t + C$$

31. $\displaystyle\int e^w(1 + e^w)^5\,dw$ $\boxed{u = 1 + e^w,\ du = e^w\,dw}$

$$= \int u^5\,du = \frac{1}{6}u^6 + C = \frac{1}{6}(1 + e^w)^6 + C$$

33. $\displaystyle\int \cot^3 4x\,dx = \int (\csc^2 4x - 1)\cot 4x\,dx = \int \csc 4x\csc 4x\cot 4x\,dx - \int \cot 4x\,dx$

$$\boxed{u = \csc 4x,\ du = -4\csc 4x\cot 4x\,dx}$$

$$= -\frac{1}{4}\int u\,du - \frac{1}{4}\ln|\sin 4x| + C = -\frac{1}{8}u^2 - \frac{1}{4}\ln|\sin 4x| + C$$

$$= -\frac{1}{8}\csc^2 4x - \frac{1}{4}\ln|\sin 4x| + C$$

35. $\displaystyle\int_0^{\pi/4}\cos^2 x\tan x\,dx = \int_0^{\pi/4}\cos x\sin x\,dx = \frac{1}{2}\int_0^{\pi/4}\sin 2x\,dx = -\frac{1}{4}\cos 2x\Big]_0^{\pi/4}$

$$= -\frac{1}{4}(0-1) = \frac{1}{4}$$

37. $\displaystyle\int \frac{\sin x}{1+\sin x}\,dx = \int \frac{(\sin x)(1-\sin x)}{(1+\sin x)(1-\sin x)}\,dx = \int \frac{\sin x - \sin^2 x}{1-\sin^2 x}\,dx$

$$= \int \left(\frac{\sin x}{\cos^2 x} - \frac{\sin^2 x}{\cos^2 x}\right)dx = \int \left[\frac{1}{\cos x}\left(\frac{\sin x}{\cos x}\right) - \tan^2 x\right]dx$$

$$= \int [\sec x\tan x - (\sec^2 x - 1)]\,dx = \sec x - \tan x + x + C$$

Alternatively, the substitution $u = \tan\dfrac{x}{2}$ leads to the equivalent solution

$$\int \frac{\sin x}{1+\sin x}\,dx = x + \frac{2}{1+\tan x/2} + C.$$

39. Write $\displaystyle\frac{1}{(x+1)(x+2)(x+3)} = \frac{A}{x+1} + \frac{B}{x+2} + \frac{C}{x+3}.$

Then $1 = A(x+2)(x+3) + B(x+1)(x+3) + C(x+1)(x+2).$

Setting $x = -1$, $x = -2$, and $x = -3$ gives $A = 1/2$, $B = -1$, and $C = 1/2$. Thus

$$\int_0^1 \frac{1}{(x+1)(x+2)(x+3)}\,dx = \frac{1}{2}\int_0^1 \frac{1}{x+1}\,dx - \int_0^1 \frac{1}{x+2}\,dx + \frac{1}{2}\int_0^1 \frac{1}{x+3}\,dx$$

$$= \frac{1}{2}\ln|x+1|\Big]_0^1 - \ln|x+2|\Big]_0^1 + \frac{1}{2}\ln|x+3|\Big]_0^1$$

$$= \frac{1}{2}(\ln 2 - \ln 1) - (\ln 3 - \ln 2) + \frac{1}{2}(\ln 4 - \ln 3) = \frac{5}{2}\ln 2 - \frac{3}{2}\ln 3.$$

41. $\displaystyle\int e^x \cos 3x\,dx$
$$\boxed{u = e^x,\ du = e^x\,dx;\quad dv = \cos 3x\,dx,\ v = \frac{1}{3}\sin 3x}$$

$$= \frac{1}{3}e^x\sin 3x - \frac{1}{3}\int e^x\sin 3x\,dx$$

$$\boxed{u = e^x,\ du = e^x\,dx;\quad dv = \sin 3x\,dx,\ v = -\frac{1}{3}\cos 3x}$$

$$= \frac{1}{3}e^x\sin 3x - \frac{1}{3}\left(-\frac{1}{3}e^x\cos 3x + \frac{1}{3}\int e^x\cos 3x\,dx\right)$$

Solving for the integral, $\displaystyle\int e^x\cos 3x\,dx = \frac{3}{10}e^x\sin 3x + \frac{1}{10}e^x\cos 3x + C.$

43. $\displaystyle\int \cos(\ln t)\,dt$ $\boxed{u = \cos(\ln t),\ du = -\dfrac{\sin(\ln t)}{t}\,dt;\quad dv = dt,\ v = t}$

$$= t\cos(\ln t) + \int \sin(\ln t)\,dt$$

$\boxed{u = \sin(\ln t),\ du = \dfrac{\cos(\ln t)}{t}\,dt;\quad dv = dt,\ v = t}$

$$= t\cos(\ln t) + t\sin(\ln t) - \int \cos(\ln t)\,dt$$

Solving for the integral, $\displaystyle\int \cos(\ln t)\,dt = \frac{1}{2}t\cos(\ln t) + \frac{1}{2}t\sin(\ln t) + C.$

45. $\displaystyle\int \cos\sqrt{x}\,dx$ $\boxed{t = \sqrt{x},\ x = t^2,\ dx = 2t\,dt}$

$$= 2\int t\cos t\,dt \quad \boxed{u = t,\ du = dt;\quad dv = \cos t\,dt,\ v = \sin t}$$

$$= 2t\sin t - 2\int \sin t\,dt = 2t\sin t + 2\cos t + C$$

$$= 2\sqrt{x}\sin\sqrt{x} + 2\cos\sqrt{x} + C$$

47. $\displaystyle\int \cos x\sin 2x\,dx = 2\int \cos^2 x\sin x\,dx \quad \boxed{u = \cos x,\ du = -\sin x\,dx}$

$$= -2\int u^2\,du = -\frac{2}{3}u^3 + C = -\frac{2}{3}\cos^3 x + C$$

49. $\displaystyle\int \sqrt{x^2 + 2x + 5}\,dx = \int \sqrt{(x+1)^2 + 4}\,dx \quad \boxed{x + 1 = 2\tan\theta,\ dx = 2\sec^2\theta\,d\theta}$

$$= \int \sqrt{4\tan^2\theta + 4}(2\sec^2\theta\,d\theta) = 4\int \sec^3\theta\,d\theta$$

$\boxed{\text{See Section 7.3, Example 5}}$

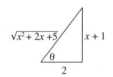

$$= 2\sec\theta\tan\theta + 2\ln|\sec\theta + \tan\theta| + C$$

$$= \frac{1}{2}(x+1)\sqrt{x^2 + 2x + 5} + 2\ln\left|\frac{\sqrt{x^2 + 2x + 5}}{2} + \frac{x+1}{2}\right| + C$$

$$= \frac{1}{2}(x+1)\sqrt{x^2 + 2x + 5} + 2\ln\left|\sqrt{x^2 + 2x + 5} + x + 1\right| + C_1$$

51. $\displaystyle\int \tan^5 x\sec^3 x\,dx = \int \tan^4 x\sec^2 x\sec x\tan x\,dx = \int (\sec^2 x - 1)^2\sec^2 x\sec x\tan x\,dx$

$\boxed{u = \sec x,\ du = \sec x\tan x\,dx}$

$$= \int (u^2 - 1)^2 u^2\,du = \int (u^6 - 2u^4 + u^2)\,du = \frac{1}{7}u^7 - \frac{2}{5}u^5 + \frac{1}{3}u^3 + C$$

$$= \frac{1}{7}\sec^7 x - \frac{2}{5}\sec^5 x + \frac{1}{3}\sec^3 x + C$$

53. $\displaystyle\int \frac{t^5}{1 + t^2}\,dt = \int \left(t^3 - t + \frac{t}{1 + t^2}\right)dt = \frac{1}{4}t^4 - \frac{1}{2}t^2 + \frac{1}{2}\ln(1 + t^2) + C$

55. $\displaystyle\int \frac{5x^3 + x^2 + 6x + 1}{(x^2 + 1)^2}\,dx = \int \frac{(5x+1)(x^2+1) + x}{(x^2+1)^2}\,dx = \int \frac{5x+1}{x^2+1}\,dx + \int \frac{x}{(x^2+1)^2}\,dx$

$$= \int \frac{5x}{x^2+1}\,dx + \int \frac{1}{x^2+1}\,dx + \int \frac{x}{(x^2+1)^2}\,dx$$

$$= \frac{5}{2}\ln(x^2+1) + \tan^{-1}x - \frac{1}{2(x^2+1)} + C$$

57. $\displaystyle\int x\sin^2 x\,dx$ $\boxed{u = x\sin x,\ du = (x\cos x + \sin x)\,dx;\quad dv = \sin x\,dx,\ v = -\cos x}$

$$= -x\sin x\cos x + \int x\cos^2 x\,dx + \int \sin x\cos x\,dx$$

$$= -\frac{1}{2}x\sin 2x + \int x(1-\sin^2 x)\,dx + \frac{1}{2}\int \sin 2x\,dx$$

$$= -\frac{1}{2}x\sin 2x + \frac{1}{2}x^2 - \int x\sin^2 x\,dx - \frac{1}{4}\cos 2x$$

Solving for the integral, $\displaystyle\int x\sin^2 x\,dx = \frac{1}{4}x^2 - \frac{1}{4}x\sin 2x - \frac{1}{8}\cos 2x + C.$

59. $\displaystyle\int e^{\sin x}\sin 2x\,dx = \int e^{\sin x}(2\sin x\cos x)\,dx = 2\int e^{\sin x}\sin x\cos x\,dx$

$$\boxed{u = \sin x,\ du = \cos x\,dx;\quad dv = e^{\sin x}\cos x,\ v = e^{\sin x}}$$

$$= 2e^{\sin x}\sin x - 2\int e^{\sin x}\cos x\,dx = 2e^{\sin x}\sin x - 2e^{\sin x} + C$$

61. $\displaystyle\int_0^{\pi/6}\frac{\cos x}{\sqrt{1+\sin x}}\,dx$ $\boxed{u = 1+\sin x,\ du = \cos x\,dx}$

$$= \int_1^{3/2}\frac{1}{\sqrt{u}}\,du = 2\sqrt{u}\,\Big]_1^{3/2} = 2\sqrt{3/2}-2 = \sqrt{6}-2$$

63. $\displaystyle\int \sinh^{-1}t\,dt$ $\boxed{u = \sinh^{-1}t,\ du = \dfrac{1}{\sqrt{t^2+1}}\,dt;\quad dv = dt,\ v = t}$

$$= t\sinh^{-1}t - \int \frac{t}{\sqrt{t^2+1}}\,dt = t\sinh^{-1}t - \sqrt{t^2+1} + C$$

65. $\displaystyle\int_3^8 \frac{1}{x\sqrt{x+1}}\,dx$ $\boxed{u^2 = x+1,\ 2u\,du = dx}$

$$= \int_2^3 \frac{2u}{(u^2-1)u}\,du = \int_2^3 \frac{2}{(u-1)(u+1)}\,du = \int_2^3 \frac{1}{u-1}\,du - \int_2^3 \frac{1}{u+1}\,du$$

$$= \ln|u-1|\,\Big]_2^3 - \ln|u+1|\,\Big]_2^3 = \ln 2 - \ln\frac{4}{3} = \ln\frac{3}{2}$$

67. $\displaystyle\int \frac{\sec^4 3u}{\cot^{12} 3u}\,du = \int \tan^{12}3u\sec^2 3u\sec^2 3u\,du = \int \tan^{12}3u(1+\tan^2 3u)\sec^2 3u\,du$

$$= \frac{1}{39}\tan^{13}3u + \frac{1}{45}\tan^{15}3u + C$$

69. $\displaystyle\int \frac{3+\sin x}{\cos^2 x}\,dx = \int 3\sec^2 x\,dx + \int \tan x\sec x\,dx = 3\tan x + \sec x + C$

71. $\displaystyle\int x(1+\ln x)^2\,dx$ $\boxed{u = (1+\ln x)^2,\ du = \dfrac{2(1+\ln x)}{x}\,dx;\quad dv = x\,dx,\ v = \dfrac{1}{x^2}}$

$$= \frac{1}{2}x^2(1+\ln x)^2 - \int x(1+\ln x)\,dx$$

$$\boxed{u = 1+\ln x,\ du = \dfrac{1}{x}\,dx;\quad dv = x\,dx,\ v = \dfrac{1}{2}x^2}$$

$$= \frac{1}{2}x^2(1+\ln x)^2 - \left[\frac{1}{2}x^2(1+\ln x) - \int \frac{1}{2}x\,dx\right]$$

$$= \frac{1}{2}x^2(1+\ln x)\ln x - \frac{1}{2}x^2(1+\ln x) + \frac{1}{4}x^2 + C$$

73. $\displaystyle\int e^x e^{e^x}\,dx$ $\boxed{u = e^x,\ du = e^x\,dx}$ $\displaystyle= \int e^u\,du = e^u + C = e^{e^x} + C$

75. $\displaystyle\int \frac{2t}{1+e^{t^2}}\,dt \qquad \boxed{u=t^2,\ du=2t\,dt}$

$$= \int \frac{1}{1+e^u}\,du = \int \frac{e^{-u}}{e^{-u}+1}\,du \qquad \boxed{v=e^{-u},\ dv=e^{-u}\,du}$$

$$= \int \frac{1}{v+1}(-dv) = -\ln|v+1|+C = -\ln(e^{-u}+1)+C = -\ln(e^{-t^2}+1)+C$$

77. $\displaystyle\int \frac{1}{\sqrt{1-(5x+2)^2}}\,dx = \frac{1}{5}\int \frac{5}{\sqrt{1-(5x+2)^2}}\,dx = \frac{1}{5}\sin^{-1}(5x+2)+C$

79. $\displaystyle\int \cos x \ln|\sin x|\,dx \qquad \boxed{u=\ln|\sin x|,\ du=\dfrac{\cos x}{\sin x}\,dx;\quad dv=\cos x\,dx,\ v=\sin x}$

$$= \sin x \ln|\sin x| - \int \cos x\,dx = \sin x \ln|\sin x| - \sin x + C$$

81. $\displaystyle\int_0^3 x(x^2-9)^{-2/3}\,dx = \lim_{t\to 3^-}\int_0^t x(x^2-9)^{-2/3}\,dx = \lim_{t\to 3^-} \left.\frac{3}{2}(x^2-9)^{1/3}\right]_0^t$

$$= \lim_{t\to 3^-}\left[\frac{3}{2}(t^2-9)^{1/3} - \frac{3}{2}(-9)^{1/3}\right] = \frac{3\sqrt[3]{9}}{2}$$

83. $\displaystyle\int_{-\infty}^0 (x+1)e^x\,dx = \lim_{s\to -\infty}\int_s^0 (x+1)e^x\,dx \qquad \boxed{u=x+1,\ du=dx;\quad dv=e^x\,dx,\ v=e^x}$

$$= \lim_{s\to -\infty}\left[\left.(x+1)e^x\right]_s^0 - \int_s^0 e^x\,dx\right] = \lim_{s\to -\infty}\left[1-(s+1)e^s - \left.e^x\right]_s^0\right]$$

$$= \lim_{s\to -\infty}\left[1-(s+1)e^s -1 + e^s\right] = \lim_{s\to -\infty}(-se^s) \qquad \boxed{\text{Let } t=-s}$$

$$= \lim_{t\to \infty} te^{-t} = \lim_{t\to \infty}\frac{t}{e^t} \overset{h}{=} \lim_{t\to \infty}\frac{1}{e^t} = 0$$

85. $\displaystyle\int_3^\infty \frac{1}{1+5x}\,dx = \lim_{t\to \infty}\int_3^t \frac{1}{1+5x}\,dx = \lim_{t\to \infty} \left.\frac{1}{5}\ln|1+5x|\right]_3^t = \lim_{t\to \infty}\left(\frac{1}{5}\ln|1+5t| - \frac{1}{5}\ln 16\right)$

The integral diverges.

87. $\displaystyle\int_0^e \ln\sqrt{x}\,dx = \lim_{s\to 0^+}\int_s^e \frac{1}{2}\ln x\,dx = \lim_{s\to 0^+} \left.\frac{1}{2}(x\ln x - x)\right]_s^e = \lim_{s\to 0^+}\left[\left(\frac{e}{2}-\frac{e}{2}\right) - \frac{1}{2}(s\ln s - s)\right]$

$$= \lim_{s\to 0^+}\left(\frac{1-\ln s}{2/s}\right) \overset{h}{=} \lim_{s\to 0^+}\frac{-1/s}{-2/s^2} = \lim_{s\to 0^+} s = 0$$

89. $\displaystyle\int_0^{\pi/2} \frac{1}{1-\cos x}\,dx = \lim_{s\to 0^+}\int_s^{\pi/2} \frac{1}{1-\cos x}\left(\frac{1+\cos x}{1+\cos x}\right)dx = \lim_{s\to 0^+}\int_s^{\pi/2}\frac{1+\cos x}{\sin^2 x}\,dx$

$$= \lim_{s\to 0^+}\int_s^{\pi/2}(\csc^2 x + \cot x \csc x)\,dx = \lim_{s\to 0^+}\left.(-\cot x - \csc x)\right]_s^{\pi/2}$$

$$= \lim_{s\to 0^+}(\cot s + \csc s - 0 - 1)$$

Since $\displaystyle\lim_{s\to 0^+}\cot s = +\infty$ and $\displaystyle\lim_{s\to 0^+}\csc s = +\infty$, the integral diverges.

91. $\displaystyle\int_0^1 \frac{1}{\sqrt{x}e^{\sqrt{x}}}\,dx = \lim_{s\to 0^+}\int_s^1 x^{-1/2}e^{-x^{1/2}}\,dx \qquad \boxed{u=-x^{1/2},\ du=-\dfrac{1}{2}x^{-1/2}\,dx}$

$$= \lim_{s\to 0^+}\int_{-\sqrt{s}}^{-1} -2e^u\,du = \lim_{s\to 0^+}\left.(-2e^u)\right]_{-\sqrt{s}}^{-1} = \lim_{s\to 0^+}(2e^{-\sqrt{s}} - 2e^{-1}) = 2 - 2e^{-1}$$

93. $\displaystyle\int_1^\infty \frac{\sqrt{x}}{(1+x)^2}\,dx = \lim_{t\to\infty}\int_1^t \frac{\sqrt{x}}{(1+x)^2}\,dx$ $\qquad \boxed{x=u^2,\ dx=2u\,du}$

$\displaystyle = \lim_{t\to\infty}\int_1^{\sqrt{t}}\frac{u}{(1+u^2)^2}(2u\,du) = 2\lim_{t\to\infty}\int_1^{\sqrt{t}}\frac{u^2}{(1+u^2)^2}\,du$

$\displaystyle = 2\lim_{t\to\infty}\int_1^{\sqrt{t}}\frac{1+u^2-1}{(1+u^2)^2}\,du = 2\lim_{t\to\infty}\int_1^{\sqrt{t}}\left[\frac{1}{1+u^2}-\frac{1}{(1+u^2)^2}\right]du$

$\displaystyle = 2\lim_{t\to\infty}\left[\tan^{-1}u\Big]_1^{\sqrt{t}} - \int_1^{\sqrt{t}}\frac{1}{(1+u^2)^2}\,du\right]$

$\displaystyle = 2\lim_{t\to\infty}\left[\tan^{-1}\sqrt{t}-\frac{\pi}{4}-\int_1^{\sqrt{t}}\frac{1}{(1+u^2)^2}\,du\right]$ $\qquad \boxed{u=\tan\theta,\ du=\sec^2\theta\,d\theta}$

$\displaystyle = 2\left[\frac{\pi}{2}-\frac{\pi}{4}-\lim_{t\to\infty}\int_{\pi/4}^{\tan^{-1}\sqrt{t}}\frac{\sec^2\theta\,d\theta}{(1+\tan^2\theta)^2}\right] = \frac{\pi}{2}-2\lim_{t\to\infty}\int_{\pi/4}^{\tan^{-1}\sqrt{t}}\frac{\sec^2\theta\,d\theta}{\sec^4\theta}$

$\displaystyle = \frac{\pi}{2}-2\lim_{t\to\infty}\int_{\pi/4}^{\tan^{-1}\sqrt{t}}\cos^2\theta\,d\theta = \frac{\pi}{2}-2\lim_{t\to\infty}\int_{\pi/4}^{\tan^{-1}\sqrt{t}}\frac{1}{2}(1+\cos2\theta)\,d\theta$

$\displaystyle = \frac{\pi}{2}-2\lim_{t\to\infty}\left(\frac{1}{2}\theta+\frac{1}{4}\sin2\theta\right)\Big]_{\pi/4}^{\tan^{-1}\sqrt{t}}$

$\displaystyle = \frac{\pi}{2}-2\lim_{t\to\infty}\left(\frac{1}{2}\theta+\frac{1}{2}\sin\theta\cos\theta\right)\Big]_{\pi/4}^{\tan^{-1}\sqrt{t}}$

$\displaystyle = \frac{\pi}{2}-2\lim_{t\to\infty}\left[\frac{1}{2}\tan^{-1}\sqrt{t}+\frac{1}{2}\left(\frac{\sqrt{t}}{\sqrt{1+t}}\right)\frac{1}{\sqrt{1+t}}-\frac{\pi}{8}-\frac{1}{4}\right]$

$\displaystyle = \frac{\pi}{2}-2\lim_{t\to\infty}\left[\frac{1}{2}\tan^{-1}\sqrt{t}+\frac{1}{2}\left(\frac{\sqrt{t}}{1+t}\right)-\frac{\pi}{8}-\frac{1}{4}\right]$

$\displaystyle = \frac{\pi}{2}-2\left[\frac{1}{2}\left(\frac{\pi}{2}\right)-\frac{\pi}{8}-\frac{1}{4}\right]-\lim_{t\to\infty}\frac{1}{1/\sqrt{t}+\sqrt{t}} = \frac{\pi}{4}+\frac{1}{2}$

95. $\displaystyle\lim_{x\to\infty}\frac{x\displaystyle\int_0^x e^{t^2}\,dt}{e^{x^2}} \overset{h}{=} \lim_{x\to\infty}\frac{xe^{x^2}+\displaystyle\int_0^x e^{t^2}\,dt}{2xe^{x^2}} = \lim_{x\to\infty}\left(\frac{1}{2}+\frac{\displaystyle\int_0^x e^{t^2}\,dt}{2xe^{x^2}}\right)$

$\displaystyle \overset{h}{=} \frac{1}{2}+\lim_{x\to\infty}\frac{e^{x^2}}{2x(2xe^{x^2})+2e^{x^2}} = \frac{1}{2}+\lim_{x\to\infty}\frac{1}{4x^2+2} = \frac{1}{2}$

97. $\displaystyle A = \int_0^\infty (e^{-x}-e^{-3x})\,dx = \lim_{t\to\infty}\int_0^t (e^{-x}-e^{-3x})\,dx$

$\displaystyle = \lim_{t\to\infty}\left(-e^{-x}+\frac{1}{3}e^{-3x}\right)\Big]_0^t = \lim_{t\to\infty}\left(\frac{1}{3}e^{-3t}-e^{-t}+1-\frac{1}{3}\right) = \frac{2}{3}$

99. (a) The horizontal asymptote is $y=1$. Using symmetry,

$$\text{Area}(R_1) = 2\int_0^\infty\left(1-\frac{x^2-1}{x^2+1}\right)dx = 2\int_0^\infty\frac{2}{x^2+1}\,dx = 2\lim_{t\to\infty}\int_0^t\frac{2}{x^2+1}\,dx$$

$$= 2\lim_{t\to\infty}2\tan^{-1}x\Big]_0^t = 2\lim_{t\to\infty}2\tan^{-1}t = 2\pi.$$

(b) If the area of R_3 were finite, then, since R_1 has finite area, the area of the infinite strip bounded by $x=0$, $y=0$, and $y=1$ would be finite. The area of the strip is infinite, so the area of R_3 must be infinite. By symmetry, the area of R_2 is infinite.

101. Using the Trapezoidal Rule,

$$W = \int_0^1 F(x)\,dx \approx \frac{1-0}{10}[0+2(50)+2(90)+2(150)+2(210)+260] = 126 \text{ joules}.$$

First-Order Differential Equations

8.1 Separable Equations

1. From $dy = \sin 5x\,dx$ we obtain $y = -\dfrac{1}{5}\cos 5x + c$.

3. $\displaystyle\int y^{-3}\,dy = \int x^{-2}\,dx$

$-\dfrac{1}{2}y^{-2} = -x^{-1} + C$

$y^{-2} = 2x^{-1} + C_1$

5. $\displaystyle\int (1 + 2y + y^2)\,dy = \int (1 + 2x + x^2)\,dx$

$y + y^2 + \dfrac{1}{3}y^3 = x + x^2 + \dfrac{1}{3}x^3 + C$

7. $\displaystyle\int \sin y\,dy = \int (x^{-2} + 5)\,dx$

$-\cos y = -x^{-1} + 5x + C$

$\cos y = \dfrac{1}{x} - 5x + C_1$

9. From $\dfrac{1}{y}dy = \dfrac{4}{x}dx$ we obtain $\ln|y| = 4\ln|x| + c$ or $y = c_1 x^4$.

11. From $e^{-2y}dy = e^{3x}dx$ we obtain $3e^{-2y} + 2e^{3x} = c$.

13. From $\left(y + 2 + \dfrac{1}{y}\right)dy = x^2 \ln x\,dx$ we obtain $\dfrac{y^2}{2} + 2y + \ln|y| = \dfrac{x^3}{3}\ln|x| - \dfrac{1}{9}x^3 + c$.

15. From $\dfrac{1}{N}dN = (te^{t+2} - 1)dt$ we obtain $\ln|N| = te^{t+2} - e^{t+2} - t + c$ or $N = c_1 e^{te^{t+2} - e^{t+2} - t}$.

17. From $\dfrac{1}{5P - P^2}dP = \left(\dfrac{1}{5P} - \dfrac{1}{5(P-5)}\right)dP = dt$ we obtain $\dfrac{1}{5}\ln|P| - \dfrac{1}{5}\ln|P - 5| = t + c$ so that

$\ln\left|\dfrac{P}{(P-5)}\right| = 5t + c_1$ or $\dfrac{P}{P-5} = c_2 e^{5t}$. Solving for P we have $P = \dfrac{5c_2 e^{5t}}{c_2 e^{5t} - 1}$

19. From $\dfrac{y-2}{y+3}dy = \dfrac{x-1}{x+4}dx$ or $\left(1 - \dfrac{5}{y+3}\right)dy = \left(1 - \dfrac{5}{x+4}\right)dx$ we obtain $y - 5\ln|y+3| = x - 5\ln|x+4| + c$ or $\left(\dfrac{x+4}{y+3}\right)^5 = c_1 e^{x-y}$.

209

21. $\int y^2 \, dy = \int x^{-2} \, dx$

$$\frac{1}{3}y^3 = -x^{-1} + C$$

$$y^3 = -3x^{-1} + C_1$$

Setting $x = 1$ and $y = 3$, we obtain $27 = -3 + C_1$ or $C_1 = 30$. Thus, $y^3 = -3x^{-1} + 30$.

23. From $\dfrac{1}{x^2 + 1} dx = 4 \, dt$ we obtain $\tan^{-1} x = 4t + c$. Using $x(\pi/4) = 1$ we find $c = -3/\pi$. The solution of the initial-value problem

is $\tan^{-1} x = 4t - \dfrac{3\pi}{4}$ or $x = \tan\left(4t - \dfrac{3\pi}{4}\right)$.

25. From $\dfrac{1}{y} dy = \dfrac{1-x}{x^2} dx = \left(\dfrac{1}{x^2} - \dfrac{1}{x}\right) dx$ we obtain $\ln|y| = -\dfrac{1}{x} - \ln|x| = c$ or $xy = c_1 e^{-1/x}$. Using $y(-1) = -1$ we find $c_1 = e^{-1}$.

The solution of the initial-value problem is $xy = e^{-1-1/x}$ or $y = e^{-(1+1/x)}$**3.**.

27. Separating variables and integrating we obtain

$$\frac{dx}{\sqrt{1-x^2}} - \frac{dy}{\sqrt{1-y^2}} = 0 \quad \text{and} \quad \sin^{-1} x - \sin^{-1} y = c.$$

Setting $x = 0$ and $y = \sqrt{3}/2$ we obtain $c = -\pi/3$. Thus, an implicit solution of the initial-value problem is $\sin^{-1} x - \sin^{-1} y = \pi/3$. Solving for y and using an addition formula from trigonometry, we get

$$y = \sin\left(\sin^{-1} x + \frac{\pi}{3}\right) = x\cos\frac{\pi}{3} + \sqrt{1-x^2}\sin\frac{\pi}{3} = \frac{x}{2} + \frac{\sqrt{3}\sqrt{1-x^2}}{2}.$$

29. Substituting $y = k$ and $\dfrac{dy}{dx} = 0$, we get $6k = 18$ or $k = 3$ so that $y = 3$.

31. Substituting $y = k$ and $\dfrac{dy}{dx} = 0$, we get $0 = k^2 - k - 20$ or $0 = (k-5)(k+4)$ so that $y = 5$ or $y = -4$.

33. Substituting $y = k$ and $\dfrac{dy}{dx} = 0$ and then solving for k, we see that there are two constant solutions: $y = 0$ and $y = 1$. Separating variables, we have

$$\frac{dy}{y^2 - y} = \frac{dx}{x} \quad \text{or} \quad \int \frac{dy}{y(y-1)} = \ln|x| + c.$$

Using partial fractions, we obtain

$$\int \left(\frac{1}{y-1} - \frac{1}{y}\right) dy = \ln|x| + c \ln|y-1| - \ln|y| = \ln|x| + c \ln\left|\frac{y-1}{xy}\right| = c\frac{y-1}{xy} = e^c = c_1.$$

Solving for y we get $y = 1/(1 - c_1 x)$.

(a) Setting $x = 0$ and $y = 1$ we have $1 = 1/(1/0)$, which is true for all values of c_1. Thus, solutions passing through $(0, 1)$ are $y = 1/(1 - c_1 x)$.

(b) Setting $x = 0$ and $y = 0$ in $y = 1/(1 - c_1 x)$ we get $0 = 1$. Thus, the only solution passing through $(0, 0)$ is $y = 0$.

(c) Setting $x = \dfrac{1}{2}$ and $y = \dfrac{1}{2}$ we have $\dfrac{1}{2} = 1/(1 - \tfrac{1}{2}c_1)$, so $c_1 = -2$ and $y = 1/(1 + 2x)$.

35. In order for $y'(x_0) = \sqrt{y + 0}$ to be defined, we must have $y + 0 \geq 0$.

37. The right side of the differential equation $\dfrac{dy}{dx} = -\dfrac{x}{y}$ is not defined for $y = 0$. At $x = \pm 5$, however, the function $y = -\sqrt{25 - x^2}$

is zero.

8.2 Linear Equations

1. For $y' - 4y = 0$, an integrating factor is $e^{-\int 4dx} = e^{-4x}$ so that $\dfrac{d}{dx}\left[e^{-4x}y\right] = 0$ and $y = ce^{4x}$ for $-\infty < x < \infty$.

3. Writing the equation as $y' + 5y = \dfrac{1}{2}$, the integrating factor is $e^{\int 5dx} = e^{5x}$. Then $\dfrac{d}{dx}\left[e^{5x}y\right] = \dfrac{1}{2}e^{5x}$, so $e^{5x}y = \dfrac{1}{10}e^{5x} + C$, and $y = \dfrac{1}{10} + Ce^{-5x}$. (This equation can also be solved by separation of variables.)

5. For $y' + y = e^{3t}$, an integrating factor is $e^{\int dt} = e^{t}$ so that $\dfrac{d}{dt}[e^{t}y] = e^{4t}$ and $y = \dfrac{1}{4}e^{3t} + ce^{-t}$ for $-\infty < t < \infty$.

7. For $y' + 3x^2 y = x^2$, an integrating factor is $e^{\int 3x^2\,dx} = e^{x^3}$ so that $\dfrac{d}{dx}\left[e^{x^3}y\right] = x^2 e^{x^3}$ and $y = \dfrac{1}{3} + ce^{-x^3}$ for $-\infty < x < \infty$.

9. For $y' + \dfrac{1}{x}y = \dfrac{1}{x^2}$, an integrating factor is $e^{\int (1/x)\,dx} = x$ so that $\dfrac{d}{dx}[xy] = \dfrac{1}{x}$ and $y = \dfrac{1}{x}\ln x + \dfrac{c}{x}$ for $0 < x < \infty$.

11. For $y' + \dfrac{e^x}{1+e^x}y = 0$, an integrating factor is $e^{\int e^x/(1+e^x)\,dx} = 1 + e^x$ so that $\dfrac{d}{dx}\left[(1+e^x)y\right] = 0$ and $y = \dfrac{c}{1+e^x}$ for $-\infty < x < \infty$.

13. For $y' - \dfrac{1}{x}y = x\sin x$ an integrating factor is $e^{-\int(1/x)dx} = \dfrac{1}{x}$ so that $\dfrac{d}{dx}\left[\dfrac{1}{x}y\right] = \sin x$ and $y = cx - x\cos x$ for $0 < x < \infty$.

15. For $y' + (\tan x)y = \sec x$, an integrating factor is $e^{\int \tan x\,dx} = \sec x$ so that $\dfrac{d}{dx}\left[(\sec x)y\right] = \sec^2 x$ and $y = \sin x + c\cos x$ for $-\dfrac{\pi}{2} < x < \dfrac{\pi}{2}$.

17. For $y' + (\cot x)y = 2\cos x$, an integrating factor is $e^{\int \cot x\,dx} = \sin x$ so that $\dfrac{d}{dx}\left[(\sin x)y\right] = 2\sin x\cos x$ and $y = \sin x + c\csc x$ for $0 < x < \pi$.

19. For $y' + \dfrac{4}{x+2}y = \dfrac{5}{(x+2)^2}$ an integrating factor is $e^{\int [4/(x+2)]dx} = (x+2)^4$ so that $\dfrac{d}{dx}\left[(x+2)^2 y\right] = 5(x+2)^2$ and $y = \dfrac{5}{3}(x+2)^{-1} + c(x+2)^{-4}$ for $-2 < x < \infty$. The entire solution is transient.

21. For $y' + \left(1 + \dfrac{2}{x}\right)y = \dfrac{e^x}{x^2}$ an integrating factor is $e^{\int [1+(2/x)]dx} = x^2 e^x$ so that $\dfrac{d}{dx}[x^2 e^x y] = e^{2x}$ and $y = \dfrac{1}{2}\dfrac{e^x}{x^2} + \dfrac{ce^{-x}}{x^2}$ for $0 < x < \infty$. The transient term is $\dfrac{ce^{-x}}{x^2}$.

23. For $y' - y = x$ and integrating factor is $e^{\int -1\,dx} = e^{-x}$ so that $\dfrac{d}{dx}[e^{-x}y] = e^{-x}x$ and $y = -x - 1 + ce^x$. Substituting $x = 0$ and $y = -4$, we have $-4 = -1 + c$ or $c = -3$ so that $y = -x - 1 - 3e^x$.

25. For $y' + \dfrac{1}{x}y = \dfrac{1}{x}e^x$, an integrating factor is $e^{\int (1/x)\,dx} = x$ so that $\dfrac{d}{dx}[xy] = e^x$ and $y = \dfrac{1}{x}e^x + \dfrac{c}{x}$ for $0 < x < \infty$. If $y(1) = 2$ then $c = 2 - e$ and $y = \dfrac{1}{x}e^x + \dfrac{2-e}{x}$.

27. For $y' - \dfrac{1}{x}y = 2x$ an integrating factor is $e^{\int -\frac{1}{x}dx} = e^{-\ln x} = \dfrac{1}{x}$ so that $\dfrac{d}{dx}\left[\dfrac{y}{x}\right] = 2$ and $y = 2x^2 + cx$. Substituting $x = 5$ and $y = 1$, we have $1 = 50 + 5c$ or $c = \dfrac{-49}{5}$ so that $y = 2x^2 - \dfrac{49}{5}x$.

29. For $x' + \dfrac{1}{t+1}x = \dfrac{\ln t}{t+1}$ and integrating factor is $e^{\int [1/(t+1)]dt} = t+1$ so that $\dfrac{d}{dt}[(t+1)x] = \ln t$ and $x = \dfrac{t}{t+1}\ln t - \dfrac{t}{t+1} + \dfrac{c}{t+1}$ for $0 < t < \infty$. If $x(1) = 10$ then $c = 21$ and $x = \dfrac{t}{t+1}\ln t - \dfrac{t}{t+1} + \dfrac{21}{t+1}$.

31. For $\dfrac{di}{dt} + \dfrac{R}{L}i = \dfrac{E}{L}$ and integrating factor is $e^{\int (R/L)dt} = e^{Rt/L}$ so that $\dfrac{d}{dt}\left[e^{Rt/L}i\right] = \dfrac{E}{L}e^{Rt/L}$ and $i = \dfrac{E}{R} + ce^{-Rt/L}$ for $-\infty < t < \infty$. If $i(0) = i_0$ then $c = i_0 - E/R$ and $i = \dfrac{E}{R} + \left(i_0 - \dfrac{E}{R}\right)e^{-Rt/L}$.

33. (a) An integrating factor for $y' - 2xy = 2$ is $e^{\int -2x\,dx} = e^{-x^2}$. Thus $\dfrac{d}{dx}[e^{-x^2}y] = 2e^{-x^2}$

$e^{-x^2}y = 2\int_0^x e^{-t^2}\,dt + c = \sqrt{\pi}\,\mathrm{erf}(x) + c.$

  (b) Using a CAS, we find $y(2) \approx 150.92$.

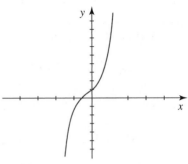

35. Note that f is discontinuous at $x = 1$. We solve the problem in two parts. For $0 \le x \le 1$, $f(x) = 1$ and an integrating factor is $e^{\int dx} = e^x$. Then $\dfrac{d}{dx}[e^x y] = e^x$, $e^x y = e^x + c$, and $y = 1 + ce^{-x}$. Since $y(0) = 0$, $c = -1$ and $y = 1 - e^{-x}$ for $0 \le x \le 1$. For $x > 1$, the differential equation is $\dfrac{dy}{dx} = -y$. By separation of variables, we obtain $y = c_1 e^{-x}$. Thus, $y = \begin{cases} 1 - e^{-x}, & 0 \le x \le 1 \\ c_1 e^{-x}, & x > 1 \end{cases}$.

In order to make y a continuous function, we require $\lim\limits_{x \to 1^-} y(x) = \lim\limits_{x \to 1^+} y(x)$ or $1 - e^{-1} = c_1 e^{-1}$. Then $c_1 = e - 1$ and $y = \begin{cases} 1 - e^{-x}, & 0 \le x \le 1 \\ (e-1)e^{-x}, & x > 1 \end{cases}$.

In the graph, f is shown in blue, and the solution of the IVP is shown in red.

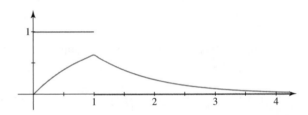

37. On the interval $(-3, 3)$ the integrating factor is

$$e^{\int x\,dx/(x^2-9)} = e^{-\int x\,dx/(9-x^2)} = e^{\frac{1}{2}\ln(9-x^2)} = \sqrt{9 - x^2}$$

and so

$$\frac{d}{dx}\left[\sqrt{9-x^2}\,y\right] = 0 \text{ and } y = \frac{c}{\sqrt{9-x^2}}.$$

39. The solution of the first equation is $x = c_1 e^{-\lambda_1 t}$. From $x(0) = x_0$, we obtain $c_1 = x_9$ and so $x = x_0 e^{-\lambda_1 t}$. The equation equation then becomes

$$\frac{dy}{dt} = -x_0 \lambda_1 e^{-\lambda_1 t} - \lambda_2 y \quad \text{or} \quad y' + \lambda_2 y = -x_0 \lambda_1 e^{-\lambda_1 t}$$

which is linear. An integrating factor is $e^{\int \lambda_2 dt} = e^{\lambda_2 t}$. Thus

$$\frac{d}{dt}\left[e^{\lambda_2 t}y\right] = -x_0 \lambda_1 e^{-\lambda_1 t} e^{\lambda_2 t} = x_0 \lambda_1 e^{(\lambda_2 - \lambda_1)t}$$

$$e^{\lambda_2 t}y = \frac{-x_0 \lambda_1}{\lambda_2 - \lambda_1}e^{-\lambda_1 t} + c_2 e^{-\lambda_2 t}.$$

From $y(0) = y_0$, we obtain

$$y_0 = \frac{-x_0 \lambda_1}{\lambda_2 - \lambda_1} + c_2 \quad \text{or} \quad c_2 = \frac{y_0 \lambda_2 - y_0 \lambda_1 + x_0 \lambda_1}{\lambda_2 - \lambda_1}.$$

The solution is then $y = \dfrac{-x_0 \lambda_1}{\lambda_2 - \lambda_1}e^{-\lambda_1 t} + \dfrac{y_0 \lambda_2 - y_0 \lambda_1 + x_0 \lambda_1}{\lambda_2 - \lambda_1}e^{-\lambda_2 t}.$

41. The new DE is $\dfrac{dx}{dy} = -x - y$ or $\dfrac{dx}{dy} + x = -y$ which is linear in the variable x and can therefore be solved.

43. (a) For $y' + \dfrac{3}{x}y = 6x$, an integrating factor is $e^{\int \frac{3}{x}dx} = e^{3\ln x} = x^3$ so that

$$\frac{d}{dx}[x^3 y] = 6x^4$$

$$x^3 y = \frac{6}{5}x^5 + c$$

$$y = \frac{6}{5}x^2 + cx^{-3}$$

(b) Substituting $x = -1$ and $y = 2$, we have

$$2 = \frac{6}{5}(-1)^2 + c(-1)^{-3}$$

$$2 = \frac{6}{5} - c$$

$$c = -\frac{4}{5}$$

The solution is $y = \dfrac{6}{5}x^3 - \dfrac{4}{5}x^{-3}$. The solution is valid for x in $(-\infty, 0)$.

(c) Substituting $x = 1$ and $y = 2$, we have

$$2 = \frac{6}{5}(1)^2 + c(1)^{-3}$$

$$2 = \frac{6}{5} + c$$

$$c = \frac{4}{5}$$

The solution is $y = \dfrac{6}{5}x^3 + \dfrac{4}{5}x^{-3}$. The solution is valid for x in $(0, \infty)$.

(d) We need $c = 0$ so that $y = \dfrac{6}{5}x^2$. Let the initial condition be $y(1) = \dfrac{6}{5}$.

8.3 Mathematical Models

1. Let $P = P(t)$ be the population at time t, and P_0 the initial population. From $\dfrac{dP}{dt} = kP$, we obtain $P = P_0 e^{kt}$. Using $P(5) = 2P_0$, we find $k = \dfrac{1}{5}\ln 2$ and $P = P_0 e^{(\ln 2)t/5}$. Setting $P(t) = 3P_0$, we have $3 = e^{(\ln 2)t/5}$, so $\ln 3 = (\ln 2)\dfrac{t}{5}$, and $t = \dfrac{5\ln 3}{\ln 2} \approx 7.9$ years. Setting $P(t) = 4P_0$, we have $4 = e^{(\ln 2)t/5}$, so $\ln 4 = (\ln 2)\dfrac{t}{5}$, and $t = 10$ years.

3. Let $P = P(t)$ be the population at time t. Then $dP/dt = kP$ and $P + ce^k t$. From $P(0) = c = 500$ we see that $P = 500e^{5t}$. Since 15% of 500 is 75, we have $P(10) = 500e^{10k} = 575$. Solving for k, we get $k = \dfrac{1}{10}\ln\dfrac{575}{500} = \dfrac{1}{10}\ln 1.15$. When $t = 30$,

$$P(30) = 500e^{(1/10)(\ln 1.15)30} = 500e^{3\ln 1.15} = 760 \text{ years.}$$

5. Let $A = A(t)$ be the amount of lead present at time t. From $dA/dt = kA$ and $A(0) = 1$ we obtain $A = e^{kt}$. Using $A(3.3) = 1/2$ we find $k = \dfrac{1}{3.3}\ln(1/2)$. When 90% of the lead has decayed, 0.1 grams will remain. Setting $A(t) = 0.1$ we have $e^{t(1/3.3)\ln(1/2)} = 0.1$, so

$$\frac{t}{3.3}\ln\frac{1}{2} = \ln 0.1 \quad \text{and} \quad t = \frac{3.3\ln 0.1}{\ln(1/2)} \approx 10.96 \text{ hours.}$$

7. Setting $N(t) = 50$ in Problem 6, we obtain $50 = 100e^{kt}$, so $kt = \ln\dfrac{1}{2}$, and $t = \dfrac{\ln 1/2}{(1/6)\ln 0.97} \approx 136.5$ hours.

9. Let $I = I(t)$ be the intensity, t the thickness, and $I(0) = I_0$. If $\dfrac{dI}{dt} = kI$ and $I(3) = 0.25I_0$, then $I = I_0 e^{kt}$, $k = \dfrac{1}{3}\ln 0.25$, and $I(15) = 0.00098I_0$.

11. Assume that $A = A_0 e^{kt}$ and $k = -0.00012378$. If $A(t) = 0.145A_0$, then $t \approx 15,600$ years.

13. Assume that $\dfrac{dT}{dt} = k(T-10)$ so that $T = 10 + ce^{kt}$. If $T(0) = 70°$ and $T(1/2) = 50°$, then $c = 60$ and $k = 2\ln\dfrac{2}{3}$ so that $T(1) = 36.67°$. If $T(t) = 15°$, then $t = 3.06$ minutes.

15. From $\dfrac{dA}{dt} = 4 - \dfrac{A}{50}$, we obtain $A = 200 + ce^{-t/50}$. If $A(0) = 30$ then $c = -170$ and $A = 200 - 170e^{-t/50}$.

17. From $\dfrac{dA}{dt} = 10 - \dfrac{A}{50}$, we obtain $A = 1000 + ce^{-t/100}$. If $A(0) = 0$ then $c = -1000$ and $A = 1000 - 1000e^{-t/100}$.

19. From $\dfrac{dA}{dt} = 10 - \dfrac{10A}{500 - (10-5)t} = 10 - \dfrac{2A}{100-t}$, we obtain $A = 1000 - 10t + c(100-t)^2$. If $A(0) = 0$, then $c = -\dfrac{1}{10}$. The tank is empty in 100 minutes. x

21. From $\dfrac{ds}{dt} = v(t)$, we have $ds = \left[\dfrac{mg}{k} + \left(v_0 - \dfrac{mg}{k}\right)e^{-kt/m}\right] dt$. Integrating, we obtain

$$s = \dfrac{mg}{k}t + \dfrac{v_0 - mg/k}{-k/m}e^{-kt/m} + C = \dfrac{mg}{k}t + \left(\dfrac{m^2g}{k^2} - \dfrac{mv_0}{k}\right)e^{-kt/m} + C.$$

Setting $s(0) = 0$, we obtain $0 = \dfrac{m^2g}{k^2} - \dfrac{mv_0}{k} + C$ or $C = -\dfrac{m^2g}{k^2} + \dfrac{mv_0}{k}$. Thus,

$$s(t) = \dfrac{mg}{k}t + \left(\dfrac{m^2g}{k^2} - \dfrac{mv_0}{k}\right)\left(e^{-kt/m} - 1\right).$$

23. From $\dfrac{dX}{dt} = A - BX$ and $X(0) = 0$, we obtain $X = \dfrac{A}{B} - \dfrac{A}{B}e^{-Bt}$ so that $X \to \dfrac{A}{B}$ as $t \to \infty$. If $X(t) = \dfrac{A}{2B}$, then $t = \dfrac{\ln 2}{B}$.

25. From $\dfrac{dE}{dt} = -\dfrac{E}{RC}$ and $E(t_1) = E_0$, we obtain $E = E_0e^{(t_1-t)/RC}$.

27. Assume $L\dfrac{di}{dt} + Ri = E(t)$, $L = 0.1$, $R = 50$, and $E(t) = 50$ so that $i = \dfrac{3}{5} + ce^{-500t}$. If $i(0) = 0$, then $c = -3/5$ and $\lim\limits_{t\to\infty} i(t) = 3/5$.

29. We note first that $P(0) = \dfrac{aP_0}{bP_0 + (a-bP_0)} = \dfrac{aP_0}{a} = P_0$ so that the initial condition is satisfied. To verify that $P(t)$ satisfies the differential equation, compute $P'(t) = \dfrac{a^2P_0(a-bP_0)e^{-at}}{(bP_0 + (a-bP_0)e^{-at})^2}$. We now compute

$$P(a-bP) = aP - bP^2 = \dfrac{a^2P_0}{bP_0 + (a-bP_0)e^{-at}} - \dfrac{a^2bP_0^2}{(bP_0 + (a-bP_0)e^{-at})^2}$$ From these calculations, we note that the differential

$$= \dfrac{a^2P_0(bP_0 + (a-bP_0)e^{-at}) - a^2bP_0^2}{(bP_0 + (a-bP_0)e^{-at})^2}$$

$$= \dfrac{a^2P_0(a-b)P_0^{-at}}{e}(bP_0 + (a-bP_0)e^{-at})^2$$

equation $\dfrac{dP}{dt} = P(a-bP)$ is satisfied.

31. $\dfrac{dx}{dt} = kx(1000 - x) = x(1000k - kx)$ Using the result from Problem 29 with $a = 1000k$, $b = k$, and $x_0 = 1$,

$$x(t) = \dfrac{1000k}{k + 999ke^{-1000kt}} = \dfrac{1000}{1 + 999e^{-1000kt}}$$

Substituting $t = 4$ and $x = 50$ yields

$$50 = \dfrac{1000}{1 + 999e^{-4000k}}$$

$$e^{-4000k} = \dfrac{\frac{1000}{50} - 1}{999} = \dfrac{19}{999}$$

$$k = \dfrac{\ln\left(\frac{19}{999}\right)}{-4000} = 0.00099$$

Hence $x(t) = \dfrac{1000}{1 + 999e^{-.99t}}$.

After 6, days, the number of infections is $x(6) = \dfrac{1000}{1 + 999e^{-99(6)}} \approx 276$.

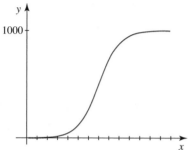

33. (a) Writing the equation as $v\,dv = -ky^{-2}\,dy$ and integrating, we obtain $\dfrac{1}{2}v^2 = \dfrac{k}{y} + C$ or $v^2 = \dfrac{2k}{y} + C_1$. Since $v(R) = v_0$, we

    have $v_0^2 = \dfrac{2k}{R} + C_1$. Thus $v^2 = \dfrac{2k}{y} + v_0^2 - \dfrac{2k}{R}$.

    (b) For the rocket to escape, v must remain positive. That is, v can never be zero; for then, the rocket stops and begins

    falling back to earth. Since $\dfrac{2k}{y}$ approaches zero as y increases, we can guarantee that v^2 will never be zero by requiring

    $v_0^2 - \dfrac{2k}{R} > 0$ or $v_0 > \sqrt{2kR}$. Since $k = gR^2$, $\dfrac{2k}{R} = 2gR$ and we take the escape velocity to be

    $$v_0 = \sqrt{2gR} \approx \sqrt{2(32\ \text{ft/s}^2)(4000\ \text{mi})} = \sqrt{\dfrac{2(32)4000}{5280}}\ \text{mi}^2/\text{s}^2$$

    $$\approx 6.96\ \text{mi/s} = 6.96(60)^2\ \text{mi/h} \approx 25,000\ \text{mi/h}.$$

35. (a) $\dfrac{dA}{\sqrt{A}(M - A)} = k\,dt$

    $\displaystyle\int \dfrac{dA}{\sqrt{A}(M - A)} = kt + c$

    With the substitution $x = \sqrt{A}$, the integral on the left becomes

    $$2\int \dfrac{1}{M - x^2}\,dx = 2\int \dfrac{1}{(\sqrt{M} + x)(\sqrt{M} - x)}\,dx$$

    $$= \int \dfrac{1}{\sqrt{M}(x + \sqrt{M})}\,dx - \int \dfrac{1}{\sqrt{M}(x - \sqrt{M})}\,dx$$

    $$= \dfrac{1}{\sqrt{M}} \ln\left| \dfrac{x + \sqrt{M}}{x - \sqrt{M}} \right|$$

    Hence

    $$\int \dfrac{dA}{\sqrt{A}(M - A)} = \dfrac{1}{\sqrt{M}} \ln\left| \dfrac{\sqrt{A} + \sqrt{M}}{\sqrt{A} - \sqrt{M}} \right| = kt + c$$

    $$\ln\left| \dfrac{\sqrt{A} + \sqrt{M}}{\sqrt{A} - \sqrt{M}} \right| = k\sqrt{M}t + c$$

    $$\dfrac{\sqrt{A} + \sqrt{M}}{\sqrt{A} - \sqrt{M}} = c_1 e^{k\sqrt{M}t}$$

    $$\sqrt{A} = \dfrac{\sqrt{M}(c_1 e^{k\sqrt{M}t} + 1)}{c_1 e^{k\sqrt{M}t} - 1}$$

    So $A(t) = \left(\dfrac{\sqrt{M}(c_1 e^{k\sqrt{M}t} + 1)}{c_1 e^{k\sqrt{mt}} - 1} \right)^2$

    (b) $\displaystyle\lim_{t \to \infty} A(t) = M$.

8.4 Solution Curves without a Solution

1.

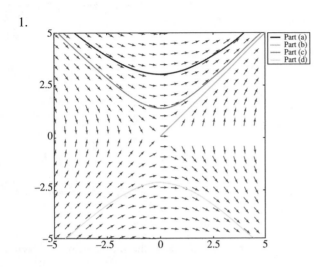

3.

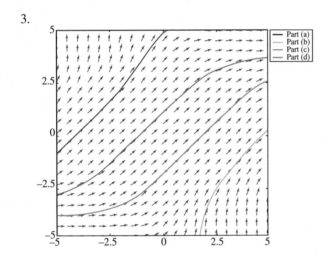

5.

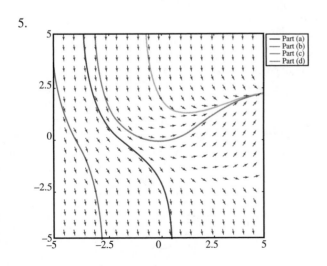

7.

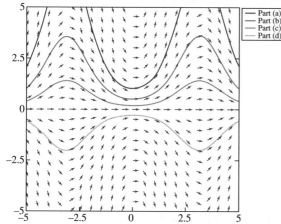

9.

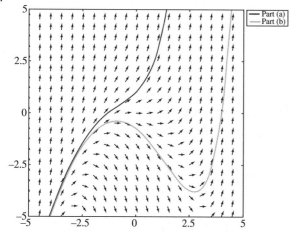

11.

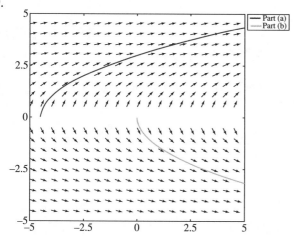

13. Solving $y^2 - 3y = y(y - 3) = 0$ we obtain the critical points 0 and 3. From the phase portrait we see that 0 is asymptotically stable (attractor) and 3 is unstable (repeller).

15. Solving $(y-2)^4 = 0$ we obtain the critical point 2. From the phase portrait we see that 2 is semi-stable.

17. Solving $y^2(4-y^2) = y^2(2-y)(2+y) = 0$ we obtain the critical points -2, 0, and 2. From the phase portrait we see that 2 is asymptotically stable, 0 is semi-stable, and -2 is unstable (repeller).

19. Solving $y\ln(y+2) = 0$ we obtain the critical points -1 and 0. From the phase portrait we see that -1 is asymptotically stable (attractor) and 0 is unstable (repeller).

21. Writing the differential equation in the form $dy/dx = y(1-y)(1+y)$ we see that critical points are located at $y = -1$, $y = 0$, and $y = 1$. The phase portrait is shown at the right.

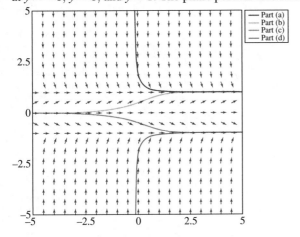

23. The critical points are 0 and c because the graph of $f(y)$ is 0 at these points. Since $f(y) > 0$ for $y < 0$ and $y > c$, the graph of the solution is increasing on $(-\infty, 0)$ and (c, ∞). Since $f(y) < 0$ for $0 < y < c$, the graph of the solution is decreasing on $(0, c)$.

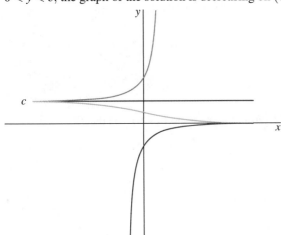

25. Writing the differential equation in the form

$$\frac{dv}{dt} = \frac{k}{m}\left(\frac{mg}{k} - v\right)$$

we see that a critical point is mg/k.
From the phase portrait we see that mg/k is an asymptotically stable critical point. Thus, $\lim_{t\to\infty} v = mg/k$.

27. From an inspection of the autonomous differential equation $\frac{di}{dt} = \frac{1}{L}(E - Ri)$
we see that $i = E/R$ is the equilibrium solution. If $i_0 < E/R$, then $E - Ri > 0$ and hence $\frac{di}{dt} > 0$. If $i_0 > E/R$, then $E - Ri < 0$ and hence $\frac{di}{dt} < 0$. Upon exaimination of the resulting phase portrait, we see that $i \to E/R$ as $t \to \infty$. Thus Ohm's Law $E = iR$ is satisfied as $t \to \infty$.

29. At points on the isocline $\frac{dy}{dx} = x^2 + y^2 = 1$, the line segments should each have a slope of 1. Isoclines of the differential equation $\frac{dy}{dx} = x + y$ have the form $x + y = c$ or $y = -x + c$. Thus, the isoclines are all lines in the plane with slope -1.

8.5 Euler's Method

1. We identify $f(x,y) = 2x - 3y + 1$. Then, for $h = 0.1$,

$$y_{n+1} = y_n + 0.1(2x_n - 3y_n + 1) = 0.2x_n + 0.7y_n + 0.1,$$

and

$$y(1.1) \approx y_1 = 0.2(1) + 0.7(5) + 0.1 = 3.8$$
$$y(1.2) \approx y_2 = 0.2(1.1) + 0.7(3.8) + 0.1 = 2.98.$$

For $h = 0.05$,

$$y_{n+1} = y_n + 0.05(2x_n - 3y_n + 1) = 0.1x_n + 0.85y_n + 0.1,$$

and

$$y(1.05) \approx y_1 = 0.1(1) + 0.85(5) + 0.1 = 4.4$$
$$y(1.1) \approx y_2 = 0.1(1.05) + 0.85(4.4) + 0.1 = 3.895$$
$$y(1.15) \approx y_3 = 0.1(1.1) + 0.85(3.895) + 0.1 = 3.47075$$
$$y(1.2) \approx y_4 = 0.1(1.15) + 0.85(3.47075) + 0.1 = 3.11514.$$

3. Separating variables and integrating, we have

$$\frac{dy}{y} = dx \text{ and } \ln|y| = x + c.$$

Thus $y = c_1 e^x$ and, using $y(0) = 1$, we find $c = 1$, so $y = e^x$ is the solution of the initial-value problem.

$h = 0.1$

x_n	y_n	Actual Value	Abs. Error	% Rel. Error
0.00	1.0000	1.0000	0.0000	0.00
0.10	1.1000	1.1052	0.0052	0.47
0.20	1.2100	1.2214	0.0114	0.93
0.30	1.3310	1.3499	0.0189	1.40
0.40	1.4641	1.4918	0.0277	1.86
0.50	1.6105	1.6487	0.0382	2.32
0.60	1.7716	1.8221	0.0506	2.77
0.70	1.9487	2.0138	0.0650	3.23
0.80	2.1436	2.2255	0.0820	3.68
0.90	2.3579	2.4596	0.1017	4.13
1.00	2.5937	2.7183	0.1245	4.58

$h = 0.05$

x_n	y_n	Actual Value	Abs. Error	% Rel. Error
0.00	1.0000	1.0000	0.0000	0.00
0.05	1.0500	1.0513	0.0013	0.12
0.10	1.1025	1.1052	0.0027	0.24
0.15	1.1576	1.1618	0.0042	0.36
0.20	1.2155	1.2214	0.0059	.48
0.25	1.2763	1.2840	0.0077	0.60
0.30	1.3401	1.3499	0.0098	0.72
0.35	1.4071	1.4191	0.01320	0.84
0.40	1.4775	1.4918	0.0144	0.96
0.45	1.5513	1.5683	0.0170	1.08
0.50	1.6289	1.6487	0.0198	1.20
0.55	1.7103	1.7333	0.0229	1.32
0.60	1.7959	1.8221	0.0263	1.44
0.65	1.8856	1.9155	0.299	1.56
0.70	1.9799	2.0138	0.0338	1.68
0.75	2.0789	2.1170	0.0381	1.80
0.80	2.1829	2.2255	0.0427	1.92
0.85	2.2920	2.3396	0.0476	2.04
0.90	2.4066	2.4596	0.0530	2.15
0.95	2.5270	2.5857	0.0588	2.27
1.00	2.6533	2.7183	0.0650	2.39

5.

$h = 0.1$

x_n	y_n
0.00	0.0000
0.10	0.1000
0.20	0.1905
0.30	0.2731
0.40	0.3492
0.50	0.4198

$h = 0.1$

x_n	y_n
0.00	0.0000
0.10	0.1000
0.20	0.1905
0.30	0.2731
0.40	0.3492
0.50	0.4198

7.

$h = 0.1$

x_n	y_n
0.00	0.5000
0.10	0.5250
0.20	0.5431
0.30	0.5548
0.40	0.5613
0.50	0.5639

$h = 0.1$

x_n	y_n
0.00	0.5000
0.05	0.5125
0.10	0.5232
0.15	0.5322
0.20	0.5395
0.25	0.5452
0.30	0.5496
0.35	0.5527
0.40	0.5547
0.45	0.5559
0.50	0.5565

9.

x_n	y_n
1.00	1.0000
1.10	1.0000
1.20	1.0191
1.30	1.0588
1.40	1.1231
1.50	1.2194

$h = 0.1$

$h = 0.1$

x_n	y_n
1.00	1.0000
1.05	1.0000
1.10	1.0049
1.15	1.0147
1.20	1.0298
1.25	1.0506
1.30	1.0775
1.35	0.1115
1.40	1.1538
1.45	1.2057
1.50	1.2696

Chapter 8 in Review

A. True/False

1. True

3. True

B. Fill In the Blanks

1. $y = x - 3x^2 + 36e^{3x} + C$

3. $e^{\int -1 dx} = e^{-x}$

5. half-life

7. $\dfrac{dP}{dt} = 0.16P, P(0) = P_0$

C. Exercises

1. Separating variables, we obtain

$$\frac{1}{y} dy = -\cot x \, dx \quad \Longrightarrow \quad \ln|y| = -\ln|\sin x| + C \quad \Longrightarrow \quad y = C_1 \csc x.$$

3. Write the equation in the form $\dfrac{dy}{dt} = -\dfrac{5}{t}y = 1$. An integrating factor is $\dfrac{1}{t^5}$, so

$$\frac{d}{dt}\left[\frac{1}{t^5}y\right] = \frac{1}{t^5} \quad \Longrightarrow \quad \frac{1}{t^5}y = -\frac{1}{4t^4} + C \quad \Longrightarrow \quad y = -\frac{1}{4}t + Ct^5.$$

5. Write the equation in the form $\dfrac{dy}{dx} + \dfrac{8x}{x^2+4}y = \dfrac{2x}{x^2+4}$. An integrating factor is $(x^2+4)^4$, so

$$\frac{d}{dx}\left[(x^2+4)^4 y\right] = 2x(x^2+4)^3 \quad \Longrightarrow \quad (x^2+4)^4 y = \frac{1}{4}(x^2+4)^4 + C$$

$$\Longrightarrow \quad y = \frac{1}{4} + C(x^2+4)^{-4}.$$

7. From $\dfrac{1}{\sqrt{1-y^2}} dy = 2x \, dx$, we have $\sin^{-1}(y) = x^2 + c$ or $y = \sin(x^2 + c)$.

9. An integrating factor is e^{-2x} so that

$$\frac{d}{dx}[e^{-2x}y] = e^{-2x}y\left(e^{3x} - e^{2x}\right) = x(e^x - 1)$$

$$e^{-2x}y = (x-1)e^x - \frac{x^2}{2} + C$$

$$y = (x-1)e^{3x} - \frac{x^2}{2}e^{2x} + Ce^{2x}$$

11. The general solution is $P = ce^{0.05t}$. Substituting $t = 0$ and $P = 1000$, we have $1000 = c$ so that $P = 1000e^{0.05t}$.

13. Write the equation in the form

$$\frac{dy}{dt} + \frac{1}{t}y = t^3 \ln t.$$

An integrating factor is $e^{\ln t} = t$, so

$$\frac{d}{dt}[ty] = t^4 \ln t$$

$$ty = -\frac{1}{25}t^5 + \frac{1}{5}t^5 \ln t + c$$

$$y = -\frac{1}{25}t^4 + \frac{1}{5}t^4 \ln t + \frac{c}{t}.$$

Substituting $y = 1$ and $y = 0$, we have $c = \frac{1}{25}$ so that $y = -\frac{1}{25}y^4 + \frac{1}{5}t^4 \ln t + \frac{1}{25t}$.

15.

$$\frac{dy}{2y + y^2} = dx$$

$$\int \frac{dy}{(2+y)y} = \int dx$$

$$\int \frac{1}{2y} - \frac{1}{2(y+2)} dy = \int dx$$

$$\frac{1}{2}\ln|y| - \frac{1}{2}\ln|y+2| = x + c$$

$$\frac{1}{2}\ln\left|\frac{y}{y+2}\right| = x + c$$

$$\ln\left|\frac{y}{y+2}\right| = 2x + c_1$$

$$\frac{y}{y+2} = c_2 e^{2x}$$

Solving for y we have $y = \frac{-2c_2 e^{2x}}{c_2 e^{2x} - 1}$. Substituting $x = 0$ and $y = 3$, we have $3 = \frac{-3c_2}{c_2 - 1}$ which yields $3c_2 - 3 = -2c_2$ or $c_2 = \frac{3}{5}$.

Thus the solution is $y = \frac{-\frac{6}{5}e^{2x}}{\frac{3}{5}e^{2x} - 1}$.

17. $\frac{dy}{1+y^2} = dx$ or $\tan^{-1}(y) = x + c$. Substituting $x = \frac{\pi}{3}$ and $y = -1$, we have

$$\tan^{-1}(-1) = \frac{\pi}{3} + c$$

$$\frac{-\pi}{4} = \frac{\pi}{3} + c$$

$$\frac{-7\pi}{12} = c.$$

Thus $\tan^{-1}(t) = x - \frac{7\pi}{12}$ or $y = \tan\left(x - \frac{7\pi}{12}\right)$.

19.

$$\frac{dy}{y^2} = -8x^2 dx$$

$$\frac{-1}{y} = -\frac{8}{3}x^3 + c$$

$$y = \frac{1}{\frac{8}{3}x^3 + c_1}$$

Substituting $x = 0$ and $y = \frac{1}{2}$, we have $\frac{1}{2} = \frac{1}{c_1}$ or $c_1 = 2$. Thus the solution is $y = \frac{1}{\frac{8}{3}x^3 + 2}$.

21.

$$\frac{dy}{dx} = \frac{2x}{3y^3}$$

$$3y^3 dy = 2x dx$$

$$\frac{3}{4}y^4 = x^2 + c$$

$$y^4 = \frac{4}{3}x^2 + c_1$$

$$y = \sqrt[4]{\frac{4}{3}x^2 + c_1}$$

Substituting $x = 0$ and $y = 2$, we have $2 = \sqrt[4]{c_1}$ or $c_1 = 16$. Thus the solution is $y = \sqrt[4]{\frac{4}{3}x^2 + 16}$.

23. Writing $\frac{dP}{dt} = kP$ as $\frac{1}{P}dP = k\,dt$ and integrating, we obtain $\ln P = kt + C$ or $P = C_1 e^{kt}$. Since $P(0) = P_0$, $P_0 = C_1$ and $P = P_0 e^{kt}$. From $P(t_1) = P_1$ and $P(t_2) = P_2$, we have $P_1 = P_0 e^{kt_1}$ and $P_2 = P_0 e^{kt_2}$ or $\frac{P_1}{P_0} = e^{kt_1}$ and $\frac{P_2}{P_0} = e^{kt_2}$. Then $e^k = (P_1/P_0)^{1/t_1}$ and $e^k = (P_2/P_0)^{1/t_2}$ or $(P_1/P_0)^{1/t_1} = (P_2/P_0)^{1/t_2}$. Thus, $(P_1/P_0)^{t_2} = (P_2/P_0)^{t_1}$.

25. (a) Write the differential equation in the form $\frac{dA}{dt} + (k_1 + k_2)A = k_1 M$. Then an integrating factor is $e^{(k_1+k_2)t}$, and

$$\frac{d}{dt}\left[e^{(k_1+k_2)t}A\right] = k_1 M e^{(k_1+k_2)t} \implies e^{(k_1+k_2)t}A = \frac{k_1 M}{k_1 + k_2}e^{(k_1+k_2)t} + C$$

$$\implies A = \frac{k_1 M}{k_1 + k_2} + Ce^{-(k_1+k_2)t}.$$

Using $A(0) = 0$, we find $C = -\frac{k_1 M}{k_1 + k_2}$ and $A = \frac{k_1 M}{k_1 + k_2}\left[1 + e^{-(k_1+k_2)t}\right]$.

(b) As $t \to \infty$, $A \to \frac{k_1 M}{k_1 + k_2}$. If $k_2 > 0$, the material will never be completely memorized.

(c)

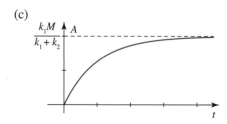

27. (a) Separating variables, we obtain $\frac{1}{P}\,dP = k\cos t\,dt$, so $\ln|P| = k\sin t + c$, and $P = c_1 e^{k\sin t}$. If $P(0) = P_0$ then $c_1 = P_0$ and $P = P_0 e^{k\sin t}$.

(b)

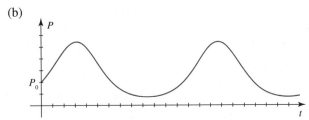

29. (a)

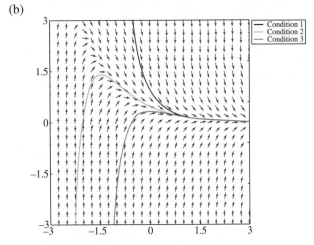

(b)

(c) As $x \to \infty$, $y(x) \to \infty$.
As $x \to -\infty$, $y(x) \to -\infty$.

31. Using a CAS we find that the zero of f occurs at approximately $y = 1.3214$. From the graph we observe that $dy/dx > 0$ for $y < 1.3214$ and $dy/dt < 0$ for $y > 1.3214$, so $y = 1.3214$ is an asymptotically stable critical point. Thus, $\lim\limits_{t \to \infty} y(x) = 1.3214$.

33. For the curve $y = x^3$, we have $\dfrac{dy}{dx} = 3x^2 = 3$ at both $(-1, -1)$ and $(1, 1)$. For the curve $= x^2 + 3y^2 = 4$, we can differentiate implicitly to get $2x + 6y\dfrac{dy}{dx} = 0$ or $\dfrac{dy}{dx} = \dfrac{-x}{3y} = \dfrac{-1}{3}$ at both $(-1, -1)$ and $(1, 1)$. Therefore, the slopes of the tangent lines are perpendicular at $(-1, -1)$ and $(1, 1)$.

Chapter 9

Sequences and Series

9.1 Sequences

In this exercise set, the symbol "$\overset{h}{=}$" is used to denote the fact that L'Hôpital's Rule was applied to obtain the equality.

1. $\dfrac{1}{3}, \dfrac{1}{5}, \dfrac{1}{7}, \dfrac{1}{9}$

3. $-1, \dfrac{1}{2}, -\dfrac{1}{3}, \dfrac{1}{4}$

5. $10, 100, 1000, 10000$

7. $2, 4, 12, 48$

9. $1, \dfrac{3}{2}, \dfrac{11}{6}, \dfrac{25}{12}$

11. Let $\varepsilon > 0$. Then $\left|\dfrac{1}{n} - 0\right| < \varepsilon$ implies $\dfrac{1}{n} < \varepsilon$ or $n > \dfrac{1}{\varepsilon}$. Take N to be the smallest integer greater than $\dfrac{1}{\varepsilon}$.

13. Let $\varepsilon > 0$. Then $\left|\dfrac{n}{n+1} - 1\right| = \left|-\dfrac{1}{n+1}\right| < \varepsilon$ implies $\dfrac{1}{n+1} < \varepsilon$ or $n > \dfrac{1}{\varepsilon} - 1$. Take N to be the smallest integer greater than $\dfrac{1}{\varepsilon} - 1$.

15. $\displaystyle\lim_{n\to\infty} \dfrac{10}{\sqrt{n+1}} = 0$

17. $\displaystyle\lim_{n\to\infty} \dfrac{1}{5n+6} = 0$

19. $\displaystyle\lim_{n\to\infty} \dfrac{3n-2}{6n+1} = \lim_{n\to\infty} \dfrac{3 - 2/n}{6 + 1/n} = \dfrac{1}{2}$

21. The terms alternate between 20 and -20. The sequence diverges.

23. $\displaystyle\lim_{n\to\infty} \dfrac{n^2-1}{2n} = \lim_{n\to\infty} \dfrac{n - 1/n}{2} = \infty$. The sequence diverges.

25. $\displaystyle\lim_{n\to\infty} ne^{-n} = \lim_{n\to\infty} \dfrac{n}{e^n} \overset{h}{=} \lim_{n\to\infty} \dfrac{1}{e^n} = 0$

27. $\displaystyle\lim_{n\to\infty} \dfrac{\sqrt{n+1}}{n} = \lim_{n\to\infty} \dfrac{\sqrt{1 + 1/n}}{\sqrt{n}} = 0$

29. Since the terms alternate between -1 and 1, the sequence diverges.

225

31. $\displaystyle\lim_{n\to\infty}\frac{\ln n}{n} \overset{h}{=} \lim_{n\to\infty}\frac{1/n}{1} = 0$

33. $\displaystyle\lim_{n\to\infty}\frac{5-2^{-n}}{7+4^{-n}} = \frac{5}{7}$

35. $\displaystyle\lim_{n\to\infty}\frac{e^n+1}{e^n} = \lim_{n\to\infty}\frac{1+1/e^n}{1} = 1$

37. $\displaystyle\lim_{n\to\infty} n\sin\left(\frac{6}{n}\right) = \lim_{n\to\infty}\frac{\sin(6/n)}{1/n} \overset{h}{=} \lim_{n\to\infty}\frac{[\cos(6/n)](-6/n^2)}{-1/n^2} = \lim_{n\to\infty} 6\cos\left(\frac{6}{n}\right) = 6$

39. $\displaystyle\lim_{n\to\infty}\frac{e^n-e^{-n}}{e^n+e^{-n}} = \lim_{n\to\infty}\frac{1-e^{-2n}}{1+e^{-2n}} = 1$

41. Let $y = x^{2/(x+1)}$. Then $\ln y = \dfrac{2}{x+1}\ln x$ and $\displaystyle\lim_{x\to\infty}\frac{2\ln x}{x+1} \overset{h}{=} \lim_{x\to\infty}\frac{2/x}{1} = 0$. Thus,

$\displaystyle\lim_{n\to\infty} n^{2/(n+1)} = e^0 = 1$.

43. $\displaystyle\lim_{n\to\infty}\ln\left(\frac{4n+1}{3n-1}\right) = \ln\left(\lim_{n\to\infty}\frac{4+1/n}{3-1/n}\right) = \ln\frac{4}{3}$

45. $\displaystyle\lim_{n\to\infty}\left(\sqrt{n+1}-\sqrt{n}\right) = \lim_{n\to\infty}\left[\left(\sqrt{n+1}-\sqrt{n}\right)\left(\frac{\sqrt{n+1}+\sqrt{n}}{\sqrt{n+1}+\sqrt{n}}\right)\right] = \lim_{n\to\infty}\frac{n+1-n}{\sqrt{n+1}+\sqrt{n}}$

$\displaystyle = \lim_{n\to\infty}\frac{1}{\sqrt{n+1}+\sqrt{n}} = 0$

47. $\left\{\dfrac{2n}{2n-1}\right\}$ $\displaystyle\lim_{n\to\infty}\frac{2n}{2n-1} = \lim_{n\to\infty}\frac{2n-1+1}{2n-1} = \lim_{n\to\infty}\left(1+\frac{1}{2n-1}\right) = 1$

49. $\{(-1)^{n+1}(2n+1)\}$

Since the terms alternate between values that are increasingly greater than and less than 0, the sequence diverges.

51. $\left\{\dfrac{2}{3^{n-1}}\right\}$ $\displaystyle\lim_{n\to\infty}\frac{2}{3^{n-1}} = 0$

53. $-\dfrac{1}{2}, -\dfrac{1}{4}, -\dfrac{1}{8}, -\dfrac{1}{16}, \ldots$

55. $3, 1, \dfrac{1}{3}, \dfrac{1}{3}, \ldots$

57. $a_{n+1} = \dfrac{1}{4}a_n + 6 \implies \displaystyle\lim_{n\to\infty} a_{n+1} = \frac{1}{4}\lim_{n\to\infty} a_n + 6 \implies L = \frac{1}{4}L + 6 \implies L = 8$

59. For $a_n = \dfrac{5^n}{n!}$, we have $a_{n+1} = \dfrac{5^{n+1}}{(n+1)!}$. Expanding, we get $a_{n+1} = \dfrac{5\cdot 5^n}{(n+1)\cdot n!} = \dfrac{5}{n+1}\left(\dfrac{5^n}{n!}\right)$. The last factor of the last term

is a_n, so $a_{n+1} = \dfrac{5}{n+1}a_n$.

61. Let $a_n = 0$, $b_n = \dfrac{\sin^2 n}{4^n}$, and $c_n = \dfrac{1}{4^n}$. Then $\displaystyle\lim_{n\to\infty} a_n = \lim_{n\to\infty} c_n = 0$, so $\displaystyle\lim_{n\to\infty}\frac{\sin^2 n}{4^n} = 0$.

63. Let $a_n = 0$, $b_n = \dfrac{\ln n}{n(n+2)}$, and $c_n = \dfrac{n}{n(n+2)} = \dfrac{1}{n+2}$. Then, for $n \geq 1$, $a_n \leq b_n \leq c_n$. Since $\displaystyle\lim_{n\to\infty} a_n = \lim_{n\to\infty} c_n = 0$, we have

$\displaystyle\lim_{n\to\infty}\frac{\ln n}{n(n+2)} = 0$.

65. Let $y = \left(1 + \dfrac{x}{t}\right)^t$. Then $\ln y = t \ln\left(1 + \dfrac{x}{t}\right)$ and using L'Hôpital's Rule,

$$\lim_{t \to \infty} \ln y = \lim_{t \to \infty} \frac{\ln(1 + x/t)}{1/t} = \lim_{t \to \infty} \frac{\dfrac{-x/t^2}{1 + x/t}}{-1/t^2} = \lim_{t \to \infty} \frac{x}{1 + x/t} = x.$$

Thus, $\displaystyle\lim_{n \to \infty} \left(1 + \frac{x}{n}\right)^n = e^x$.

67. Let a_n be the height of the ball on the n-th bounce. Then $a_0 = 15$, $a_1 = \dfrac{2}{3}(15) = 10$, $a_2 = \dfrac{2}{3}(10) = \dfrac{20}{3}$, $a_3 = \dfrac{2}{3}\left(\dfrac{20}{3}\right) = \dfrac{40}{9}$ ft,

$a_n = 15\left(\dfrac{2}{3}\right)^n$.

69. $A_1 = 15$, $A_2 = 15(0.2) + 15 = 18$, $A_3 = 18(0.2) + 15 = 18.6$, $A_4 = 18.6(0.2) + 15 = 18.72$, $A_5 = 18.72(0.2) + 15 = 18.744$, $A_6 = 18.744(0.2) + 15 = 18.7488$

71. Parents: 2; grandparents: $2 \cdot 2 = 4$; great-grandparents: $2 \cdot 4 = 8$; great-great-grandparents: $2 \cdot 8 = 16$; great-great-great-grandparents: $2 \cdot 16 = 32$

73. (a) $a_{n+1} = 1 + \dfrac{1}{1 + a_n}$

　　(b) $a_5 = 1 + \dfrac{1}{2 + \dfrac{1}{2 + \dfrac{1}{2 + \dfrac{1}{2}}}} \qquad a_6 = 1 + \dfrac{1}{2 + \dfrac{1}{2 + \dfrac{1}{2 + \dfrac{1}{2}}}}$

　　(c) Letting $L = \displaystyle\lim_{n \to \infty} a_n = \lim_{n \to \infty} a_{n+1}$, we have $L = 1 + \dfrac{1}{1 + L}$, $L^2 + L = 1 + L + 1$, and $L^2 = 2$. Thus, $L = \sqrt{2} = \displaystyle\lim_{n \to \infty} a_n$, and the sequence converges to $\sqrt{2}$.

75. Since $\{a_n\}$ converges, then $\displaystyle\lim_{n \to \infty} a_n$ is some value L. Since the limit of a product is the same as the product of the factors' limits, we have $\displaystyle\lim_{n \to \infty} a_n^2 = \lim_{n \to \infty}(a_n \cdot a_n) = \left(\lim_{n \to \infty} a_n\right)\left(\lim_{n \to \infty} a_n\right) = L^2$. Thus, $\{a_n^2\}$ actually does converge.

77. (a) $P_1 = 3$, $P_2 = 3\left(\dfrac{4}{3}\right)$, $P_3 = 3\left(\dfrac{4}{3}\right)^2$, $P_4 = 3\left(\dfrac{4}{3}\right)^3$

　　(b) $P_n = 3\left(\dfrac{4}{3}\right)^{n-1}$

　　(c) $\displaystyle\lim_{n \to \infty} P_n = \lim_{n \to \infty}\left[3\left(\dfrac{4}{3}\right)^{n-1}\right] = \infty$

79. $2, 3, 5, 8, 13, \cdots$

The recursion formula matches the pattern in Problem 78. Not surprisingly, the resulting sequence is called the *Fibonacci sequence*.

9.2　Monotonic Sequences

1. $a_{n+1} - a_n = \dfrac{n+1}{3n+4} - \dfrac{n}{3n+1} = \dfrac{1}{(3n+4)(3n+1)} > 0$. The sequence is increasing.

3. $a_1 = -1$, $a_2 = \sqrt{2}$, $a_3 = -\sqrt{3}$. The sequence is not monotonic.

5. Let $f(x) = \dfrac{e^x}{x}$. Then $f'(x) = \dfrac{(x-1)e^x}{x^2} > 0$ for $x > 1$. The sequence is increasing.

7. $a_1 = 2$, $a_2 = 2$, $\dfrac{a_{n+1}}{a_n} = \dfrac{2^{n+1}/(n+1)!}{2^n/n!} = \dfrac{2}{n+1} < 1$ for $n > 1$.

The sequence is nonincreasing.

9. Let $f(x) = x + \dfrac{1}{x}$. Then $f'(x) = 1 - \dfrac{1}{x^2} > 0$ for $x > 1$. The sequence is increasing.

11. Since $n < \pi$ for $n = 1, 2, 3$, we have $\sin n > 0$ for $n = 1, 2, 3$. Since $\pi < n < 2\pi$ for $n = 4, 5$, we have $\sin n < 0$ for $n = 4, 5$. Thus, $a_3 > 0$, $a_4 < 0$, and $a_5 > 0$. The sequence is not monotonic.

13. Since $a_{n+1} - a_n = \dfrac{4n+3}{5n+7} - \dfrac{4n-1}{5n+2} = \dfrac{13}{(5n+7)(5n+2)} > 0$, the sequence is monotonic. Using $\dfrac{4n-1}{5n+2} > 0$ and $\dfrac{4n-1}{5n+2} < \dfrac{4n}{5n+2} < \dfrac{4n}{5n} = \dfrac{4}{5}$, we see that the sequence is bounded. Thus, the sequence converges.

15. Since $\dfrac{a_{n+1}}{a_n} = \dfrac{3^{n+1}/(1+3^{n+1})}{3^n/(1+3^n)} = \dfrac{3+3^{n+1}}{1+3^{n+1}} = 1 + \dfrac{2}{1+3^{n+1}} > 1$, the sequence is monotonic. Using $\dfrac{3^n}{1+3^n} > 0$ and $\dfrac{3^n}{1+3^n} = 1 - \dfrac{1}{1+3^n} < 1$, we see that the sequence is bounded. Thus, the sequence converges.

17. Let $f(x) = e^{1/x}$. Then $f'(x) = \dfrac{-e^{1/x}}{x^2} < 0$ and the sequence is monotonic. Using $e^{1/n} > 0$ and $e^{1/n} \leq e$ (since the sequence is decreasing), we see that the sequence is bounded. Thus, the sequence converges.

19. Since $\dfrac{a_{n+1}}{a_n} = \dfrac{(n+1)!/[1 \cdot 3 \cdot 5 \cdots (2n+1)]}{n!/[1 \cdot 3 \cdot 5 \cdots (2n-1)]} = \dfrac{n+1}{2n+1} < 1$, the sequence is monotonic. Using $\dfrac{n!}{1 \cdot 3 \cdot 5 \cdots (2n-1)} > 0$ and $\dfrac{n!}{1 \cdot 3 \cdot 5 \cdots (2n-1)} \leq 1$ (since the sequence is decreasing), we see that the sequence is bounded. Thus, the sequence converges.

21. Let $f(x) = \tan^{-1} x$. Then $f'(x) = \dfrac{1}{1+x^2} > 0$ and the sequence is monotonic. Since $|\tan^{-1} n| < \dfrac{\pi}{2}$, we see that the sequence is bounded. Thus, the sequence converges.

23. The sequence is $\{(0.8)^n\}$. Since $\dfrac{a_{n+1}}{a_n} = \dfrac{(0.8)^{n+1}}{(0.8)^n} = 0.8 < 1$, the sequence is monotonic. Using $(0.8)^n > 0$ and $(0.8)^n \leq 0.8$ (since the sequence is decreasing), we see that the sequence is bounded. Thus, the sequence converges.

25. $a_{n+1} = \dfrac{1}{2}a_n + 5$, $a_1 = 1$. We will show that $a_n < 10$ for all n.

For $n = 1$, we have $a_2 = \dfrac{11}{2} < 10$. Assume that $a_k < 10$. Then $a_{k+1} = \dfrac{1}{2}a_k + 5 < \dfrac{1}{2}(10) + 5 = 10$; that is, $a_{k+1} < 10$ whenever $a_k < 10$. The sequence is bounded because $0 < a_n < 10$.

Next, we will show that the sequence $\{a_n\}$ is monotonic. Because $a_n < 10$, necessarily $\dfrac{1}{2}a_n < \dfrac{1}{2} \cdot 10 = 5$. Therefore, from the recursion formula,

$$a_{n+1} = \dfrac{1}{2}a_n + 5 > \dfrac{1}{2}a_n + \dfrac{1}{2}a_n = a_n.$$

This shows that $a_{n+1} > a_n$ for all n, and so the sequence is increasing.

Since $\{a_n\}$ is bounded and monotonic, it follows from Theorem 9.2.1 that the sequence converges. Because we must have $\lim_{n \to \infty} a_n = L$ and $\lim_{n \to \infty} a_{n+1} = L$, the limit of the sequence can be determined from the recursion formula:

$$\lim_{n \to \infty} a_{n+1} = \dfrac{1}{2}\lim_{n \to \infty} a_n + 5 \quad \Longrightarrow \quad L = \dfrac{1}{2}L + 5 \quad \Longrightarrow \quad L = 10.$$

27. $a_{n+1} = \sqrt{7a_n}$, $a_1 = \sqrt{7}$. Now $0 < a_n < 7 \Longrightarrow \sqrt{a_n} < \sqrt{7}$, and so

$$a_{n+1} = \sqrt{7a_n} = \sqrt{7}\sqrt{a_n} > \sqrt{a_n}\sqrt{a_n} = a_n.$$

Thus, $a_{n+1} > a_n$ for all n. The sequence is therefore monotonic (increasing) and bounded. By Theorem 9.2.1, the sequence converges. From

$$\lim_{n\to\infty} a_{n+1} = \lim_{n\to\infty} \sqrt{7a_n} \quad\Longrightarrow\quad L = \sqrt{7L} \quad\Longrightarrow\quad L^2 = 7L \quad\Longrightarrow\quad L(L-7) = 0,$$

we have $L = 7$.

29. (a) If $\lim_{n\to\infty} p_n = L$, then $\lim_{n\to\infty} p_{n+1} = L$ and $\lim_{n\to\infty} \dfrac{bp_n}{a+p_n} = \dfrac{bL}{a+L}$. Thus, $L = \dfrac{bL}{a+L}$ or $(a+L-b)L = 0$ and $L = 0$ or $L = b-a$.

(b) Since $p_n > 0$, $p_{n+1} = \dfrac{bp_n}{a+p_n} = \dfrac{b}{a+p_n} p_n < \dfrac{b}{a} p_n$.

(c) If $a > b$ then $\dfrac{b}{a} < 1$ and by part (b), $p_{n+1} < p_n$. The sequence is thus monotonically decreasing. Now $p_1 < \dfrac{b}{a} p_0$ implies $p_2 < \dfrac{b}{a} p_1 < \dfrac{b}{a}\left(\dfrac{b}{a} p_0\right) = \left(\dfrac{b}{a}\right)^2 p_0$, which in turn implies $p_3 < \dfrac{b}{a} p_2 < \dfrac{b}{a}\left[\left(\dfrac{b}{a}\right)^2 p_0\right] = \left(\dfrac{b}{a}\right)^3 p_0$. In general, $p_{n+1} < \left(\dfrac{b}{a}\right)^{n+1} p_0$. Since $\dfrac{b}{a} < 1$, $\lim_{n\to\infty} p_n = \lim_{n\to\infty} p_{n+1} \le \lim_{n\to\infty} \left(\dfrac{b}{a}\right)^{n+1} p_0 = 0$. Since $p_n > 0$ for all n, $\lim_{n\to\infty} p_n = 0$.

(d) We first note that

$$|b-a-p_{n+1}| = \left|b-a-\frac{bp_n}{a+p_n}\right| = \left|\frac{ab+bp_n-a^2-ap_n-bp_n}{a+p_n}\right| = \frac{a}{a+p_n}|b-a-p_n|.$$

Since $0 < \dfrac{a}{a+p_n} < 1$, the distance from p_{n+1} to $b-a$ is less than the distance from p_n to $b-a$. This means that p_{n+1} is between $b-a$ and p_n. Thus, if $0 < p_0 < b-a$, the sequence $\{p_n\}$ is increasing and bounded above by $b-a$. If $0 < b-a < p_0$, the sequence $\{p_n\}$ is decreasing and bounded below by $b-a$. In either case, it follows from part (a) that the sequence converges to $b-a$.

31. Since $\{a_n\}$ is convergent, it follows from Definition 9.1.2 that there exists an N such that $|a_n - L| < 1$ whenever $n > N$. Adding $|L|$ to both sides, we have $|a_n - L| + |L| < 1 + |L|$. By the triangle inequality, $|(a_n - L) + L| = |a_n| \le |a_n - L| + |L|$, and so $|a_n| \le |a_n - L| + |L| < 1 + |L|$ for all $n > N$. For $n \le N$, we have a finite set of numbers that therefore has a maximum value M and minimum value m; thus, $\{a_n\}$ is bounded.

33. Note that $a_1 = 1$ and assume $n \ge 2$. Then the area under the graph of $y = \dfrac{1}{x}$ on $[1,n]$ is $A = \displaystyle\int_1^n \frac{1}{x}\,dx = \ln x\Big]_1^n = \ln n$.

Partitioning $[1,n]$ at 1, 2, 3, $\ldots$, n, the upper sum is $U = 1 + \dfrac{1}{2} + \cdots + \dfrac{1}{n-1}$ and the lower sum is $L = \dfrac{1}{2} + \dfrac{1}{3} + \cdots + \dfrac{1}{n}$. Since $L < A < U$, we have, for $n \ge 2$, $\dfrac{1}{2} + \dfrac{1}{3} + \cdots + \dfrac{1}{n} < \ln n < 1 + \dfrac{1}{2} + \cdots + \dfrac{1}{n-1}$. Now, $a_n = 1 + \dfrac{1}{2} + \cdots + \dfrac{1}{n} - \ln n$, so $\dfrac{1}{2} + \dfrac{1}{3} + \cdots + \dfrac{1}{n} = a_n - 1 + \ln n$ and $1 + \dfrac{1}{2} + \cdots + \dfrac{1}{n-1} = a_n - \dfrac{1}{n} + \ln n$. Thus, for $n \ge 2$, $a_n - 1 + \ln n < \ln n < a_n - \dfrac{1}{n} + \ln n$ and $a_n - 1 < 0 < a_n - \dfrac{1}{n}$, or $a_n < 1$ and $a_n > \dfrac{1}{n} > 0$. Since $a_1 = 1$, the sequence is bounded below by 0 and above by 1. To see that the sequence is monotonic, note that $a_n - 1 + \ln n < \ln n$ implies $a_{n+1} - 1 + \ln(n+1) < \ln(n+1)$. Subtracting, we have $[a_{n+1} - 1 + \ln(n+1)] - (a_n - 1 + \ln n) < \ln(n+1) - \ln n$ or $a_{n+1} - a_n < 0$. Since the sequence is bounded and monotonic, it is convergent.

9.3 Series

1. $3 + \dfrac{5}{2} + \dfrac{7}{3} + \dfrac{9}{4} + \cdots$

3. $\dfrac{1}{2} - \dfrac{1}{6} + \dfrac{1}{12} - \dfrac{1}{20} + \cdots$

5. $1 + 2 + \dfrac{3}{2} + \dfrac{2}{3} + \cdots$

7. $2 + \dfrac{8}{3} + \dfrac{16}{5} + \dfrac{128}{35} + \cdots$

9. $-\dfrac{1}{7} + \dfrac{1}{9} - \dfrac{1}{11} + \dfrac{1}{13} - \cdots$

11. Write $a_k = \dfrac{1}{k} - \dfrac{1}{k+1}$. Then $S_n = \left(1 - \dfrac{1}{2}\right) + \left(\dfrac{1}{2} - \dfrac{1}{3}\right) + \cdots + \left(\dfrac{1}{n} - \dfrac{1}{n+1}\right) = 1 - \dfrac{1}{n+1}$ and

$\displaystyle\sum_{k=1}^{\infty} \dfrac{1}{k(k+1)} = \lim_{n \to \infty} S_n = 1.$

13. Write $a_k = \dfrac{1/2}{2k-1} - \dfrac{1/2}{2k+1}$. Then

$$S_n = \dfrac{1}{2}\left[\left(1 - \dfrac{1}{3}\right) + \left(\dfrac{1}{3} - \dfrac{1}{5}\right) + \cdots + \left(\dfrac{1}{2n-1} - \dfrac{1}{2n+1}\right)\right] = \dfrac{1}{2}\left(1 - \dfrac{1}{2n+1}\right)$$

and $\displaystyle\sum_{k=1}^{\infty} \dfrac{1}{4k^2 - 1} = \lim_{n \to \infty} S_n = \dfrac{1}{2}.$

15. Identify $r = \dfrac{1}{5}$ and $a = 3$. The series converges to $\dfrac{3}{1 - 1/5} = \dfrac{15}{4}.$

17. Identify $r = -\dfrac{1}{2}$ and $a = 1$. The series converges to $\dfrac{1}{1 + 1/2} = \dfrac{2}{3}.$

19. The common ratio is $\dfrac{5}{4} > 1$. The series diverges.

21. Identify $r = 0.9$ and $a = 900$. The series converges to $\dfrac{900}{1 - 0.9} = 9000.$

23. Identify $r = \dfrac{1}{\sqrt{3} - \sqrt{2}} > 1$. The series diverges.

25. $0.222\ldots = 0.2 + 0.02 + 0.002 + \cdots$. Identify $r = 0.1$ and $a = 0.2$. Then $0.222\ldots = \dfrac{0.2}{1 - 0.1} = \dfrac{2}{9}.$

27. $0.616161\ldots = 0.61 + 0.0061 + 0.000061 + \cdots$. Identify $r = 0.01$ and $a = 0.61$. Then $0.616161\ldots = \dfrac{0.61}{1 - 0.01} = \dfrac{61}{99}.$

29. $1.314314\ldots = 1 + (0.314 + 0.000314 + \cdots)$. Identify $r = 0.01$ and $a = 0.314$. Then $1.314314\ldots = 1 + \dfrac{0.314}{1 - 0.001} = 1 + \dfrac{314}{999} = \dfrac{1313}{999}.$

31. $\displaystyle\sum_{k=1}^{\infty}\left[\left(\dfrac{1}{3}\right)^{k-1} + \left(\dfrac{1}{4}\right)^{k-1}\right] = \sum_{k=1}^{\infty}\left(\dfrac{1}{3}\right)^{k-1} + \sum_{k=1}^{\infty}\left(\dfrac{1}{4}\right)^{k-1} = \dfrac{1}{1 - 1/3} + \dfrac{1}{1 - 1/4} = \dfrac{3}{2} + \dfrac{4}{3} = \dfrac{17}{6}$

33. $\displaystyle\lim_{k \to \infty} a_k = \lim_{k \to \infty} 10 \neq 0$, so the series diverges.

35. $\displaystyle\lim_{k \to \infty} a_k = \lim_{k \to \infty} \dfrac{k}{2k+1} = \lim_{k \to \infty} \dfrac{1}{2 + 1/k} = \dfrac{1}{2} \neq 0$, so the series diverges.

37. $\displaystyle\lim_{k \to \infty} a_k = \lim_{k \to \infty} (-1)^k$ does not exist, so the series diverges.

39. $10 \displaystyle\sum_{k=1}^{\infty} \dfrac{1}{k}$ diverges because the harmonic series diverges. Thus $\displaystyle\sum_{k=1}^{\infty} \dfrac{10}{k}$ diverges.

41. Since $\displaystyle\sum_{k=1}^{\infty} \dfrac{1}{2^{k-1}}$ is a geometric series with $r = \dfrac{1}{2}$, it converges. Since $\displaystyle\sum_{k=1}^{\infty} \dfrac{1}{k}$ is the harmonic series, it diverges. Thus, $\displaystyle\sum_{k=1}^{\infty}\left(\dfrac{1}{2^{k-1}} + \dfrac{1}{k}\right)$ diverges.

43. This is a geometric series with $r = \dfrac{x}{2}$ and will converge for $\left|\dfrac{x}{2}\right| < 1$ or $|x| < 2$.

45. This is a geometric series with $r = x+1$ and will converge for $|x+1| < 1$ or $-2 < x < 0$.

47. The total distance is $15 + 2(15)\left(\dfrac{2}{3}\right) + 2(15)\left(\dfrac{2}{3}\right)^2 + \cdots = 15 + \sum\limits_{k=1}^{\infty} 30\left(\dfrac{2}{3}\right)^k$. The sum is a geometric series with $a = 20$ and

$r = \dfrac{2}{3}$, so $\sum\limits_{k=1}^{\infty} 30\left(\dfrac{2}{3}\right)^k = \dfrac{20}{1-2/3} = 60$. The total distance is $15 + 60 = 75$ ft.

49. $N_0 + N_0 s + N_0 s^2 + \cdots = \sum\limits_{k=0}^{\infty} N_0 s^k = \dfrac{N_0}{1-s}$. Solving $10{,}000 = \dfrac{N_0}{1-0.9}$, we obtain $N_0 = 1000$.

51. The total amount of the drug immediately after the n-th dose is $A_n = 15 + 15(0.2) + 15(0.2)^2 + \cdots + 15(0.2)^{n-1}$. As $n \to \infty$, the total accumulation of the drug will be

$$\lim_{n\to\infty} A_n = \lim_{n\to\infty} \sum_{k=1}^{n} 15(0.2)^{k-1} = \sum_{k=1}^{\infty} 15(0.2)^{k-1} = \frac{15}{1-0.2} = \frac{75}{4} = 18.75 \text{ mg.}$$

53. By Theorem 9.3.3 in the text, if $\lim\limits_{n\to\infty} a_n \neq 0$, then the series $\sum\limits_{k=1}^{\infty} a_k$ diverges.

55. The series $\sum\limits_{k=1}^{\infty} k$ and $\sum\limits_{k=1}^{\infty} (-k)$ both diverge, but their sum $\sum\limits_{k=1}^{\infty} (k-k) = \sum\limits_{k=1}^{\infty} 0 = 0$ converges.

57. $\dfrac{1+9}{25} + \dfrac{1+27}{125} + \dfrac{1+81}{625} + \cdots = \dfrac{1+3^2}{5^2} + \dfrac{1+3^3}{5^3} + \dfrac{1+3^4}{5^4} + \cdots$

$= \dfrac{1}{5^2} + \left(\dfrac{3}{5}\right)^2 + \dfrac{1}{5^3} + \left(\dfrac{3}{5}\right)^3 + \dfrac{1}{5^4} + \left(\dfrac{3}{5}\right)^4 + \cdots$

$= \dfrac{1}{5^2} + \dfrac{1}{5^3} + \dfrac{1}{5^4} + \cdots + \left(\dfrac{3}{5}\right)^2 + \left(\dfrac{3}{5}\right)^3 + \cdots$

$= \left[\left(\sum\limits_{k=1}^{\infty} \dfrac{1}{5^{k-1}}\right) - 1 - \dfrac{1}{5}\right] + \left[\left(\sum\limits_{k=1}^{\infty} \dfrac{3^{k-1}}{5^{k-1}}\right) - 1 - \dfrac{3}{5}\right]$

$= \sum\limits_{k=1}^{\infty} \dfrac{1}{5^{k-1}} + \sum\limits_{k=1}^{\infty} \left(\dfrac{3}{5}\right)^{k-1} - 1 - 1 - \dfrac{1}{5} - \dfrac{3}{5}$

$= \sum\limits_{k=1}^{\infty} \dfrac{1}{5^{k-1}} + \sum\limits_{k=1}^{\infty} \left(\dfrac{3}{5}\right)^{k-1} - 2 - \dfrac{4}{5}$

$= \dfrac{1}{1-1/5} + \dfrac{1}{1-3/5} - 2 - \dfrac{4}{5} = \dfrac{19}{20}$

59. $\sum\limits_{k=0}^{n} \tan^k x$ is a geometric series with $a = 1$ and $r = \tan x$. The infinite series $\sum\limits_{k=0}^{\infty} \tan^k x$ will converge to $\dfrac{1}{1-\tan x}$ when $|\tan x| < 1$ or $|x| < \dfrac{\pi}{4}$. Thus, $\lim\limits_{n\to\infty}\left(\dfrac{1}{1-\tan x} - \sum\limits_{k=0}^{n} \tan^k x\right) = 0$ for $|x| < \dfrac{\pi}{4}$.

61. The general term of the series is $a_n = \sum\limits_{k=1}^{n} \dfrac{1}{k}$. Since $\lim\limits_{n\to\infty} a_n = \sum\limits_{k=1}^{\infty} \dfrac{1}{k}$ and the harmonic series diverges, then by Theorem 9.3.3, the series $\sum\limits_{n=1}^{\infty} a_n = \sum\limits_{n=1}^{\infty} \left(\sum\limits_{k=1}^{n} \dfrac{1}{k}\right)$ diverges.

63. (a) The partial sum $1 + \dfrac{1}{2} + \dfrac{1}{3} + \dfrac{1}{4} + \cdots + \dfrac{1}{n}, n > 1$ can be represented as an area. The graph on the following page shows an area of $1 + \dfrac{1}{2}$ on $[0,2]$, $1 + \dfrac{1}{2} + \dfrac{1}{3}$ on $[0,3]$, and so on:

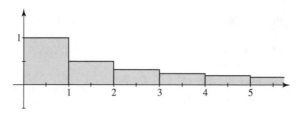

In addition, the area under the graph of $f(x) = \dfrac{1}{x}$ from $x = 1$ up to any $n > 1$ is $\displaystyle\int_1^n \dfrac{1}{x}\,dx = \ln x\Big]_1^n = \ln n - \ln 1 = \ln n$.

Shifting the partial sum areas to the right allows us to compare $\displaystyle\int_1^{n+1} \dfrac{1}{x}\,dx = \ln(n+1)$ with their corresponding partial

sums. This shows that $\ln(n+1) < S_n = 1 + \dfrac{1}{2} + \dfrac{1}{3} + \dfrac{1}{4} + \cdots + \dfrac{1}{n}$:

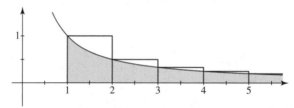

Similarly, the graphs can be aligned to show that $1 + \displaystyle\int_1^n \dfrac{1}{x}\,dx = 1 + \ln n > S_n = 1 + \dfrac{1}{2} + \dfrac{1}{3} + \dfrac{1}{4} + \cdots + \dfrac{1}{n}$. Recall that $n > 1$ is stipulated, so the area on $[0,1]$ consists solely of the 1×1 square (thus corresponding to the first term in $1 + \ln n$):

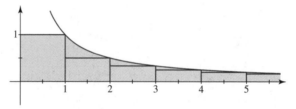

This yields the overall inequality $\quad \ln(n+1) < 1 + \dfrac{1}{2} + \dfrac{1}{3} + \dfrac{1}{4} + \cdots + \dfrac{1}{n} < 1 + \ln n$.

(b) Using the inequality in part (a), we know that $S_n \geq 10$ when $\ln(n+1) \geq 10$. Solving $\ln(n+1) = 10$, we get $n + 1 = e^{10}$, and $n = e^{10} - 1$. Thus, $S_n \geq 10$ for $n > e^{10} - 1 \approx 22026$. Using a calculator yields $\displaystyle\sum_{k=1}^{22026} \dfrac{1}{k} \approx 10.5772$ (it should be noted that the smallest n for which $S_n \geq 10$ is 12367). Similarly, to estimate the value of n for which $S_n \geq 100$, we solve $\ln(n+1) = 100$, getting $n = e^{100} - 1 \approx 2.6811 \times 10^{43}$ (the smallest n for which $S_n \geq 100$ is approximately 1.509×10^{43}).

65. (a) We observe $L_1 = d$, $L_2 = d + L_1 - pL_1 = d + d(1 - p)$, $L_3 = d + L_2 - pL_2 = d + (1 - p)L_2 = d + d(1 - p) + d(1 - p)^2$. In general, $L_n = d + d(1 - p) + d(1 - p)^2 + \cdots + d(1 - p)^{n-1}$. This is a geometric series with first term d and common ratio $(1 - p)$. Thus

$$L_n = \frac{d[1 - (1 - p)^n]}{1 - (1 - p)} = \frac{d}{p}[1 - (1 - p)^n].$$

Since $0 < p < 1$, $\displaystyle\lim_{n \to \infty} L_n = d/p$.

(b) With $d = 1.4$ and $p = 0.009$, $\displaystyle\lim_{n \to \infty} L_n = 1.4/0.009 \approx 155.56 \approx 155.56$ mg. To determine when the various symptoms occur, we solve $L_n = \dfrac{d}{p}[1 - (1 - p)^n]$ for n, obtaining $n = \dfrac{\ln(1 - pL_n/d)}{\ln(1 - p)}$. With $d = 1.4$ and $p = 0.009$, this becomes $n = \dfrac{\ln(1 - 9L_n/1400)}{\ln 0.991}$.

$$\begin{aligned}
\text{parasthesia:} \quad & L_n = 25, \quad n \approx 19.4 \\
\text{ataxia:} \quad & L_n = 55, \quad n \approx 48.3 \\
\text{dysarthia:} \quad & L_n = 90, \quad n \approx 95.6
\end{aligned}$$

Neither deafness nor death can occur at these values of d and p.

(c) Solving $200 = \dfrac{d}{0.009}[1 - (1 - 0.009)^{100}]$ for d, we obtain $d \approx 3.02$ mg.

67. This exercise involves a research report, and thus a preset solution is not applicable. It should be noted, however, that the series of the reciprocal of primes (i.e., the *harmonic series of primes*) does diverge, with multiple proofs available in the literature.

69. $AP_1 + P_1P_2 + P_2P_3 + P_3P_4 + P_4P_5 + P_5P_6 + \cdots$

$$= \sin 30° + (\cos 30°)\sin 30° + (\cos 30°)^2 \sin 30° + (\cos 30°)^3 \sin 30° + \cdots$$

$$= \frac{1}{2} + \left(\frac{\sqrt{3}}{2}\right)\frac{1}{2} + \left(\frac{\sqrt{3}}{2}\right)^2 \frac{1}{2} + \left(\frac{\sqrt{3}}{2}\right)^3 \frac{1}{2} + \cdots$$

$$= \frac{1}{2}\left[1 + \frac{\sqrt{3}}{2} + \left(\frac{\sqrt{3}}{2}\right)^2 + \left(\frac{\sqrt{3}}{2}\right)^3 + \cdots\right]$$

$$= \frac{1}{2}\left(\frac{1}{1 - \sqrt{3}/2}\right) = \frac{1}{2 - \sqrt{3}} = 2 + \sqrt{3}$$

71. (a) The overhangs from the edge of the table are:

$$d_2 = \frac{L}{2} + \frac{L}{4} = \frac{L}{2}\left(1 + \frac{1}{2}\right) = \frac{L}{2}H_2 = \frac{1}{2}\left(\frac{3}{2}\right) = 0.75$$

$$d_3 = \frac{L}{2} + \frac{L}{4} + \frac{L}{6} = \frac{L}{2}\left(1 + \frac{1}{2} + \frac{1}{3}\right) = \frac{L}{2}H_3 = \frac{1}{2}\left(\frac{11}{6}\right) \approx 0.917$$

$$d_4 = \frac{L}{2} + \frac{L}{4} + \frac{L}{6} + \frac{L}{8} = \frac{L}{2}\left(1 + \frac{1}{2} + \frac{1}{3} + \frac{1}{4}\right) = \frac{L}{2}H_4 = \frac{1}{2}\left(\frac{25}{12}\right) \approx 1.0417,$$

where $H_n = \sum_{k=1}^{n} \frac{1}{k}$ are the harmonic numbers. If m denotes the mass of a book and x_k denotes the x-coordinate of the center of mass of the kth book, then the centers of mass are defined by $\bar{x}_n = \dfrac{mx_1 + mx_2 + \cdots + mx_n}{nm} = \dfrac{1}{n}\sum_{k=1}^{n} x_k$.

Therefore,

$$\bar{x}_2 = \frac{\dfrac{L}{2} + \left(\dfrac{L}{2} + \dfrac{L}{2}\right)}{2} = \frac{L\left(\dfrac{2}{2} + \dfrac{1}{2}\right)}{2} = \frac{3}{4}L$$

$$\bar{x}_3 = \frac{\dfrac{L}{2} + \left(\dfrac{L}{2} + \dfrac{L}{4}\right) + \left(\dfrac{L}{2} + \dfrac{L}{4} + \dfrac{L}{2}\right)}{3} = \frac{L\left(\dfrac{3}{2} + \dfrac{2}{4} + \dfrac{1}{2}\right)}{3} = \frac{5}{6}L$$

$$\bar{x}_4 = \frac{\dfrac{L}{2} + \left(\dfrac{L}{2} + \dfrac{L}{6}\right) + \left(\dfrac{L}{2} + \dfrac{L}{6} + \dfrac{L}{4}\right) + \left(\dfrac{L}{2} + \dfrac{L}{6} + \dfrac{L}{4} + \dfrac{L}{2}\right)}{4}$$

$$= \frac{L\left(\dfrac{4}{2} + \dfrac{3}{6} + \dfrac{2}{4} + \dfrac{1}{2}\right)}{4} = \frac{7}{8}L.$$

In other words, the center of mass for each stack of books is at the edge of the table.

(b) $d_4 = \dfrac{1}{2}H_4 = \dfrac{1}{2}\left(\dfrac{25}{12}\right) \approx 1.0417 > 1$ means that the fourth book is completely beyond the edge of the table.

(c) The overhang of n books from the edge of the table is

$$d_n = \frac{L}{2} + \frac{L}{4} + \frac{L}{6} + \cdots + \frac{L}{2(n-1)} + \frac{L}{2n} = \frac{L}{2}\left(1 + \frac{1}{2} + \frac{1}{3} + \cdots + \frac{1}{n-1} + \frac{1}{n}\right) = \frac{L}{2}H_n,$$

where $H_n = \sum_{k=1}^{n} \frac{1}{k}$ are the harmonic numbers. The x-coordinate of the center of mass is

$$\bar{x}_n = \frac{1}{n}\left\{ \frac{L}{2} + \left[\frac{L}{2} + \frac{L}{2(n-1)}\right] + \left[\frac{L}{2} + \frac{L}{2(n-1)} + \frac{L}{2(n-2)}\right] + \cdots \right.$$

$$\left. \cdots + \left[\frac{L}{2} + \frac{L}{2(n-1)} + \frac{L}{2(n-2)} + \cdots + \frac{L}{4} + \frac{L}{2} + \frac{L}{2}\right] \right\}$$

$$= \frac{L}{n}\left[\frac{n}{2} + \frac{n-1}{2(n-1)} + \frac{n-2}{2(n-2)} + \cdots + \frac{3}{6} + \frac{2}{4} + \frac{1}{2}\right]$$

$$= \frac{L}{n}\left[\frac{n}{2} + (n-1)\frac{1}{2}\right] = \frac{L}{n}\left(\frac{2n-1}{2}\right) = L\left(\frac{2n-1}{2n}\right) = L - \frac{L}{2n}$$

Since the overhang of the first (or bottom) book in the stack from the edge of the table is $\frac{L}{2n}$, $\bar{x}_n$ is the distance to the edge of the table. That is, the center of mass of n books is again at the edge of the table.

(d) For $n = 30$ and $n = 31$, *Mathematica* gives

$$d_{30} = \frac{9304682830147}{465817912560} \approx 1.99749L$$

$$d_{31} = \frac{L}{2}\left(\frac{290774257297357}{72201776446800}\right) \approx 2.01362L,$$

which means that for a stack of 31 books, the overhang of the top book from the edge of the table is over twice the length of the book.

(e) There is no theoretical limit to the number of books that can be stacked in this manner because the overhang $d_n = \frac{L}{2}H_n$ for large n behaves as the divergent harmonic series. Namely, $d_n \to \infty$ as $n \to \infty$.

9.4 Integral Test

1. The function $f(x) = \frac{1}{x^{1.1}}$ is continuous and decreasing on $[1, \infty)$. Since

$$\int_1^\infty \frac{1}{x^{1.1}}\,dx = \lim_{t\to\infty}\int_1^t x^{-1.1}\,dx = \lim_{t\to\infty}\left(-\frac{1}{0.1}x^{-0.1}\right)\Big|_1^t = -10(0-1) = 10,$$

the integral converges and $\sum_{k=1}^{\infty} \frac{1}{k^{1.1}}$ converges.

3. Rewriting the series, we have

$$1 + \frac{1}{2\sqrt{2}} + \frac{1}{3\sqrt{3}} + \cdots = \sum_{k=1}^{\infty} \frac{1}{x\sqrt{x}} = \sum_{k=1}^{\infty} \frac{1}{x^{3/2}}.$$

The function $f(x) = \frac{1}{x^{3/2}}$ is continuous and decreasing on $[1, \infty)$. Since

$$\int_1^\infty \frac{1}{x^{3/2}}\,dx = \lim_{t\to\infty}\int_1^t x^{-3/2}\,dx = \lim_{t\to\infty}\left(-2x^{-1/2}\right)\Big|_1^t = -2(0-1) = 2,$$

the integral converges and the series $1 + \frac{1}{2\sqrt{2}} + \frac{1}{3\sqrt{3}} + \cdots$ converges.

5. The function $f(x) = \frac{1}{2x+7}$ is continuous and decreasing on $[1, \infty)$. Since

$$\int_1^\infty \frac{1}{2x+7}\,dx = \lim_{t\to\infty}\int_1^t \frac{1}{2x+7}\,dx = \lim_{t\to\infty}\left(\frac{1}{2}\ln(2x+7)\right)\Big|_1^t = \infty,$$

the integral diverges and $\sum_{k=1}^{\infty} \frac{1}{2k+7}$ diverges.

7. The function $f(x) = \dfrac{1}{1+5x^2}$ is continuous and decreasing on $[1,\infty)$. Since

$$\int_1^\infty \frac{1}{1+5x^2}\,dx = \lim_{t\to\infty}\int_1^t \frac{1}{1+5x^2}\,dx = \lim_{t\to\infty}\left(\frac{\sqrt{5}}{5}\tan^{-1}(\sqrt{5}x)\right)\Bigg|_1^t$$
$$= \frac{\sqrt{5}}{5}\left(\frac{\pi}{2} - \tan^{-1}(\sqrt{5})\right),$$

the integral converges and $\displaystyle\sum_{k=1}^\infty \frac{1}{1+5k^2}$ converges.

9. Using the limit comparison test with $a_n = ne^{-n^2}$ and $b_n = 1/n^2$, we have (using L'Hôpital's Rule)

$$\lim_{n\to\infty}\frac{a_n}{b_n} = \lim_{n\to\infty}\frac{n/e^{n^2}}{1/n^2} = \lim_{n\to\infty}\frac{n^3}{e^{n^2}} \overset{h}{=} \lim_{n\to\infty}\frac{3n^2}{2ne^{n^2}} = \lim_{n\to\infty}\frac{3n}{2e^{n^2}} \overset{h}{=} \lim_{n\to\infty}\frac{3}{4ne^{n^2}} = 0.$$

Since $\displaystyle\sum_{n=1}^\infty b_n$ is a p-series with $p = 2 > 1$, it converges and $\displaystyle\sum_{n=1}^\infty ke^{-k^2}$ converges.

11. The function $f(x) = \dfrac{x}{e^x}$ is continuous and decreasing on $[1,\infty)$. Since

$$\int_1^\infty \frac{x}{e^x}\,dx = \lim_{t\to\infty}\int_1^t \frac{x}{e^x}\,dx$$
$$= \lim_{t\to\infty}\left((-x-1)e^{-x}\right)\Big|_1^t = \frac{2}{e},$$

the integral converges and $\displaystyle\sum_{k=1}^\infty \frac{k}{e^k}$ converges.

13. The function $f(x) = \dfrac{1}{x\ln x}$ is continuous and decreasing on $[2,\infty)$. Since

$$\int_2^\infty \frac{1}{x\ln x}\,dx = \lim_{t\to\infty}\int_2^t \frac{1}{x\ln x}\,dx = \lim_{t\to\infty}\ln(\ln x)\Big]_2^t = \lim_{t\to\infty}[\ln(\ln t) - \ln(\ln 2)] = \infty,$$

the integral diverges and $\displaystyle\sum_{k=2}^\infty \frac{1}{k\ln k}$ diverges.

15. The function $f(x) = \dfrac{10}{x(\ln x)^2}$ is continuous and decreasing on $[2,\infty)$. Since

$$\int_2^\infty \frac{10}{x(\ln x)^2}\,dx = 10\lim_{t\to\infty}\int_2^t \frac{(\ln x)^{-2}}{x}\,dx = 10\lim_{t\to\infty}\left[-(\ln x)^{-1}\right]\Big]_2^t$$
$$= 10\lim_{t\to\infty}\left[(\ln 2)^{-1} - (\ln t)^{-1}\right] = 10/\ln 2,$$

the integral converges and $\displaystyle\sum_{k=2}^\infty \frac{10}{k(\ln k)^2}$ converges.

17. The function $f(x) = \dfrac{\arctan x}{1+x^2}$ is continuous and decreasing on $[1,\infty)$. Since

$$\int_1^\infty \frac{\arctan x}{1+x^2}\,dx = \lim_{t\to\infty}\int_1^t \frac{\arctan x}{1+x^2}\,dx = \lim_{t\to\infty}\frac{(\arctan x)^2}{2}\Bigg]_1^t = \lim_{t\to\infty}\left[\frac{(\arctan t)^2}{2} - \frac{\pi^2}{32}\right]$$
$$= \frac{\pi^2}{8} - \frac{\pi^2}{32} = \frac{3\pi^2}{32},$$

the integral converges and $\displaystyle\sum_{k=1}^\infty \frac{\arctan k}{1+k^2}$ converges.

19. The function $f(x) = \dfrac{1}{1+\sqrt{x}}$ is continuous and decreasing on $[1,\infty)$. Since

$$\int_1^\infty \frac{1}{1+\sqrt{x}}\,dx = \lim_{t\to\infty}\int_1^t \frac{1}{1+\sqrt{x}}\,dx$$
$$= \lim_{t\to\infty}\left(2\sqrt{x+1}\right)\Big|_1^t = \infty,$$

the integral diverges and $\displaystyle\sum_{k=1}^\infty \frac{1}{\sqrt{1+k}}$ diverges.

21. The function $f(x) = \dfrac{x}{(x^2+1)^3}$ is continuous and decreasing on $[1,\infty)$. Since

$$\int_1^\infty \frac{x}{(x^2+1)^3}\,dx = \lim_{t\to\infty}\int_1^t \frac{x}{(x^2+1)^3}\,dx$$
$$= \lim_{t\to\infty}\left(\frac{-1}{4(x^2+1)^2}\right)\Big|_1^t = \frac{1}{16},$$

the integral converges and $\displaystyle\sum_{k=1}^\infty \frac{k}{(k^2+1)^3}$ diverges.

23. Since $\displaystyle\lim_{x\to\infty} x\sin\frac{1}{x} = \lim_{t\to 0}\frac{\sin t}{t} = 1$, the series diverges by the n-th term test.

25. The function $f(x) = \dfrac{1}{x(x+1)}$ is continuous and decreasing on $[1,\infty)$. Since

$$\int_1^\infty \frac{1}{x(x+1)}\,dx = \lim_{t\to\infty}\int_1^t \frac{1}{x(x+1)}\,dx$$
$$= \lim_{t\to\infty}\ln\left(\frac{x}{x+1}\right)\Big|_1^t = 0 - \ln\left(\frac{1}{2}\right) = \ln(2),$$

the integral converges and $\displaystyle\sum_{k=1}^\infty \frac{1}{k(k+1)}$ converges.

27. The function $f(x) = \dfrac{1}{(x+1)(x+7)}$ is continuous and decreasing on $[1,\infty)$. Since

$$\int_1^\infty \frac{1}{(x+1)(x+7)}\,dx = \lim_{t\to\infty}\int_1^t \frac{1}{(x+1)(x+7)}\,dx$$
$$= \lim_{t\to\infty}\left(\frac{1}{6}\ln\left(\frac{x+1}{x+7}\right)\right)\Big|_1^t$$
$$= 0 - \frac{1}{6}\ln\left(\frac{2}{8}\right) = \frac{1}{6}\ln(4),$$

the integral converges and $\displaystyle\sum_{k=1}^\infty \frac{1}{(k+1)(k+7)}$ converges.

29. The function $f(x) = \dfrac{2}{e^x+e^{-x}}$ is continuous and decreasing on $[1,\infty)$. Since

$$\int_1^\infty \frac{2}{e^x+e^{-x}}\,dx = \lim_{t\to\infty}\int_1^t \frac{2}{e^x+e^{-x}}\,dx$$
$$= \lim_{t\to\infty} 2\tan^{-1}(e^x)\Big|_1^t = \pi - 2\tan^{-1}(e),$$

the integral converges and $\displaystyle\sum_{k=1}^\infty \frac{2}{e^k+e^{-k}}$ converges.

31. Since $\displaystyle\sum_{k=1}^{\infty}\frac{2}{k}$ diverges while $\displaystyle\sum_{k=1}^{\infty}\frac{3}{k^2}$ converges, the entire series $\displaystyle\sum_{k=1}^{\infty}\frac{2}{k}+\frac{3}{k^2}$ diverges.

33. $\displaystyle\sum_{k=1}^{\infty}\frac{1}{k^2}$ donverges by the p-series test, $\displaystyle\sum_{k=1}^{\infty}\frac{1}{2^k}$ is a geometric series with $r=\frac{1}{2}<1$. Therefore, $\displaystyle\sum_{k=1}^{\infty}\frac{1}{k^2}+\frac{1}{2^k}$ converges.

35. For $\displaystyle\sum_{k=2}^{\infty}\frac{1}{k(\ln k)^p}$, we have

$$\int_2^{\infty}(\ln x)^{-p}\left(\frac{1}{x}dx\right)=\lim_{b\to\infty}(\ln x)^{-p+1}\Big]_2^b=\lim_{b\to\infty}\left[\frac{1}{(\ln b)^{p-1}}-\frac{1}{(\ln 2)^{p-1}}\right].$$

For $p>1$, this limit converges to $-\dfrac{1}{(\ln 2)^{p-1}}$. For $p=1$, $\displaystyle\int_2^{\infty}\frac{1}{\ln x}\left(\frac{1}{x}\right)dx=\lim_{b\to\infty}[\ln(\ln b)-\ln(\ln 2)]=\infty$. For $p<1$, the integral also diverges.

37. For $p\geq 0$, $f(x)=x^p\ln x$ is not decreasing. So for $p<0$, integration by parts gives

$$\sum_{k=2}^{\infty}k^p\ln k\implies\int_2^{\infty}x^p\ln x\,dx=\lim_{b\to\infty}\frac{x^{p+1}}{(p+1)^2}[1-(p+1)\ln x]\Big]_2^b$$

$$=\lim_{b\to\infty}\left\{\frac{b^{p+1}}{(p+1)^2}[1-(p+1)\ln b]-\frac{2^{p+1}}{(p+1)^2}[1-(p+1)\ln 2]\right\}$$

$$=-\frac{2^{p+1}}{(p+1)^2}[1-(p+1)\ln 2]\text{ for }p+1<0\text{ or }p<-1.$$

The integral converges for $p<-1$ and diverges for $-1\leq p<0$ and for $p\geq 0$ (that is, the integral diverges for $p\geq -1$).

39. The function $f(x)=\dfrac{1}{1+x^2}$ is continuous, positive, and decreasing on $[1,\infty)$. Using the result from Problem 38, we have

$$\int_1^{n+1}\frac{1}{1+x^2}dx\leq\sum_{k=1}^{n}\frac{1}{1+k^2}\leq\frac{1}{2}+\int_1^n\frac{1}{1+x^2}dx$$

Letting $n\to\infty$, we have

$$\int_1^{\infty}\frac{1}{1+x^2}dx\leq\sum_{k=1}^{\infty}\frac{1}{1+k^2}\leq\frac{1}{2}+\int_1^{\infty}\frac{1}{1+x^2}dx$$

After integrating, this becomes

$$\frac{\pi}{4}\leq\sum_{k=1}^{\infty}\frac{1}{1+k^2}\leq\frac{1}{2}+\frac{\pi}{4}$$

41. Since f is decreasing and $f(k)=a_k$, we have $f(x)\leq f(k)=a_k$ on $[k,k+1]$. This implies $\int_k^{k+1}f(x)dx\leq\int_k^{k+1}a_k dx=a_k$. Therefore, $\displaystyle\int_{n+1}^{n+p}f(x)dx=\sum_{k=n+1}^{n+p-1}\int_k^{k+1}f(x)dx\leq\sum_{k=n+1}^{n+p-1}a_k$. Since this is true for every integer p, we can let $p\to\infty$ to get $\displaystyle\int_{n+1}^{\infty}f(x)dx\leq\sum_{k=n+1}^{\infty}a_k$ or $\displaystyle\int_{n+1}^{\infty}f(x)dx\leq R_n$. Also, note that $f(x)\geq f(k+1)=a_{k+1}$ on $[k,k+1]$. This implies $\int_k^{k+1}f(x)dx\geq\int_k^{k+1}a_{k+1}dx=a_{k+1}$. Therefore,

$$\sum_{k=n+1}^{n+p}a_k=\sum_{k=n}^{n+p-1}a_{k+1}\leq\sum_{k=n}^{n+p-1}\int_k^{k+1}f(x)dx=\int_n^{n+p}f(x)dx.$$

Since this is true for every integer p, we can let $p\to\infty$ to get $\displaystyle\sum_{k=n+1}^{\infty}a_k\leq\int_n^{\infty}f(x)dx$ or $R_n\leq\int_n^{\infty}f(x)dx$. Hence $\int_{n+1}^{\infty}f(x)dx\leq R_n\leq\int_n^{\infty}f(x)dx$.

9.5 Comparison Tests

1. Since $\dfrac{1}{(k+1)(k+2)} < \dfrac{1}{k^2}$, the series converges by comparison with the p-series $\displaystyle\sum_{k=1}^{\infty} \dfrac{1}{k^2}$.

3. Since $\dfrac{1}{\sqrt{k}-1} \geq \dfrac{1}{\sqrt{k}}$ for $k \geq 2$, the series diverges by comparison with the p-series $\displaystyle\sum_{k=1}^{\infty} \dfrac{1}{k^{1/2}}$.

5. Since $\dfrac{1}{\ln k} > \dfrac{1}{k}$ for $k \geq 2$, the series diverges by comparison with the harmonic series $\displaystyle\sum_{k=1}^{\infty} \dfrac{1}{k}$.

7. Since $\dfrac{1+3^k}{2^k} = \dfrac{1}{2^k} + \left(\dfrac{3}{2}\right)^k > 1$ for all $k \geq 1$ the series diverges by comparison with the series $\displaystyle\sum_{k=1}^{\infty} 1$.

9. Since $\dfrac{2+\sin k}{\sqrt[3]{k^4+1}} < \dfrac{3}{k^{4/3}}$, the series converges by comparison with the p-series $\displaystyle\sum_{k=1}^{\infty} \dfrac{3}{k^{4/3}}$.

11. Since $j+e^{-j} < j+9$, then $\dfrac{j+e^{-j}}{5^j(j+9)} < \dfrac{1}{5^j}$, and the series converges by comparison with the geometric series $\displaystyle\sum_{j=1}^{\infty} \left(\dfrac{1}{5}\right)^j$.

13. Since $\dfrac{\sqrt{k+1}-\sqrt{k}}{k} = \dfrac{1}{k(\sqrt{k+1}+\sqrt{k})} \leq \dfrac{1}{k(\sqrt{k}+\sqrt{k})} = \dfrac{1}{2k^{3/2}}$, the series converges by comparison with the p-series $\dfrac{1}{2}\displaystyle\sum_{k=1}^{\infty} \dfrac{1}{k^{3/2}}$.

15. Using the limit comparison test with $a_n = \dfrac{1}{2n+7}$ and $b_n = \dfrac{1}{n}$, we have

$$\lim_{n\to\infty} \frac{a_n}{b_n} = \lim_{n\to\infty} \frac{1/(2n+7)}{1/n} = \lim_{n\to\infty} \frac{n}{2n+7} = \frac{1}{2}.$$

Since $\displaystyle\sum_{n=1}^{\infty} b_n$ diverges, $\displaystyle\sum_{n=1}^{\infty} \dfrac{1}{2n+7}$ diverges.

17. Using the limit comparison test with $a_n = \dfrac{1}{n\sqrt{n^2-1}}$ and $b_n = \dfrac{1}{n^2}$, we have

$$\lim_{n\to\infty} \frac{a_n}{b_n} = \lim_{n\to\infty} \frac{1/n\sqrt{n^2-1}}{1/n^2} = \lim_{n\to\infty} \frac{n^2}{n\sqrt{n^2-1}} = \lim_{n\to\infty} \frac{1}{\sqrt{1-1/n^2}} = 1.$$

Since $\displaystyle\sum_{n=1}^{\infty} b_n$ is a p-series with $p = 2 > 1$, it converges and $\displaystyle\sum_{n=2}^{\infty} \dfrac{1}{n\sqrt{n^2-1}}$ converges.

19. Using the limit comparison test with $a_n = \dfrac{n^2-n+2}{3n^5+n^2}$ and $b_n = \dfrac{1}{n^3}$, we have

$$\lim_{n\to\infty} \frac{a_n}{b_n} = \lim_{n\to\infty} \frac{(n^2-n+2)/(3n^5+n^2)}{1/n^3} = \lim_{n\to\infty} \frac{n^5-n^4+2n^3}{3n^5+n^2} = \lim_{n\to\infty} \frac{1-1/n+2/n^2}{3+1/n^3} = \frac{1}{3}.$$

Since $\displaystyle\sum_{n=1}^{\infty} b_n$ is a p-series with $p = 3 > 1$, it converges and $\displaystyle\sum_{n=1}^{\infty} \dfrac{n^2-n+2}{3n^5+n^2}$ converges.

21. Using the Limit Comparison Test with $a_k = \dfrac{\sqrt{k+1}}{\sqrt[3]{64k^9+40}}$ and $b_k = \dfrac{\sqrt{k}}{\sqrt[3]{64k^9}}$, we have

$$\lim_{k\to\infty}\frac{a_k}{b_k} = \lim_{k\to\infty}\frac{\left(\dfrac{\sqrt{k+1}}{\sqrt[3]{64k^9+40}}\right)}{\left(\dfrac{\sqrt{k}}{\sqrt[3]{64k^9}}\right)}$$

$$= \lim_{k\to\infty}\frac{\sqrt{k+1}}{\sqrt{k}}\cdot\frac{\sqrt[3]{64k^9}}{\sqrt[3]{64k^9+40}}$$

$$= \lim_{k\to\infty}\frac{\sqrt{k+1}}{\sqrt{k}}\cdot\lim_{k\to\infty}\frac{\sqrt[3]{64k^9}}{\sqrt[3]{64k^9+40}} = 1\cdot 1 = 1.$$

Since $\displaystyle\sum_{k=1}^{\infty}b_k = \sum_{k=1}^{\infty}\frac{\sqrt{k}}{\sqrt[3]{64k^9}} = \sum_{k=1}^{\infty}\frac{1}{4k^{5/2}}$ is a p-series with $p = 5/2 > 1$, it converges and $\displaystyle\sum_{k=1}^{\infty}\frac{\sqrt{k+1}}{\sqrt[3]{64k^9+40}}$ converges.

23. Using the limit comparison test with $a_n = \dfrac{n+\ln n}{n^3+2n-1}$ and $b_n = \dfrac{1}{n^2}$, we have

$$\lim_{n\to\infty}\frac{a_n}{b_n} = \lim_{n\to\infty}\frac{(n+\ln n)/(n^3+2n-1)}{1/n^2} = \lim_{n\to\infty}\frac{n^3+n^2\ln n}{n^3+2n-1} = \lim_{n\to\infty}\frac{1+(\ln n)/n}{1+2/n-1/n} = 1.$$

(By L'Hôpital's Rule, $\displaystyle\lim_{n\to\infty}\frac{\ln n}{n} = \lim_{n\to\infty}\frac{1/n}{1} = 0$.) Since $\displaystyle\sum_{n=1}^{\infty}\frac{1}{n^2}$ is a p-series with $p = 2 > 1$, it converges and $\displaystyle\sum_{n=2}^{\infty}\frac{k+\ln k}{k^3+2k-1}$ converges.

25. Using the limit comparison test with $a_n = \sin\dfrac{1}{n}$ and $b_n = \dfrac{1}{n}$, we have

$$\lim_{n\to\infty}\frac{a_n}{b_n} = \lim_{n\to\infty}\frac{\sin(1/n)}{1/n} \overset{h}{=} \lim_{n\to\infty}\frac{\cos(1/n)(-1/n^2)}{-1/n^2} = \lim_{n\to\infty}\cos\frac{1}{n} = 1.$$

Since $\displaystyle\sum_{n=1}^{\infty}\frac{1}{n}$ diverges, the series $\displaystyle\sum_{k=1}^{\infty}\sin\frac{1}{k}$ diverges.

27. Using the limit comparison test with $a_n = \left(\dfrac{1}{2}+\dfrac{1}{2n}\right)^n$ and $b_n = \left(\dfrac{1}{2}\right)^n$, we have

$$\lim_{n\to\infty}\frac{a_n}{b_n} = \lim_{n\to\infty}\frac{\left(\dfrac{1}{2}+\dfrac{1}{2n}\right)^n}{(1/2)^n} = \lim_{n\to\infty}\left(1+\frac{1}{n}\right)^n = e.$$

Since $\displaystyle\sum_{n=1}^{\infty}b_n$ is a geometric series with $r = 1/2 < 1$, it converges and the series $\displaystyle\sum_{k=1}^{\infty}\left(\frac{1}{2}+\frac{1}{2k}\right)^k$ converges.

29. Since $\displaystyle\lim_{k\to\infty}\frac{k}{100\sqrt{k^2+1}} = \frac{1}{100}$, the series diverges by the n-th term test.

31. Since $\displaystyle\lim_{k\to\infty}\ln\left(5+\frac{k}{5}\right) = \infty$, the series diverges by the n-th term test.

33. The function $f(x) = \dfrac{x}{(x^2+1)^2}$ is continuous and decreasing on $[1,\infty)$. Since

$$\int_1^{\infty}\frac{x}{(x^2+1)^2}\,dx = \lim_{t\to\infty}\int_1^t\frac{x}{(x^2+1)^2}\,dx = -\frac{1}{2}\lim_{t\to\infty}\frac{1}{x^2+1}\Big]_1^t = -\frac{1}{2}\lim_{t\to\infty}\left(\frac{1}{t^2+1}-\frac{1}{2}\right) = \frac{1}{4},$$

the integral converges and $\displaystyle\sum_{k=1}^{\infty}\frac{k}{(k^2+1)^2}$ converges.

(The direct comparison test and limit comparison test can also be used.)

35. Since $\dfrac{1}{9+\sin^2 k} > \dfrac{1}{10}$ for $k \geq 1$, the series diverges by comparison with the series $\displaystyle\sum_{k=1}^{\infty} \dfrac{1}{10}$.

37. Since $\dfrac{2}{2+k2^k} < \dfrac{2}{k2^k} < \dfrac{2}{2^k} = \dfrac{1}{2^{k-1}}$, the series converges by comparison with the geometric series $\displaystyle\sum_{k=1}^{\infty} \dfrac{1}{2^{k-1}}$.

39. Since $\ln\left(1+\dfrac{1}{k}\right) > \ln\dfrac{1}{k} = -\ln k$, the series diverges by comparison with the series $-\displaystyle\sum_{k=2}^{\infty} \ln k$.

41. Since $\sum a_k$ converges, then $\displaystyle\lim_{k\to\infty} a_k = 0$. Therefore, for a sufficiently large n and $k \geq n$, we can say that $0 < a_k < 1$ and so $a_k^2 < a_k$. Thus, $\sum a_k^2$ converges by the direct comparison test.

43. The statement is false. The condition of a positive-term series is missing: as a counterexample, consider the convergent series $\sum b_k = \sum 0 = 0$ and the divergent series $a_k = -\displaystyle\sum \dfrac{1}{k}$.

45. The series $\displaystyle\sum_{k=1}^{\infty} \dfrac{1}{k^{1+1/k}}$ diverges by the limit comparison test with $\displaystyle\sum_{k=1}^{\infty} \dfrac{1}{k}$:

$$\lim_{n\to\infty} \frac{a_n}{b_n} = \lim_{n\to\infty} \frac{1/n^{1+1/n}}{1/n} = \lim_{n\to\infty} \frac{1}{\sqrt[n]{n}} = 1.$$

47. Of the set of integers a_i used in the decimal representation of the number, let a_B be the biggest integer. Then for the series $\displaystyle\sum_{k=1}^{\infty} \dfrac{a_k}{10^k}$, we have $\dfrac{a_k}{10^k} \leq \dfrac{a_B}{10^k}$ for all k. Since $\displaystyle\sum_{k=1}^{\infty} \dfrac{a_B}{10^k} = a_B \displaystyle\sum_{k=1}^{\infty} \dfrac{1}{10^k}$ and $\displaystyle\sum_{k=1}^{\infty} \dfrac{1}{10^k} = \displaystyle\sum_{k=1}^{\infty} \left(\dfrac{1}{10}\right)^k$ is a convergent geometric series, then $\displaystyle\sum_{k=1}^{\infty} \dfrac{a_k}{10^k}$ converges by the direct comparison test.

9.6   Ratio and Root Tests

1. Since $\displaystyle\lim_{n\to\infty} \dfrac{a_{n+1}}{a_n} = \lim_{n\to\infty} \dfrac{1/(n+1)!}{1/n!} = \lim_{n\to\infty} \dfrac{n!}{(n+1)!} = \lim_{n\to\infty} \dfrac{1}{n+1} = 0 < 1$, the series converges by the ratio test.

3. Since $\displaystyle\lim_{n\to\infty} \dfrac{a_{n+1}}{a_n} = \lim_{n\to\infty} \dfrac{(n+1)!/1000^{n+1}}{n!/1000^n} = \lim_{n\to\infty} \dfrac{n+1}{1000} = \infty$, the series diverges by the ratio test.

5. Since $\displaystyle\lim_{n\to\infty} \dfrac{a_{n+1}}{a_n} = \lim_{n\to\infty} \dfrac{(n+1)^{10}/(1.1)^{n+1}}{n^{10}/(1.1)^n} = \lim_{n\to\infty} \left(1+\dfrac{1}{n}\right)^{10} \dfrac{1}{1.1} = \dfrac{1}{1.1} < 1$, the series converges by the ratio test.

7. Since $\displaystyle\lim_{n\to\infty} \dfrac{a_{n+1}}{a_n} = \lim_{n\to\infty} \dfrac{4^n/(n+1)3^{n-1}}{4^{n-1}/n3^{n-2}} = \lim_{n\to\infty} \dfrac{4}{3}\left(\dfrac{n}{n+1}\right) = \dfrac{4}{3} > 1$, the series diverges by the ratio test.

9. Since

$$\lim_{n\to\infty} \frac{a_{n+1}}{a_n} = \lim_{n\to\infty} \frac{(n+1)!/(2n+2)!}{n!/(2n)!} = \lim_{n\to\infty} \frac{n+1}{(2n+1)(2n+2)} = \lim_{n\to\infty} \frac{n+1}{4n^2+6n+2} = 0 < 1,$$

the series converges by the ratio test.

11. Since

$$\lim_{n\to\infty} \frac{a_{n+1}}{a_n} = \lim_{n\to\infty} \frac{99^{n+1}((n+1)^3+1)}{99^n(n^3+1)} \cdot \frac{n^2 100^n}{(n+1)^2 100^{n+1}}$$

$$= \lim_{n\to\infty} \left(\frac{99}{100}\right) \cdot \left(\frac{(n+1)^3+1}{n^3+1}\right) \cdot \left(\frac{n^2}{(n+1)^2}\right)$$

$$= \left(\frac{99}{100}\right)\left(\lim_{n\to\infty} \frac{(n+1)^3+1}{n^3+1}\right)\left(\lim_{n\to\infty} \frac{n^2}{(n+1)^2}\right)$$

$$= \frac{99}{100} \cdot 1 \cdot 1 = \frac{99}{100} < 1,$$

the series converges by the ratio test.

13. Since

$$\lim_{n\to\infty} \frac{a_{n+1}}{a_n} = \lim_{n\to\infty} \frac{5^{n+1}}{5^n} \cdot \frac{n^n}{(n+1)^{n+1}}$$

$$\leq \lim_{n\to\infty} 5 \cdot \frac{n^n}{n^{n+1}}$$

$$= \lim_{n\to\infty} \frac{5}{n} = 0 < 1,$$

the series converges by the ratio test.

15. Since $\lim_{k\to\infty} \dfrac{a_{k+1}}{k_n} = \lim_{k\to\infty} \dfrac{1 \cdot 3 \cdot 5 \cdots (2k+1)/(k+1)!}{1 \cdot 3 \cdot 5 \cdots (2k-1)/k!} = \lim_{k\to\infty} \dfrac{2k+1}{k+1} = 2 > 1$, the series diverges by the ratio test.

17. Since $\lim_{n\to\infty} \left(\dfrac{1}{n^n}\right)^{1/n} = \lim_{n\to\infty} \dfrac{1}{n} = 0 < 1$, the series converges by the root test.

19. Since $\lim_{n\to\infty} \left[\left(\dfrac{n}{\ln n}\right)^n\right]^{1/n} = \lim_{n\to\infty} \dfrac{n}{\ln n} = \infty$, the series diverges by the root test.

21. Since $\lim_{n\to\infty} \left[\left(\dfrac{n}{n+1}\right)^{n^2}\right]^{1/n} = \lim_{n\to\infty} \left(\dfrac{n}{n+1}\right)^n = \lim_{n\to\infty} \dfrac{1}{\left(\dfrac{n+1}{n}\right)^n} = \lim_{n\to\infty} \dfrac{1}{\left(1+\dfrac{1}{n}\right)^n} = \dfrac{1}{e} < 1$, the series converges by the root test.

23. Since $\lim_{n\to\infty} \left(\dfrac{6^{2n+1}}{n^n}\right)^{1/n} = \lim_{n\to\infty} \dfrac{6^{2+1/n}}{n} = 0 < 1$, the series converges by the root test.

25. Since $\dfrac{k^2+k}{k^3+2k+1} \geq \dfrac{k^2}{k^3+2k+1} \geq \dfrac{k^2}{k^3+2k^3+k^3} = \dfrac{1}{4k}$ for $k \geq 1$, the series diverges by comparison with the series $\dfrac{1}{4}\sum_{k=1}^{\infty} \dfrac{1}{k}$.

27. Since $\dfrac{e^{1/n}}{n^2} \leq \dfrac{e}{n^2}$ for $n \geq 1$, the series converges by the comparison with the series $\sum_{n=1}^{\infty} \dfrac{e}{n^2}$.

29. Since $\lim_{n\to\infty} \dfrac{a_{n+1}}{a_n} = \lim_{n\to\infty} \dfrac{(5^{n+1}(n+1)!}{5^n \cdot n!} \cdot \dfrac{(n+1)!}{(n+2)!}$

$$= \lim_{n\to\infty} \dfrac{5(n+1)}{n+2} = 5 > 1,$$

the series diverges by the ratio test.

31. Since $\dfrac{2^k}{3^k+4^k} \leq \dfrac{2^k}{3^k} = \left(\dfrac{2}{3}\right)^k$ for $k \geq 0$, the series converges by comparison with $\sum_{k=0}^{\infty} \left(\dfrac{2}{3}\right)^k$.

33. Applying the ratio test, we have $\lim_{n\to\infty} \dfrac{a_{n+1}}{a_n} = \lim_{n\to\infty} \dfrac{(n+1)p^{n+1}}{np^n} = \lim_{n\to\infty} \left(1+\dfrac{1}{n}\right)p = p$. Thus, the series converges for $0 \leq p < 1$

and diverges for $p > 1$. For $p = 1$, the series is $\sum_{k=1}^{\infty} k$, which diverges.

35. Applying the ratio test, we have $\lim_{n\to\infty} \dfrac{a_{n+1}}{a_n} = \lim_{n\to\infty} \dfrac{(n+1)^p/(n+1)!}{n^p/n!} = \lim_{n\to\infty} \left(\dfrac{n+1}{n}\right)^p \dfrac{1}{n+1} = 0$. The series converges for all real values of p.

37. (a) $F_n + F_{n-1} = \dfrac{1}{\sqrt{5}}\left(\dfrac{1+\sqrt{5}}{2}\right)^n - \dfrac{1}{\sqrt{5}}\left(\dfrac{1-\sqrt{5}}{2}\right)^n + \dfrac{1}{\sqrt{5}}\left(\dfrac{1+\sqrt{5}}{2}\right)^{n-1} - \dfrac{1}{\sqrt{5}}\left(\dfrac{1-\sqrt{5}}{2}\right)^{n-1}$

$\qquad = \dfrac{1}{\sqrt{5}}\left(\dfrac{1+\sqrt{5}}{2}\right)^{n-1}\left(\dfrac{1+\sqrt{5}}{2}+1\right) - \dfrac{1}{\sqrt{5}}\left(\dfrac{1-\sqrt{5}}{2}\right)^{n-1}\left(\dfrac{1-\sqrt{5}}{2}+1\right)$

$\qquad = \dfrac{1}{\sqrt{5}}\left(\dfrac{1+\sqrt{5}}{2}\right)^{n-1}\left(\dfrac{1+\sqrt{5}}{2}\right)^2 - \dfrac{1}{\sqrt{5}}\left(\dfrac{1-\sqrt{5}}{2}\right)^{n-1}\left(\dfrac{1-\sqrt{5}}{2}\right)^2$

$\qquad = \dfrac{1}{\sqrt{5}}\left(\dfrac{1+\sqrt{5}}{2}\right)^{n-1} - \dfrac{1}{\sqrt{5}}\left(\dfrac{1-\sqrt{5}}{2}\right)^{n-1} = F_{n+1}$

(b) $F_1 = \dfrac{1}{\sqrt{5}}\left(\dfrac{1+\sqrt{5}}{2}\right) - \dfrac{1}{\sqrt{5}}\left(\dfrac{1-\sqrt{5}}{2}\right) = 1$

$\quad F_2 = \dfrac{1}{\sqrt{5}}\left(\dfrac{1+\sqrt{5}}{2}\right)^2 - \dfrac{1}{\sqrt{5}}\left(\dfrac{1-\sqrt{5}}{2}\right)^2 = 1$

$\quad F_3 = \dfrac{1}{\sqrt{5}}\left(\dfrac{1+\sqrt{5}}{2}\right)^3 - \dfrac{1}{\sqrt{5}}\left(\dfrac{1-\sqrt{5}}{2}\right)^3 = 2$

$\quad F_4 = \dfrac{1}{\sqrt{5}}\left(\dfrac{1+\sqrt{5}}{2}\right)^4 - \dfrac{1}{\sqrt{5}}\left(\dfrac{1-\sqrt{5}}{2}\right)^4 = 3$

$\quad F_5 = \dfrac{1}{\sqrt{5}}\left(\dfrac{1+\sqrt{5}}{2}\right)^5 - \dfrac{1}{\sqrt{5}}\left(\dfrac{1-\sqrt{5}}{2}\right)^5 = 5$

39. Applying the ration test to $\displaystyle\sum_{n-1}^{\infty}\dfrac{1}{F_n}$, we have

$$\lim_{n\to\infty}\dfrac{a_{n+1}}{a_n} = \lim_{n\to\infty}\dfrac{1/F_{n+1}}{1/F_n} = \lim_{n\to\infty}\dfrac{F_n}{F_{n+1}} = \dfrac{2}{1+\sqrt{5}} < 1.$$

Thus, the series $\displaystyle\sum_{n=1}^{\infty}\dfrac{1}{F_n}$ converges.

9.7   Alternating Series

1. Since $a_{k+1} = \dfrac{1}{k+3} < \dfrac{1}{k+2} = a_k$ and $\displaystyle\lim_{k\to\infty}\dfrac{1}{k+2} = 0$, the series converges.

3. Since $\displaystyle\lim_{k\to\infty}\dfrac{k}{k+1} = 1 \neq 0$, the series diverges.

5. Let $f(x) = \dfrac{x^2+2}{x^3}$. Then $f'(x) = -\dfrac{x^2+6}{x^4} < 0$ for $x \geq 1$ and $a_{k+1} < a_k$. Since $\displaystyle\lim_{k\to\infty}\dfrac{k^2+2}{k^3} = \lim_{k\to\infty}\dfrac{1/k+2/k^3}{1} = 0$, the series converges.

7. Since $a_{k+1} = \dfrac{1}{k+1} + \dfrac{1}{3^{k+1}} < \dfrac{1}{k} + \dfrac{1}{3^k} = a_k$ and $\displaystyle\lim_{k\to\infty}\left(\dfrac{1}{k}+\dfrac{1}{3^k}\right) = 0$, the series converges.

9. Let $f(x) = \dfrac{4\sqrt{x}}{2x+1}$. Then $f'(x) = \dfrac{2-4x}{\sqrt{x}(2x+1)^2} < 0$ for $x \geq 1$ and $a_{n+1} < a_n$. Since $\displaystyle\lim_{n\to\infty}\dfrac{4\sqrt{n}}{2n+1} = \lim_{n\to\infty}\dfrac{4}{2\sqrt{n}+1/\sqrt{n}} = 0$, the series converges.

11. Note that $\cos n\pi = (-1)^n$. Let $f(x) = \dfrac{\sqrt{x+1}}{x+2}$. Then $f'(x) = \dfrac{-2x}{2\sqrt{x+1}(x+2)^2} < 0$ for $x \geq 2$ and $a_{n+1} < a_n$. Since $\displaystyle\lim_{n\to\infty}\dfrac{\sqrt{n+1}}{n+2} = \lim_{n\to\infty}\dfrac{\sqrt{1+1/n}}{\sqrt{n}+2/\sqrt{n}} = 0$, the series converges.

13. Using L'Hôpital's Rule, $\lim\limits_{x\to\infty} \dfrac{x}{\ln x} = \lim\limits_{x\to\infty} \dfrac{1}{1/x} = \lim\limits_{x\to\infty} x \neq 0$, and the series diverges.

15. Apply the limit comparison test to $\displaystyle\sum_{k=1}^{\infty} \dfrac{1}{2k+1}$ with $a_k = \dfrac{1}{2k+1}$ and $b_k = \dfrac{1}{k}$:

$$\lim_{k\to\infty} \frac{a_k}{b_k} = \lim_{k\to\infty} \frac{1/(2k+1)}{1/k} = \lim_{k\to\infty} \frac{k}{2k+1} = \frac{1}{2}.$$

Since $\displaystyle\sum_{k=1}^{\infty} b_k$ diverges, the given series is not absolutely convergent. Since $a_{k+1} = \dfrac{1}{2k+3} < \dfrac{1}{2k+1} = a_k$ and $\lim\limits_{k\to\infty} \dfrac{1}{2k+1} = 0$, the series is conditionally convergent.

17. Since $\displaystyle\sum_{k=1}^{\infty} \left(\dfrac{2}{3}\right)^k$ is a geometric series with $r = \dfrac{2}{3} < 1$, the series is absolutely convergent.

19. Since $\lim\limits_{k\to\infty} \left|\dfrac{a_{k+1}}{a_k}\right| = \lim\limits_{k\to\infty} \dfrac{(k+1)/5^{k+1}}{k/5^k} = \lim\limits_{k\to\infty} \dfrac{k+1}{5k} = \dfrac{1}{5} < 1$, the series is absolutely convergent by the ratio test.

21. Since $\lim\limits_{k\to\infty} \left|\dfrac{a_{k+1}}{a_k}\right| = \lim\limits_{k\to\infty} \dfrac{1/(k+1)!}{1/k!} = \lim\limits_{k\to\infty} \dfrac{1}{k+1} = 0 < 1$, the series is absolutely convergent by the ratio test.

23. Since $\lim\limits_{k\to\infty} \left|\dfrac{a_{k+1}}{a_k}\right| = \lim\limits_{k\to\infty} \dfrac{(k+1)!/100^{k+1}}{k!/100^k} = \lim\limits_{k\to\infty} \dfrac{k+1}{100} = \infty$, the series is divergent by the ratio test.

25. Apply the limit comparison test to $\displaystyle\sum_{k=1}^{\infty} \dfrac{k}{1+k^2}$ with $a_k = \dfrac{k}{1+k^2}$ and $b_k = \dfrac{1}{k}$: $\lim\limits_{k\to\infty} \dfrac{a_k}{b_k} = \lim\limits_{k\to\infty} \dfrac{k/(1+k^2)}{1/k} = \lim\limits_{k\to\infty} \dfrac{k^2}{1+k^2} = 1$. Since

$\displaystyle\sum_{k=1}^{\infty} b_k$ diverges, the given series is not absolutely convergent. Let $f(x) = \dfrac{x}{1+x^2}$. Then $f'(x) = \dfrac{1-x^2}{(1+x^2)^2} < 0$ for $x > 1$ and

$a_{k+1} < a_k$. Also, $\lim\limits_{k\to\infty} a_k = \lim\limits_{k\to\infty} \dfrac{k}{1+k^2} = 0$, so the series is conditionally convergent.

27. Since $\cos k\pi = (-1)^k$ and $\lim\limits_{k\to\infty} (-1)^k$ is not 0, the series diverges by the n-th term test.

29. Apply the limit comparison test to $\displaystyle\sum_{k=1}^{\infty} \sin\left(\dfrac{1}{k}\right)$ with $a_k = \sin\left(\dfrac{1}{k}\right)$ and $b_k = \dfrac{1}{k}$:

$$\lim_{k\to\infty} \frac{a_k}{b_k} = \lim_{k\to\infty} \frac{\sin(1/k)}{1/k} \overset{h}{=} \lim_{k\to\infty} \frac{(-1/k^2)\cos(1/k)}{-1/k^2} = \lim_{k\to\infty} \cos(1/k) = 1.$$

Since $\displaystyle\sum_{k=1}^{\infty} b_k$ diverges, the given series is not absolutely convergent. Let $f(x) = \sin\left(\dfrac{1}{x}\right)$. Then $f'(x) = \left(-\dfrac{1}{x^2}\right)\cos\left(\dfrac{1}{x}\right) < 0$

for $x \geq 1$ and $a_{k+1} < a_k$. Since $\lim\limits_{k\to\infty} a_k = \lim\limits_{k\to\infty} \sin\left(\dfrac{1}{k}\right) = 0$, the original series is conditionally convergent.

31. Since $\dfrac{1}{k+1} - \dfrac{1}{k} = -\dfrac{1}{k^2+k}$, the series can be written as $\displaystyle\sum_{k=1}^{\infty} (-1)^{k+1} \dfrac{1}{k^2+k}$. Now, $\dfrac{1}{k^2+k} < \dfrac{1}{k^2}$, so the series is absolutely

convergent by comparison with the p-series $\displaystyle\sum_{k=1}^{\infty} \dfrac{1}{k^2}$.

33. Since $\lim\limits_{k\to\infty} \left(\dfrac{2k}{k+50}\right)^k = \lim\limits_{k\to\infty} \left(\dfrac{2}{1+50/k}\right)^k = \infty$, the series diverges by the n-th term test.

35. We must have $a_{n+1} = \dfrac{1}{(2n+1)!} < 0.000005$. Taking $n = 4$ we have $a_5 = \dfrac{1}{9!} \approx 0.000003 < 0.000005$. Thus, $S_4 = \dfrac{1}{1!} - \dfrac{1}{3!} + \dfrac{1}{5!} - \dfrac{1}{7!} \approx 0.84147$ has the desired accuracy.

37. We must have $a_{n+1} = \dfrac{1}{(n+1)^3} < 0.005$. Taking $n = 5$ we have $a_6 = \dfrac{1}{6^3} \approx 0.0046 < 0.005$. Thus, S_5 has the desired accuracy.

39. We must have $a_{n+1} = \dfrac{1}{4^{n+1}} < 0.001$. Taking $n = 4$ we have $a_5 = \dfrac{1}{4^5} \approx 0.00098 < 0.001$. Thus, $S_4 = 1 - \dfrac{1}{4^2} + \dfrac{1}{4^3} - \dfrac{1}{4^4} \approx 0.9492$ has the desired accuracy.

41. The error will be less than $a_{101} = \dfrac{1}{101} \approx 0.009901$.

43. This is not an alternating series since, for $k = 1$ to $k = 5$, the terms are positive, while for $k = 7$ to $k = 11$, the terms are negative. Since $|a_{k+1}| \leq \dfrac{1}{k^2}$, the series is absolutely convergent by comparison with the p-series $\displaystyle\sum_{k=1}^{\infty} \dfrac{1}{k^2}$. Hence, the series is convergent.

45. This is not an alternating series. Since $|a_k| = \dfrac{1}{2^{k-1}}$, $\displaystyle\sum_{k=1}^{\infty} |a_k|$ is a geometric series with $r = \dfrac{1}{2} < 1$, and the original series is absolutely convergent.

47. The terms of the series do not satisfy $|a_{n+1}| \leq |a_n|$ since $|a_5| = \dfrac{2}{3} > \dfrac{1}{2} = |a_4|$. Grouping pairs of terms, we obtain the harmonic series $1 + \dfrac{1}{2} + \dfrac{1}{3} + \cdots$. Thus, the sequence of partial sums $\{S_{2n}\}$ is the same as the sequence of partial sums for the harmonic series. Since the latter sequence diverges, so does $\{S_{2n}\}$. Finally, if $\{S_{2n}\}$ diverges, so must $\{S_n\}$. Thus, the original series diverges.

49. The terms do not approach 0, so the series diverges.

51. All terms of the series after the first are 0, so the series converges.

53. The statement is true because a positive-term series $\sum a_k$ is the same as the series of its terms' absolute values $\sum |a_k|$. If this series is convergent, then it is also absolutely convergent, and so, as stated in the discussion, its terms can be rearranged in any manner and the resulting series will converge to the same number as the original series.

55. Let $\displaystyle\sum_{n=1}^{\infty} a_n = 1 - \dfrac{1}{2} + \dfrac{1}{3} - \dfrac{1}{4} + \dfrac{1}{5} - \dfrac{1}{6} + \cdots$, and from Problem 54 let $\displaystyle\sum_{n=1}^{\infty} b_n = 0 + \dfrac{1}{2} + 0 - \dfrac{1}{4} + 0 + \dfrac{1}{6} + \cdots$. Then by Theorem 9.3.5 in the text,

$$\frac{3}{2}S = S + \frac{1}{2}S = \sum_{n=1}^{\infty}(a_n + b_n) = 1 + 0 + \frac{1}{3} - \frac{1}{2} + \frac{1}{5} + 0 + \frac{1}{7} - \frac{1}{4} + \cdots$$

$$= 1 + \frac{1}{3} - \frac{1}{2} + \frac{1}{5} + \frac{1}{7} - \frac{1}{4} + \cdots.$$

57. Since $\displaystyle\sum_{k=1}^{\infty} a_k$ is absolutely convergent, $\displaystyle\lim_{n \to \infty} a_k = 0$. Thus, for n sufficiently large, $|a_n| < 1$ and $a_n^2 < |a_n|$. Therefore, $\displaystyle\sum_{k=1}^{\infty} a_k^2$ converges by the comparison test.

59. The alternating harmonic series converges (see Example 2 in this section) and the series consisting of its terms' squares, $\displaystyle\sum_{k=1}^{\infty} \dfrac{1}{x^2}$, is a p-series with $p = 2 > 1$, which also converges.

61. $e^{-x}\sin x + e^{-2x}\sin 2x + e^{-3x}\sin 3x + \cdots$ can be written as $\displaystyle\sum_{k=1}^{\infty} e^{-kx}\sin kx$, and for $x > 0$, $|e^{-kx}\sin kx| \leq |e^{-kx}| = (e^{-x})^k$. Now, $\displaystyle\sum_{k=1}^{\infty}(e^{-x})^k$ is a convergent geometric series since $e^{-x} < 1$ for $x > 0$. Thus, $\displaystyle\sum_{k=1}^{\infty} |e^{-kx}\sin kx|$ converges by the direct comparison test for all positive x, and so $\displaystyle\sum_{k=1}^{\infty} e^{-kx}\sin kx$ must also converge for all positive x.

9.8 Power Series

1. $\lim\limits_{n\to\infty}\left|\dfrac{a_{n+1}}{a_n}\right| = \lim\limits_{n\to\infty}\left|\dfrac{x^{n+1}/(n+1)}{x^n/n}\right| = \lim\limits_{n\to\infty}\dfrac{n}{n+1}|x| = |x|$

The series is absolutely convergent on $(-1,1)$. At $x=-1$, the series $\sum\limits_{k=1}^{\infty}\dfrac{1}{k}$ is the harmonic series which diverges. At $x=1$, the

series $\sum\limits_{k=1}^{\infty}\dfrac{(-1)^k}{k}$ converges by the alternating series test. Thus, the given series converges on $(-1,1]$.

3. $\lim\limits_{n\to\infty}\left|\dfrac{a_{n+1}}{a_n}\right| = \lim\limits_{n\to\infty}\left|\dfrac{2^{n+1}x^{n+1}/(n+1)}{2^n x^n/n}\right| = \lim\limits_{n\to\infty}\dfrac{2n}{n+1}|x| = 2|x|$

The series is absolutely convergent for $2|x|<1$ or $|x|<1/2$. At $x=-1/2$, the series $\sum\limits_{k=1}^{\infty}\dfrac{(-1)^k}{k}$ converges by the alternating

series test. At $x=1/2$, the series $\sum\limits_{k=1}^{\infty}\dfrac{1}{k}$ is the harmonic series which diverges. Thus, the given series converges on $[-1/2,1/2)$.

5. $\lim\limits_{n\to\infty}\left|\dfrac{a_{n+1}}{a_n}\right| = \lim\limits_{n\to\infty}\left|\dfrac{(x-3)^{n+1}/(n+1)^3}{(x-3)^n/n^3}\right| = \lim\limits_{n\to\infty}\left(\dfrac{n}{n+1}\right)^3|x-3| = |x-3|$

The series is absolutely convergent for $|x-3|<1$ or on $(2,4)$. At $x=2$, the series $\sum\limits_{k=1}^{\infty}\dfrac{(-1)^k}{k^3}$ converges by the alternating

series test. At $x=4$, the series $\sum\limits_{k=1}^{\infty}\dfrac{1}{k^3}$ is a convergent p-series. Thus, the given series converges on $[2,4]$.

7. $\lim\limits_{n\to\infty}\left|\dfrac{a_{n+1}}{a_n}\right| = \lim\limits_{n\to\infty}\left|\dfrac{(x-5)^{n+1}/10^{n+1}}{(x-5)^n/10^n}\right| = \lim\limits_{n\to\infty}\dfrac{1}{10}|x-5| = \dfrac{1}{10}|x-5|$

The series is absolutely convergent for $\dfrac{1}{10}|x-5|<1$, $|x-5|<10$, or on $(-5,15)$. At $x=-5$, the series $\sum\limits_{k=1}^{\infty}\dfrac{(-1)^k(-10)^k}{10^k} = \sum\limits_{k=1}^{\infty}1$

diverges by the n-th term test. At $x=15$, the series $\sum\limits_{k=1}^{\infty}\dfrac{(-1)^k 10^k}{10^k} = \sum\limits_{k=1}^{\infty}(-1)^k$ diverges by the n-th term test. Thus, the series

converges on $(-5,15)$.

9. $\lim\limits_{n\to\infty}\left|\dfrac{a_{n+1}}{a_n}\right| = \lim\limits_{n\to\infty}\left|\dfrac{(n+1)!2^{n+1}x^{n+1}}{n!2^n x^n}\right| = \lim\limits_{n\to\infty}2(n+1)|x| = \infty,\ x\neq 0$ The series converges only at $x=0$.

11. $\lim\limits_{n\to\infty}\left|\dfrac{a_{n+1}}{a_n}\right| = \lim\limits_{n\to\infty}\left|\dfrac{(3x-1)^{n+1}/[(n+1)^2+(n+1)]}{(3x-1)^n/(n^2+n)}\right| = \lim\limits_{n\to\infty}\dfrac{n^2+n}{n^2+3n+2}|3x-1| = |3x-1|$

The series is absolutely convergent for $|3x-1|<1$ or on $(0,2/3)$. At $x=0$, the series $\sum\limits_{k=1}^{\infty}\dfrac{(-1)^k}{k^2+k}$ converges by the alternating

series test. At $x=2/3$, the series $\sum\limits_{k=1}^{\infty}\dfrac{1}{k^2+k}$ converges by comparison with the p-series $\sum\limits_{k=1}^{\infty}\dfrac{1}{k^2}$. Thus, the given series converges

on $[0,2/3]$.

13. $\lim\limits_{n\to\infty}\left|\dfrac{a_{n+1}}{a_n}\right| = \lim\limits_{n\to\infty}\left|\dfrac{x^{n+1}/\ln(n+1)}{x^n/\ln n}\right| = \lim\limits_{n\to\infty}\dfrac{\ln n}{\ln(n+1)}|x|$

By L'Hôpital's Rule, $\lim\limits_{n\to\infty}\dfrac{\ln n}{\ln(n+1)} \overset{h}{=} \lim\limits_{n\to\infty}\dfrac{1/n}{1/(n+1)} = \lim\limits_{n\to\infty}\dfrac{n+1}{n} = 1$. Thus, $\lim\limits_{n\to\infty}\left|\dfrac{a_{n+1}}{a_n}\right| = |x|$. The series is absolutely

convergent on $(-1,1)$. At $x=-1$, the series $\sum\limits_{k=2}^{\infty}\dfrac{(-1)^k}{\ln k}$ converges by the alternating series test. At $x=1$, the series $\sum\limits_{k=2}^{\infty}\dfrac{1}{\ln k}$

diverges by comparison with $\sum\limits_{k=2}^{\infty}\dfrac{1}{k}$. Thus, the given series converges on $[-1,1)$.

15. $\lim\limits_{n\to\infty}\left|\dfrac{a_{n+1}}{a_n}\right| = \lim\limits_{n\to\infty}\left|\dfrac{(n+1)^2(x+7)^{n+1}/3^{2n+2}}{n^2(x+7)^n/3^{2n}}\right| = \lim\limits_{n\to\infty}\dfrac{1}{9}\left(\dfrac{n+1}{n}\right)^2|3x+7| = \dfrac{1}{9}|x+7|$

The series is absolutely convergent for $\frac{1}{9}|x+7| < 1$, $|x+7| < 9$, or on $(-16,2)$. At $x = -16$, the series $\sum_{k=1}^{\infty} \frac{(-9)^k k^2}{9^k} = \sum_{k=1}^{\infty} (-1)^k k^2$ diverges by the n-th term test. At $x = 2$, the series $\sum_{k=1}^{\infty} \frac{9^k k^2}{9^k} = \sum_{k=1}^{\infty} k^2$ diverges by the n-th term test. Thus, the given series converges on $(-16,2)$.

17. Write the series as $\sum_{k=1}^{\infty} \left(\frac{32}{75}\right)^k x^k$. Then

$$\lim_{n\to\infty} \left|\frac{a_{n+1}}{a_n}\right| = \lim_{n\to\infty} \left|\frac{(35/75)^{n+1} x^{n+1}}{(32/75)^n x^n}\right| = \lim_{n\to\infty} \left|\frac{32}{75} x\right| = \frac{32}{75}|x|.$$

The series is absolutely convergent for $\frac{32}{75}|x| < 1$, or on $(-75/32, 75/32)$. At $x = -75/32$ the series $\sum_{k=1}^{\infty}(-1)^k$ diverges by the n-th term test. At $x = 75/32$ the series $\sum_{k=1}^{\infty} 1$ diverges by the n-th term test. Thus, the given series converges on $(-75/32, 75/32)$

19. $\lim_{n\to\infty}\left|\frac{a_{n+1}}{a_n}\right| = \lim_{n\to\infty}\left|\frac{3^{n+1}(x-1)^{n+1}/(n+2)(n+3)}{3^n(x-1)^n/(n+1)(n+2)}\right| = \lim_{n\to\infty} 3\left(\frac{n+1}{n+3}\right)|x-1| = 3|x-1|$
The series is absolutely convergent for $3|x-1| < 1$, $|x-1| < 1/3$, or on $(2/3, 4/3)$. At $x = 2/3$, the series $\sum_{k=0}^{\infty}\frac{(-3)^k}{(k+1)(k+2)}\left(-\frac{1}{3}\right)^k = \sum_{k=0}^{\infty}\frac{1}{(k+1)(k+2)}$ converges by comparison with the p-series $\sum_{k=1}^{\infty}\frac{1}{k^2}$. At $x = 4/3$, the series $\sum_{k=0}^{\infty}\frac{(-3)^k}{(k+1)(k+2)}\left(\frac{1}{3}\right)^k = \sum_{k=0}^{\infty}\frac{(-1)^k}{(k+1)(k+2)}$ converges by the alternating series test. Thus, the given series converges on $[2/3, 4/3]$.

21. $\lim_{n\to\infty}\left|\frac{a_{n+1}}{a_n}\right| = \lim_{n\to\infty}\left|\frac{(x-2)^{n+1}/(n+1)!(n+1)!3^{n+1}}{(x-2)^n/n!n!3^n}\right| = \lim_{n\to\infty}\frac{1}{3(n+1)^2}|x-2| = 0$. The series converges on $(-\infty,\infty)$.

23. $\lim_{n\to\infty}\left|\frac{a_{n+1}}{a_n}\right| = \lim_{n\to\infty}\left|\frac{x^{2(n+1)+1}/9^{n+1}}{x^{2n+1}/9^n}\right| = \lim_{n\to\infty}\frac{1}{9}x^2 = \frac{1}{9}x^2$

The series is absolutely convergent for $\frac{1}{9}x^2 < 1$ or on $(-3,3)$. At $x = -3$ the series $\sum_{k=0}^{\infty}(-1)^k(-3)$ diverges by the n-th term test. At $x = 3$ the series $\sum_{k=0}^{\infty}(-1)^k 3$ diverges by the n-th term test. Thus, the given series converges on $(-3,3)$.

25. $\lim_{n\to\infty}|a_n|^{1/n} = \lim_{n\to\infty}\left|\frac{x^n}{(\ln n)^n}\right|^{1/n} = \lim_{n\to\infty}\frac{|x|}{\ln n} = 0$. The series is absolutely convergent on $(-\infty,\infty)$.

27. $\lim_{n\to\infty}|a_n|^{1/n} = \lim_{n\to\infty}\left|\left(\frac{4^n}{3}(x+3)^n\right)\right|^{1/n} = \lim_{n\to\infty}\frac{4}{3}|x+3| = \frac{4}{3}|x+3|$
The series is absolutely convergent for $\frac{4}{3}|x+3| < 1$, $|x+3| < \frac{3}{4}$, or on $(-15/4, -9/4)$. At $x = -15/4$, the series $\sum_{k=1}^{\infty}\left(\frac{4}{3}\right)^k\left(-\frac{3}{4}\right)^k = \sum_{k=1}^{\infty}(-1)^k$ is divergent by the n-th term test. At $x = -9/4$, the series $\sum_{k=1}^{\infty}\left(\frac{4}{3}\right)^k\left(\frac{3}{4}\right)^k = \sum_{k=1}^{\infty}1^k$ is divergent by the n-th term test. Thus, the given series converges on $(-15/4, -9/4)$.

29. $\lim_{n\to\infty}\left|\frac{a_{n+1}}{a_n}\right| = \lim_{n\to\infty}\left|\frac{\dfrac{(n+1)!(x/2)^{n+1}}{1\cdot 3\cdot 5\cdots(2n-1)(2n+1)}}{\dfrac{n!(x/2)^n}{1\cdot 3\cdot 5\cdots(2n-1)}}\right| = \lim_{n\to\infty}\frac{n+1}{2n+1}\left|\frac{x}{2}\right| = \frac{1}{4}|x|$
The series has radius of convergence 4.

31. $\sum_{k=1}^{\infty}\left(\frac{1}{x}\right)^k$ is a geometric series with common ratio $r = 1/x$. It converges for $|1/x| < 1$ or $x > 1$.

33. This is a geometric series with common ratio $r = \dfrac{x+1}{x}$ and will converge for $\left| \dfrac{x+1}{x} \right| < 1$ or $|x+1| < |x|$. If $x \geq 0$, this means that $x+1 < x$ which has no solution. For $x < 0$, $|x| = -x$ and the inequality can be written as $|x+1| < -x$ or $x < x+1 < -x$. Since $x < x+1$ is valid for all x, we have $x+1 < -x$, $2x < -1$, or $x < -1/2$. Thus, the given series converges on $(-\infty, -1/2)$.

35. Applying the root test, we have $\displaystyle\lim_{n\to\infty} |a_n|^{1/n} = \lim_{n\to\infty} \left| \left(\frac{x^2+2}{6} \right)^{n^2} \right|^{1/n} = \lim_{n\to\infty} \left(\frac{x^2+2}{6} \right)^n < 1$. Setting $\left(\dfrac{x^2+2}{6} \right)^n < 1$, we obtain

$\dfrac{x^2+2}{6} < 1$ or $x^2 < 4$. Thus, the series converges on $(-2,2)$. At $x = \pm 2$, the series $\displaystyle\sum_{k=0}^{\infty} 1^{k^2}$ diverges by the n-th term test. Therefore, the given series converges on $(-2,2)$.

37. This is a geometric series with common ratio $r = x^x$ and will converge for $|e^x| = e^x < 1$ or $x < 0$. Thus, the series converges on $(-\infty, 0)$.

39. This is a geometric series with $r = \dfrac{2}{\sqrt{3}} \sin x$ and will converge for $\left| \dfrac{2}{\sqrt{3}} \sin x \right| < 1$ or $|\sin x| < \sqrt{3}/2$. On $[0, 2\pi]$ this will be on $[0, \pi/3) \cup (2\pi/3, 4\pi/3) \cup (5\pi/3, 2\pi]$.

41. (a) $\displaystyle\lim_{n\to\infty} \frac{|a_{n+1}|}{|a_n|} = \lim_{n\to\infty} \frac{x^{2(n+1)}}{x^{2n}} \cdot \frac{2^{2n}(n!)^2}{2^{2(n+1)}[(n+1)!]^2}$ for all x values. Therefore, the interval of convergence is $(-\infty, \infty)$ and hence

$= \displaystyle\lim_{n\to\infty} x^2 \cdot \frac{1}{4(n+1)^2} = 0 < 1$

the domain is $(-\infty, \infty)$.

(b)

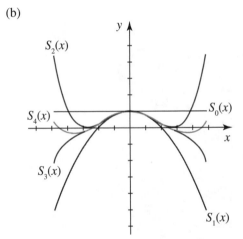

(c)

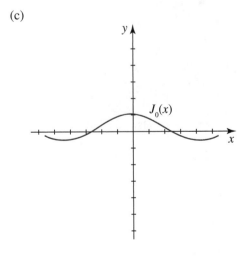

9.9 Representing Functions by Power Series

In this section we will use the fact that the geometric series $\sum_{k=0}^{\infty} ar^k$ converges to $\dfrac{a}{1-r}$ for $r < 1$.

1. $\dfrac{1}{3-x} = \dfrac{1}{3} \cdot \dfrac{1}{1-\frac{x}{3}} = \dfrac{1}{3}\sum_{k=0}^{\infty}\left(\dfrac{x}{3}\right)^k$ for $\left|\dfrac{x}{3}\right| < 1$ or x in $(-3,3)$. At $x = -3$ and $x = 3$, the series diverges by the n-th term test. Thus, the interval of convergence is $(-3,3)$.

3. $\dfrac{1}{2+x} = \sum_{k=0}^{\infty}(-1)^k (2x)^k$ for $|2x| < 1$ or x in $\left(-\dfrac{1}{2},\dfrac{1}{2}\right)$. At $x = -\dfrac{1}{2}$ and $x = \dfrac{1}{2}$, the series diverges by the n-th term test. Thus, the interval of convergence is $\left(-\dfrac{1}{2},\dfrac{1}{2}\right)$.

5. Identify $a = 1$ and $r = -x^2$. Then, $\dfrac{1}{1+x^2} = \sum_{k=0}^{\infty}(-x^2)^k = \sum_{k=0}^{\infty}(-1)^k x^{2k}$ for $x^2 < 1$ or x in $(-1,1)$.

   At $x = -1$ and $x = 1$, the series diverges by the n-th term test. Thus, the interval of convergence is $(-1,1)$.

7. $\dfrac{1}{4+x^2} = \dfrac{1}{4} \cdot \dfrac{1}{1+\left(\frac{x}{2}\right)^2} = \dfrac{1}{4}\sum_{k=0}^{\infty}(-1)^k \left(\dfrac{x}{2}\right)^{2k}$ for $\left|\left(\dfrac{x}{2}\right)^2\right| < 1$ or x in $(-2,2)$. At $x = -2$ and $x = 2$, the series diverges by the n-th term test. Thus, the interval of convergence is $(-2,2)$.

9. $\dfrac{1}{(3-x)^2} = -\dfrac{d}{dx}\left(\dfrac{1}{3-x}\right) = -\dfrac{d}{dx}\left(\dfrac{1}{3}\sum_{k=0}^{\infty}\left(\dfrac{x}{3}\right)^k\right) = -\dfrac{1}{3}\sum_{k=1}^{\infty}k\left(\dfrac{x}{3}\right)^{k-1}$ The integral of convergence remains $(-3,3)$.

11. $\dfrac{1}{(5+2x)^3} = \dfrac{1}{8}\dfrac{d^2}{dx^2}\left(\dfrac{1}{5+2x}\right) = \dfrac{1}{8}\dfrac{d^2}{dx^2}\left(\dfrac{1}{5}\sum_{k=0}^{\infty}(-1)^k\left(\dfrac{2x}{5}\right)^k\right)$

    $= \dfrac{1}{250}\sum_{k=2}^{\infty}k(k-1)(-1)^k\dfrac{2x^{k-2}}{5}$.

    The integral of convergence remains $\left(-\dfrac{5}{2},\dfrac{5}{2}\right)$.

13. $\dfrac{x}{(1+x^2)^2} = -\dfrac{1}{2}\dfrac{d}{dx}\left(\dfrac{1}{1+x^2}\right) = -\dfrac{1}{2}\dfrac{d}{dx}\left(\sum_{k=0}^{\infty}(-1)^k x^{2k}\right)$

    $= -\dfrac{1}{2}\sum_{k=1}^{\infty}2k(-1)^k x^{2k-1}$.

    The integral of convergence remains $(-1,1)$.

15. Using Problem 5,

$$\tan^{-1}x = \int_0^x \dfrac{1}{1+t^2} = \int_0^x \sum_{k=0}^{\infty}(-1)^k t^{2k}\,dt = \sum_{k=0}^{\infty}(-1)^k \int_0^x t^{2k}\,dt = \sum_{k=0}^{\infty}(-1)^k \dfrac{x^{2k+1}}{2k+1}$$

for x in $(-1,1)$. At $x = -1$ and $x = 1$, the series converges by the alternating series test. Thus, the interval of convergence is $[-1,1]$.

17. Using Problem 5, $\dfrac{1}{1+x^2} = \sum_{k=0}^{\infty}(-1)^k x^{2k}$ and

$$\dfrac{2x}{1+x^2} = 2x\sum_{k=0}^{\infty}(-1)^k x^{2k} = \sum_{k=0}^{\infty}2(-1)^k x^{2k+1}.$$

Then

$$\ln(1+x^2) = \int_0^x \frac{2t}{1+t^2}\,dt = \int_0^x \sum_{k=0}^{\infty} 2(-1)^k t^{2k+1}\,dt = 2\sum_{k=0}^{\infty}(-1)^k \int_0^x t^{2k+1}\,dt$$

$$= 2\sum_{k=0}^{\infty}(-1)^k \frac{x^{2k+2}}{2k+2} = \sum_{k=0}^{\infty}\frac{(-1)^k}{k+1}x^{2k+2}$$

for $x^2 < 1$ or x in $(-1,1)$. At $x = -1$ and $x = 1$, the series converges by the alternating series test. Thus, the interval of convergence is $[-1,1]$.

19. Writing $\dfrac{1}{4+x} = \dfrac{1}{4}\cdot\dfrac{1}{1+x/4}$, we have

$$\ln(4+x) - \ln 4 = \int_0^x \frac{dt}{4+t} = \int_0^x \frac{1}{4}\cdot\frac{dt}{1+t/4} = \frac{1}{4}\int_0^x \sum_{k=0}^{\infty}(-t/4)^k\,dt$$

$$= \frac{1}{4}\sum_{k=0}^{\infty}(-1)^k \int_0^x \frac{t^k}{4^k}\,dt = \frac{1}{4}\sum_{k=0}^{\infty}\frac{(-1)^k}{k+1}\cdot\frac{x^{k+1}}{4^k}.$$

Thus $\ln(4+x) = \ln 4 + \displaystyle\sum_{k=0}^{\infty}\frac{(-1)^k}{(k+1)4^{k+1}}x^{k+1}$. This series converges for $|x/4| < 1$, $|x| < 4$, or on $(-4,4)$. At $x = -4$, the series diverges since it is the negative harmonic series. At $x = 4$, the series converges by the alternating series test. Thus, the interval of convergence is $(-4,4]$.

21. $\dfrac{1-x}{1+2x} = (1-x)\cdot\left(\dfrac{1}{1+2x}\right)(1-x)\cdot\displaystyle\sum_{k=0}^{\infty}(-1)^k(2x)^k$

$$= \sum_{k=0}^{\infty}(-1)^k x^k(2x)^k - \frac{1}{2}(-1)^k(2x)^{k+1}$$

$$= 1 + \frac{3}{2}\sum_{k=1}^{\infty}(-1)^k x^k(2x)^k$$

which converges for x in $\left(-\dfrac{1}{2},\dfrac{1}{2}\right)$ and diverges at $x = -\dfrac{1}{2}$ and $x = \dfrac{1}{2}$. Thus, the interval of convergence is $\left(-\dfrac{1}{2},\dfrac{1}{2}\right)$.

23. $\dfrac{x^2}{(1-x)^3} = x^2 \cdot \dfrac{1}{(1+x)^3} = x^2 \cdot \dfrac{1}{2}\dfrac{d^2}{dx^2}\left[\dfrac{1}{(1+x)}\right]$

$$= x^2 \cdot \frac{1}{2}\cdot\sum_{k=2}^{\infty}k(k-1)(-1)^k x^{k-2}$$

$$= \frac{1}{2}\cdot\sum_{k=2}^{\infty}k(k-1)(-1)^k x^k$$

which converges for x in $(-1,1)$ and diverges at $x = -1$ and $x = 1$. Thus, the interval of convergence is $(-1,1)$.

25. $s\ln(1+x^2) = x\cdot 2\displaystyle\int \frac{x}{1+x^2}\,dx$

$$= x\cdot 2\int\left(\sum_{k=0}^{\infty}(-1)^k x^{2k+1}\right)dx$$

$$= x\cdot 2\sum_{k=0}^{\infty}\frac{(-1)^k x^{2k+2}}{2k+2} = \sum_{k=0}^{\infty}\frac{(-1)^k x^{2k+3}}{k+1}$$

which converges for x in $(-1,1)$. At $x = -1$, we have the series $\displaystyle\sum_{k=0}^{\infty}\frac{(-1)^{3k+3}}{k+1}$ which converges. At $x = 1$, we have the series $\displaystyle\sum_{k=0}^{\infty}\frac{(-1)^k}{k+1}$ which converges. Thus, the interval of convergence is $[-1,1]$.

27. Since $\tan^{-1} t = \sum_{k=0}^{\infty} \frac{(-1)^k t^{2k+1}}{2k+1}$, we have

$$\int_0^x \tan^{-1} t\, dt = \int_0^x \left(\sum_{k=0}^{\infty} \frac{(-1)^k t^{2k+1}}{2k+1} \right) dt$$

$$= \sum_{k=0}^{\infty} \left(\int_0^x \frac{(-1)^k t^{2k+1}}{2k+1}\, dt \right)$$

$$= \sum_{k=0}^{\infty} \left(\frac{(-1)^k t^{2k+2}}{(2k+1)(2k+2)} \right) \Big|_0^x$$

$$= \sum_{k=0}^{\infty} \frac{(-1)^k x^{2k+2}}{(2k+1)(2k+2)}$$

which converges for x in $(-1,1)$. At $x=-1$, we have the series $\sum_{k=0}^{\infty} \frac{(-1)^{3k+2}}{(2k+1)(2k+2)}$ which converges. At $x=1$, we have the

series $\sum_{k=0}^{\infty} \frac{(-1)^k}{(2k+1)(2k+2)}$ which converges. Thus, the interval of convergence is $[-1,1]$.

29. $\dfrac{1}{1-x} = \dfrac{1}{-5-(x-6)} = -\dfrac{1}{5} \cdot \dfrac{1}{1+\left(\frac{x+6}{5}\right)}$

$$= -\frac{1}{5} \sum_{k=0}^{\infty} (-1)^k \left(\frac{x-6}{5} \right)^k$$

which converges for $\left| \dfrac{x-6}{5} \right| < 1$ or $|x-6| < 5$. Since the series diverges at $x=11$ and $x=1$, the interval of convergence is
$(1,11)$.

31. $\dfrac{x}{2+x} = \dfrac{(x+2)-2}{x+2} = 1 - \dfrac{2}{x+2}$

$$= 1 - \frac{2}{1+(x+1)} = 1 - 2\sum_{k=0}^{\infty} (-1)^k (x+1)^k$$

$$= -1 - 2\sum_{k=1}^{\infty} (-1)^k (x+1)^k = -1 + 2\sum_{k=0}^{\infty} (-1)^k (x+1)^{k+1}$$

which converges for $|x+1| < 1$. Since the series diverges at $x=-2$ and $x=0$, the interval of convergence is $(-2,0)$.

33. $\dfrac{7x}{x^2+x-12} = \dfrac{4}{x+4} + \dfrac{3}{x-3}$

$$= \frac{1}{1+\left(\frac{x}{4}\right)} - \frac{1}{1-\left(\frac{x}{3}\right)}$$

$$= \sum_{k=0}^{\infty} (-1)^k \left(\frac{x}{4} \right)^k - \sum_{k=0}^{\infty} \left(\frac{x}{3} \right)^k$$

$$= \sum_{k=0}^{\infty} \left[\frac{(-1)^k}{4^k} - \frac{1}{3^k} \right] x^k$$

which converges for $(-3,3)$. Since the series diverges at $x=-3$ and $x=3$, the interval of convergence is $(-3,3)$.

35. $\dfrac{1}{(2-x)(1-x)} = \dfrac{1}{2-x} \cdot \dfrac{1}{1-x} = \left[\dfrac{1}{2} \sum_{k=0}^{\infty} \left(\dfrac{x}{2} \right)^k \right] \cdot \left[\sum_{k=0}^{\infty} x^k \right]$

$$= \frac{1}{2} \left[1 + \frac{x}{2} + \frac{x^2}{4} + \frac{x^3}{8} + \cdots \right] \cdot [1 + x + x^2 + x^3 + \cdots]$$

$$= \frac{1}{2} + \frac{3}{4}x + \frac{7}{8}x^2 + \frac{15}{16}x^3 + \cdots$$

37. Writing $f(x) = \sum_{k=1}^{\infty} (-1)^{k+1} \frac{x^k}{k3^k}$ and applying the ratio test, we have

$$\lim_{n\to\infty} \left| \frac{a_{n+1}}{a_n} \right| = \lim_{n\to\infty} \left| \frac{x^{n+1}/(n+1)3^{n+1}}{x^n/n3^n} \right| = \lim_{n\to\infty} \frac{1}{3} \left(\frac{n}{n+1} \right) |x| = \frac{1}{3}|x|.$$

The series is absolutely convergent for $\frac{1}{3}|x| < 1$, $|x| < 3$, or on $(-3,3)$. At $x = -3$, the series is divergent since it is the negative harmonic series. At $x = 3$, the series converges by the alternating series test. Thus, the domain of the function is $(-3,3]$.

39. Using Example 7 in the text with $x = 0.1$,

$$\ln 1.1 = 0.1 - \frac{(0.1)^2}{2} + \frac{(0.1)^3}{3} - \frac{(0.1)^4}{4} + \frac{(0.1)^5}{5} - \cdots$$

$$\approx 0.1 - 0.005 + 0.00033 - 0.00003 + \cdots \approx 0.0953.$$

41. Identify $a = 1$ and $r = -x^3$. Then $\frac{1}{1+x^3} = \sum_{k=0}^{\infty} (-x^3)^k = \sum_{k=0}^{\infty} (-1)^k x^{3k}$ and

$$\int_0^x \frac{dt}{1+t^3} = \int_0^x \sum_{k=0}^{\infty} (-1)^k t^{3k} dt = \sum_{k=0}^{\infty} (-1)^k \int_0^x t^{3k} dt == \sum_{k=0}^{\infty} (-1)^k \frac{x^{3k+1}}{3k+1}.$$

Letting $x = 1/2$, we have

$$\int_0^{1/2} \frac{dx}{1+x^3} = \sum_{k=0}^{\infty} (-1)^k \frac{(1/2)^{3k+1}}{3k+1} = \sum_{k=0}^{\infty} (-1)^k \frac{1}{2^{3k+1}(3k+1)}$$

$$= \frac{1}{2} - \frac{1}{2^4(4)} + \frac{1}{2^7(7)} - \frac{1}{2^{11}(11)} + \cdots$$

$$\approx 0.5 - 0.01563 + 0.00112 - 0.00004 + \cdots \approx 0.4854.$$

43. Using Problem 15, $x\tan^{-1} x = x \sum_{k=0}^{\infty} (-1)^k \frac{x^{2k+1}}{2k+1} = \sum_{k=0}^{\infty} (-1)^k \frac{x^{2k+2}}{2k+1}$, so

$$\int_0^x t\tan^{-1} t\, dt = \int_0^x \sum_{k=0}^{\infty} (-1)^k \frac{t^{2k+2}}{2k+1} dt = \sum_0^{\infty} \frac{(-1)^k}{2k+1} \int_0^x t^{2k+2} dt = \sum_0^{\infty} \frac{(-1)^k}{2k+1} \frac{x^{2k+3}}{2k+3}.$$

Letting $x = 0.3$, we have

$$\int_0^{0.3} x\tan^{-1} x\, dx = \sum_0^{\infty} \frac{(-1)^k}{2k+1} \frac{(0.3)^{2k+3}}{2k+3} = \frac{(0.3)^3}{1 \cdot 3} - \frac{(0.3)^5}{3 \cdot 5} + \frac{(0.3)^7}{5 \cdot 7} - \cdot$$

$$\approx 0.0090 - 0.0016 + 0.00001 - \cdot \approx 0.008.$$

45. Using $\tan^{-1} x = \sum_0^{\infty} (-1)^k \frac{x^{2k+1}}{2k+1}$, we have $\frac{\pi}{4} = \tan^{-1} 1 = \sum_0^{\infty} \frac{(-1)^k}{2k+1} = 1 - \frac{1}{3} + \frac{1}{5} - \frac{1}{7} + \cdots$

47. $y' = \sum_{k=1}^{\infty}(-1)^{k+1}x^{k-1} = -\sum_{k=1}^{\infty}(-1)^{k}x^{k-1}$ $y'' = \sum_{k=2}^{\infty}(k-1)(-1)^{k+1}x^{k-2}$

$$(x+1)y'' = \sum_{k=2}^{\infty}(k-1)(-1)^{k+1}x^{k-1} + \sum_{k=2}^{\infty}(k-1)(-1)^{k+1}x^{k-2}$$

$$= \sum_{k=1}^{\infty}k(-1)^{k+2}x^{k} + \sum_{k=1}^{\infty}k(-1)^{k+2}x^{k-1}$$

$$= \sum_{k=1}^{\infty}k(-1)^{k}x^{k} + \sum_{k=1}^{\infty}(-1)^{k}x^{k-1}$$

$$(x+1)y'' + y' = \sum_{k=1}^{\infty}k(-1)^{k}x^{k} + \sum_{k=1}^{\infty}k(-1)^{k}x^{k-1} - \sum_{k=1}^{\infty}(-1)^{k}x^{k-1}$$

$$= \sum_{k=1}^{\infty}k(-1)^{k}x^{k} + \sum_{k=1}^{\infty}(k-1)(-1)^{k}x^{k-1}$$

$$= \sum_{k=1}^{\infty}k(-1)^{k}x^{k} + \sum_{k=2}^{\infty}(k-1)(-1)^{k}x^{k-1}$$

$$= \sum_{k=1}^{\infty}k(-1)^{k}x^{k} - \sum_{k=1}^{\infty}k(-1)^{k}x^{k} = 0$$

49. (a) $f'(x) = \sum_{k=1}^{\infty}\frac{x^{k-1}}{(k-1)!} = \sum_{k=0}^{\infty}\frac{x^{k}}{k!} = f(x)$

(b) e^{x}

9.10 Taylor Series

1.

$$f(x) = \frac{1}{2-x}, \qquad f(0) = \frac{1}{2}$$

$$f'(x) = \frac{1}{(2-x)^{2}}, \qquad f'(0) = \frac{1}{2^{2}}$$

$$f''(x) = \frac{2}{(2-x)^{3}}, \qquad f''(0) = \frac{2}{2^{3}}$$

$$f'''(x) = \frac{2\cdot3}{(2-x)^{4}}, \qquad f'''(0) = \frac{3!}{2^{4}}$$

$$\vdots$$

$$f^{(k)}(x) = \frac{k!}{(2-x)^{k+1}}, \qquad f^{k}(0) = \frac{k!}{2^{k+1}}$$

The Maclaurin series is $\displaystyle\sum_{k=0}^{\infty}\frac{k!/2^{k+1}}{k!}x^{k} = \sum_{k=0}^{\infty}\frac{1}{2^{k+1}}x^{k}.$

3.

$$f(x) = \ln(1+x), \qquad\qquad f(0) = 0$$

$$f'(x) = \frac{1}{1+x}, \qquad\qquad f'(0) = 1$$

$$f''(x) = -\frac{1}{(1+x)^2}, \qquad\qquad f''(0) = -1$$

$$f'''(x) = \frac{2}{(1+x)^3}, \qquad\qquad f'''(0) = 2$$

$$f^{(4)}(x) = -\frac{2 \cdot 3}{(1+x)^4}, \qquad\qquad f^{(4)}(0) = -3!$$

$$\vdots$$

$$f^{(k)}(x) = (-1)^{k-1}\frac{(k-1)!}{(1+x)^k}, \qquad f^{(k)}(0) = (-1)^{k-1}(k-1)!$$

The Maclaurin series is $\displaystyle\sum_{k=1}^{\infty} \frac{(-1)^{k-1}(k-1)!}{k!}x^k = \sum_{k=1}^{\infty} \frac{(-1)^{k-1}(k-1)!}{k!}x^k = \sum_{k=1}^{\infty} \frac{(-1)^{k-1}}{k}x^k.$

5.

$$f(x) = \sin x, \qquad\qquad f(0) = 0$$

$$f'(x) = \cos x, \qquad\qquad f'(0) = 1$$

$$f''(x) = -\sin x, \qquad\qquad f''(0) = 0$$

$$f'''(x) = -\cos x, \qquad\qquad f'''(0) = -1$$

$$\vdots$$

$$f^{(2k+1)}(x) = (-1)^k \cos x, \qquad f^{(2k+1)}(0) = (-1)^k$$

The Maclaurin series is $\displaystyle\sum_{k=1}^{\infty} \frac{(-1)^k}{(2k+1)!}x^{2k+1}.$

7.

$$f(x) = e^x, \qquad f(0) = 1$$

$$f'(x) = e^x, \qquad f'(0) = 1$$

$$\vdots$$

$$f^{(2k)}(x) = e^x, \qquad f^{(k)}(0) = 1$$

The Maclaurin series is $\displaystyle\sum_{k=0}^{\infty} \frac{1}{k!}x^k.$

9.

$$f(x) = \sinh x, \qquad\qquad f(0) = 0$$

$$f'(x) = \cosh x, \qquad\qquad f'(0) = 1$$

$$f''(x) = \sinh x, \qquad\qquad f''(0) = 0$$

$$\vdots$$

$$f^{(2k+1)}(x) = \cosh x, \qquad f^{(2k+1)}(0) = 1$$

The Maclaurin series is $\displaystyle\sum_{k=0}^{\infty} \frac{1}{(2k+1)!}x^{2k+1}.$

11.

$$f(x) = \tan x, \qquad\qquad\qquad\qquad\qquad\qquad\qquad f(0) = 0$$

$$f'(x) = \sec^2 x = 1 + \tan^2 x, \qquad\qquad\qquad\qquad\qquad f'(0) = 1$$

$$f''(x) = 2\tan x(1 + \tan^2 x) = 2\tan x + 2\tan^3 x, \qquad\qquad f''(0) = 0$$

$$f'''(x) = 2 + 8\tan^2 x + 6\tan^4 x, \qquad\qquad\qquad\qquad f'''(0) = 2$$

$$f^{(4)}(x) = 16\tan x + 40\tan^3 x + 24\tan^5 x, \qquad\qquad\qquad f^{(4)}(0) = 0$$

$$f^{(5)}(x) = 16 + 136\tan^2 x + 240\tan^4 x + 120\tan^6 x, \qquad\quad f^{(5)}(0) = 16$$

$$f^{(6)}(x) = 276\tan x + 1246\tan^3 x + 1680\tan^5 x + 720\tan^7 x, \quad f^{(6)}(0) = 0$$

$$f^{(7)}(x) = 276 + 4014\tan^2 x + 12,138\tan^4 x + 13,440\tan^6 x + 5040\tan^8 x, \qquad f^{(7)}(0) = 276$$

The Maclaurin series is $x + \dfrac{2}{3!}x^3 + \dfrac{16}{5!}x^5 + \dfrac{272}{7!}x^7 + \cdots = x + \dfrac{1}{3}x^3 + \dfrac{2}{15}x^5 + \dfrac{17}{315}x^7 + \cdots$

13.

$$f(x) = \frac{1}{1+x}, \qquad\qquad\qquad f(4) = \frac{1}{5}$$

$$f'(x) = -\frac{1}{(1+x)^2}, \qquad\qquad f'(4) = -\frac{1}{5^2}$$

$$f''(x) = \frac{2}{(1+x)^3}, \qquad\qquad f''(4) = \frac{3}{5^3}$$

$$f'''(x) = -\frac{2\cdot 3}{(1+x)^4}, \qquad\qquad f'''(4) = -\frac{3!}{5^4}$$

$$\vdots$$

$$f^{(2k+1)}(x) = \frac{(-1)^k k!}{(1+x)^{k+1}}, \qquad f^{(2k+1)}(4) = \frac{(-1)^k k!}{5^{k+1}}$$

The Taylor series is $\displaystyle\sum_{k=0}^{\infty} \frac{(-1)^k k!/5^{k+1}}{k!}(x-4)^k = \sum_{k=0}^{\infty} \frac{(-1)^k}{5^{k+1}}(x-4)^k.$

15.

$$f(x) = \frac{1}{x}, \qquad\qquad f(1) = 1$$

$$f'(x) = -\frac{1}{x^2}, \qquad\qquad f'(1) = -1$$

$$f''(x) = \frac{2}{x^3}, \qquad\qquad f''(1) = 2$$

$$f'''(x) = -\frac{6}{x^4}, \qquad\qquad f'''(1) = -6$$

$$\vdots$$

$$f^n(x) = \frac{(-1)^n n!}{x^{n+1}} \qquad f^{(n)}(1) = (-1)^n n!$$

The Taylor series is

$$\sum_{k=0}^{\infty} \frac{f(k)(1)}{k!}(x-1)^k = \sum_{k=0}^{\infty} \frac{(-1)^k k!}{k!}(x-1)^k = \sum_{k=0}^{\infty}(-1)^k(x-1)^k$$

17.

$$f(x) = \sin x, \qquad\qquad f(\pi/4) = \sqrt{2}/2$$

$$f'(x) = \cos x, \qquad\qquad f'(\pi/4) = \sqrt{2}/2$$

$$f''(x) = -\sin x, \qquad\qquad f''(\pi/4) = -\sqrt{2}/2$$

$$f'''(x) = -\cos x, \qquad\qquad f'''(\pi/4) = -\sqrt{2}/2$$

The Taylor series is $\dfrac{\sqrt{2}}{2} + \dfrac{\sqrt{2}}{2}\left(x - \dfrac{\pi}{4}\right) - \dfrac{\sqrt{2}}{2 \cdot 2!}\left(x - \dfrac{\pi}{4}\right)^2 - \dfrac{\sqrt{2}}{2 \cdot 3!}\left(x - \dfrac{\pi}{4}\right)^3 + \cdots$

19.

$$f(x) = \cos x, \qquad\qquad f(\pi/3) = 1/2$$

$$f'(x) = -\sin x, \qquad\qquad f'(\pi/3) = -\sqrt{3}/2$$

$$f''(x) = -\cos x, \qquad\qquad f''(\pi/3) = -1/2$$

$$f'''(x) = \sin x, \qquad\qquad f'''(\pi/3) = \sqrt{3}/2$$

The Taylor series is $\dfrac{1}{2} - \dfrac{\sqrt{3}}{2}\left(x - \dfrac{\pi}{3}\right) - \dfrac{1}{2 \cdot 2!}\left(x - \dfrac{\pi}{3}\right)^2 + \dfrac{\sqrt{3}}{2 \cdot 3!}\left(x - \dfrac{\pi}{3}\right)^3 + \cdots.$

21.

$$f(x) = e^x, \qquad\qquad f(1) = e$$

$$f'(x) = e^x, \qquad\qquad f'(1) = e$$

$$\vdots$$

$$f^{(k)}(x) = e^x, \qquad\qquad f^{(k)}(1) = e$$

The Taylor series is $\displaystyle\sum_{k=0}^{\infty} \dfrac{e}{k!}(x-1)^k.$

23.

$$f(x) = \ln x, \qquad\qquad\qquad f(2) = 0$$

$$f'(x) = \frac{1}{x}, \qquad\qquad\qquad f'(2) = \frac{1}{2}$$

$$f''(x) = -\frac{1}{x^2}, \qquad\qquad\qquad f''(2) = -\frac{1}{2^2}$$

$$f'''(x) = \frac{2}{x^3}, \qquad\qquad\qquad f'''(2) = \frac{2}{2^3}$$

$$f^{(4)}(x) = -\frac{2 \cdot 3}{x^4}, \qquad\qquad\qquad f^{(4)}(2) = -\frac{3!}{2^4}$$

$$\vdots$$

$$f^{(k)}(x) = (-1)^{k-1}\frac{(k-1)!}{x^k}, \qquad f^{(k)}(2) = (-1)^{k-1}\frac{(k-1)!}{2^k}$$

The Taylor series is

$$\ln 2 + \sum_{k=1}^{\infty} \frac{(-1)^{k-1}(k-1)!/2^k}{k!}(x-2)^k = \ln 2 + \sum_{k=1}^{\infty} \frac{(-1)^{k-1}}{k2^k}(x-2)^k.$$

25. Substituting x^2 for x in Problem 8, we have $\displaystyle\sum_{k=0}^{\infty} \frac{(-1)^k}{k!}(x^2)^k = \sum_{k=0}^{\infty} \frac{(-1)^k}{k!}x^{2k}.$

27. Multiplying by x in Example 3 in the text, we have $\displaystyle x\sum_{k=0}^{\infty} \frac{(-1)^k}{(2k)!}x^{2k} = \sum_{k=0}^{\infty} \frac{(-1)^k}{(2k)!}x^{2k+1}.$

29. Substituting $-x$ for x in Problem 3, we have $\displaystyle\sum_{k=1}^{\infty}\frac{(-1)^{k-1}}{k}(-x)^k = \sum_{k=1}^{\infty}\frac{-1}{k}x^k$.

31. Using Problem 11 and the fact that $\sec^2 x = \dfrac{d}{dx}\tan x$, we have

$$\frac{d}{dx}\left(x + \frac{1}{3}x^3 + \frac{2}{15}x^5 + \frac{17}{315}x^7 + \cdots\right) = 1 + x^2 + \frac{2}{3}x^4 + \frac{17}{45}x^6 + \cdots.$$

33. $\displaystyle\lim_{x\to 0}\frac{x^3}{x - \sin x} = \lim_{x\to 0}\frac{x^3}{x - \left(x - \dfrac{x^3}{6} + \dfrac{x^5}{120} - \cdots\right)}$

$$= \lim_{x\to 0}\frac{x^3}{\dfrac{x^3}{6} - \dfrac{x^5}{120} + \cdots}$$

$$= \lim_{x\to 0}\frac{1}{\dfrac{1}{6} - \dfrac{x^2}{120} + \cdots}$$

$$= \frac{1}{\dfrac{1}{6}} = 6$$

35. $\cosh x = \dfrac{e^x + e^{-x}}{2}$

$$= \frac{\left(1 + x + \frac{x^2}{2!} + \frac{x^3}{3!}\right) + \left(1 - x + \frac{x^2}{2!} - \frac{x^3}{3!}\right)}{2}$$

$$= \frac{2\left(1 + \frac{x^2}{2!} + \frac{x^4}{4!} + \cdots\right)}{2}$$

$$= \sum_{k=0}^{\infty}\frac{x^{2k}}{(2k)!}$$

37. $\dfrac{e^x}{1-x} = e^x \cdot \left(\dfrac{1}{1-x}\right)$

$$= \left(1 + x + \frac{x^2}{2!} + \frac{x^3}{3!} + \frac{x^4}{4!} + \cdots\right) \cdot \left(1 + x + x^2 + x^3 + x^4 + \cdots\right)$$

$$= 1 + 2x + \frac{5}{2}x^2 + \frac{8}{3}x^3 + \frac{65}{24}x^4 + \cdots$$

39. $\dfrac{e^x}{\cos x} = \dfrac{1 + x + \frac{x^2}{2} + \frac{x^3}{3!} + \cdots}{1 - \frac{x^2}{2} + \frac{x^4}{4!} - \cdots}$

$$= 1 + x + x^2 + \frac{2x^3}{3} + \frac{x^4}{2} + \cdots$$

41. $\displaystyle\int_0^{-1} e^{-x^2}\,dx = \int_0^{-1} 1 + (-x^2) + \frac{(-x^2)^2}{2} + \frac{(-x^2)^3}{3!} + \cdots\,dx$

$$= \int_0^{-1}\left(1 - x^2 + \frac{x^4}{2} - \frac{x^6}{6} + \cdots\right)dx$$

$$= \left(x - \frac{x^3}{3} + \frac{x^5}{10} - \frac{x^7}{42}\right)\Big|_0^1$$

$$= 1 - \frac{1}{3} + \frac{1}{10} - \frac{1}{42} + \cdots$$

43. Using $\tan^{-1} x = \sum_{k=0}^{\infty} (-1)^k \dfrac{x^{2k+1}}{2k+1}$, we have

$$\frac{\pi}{4} = \tan^{-1} 1 = \sum_{k=0}^{\infty} \frac{(-1)^k}{2k+1} = 1 - \frac{1}{3} + \frac{1}{5} - \frac{1}{7} + \cdots$$

45. Using $\cos x = 1 - \dfrac{x^2}{2!} + \dfrac{x^4}{4!} - \dfrac{x^6}{6!} + \cdots$, we have

$$\cos \pi = 1 - \frac{\pi^2}{2!} + \frac{\pi^4}{4!} - \frac{\pi^6}{6!} + \cdots$$

47. Using Problem 17 and the fact that $46° \approx 0.802851456$ radians, we have

$$\sin 46° \approx= \frac{\sqrt{2}}{2} + \frac{\sqrt{2}}{2}\left(0.802851456 - \frac{\pi}{4}\right) - \frac{\sqrt{2}}{4}\left(0.802851456 - \frac{\pi}{4}\right)^2 \approx 0.719340424.$$

Now, for $f(x) = \sin x$, $f'''(x) = -\cos x$ and $|R_2(x)| = \dfrac{|\cos c|}{3!}|x - \pi/4|^3 < \dfrac{|x - \pi/4|^3}{3!}$.

Thus, $|R_2(0.802851456)| = \dfrac{|0.802851456 - \pi/4|^3}{6} < 0.000001$.

49. Using Problem 7, we have $e^{0.3} \approx 1 + 0.3 + \dfrac{(0.3)^2}{2} + \dfrac{(0.3)^3}{3!} + \dfrac{(0.3)^4}{4!} \approx 1.349837500$. Now, $|R_4(x)| = \dfrac{e^c}{5!}|x|^5 < \dfrac{3|x|^5}{5!}$ since

$0 < c < 1$ and $e^c < e < 3$. Thus, $|R_4(0.3)| < \dfrac{3|0.3|^5}{5!} < 0.0001$.

51. We use $|f^{(n+1)}(x)| = \begin{cases} |\cos x|, & n \text{ even} \\ |\sin x|, & n \text{ odd} \end{cases}$. Since $|\cos x| \le 1$ and $|\sin x| \le 1$, $|R_n(x)| = \dfrac{|f^{(n+1)}(c)|}{(n+1)!}|x|^{n+1} \le \dfrac{|x|^{n+1}}{(n+1)!}$ and

$\lim\limits_{n \to \infty} |R_n(x)| = 0$. Thus, the series represents $\sin x$ for all x.

53. We use $R_n(x) = \begin{cases} \dfrac{\sinh c}{(n+1)!}x^{n+1}, & n \text{ odd} \\ \dfrac{\cosh c}{(n+1)!}x^{n+1}, & n \text{ even} \end{cases}$ for c between 0 and x. Now, $|\sinh c| < |\sinh x|$ and $\cosh c < \cosh x$ for c between 0

and x, so for n odd, $|R_n(x)| = \dfrac{|\sinh c|}{(n+1)!}|x|^{n+1} < \dfrac{|\sinh x||x|^{n+1}}{(n+1)!}$, and for n even, $|R_n(x)| = \dfrac{\cosh c}{(n+1)!}|x|^{n+1} < \dfrac{(\cosh x)|x|^{n+1}}{(n+1)!}$. Thus,

$\lim\limits_{n \to \infty} |R_n(x)| = 0$ and the series represents $\sinh x$ for all x.

55. (a) From Figure 9.10.3, $L = Rx$ and $\sec x = \dfrac{R+y}{R}$. This gives $y = R \sec x - R = R \sec\left(\dfrac{L}{R}\right) - R$

(b) We need to find the Maclaurin series for $f(x) = \sec x$. We compute

$f(x) = \sec x,$ $\qquad\qquad\qquad\qquad f(0) = 1$

$f'(x) = \sec x \tan x,$ $\qquad\qquad\qquad f'(0) = 0$

$f''(x) = 2\sec^3 x - \sec x,$ $\qquad\qquad f''(0) = 1$

$f'''(x) = 6\sec^3 x \tan x - \sec x \tan x,$ $\qquad f'''(0) = 0$

$f^{(4)} = 6\sec^5 x + 18\sec^3 \tan^2 x - \sec^3 x - \sec x \tan x,$ $\qquad f^{(4)}(0) = 5$

Therefore, $\sec x = 1 + \dfrac{x^2}{2} + \cdots$.

Approximating $\sec x$ by $1 + \dfrac{x^2}{2}$, we have

$$y = R\left(\sec\left(\frac{L}{R}\right) - 1\right) = R\left(1 + \frac{(L/R)^2}{2} - 1\right) = R\left(\frac{L^2}{2r^2}\right) = \frac{L^2}{2R}$$

(c) Using $y \approx \dfrac{L^2}{2R}$, we have

$$y = \frac{(5280)^2}{2(4000)(5280)} \text{ ft} \approx 0.66 \text{ ft} = 7.92 \text{ in}$$

(d) Approximating $\sec x$ by $1 + \dfrac{x^2}{2} + \dfrac{5}{24}x^4$, we have

$$y = R\left(1 + \frac{(L/R)^2}{2} + \frac{5}{24}(L/R)^4 - 1\right)$$

$$= \frac{L^2}{2R} + \frac{5L^4}{24R^3}.$$

57. $\sin^2 x = (\sin x)(\sin x)$

$$= \left(x - \frac{x^3}{3!} + \frac{x^5}{5!} - \frac{x^7}{7!} + \cdots\right)\left(x - \frac{x^3}{3!} + \frac{x^5}{5!} - \frac{x^7}{7!} + \cdots\right)$$

$$= x^2 - \frac{x^4}{3} + \frac{2x^6}{45} - \cdots$$

Also,

$\sin^2 x = 1 - \cos^2 x - 1 - (\cos x)(\cos x)$

$$= 1 - \left(1 - \frac{x^2}{2!} + \frac{x^4}{4!} - \cdots\right)\left(1 - \frac{x^2}{2!} + \frac{x^4}{4!} - \cdots\right)$$

$$= 1 - \left(1 - x^2 + \frac{x^4}{3} - \frac{2x^6}{45} + \cdots\right)$$

$$= x^2 - \frac{x^4}{3} + \frac{2x^6}{45} - \cdots$$

59. Using $e^x = 1 + x + \dfrac{x^2}{2} + \dfrac{x^3}{3!} + \cdots$, we have

$(x+1)^2 e^x = (x+1)^2 e^{x+1-1} = e^{-1}(x+1)^2 e^{x+1}$

$$= e^{-1}(x+1)^2\left[1 + (x+1) + \frac{(x+1)^2}{2} + \frac{(x+1)^3}{3!} + \cdots\right]$$

$$= e^{-1}\left[(x+1)^2 + (x+1)^3 + \frac{(x+1)^4}{2} + \frac{(x+1)^5}{3!} + \cdots\right]$$

$$= \sum_{k=0}^{\infty} \frac{e^{-1}(x+1)^{k+2}}{k!}$$

61. $\cos x$ is an even function while $\sin x$ is an odd function. From (18), (19), and (20), we see that $\tan^{-1} x$ is an odd function, $\cosh x$ is an even function, and $\sinh x$ is an odd function.

9.11 Binomial Series

1. With $r = 1/3$, for $|x| < 1$,

$$\sqrt[3]{1+x} = 1 + \frac{1}{3}x + \frac{\frac{1}{3}\left(\frac{1}{3}-1\right)}{2!}x^2 + \frac{\frac{1}{3}\left(\frac{1}{3}-1\right)\left(\frac{1}{3}-2\right)}{3!}x^3 + \cdots$$

$$= 1 + \frac{1}{3}x - \frac{2}{3^2 \cdot 2!}x^2 + \frac{2 \cdot 5}{3^3 \cdot 3!}x^3 - \cdots.$$

3. With $r = 1/2$, for $|x/9| < 1$ or $|x| < 9$,

$$\sqrt{9-x} = 3\sqrt{1-x/9} = 3\left[1 + \frac{1}{2}\left(-\frac{x}{9}\right) + \frac{\frac{1}{2}\left(\frac{1}{2}-1\right)}{2!}\left(-\frac{x}{9}\right)^2 + \frac{\frac{1}{2}\left(\frac{1}{2}-1\right)\left(\frac{1}{2}-2\right)}{3!}\left(-\frac{x}{9}\right)^3 + \cdots\right]$$

$$= 3 - \frac{3}{2 \cdot 9}x - \frac{3 \cdot 1}{2^2 \cdot 2! \cdot 9^2}x^2 - \frac{3 \cdot 1 \cdot 3}{2^3 \cdot 3! \cdot 9^3}x^3 - \cdots.$$

5. With $r = -1/2$, for $|x| < 1$,

$$\frac{1}{\sqrt{1+x^2}} = 1 + \left(-\frac{1}{2}\right)(x^2) + \frac{-\frac{1}{2}\left(-\frac{1}{2}-1\right)}{2!}(x^2)^2 + \frac{-\frac{1}{2}\left(-\frac{1}{2}-1\right)\left(-\frac{1}{2}-2\right)}{3!}(x^2)^3 + \cdots$$

$$= 1 - \frac{1}{2}x - \frac{3}{2^2 \cdot 2!}x^4 - \frac{3 \cdot 5}{2^3 \cdot 3!}x^6 - \cdots.$$

7. With $r = 3/2$, for $|x/4| < 1$ or $|x| < 4$,

$$(4+x)^{3/2} = 8(1+x/4)^{3/2} = 8\left[1 + \frac{3}{2}\left(\frac{x}{4}\right) + \frac{\frac{3}{2}\left(\frac{3}{2}-1\right)}{2!}\left(\frac{x}{4}\right)^2 + \frac{\frac{3}{2}\left(\frac{3}{2}-1\right)\left(\frac{3}{2}-2\right)}{3!}\left(\frac{x}{4}\right)^3 + \cdots\right]$$

$$= 8 + \frac{8 \cdot 3}{2 \cdot 4}x + \frac{8 \cdot 3 \cdot 1}{3^2 \cdot 2! \cdot 4^2}x^2 + \frac{8 \cdot 3 \cdot 1 \cdot (-1)}{2^3 \cdot 3! \cdot 4^3}x^3 + \cdots.$$

9. With $r = -2$, for $|x/2| < 1$ or $|x| < 2$,

$$\frac{x}{(2+x)^2} = \frac{x}{4}(1+x/2)^{-2}$$

$$= \frac{x}{4}\left[1 - 2\left(\frac{x}{2}\right) + \frac{-2(-2-1)}{2!}\left(\frac{x}{2}\right)^2 + \frac{-2(-2-1)(-2-2)}{3!}\left(\frac{x}{2}\right)^3 + \cdots\right]$$

$$= \frac{1}{4}x - \frac{1}{4}x^2 + \frac{2 \cdot 3}{4 \cdot 2! \cdot 2^2}x^3 - \frac{2 \cdot 3 \cdot 4}{4 \cdot 3! \cdot 2^3}x^4 + \cdots.$$

11. See Problem 1. Since the series is alternating on $(0,1)$, by Theorem 9.7.2 the approximation to the sum using $S_2 = 1 + \frac{x}{3}$ is accurate within $a_3 = \frac{1}{9}x^2$.

13. With $r = -1/2$,

$$\sin^{-1}x = \int_0^x (1-t^2)^{(-1/2)}\,dt$$

$$= \int_0^x \left[1 - \frac{1}{2}(-t^2) + \frac{-\frac{1}{2}\left(-\frac{1}{2}-1\right)}{2!}(-t^2)^2 + \frac{-\frac{1}{2}\left(-\frac{1}{2}-1\right)\left(-\frac{1}{2}-2\right)}{2!}(-t^2)^3 + \cdots\right]dt$$

$$= \int_0^x \left[1 + \frac{1}{2}t^2 + \frac{3}{2^2 \cdot 2!}t^4 + \frac{3 \cdot 5}{2^3 \cdot 3!} + \cdots\right]dt$$

$$= x + \frac{1}{2 \cdot 3}x^3 + \frac{3}{2^2 \cdot 2! \cdot 5}x^5 + \frac{3 \cdot 5}{2^3 \cdot 3! \cdot 7}x^7 + \cdots$$

$$= x + \sum_{k=1}^{\infty} \frac{1 \cdot 3 \cdot 5 \cdots (2k-1)}{2^k k!(2k+1)}x^{2k+1}.$$

15. $y' = \frac{8d}{l^2}x$. Using the formula for arc length and $r = 1/2$, we have

$$s = 2\int_0^{l/2} \sqrt{1 + \frac{64d^2}{l^4}x^2}\,dx = 2\int_0^{l/2}\left[1 + \frac{1}{2} \cdot \frac{64d^2}{l^4}x^2 + \frac{\frac{1}{2}\left(\frac{1}{2}-1\right)}{2!}\left(\frac{64d^2}{l^4}x^2\right)^2 + \cdots\right]dx$$

$$= 2\int_0^{l/2}\left[1 + \frac{32d^2}{l^4}x^2 - \frac{64^2d^4}{8l^8}x^4 + \cdots\right]dx = 2\left[x + \frac{32d^2}{3l^4}x^3 - \frac{64^2d^4}{5 \cdot 8l^8}x^5 + \cdots\right]\Big|_0^{l/2}$$

$$= l + \frac{8d^2}{3l} - \frac{32d^4}{5l^3} + \cdots.$$

17. From Theorem 9.11.1 in the text with $r = -1/2$, we have

$$(1 - 2xr + r^2)^{-1/2} = [1 + r(r - 2x)]^{-1/2} = 1 - \frac{1}{2}r(r - 2x) + \frac{-\frac{1}{2}\left(-\frac{1}{2} - 1\right)}{2!}r^2(r - 2x)^2 + \cdots$$

$$= 1 - \frac{1}{2}r^2 + xr + \frac{3}{8}(r^4 - 4xr^3 + 4x^2r^2) + \cdots = 1 + xr + \left(\frac{3}{2}x^2 - \frac{1}{2}\right)r^2 + \cdots$$

Thus, $P_0 = 1$ $P_1 = x$, and $P_2 = \frac{3}{2}x^2 - \frac{1}{2}$.

19. $(1 + x)^{1/2} = [2 + (x - 1)]^{1/2} = \sqrt{2}\left(1 + \frac{x - 1}{2}\right)^{1/2}$

$$= \sqrt{2}\left[1 + \left(\frac{1}{2}\right)\frac{x - 1}{2} + \frac{\left(\frac{1}{2}\right)\left(\frac{1}{2} - 1\right)}{2!}\frac{(x - 1)^2}{2^2} + \frac{\left(\frac{1}{2}\right)\left(\frac{1}{2} - 1\right)\left(\frac{1}{2} - 2\right)}{3!}\frac{(x - 1)^4}{2^4} + \cdots\right]$$

$$= \sqrt{2} + \frac{\sqrt{2}}{2^2}(x - 1) - \frac{\sqrt{2}}{2^2 2!}(x - 1)^2 + \frac{\sqrt{2} \cdot 1 \cdot 3}{2^6 \cdot 3!}(x - 1)^3 + \cdots$$

Chapter 9 in Review

A. True/False

1. False; since $|a_n| = \frac{n}{2n + 1} \to \frac{1}{2}$, the series diverges by the n-th term test.

3. False; $\{(-1)^n/n\}$ is convergent and not monotonic.

5. True; since $a_{n+1}/a_n \geq 1$, $a_{n+1} \geq a_n$ and $\{a_n\}$ is a bounded monotonic sequence.

7. False; $\{1/n\}$ converges, but $\sum_{k=1}^{\infty} 1/k$ is the divergent harmonic series.

9. True; if this were false, the series would diverge by the n-th term test.

11. False; let $a_k = 1/k$.

13. True

15. False; let $a_k = (-1)^k/k$.

17. True

19. False; the ratio test is inconclusive in this case.

21. False; $\sum_{k=1}^{\infty} \frac{1}{k}x^k$ converges at $x = -1$ but is not absolutely convergent there.

23. False; let $c_k = -1/k$.

25. False; the integral test simply indicates that the series converges.

27. True; $\ln x$ is not differentiable at $x = 0$.

29. True

B. Fill In the Blanks

1. 20, 9, 4/5, 16

3. 4, since $a_5 = 1/10^5 = 0.00001 < 0.00005$.

5. $\dfrac{n}{9}; \dfrac{22}{9}$

$$0.nnnn\ldots = \frac{n}{10}\left(1+\frac{1}{10}+\frac{1}{10^2}+\ldots\right)$$

$$= \frac{n}{10}\left(\frac{1}{1-\dfrac{1}{10}}\right) = \frac{n}{10}\left(\frac{1}{\frac{9}{10}}\right) = \frac{n}{9}$$

$$2.444\ldots = 2+.044\ldots = 2+\frac{4}{9} = \frac{22}{9}$$

7. e^x

9. $x < -5$ and $x > 5$.

11. The series converges absolutely for x in $(-1,1)$. At $x=-1$, the series diverges. At $x=1$, the series converges conditionally. Thus, the interval of convergence is $(-1,1]$.

C. Exercises

1. Since $\dfrac{k}{(k^2+1)^2} < \dfrac{1}{k^3}$, the series converges by comparison with the p-series $\displaystyle\sum_{k=1}^{\infty}\frac{1}{k^p}$.

3. This is a geometric series with $r = \dfrac{1}{\pi} < 1$. Thus, the series converges.

5. Since $\ln k < k$, $\dfrac{\sqrt{k}\ln k}{k^4+4} < \dfrac{1}{k^{5/2}}$ and the series converges by comparison with the p-series $\displaystyle\sum_{k=1}^{\infty}\frac{1}{k^{5/2}}$

7. Since $\dfrac{k}{\sqrt[3]{k^6-4k}} > \dfrac{k}{k^2} = \dfrac{1}{k}$, the series diverges by comparison with the harmonic series.

9. All of the odd-numbered terms of this series are 0. We may thus express the series as $\displaystyle\sum_{k=1}^{\infty}\frac{2}{\sqrt{2k}}$ or $\displaystyle\sum_{k=1}^{\infty}\frac{1}{\sqrt{k}}$. This is a divergent p-series.

11. Since $\dfrac{1}{3k^2+4k+6} < \dfrac{1}{k^2}$, the series converges by comparison with the p-series $\displaystyle\sum_{k=1}^{\infty}\frac{1}{k^2}$.

13. $\displaystyle\sum_{k=1}^{\infty}\frac{(-1)^{k-1}+3}{(1.01)^{k-1}} = \sum_{k=1}^{\infty}\left(-\frac{1}{1.01}\right)^{k-1} + 3\sum_{k=1}^{\infty}\left(\frac{1}{1.01}\right)^{k-1} = \frac{1}{1-(-1/1.01)} + \frac{3}{1-(1/1.01)}$

$$= \frac{1.01}{1.01+1} + \frac{3.03}{1.01-1} = \frac{101}{201} + 303 = \frac{61,004}{201}$$

15. $\displaystyle\lim_{n\to\infty}\left|\frac{a_{n+1}}{a_n}\right| \lim_{n\to\infty}\left|\frac{3^{n+1}x^{n+1}/(n+1)^3}{3^n x^n/n^3}\right| = \lim_{n\to\infty} 3\left(\frac{n}{n+1}\right)^3 |x| = 3|x|$

The series is absolutely convergent for $3|x| < 1$ or on $(-1/3,1/3)$. At $x=-1/3$, the series $\displaystyle\sum_{k=1}^{\infty}\frac{(-1)^k}{k^3}$ converges by the alternating series test. At $x=1/3$, the series $\displaystyle\sum_{k=1}^{\infty}\frac{1}{k^3}$ is a convergent p-series. Thus, the given series converges on $[-1/3,1/3]$.

17. $\displaystyle\lim_{n\to\infty}\left|\frac{a_{n+1}}{a_n}\right| \lim_{n\to\infty}\left|\frac{(n+1)!(x+5)^{n+1}}{n!(x+5)^n}\right| = \lim_{n\to\infty}(n+1)|x+5| = \infty$ for $x \ne -5$. Thus, the series converges only for $x=-5$.

19. $\displaystyle\lim_{n\to\infty}\left|\frac{a_{n+1}}{a_n}\right| = \lim_{n\to\infty}\left|\frac{\dfrac{2\cdot5\cdots(3n+2)x^{n+1}}{3\cdot7\cdots(4n+3)}}{\dfrac{2\cdot5\cdots(3n-1)x^n}{3\cdot7\cdots(4n-1)}}\right| = \lim_{n\to\infty}\frac{3n+2}{4n+3}|x| = \frac{3}{4}|x$

The series converges for $\dfrac{3}{4}|x| < 1$ or $|x| < \dfrac{4}{3}$. Thus, the radius of convergence is $4/3$.

21. $\dfrac{1}{\alpha} + \dfrac{1}{\alpha^2} + \dfrac{1}{\alpha^3} + \cdots = \left(1 + \dfrac{1}{\alpha} + \dfrac{1}{\alpha^2} + \cdots\right) - 1$

$$= \dfrac{1}{1 - \left(\frac{1}{\alpha}\right)} - 1 = \dfrac{\alpha}{\alpha - 1} - 1 = \dfrac{1}{\alpha - 1}$$

23. Using a binomial series expansion,

$$\dfrac{1}{\sqrt[3]{1+x^5}} = (1+x^5)^{-1/3} = 1 - \dfrac{1}{3}x^5 + \dfrac{\left(-\frac{1}{3}\right)\left(-\frac{1}{3}-1\right)}{2!}(x^5)^2 + \cdots = 1 - \dfrac{1}{3}x^5 + \dfrac{2}{9}x^{10} - \cdots.$$

25. Using the Maclaurin series for $\sin x$,

$$\sin x \cos x = \dfrac{1}{2}\sin 2x = \dfrac{1}{2}\left[2x - \dfrac{(2x)^3}{3!} + \dfrac{(2x)^5}{5!} - \cdots\right] = x - \dfrac{2}{3}x^3 + \dfrac{2}{15}x^5 - \cdots.$$

27.

$f(x) = \cos x,$ $\qquad\qquad\qquad f(\pi/2) = 0$

$f'(x) = -\sin x,$ $\qquad\qquad\qquad f'(\pi/2) = -1$

$f''(x) = -\cos x,$ $\qquad\qquad\qquad f''(\pi/2) = 0$

$f'''(x) = \sin x,$ $\qquad\qquad\qquad f'''(\pi/2) = 1$

$\vdots$

$f^{(2k+1)}(x) = (-1)^{k+1}\sin x,$ $\qquad f^{(2k+1)}(\pi/2) = (-1)^{k+1}$

The Taylor series is $\displaystyle\sum_{k=0}^{\infty} \dfrac{(-1)^{k+1}}{(2k+1)!}\left(x - \dfrac{\pi}{2}\right)^{2k+1}.$

29. $3\left(\dfrac{2}{3}\right) + 2\left(\dfrac{2}{3}\right) + \dfrac{4}{3}\left(\dfrac{2}{3}\right) + \dfrac{8}{9}\left(\dfrac{2}{3}\right) + \dfrac{16}{27}\left(\dfrac{2}{3}\right) + \cdots$

$$= 3\left(\dfrac{2}{3}\right) + 3\left(\dfrac{2}{3}\right)^2 + 3\left(\dfrac{2}{3}\right)^3 + 3\left(\dfrac{2}{3}\right)^4 + 3\left(\dfrac{2}{3}\right)^5 + \cdots = \dfrac{2}{1 - 2/3} = 6 \text{ million dollars.}$$

Chapter 10

Conics and Polar Coordinates

10.1   Conic Sections

1. vertex: $(0,0)$
    focus: $(1,0)$
    directrix: $x = -1$
    axis: $y = 0$

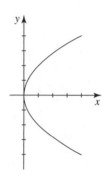

3. vertex: $(0,0)$
    focus: $(0,-4)$
    directrix: $y = 4$
    axis: $x = 0$

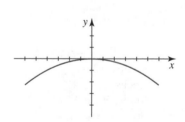

5. vertex: $(0,1)$
    focus: $(4,1)$
    directrix: $x = -4$
    axis: $y = 1$

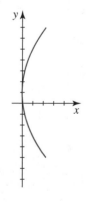

7. vertex: $(-5,-1)$
   focus: $(-5,2)$
   directrix: $y=0$
   axis: $x=-5$

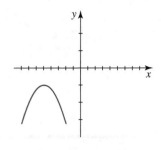

9. $(y+6)^2 = 4(x+5)$
   vertex: $(-5,-6)$
   focus: $(-4,-6)$
   directrix: $x=-6$
   axis: $y=-6$

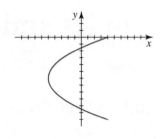

11. $\left(x+\frac{5}{2}\right)^2 = \frac{1}{4}(y-1)$
    vertex: $(-5/2,-1)$
    focus: $(-5/2,-15/16)$
    directrix: $y=-17/16$
    axis: $x=--5/2$

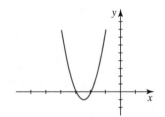

13. $(y-4)^2 = -2(x+3)$
    vertex: $(3,4)$
    focus: $(5/2,4)$
    directrix: $x=7/2$
    axis: $y=4$

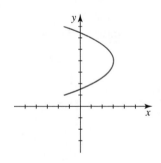

15. $x^2 = 28$

17. $y^2 = 10x$

19. The parabola is of the form $(y-k)^2 = 4p(x-h)$ with $(h,k) = (-2,-7)$ and $p=3$. Thus the equation is $(y+7)^2 = 12(x+2)$.

21. The parabola is of the form $x^2 = 4py$ with $(-2)^2 = 4p(8)$. Thus $p=\frac{1}{8}$ and the equation is $x^2 = \frac{1}{2}y$.

23. To find the x-intercept set $y=0$. Solving $4^2 = 4(x+1)$ gives $x=3$. The x-intercept is $(3,0)$. To find the y-intercept set $x=0$.
    Solving $(y+4)^2 = 4$ gives $y=-4\pm2$. The y-intercepts are $(0,-2)$ and $(0,-6)$.

25. center: $(0,0)$
   foci: $(0,\pm\sqrt{15})$
   vertices: $(0,\pm4)$
   endpoints of the minor axis: $(\pm1,0)$
   eccentricity: $\dfrac{\sqrt{15}}{4}$

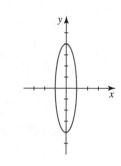

27. $\dfrac{x^2}{16}+\dfrac{y^2}{9}=1$ center: $(0,0)$
   foci: $(\pm\sqrt{7},0)$
   vertices: $(\pm4,0)$
   endpoints of the minor axis: $(0,\pm3)$
   eccentricity: $\dfrac{\sqrt{7}}{4}$

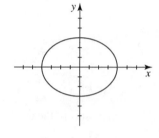

29. center: $(1,3)$
   foci: $(1\pm\sqrt{13},3)$
   vertices: $(-6,3),(8,3)$
   endpoints of the minor axis: $(1,-3),(1,9)$
   eccentricity: $\dfrac{\sqrt{13}}{7}$

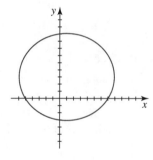

31. center: $(-5,-2)$
   foci: $(-5,-2\pm\sqrt{15})$
   vertices: $(-5,-6),(-5,2)$
   endpoints of the minor axis: $(-6,-2),(-4,-2)$
   eccentricity: $\dfrac{\sqrt{15}}{4}$

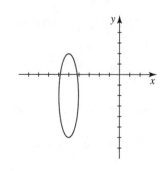

33. $x^2 + \dfrac{\left(y+\frac{1}{2}\right)^2}{4} = 1$
    center: $(0, -1/2)$
    foci: $(0, -1/2 \pm \sqrt{3})$
    vertices: $(0, -5/2), (0, 3/2)$
    endpoints of the minor axis: $(-1, -1/2), (1, -1/2)$
    eccentricity: $\dfrac{\sqrt{3}}{2}$

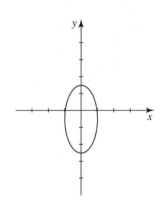

35. $\dfrac{(x-7)^2}{9} + \dfrac{(y+1)^2}{25} = 1$
    center: $(2, -1)$
    foci: $(2, -5), (2, 3)$
    vertices: $(2, -6), (2, 4)$
    endpoints of the minor axis: $(-1, -1), (5, -1)$
    eccentricity: $\dfrac{4}{5}$

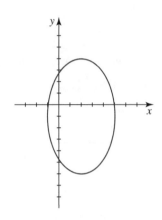

37. $\dfrac{x^2}{9} + \dfrac{(y+3)^2}{3} = 1$
    center: $(0, -3)$
    foci: $(\pm\sqrt{6}, -3)$
    vertices: $(\pm3, -3)$
    endpoints of the minor axis: $(0, -3 \pm \sqrt{3})$
    eccentricity: $\dfrac{\sqrt{6}}{3}$

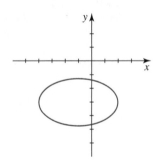

39. The center is $(0,0)$ with the x-axis as the major axis. $a = 5$ and $c = 3$, so $b = 4$. Thus the equation is $\dfrac{x^2}{25} + \dfrac{y^2}{16} = 1$.

41. The center is $(1, -3)$ with the x-axis as the major axis. $a = 4$ and $b = 2$. Thus the equation is $\dfrac{(x-1)^2}{16} + \dfrac{(y+3)^2}{4} = 1$.

43. The center is $(0,0)$ with the x-axis as the major axis. $c = \sqrt{2}$ and $b = 3$, so $a = \sqrt{11}$. Thus the equation is $\dfrac{x^2}{11} + \dfrac{y^2}{9} = 1$.

45. The center is $(0,0)$ with the y-axis as the major axis. $c = 3$ thus $9 = a^2 - b^2$ and $a = \sqrt{9 + b^2}$. Thus the equation is of the form $\dfrac{x^2}{b^2} + \dfrac{y^2}{9 + b^2} = 1$. The ellipse passes through the point $(-1, 2\sqrt{2})$, thus $\dfrac{(-1)^2}{b^2} + \dfrac{(2\sqrt{2})^2}{9 + b^2} = 1$. Solving this for b, we obtain $b = \sqrt{3}$. Thus $a^2 = 12$ and the equation is $\dfrac{x^2}{3} + \dfrac{y^2}{12} = 1$.

47. The y-axis as the major axis with $c = 3$ and $a = 4$. Thus $b = \sqrt{7}$ and the equation of the ellipse is $\dfrac{(x-1)^2}{7} + \dfrac{(y-3)^2}{16} = 1$

49. center: $(0,0)$
    foci: $(\pm\sqrt{41},0)$
    vertices: $(\pm4,0)$
    asymptotes: $y=\pm\frac{5}{4}x$
    eccentricity: $\dfrac{\sqrt{41}}{4}$

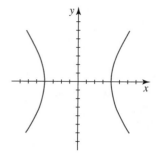

51. $\dfrac{y^2}{20}-\dfrac{x^2}{4}=1$ center: $(0,0)$
    foci: $(0,\pm2\sqrt{6})$
    vertices: $(0,\pm2\sqrt{5})$
    asymptotes: $y=\pm\sqrt{5}x$
    eccentricity: $\sqrt{\dfrac{6}{5}}$

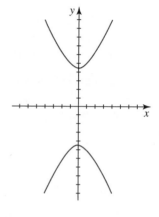

53. center: $(5,-1)$
    foci: $(5\pm\sqrt{53},-1)$
    vertices: $(3,-1),(7,-1)$
    asymptotes: $y=-1\pm\frac{7}{2}(x-5)$
    eccentricity: $\dfrac{\sqrt{53}}{2}$

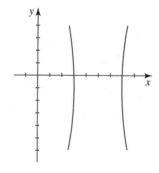

55. center: $(0,4)$
    foci: $(0,4\pm\sqrt{37})$
    vertices: $(0,-2),(0,10)$
    asymptotes: $y=4\pm6x$
    eccentricity: $\dfrac{\sqrt{37}}{6}$

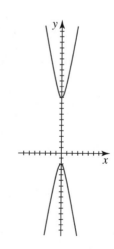

57. $\dfrac{(x-3)^2}{5} - \dfrac{(y-1)^2}{25} = 1$ center: $(3,1)$
foci: $(3 \pm \sqrt{30}, 1)$
vertices: $(3 \pm \sqrt{5}, 1)$
asymptotes: $y = 1 \pm \sqrt{5}(x-3)$
eccentricity: $\sqrt{6}$

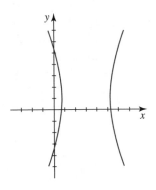

59. $\dfrac{(x-2)^2}{6} - \dfrac{(y-1)^2}{5} = 1$ center: $(2,1)$
foci: $(2 \pm \sqrt{11}, 1)$
vertices: $(2 \pm \sqrt{6}, 1)$
asymptotes: $y = 1 \pm \sqrt{\dfrac{5}{6}}(x-2)$
eccentricity: $\sqrt{\dfrac{11}{6}}$

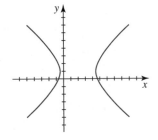

61. $(y-3)^2 - \dfrac{(x-1)^2}{1/4} = 1$ center: $(1,3)$
foci: $(1, 3 \pm \sqrt{5}/2)$
vertices: $(1,2), (1,4)$
asymptotes: $y = 3 \pm 2(x-1)$
eccentricity: $\dfrac{\sqrt{5}}{2}$

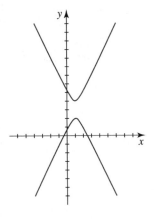

63. The center is $(0,0)$ with the y-axis as the transverse axis. $c = 4$ and $a = 2$, thus $b = \sqrt{12}$. The equation is $\dfrac{y^2}{4} - \dfrac{x^2}{12} = 1$

65. The center is $(1,-3)$ with the y-axis as the transverse axis. $c = 3$ and $a = 2$, thus $b = \sqrt{5}$. The equation is $\dfrac{(y+3)^2}{4} - \dfrac{(x-1)^2}{5} = 1$

67. The center is $(-1,3)$ with the y-axis as the transverse axis. $a = 1$ and the equation is of the form $\dfrac{(y-3)^2}{1} - \dfrac{(x+1)^2}{b^2} = 1$. The hyperbola passes through the point $(-5, 3 + \sqrt{5})$ thus $(3 + \sqrt{5} - 3)^2 - \dfrac{(-5+1)^2}{b^2} = 1$. Thus $b^2 = 4$ and the equation is $(y-3)^2 - \dfrac{(x+1)^2}{4} = 1$.

69. The center is $(2,4)$ with the y-axis as the transverse axis and $a = 1$. After solving the asymptote given in the problem for y, we obtain $y = \dfrac{x+6}{2} = \dfrac{x}{2} + 3$. The equation of the hyperbola is of the form $(y-4)^2 - \dfrac{(x-2)^2}{b^2} = 1$. The asymptote equations for

this hyperbola are $y - 4 = \dfrac{x-2}{b}$ and $y - 4 = \dfrac{-x+2}{b}$ (these are also equivalent to $y = \dfrac{x}{b} + \left(4 - \dfrac{2}{b}\right)$ and $y = -\dfrac{x}{b} + \left(4 + \dfrac{2}{b}\right)$).

Letting b equal 2 or -2 will yield one asymptote with the equation $y = \dfrac{x}{2} + 3$. In either case, the equation of the hyperbola is

$$(y-4)^2 - \frac{(x-2)^2}{4} = 1.$$

71. We place the coordinate axes so that the origin is at the vertex of the parabola. The point $(2,2)$ lies on the parabola. Thus the equation is $x^2 = 2y$ with $p = 1/2$. The focus of this parabola occurs at the point $(0, 1/2)$. Thus the light source is 6 inches from the vertex.

73. We place the coordinate axes so that the origin is at the vertex of the parabola. The parabola is of the form $x^2 = 4py$ and contains the point $(20, 1)$. Thus the equation of the parabola is $x^2 = 400y$. The towers are located at $x = 175$ and $x = -175$. Hence the height of the towers is found by solving $(175)^2 = 400y$. Solving this equation yields $y = 76.5625$. Therefore the towers are 76.5625 ft above the road.

75. We place the coordinate axes so that the origin is at end of the pipe with the parabola in Quadrants 3 and 4. The equation is of the form $x^2 = 4py$ and the point $(4, -2)$ lies on the parabola. Therefore the equation is $x^2 = -8y$. The water hits the ground at $y = -20$. The point on the parabola with y-value -20 is found by solving $x^2 = -8(-20)$. This point is $x = 12.65$. Thus the water hits the ground 12.65 m from the point on the ground directly beneath the end of the pipe.

77. Taking the center of the ellipse to be at the origin, we have $a = 3.6 \times 10^7$ and $b = 3.52 \times 10^7$. Since $c^2 = a^2 - b^2$, $c^2 = 12.96 \times 10^{14} - 12.3904 \times 10^{14} = 0.5696 \times 10^{14}$ and $c \approx 0.75 \times 10^7$. The perihelion or least distance is $a - c \approx 2.85 \times 10^7$ miles or 28.5 million miles. And the aphelion or greatest distance is $a + c \approx 4.35 \times 10^7$ miles or 43.5 million miles.

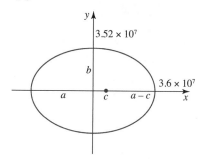

79. From $a = 1.67 \times 10^9$ and 4.25×10^8 we obtain

$$c^2 = a^2 - b^2 = 2.7889 \times 10^{18} - 18.0625 \times 10^6 = 2.78.89 \times 10^{16} - 18.0625 \times 10^{16}$$
$$= 260.8275 \times 10^{16} = 2.608275 \times 10^{18}.$$

Then $c \approx 1.615 \times 10^9$ and the eccentricity is $\dfrac{c}{a} \approx \dfrac{1.615 \times 10^9}{1.67 \times 10^9} \approx 0.967.$

81. We place the coordinate axes so that the origin is at the point midway across the base. Thus $a = 5$ and $b = 15$ so the equation of the doorway is $y = \sqrt{15^2 \left(1 - \dfrac{x^2}{25}\right)}$. The height of the doorway at a point on the base 4 ft from the center is

$$y = \sqrt{15^2 \left(1 - \frac{3^2}{25}\right)} = 12 \text{ ft.}$$

83.

	ellipse	shifted ellipse
center	$(0,1)$	$(4,1)$
vertices	$(-2,1), (2,1), (0,-2), (0,4)$	$(2.1), (6.1), (4.-2), (4.4)$
foci	$(0, 1 - \sqrt{5}), (0, 1 + \sqrt{5})$	$(4, 1 - \sqrt{5}), (4, 1 + \sqrt{5})$

85. (a) $\dfrac{y^2}{144} - \dfrac{x^2}{25} = 1$

(b) Conjugate hyperbolas have the same asymptotes and do not intersect.

87. Since $a = 4$ and $b = \sqrt{20}$, we have $c^2 = a^2 + b^2 = 16 + 20 = 36$ and hence $c = 6$. Thus the foci occur at $F_1 = (-6, 0)$ and $F_2 = (6, 0)$. The line joining $(-6, -5)$ and F_2 is given by $y = \dfrac{5}{12}x - \dfrac{5}{2}$. The ray of light travels southwest along this line.

10.2 Parametric Equations

1.

t	-3	-2	-1	0	1	2	3
x	-5	-3	-1	1	3	5	7
y	6	2	0	0	2	6	12

3.

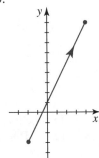

5.

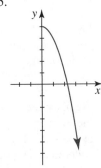

7.

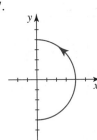

9.

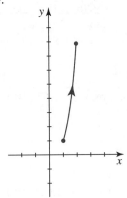

11. $y = (t^2)^2 + 3t^2 - 1 = x^2 + 3x - 1; \quad y = x^2 + 3x - 1, \quad x \geq 0$

13. $x = \cos 2t = \cos^2 t - \sin^2 t = 1 - 2\sin^2 t = 1 - 2y^2; \quad y = 1 - 2y^2, \quad -1 \leq y \leq 1$

15. $t = x^{1/3}; \quad y = 3\ln x^{1/3}; \quad y = \ln x, \quad x > 0$

17.

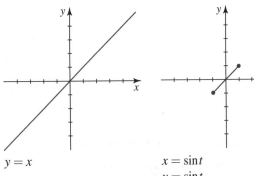

$y = x$

$x = \sin t$
$y = \sin t$

19.

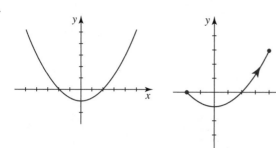

$y = \dfrac{x^2}{4} - 1$

$x = 2t$
$y = t^2 - 1$
$-1 \le t \le 2$

21.

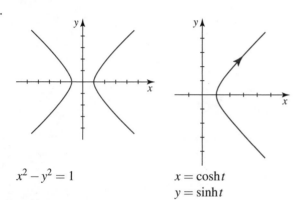

$x^2 - y^2 = 1$

$x = \cosh t$
$y = \sinh t$

23.

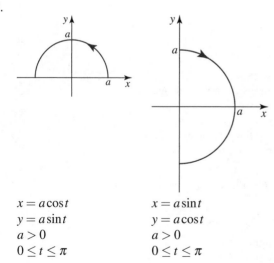

$x = a\cos t$
$y = a\sin t$
$a > 0$
$0 \le t \le \pi$

$x = a\sin t$
$y = a\cos t$
$a > 0$
$0 \le t \le \pi$

25.

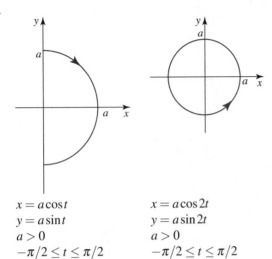

$x = a\cos t$
$y = a\sin t$
$a > 0$
$-\pi/2 \le t \le \pi/2$

$x = a\cos 2t$
$y = a\sin 2t$
$a > 0$
$-\pi/2 \le t \le \pi/2$

27.

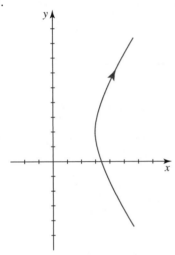

29. This is the same as $x = 1/y$ or $xy = 1$. The graphs are the same.

31. Since $|\cos t| \ge 1$, x can never be 2. But $(2, 1/2)$ is on $xy = 1$, so the graphs are not the same.

33. Since $e^{-2t} > 0$ for all t, x can never be -1. But $(-1, -1)$ is on $xy = 1$, so the graphs are not the same.

35. From $\sin\phi = \frac{y}{L}$ we have $t = L\sin\phi$. Since (x, y) is on the circle $x^2 + y^2 = r^2$, $x = \pm\sqrt{r^2 - y^2} = \pm\sqrt{r^2 - l^2\sin^2\phi}$.

37. From the figure we see that $\beta = \theta - \pi/2$ and $\alpha = \beta = \theta - \pi/2$. The length of the line segment from R to P is equal to the arc of the circle subtended by θ; that is $a\theta$. Now, $x = a\theta\sin\alpha = a\theta\sin(\theta - \pi/2) = -a\theta\cos\theta$, $t = a\cos\beta = a\cos(\theta - \pi/2) = a\sin\theta$, $b = a\sin\beta = a\sin(\theta - \pi/2) = -a\cos\theta$, and $c = a\theta\cos\alpha = a\theta\cos(\theta - \pi/2) = a\theta\sin\theta$. Thus

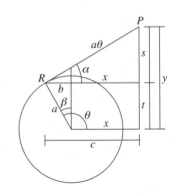

$$x = c - b = a\theta\sin\theta - (-a\cos\theta) = a(\cos\theta + \theta\sin\theta)$$
$$y = s + t = -a\theta\cos\theta = a(\sin\theta - \theta\cos\theta).$$

39. (a) When $b = 4a$, the equations become $x = 3a\cos\theta + a\cos 3\theta$, $y = 3a\sin\theta - a\sin 3\theta$. Using the identities $\cos 3\theta = 4\cos^3\theta - 3\cos\theta$ and $\sin 3\theta = 3\sin\theta - 4\sin^3\theta$, the parametric equations of the hypocycloid of four cusps become $x = 4a\cos^3\theta = b\cos^3\theta$, $y = 4a\sin^3\theta = b\sin^3\theta$.

    (b)

    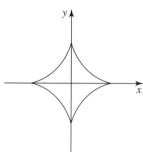

    (c) Writing $x^{2/3} = b^{2/3}\cos^2\theta$, $y^{2/3} = b^{2/3}\sin^2\theta$ we obtain $x^{2/3} + y^{2/3} = b^{2/3}$.

41. (a) When $b = 3a$, the equations become $x = 4a\cos\theta - a\cos 4\theta$, $y = 4a\sin\theta - a\sin 4\theta$.

    (b)

    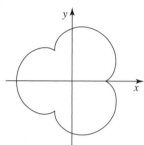

43.

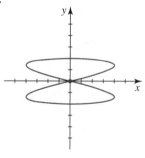

45.

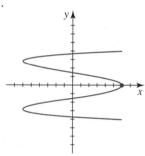

47.

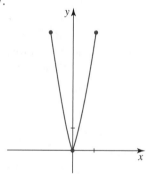

49. Using the equation from x to solve for t, we have $t = \dfrac{x - x_1}{x_2 - x_1}$. Plugging this into the equation for y yields

$$y = y_1 + (y_2 - y_1)\left(\frac{x - x_1}{x_2 - x_1}\right)$$

$$= y_1 + \left(\frac{y_2 - y_1}{x_2 - x_1}\right)(x - x_1)$$

which is the equation of a line joining (x_1, y_1) and (x_2, y_2). When $0 \leq t \leq 1$, we get the line segment with endpoints (x_1, y_1) and (x_2, y_2).

51. If the launch point is designated as the origin, then the equations describing the skier's motion from launch until landing are given by

$$x = 75t \quad \text{and} \quad y = -16t^2$$

where $t = 0$ at the moment of launch. At the moment of impact, we have

$$\tan 33° = \frac{|y|}{|x|} = \frac{16}{15}t.$$

Thus, $t = \dfrac{75}{16}\tan 33° \approx 3.044$, and therefore $x \approx 228.3$ ft, $y \approx -148.25$ ft.

10.3 Calculus and Parametric Equations

1. $\dfrac{dx}{dt} = 3t^2 - 2t; \quad \dfrac{dy}{dt} = 2t + 5; \quad \dfrac{dy}{dx} = \dfrac{2t+5}{3t^2 - 2t}; \quad \dfrac{dy}{dx}\Big|_{t=-1} = \dfrac{3}{5}$

3. $\dfrac{dx}{dt} = \dfrac{t}{\sqrt{t^2+1}}; \quad \dfrac{dy}{dt} = 4t^3; \quad \dfrac{dy}{dx} = \dfrac{4t^3}{t/\sqrt{t^2+1}} = 4t^2\sqrt{t^2+1}; \quad \dfrac{dy}{dx}\Big|_{t=\sqrt{3}} = 4(3)\sqrt{4} = 24$

5. $\dfrac{dx}{d\theta} = 2\cos\theta(-\sin\theta); \quad \dfrac{dy}{d\theta}\cos\theta; \quad \dfrac{dy}{dx} = \dfrac{\cos\theta}{-2\sin\theta\cos\theta} = -\dfrac{1}{2\sin\theta}$

   $\dfrac{dy}{dx}\Big|_{\theta=\pi/6} = -\dfrac{1}{2(1/2)} = -1$

7. $\dfrac{dy}{dx} = \dfrac{12t}{3t^2+3} = \dfrac{4t}{t^2+1}.$

   At $t = -1$ we observe $x = -4$, $y = 7$, and $m = dy/dx = -2$. The tangent line is $y = -2x - 1$.

9. $\dfrac{dy}{dt} = \dfrac{2t}{(2t+1)}.$ At $(2,4)$, $t = -2$ and $m = dy/dx = 4/3$. The tangent line is $y = \dfrac{4}{3}x + \dfrac{4}{3}.$

11. $\dfrac{dy}{dx} = \dfrac{-2\sin t}{8\cos 2t}.$ When $y = 1$, $\cos t = 1/2$ and $t = \pi/3$ or $5\pi/3$. For $t = \pi/3$, $x = 4\sin(2\pi/3) = 4(\sqrt{3}/2) = 2\sqrt{3}$, and $m = \dfrac{dy}{dx} = \dfrac{-2\sin(\pi/3)}{8\cos(2\pi/3)} = \dfrac{\sqrt{3}}{4}.$

13. $\dfrac{dx}{dt} = 2; \quad \dfrac{dy}{dt} = 2t - 4; \quad \dfrac{dy}{dx} = \dfrac{2t-4}{2} = t - 2$

   We want $t - 2 = 3$. Then $t = 5$ and the point of tangency is $(5,8)$. The equation of the tangent line is $y - 8 = 3(x-5)$ or $y = 3x - 7.$

15. $\dfrac{dx}{dt} = 3t^2 - 1; \quad \dfrac{dy}{dt} = 2t; \quad \dfrac{dy}{dx} = \dfrac{2t}{3t^2-1}.$ The tangent line is horizontal when $2t = 0$ or $t = 0$, and vertical when $3t^2 - 1 = 0$ or $t = \pm 1/\sqrt{3}$. Thus, there is a horizontal tangent at $(0,0)$ and vertical at $(-2/3\sqrt{3}, 1/3)$ and $(2/3\sqrt{3}, 1/3)$.

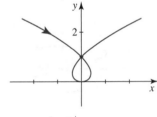

17. $\dfrac{dx}{dt} = 1; \quad \dfrac{dy}{dt} = 3t^2 - 6t; \quad \dfrac{dy}{dx} = 3t^2 - 6t.$ The tangent line is horizontal when $3t^2 - 6t = 3t(t-2) = 0$ or $t = 0, 2$. Thus, there are horizontal tangents at $(-1, 0)$ and $(1, -4)$. There are no vertical tangents.

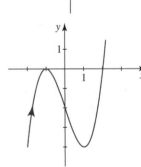

19. $\dfrac{dy}{dx} = \dfrac{18t^2}{6t} = 3t; \quad \dfrac{d^2y}{dx^2} = \dfrac{3}{6t} = \dfrac{1}{2t}; \quad \dfrac{d^3y}{dx^3} = \dfrac{-1/2t^2}{6t} = -\dfrac{1}{12t^3}$

21. $\dfrac{dy}{dx} = \dfrac{2e^{2t} + 3e^{3t}}{-e^{-t}} = -2e^{3t} - 3e^{4t}; \quad \dfrac{d^2y}{dx^2} = \dfrac{-6e^{3t} - 12e^{4t}}{-e^{-t}} = 6e^{4t} + 12e^{5t}$

   $\dfrac{d^3y}{dx^3} = \dfrac{24e^{4t} + 60e^{5t}}{-e^{-t}} = -24e^{5t} - 60e^{6t}$

23. Using Problem 16, $\dfrac{d^2y}{dx^2} = \dfrac{dy'/dt}{dx/dt} = \dfrac{(32-16t)/3t^3}{3t^2/8} = \dfrac{256 - 128t}{9t^5} = \dfrac{128}{9}\left(\dfrac{2-t}{t^5}\right).$ Then d^2y/dx^2 is 0 when $t = 2$ and undefined when $t = 0$. The graph is concave downward on $(-\infty, 0)$ and $(2, \infty)$ and concave upward on $(0, 2)$.

t		0		2	
y''	−	und	+	0	−

25. $x'(t) = 5t^2, \quad y'(t) = 12t^2; \quad s = \int_0^2 \sqrt{25t^4 + 144t^4} \, dt = 13 \int_0^2 t^2 \, dt = \frac{13}{3} t^3 \Big|_0^2 = \frac{104}{3}$

27. $x'(t) = e^t \cos t + e^t \sin t; \quad y'(t) = -e^t \sin t + e^t \cos t$

$s = \int_0^\pi [e^{2t}(\cos^2 t + 2\sin t \cos t + \sin^2 t) + e^{2t}(\sin^2 t - 2\sin t \cos t + \cos^2 t)]^{1/2} dt$

$= \int_0^\pi e^t (2)^{1/2} dt = \sqrt{2} e^t \Big|_0^\pi = \sqrt{2}(e^\pi - 1)$

29. $x'(\theta) = -3b\cos^2\theta \sin\theta; \quad y'(\theta) = 3b\sin^2\theta\cos\theta$

$s = \int_0^{\pi/2} (9b^2\cos^4\theta\sin^2\theta + 9b^2\sin^4\theta\cos^2\theta)^{1/2} d\theta = 3|b| \int_0^{\pi/2} \sin\theta\cos\theta\sqrt{\cos^2\theta + \sin^2\theta} \, d\theta$

$= 3|b| \int_0^{\pi/2} \frac{1}{2}\sin 2\theta \, d\theta = -\frac{3}{4}|b|\cos 2\theta \Big|_0^{\pi/2} = -\frac{3}{4}|b|(-1-1) = \frac{3}{2}|b|$

31. (a) Setting $x = 0$ we have $t^2 - 4t - 2 = 0$ which implies $t = 2 \pm \sqrt{6}$. When $t = 2 - \sqrt{6}$, $y \approx -0.6551$. When $t = 2 + \sqrt{6}$, $y \approx 1390.66$.

(b) Using Newton's Method to solve $t^5 - 4t^3 - 1 = 0$ we obtain $t \approx -1.96687, \ -0.654175, \ 2.02968$ with corresponding x values 9.73606, 1.04465, −5.99912.

33. From Example 7 in Section 10.2, $f(\theta) = a(\theta - \sin\theta)$ and $g(\theta) = a(1 - \cos\theta)$ for $0 \le \theta \le 2\pi$. Then $f'(\theta) = a(1 - \cos\theta)$, and using symmetry,

$$A = 2\int_0^\pi a^2(1-\cos\theta)^2 d\theta = 2a^2 \int_0^\pi (1 - 2\cos\theta + \cos^2\theta) d\theta$$

$$= 2a^2 \int_0^\pi \left(1 - 2\theta + \frac{1}{2} + \frac{1}{2}\cos 2\theta\right) d\theta = 2a^2 \left(\frac{3}{2}\theta - 2\sin\theta + \frac{1}{4}\sin 2\theta\right)\Big|_0^\pi$$

$$= 2a^2 \left(\frac{3}{2}\pi\right) = 3(\pi a^2).$$

Thus, the area under an arch of the cycloid is three times the area of the circle.

10.4 Polar Coordinate System

1.

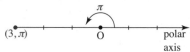

3.

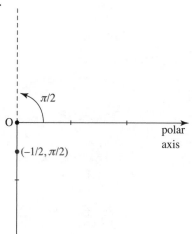

5.

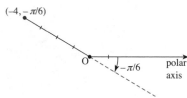

7. **(a)** $\left(2,-\frac{5\pi}{4}\right)$ **(b)** $\left(2,\frac{11\pi}{4}\right)$ **(c)** $\left(-2,\frac{7\pi}{4}\right)$ **(d)** $\left(-2,-\frac{\pi}{4}\right)$

9. **(a)** $\left(4,-\frac{5\pi}{3}\right)$ **(b)** $\left(4,\frac{7\pi}{3}\right)$ **(c)** $\left(-4,\frac{4\pi}{3}\right)$ **(d)** $\left(-4,-\frac{2\pi}{3}\right)$

11. **(a)** $\left(1,-\frac{11\pi}{6}\right)$ **(b)** $\left(1,\frac{13\pi}{6}\right)$ **(c)** $\left(-1,\frac{7\pi}{6}\right)$ **(d)** $\left(-1,-\frac{5\pi}{6}\right)$

13. With $r=1/2$ and $\theta=2\pi/3$ we have $x=1/2\cos 2\pi/3=1/2(-1/2)=-1/4$, $y=1/2\sin 2\pi/3=1/2(\sqrt{3}/2)=\sqrt{3}/4$. The point is $\left(-\dfrac{1}{4},\dfrac{\sqrt{3}}{4}\right)$ in rectangular coordinates.

15. With $r=-6$ and $\theta=-\pi/3$ we have $x=-6\cos(-\pi/3)=-6(1/2)=-3$, $y=-6\sin(-\pi/3)=-6(-\sqrt{3}/2)=3\sqrt{3}$. The point is $(-3,3\sqrt{3})$ in rectangular coordinates.

17. With $r=4$ and $\theta=5\pi/4$ we have $x=4\cos 5\pi/4=4(-\sqrt{2}/2)=-2\sqrt{2}$, $y=4\sin 5\pi/4=4(-\sqrt{2}/2)=-2\sqrt{2}$. The point is $\left(-2\sqrt{2},-2\sqrt{2}\right)$ in rectangular coordinates.

19. With $x=-2$ and $y=-2$ we have $r^2=8$ and $\tan\theta=1$.
    (a) $\left(2\sqrt{2},-\frac{3\pi}{4}\right)$ **(b)** $\left(-2\sqrt{2},\frac{\pi}{4}\right)$

21. With $x=1$ and $y=-\sqrt{3}$ we have $r^2=4$ and $\tan\theta=-\sqrt{3}$.
    (a) $\left(2,-\frac{\pi}{3}\right)$ **(b)** $\left(-2,\frac{3\pi}{3}\right)$

23. With $x=7$ and $y=0$ we have $r^2=49$ and $\tan\theta=0$.
    (a) $(7,0)$ **(b)** $(-7,\pi)$ or $(-7,-\pi)$

25.

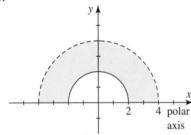

27.

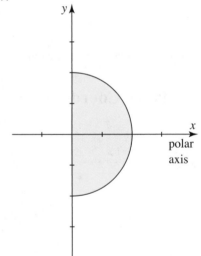

29.

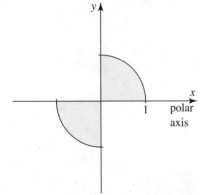

In Problems 31–40, we use $x = r\cos\theta$ and $y = r\sin\theta$.

31. $r\sin\theta = 5$; $r = 5\csc\theta$

33. $r\sin\theta = 7r\cos\theta$; $\tan\theta = 7$; $\theta = \tan^{-1}7$

35. $r^2\sin^2\theta = -4r\cos\theta$; $r^2(1 - \cos^2\theta) + 4r\cos\theta = 0$; $r^2 - (r^2\cos^2\theta - 4r\cos\theta + 4) = 0$; $r^2 - (r\cos\theta - 2)^2 = 0$; $[r - (r\cos\theta - 2)][r + (r\cos\theta) - 2] = 0$
Solving for r, we obtain $r = \dfrac{-2}{(1 - \cos\theta)}$ or $r = \dfrac{2}{(1 + \cos\theta)}$. Since replacement of (r, θ) by $(-r, \theta + \pi)$ in the first equation gives the second equation, we take the polar equation to be $r = \dfrac{2}{1 + \cos\theta}$.

37. $r^2 = 36$. Since $r = -6$ has the same graph as $r = 6$, we take the equation to be $r = 6$.

39. $r^2 + r\cos\theta = \sqrt{r^2} = \pm r$; $r(r + \cos\theta \mp 1) = 0$. Solving for r, we obtain $r = 0$ or $r = \pm 1 - \cos\theta$. Since replacement of (r, θ) in $r = -1 - \cos\theta$ by $(-r, \theta + \pi)$ gives $r = 1 - \cos\theta$, and since $\theta = 0$ gives $r = 0$, we take the polar equation to be $r = 1 - \cos\theta$.

In Problems 41–52, we use $r^2 = x^2 + y^2$, $r\cos\theta = x$, $r\sin\theta = y$, and $\tan\theta = y/x$.

41. $r\cos\theta = 2$; $x = 2$

43. $r = 12\sin\theta\cos\theta$; $r^3 = 12r\sin\theta r\cos\theta$; $(x^2 + y^2)^{3/2} = 12xy$; $(x^2 + y^2)^3 = 144x^2y^2$

45. $r^2 = 8\sin\theta\cos\theta$; $r^4 = 8r\sin\theta r\cos\theta$; $(x^2 + y^2)^2 = 8xy$

47. $r^2 + 5r\sin\theta = 0$; $x^2 + y^2 + 5y = 0$

49. $r + 3r\cos\theta = 2$; $(x^2 + y^2)^{1/2} + 3x = 2$; $(x^2 + y^2)^{1/2} = 2 - 3x$; $x^2 + y^2 = 4 - 12x + 9x^2$; $8x^2 - y^2 - 12x + 4 = 0$

51. $3r\cos\theta + 8r\sin\theta = 5$; $3x + 8y = 5$

53. $\sqrt{(x_2 - x_1)^2 + (y_2 - y_1)^2}$
$= \sqrt{(r_2\cos\theta_2 - r_1\cos\theta_1)^2 + (r_2\sin\theta_2 - r_1\sin\theta_1)^2}$
$= \sqrt{r_2^2\cos^2\theta_2 - 2r_2r_1\cos\theta_2\cos\theta_1 + r_1^2\cos^2\theta_1 + r_2^2\sin^2\theta_2 - 2r_2r_2\sin\theta_2\sin\theta_1 + r_1^2\sin^2\theta_1}$
$= \sqrt{r_2^2 + r_1^2 - 2r_1r_2(\cos\theta_2\cos\theta_1 + \sin\theta_2\sin\theta_1)}$
$= \sqrt{r_2^2 + r_1^2 - 2r_1r_2\cos(\theta_2 - \theta_1)}$

55. Solutions of $f(\theta) = 0$ are θ values at which the graph of $r = f(\theta)$ passes through the origin.

10.5 Graphs of Polar Equations

1.

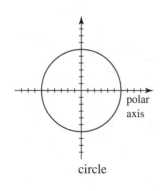

circle

3.

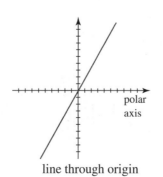

line through origin

5.

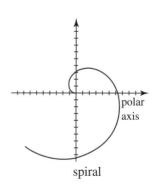

spiral

7.

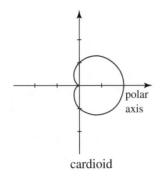

cardioid

9.

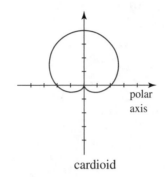

cardioid

11.

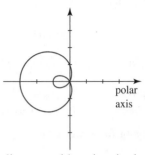

limaçon with an interior loop

13.

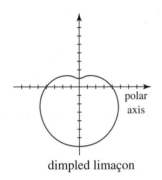

dimpled limaçon

15.

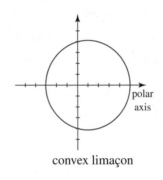

convex limaçon

17.

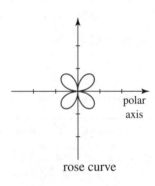

rose curve

19.

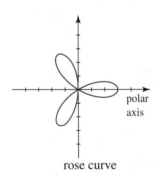

rose curve

21.

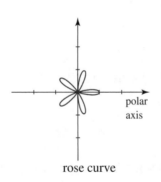

rose curve

23.

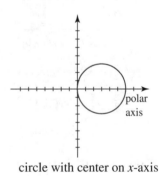

circle with center on x-axis

25.

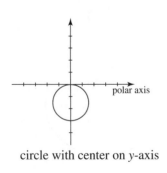

circle with center on y-axis

27.

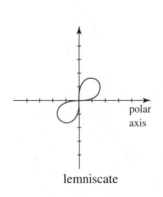

lemniscate

29.

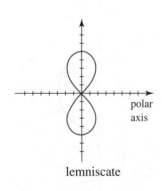

lemniscate

31.

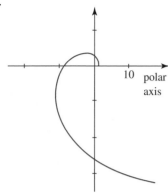

33. $r = 2.5$

35. $r = 4 - 3\cos\theta$ s

37. $r = 2\cos 4\theta$

39. Solving $4\sin\theta = 2$ we have $\sin\theta = 1/2$ and $\theta = \pi/6$ and $5\pi/6$. The points of intersection are $(2, \pi/6)$ and $(2, 5\pi/6)$.

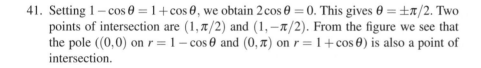

41. Setting $1 - \cos\theta = 1 + \cos\theta$, we obtain $2\cos\theta = 0$. This gives $\theta = \pm\pi/2$. Two points of intersection are $(1, \pi/2)$ and $(1, -\pi/2)$. From the figure we see that the pole $((0,0)$ on $r = 1 - \cos\theta$ and $(0, \pi)$ on $r = 1 + \cos\theta)$ is also a point of intersection.

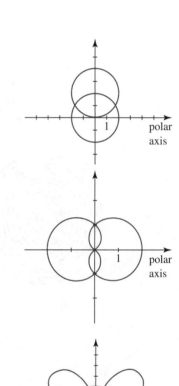

43. Setting $6\sin 2\theta = 3$, we obtain $\sin 2\theta = 1/2$. Then $2\theta = \pi/6, 5\pi/6, 13\pi/6$, and $17\pi/6$. This gives the points of intersection $(3, \pi/12), (3, 5\pi/12), (3, 13\pi/12)$, and $(3, 17\pi/12)$. Writing the second equation in the form $-r = 6\sin 2(\theta + \pi)$, we obtain $r = -6\sin 2\theta$. Setting $-6\sin 2\theta = 3$, we obtain $\sin 2\theta = -1/2$. Then $2\theta = -\pi/6, -5\pi/6, -13\pi/6$, and $-17\pi/6$. This gives the points of intersection $(3, -\pi/12), (3, -5\pi/12), (3, -13\pi/12)$, and $(3, -17\pi/12)$.

45. Setting $4\sin\theta\cos^2\theta = \sin\theta$ we obtain $\sin\theta(4\cos^2\theta - 1) = 0$. This gives $\theta = 0$, $\theta = \pi/3$, and $\theta = 2\pi/3$. The points of intersection are $(0,0), (\sqrt{3}/2, \pi/3)$, and $(\sqrt{3}/2, 2\pi/3)$.

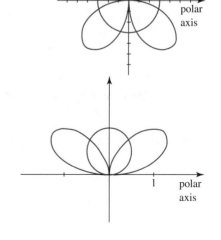

47.

(a)

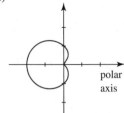

(b)

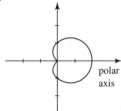

(c)

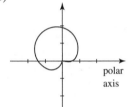

(d)

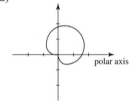

49. (d)

51. (b)

53.

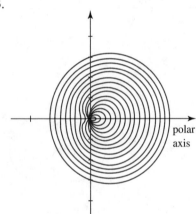

55.

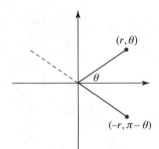

Symmetric with respect to the
x-axis.

57.

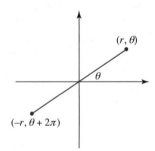

Symmetric with respect to the
origin.

59. Symmetric with respect to the *x*-axis.

61. (a) The graphs are identical.

(b) The graphs are identical.

10.6   Calculus in Polar Coordinates

In this exercise set we make frequent use of the formulas

$$\int \sin^2 \theta \, d\theta = \frac{1}{2}\theta - \frac{1}{4}\sin 2\theta + C, \quad \int \cos^2 \theta \, d\theta = \frac{1}{2}\theta + \frac{1}{4}\sin 2\theta + C.$$

1. $\dfrac{dy}{d\theta} = \theta \cos\theta + \sin\theta$

   $\dfrac{dx}{d\theta} = -\theta \sin\theta + \cos\theta$

   $\dfrac{dy}{dx} = \dfrac{\theta \cos\theta + \sin\theta}{-\theta \sin\theta + \cos\theta} = -\dfrac{2}{\pi}$ at $\theta = \frac{\pi}{2}$.

3. At $\theta = \frac{\pi}{3}$,

   $\dfrac{dy}{d\theta} = (4 - 2\sin\theta)\cos\theta + (-2\cos\theta)\sin\theta$

   $\qquad = 4 - 4\sin\theta\cos\theta = 4 - 4\left(\dfrac{\sqrt{3}}{2}\right)\left(\dfrac{1}{2}\right) = 4 - \sqrt{3}$

   $\dfrac{dx}{d\theta} = -(4 - 2\sin\theta)\sin\theta + (-2\cos\theta)\cos\theta$

   $\qquad = -4 + 2\sin^2\theta - 2\cos^2\theta = -4 + 2\left(\dfrac{3}{4}\right) - 2\left(\dfrac{1}{4}\right)$

   $\qquad = -4 + \dfrac{3}{2} - \dfrac{1}{2} = -3$

   $\dfrac{dy}{dx} = \dfrac{4 - \sqrt{3}}{-3}$

5. At $\theta = \pi/6$,

   $\dfrac{dy}{d\theta} = \sin\theta\cos\theta + \cos\theta\sin\theta$

   $\qquad = 2\cos\theta\sin\theta = 2\left(\dfrac{\sqrt{3}}{2}\right)\left(\dfrac{1}{2}\right) = \dfrac{\sqrt{3}}{2}$

   $\dfrac{dx}{d\theta} = -\sin^2\theta + \cos^2\theta = -\left(\dfrac{1}{4}\right) + \left(\dfrac{3}{4}\right) = \dfrac{1}{2}$

   $\dfrac{dy}{dx} = \sqrt{3}$

7. $\dfrac{dy}{d\theta} = (2 + 2\cos\theta)\cos\theta + (-2\sin\theta)\sin\theta$

   $\qquad = 2(1 + \cos^2\theta - \sin^2\theta)$

   $\dfrac{dx}{d\theta} = -(2 + 2\cos\theta)\sin\theta + (-2\sin\theta)\cos\theta$

   $\qquad = -2 - 4\cos\theta\sin\theta$

   $\dfrac{dy}{dx} = \dfrac{2(1 + \cos^2\theta - \sin^2\theta)}{2(-2 - 2\cos\theta\sin\theta)} = \dfrac{1 + \cos^2\theta - \sin^2\theta}{-1 - 2\cos\theta\sin\theta}$

   If the tangent line is horizontal, we must have

   $$1 + \cos^2\theta - \sin^2\theta = 0$$

   which requires $\sin\theta = \pm 1$ and thus $\theta = \frac{\pi}{2}$ or $\theta = \frac{3\pi}{2}$. Hence, the polar coordinates of points on the graph with horizontal tangents are $(2, \pi/2)$ and $(2, 3\pi/2)$. If the tangent line is vertical, we must have

   $$-1 - 2\cos\theta\sin\theta = 0 \quad \text{or} \quad \cos\theta\sin\theta = -\dfrac{1}{2}$$

   which occurs at $\theta = \frac{3\pi}{4}$ or $\theta = \frac{7\pi}{4}$. Hence the polar coordinates of points on the graph with vertical tangents are $(2 - \sqrt{3}, 3\pi/4)$ and $(2 + \sqrt{3}, 7\pi/4)$.

9. $\dfrac{dy}{d\theta} = (4\cos 3\theta)(\cos\theta) + (-12\sin 3\theta)(\sin\theta)$

$\dfrac{dx}{d\theta} = -(4\cos 3\theta)(\sin\theta) + (-12\sin 3\theta)(\cos\theta)$

The points on the graph correspond to $\theta = \frac{\pi}{3}$ and $\theta = \frac{2\pi}{3}$. At $\theta = \frac{\pi}{3}$, we have

$$\frac{dy}{dx} = \frac{(-4)\left(\frac{1}{2}\right) + (0)\left(\frac{\sqrt{3}}{2}\right)}{-(-4)\left(\frac{\sqrt{3}}{2}\right) + (0)\left(\frac{1}{2}\right)}$$

$$= \frac{-2}{2\sqrt{3}} = -\frac{\sqrt{3}}{3}$$

and the rectangular coordinates of the point are $(-2, -2\sqrt{3})$. Hence, the equation of the tangent line is $y = -2\sqrt{3} - \dfrac{\sqrt{3}}{3}(x+2)$.
At $\theta = \frac{2\pi}{3}$, we have

$$\frac{dy}{dx} = \frac{4\left(-\frac{1}{2}\right) + (0)\left(\frac{\sqrt{3}}{2}\right)}{-(4)\left(\frac{\sqrt{3}}{2}\right) + (0)\left(-\frac{1}{2}\right)}$$

$$= \frac{2}{2\sqrt{3}} = \frac{\sqrt{3}}{3}$$

and the rectangular coordinates of the point are $(-2, 2\sqrt{3})$. Hence, the equation of the tangent line is $y = 2\sqrt{3} + \dfrac{\sqrt{3}}{3}(x+2)$.

11. $r = 0$ when $\sin\theta = 0$ which occurs at $\theta = 0$ and $\theta = \pi$. $\dfrac{dr}{d\theta} = -2\cos\theta \neq 0$ at either $\theta = 0$ or $\theta = \pi$. Therefore, $\theta = 0$ and $\theta = \pi$ define tangent lines to the graph at the origin.

13. $r = 0$ when $\sin\theta = -\dfrac{1}{\sqrt{2}} = -\dfrac{\sqrt{2}}{2}$ which occurs at $\theta = \dfrac{5\pi}{4}$ and $\theta = \dfrac{7\pi}{4}$. $\dfrac{dr}{d\theta} = \sqrt{2}\cos\theta \neq 0$ at $\theta = \dfrac{5\pi}{4}$ or $\theta = \dfrac{7\pi}{4}$. Therefore, $\theta = \dfrac{5\pi}{4}$ and $\theta = \dfrac{7\pi}{4}$ define tangent lines to the graph at the origin.

15. $r = 0$ when $\cos 5\theta = 0$ which occurs at $\theta = \frac{\pi}{10}, \frac{3\pi}{10}, \frac{5\pi}{10}, \frac{7\pi}{10}, \frac{9\pi}{10}, \frac{11\pi}{10}, \frac{13\pi}{10}, \frac{15\pi}{10}, \frac{17\pi}{10}$, and $\frac{19\pi}{10}$. $\dfrac{dr}{d\theta} = -5\sin 5\theta \neq 0$ at any of these θ values. Therefore, $\theta = \dfrac{n\pi}{10}$ defines a tangent line to the graph at the origin for $n = 1, 3, 5, \ldots, 19$.

17. $A = \dfrac{1}{2}\displaystyle\int_0^\pi 4\sin^2\theta\, d\theta = \left(\theta - \dfrac{1}{2}\sin 2\theta\right)\Big|_0^\pi = \pi$

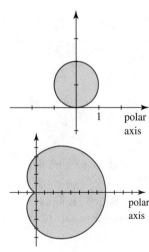

1   polar axis

19. $A = \dfrac{1}{2}\displaystyle\int_0^{2\pi}(4 + 4\cos\theta)^2 d\theta = 8\displaystyle\int_0^{2\pi}(1 + 2\cos\theta + \cos^2\theta)d\theta$

$\quad = (8\theta + 16\sin\theta + 4\theta + 2\sin 2\theta)\big|_0^{2\pi} = 24\pi$

polar axis

21. $A = \dfrac{1}{2} \displaystyle\int_0^{2\pi} (3 + 2\sin\theta)^2 d\theta = \dfrac{1}{2} \displaystyle\int_0^{2\pi} (9 + 12\sin\theta + 4\sin^2\theta) d\theta$

$= \left(\dfrac{9}{2}\theta - 6\cos\theta + \theta - \dfrac{1}{2}\sin 2\theta \right) \Big|_0^{2\pi} = 11\pi$

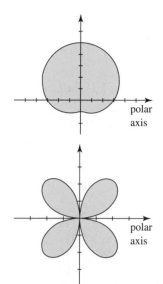

23. $A = \dfrac{1}{2} \displaystyle\int_0^{2\pi} 9\sin^2 2\theta\, d\theta$ $\boxed{u = 2\theta \ \ du = 2d\theta}$

$= \dfrac{9}{4} \displaystyle\int_0^{4\pi} \sin^2 u\, du = \left(\dfrac{9}{8}u - \dfrac{9}{16}\sin 2u \right) \Big|_0^{4\pi} = \dfrac{9}{2}\pi$

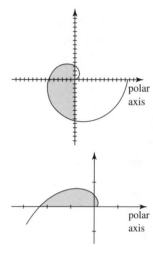

25. $A = \dfrac{1}{2} \displaystyle\int_0^{3\pi/2} 4\theta^2 d\theta = \dfrac{2}{3}\theta^3 \Big|_0^{3\pi/2} = \dfrac{29}{4}\pi^3$

27. $A = \dfrac{1}{2} \displaystyle\int_0^{\pi} e^{2\theta} d\theta = \dfrac{1}{4}e^{2\theta} \Big|_0^{\pi} = \dfrac{1}{4}e^{2\theta} - \dfrac{1}{4}$

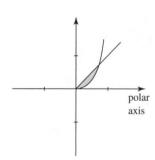

29. $A = \dfrac{1}{2} \displaystyle\int_0^{\pi/4} \tan^2\theta\, d\theta = \dfrac{1}{2} \displaystyle\int_0^{\pi/4} (\sec^2\theta - 1) d\theta$

$= \left(\dfrac{1}{2}\tan\theta - \dfrac{1}{2}\theta \right) \Big|_0^{\pi/4} = \left(\dfrac{1}{2} - \dfrac{\pi}{8} \right) - 0 = \dfrac{4 - \pi}{8}$

31. $A = \dfrac{1}{2} \displaystyle\int_{2\pi/3}^{4\pi/3} (1 + 2\cos\theta)^2 d\theta$

$= \dfrac{1}{2} \displaystyle\int_{2\pi/3}^{4\pi/3} (1 + 4\cos\theta + 4\cos^2\theta) d\theta$

$= \dfrac{1}{2} \left[\theta + 4\sin\theta + 4\left(\dfrac{1}{2}\theta + \dfrac{1}{4}\sin 2\theta \right) \right] \Big|_{2\pi/3}^{4\pi/3}$

$= \dfrac{2\pi - 3\sqrt{3}}{2}$

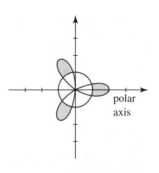

33. Solving $2\cos 3\theta = 1$ in the first quadrant, we obtain $\cos 3\theta = 1/2$, $3\theta = \pi/3$, and $\theta = \pi/9$. Using symmetry,

$$A = 6\left[\int_0^{\pi/9}(4\cos^2 3\theta - 1)d\theta\right] \quad \boxed{u = 3\theta, \ du = 3d\theta}$$

$$= 4\int_0^{\pi/3}(\cos^2 u - 1)du = (2u + \sin 2u - u)\Big|_0^{\pi/3} = \frac{\pi}{3} + \frac{\sqrt{3}}{2}.$$

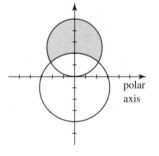

35. Solving $5\sin\theta = 3 - \sin\theta$ in the first quadrant, we obtain $\sin\theta = 1/2$ and $\theta = \pi/6$. Using symmetry,

$$A = 2\left[\frac{1}{2}\int_{\pi/6}^{\pi/2}(25\sin^2\theta - (3-\sin\theta)^2)d\theta\right]$$

$$= \int_{\pi/6}^{\pi/2}(24\sin^2\theta + 6\sin\theta - 9)d\theta$$

$$= (12\theta - 6\sin 2\theta - 6\cos\theta - 9\theta)\Big|_{\pi/6}^{\pi/2}$$

$$= \frac{3\pi}{2} - \left(\frac{\pi}{2} - 3\sqrt{3} - 3\sqrt{3}\right) = \pi + 6\sqrt{3}.$$

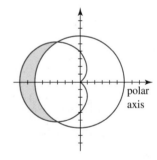

37. Solving $4 - 4\cos\theta = 6$ in the second quadrant, we obtain $\cos\theta = -1/2$ and $\theta = 2\pi/3$. Using symmetry,

$$A = 2\left[\frac{1}{2}\int_{2\pi/3}^{\pi}\left((4-4\cos\theta)^2 - 36\right)\right]d\theta$$

$$= \int_{2\pi/3}^{\pi}(16\cos^2\theta - 32\cos\theta - 20)d\theta$$

$$= (8\theta + 4\sin 2\theta - 32\sin\theta - 20\theta)\Big|_{2\pi/3}^{\pi}$$

$$= -12\pi - (8\pi - 2\sqrt{3} - 16\sqrt{3}) = 18\sqrt{3} - 4\pi.$$

39. $\dfrac{dr}{d\theta} = 0$ so we have $L = \int_0^{2\pi}\sqrt{3^2}d\theta = \int_0^{2\pi} 3d\theta = 6\pi$

41. $\dfrac{dr}{d\theta} = \dfrac{1}{2}e^{\theta/2}; \quad \left(\dfrac{dr}{d\theta}\right)^2 + r^2 = \dfrac{1}{4}e^\theta + e^\theta = \dfrac{5}{4}e^\theta; \quad s = \dfrac{\sqrt{5}}{2}\int_0^4 e^{\theta/2}d\theta = \sqrt{5}e^{\theta/2}\Big|_0^4 = \sqrt{4}(e^2 - 1)$

43. $\dfrac{dr}{d\theta} = 3\sin\theta$ so we have

$$L = \int_0^{2\pi} \sqrt{(3 - 3\cos\theta)^2 + (3\sin\theta)^2}\,d\theta$$

$$= \int_0^{2\pi} \sqrt{9 - 18\cos\theta + 9\cos^2\theta + 9\sin^2\theta}\,d\theta$$

$$= \int_0^{2\pi} \sqrt{18 - 18\cos\theta}\,d\theta$$

$$= \sqrt{18} \int_0^{2\pi} \sqrt{1 - \cos\theta}\,d\theta$$

$$\boxed{\cos^2\left(\frac{\theta}{2}\right) = \frac{1}{2}(1 + \cos\theta) \longrightarrow 1 - \cos\theta = 2\sin^2\left(\frac{\theta}{2}\right)}$$

$$= \sqrt{18} \int_0^{2\pi} \sqrt{2\sin^2\left(\frac{\theta}{2}\right)}\,d\theta$$

$$= 6\int_0^{2\pi} \sin\left(\frac{\theta}{2}\right)d\theta \quad \boxed{u = \frac{\theta}{2},\ du = 2d\theta}$$

$$= 12\int_0^{\pi} \sin u\,du$$

$$= 12\left(-\cos u\right)\big|_0^{\pi}$$

$$= 24$$

45. (a) The lemniscate $r^2 = 9\cos 2\theta$ is only defined for $-\frac{\pi}{4} \le \theta \le \frac{\pi}{4}$ and $\frac{3\pi}{4} \le \theta \le \frac{5\pi}{4}$.

(b) $A = \dfrac{1}{2}\displaystyle\int_{-\pi/4}^{\pi/4} 9\cos 2\theta\,d\theta + \dfrac{1}{2}\displaystyle\int_{3\pi/4}^{5\pi/4} 9\cos 2\theta\,d\theta$

$$\boxed{u = 2\theta,\ du = 2d\theta}$$

$$= \frac{9}{4}\left(\sin u\right)\big|_{-\pi/2}^{\pi/2} + \frac{9}{4}\left(\sin u\right)\big|_{3\pi/2}^{5\pi/2}$$

$$= \frac{9}{2} + \frac{9}{2} = 9$$

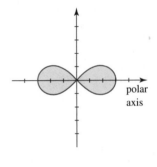

47. To obtain the area, we can compute the area of half of one of the petals and then use symmetry, multiplying by 16.

$$\text{area of a half-petal} = \frac{1}{2}\int_0^{\pi/8} \sin^2 2\theta\,d\theta \quad \boxed{u = 2\theta,\ du = 2d\theta}$$

$$= \frac{1}{4}\int_0^{\pi/4} \sin^2 u\,du$$

$$= \frac{1}{4}\left(\frac{1}{2}u - \frac{1}{4}\sin 2u\right)\Big|_0^{\pi/4}$$

$$= \frac{\pi - 2}{32}$$

$$\text{Total area} = 16 \cdot \left(\frac{\pi - 2}{32}\right) = \frac{\pi - 2}{2}$$

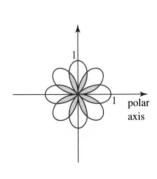

49. No; Let A_1 denote the area of the graph of $r = 2(1 + \cos\theta)$ and let A_2 denote the area of the graph of $r = 1 + \cos\theta$. Then

$$A = \frac{1}{2} \int_0^{2\pi} 4(1 + \cos\theta)^2 d\theta = 4 \left[\frac{1}{2} \int_0^{2\pi} (1 + \cos\theta)^2 d\theta \right]$$

$$= 4A_1$$

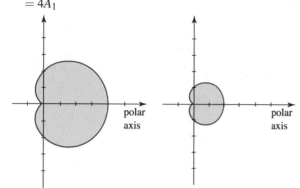

51. From $L = mr^3 d\theta = dt$ we obtain $r^2 d\theta = Ldt/m$. Then

$$A = \frac{1}{2} \int_{\theta_1}^{\theta_2} r^2 d\theta = \frac{1}{2} \int_a^b \frac{L}{m} dt = \frac{L}{2m}(b - a).$$

10.7 Conic Sections in Polar Coordinates

1. Identifying $e = 1$, the graph is a parabola.

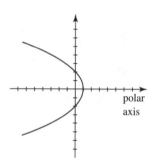

3. Writing $r = \dfrac{15/4}{1 - (1/4)\cos\theta}$, we identify $e = 1/4$. The graph is an ellipse.

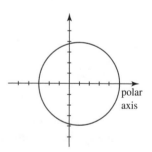

5. Identifying $e = 2$, the graph is a hyperbola.

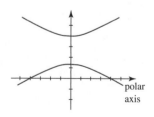

7. Writing $r = \dfrac{6}{1+2\cos\theta}$, we identify $e = 2$. The graph is a hyperbola.

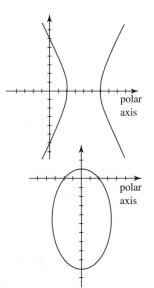

9. Writing $r = \dfrac{2}{1+(4/5)\sin\theta}$, we identify $e = 4/5$. The graph is an ellipse.

11. From $r = \dfrac{6}{1+2\sin\theta}$, we have $e = 2$. Converting to a rectangular equation, we get

$$r = \frac{6}{1+2\sin\theta}$$
$$r+2r\sin\theta = 6$$
$$r = 6-2r\sin\theta$$
$$\sqrt{x^2+y^2} = 6-2y$$
$$x^2+y^2 = 36-24y+4y^2$$
$$\frac{(y-4)^2}{4} - \frac{x^2}{12} = 1$$

with $a = 2$ and $b = \sqrt{12}$ so $c^2 = a^2 + b^2 = 4+12 = 16$ so $c = 4$. Thus $e = \dfrac{c}{a} = \dfrac{4}{2} = 2$.

13. From $r = \dfrac{12}{3-2\cos\theta} = \dfrac{4}{1-\frac{2}{3}\cos\theta}$, we have $e = \frac{2}{3}$. Converting to a rectangular equation, we get

$$r = \frac{12}{3-2\cos\theta}$$
$$3r = 12+2r\cos\theta$$
$$3\sqrt{x^2+y^2} = 12+2x$$
$$\frac{\left(x-\frac{24}{5}\right)^2}{\frac{1296}{25}} + \frac{y^2}{\frac{144}{5}} = 1$$

with $a = \frac{36}{5}$ and $b = \frac{12}{\sqrt{5}}$ so $c^2 = a^2 - b^2 = \frac{1296}{25} - \frac{144}{5} = \frac{576}{25}$. Thus $c = \frac{24}{5}$ so $e = \frac{c}{a} = \frac{25/4}{36/5} = \frac{2}{3}$.

15. Since $e = 1$, the conic is a parabola. The directrix is 3 units to the right of the focus and perpendicular to the x-axis. Therefore, $r = \dfrac{3}{1+\cos\theta}$.

17. Since $e = \frac{2}{3}$, the conic is an ellipse. The directrix is 2 units below the focus and parallel to the x-axis. Therefore, $r = \dfrac{\frac{4}{3}}{1-\frac{2}{3}\sin\theta}$.

19. Since $e = 2$, the conic is a hyperbola. The directrix is 6 units to the right of the focus and perpendicular to the x-axis. Therefore, $r = \dfrac{12}{1+2\cos\theta}$.

21. $r = \dfrac{3}{1+\cos\left(\theta+\frac{2\pi}{3}\right)}$

23. Since the vertex is $\frac{3}{2}$ units below the focus, the directrix must be 3 units below the focus and parallel to the x-axis. Therefore,
$$r = \frac{3}{1 - \sin\theta}.$$

25. Since the vertex is $\frac{1}{2}$ units to the left of the focus, the directrix must be 1 units to the left of the focus and perpendicular to the x-axis. Therefore, $r = \dfrac{1}{1 - \cos\theta}.$

27. Since the vertex is $\frac{1}{4}$ units below the focus, the directrix must be $\frac{1}{2}$ units below the focus and parallel to the x−axis. Therefore,
$$r = \frac{1/2}{1 - \sin\theta}.$$

29. This is the parabola $r = \dfrac{4}{1 + \cos\theta}$ rotated counterclockwise by $\pi/4$. The original parabola had its focus at the origin and its directrix at $x = 4$. The original vertex therefore had polar coordinates $(2, 0)$. After rotation, the vertex is located at $(2, \pi/4)$.

31. This is the ellipse $r = \dfrac{10}{2 - \sin\theta} = \dfrac{5}{2 - \frac{1}{2}\sin\theta}$ rotated clockwise by $\pi/6$. The original ellipse had vertices at $\theta = \pi/2$ and $\theta = 3\pi/2$. The polar coorinates of the vertices were $(10, \pi/2)$ and $(10.3, 3\pi/2)$. After rotation, the vertices are located at $(10.\pi/3)$ and $(10/3, 4\pi/3)$.

33. Identifying $r_a = 12,000$ and $e = 0.2$, we have from (7) in the text $0.2 = \dfrac{12,000 - r_p}{12,000 + r_p}$. Solving for r_p, we obtain $r_p = 8,000$ km.

35. The equation of the orbit is $r = \dfrac{ep}{(1 - e\cos\theta)}$. From (7) in the text,

$$e = \frac{1.5 \times 10^8 - 1.47 \times 10^8}{1.52 \times 10^8 + 1.47 \times 10^8} = \frac{5}{299} \approx 1.67 \times 10^{-2}.$$

When $\theta = 0$, $r = r_a = 1.52 \times 10^8 = \dfrac{ep}{(1 - 1.67 \times 10^{-2})}$. Thus $ep \approx 1.52 \times 10^8 - 2.52 \times 10^6 \approx 1.49 \times 10^8$ and the equation of the orbit is $r = \dfrac{(1.49 \times 10^8)}{(1 - 1.67 \times 10^{-2}\cos\theta)}.$

37.

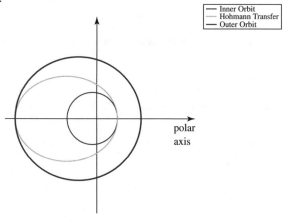

39.

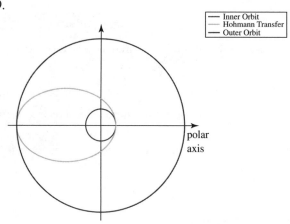

41.

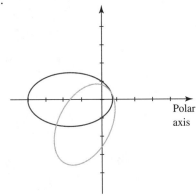

Polar
axis

43. Continuing the development in the paragraph following **(3)** in the text, we see that $r = e(d + r\cos\theta)$ yields $(1 - e^2)x^2 - 2e^2dx + y^2 = e^2d^2$ which in turn gives

$$x^2 - \frac{2e^2d}{1-e^2}x + \frac{y^2}{1-e^2} = \frac{e^2d^2}{1-e^2}$$

$$x^2 - \frac{2e^2d}{1-e^2}x + \left(\frac{e^2d}{1-e^2}\right)^2 + \frac{y^2}{1-e^2} = \frac{e^2d^2}{1-e^2} + \left(\frac{e^2d}{1-e^2}\right)^2$$

$$\left(x - \frac{e^2d}{1-e^2}\right)^2 + \frac{y^2}{1-e^2} = \frac{e^2d^2(1-e^2)}{(1-e^2)^2} + \frac{e^4d^2}{(1-e^2)^2} = \frac{1}{(1-e^2)^2}$$

$$\frac{\left(x - \frac{e^2d}{1-e^2}\right)^2}{\left[\frac{1}{(1-e^2)^2}\right]} + \frac{y^2}{\left[\frac{1}{1-e^2}\right]} = 1,$$

When $0 < e < 1$, both the denominators are positive and the denominator of the fraction involving the x term is smaller. Therefore, this is in the standard form for an ellipse with center and foci on the x-axis. When $e > 1$, the first denominator is positive while the second is negative. Therefore, this is in the standard form for a hyperbola with center and foci on the x-axis.

Chapter 10 in Review

A. True/False

1. True

3. True

5. True

7. False; $(-r, \theta)$ and $(r, \theta + \pi)$ are the same point.

9. True; solving $x = t^2 + t - 12 = (t-3)(t+4) = 0$ we obtain $t = 3$ and $t = 4$. Since $3^3 - 7(3) = 27 - 21 = 6$, the graph intersects the y-axis at $(0, 6)$.

11. True

13. False; the same point can be expressed as $(-4, \pi/2)$, which does satisfy the equation.

15. True

17. True

19. False; if $r < 0$, the point (r, θ) is in the same quadrant as the terminal side of $\theta + \pi$.

21. True

23. True

25. False

B. Fill In the Blanks

1. $4p = 1/2$, $p = 1/8$. The focus is $(0, 1/8)$.

3. The center is $(0, 2)$.

5. $4p = 8$, $p = 2$. The directrix is $y = -3 - 2 = -5$.

7. Completing the square, $y + 10 = (x+2)^2$. The vertex is $(-2, -10)$.

9. $a^2 = 4$. The endpoints of the transverse axis are the vertices $(4 \pm 2, -1)$ or $(2, -1)$ and $(6, -1)$.

11. Completing the square, we have $25(x^2 - 8x + 16) + (y^2 + 6y + 9) = -384 + 409 = 25$ or $(x - 4)^2 + (y+3)^2/25 = 1$. The center of the ellipse is at $(4, -3)$.

13. Setting $x = 0$ and solving, we have $y^2 - 4 = 1$, $y^2 = 5$, or $y = \pm\sqrt{5}$. The y-intercepts are $\pm\sqrt{5}$.

15. line

17. circle

19. $r = 0$ at $\theta = 0$, $\pi/3$, $2\pi/3$, π, $4\pi/3$, and $5\pi/3$. $\dfrac{dr}{d\theta} = 3\cos 3\theta \neq 0$ at any of the θ values mentioned. Thus, the polar equation $\theta = \dfrac{n\pi}{3}$ defines a tangnet to the graph at the origin for $n = 0, \dots, 5$.

21. The focus is the origin. The directrix is 10 units below the origin and parallel to the x-axis. Therefore, the vertex is 5 units below the origin at $(0, -5)$.

C. Exercises

1. $\dfrac{dy}{dx} = \dfrac{\sin t}{1 - \cos t}$. At $t = \pi/2$, $\dfrac{dy}{dx} = \dfrac{(\sqrt{3}/2)}{(1/2)} = \sqrt{3}$. The slope of the normal line is $\frac{-1}{\sqrt{3}}$ and its equation is

$$y - \frac{1}{2} = -\frac{1}{\sqrt{3}}\left[x - \left(\frac{\pi}{3} - \frac{\sqrt{3}}{2}\right)\right] \quad \text{or} \quad y = -\frac{1}{\sqrt{3}}x + \frac{\pi}{3\sqrt{3}} = -\frac{\sqrt{3}}{3}x + \frac{\sqrt{3}\pi}{9}.$$

3. The slope of the line $6x + y = 8$ is -6 and $\dfrac{dy}{dx} = \dfrac{(3t^2 - 18t)}{2t} = \frac{3t}{2} - 9$, $t \neq 0$. Solving $\frac{3t}{2} - 9 = -6$ we obtain $t = 2$. Since $x(2) = 8$ and $y(2) = -26$, the point on the graph is $(8, -26)$.

5. (a) Since $4x^2(1 - x^2) = y^2 \geq 0$, $1 - x^2 \geq 0$, $x^2 \leq 1$, and $|x| \leq 1$.

   (b) Letting $x = \sin t$, we have $y^2 = 4\sin^2 t(1 - \sin^2 t) = 4\sin^2 t \cos^2 t = \sin^2 2t$. Parametric equations for the curve are $x = \sin t$, $y = \sin 2t$, for $0 \leq t \leq 2\pi$.

   (c) $\dfrac{dy}{dx} = \dfrac{(2\cos 2t)}{\cos t}$. The tangent line is horizontal when $\cos 2t = 0$ or $t = \pi/4$, $3\pi/4$, $5\pi/4$, $7\pi/4$. The points on the graph are $(\sqrt{2}/2, 1)$, $(\sqrt{2}/2, -1)$, $(-\sqrt{2}/2, 1)$ and $(-\sqrt{2}/2, -1)$.

   (d)

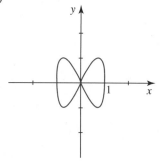

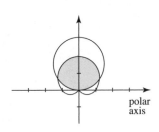

polar
axis

7. Solving $3\sin\theta = 1 + \sin\theta$ in the first quadrant, we have $\sin\theta = 1/2$ and $\theta = \pi/6$. Using symmetry,

$$A = 2\left[\frac{1}{2}\int_0^{\pi/6} 9\sin^2\theta\, d\theta + \frac{1}{2}\int_{\pi/6}^{\pi/2} (1+\sin\theta)^2 d\theta\right]$$

$$= \left(\frac{9}{2}\theta - \frac{9}{4}\sin 2\theta\right)\Bigg|_0^{\pi/6} + \left(\theta - 2\cos\theta + \frac{1}{2}\theta - \frac{1}{4}\sin 2\theta\right)\Bigg|_{\pi/6}^{\pi/2}$$

$$= \left(\frac{3\pi}{4} - \frac{9\sqrt{3}}{8}\right) + \left[\frac{3\pi}{4}\left(\frac{\pi}{4} - \sqrt{3} - \frac{\sqrt{3}}{8}\right)\right] = \frac{5\pi}{4}.$$

9. From $x = 2\sin 2\theta\cos\theta$, $y = 2\sin 2\theta\sin\theta$, we find

$$\frac{dy}{dx} = \frac{2\sin 2\theta\cos\theta + 4\cos 2\theta\sin\theta}{-2\sin 2\theta\sin\theta + 4\cos 2\theta\cos\theta} \quad\text{and}\quad \frac{dy}{dx}\Bigg|_{\theta=\pi/4} = \frac{2(1)\left(\dfrac{\sqrt{2}}{2}\right) + 4(0)\left(\dfrac{\sqrt{2}}{2}\right)}{-2(1)\left(\dfrac{\sqrt{2}}{2}\right) + 4(0)\left(\dfrac{\sqrt{2}}{2}\right)} = -1.$$

(a) At $\theta = \pi/4$, $x = \sqrt{2}$ and $y = \sqrt{2}$. The Cartesian equation of the tangent line is $y - \sqrt{2} = -1(x - \sqrt{2})$ or $y = -x + 2\sqrt{2}$.

(b) Using $x = r\cos\theta$ and $y = r\sin\theta$ in (a), we obtain $r\sin\theta = -r\cos\theta = 2\sqrt{2}$ or $r = \dfrac{2\sqrt{2}}{\sin\theta + \cos\theta}$.

11. Multiplying both sides of the equation by r, we have $r^2 = r\cos\theta + r\sin\theta$. The corresponding Cartesian equation is $x^2 + y^2 = x + y$.

13. $2(r\cos\theta)(r\sin\theta) = 5$

$$r^2 = \frac{5}{2\cos\theta\sin\theta}$$

15. By writing the equation as $r\cos\theta = -1$ we see that the line is $x = -2$. The curve is then a parabola with axis along the x-axis, directrix $x = -1$, and focus at the origin. Since the vertex is at $(-1/2, 0)$, $p = 1$ and the equation is $r = \dfrac{1}{(1 - \cos\theta)}$.

17. $r = 3\sin(10\theta)$

19. The form of the equation is $\dfrac{y^2}{100} - \dfrac{x^2}{b^2} = 1$. The asymptotes for the hyperbola are $\dfrac{y^2}{100} - \dfrac{x^2}{b^2} = 0$ or $by = \pm 10x$. Since the given asymptotes are $3y = \pm 5x$, we have the proportion $\dfrac{b}{3} = \dfrac{10}{5}$. Thus, $b = 6$ and the equation of the hyperbola is $\dfrac{y^2}{100} - \dfrac{x^2}{36} = 1$.

21. Substituting $y = tx$ into $x^3 + y^3 = 3axy$ we obtain

$$x^3 + y^3x = 3atx^2 \implies (1 + t^3)x = 3at \implies x = \frac{3at}{1 + t^3},$$

and $y = tx = \dfrac{3at^2}{(1 + t^3)}$.

23. (a) From $x = r\cos\theta$ and $y = r\sin\theta$ we obtain $r^3(\cos^3\theta + \sin^3\theta) = 3ar^2\cos\theta\sin\theta$ or $r = \dfrac{(3a\cos\theta\sin\theta)}{(\cos^3\theta + \sin^3\theta)}$.

(b) The loop is formed from $\theta = 0$ to $\theta = \pi/2$, so the area is

$$A = \frac{1}{2} \int_0^{\pi/2} \frac{9a^2 \cos^2 \theta \sin^2 \theta}{(\cos^3 \theta + \sin^\theta)} d\theta = \frac{9a^2}{2} \int_0^{\pi/2} \frac{\cos^2 \theta \sin^2 \theta}{(1 + \tan^3 \theta)^2 \cos^6 \theta} d\theta$$

$$= \frac{9a^2}{2} \int_0^{\pi/2} \frac{\sin^2 \theta}{(1 + \tan^3 \theta)^2 \cos^4 \theta} d\theta = \frac{9a^2}{2} \int_0^{\pi/2} \frac{\tan^2 \theta}{(1 + \tan^3 \theta) \cos^2 \theta} d\theta$$

$$= \frac{9a^2}{2} \int_0^{\pi/2} \frac{\tan^2 \theta \sec^3 \theta}{(1 + \tan^3 \theta)\theta} d\theta \quad \boxed{u = \tan \theta, \ du = \sec^2 \theta d\theta}$$

$$= \frac{9a^2}{2} \int_0^{\infty} \frac{u^2}{(1 + u^3)^2} du = \frac{9a^2}{2} \left[-\frac{1}{3} \left(\frac{1}{1 + u^3} \right) \right] \Big|_0^{\infty} = -\frac{3a^2}{2} \lim_{t \to \infty} \left(\frac{1}{1 + t^3} - 1 \right) = \frac{3a^2}{2}.$$

25. Using symmetry

$$A = 2 \left(\frac{1}{2} \right) \int_0^{\pi/2} \left(2 \sin \frac{\theta}{3} \right)^2 d\theta = 4 \int_0^{\pi/2} \frac{1}{2} \left(1 - \cos \frac{2\theta}{3} \right) d\theta$$

$$= 2 \left(\theta - \frac{3}{2} \sin \frac{2\theta}{3} \right) \Big|_0^{\pi/2} = 2 \left(\frac{\pi}{2} - \frac{3}{2} \frac{\sqrt{3}}{2} \right) = \pi - \frac{3\sqrt{3}}{2}.$$

27. (a) $r = 2 \cos \left(\theta - \frac{\pi}{4} \right)$

(b) Note that $r = 2 \cos \theta$ defines a circle of radius 1 centered at $(1, 0)$. A rotation of $\pi/4$ puts the center at $\left(\frac{\sqrt{2}}{2}, \frac{\sqrt{2}}{2} \right)$. The new rectangular equation is therefore $\left(x - \frac{\sqrt{2}}{2} \right)^2 + \left(y - \frac{\sqrt{2}}{2} \right)^2 = 1$

29. Taking the center of the ellipse to be at the origin, we have $a = 5 \times 10^8$ and $b = 3 \times 10^8$, since $c^2 = a^2 - b^2$, $c^2 = 1.6 \times 10^{17}$ and $c = 4 \times 10^8$. The minimum distance is $a - c = 10^8$m and the maximum distance is $a + c = 9 \times 10^8$m.

Vectors and 3-Space

11.1 Vectors in 2-Space

1. (a) $6\mathbf{i} + 12\mathbf{j}$ (b) $\mathbf{i} + 8\mathbf{j}$ (c) $3\mathbf{i}$ (d) $\sqrt{65}$ (e) 3

3. (a) $\langle 12, 0 \rangle$ (b) $\langle 4, -5 \rangle$ (c) $\langle 4, 5 \rangle$ (d) $\sqrt{41}$ (e) $\sqrt{41}$

5. (a) $-9\mathbf{i} + 6\mathbf{j}$ (b) $-3\mathbf{i} + 9\mathbf{j}$ (c) $-3\mathbf{i} - 5\mathbf{j}$ (d) $3\sqrt{10}$ (e) $\sqrt{34}$

7. (a) $-6\mathbf{i} + 27\mathbf{i}$ (b) $\mathbf{0}$ (c) $-4\mathbf{i} + 18\mathbf{j}$ (d) 0 (e) $2\sqrt{85}$

9. (a) $\langle 4, -12 \rangle - \langle -2, 2 \rangle = \langle 6, -14 \rangle$ (b) $\langle -3, 9 \rangle - \langle -5, 5 \rangle = \langle 2, 4 \rangle$

11. (a) $(4\mathbf{i} - 4\mathbf{j}) - (-6\mathbf{i} + 8\mathbf{j}) = 10\mathbf{i} - 12\mathbf{j}$ (b) $(-3\mathbf{i} + 3\mathbf{j}) - (-15\mathbf{i} + 20\mathbf{j}) = 12\mathbf{i} - 17\mathbf{j}$

13. (a) $\langle 16, 40 \rangle - \langle -4, -12 \rangle = \langle 20, 52 \rangle$ (b) $\langle -12, -30 \rangle - \langle -10, -30 \rangle = \langle -2, 0 \rangle$

15.

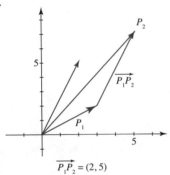

$\overrightarrow{P_1P_2} = (2, 5)$

17.

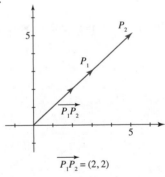

$\overrightarrow{P_1P_2} = (2, 2)$

19. Since $\overrightarrow{P_1P_2} = \overrightarrow{OP_2} - \overrightarrow{OP_1}$, $\overrightarrow{OP_2} = \overrightarrow{P_1P_2} + \overrightarrow{OP_1} = (4\mathbf{i} + 8\mathbf{j}) + (-3\mathbf{i} + 10\mathbf{j}) = \mathbf{i} + 18\mathbf{j}$, and the terminal point is $(1, 18)$

21. $a(= -\mathbf{a})$, $b = (-\frac{1}{4}\mathbf{a})$, $c(= \frac{5}{2}\mathbf{a})$, $e(= 2\mathbf{a})$, and $f(= -\frac{1}{2}\mathbf{a})$ are parallel to $\mathbf{a}$.

23. $\langle 6, 15 \rangle$

25. $|\mathbf{a}| = \sqrt{4 + 4} = 2\sqrt{2};$ (a) $\mathbf{u} = \dfrac{1}{2\sqrt{2}}\langle 2, 2 \rangle = \left\langle \dfrac{1}{\sqrt{2}}, \dfrac{1}{\sqrt{2}} \right\rangle;$ (b) $-\mathbf{u} = \left\langle -\dfrac{1}{\sqrt{2}}, -\dfrac{1}{\sqrt{2}} \right\rangle$

27. $|\mathbf{a}| = 5;$ (a) $\mathbf{u} = \dfrac{1}{5}\langle 0, -5 \rangle = \langle 0, -1 \rangle;$ (b) $-\mathbf{u} = \langle 0, 1 \rangle$

29. $|\mathbf{a} + \mathbf{b}| = |\langle 5, 12 \rangle| = \sqrt{25 + 144} = 13;$ $\mathbf{u} = \frac{1}{13}\langle 5, 12 \rangle = \left\langle \frac{5}{13}, \frac{12}{13} \right\rangle$

31. $|\mathbf{a}| = \sqrt{9 + 49} = \sqrt{58};$ $\mathbf{b} = 2\left((\frac{1}{\sqrt{58}})3\mathbf{i} + 7\mathbf{j} \right) = \frac{6}{\sqrt{58}}\mathbf{i} + \frac{14}{\sqrt{58}}\mathbf{j}$

33. $-\frac{3}{4}\mathbf{a} = \langle -3 - 15/2 \rangle$

35.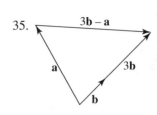

37. $\mathbf{x} = -(\mathbf{a}+\mathbf{b}) = -\mathbf{a} - \mathbf{b}$

39.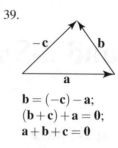

$\mathbf{b} = (-\mathbf{c}) - \mathbf{a};$
$(\mathbf{b}+\mathbf{c}) + \mathbf{a} = \mathbf{0};$
$\mathbf{a} + \mathbf{b} + \mathbf{c} = \mathbf{0}$

41. From $2\mathbf{i} + 3\mathbf{j} = k_1\mathbf{b} + k_2\mathbf{c} = k_1(\mathbf{i}+\mathbf{j}) + k_2(\mathbf{i}-\mathbf{j}) = (k_1+k_2)\mathbf{i} + (k_1-k_2)\mathbf{j}$ we obtain the system of equations $k_1 + k_2 = 2,\ k_1 - k_2 = 3$. Solving, we find $k_1 = \frac{5}{2}$ and $k_2 = -\frac{1}{2}$. Then $\mathbf{a} = \frac{5}{2}\mathbf{b} - \frac{1}{2}\mathbf{c}$.

43. From $y' = \frac{1}{2}x$ we see that the slope of the tangent line at $(2,2)$ is 1. A vector with slope 1 is $\mathbf{i}+\mathbf{j}$. A unit vector is $(\mathbf{i}+\mathbf{j})/|\mathbf{i}+\mathbf{j}| = (\mathbf{i}+\mathbf{j})/\sqrt{2} = \frac{1}{\sqrt{2}}\mathbf{i} + \frac{1}{\sqrt{2}}\mathbf{j}$. Another unit vector tangent to the curve is $-\frac{1}{\sqrt{2}}\mathbf{i} - \frac{1}{\sqrt{2}}\mathbf{j}$.

45.   (a) Since the shortest distance between two point is a straight line,
  $|\mathbf{a}+\mathbf{b}| \le |\mathbf{a}| + |\mathbf{b}|$.

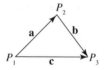

  (b) When P_2 lies on the line segment between P_1 and P_3, $|\mathbf{a}+\mathbf{b}| = |\mathbf{a}| + |\mathbf{b}|$.

47.   (a) Since $\mathbf{F}_f = -\mathbf{F}_g$, $|\mathbf{F}_g| = |\mathbf{F}_f| = \mu|\mathbf{F}_n|$ and $\tan\theta = |\mathbf{F}_g|/|\mathbf{F}_n| = \mu|\mathbf{F}_n|/|\mathbf{F}_n| = \mu$

  (b) $\theta = \arctan 0.6 \approx 31°$

49. Since $\mathbf{F}_2 = 200(\mathbf{i}+\mathbf{j})/\sqrt{2} = 100\sqrt{2}\mathbf{i} + 100\sqrt{2}\mathbf{j}$, $\mathbf{F}_3) = \mathbf{F}_2\mathbf{F}_1 = (100\sqrt{2} - 200)\mathbf{i} + 100\sqrt{2}\mathbf{j}$ and $|\mathbf{F}_3| = \sqrt{(100\sqrt{2} - 200)^2 + (100\sqrt{2})^2}$ $200\sqrt{2 - \sqrt{2}} \approx 153$ lb.

51.

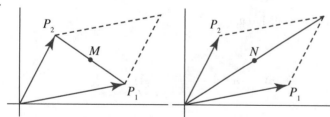

Place one corner of the parallelogram at the origin, and let two adjacent sides be $\overrightarrow{OP_1}$ and $\overrightarrow{OP_2}$. Let M be the midpoint of the diagonal connecting P_1 and P_2 and N be the midpoint of the other diagonal. By Problem 37, $\overrightarrow{OM} = \frac{1}{2}(\overrightarrow{OP_1} + \overrightarrow{OP_2})$. Since $\overrightarrow{OP_1} + \overrightarrow{OP_2}$ is the main diagonal of the parallelogram and N is its midpoint, $\overrightarrow{ON} = \frac{1}{2}(\overrightarrow{OP_1} + \overrightarrow{OP_2})$. Thus, $\overrightarrow{OM} = \overrightarrow{ON}$ and the diagonals bisect each other.

11.2   3-Space and Vectors

1–6.

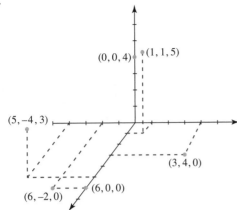

7. A plane is perpendicular to the z-axis, 5 units above the xy-plane.

9. A line perpendicular to the xy-plane at $(2,3,0)$. 0

11. $(2,0,0)$, $(2,5,0)$, $(2,0,8)$, $(2,5,8)$, $(0,5,0)$, $(0,5,8)$, $(0,0,8)$, $(0,0,0)$

13. (a) xy-plane: $(-2,5,0)$, xz-plane: $(-2,0,4)$, yz-plane: $(0,5,4)$;

(b) $(-2,5,-2)$

(c) Since the shortest distance between a point and a plane is a perpendicular line, the point in the plane $x = 3$ is $(3,5,4)$.

15. The union of the planes $x = 0$, $y = 0$, and $z = 0$.

17. The point $(-1,2,-3)$.

19. The union of the planes $z = 5$ and $z = -5$.

21. $d = \sqrt{(3-6)^2 + (-1-4)^2 + (2-8)^2} = \sqrt{70}$

23. (a)   7;   (b) $d = \sqrt{(-3)^2 + (-4)^2} = 5$

25. $d(P_1,P_2) = \sqrt{(3^2 + 6^2 + (-6)^2} = 9$; $d(P_1,P_3) = \sqrt{2^2 + 1^2 + 2^2} = 3$
$d(P_2,P_3) = \sqrt{(2-3)^2 + (1-6)^2 + (2-(-6))^2} = \sqrt{90}$. The triangle is a right triangle.

27. $d(P_1,P_2) = \sqrt{(4-1)^2 + (1-2)^2 + (3-3)^2} = \sqrt{10}$
$d(P_1,P_3) = \sqrt{(4-1)^2 + (6-2)^2 + (4-3)^2} = \sqrt{26}$
$d(P_2,P_3) = \sqrt{(4-4)^2 + (6-1)^2 + (4-3)^2} = \sqrt{26}$; The triangle is an isosceles triangle.

29. $d(P_1,P_2) = \sqrt{(-2-1)^2 + (-2-2)^2 + (-3-0)^2} = \sqrt{34}$
$d(P_1,P_3) = \sqrt{(7-1)^2 + (10-2)^2 + (6-0)^2} = 2\sqrt{34}$
$d(P_2,P_3) = \sqrt{(7-(-2))^2 + (10-(-2))^2 + (6-(-3))^2} = 3\sqrt{34}$
Since $d(P_1,P_2) + d(P_1,P_3) = d(P_2,P_3)$, the points P_1, P_2, and P_3 are collinear.

31. $d(P_1,P_2) = \sqrt{((-4)-1)^2 + ((-3)-0)^2 + (5-4)^2} = \sqrt{35}$
$d(P_1,P_3) = \sqrt{((-7)-1)^2 + ((-4)-0)^2 + (8-4)^2} = \sqrt{96}$
$d(P_2,P_3) = \sqrt{((-7)-(-4))^2 + ((-4)-(-3))^2 + (8-5)^2} = \sqrt{19}$
Since adding any two of the above distances will not result in the third, the points cannot be collinear.

33. $\sqrt{(2-x)^2 + (1-2)^2 + (1-3)^2} = \sqrt{21} \longrightarrow x^2 - 4x + 9 = 21 \longrightarrow x^2 - 4x + 4 = 16 \longrightarrow (x-2)^2 = 16 \longrightarrow x = 2 + -4$ or $x = 6, -2$

35. $\left(\dfrac{1+7}{2}, \dfrac{3+(-2)}{2}, \dfrac{1/2+5/2}{2}\right) = (4, 1/2, 3/2)$

37. $(x_1+2)/2 = -1$, $x_1 = -4$, $(y_1+3)/2 = -4$, $y_1 = -11$; $(z_1+6)/2 = 8$, $z_1 = 10$
The coordinates of P_1 are $(-4,-11,10)$.

39. $\overrightarrow{P_1P_2} = \langle -3,-6,1 \rangle$

41. $\overrightarrow{P_1P_2} = \langle 2,1,1 \rangle$

43.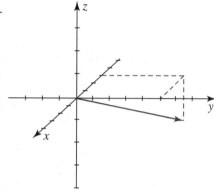

45.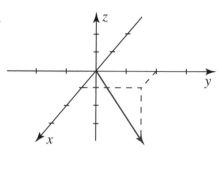

47. Since the **k** component is zero, while the **i** and **j** components are nonzero, the vector lies in the xy-plane.

49. Since the vector is a scalar multiple of **k**, the vector lies on the z-axis.

51. $\mathbf{a} + (\mathbf{b}+\mathbf{c}) = \langle 2,4,12 \rangle$

53. $\mathbf{b} + 2(\mathbf{a}-3\mathbf{c}) = \langle -1,1,1 \rangle + 2\langle -5,-21,-25 \rangle = \langle -11,-41,-49 \rangle$

55. $|\mathbf{a}+\mathbf{c}| = |\langle 3,3,11 \rangle| = \sqrt{9+9+121} = \sqrt{139}$

57. $\left| \dfrac{a}{|a|} \right| + 5 \left| \dfrac{\mathbf{b}}{|\mathbf{b}|} \right| = \dfrac{1}{|\mathbf{b}|}|\mathbf{b}| = 1+5 = 6$

59. $|\mathbf{a}| = \sqrt{100+25+100} = 15$; $\mathbf{u} = -\dfrac{1}{15}\langle 10,-5,10 \rangle = \langle -2/3, 1/3, -2/3 \rangle$

61. $\mathbf{b} = 4\mathbf{a} = 4\mathbf{i} - 4\mathbf{j} + 4\mathbf{k}$

63.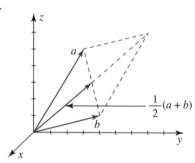

65. $x_P = 1$, $y_P = \cos 30° + \sin 30° = \dfrac{\sqrt{3}}{2} + \dfrac{1}{2}(\sqrt{3}+1)$,

$x_P = -\sin 30° + \cos 30° = -\dfrac{1}{2} + \dfrac{\sqrt{3}}{2} = \dfrac{1}{2}(\sqrt{3}-1)$,

$x_R = \cos 45° - \dfrac{1}{2}(\sqrt{3}-1)\sin 45° = \dfrac{\sqrt{2}}{2} - \dfrac{1}{2}(\sqrt{3}-1)\dfrac{\sqrt{2}}{2} = \dfrac{1}{4}(3\sqrt{2}-\sqrt{6})$, $y_r = \dfrac{1}{2}(\sqrt{3}+1)$,

$z_r = \sin 45° + \dfrac{1}{2}(\sqrt{3}-1)\cos 45° = \dfrac{\sqrt{2}}{2} + \dfrac{1}{2}(\sqrt{3}-1)\dfrac{\sqrt{2}}{2} = \dfrac{1}{4}(\sqrt{2}+\sqrt{6})$,

$x_S = \dfrac{1}{4}(3\sqrt{2}-\sqrt{6})\cos 60° + \dfrac{1}{2}(\sqrt{3}+1)\sin 60° = \dfrac{1}{4}(3\sqrt{2}-\sqrt{6})\dfrac{1}{2} + \dfrac{1}{2}(\sqrt{3}+1)\dfrac{\sqrt{3}}{2}$

$= \dfrac{1}{8}(3\sqrt{2}-\sqrt{6}+6+2\sqrt{3})$,

$$y_S = -\frac{1}{4}(3\sqrt{2}-\sqrt{6})\sin 60° + \frac{1}{2}(\sqrt{3}+1)\cos 60° = -\frac{1}{4}(3\sqrt{2}-\sqrt{6})\frac{\sqrt{3}}{2} + \frac{1}{2}(\sqrt{3}+1)\frac{1}{2}$$

$$= \frac{1}{8}(-3\sqrt{6}+3\sqrt{2}+2\sqrt{3}+2),$$

$$z_S = \frac{1}{4}(\sqrt{2}+\sqrt{6})$$

Thus, $x_S \approx 1.4072$, $y_S \approx 0.2948$, $z_S \approx 0.9659$.

11.3 Dot Product

1. $\mathbf{a}\cdot\mathbf{b} = 2(-1)+(-3)2+4(5) = 12$

3. $\mathbf{a}\cdot\mathbf{c} = 2(3)+(-3)6+4(-1) = -16$

5. $\mathbf{a}\cdot(4\mathbf{b}) = 2(-4)+(-3)8+4(20) = 48$

7. $\mathbf{a}\cdot\mathbf{a} = 2^2+(-3)^2+4^2 = 29$

9. $\mathbf{a}\cdot(\mathbf{a}+\mathbf{b}+\mathbf{c}) = 2(4)+(-3)5+4(8) = 25$

11. $\left(\dfrac{\mathbf{a}\cdot\mathbf{b}}{\mathbf{b}\cdot\mathbf{b}}\right)\mathbf{b} = \left[\dfrac{2(-1)+(-3)2+4(5)}{(-1)^2+2^2+5^2}\right]\langle -1,2,5\rangle = \dfrac{12}{30}\langle -1,2,5\rangle = \langle -2/5, 4/5, 2\rangle$

13. $\mathbf{a}\cdot\mathbf{b} = 10(5)\cos(\pi/4) = 25\sqrt{2}$

15. $\mathbf{a}\cdot\mathbf{b} = |\mathbf{a}||\mathbf{b}|\cos\theta = (2)(3)\cos(2\pi/3) = 6(-1/2) = -3$

17. $\sqrt{5}\,\mathbf{a}\cdot\mathbf{b} = 3(2)+(-1)2 = 4$; $|\mathbf{a}| = \sqrt{10}$, $|\mathbf{b}| = 2\sqrt{2}$
$$\cos\theta = \frac{4}{(\sqrt{10})(2\sqrt{2})} = \frac{1}{\sqrt{5}} \longrightarrow \theta = \arccos\frac{1}{\sqrt{5}} \approx 1.11 \text{ rad} \approx 63.45°$$

19. $\sqrt{5}\,\mathbf{a}\cdot\mathbf{b} = 2(-1)+4(-1)+0(4) = -6$; $|\mathbf{a}| = 2\sqrt{5}$, $|\mathbf{b}| = 3\sqrt{2}$
$$\cos\theta = \frac{-6}{(2\sqrt{5})(3\sqrt{2})} = -\frac{1}{\sqrt{10}} \longrightarrow \theta = \arccos(-1/\sqrt{10}) \approx 1.89 \text{ rad} \approx 108.43°$$

21. a and f, b and e, c and d

23. Solving the system of equations $3x_1+y_1-1=0$, $-3x_1+2y_1+2=0$ gives $x_1=4/9$ and $y_1=-1/3$. Thus, $\mathbf{v} = \langle 4/9, -1/3, 1\rangle$.

25. Since
$$\mathbf{c}\cdot\mathbf{a} = \left(\mathbf{b} - \frac{\mathbf{a}\cdot\mathbf{b}}{|\mathbf{a}|^2}\mathbf{a}\right)\cdot\mathbf{a} = \mathbf{a}\cdot\mathbf{a} - \frac{\mathbf{a}\cdot\mathbf{b}}{|\mathbf{a}|^2}(\mathbf{a}\cdot\mathbf{a}) = \mathbf{b}\cdot\mathbf{a} - \frac{\mathbf{a}\cdot\mathbf{b}}{|\mathbf{a}|^2}|\mathbf{a}|^2 = \mathbf{b}\cdot\mathbf{a} - \mathbf{a}\cdot\mathbf{b} = 0,$$

the vectors $\mathbf{c}$ and $\mathbf{a}$ are orthogonal.

27. $|\mathbf{a}| = \sqrt{14}$; $\cos\alpha = 1/\sqrt{14}$, $\alpha \approx 74.50°$; $\cos\beta = 2/\sqrt{14}$, $\beta \approx 57.69°$; $\cos\gamma = 3/\sqrt{14}$, $\gamma = 36.70°$

29. $|\mathbf{a}| = 2$; $\cos\alpha = 1/2$, $\alpha \approx 60°$; $\cos\beta = 0$, $\beta \approx 90°$; $\cos\gamma = -\sqrt{3}/2$, $\gamma = 150°$

31. Let θ be the angle between $\overrightarrow{AD}$ and $\overrightarrow{AB}$ and a be the length of an edge of an edge of the cube. Then $\overrightarrow{AD} = a\mathbf{i}+a\mathbf{j}+a\mathbf{k}$, $\overrightarrow{AB} = a\mathbf{i}$ and
$$\cos\theta\,\frac{\overrightarrow{AD}\cdot\overrightarrow{AB}}{|\overrightarrow{AD}||\overrightarrow{AB}|} = \frac{a^2}{\sqrt{3a^2}\sqrt{a^2}} = \frac{1}{\sqrt{3}}$$

so $\theta \approx 0.955317$ radian or $54.7356°$. Letting ϕ be the angle between $\overrightarrow{AD}$ and $\overrightarrow{AC}$ and noting that $\overrightarrow{AC} = a\mathbf{i}+a\mathbf{j}$ we have
$$\cos\theta\,\frac{\overrightarrow{AD}\cdot\overrightarrow{AC}}{|\overrightarrow{AD}||\overrightarrow{AC}|} = \frac{a^2+a^2}{\sqrt{3a^2}\sqrt{2a^2}} = \sqrt{\frac{2}{3}}$$

so $\phi \approx 0.61548$ radian or $35.2644°$

33. $\text{comp}_{\mathbf{b}}\mathbf{a} = \mathbf{a} \cdot \mathbf{b}/|\mathbf{b}| = \langle 1, -1, 3 \rangle \cdot \langle 2, 6, 3 \rangle/7 = 5/7$

35. $\mathbf{b} - \mathbf{a} = \langle 1, 7, 0 \rangle$; $\text{comp}_{\mathbf{a}}(\mathbf{b} - \mathbf{a}) = (\mathbf{b} - \mathbf{a}) \cdot \mathbf{a}/|\mathbf{a}| = \langle 1, 7, 0 \rangle \cdot \langle 1, -1, 3 \rangle/\sqrt{11} = -6/\sqrt{11}$

37. $\overrightarrow{OP} = 3\mathbf{i} + 10\mathbf{j}$; $|\overrightarrow{OP}| = \sqrt{109}$; $\text{comp}_{\overrightarrow{OP}}\mathbf{a} = \mathbf{a} \cdot \overrightarrow{OP}/|\overrightarrow{OP}| = (4\mathbf{i} + 6\mathbf{j}) \cdot (3\mathbf{i} + 10\mathbf{j})/\sqrt{109} = 72/\sqrt{109}$

39. (a) $\text{comp}_{\mathbf{b}}\mathbf{a} = \mathbf{a} \cdot \mathbf{b}/|\mathbf{b}| = (-5\mathbf{i} + 5\mathbf{j}) \cdot (-3\mathbf{i} + 4\mathbf{j})/5 = 7$
   $\text{proj}_{\mathbf{b}}\mathbf{a} = (\text{comp}_{\mathbf{b}}\mathbf{a})\mathbf{b}/|\mathbf{b}| = 7(-3\mathbf{i} + 4\mathbf{j})/5 = -\frac{21}{5}\mathbf{i} + \frac{28}{5}\mathbf{j}$

   (b) $\text{proj}_{\mathbf{b}\perp}\mathbf{a} = \mathbf{a} - \text{proj}_{\mathbf{b}}\mathbf{a} = (-5\mathbf{i} + 5\mathbf{j}) - (-\frac{21}{5} + \frac{28}{5}\mathbf{j}) = -\frac{4}{5}\mathbf{i} - \frac{3}{5}\mathbf{j}$

41. (a) $\text{comp}_{\mathbf{b}}\mathbf{a} = \mathbf{a} \cdot \mathbf{b}/|\mathbf{b}| = (-\mathbf{i} - 2\mathbf{j} + 7\mathbf{k}) \cdot (6\mathbf{i} - 3\mathbf{j} - 2\mathbf{k})/7 = -2$
   $\text{proj}_{\mathbf{b}}\mathbf{a} = (\text{comp}_{\mathbf{b}}\mathbf{a})\mathbf{b}/|\mathbf{b}| = -2(6\mathbf{i} - 3\mathbf{j} - 2\mathbf{k})/7 = -\frac{12}{7}\mathbf{i} + \frac{6}{7}\mathbf{j} + \frac{4}{7}\mathbf{k}$

   (b) $\text{proj}_{\mathbf{b}\perp}\mathbf{a} = \mathbf{a} - \text{proj}_{\mathbf{b}}\mathbf{a} = (-\mathbf{i} - 2\mathbf{j} + 7\mathbf{k}) - (-\frac{12}{7}\mathbf{i} - \frac{6}{7}\mathbf{j} + \frac{4}{7}\mathbf{k}) = \frac{5}{7}\mathbf{i} - \frac{20}{7}\mathbf{j} + \frac{45}{7}\mathbf{k}$

43. $\mathbf{a} + \mathbf{b} = 3\mathbf{i} + 4\mathbf{j}$; $|\mathbf{a} + \mathbf{b}| = 5$; $\text{comp}_{\mathbf{a}+\mathbf{b}}\mathbf{a} = \mathbf{a} \cdot (\mathbf{a} + \mathbf{b})/|\mathbf{a} + \mathbf{b}| = (4\mathbf{i} + 3\mathbf{j}) \cdot (3\mathbf{i} + 4\mathbf{j})/5 = 24/5$; $\text{proj}_{(\mathbf{a}+\mathbf{b})}\mathbf{a} = (\text{comp}_{(\mathbf{a}+\mathbf{b})}\mathbf{a})(\mathbf{a} + \mathbf{b})/|\mathbf{a} + \mathbf{b}| = \frac{24}{5}(3\mathbf{i} + 4\mathbf{j})/5 = \frac{72}{25}\mathbf{i} + \frac{96}{25}\mathbf{j}$

45. We identify $\mathbf{F} = 29$, $\theta = 60°$ and $|\mathbf{d}| = 100$. Then $W = |\mathbf{F}||\mathbf{d}| \cos\theta = 20(100)(\frac{1}{2}) = 1000$ ft-lb.

47. We identify $\mathbf{d} = -\mathbf{i} + 3\mathbf{j} + 8\mathbf{k}$. Then $W = \mathbf{F} \cdot \mathbf{d} = \langle 4, 3, 5 \rangle \cdot \langle -1, 3, 8 \rangle = 45$ N-m.

49. Using $\mathbf{d} = 6\mathbf{i} + 2\mathbf{j}$ and $\mathbf{F} = 3\left(\frac{3}{5}\mathbf{i} + \frac{4}{5}\mathbf{j}\right)$, $W = \mathbf{F} \cdot \mathbf{d} = \left\langle \frac{9}{5}, \frac{12}{5} \right\rangle \cdot \langle 6, 2 \rangle = \frac{78}{5}$ ft-lb.

51. If $\mathbf{a}$ and $\mathbf{b}$ are orthogonal, then $\mathbf{a} \cdot \mathbf{b} = 0$ and

$$\cos\alpha_1 \cos\alpha_2 + \cos\beta_1 \cos\beta_2 + \cos\gamma_1 \cos\gamma_2 = \frac{a_1}{\mathbf{a}}\frac{b_1}{\mathbf{b}} + \frac{a_2}{\mathbf{a}}\frac{b_2}{\mathbf{b}} + \frac{a_3}{\mathbf{a}}\frac{b_3}{\mathbf{b}}$$
$$= \frac{1}{|\mathbf{a}||\mathbf{b}|}(a_1b_1 + a_2b_2a_3b_3) = \frac{1}{|\mathbf{a}||\mathbf{b}|}(\mathbf{a} \cdot \mathbf{b}) = 0.$$

53. For the following, let $\mathbf{a} = \langle a_1, a_2, a_3 \rangle$, $\mathbf{b} = \langle b_1, b_2, b_3 \rangle$, and let k be any scalar.
   Proof of (i): If $\mathbf{a} = \mathbf{0} = \langle 0, 0, 0 \rangle$, then $\mathbf{a} \cdot \mathbf{b} = (0)b_1 + (0)b_2 + (0)b_3 = 0$. Similarly, if $\mathbf{b} = \mathbf{0}$, then $\mathbf{a} \cdot \mathbf{b} = a_1(0) + a_2(0) + a_3(0) = 0$
   Proof of (ii): Using the Commutative Property of real numbers, we have

$$\mathbf{a} \cdot \mathbf{b} = a_1b_1 + a_2b_2 + a_3b_3$$
$$= b_1a_1 + b_2a_2 + b_3a_3$$
$$= \mathbf{b} \cdot \mathbf{a}$$

Proof of (iv): $\mathbf{a} \cdot (k\mathbf{b}) = \langle a_1, a_2, a_3 \rangle \cdot \langle kb_2, kb_2, kb_3 \rangle$
$$= a_1kb_1 + a_2kb_2 + a_3kb_3$$
$$= k(a_1b_1 + a_2b_2 + a_3b_3)$$
$$= k(\mathbf{a} \cdot \mathbf{b})$$
$$= \langle la_1, ka_2, ka_3 \rangle \cdot \langle b_1, b_2, b_3 \rangle$$
$$= ka_1b_1 + ka_2b_2 + ka_3b_3$$
$$= k(a_1b_1 + a_2b_2 + a_3b_3)$$
$$= k(\mathbf{a} \cdot \mathbf{b})$$

Therefore, $\mathbf{a} \cdot (k\mathbf{b}) = (k\mathbf{a}) \cdot \mathbf{b} = k(\mathbf{a} \cdot \mathbf{b})$
Proof of (v): Since $x^2 \geq 0$ for any real number x, we have $\mathbf{a} \cdot \mathbf{a} = a_1^2 + a_2^2 + a_3^2 \geq 0$.

55. $|\mathbf{a} + \mathbf{b}|^2 = (\mathbf{a} + \mathbf{b}) \cdot (\mathbf{a} + \mathbf{b}) = \mathbf{a} \cdot \mathbf{a} + 2\mathbf{a} \cdot \mathbf{b} + \mathbf{b} \cdot \mathbf{b} = |\mathbf{a}|^2 + 2\mathbf{a} \cdot \mathbf{b} + |\mathbf{b}|^2$

$\leq |\mathbf{a}|^2 + 2|\mathbf{a} \cdot \mathbf{b}| + |\mathbf{b}|^2$ $\boxed{\text{since } x \leq |x|}$

$\leq |\mathbf{a}|^2 + 2|\mathbf{a}||\mathbf{b}| + |\mathbf{b}|^2 = (|\mathbf{a}| + |\mathbf{b}|)^2$ $\boxed{\text{by Problem 54}}$

Thus, since $|\mathbf{a} + \mathbf{b}|$ and $|\mathbf{a}| + |\mathbf{b}|$ are positive, $|\mathbf{a} + \mathbf{b}| \leq |\mathbf{a}| + |\mathbf{b}|$.

57. Let θ be the angle between $\mathbf{n}$ and $\overrightarrow{P_2P_1}$. Then

$$d = \|\overrightarrow{P_1P_2}|\cos\theta| = \frac{|\mathbf{n}\cdot\overrightarrow{P_2P_1}|}{|\mathbf{n}|} = \frac{|\langle a,b\rangle\cdot\langle x_1-x_2,y_1-y_2\rangle|}{\sqrt{a^2+b^2}} = \frac{|ax_1-ax_2+by_1=by_2|}{\sqrt{a^2+b^2}}$$

$$= \frac{|ax_1+by_1-(ax_2+by_2)|}{\sqrt{a^2+b^2}} = \frac{|ax_1+by_1-(-c)|}{\sqrt{a^2+b^2}} = \frac{|ax_1+by_1+c|}{\sqrt{a^2+b^2}}.$$

11.4 Cross Product

1. $\mathbf{a}\times\mathbf{b} = \begin{vmatrix} \mathbf{i} & \mathbf{j} & \mathbf{k} \\ 1 & -1 & 0 \\ 0 & 3 & 5 \end{vmatrix} = \begin{vmatrix} -1 & 0 \\ 3 & 5 \end{vmatrix}\mathbf{i} - \begin{vmatrix} 1 & 0 \\ 0 & 5 \end{vmatrix}\mathbf{j} + \begin{vmatrix} 1 & -1 \\ 0 & 3 \end{vmatrix}\mathbf{k} = -5\mathbf{i} - 5\mathbf{j} + 3\mathbf{k}$

3. $\mathbf{a}\times\mathbf{b} = \begin{vmatrix} \mathbf{i} & \mathbf{j} & \mathbf{k} \\ 1 & -3 & 1 \\ 2 & 0 & 4 \end{vmatrix} = \begin{vmatrix} -3 & 1 \\ 0 & 4 \end{vmatrix}\mathbf{i} - \begin{vmatrix} 1 & 1 \\ 2 & 4 \end{vmatrix}\mathbf{j} + \begin{vmatrix} 1 & -3 \\ 2 & 0 \end{vmatrix}\mathbf{k} = \langle -12,-2,6\rangle$

5. $\mathbf{a}\times\mathbf{b} = \begin{vmatrix} \mathbf{i} & \mathbf{j} & \mathbf{k} \\ 2 & -1 & 2 \\ -1 & 3 & -1 \end{vmatrix} = \begin{vmatrix} -1 & 2 \\ 3 & -1 \end{vmatrix}\mathbf{i} - \begin{vmatrix} 2 & 2 \\ -1 & -1 \end{vmatrix}\mathbf{j} + \begin{vmatrix} 2 & -1 \\ -1 & 3 \end{vmatrix}\mathbf{k} = -5\mathbf{i} + 5\mathbf{k}$

7. $\mathbf{a}\times\mathbf{b} = \begin{vmatrix} \mathbf{i} & \mathbf{j} & \mathbf{k} \\ 1/2 & 0 & 1/2 \\ 4 & 6 & 0 \end{vmatrix} = \begin{vmatrix} 0 & 1/2 \\ 6 & 0 \end{vmatrix}\mathbf{i} - \begin{vmatrix} 1/2 & 1/2 \\ 4 & 0 \end{vmatrix}\mathbf{j} + \begin{vmatrix} 1/2 & 0 \\ 4 & 6 \end{vmatrix}\mathbf{k} = \langle -3,2,3\rangle$

9. $\mathbf{a}\times\mathbf{b} = \begin{vmatrix} \mathbf{i} & \mathbf{j} & \mathbf{k} \\ 2 & 2 & -4 \\ -3 & -3 & 6 \end{vmatrix} = \begin{vmatrix} 2 & -4 \\ -3 & 6 \end{vmatrix}\mathbf{i} - \begin{vmatrix} 2 & -4 \\ -3 & 6 \end{vmatrix}\mathbf{j} + \begin{vmatrix} 2 & 2 \\ -3 & -3 \end{vmatrix}\mathbf{k} = \langle 0,0,0\rangle$

11. $\overrightarrow{P_1P_2} = (-2,2,-4);\ \overrightarrow{P_1P_3} = (-3,1,1)$

$\overrightarrow{P_1P_2}\times\overrightarrow{P_1P_3} = \begin{vmatrix} \mathbf{i} & \mathbf{j} & \mathbf{k} \\ -2 & 2 & -4 \\ -3 & 1 & 1 \end{vmatrix} = \begin{vmatrix} 2 & -4 \\ 1 & 1 \end{vmatrix}\mathbf{i} - \begin{vmatrix} -2 & -4 \\ -3 & 1 \end{vmatrix}\mathbf{j} + \begin{vmatrix} -2 & 2 \\ -3 & 1 \end{vmatrix}\mathbf{k} = 6\mathbf{i} + 14\mathbf{j} + 4\mathbf{k}$

13. $\mathbf{a}\times\mathbf{b} = \begin{vmatrix} \mathbf{i} & \mathbf{j} & \mathbf{k} \\ 2 & 7 & -4 \\ 1 & 1 & -1 \end{vmatrix} = \begin{vmatrix} 7 & -4 \\ 1 & -1 \end{vmatrix}\mathbf{i} - \begin{vmatrix} 2 & -4 \\ 1 & -1 \end{vmatrix}\mathbf{j} + \begin{vmatrix} 2 & 7 \\ 1 & 1 \end{vmatrix}\mathbf{k} = -3\mathbf{i} - 2\mathbf{j} - 5\mathbf{k}$

is perpendicular to both $\mathbf{a}$ and $\mathbf{b}$.

15. $\mathbf{a}\times\mathbf{b} = \begin{vmatrix} \mathbf{i} & \mathbf{j} & \mathbf{k} \\ 5 & -2 & 1 \\ 2 & 0 & -7 \end{vmatrix} = \begin{vmatrix} -2 & 1 \\ 0 & -7 \end{vmatrix}\mathbf{i} - \begin{vmatrix} 5 & 1 \\ 2 & -7 \end{vmatrix}\mathbf{j} + \begin{vmatrix} 5 & -2 \\ 2 & 0 \end{vmatrix}\mathbf{k} = \langle 14,37,4\rangle$

$\mathbf{a}\cdot(\mathbf{a}\times\mathbf{b}) = \langle 5,-2,1\rangle\cdot\langle 14,37,4\rangle = 70 - 74 + 4 = 0$

$\mathbf{b}\cdot(\mathbf{a}\times\mathbf{b}) = \langle 2,0,-7\rangle\cdot\langle 14,37,4\rangle = 28 + 0 - 28 = 0$

17. (a) $\mathbf{b}\times\mathbf{c} = \begin{vmatrix} \mathbf{i} & \mathbf{j} & \mathbf{k} \\ 2 & 1 & 1 \\ 3 & 1 & 1 \end{vmatrix} = \begin{vmatrix} 1 & 1 \\ 1 & 1 \end{vmatrix}\mathbf{i} - \begin{vmatrix} 2 & 1 \\ 3 & 1 \end{vmatrix}\mathbf{j} + \begin{vmatrix} 2 & 1 \\ 3 & 1 \end{vmatrix}\mathbf{k} = \mathbf{j} - \mathbf{k}$

$\mathbf{a}\times(\mathbf{b}\times\mathbf{c}) = \begin{vmatrix} \mathbf{i} & \mathbf{j} & \mathbf{k} \\ 1 & -1 & 2 \\ 0 & 1 & -1 \end{vmatrix} = \begin{vmatrix} -1 & 2 \\ 1 & -1 \end{vmatrix}\mathbf{i} - \begin{vmatrix} 1 & 2 \\ 0 & -1 \end{vmatrix}\mathbf{j} + \begin{vmatrix} 1 & -1 \\ 0 & 1 \end{vmatrix}\mathbf{k} = -\mathbf{i} + \mathbf{j} + \mathbf{k}$

(b) $\mathbf{a}\cdot\mathbf{c} = (\mathbf{i}-\mathbf{j}+2\mathbf{k})\cdot(3\mathbf{i}+\mathbf{j}+\mathbf{k}) = 4;\ (\mathbf{a}\cdot\mathbf{c})\mathbf{b} = 4(2\mathbf{i}+\mathbf{j}+\mathbf{k}) = 8\mathbf{i}+4\mathbf{j}+4\mathbf{k}$

$\mathbf{a}\cdot\mathbf{b} = (\mathbf{i}-\mathbf{j}+2\mathbf{k})\cdot(2\mathbf{i}+\mathbf{j}+\mathbf{k}) = 3;\ (\mathbf{a}\cdot\mathbf{b})\mathbf{c} = 3(3\mathbf{i}+\mathbf{j}+\mathbf{k}) = 9\mathbf{i}+3\mathbf{j}+3\mathbf{k}$

$\mathbf{a}\times(\mathbf{b}\times\mathbf{c}) = (\mathbf{a}\cdot\mathbf{c})\mathbf{b} - (\mathbf{a}\cdot\mathbf{b})\mathbf{c} = (8\mathbf{i}+4\mathbf{j}+4\mathbf{k}) - (9\mathbf{i}+3\mathbf{j}+3\mathbf{k}) = -\mathbf{i}+\mathbf{j}+\mathbf{k}$

19. $(2\mathbf{i})\times\mathbf{j} = 2(\mathbf{i}\times\mathbf{j}) = 2\mathbf{k}$

21. $\mathbf{k}\times(2\mathbf{i}-\mathbf{j}) = \mathbf{k}\times(2\mathbf{i}) + \mathbf{k}\times(-\mathbf{j}) = 2(\mathbf{k}\times\mathbf{i}) - (\mathbf{k}\times\mathbf{j}) = 2\mathbf{j} - (-\mathbf{i}) = \mathbf{i}+2\mathbf{j}$

23. $[(2\mathbf{k}) \times (3\mathbf{j})] \times (4\mathbf{j}) = [2 \cdot 3(\mathbf{k} \times \mathbf{j}) \times (4\mathbf{j})] = 6(-\mathbf{i}) \times 4\mathbf{j} = (-6)(4)(\mathbf{i} \times \mathbf{j}) = -24\mathbf{k}$

25. $(\mathbf{i} + \mathbf{j}) \times (\mathbf{i} + 5\mathbf{k}) = [(\mathbf{i} + \mathbf{j}) \times \mathbf{i}] + [(\mathbf{i} + \mathbf{j}) \times 5\mathbf{k}] = (\mathbf{i} \times \mathbf{i}) + (\mathbf{j} \times \mathbf{i}) + (\mathbf{i} \times 5\mathbf{k}) + (\mathbf{j} \times 5\mathbf{k})$
$$= -\mathbf{k} + 5(-\mathbf{j}) + 5\mathbf{i} = 5\mathbf{i} - 5\mathbf{j} - \mathbf{k}$$

27. $\mathbf{k} \cdot (\mathbf{j} \times \mathbf{k}) = \mathbf{k} \cdot \mathbf{i} = 0$

29. $|4\mathbf{j} - 5(\mathbf{i} \times \mathbf{j})| = |4\mathbf{j} - 5\mathbf{k}| = \sqrt{41}$

31. $\mathbf{i} \times (\mathbf{i} \times \mathbf{j}) = \mathbf{i} \times \mathbf{k} = -\mathbf{j}$

33. $(\mathbf{i} \times \mathbf{i}) \times \mathbf{j} = \mathbf{0} \times \mathbf{j} = \mathbf{0}$

35. $2\mathbf{j} \cdot [\mathbf{i} \times (\mathbf{j} - 3\mathbf{k})] = 2\mathbf{j} \cdot [(\mathbf{i} \times \mathbf{j}) + (\mathbf{i} \times (-3\mathbf{k}))] = 2\mathbf{j} \cdot [\mathbf{k} + 3(\mathbf{k} \times \mathbf{i})] = 2\mathbf{j} \cdot (\mathbf{k} + 3\mathbf{j}) = 2\mathbf{j} \cdot \mathbf{k} + 2\mathbf{j} \cdot 3\mathbf{j}$
$$= 2(\mathbf{j} \cdot \mathbf{k}) + 6(\mathbf{j} \cdot \mathbf{j}) = 2(0) + 6(1) = 6$$

37. $\mathbf{a} \times (3\mathbf{b}) = 3(\mathbf{a} \times \mathbf{b}) = 3(4\mathbf{i} - 3\mathbf{j} + 6\mathbf{k}) = 12\mathbf{i} - 9\mathbf{j} + 18\mathbf{k}$

39. $(-\mathbf{a}) \times \mathbf{b} = -(\mathbf{a} \times \mathbf{b}) = -4\mathbf{i} + 3\mathbf{j} - 6\mathbf{k}$

41. $(\mathbf{a} \times \mathbf{b}) \times \mathbf{c} = \begin{vmatrix} \mathbf{i} & \mathbf{j} & \mathbf{k} \\ 4 & -3 & 6 \\ 2 & 4 & -1 \end{vmatrix} = \begin{vmatrix} -3 & 6 \\ 4 & -1 \end{vmatrix} \mathbf{i} - \begin{vmatrix} 4 & 6 \\ 2 & -1 \end{vmatrix} \mathbf{j} + \begin{vmatrix} 4 & -3 \\ 2 & 4 \end{vmatrix} \mathbf{k} = -21\mathbf{i} + 16\mathbf{j} + 22\mathbf{k}$

43. $\mathbf{a} \cdot (\mathbf{b} \times \mathbf{c}) = (\mathbf{a} \times \mathbf{b}) \cdot \mathbf{c} = 4(2) + (-3)4 + 6(-1) = -10$

45. (a) Let $A = (1, 3, 0), B = (2, 0, 0), C = (0, 0, 4)$, and $D = (1, -3, 4)$. Then $\overrightarrow{AB} = \mathbf{i} - 3\mathbf{j}$,
$\overrightarrow{AC} = -\mathbf{i} - 3\mathbf{j} + 4\mathbf{k}$, $\overrightarrow{CD} = \mathbf{i} - 3\mathbf{j}$, and $\overrightarrow{BD} = -\mathbf{i} - 3\mathbf{j} + 4\mathbf{k}$. Since $\overrightarrow{AB} = \overrightarrow{CD}$ and $\overrightarrow{AC} = \overrightarrow{BD}$, the quadrilateral is a parallelogram.

(b) Computing
$$\overrightarrow{AB} \times \overrightarrow{AC} = \begin{vmatrix} \mathbf{i} & \mathbf{j} & \mathbf{k} \\ 1 & -3 & 0 \\ -1 & -3 & 4 \end{vmatrix} = -12\mathbf{i} - 4\mathbf{j} - 6\mathbf{k}$$

we find that the area is $|-12\mathbf{i} - 4\mathbf{j} - 6\mathbf{k}| = \sqrt{144 + 16 + 36} = 14$.

47. $\overrightarrow{P_1P_2} = \mathbf{j}$; $\overrightarrow{P_2P_3} = -\mathbf{j} + \mathbf{k}$
$$\overrightarrow{P_1P_2} \times \overrightarrow{P_2P_3} = \begin{vmatrix} \mathbf{i} & \mathbf{j} & \mathbf{k} \\ 0 & 1 & 0 \\ 0 & -1 & 1 \end{vmatrix} = \begin{vmatrix} 1 & 0 \\ -1 & 1 \end{vmatrix} \mathbf{i} - \begin{vmatrix} 0 & 0 \\ 0 & 1 \end{vmatrix} \mathbf{j} + \begin{vmatrix} 0 & 1 \\ 0 & -1 \end{vmatrix} \mathbf{k} = \mathbf{i}$$
$A = \frac{1}{2}|\mathbf{i}| = \frac{1}{2}$ sq. unit

49. $\overrightarrow{P_1P_2} = -3\mathbf{j} - \mathbf{k}$; $\overrightarrow{P_2P_3} = -2\mathbf{i} - \mathbf{k}$
$$\overrightarrow{P_1P_2} \times \overrightarrow{P_2P_3} = \begin{vmatrix} \mathbf{i} & \mathbf{j} & \mathbf{k} \\ 0 & -3 & -1 \\ -2 & 0 & -1 \end{vmatrix} = \begin{vmatrix} -3 & -1 \\ 0 & -1 \end{vmatrix} \mathbf{i} - \begin{vmatrix} 0 & -1 \\ -2 & -1 \end{vmatrix} \mathbf{j} + \begin{vmatrix} 0 & -3 \\ -2 & 0 \end{vmatrix} \mathbf{k} = 3\mathbf{i} + 2\mathbf{j} - 6\mathbf{k}$$
$A = \frac{1}{2}|3\mathbf{i} + 2\mathbf{j} - 6\mathbf{k}| = \frac{7}{2}$ sq. units

51. $\mathbf{b} \times \mathbf{c} = \begin{vmatrix} \mathbf{i} & \mathbf{j} & \mathbf{k} \\ -1 & 4 & 0 \\ 2 & 2 & 2 \end{vmatrix} = \begin{vmatrix} 4 & 0 \\ 2 & 2 \end{vmatrix} \mathbf{i} - \begin{vmatrix} -1 & 0 \\ 2 & 2 \end{vmatrix} \mathbf{j} + \begin{vmatrix} -1 & 4 \\ 2 & 2 \end{vmatrix} \mathbf{k} = 8\mathbf{i} + 2\mathbf{j} - 10\mathbf{k}$
$\mathbf{v} = |\mathbf{a} \cdot (\mathbf{b} \times \mathbf{c})| = |(\mathbf{i} + \mathbf{j}) \cdot (8\mathbf{i} + 2\mathbf{j} - 10\mathbf{k})| = |8 + 2 + 0| = 10$ cu. units

53. $\mathbf{b} \times \mathbf{c} = \begin{vmatrix} \mathbf{i} & \mathbf{j} & \mathbf{k} \\ -2 & 6 & -6 \\ \frac{5}{2} & 3 & \frac{1}{2} \end{vmatrix} = \begin{vmatrix} 6 & -6 \\ 3 & \frac{1}{2} \end{vmatrix} \mathbf{i} - \begin{vmatrix} -2 & -6 \\ \frac{5}{2} & \frac{1}{2} \end{vmatrix} \mathbf{j} + \begin{vmatrix} -2 & 6 \\ \frac{5}{2} & 3 \end{vmatrix} \mathbf{k} = 21\mathbf{i} - 14\mathbf{j} - 21\mathbf{k}$
$\mathbf{a} \cdot (\mathbf{b} \times \mathbf{c}) = (4\mathbf{i} + 6\mathbf{j}) \cdot (21\mathbf{i} - 14\mathbf{j} - 21\mathbf{k}) = 84 - 84 + 0 = 0$. The vectors are coplanar.

55. The four points will be coplanar if the three vectors $\overrightarrow{P_1P_2} = \langle 3,-1,-1 \rangle$, $\overrightarrow{P_2P_3} = \langle -3,-5,13 \rangle$, and $\overrightarrow{P_3P_4} = \langle -8,7,-6 \rangle$ are coplanar.

$$\overrightarrow{P_2P_3} \times \overrightarrow{P_3P_4} = \begin{vmatrix} \mathbf{i} & \mathbf{j} & \mathbf{k} \\ -3 & -5 & 13 \\ -8 & 7 & -6 \end{vmatrix} = \begin{vmatrix} -5 & 13 \\ 7 & -6 \end{vmatrix} \mathbf{i} - \begin{vmatrix} -3 & 13 \\ -8 & -6 \end{vmatrix} \mathbf{j} + \begin{vmatrix} -3 & -5 \\ -8 & 7 \end{vmatrix} \mathbf{k} = \langle -61,-122,-61 \rangle$$

$$\overrightarrow{P_1P_2} \cdot (\overrightarrow{P_2P_3} \times \overrightarrow{P_3P_4}) = \langle 3,-1,-1 \rangle \cdot \langle -61,-122,-61 \rangle = -183 + 122 + 61 = 0$$

The four points are coplanar.

57. (a) Since $\theta = 90°$, $|\mathbf{a} \times \mathbf{b}| = |\mathbf{a}||\mathbf{b}||\sin 90°| = 6.4(5) = 32$.

(b) The direction of $\mathbf{a} \times \mathbf{b}$ is into the fourth quadrant of the xy-plane or to the left of the plane determined by $\mathbf{a}$ and $\mathbf{b}$ as shown in Figure 11.4.9 in the text. It makes an angle of $30°$ with the positive x-axis.

(c) We identify $\mathbf{n} = (\sqrt{3}\mathbf{i} - \mathbf{j})/2$. Then $\mathbf{a} \times \mathbf{b} = 32\mathbf{n} = 16\sqrt{3}\mathbf{i} - 16\mathbf{j}$.

59. $\mathbf{b} \times \mathbf{c} = \begin{vmatrix} \mathbf{i} & \mathbf{j} & \mathbf{k} \\ 4 & 5 & 6 \\ 7 & 8 & 3 \end{vmatrix} = \begin{vmatrix} 5 & 6 \\ 8 & 3 \end{vmatrix} \mathbf{i} - \begin{vmatrix} 4 & 6 \\ 7 & 3 \end{vmatrix} \mathbf{j} + \begin{vmatrix} 4 & 5 \\ 7 & 8 \end{vmatrix} \mathbf{k} = \langle -30, 30, -3 \rangle$

$\mathbf{a} \times \mathbf{b} = \begin{vmatrix} \mathbf{i} & \mathbf{j} & \mathbf{k} \\ 1 & 2 & 3 \\ 4 & 5 & 6 \end{vmatrix} = \begin{vmatrix} 1 & 3 \\ 5 & 6 \end{vmatrix} \mathbf{i} - \begin{vmatrix} 1 & 3 \\ 4 & 6 \end{vmatrix} \mathbf{j} + \begin{vmatrix} 1 & 2 \\ 4 & 5 \end{vmatrix} \mathbf{k} = \langle -3, 6, -3 \rangle$

$\mathbf{a} \times (\mathbf{b} \times \mathbf{c}) = \begin{vmatrix} \mathbf{i} & \mathbf{j} & \mathbf{k} \\ 1 & 2 & 3 \\ -33 & 30 & -3 \end{vmatrix} = \begin{vmatrix} 2 & 3 \\ 30 & -3 \end{vmatrix} \mathbf{i} - \begin{vmatrix} 1 & 3 \\ -33 & -3 \end{vmatrix} \mathbf{j} + \begin{vmatrix} 1 & 2 \\ -30 & 30 \end{vmatrix} \mathbf{k} = \langle -96, -93, 96 \rangle$

$(\mathbf{a} \times \mathbf{b}) \times \mathbf{c} = \begin{vmatrix} \mathbf{i} & \mathbf{j} & \mathbf{k} \\ -3 & 6 & -3 \\ 7 & 8 & 3 \end{vmatrix} = \begin{vmatrix} 6 & -3 \\ 8 & 3 \end{vmatrix} \mathbf{i} - \begin{vmatrix} -3 & -3 \\ 7 & 3 \end{vmatrix} \mathbf{j} + \begin{vmatrix} -3 & 6 \\ 7 & 8 \end{vmatrix} \mathbf{k} = \langle 42, -12, -66 \rangle$

Therefore, $\mathbf{a} \times (\mathbf{b} \times \mathbf{c}) \neq (\mathbf{a} \times \mathbf{b}) \times \mathbf{c}$

61. Using equation 9 in the text,

$$\mathbf{a} \cdot (\mathbf{b} \times \mathbf{c}) = \begin{vmatrix} a_1 & a_2 & a_3 \\ b_1 & b_2 & b_3 \\ c_1 & c_2 & c_3 \end{vmatrix} \text{ and } (\mathbf{a} \times \mathbf{b}) \cdot \mathbf{c} = \mathbf{c} \cdot (\mathbf{a} \times \mathbf{b}) = \begin{vmatrix} c_1 & c_2 & c_3 \\ a_1 & a_2 & a_3 \\ b_1 & b_2 & b_3 \end{vmatrix}.$$

The second determinant can be obtained from the first by an interchange of the second and third rows followed by an interchange of the new first and second rows. Using Property (iii) of determinates in Appendix I in the text, we see that $\mathbf{a} \cdot (\mathbf{b} \times \mathbf{c}) = (\mathbf{a} \times \mathbf{b}) \cdot \mathbf{c}$.

63. $\mathbf{a} \times (\mathbf{b} \times \mathbf{c}) + \mathbf{b} \times (\mathbf{c} \times \mathbf{a}) + \mathbf{c} \times (\mathbf{a} \times \mathbf{b})$

$= (\mathbf{a} \cdot \mathbf{c})\mathbf{b} - (\mathbf{a} \cdot \mathbf{b})\mathbf{c} - (\mathbf{b} \cdot \mathbf{c})\mathbf{a} + (\mathbf{c} \cdot \mathbf{b})\mathbf{a} - (\mathbf{c} \cdot \mathbf{a})\mathbf{b}$

$= [(\mathbf{a} \cdot \mathbf{c})\mathbf{b} - (\mathbf{c} \cdot \mathbf{a})\mathbf{b}] + [(\mathbf{b} \cdot \mathbf{a})\mathbf{c} - (\mathbf{a} \cdot \mathbf{b})\mathbf{c}] + [(\mathbf{c} \cdot \mathbf{b})\mathbf{a} - (\mathbf{b} \cdot \mathbf{c})\mathbf{a}] = \mathbf{0}$

65. (a) We first note that $\mathbf{a} \times \mathbf{b} = \mathbf{k}$, $\mathbf{b} \times \mathbf{c} = \frac{1}{2}(\mathbf{i} - \mathbf{k})$, $\mathbf{c} \times \mathbf{a} = \frac{1}{2}(\mathbf{j} - \mathbf{k})$, $\mathbf{a} \cdot (\mathbf{b} \times \mathbf{c}) = \frac{1}{2}$, $\mathbf{b} \cdot (\mathbf{c} \times \mathbf{a}) = \frac{1}{2}$, and $\mathbf{c} \cdot (\mathbf{a} \times \mathbf{b}) = \frac{1}{2}$. Then

$$A = \frac{\frac{1}{2}(\mathbf{i} - \mathbf{k})}{\frac{1}{2}} = \mathbf{i} - \mathbf{k}, \quad B = \frac{\frac{1}{2}(\mathbf{j} - \mathbf{k})}{\frac{1}{2}} = \mathbf{j} - \mathbf{k}, \quad \text{and } C = \frac{\mathbf{k}}{\frac{1}{2}} = 2\mathbf{k}.$$

(b) We need to compute $A \cdot (B \times C)$. Using the formula from Problem 62 we have

$$B \times C = \frac{(\mathbf{c} \times \mathbf{a}) \times (\mathbf{a} \times \mathbf{b})}{[\mathbf{b} \cdot (\mathbf{c} \times \mathbf{a})][\mathbf{c} \times (\mathbf{a} \times \mathbf{b})]} = \frac{([\mathbf{c} \times \mathbf{a}] \cdot \mathbf{b})\mathbf{a} - [(\mathbf{c} \times \mathbf{a}) \cdot \mathbf{a}]\mathbf{b}}{[\mathbf{b} \cdot (\mathbf{c} \times \mathbf{a})][\mathbf{c} \times (\mathbf{a} \times \mathbf{b})]}$$

$$= \frac{\mathbf{a}}{\mathbf{c} \cdot (\mathbf{a} \times \mathbf{b}} \quad \boxed{\text{since } (\mathbf{c} \times \mathbf{a}) \cdot \mathbf{a} = 0.}$$

Then

$$A \cdot (B \times C) = \frac{\mathbf{b} \times \mathbf{c}}{\mathbf{a} \cdot (\mathbf{b} \times \mathbf{c})} \cdot \frac{\mathbf{a}}{\mathbf{c} \cdot (\mathbf{a} \times \mathbf{b})} = \frac{1}{\mathbf{c} \cdot (\mathbf{a} \times \mathbf{b})}$$

and the volume of the unit cell of the reciprocal lattice is the reciprocal of the volume of the unit cell of the original lattice.

11.5 Lines in 3-Space

1. $\langle x, y, z \rangle = \langle 4, 6, -7 \rangle + t \langle 3, \frac{1}{2}, -\frac{3}{2} \rangle$

3. $\langle x, y, z \rangle = \langle 0, 0, 0 \rangle + t \langle 5, 9, 4 \rangle$ The equation of a line through P_1 and P_2 is 3-space with $\mathbf{r}_1 = \overrightarrow{OP_1}$ and $\mathbf{r}_2 = \overrightarrow{OP_2}$ can be expressed as $\mathbf{r} = \mathbf{r}_1 + t(k\mathbf{a})$ or $\mathbf{r} = \mathbf{r}_2 + t(k\mathbf{a})$ where $\mathbf{a} = \mathbf{r}_2 - \mathbf{r}_1$ and k is any non-zero scalar. Thus, the form of the equation of a line is not unique. (See the alternative solution to Problem 5.)

5. $\mathbf{a} = \langle 1 - 3, 2 - 5, 1 - (-2) \rangle = \langle -2, -3, 3 \rangle$; $\quad \langle x, y, z \rangle = \langle 1, 2, 1 \rangle + t \langle -2, -3, 3 \rangle$

   Alternate Solution: $\mathbf{a} = \langle -31, 5 - 2, -2 - 1 \rangle = \langle 2, 3, -3 \rangle$; $\quad \langle x, y, z \rangle = \langle 3, 5, -2 \rangle + t \langle 2, 3, -3 \rangle$

7. $\mathbf{a} = \langle 1/2 - (-3/2), -1/2 - 5/2, 1 - (-1/2) \rangle = \langle 2, -3, 3/2 \rangle$;
   $\langle x, y, z \rangle = \langle 1/2, -1/2, 1 \rangle + t \langle 2, -3, 3/2 \rangle$

9. $\mathbf{a} = \langle 1 - (-4), 1 - 1, -1 - (-1) \rangle = \langle 5, 0, 0 \rangle$; $\quad \langle x, y, z \rangle = \langle 1, 1, -1 \rangle + t \langle 5, 0, 0 \rangle$

11. $\mathbf{a} = \langle 2 - 6, 3 - (-1), 5 - 8 \rangle = \langle -4, 4, -3 \rangle$; $x = 2 - 4t$, $y = 3 + 4t$, $z = 5 - 3t$

13. $\mathbf{a} = \langle 1 - 3, 0 - (-2), 0 - (-7) \rangle = \langle -2, 2, 7 \rangle$; $x = 1 - 2t$, $y = 2t$, $z = 7t$

15. $\mathbf{a} = \langle 4 - (-), 1/2 - (-1/4), 1/3 - 1/6 \rangle = \langle 10, 3/4, 1/6 \rangle$; $x = 4 + 10t$, $y = \dfrac{1}{2} + \dfrac{3}{4}t$, $z = \dfrac{1}{3} + \dfrac{1}{6}t$

17. $a_1 = 10 - 1 = 9$, $a_2 = 14 - 4 = 10$, $a_3 = -2 - (-9) = 7$; $\dfrac{x - 10}{9} = \dfrac{y - 14}{10} = \dfrac{z + 2}{7}$

19. $a_1 = -7 - 4 = -11$, $a_2 = 2 - 2 = 0$, $a_3 = 5 - 1 = 4$; $\dfrac{x - 7}{-11} = \dfrac{z - 5}{4}$, $y = 2$

21. $a_1 = 5 - 5 = 0$, $a_2 = 10 - 1 = 9$, $a_3 = -2 - (-14) = 12$; $x = 5$, $\dfrac{y - 10}{9} = \dfrac{z + 2}{12}$

23. Writing the given line in the form $x/2 = (y - 1)/(-3) = (z - 5)/6$, we see that a direction vector is $\langle 2, -3, 6 \rangle$. Parametric equations for the lines are $x = 6 + 2t$, $y = 4 - 3t$, $z = -2 + 6t$.

25. A direction vector parallel to both the xy- and xy-planes is $\mathbf{i} = \langle 1, 0, 0 \rangle$. Parametric equations for the line are $x = 2 + t$, $y = -2$, $z = 15$.

27. Both lines go through the points $(0, 0, 0)$ and $(6, 6, 6)$. Since two points determine a line, the lines are the same.

29. (a) Equating the x components, we have $x = 3 + 2t, = -7$, which gives $t = \dfrac{-7 - 3}{2} = -5$. We can check our work by plugging this value of t into the y and z components to get $y = 4 - (-5) = 9$ and $z = -1 + 6(-5) = -31$

    (b) Equating the x components, we have $x = 5 - x = -7$ which gives $s = 5 + 7 = 12$. We can check our work by plugging this value of s into the y and z components to get $y = 3 + \dfrac{1}{2}(12) = 9$ and $z = 5 - 3(12) = -31$

31. In the xy-plane, $z = 9 + 3t = 0$ and $t = -3$. Then $x = 4 - 2(-3) = 10$ and $y = 1 + 2(-3) = -5$. The point is $(10, -5, 0)$. In the xz-plane, $y = 1 + 2t = 0$ and $t = -1/2$. Then $x = 4 - 2(-1/2) = 5$ and $z = 9 + 3(-1/2) = 15/2$. The point is $(5, 0, 15/2)$. In the yz-plane, $x = 4 - 2t = 0$ and $t = 2$. Then $y = 1 + 2(2) = 5$ and $z = 9 + 3(2) = 15$. The point is $(0, 5, 15)$.

33. Solving the system $4 + t = 6 + 2s$, $5 + t = 11 + 4s$, $-1 + 2t = -3 + s$, or $t - 2s = 2$, $t - 4s = 6$, $2t - s = -2$ yields $s = -2$ and $t = -2$ in all three equations. Thus, the lines intersect at the point $x = 4 + (-2) = 2$, $y = 5 + (-2) = 3$, $z = -1 + 2(-2) = -5$, or $(2, 3, -5)$.

35. The system of equations $2 - t = 4 + s$, $3 + t = 1 + s$, $1 + t = 1 - s$, or $t + s = -2$, $t - s = -2$, $t + s = 0$ has no solution since $-2 \neq 0$. Thus, the lines do not intersect.

37. Using the first two points, we determine the line $x = 4 + 6t$, $y = 3 + 12t$, $z = -5 - 6t$. Letting $t = -5/6$ we see that $(-1, -7, 0)$ is on the line. Thus, the points lie on the same line.

39. A direction vector for the line is $\langle 6 - 2, -1 - 5, 3 - 9 \rangle = \langle 4, -6, -6 \rangle$. Thus, parametric equations for the line segment are $x = 2 + 4t$, $y = 5 - 6t$, $z = 9 - 6t$, where $0 \leq t \leq 1$.

41. $\mathbf{a} = \langle -1, 2, -2 \rangle$, $\mathbf{b} = \langle 2, 3, -6 \rangle$, $\mathbf{a} \cdot \mathbf{b} = 16$, $|\mathbf{a}| = 3$, $|\mathbf{b}| = 7$; $\cos\theta = \dfrac{\mathbf{a} \cdot \mathbf{b}}{|\mathbf{a}||\mathbf{b}|} = \dfrac{16}{3 \cdot 7}$; $\theta = \arccos\dfrac{16}{21} \approx 40.37°$

43. A direction vector perpendicular to the given lines will be $\langle 1, 1, 1 \rangle \times \langle -1, 1, -5 \rangle = \langle -6, 3, 3 \rangle$. Equations of the lines are $x = 4 - 6t$, $y = 1 + 3t$, $z = 6 + 3t$.

45. In the system $-3 + t = 4 + s$, $7 + 3t = 8 - 2s$, $5 + 2t = 10 - 4s$, or $t - s = 7$, $3t + 2s = 1$, $2t + 4s = 5$, the first and second equations have solution $t = 3$ and $s = -4$. Substituting into the third equation, we find $2(3) + 4(-4) = 6 - 16 = -10 \neq 5$. The direction vectors of the lines are $\langle 1, 3, 2 \rangle$ and $\langle 1, -2, -4 \rangle$, so the lines are not parallel. Thus, the lines are skew.

47. The vector $(\overrightarrow{P_1P_2} \times \overrightarrow{P_3P_4})/|\overrightarrow{P_1P_2} \times \overrightarrow{P_3P_4}|$ is a unit vector perpendicular to the two planes. To find the shortest distance between the planes we compute the absolute value of the component of $\overrightarrow{P_1P_3}$ on this unit vector. Then

$$d = \left| \overrightarrow{P_1P_3} \cdot \frac{\overrightarrow{P_1P_2} \times \overrightarrow{P_3P_4}}{|\overrightarrow{P_1P_2} \times \overrightarrow{P_3P_4}|} \right| = \frac{|\overrightarrow{P_1P_3} \cdot \overrightarrow{P_1P_2} \times \overrightarrow{P_3P_4}|}{|\overrightarrow{P_1P_2} \times \overrightarrow{P_3P_4}|}$$

11.6 Planes

1. $2(x - 5) - 3(y - 1) + 4(z - 3) = 0$; $2x - 3y + 4z = 19$

3. $-5(x - 6) + 0(y - 10) + 3(z + 7) = 0$; $-5x + 3z = -51$

5. $6(x - 1/2) + 8(y - 3/4) - 4(z - 1/2) = 0$; $6x + 8y - 4z = 11$

7. From the points $(3, 5, 2)$ and $(2, 3, 1)$ we obtain the vector $\mathbf{u} = \mathbf{i} + 2\mathbf{j} + \mathbf{k}$. From the points $(2, 3, 1)$ and $(-1, -1, 4)$ we obtain the vector $\mathbf{v} = 3\mathbf{i} + 4\mathbf{j} - 3\mathbf{k}$. From the points $(-1, -1, 4)$ and (x, y, z) we obtain the vector $\mathbf{w} = (x + 1)\mathbf{i} + (y + 1)\mathbf{j} + (z - 4)\mathbf{k}$. Then, a normal vector is

$$\mathbf{u} \times \mathbf{v} = \begin{vmatrix} \mathbf{i} & \mathbf{j} & \mathbf{k} \\ 1 & 2 & 1 \\ 3 & 4 & -3 \end{vmatrix} = -10\mathbf{i} + 6\mathbf{j} - 2\mathbf{k}$$

A vector equation of the plane is $-10(x + 1) + 6(y + 1) - 2(z - 4) = 0$ or $5x - 3y + z = 2$.

9. From the points $(0, 0, 0)$ and $(1, 1, 1)$ we obtain the vector $\mathbf{u} = \mathbf{i} + \mathbf{j} + \mathbf{k}$. From the points $(1, 1, 1)$ and $(3, 2, -1)$ we obtain the vector $\mathbf{v} = 2\mathbf{i} + \mathbf{j} - 2\mathbf{k}$. From the points $(3, 2, -1)$ and (x, y, z) we obtain the vector $\mathbf{w} = (x - 3)\mathbf{i} + (y - 2)\mathbf{j} + (z + 1)\mathbf{k}$. Then, a normal vector is

$$\mathbf{u} \times \mathbf{v} = \begin{vmatrix} \mathbf{i} & \mathbf{j} & \mathbf{k} \\ 1 & 1 & 1 \\ 2 & 1 & -2 \end{vmatrix} = -3\mathbf{i} + 4\mathbf{j} - \mathbf{k}$$

A vector equation of the plane is $-3(x - 3) + 4(y - 2) - (z + 1) = 0$ or $-3x + 4y - z = 0$.

11. From the points $(1, 2, -1)$ and $(4, 3, 1)$ we obtain the vector $\mathbf{u} = 3\mathbf{i} + \mathbf{j} + 2\mathbf{k}$. From the points $(4, 3, 1)$ and $(7, 4, 3)$ we obtain the vector $\mathbf{v} = 3\mathbf{i} + \mathbf{j} + 2\mathbf{k}$. From the points $(7, 4, 3)$ and (x, y, z) we obtain the vector $\mathbf{w} = (x - 7)\mathbf{i} + (y - 4)\mathbf{j} + (z - 3)\mathbf{k}$. Since $\mathbf{u} \times \mathbf{v} = \mathbf{0}$, the points are collinear.

13. A normal vector to $x + y - 4z = 1$ is $\langle 1, 1, -4 \rangle$. The equation of the parallel plane is $(x - 2) + (y - 3) - 4(z + 5) = 0$ or $x + y - 4z = 25$.

15. A normal vector to the xy-plane is $\langle 0, 0, 1 \rangle$. The equation of the parallel plane is $z - 12 = 0$ or $z = 12$.

17. Direction vectors of the lines are $\langle 3, -1, 1 \rangle$ and $\langle 4, 2, 1 \rangle$. A normal vector to the plane is $\langle 3, -1, 1 \rangle \times \langle 4, 2, 1 \rangle = \langle -3, 1, 10 \rangle$. A point on the first line, and thus in the plane, is $(1, 1, 2)$. The equation of the plane is $-3(x - 1) + (y - 1) + 10(z - 2) = 0$ or $-3x + y + 10z = 18$.

19. A direction vector for the two lines is $\langle 1, 2, 1 \rangle$. Points on the lines are $(1, 1, 3)$ and $(3, 0, -2)$. Thus, another vector parallel to the plane is $\langle 1 - 3, 1 - 0, 3 + 2 \rangle = \langle -2, 1, 5 \rangle$. A normal vector to the plane is $\langle 1, 2, 1 \rangle \times \langle -2, 1, 5 \rangle = \langle 9, -7, 5 \rangle$. Using the point $(3, 0, -2)$ in the plane, the equation of the plane is $9(x - 3) - 7(y - 0) + 5(z + 2) = 0$ or $9x - 7y + 5z = 17$.

21. A direction vector for the line, and hence a normal vector for the plane, is $\langle -3,1,-1/2 \rangle$. The equation of the plane is $-3(x-2)+(y-4)-\frac{1}{2}(z-8)=0$ or $-3x+y-\frac{1}{2}z=-6$.

23. Normal vectors to the plane are **(a)** $\langle 2,-1,3 \rangle$, **(b)** $\langle 1,2,2 \rangle$, **(c)** $\langle 1,1,-3/2 \rangle$, **(d)** $\langle -5,2,4 \rangle$, **(e)** $\langle -8,-8,12 \rangle$, **(f)** $\langle -2,1,-2 \rangle$. Parallel planes are **(c)** and **(e)**, and **(a)** and **(f)**. Perpendicular planes are **(a)** and **(d)**, **(b)** and **(c)**, **(b)** and **(e)**, and **(d)** and **(f)**.

25. A direction vector of the line is $\langle -6,9,3 \rangle$, and the normal vectors of the plane are **(a)** $\langle 4,1,2 \rangle$, **(b)** $\langle 2,-3,1 \rangle$, **(c)** $\langle 10,-15,-5 \rangle$, **(d)** $\langle -4,6,2 \rangle$. Vectors **(c)** and **(d)** are multiples of the direction vector and hence the corresponding planes are perpendicular to the line.

27. Letting $z=t$ in both equations and solving $5x-4y=8+9t$, $x+4y=4-3t$, we obtain $x=2+t$, $y=\frac{1}{2}-t$, $z=t$.

29. Letting $z=t$ in both equations and solving $4x-2y=1+t$, $x+y=1-2t$, we obtain $x=\frac{1}{2}-\frac{1}{2}t$, $y=\frac{1}{2}-\frac{3}{2}t$, $z=t$.

31. Substituting the parametric equations into the equation of the plane, we obtain $2(1+2t)23(2-t)+2(-3t)=-7$ or $t=-3$. Letting $t=-3$ in the equation of the line, we obtain the point of intersection $(-5,5,9)$.

33. Substituting the parametric equations into the equation of the plane, we obtain $1+2-(1+t)=8$ or $t=-6$. Letting $t=-6$ in the equation of the line, we obtain the point of intersection $(1,2,-5)$. In Problems 35 and 26, the cross product of the normal vectors to the two planes will be a vector parallel to both planes, and hence a direction vector for a line parallel to the two planes.

35. Normal vectors are $\langle 1,1,-4 \rangle$ and $\langle 2,-1,1 \rangle$. A direction vector is

$$\langle 1,1,-4 \rangle \times \langle 2,-1,1 \rangle = \langle -3,-9,-3 \rangle = -3\langle 1,3,1 \rangle.$$

Equations of the line are $x=5+t$, $y=6+3t$, $z=-12+t$.

In Problems 37 and 38, the cross product of the direction vector of the line with the normal vector of the given plane will be a normal vector to the desired plane.

37. A direction vector of the line is $\langle 3,-1,5 \rangle$ and a normal vector to the given plane is $\langle 1,1,1 \rangle$. A normal vector to the desired plane is $\langle 3,-1,5 \rangle \times \langle 1,1,1 \rangle = \langle -6,2,4 \rangle$. A point on the line, and hence in the plane is $\langle 4,0,1 \rangle$. The equation of the plane is $-6(x-4)+2(y-0)+4(z-1)=0$ or $3x-y-2z=10$.

39.

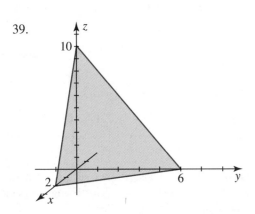

41.

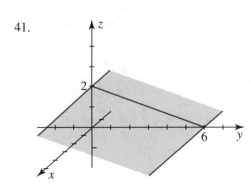

43.

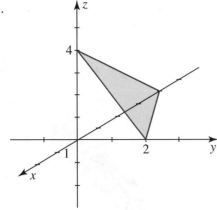

45. (a) A direction vector for the line is $\mathbf{a} = -2\mathbf{i} + \mathbf{j} - \mathbf{k}$ and a normal vector for the plane is $\mathbf{n} = \mathbf{i} + \mathbf{j} - \mathbf{k}$. Since $\mathbf{a} \cdot \mathbf{n} = -2 + 1 + 1 = 0$, the line is perpendicular to $\mathbf{n}$ and thus parallel to the plane. Since $(0, 0, 0)$ is on the line and $(0, 0, -1)$ is in the plane, the line is above the plane.

   (b) A normal vector for the plane is $\mathbf{n} - 3\mathbf{i} - 4\mathbf{j} + 2\mathbf{k}$. Since $\mathbf{a} \cdot \mathbf{n} = 6 - 4 - 2 = 0$, the line is parallel to the plane. Since $(0, 0, 0)$ is one the line and $(0, 0, 4)$ is in the plane, the line is below the plane.

47. Using, Problem 46, $D = \dfrac{|1(2) - 3(1) + 1(4) - 6|}{\sqrt{1 + 9 + 1}} = \dfrac{3}{\sqrt{11}}.$

49. Normal vectors are $\langle 1, -3, 2 \rangle$ and $\langle -1, 1, 1 \rangle$. Then

$$\cos\theta = \frac{\langle 1, -3, 2 \rangle \cdot \langle -1, 1, 1 \rangle}{|\langle 1, -3, 2 \rangle||\langle -1, 1, 1 \rangle|} = \frac{-2}{\sqrt{14}\sqrt{3}} = -\frac{2}{\sqrt{42}}$$

and $\theta = \arccos(-2/\sqrt{42}) \approx 107.98°$

51. Let the bottoms of the table legs be represented by points in 3-space. The rocking of a four-legged table occurs when these four points are not coplanar. Hence, not all four legs can rest on the plane of the floor simultaneously.
However, a three-legged table cannot have this problem. Given any three points in space, a plane can be found passing through them. Therefore, the bottoms of the legs in a three-legged table are coplanar. This implies that they will all rest on the plane of the floor, even if the legs are of uneven lengths.

53. (a) The plane should pass through the midpoint of the line segment joining $(1, -2, 3)$ and $(2, 5, -2)$. This is given in Problem 11.2.64 as $M = \left(\dfrac{1+2}{2}, \dfrac{-2+5}{2}, \dfrac{3-1}{2} \right) = \left(\dfrac{3}{2}, \dfrac{3}{2}, 1 \right)$. The vector joining $(1, -2, 3)$ and $(2, 5, -1)$ should be perpendicular to the plane. This vector is $\mathbf{n} = \langle 1, 7, -4 \rangle$. Using the point $(3/2, 3/2, 1)$ and the normal vector $\mathbf{n}$, the equation of the plane is given by $z + 7y - 4z = 8$.

   (b) The distance from the plane to either of the two points is equal to half the length of the line segment joining the two points. This is given by $\dfrac{1}{2}\sqrt{1^2 + 7^2 + (-4)^2} = \dfrac{1}{2}\sqrt{66}$

11.7 Cylinders and Spheres

1.

3.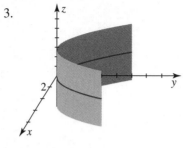

5.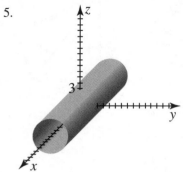

7.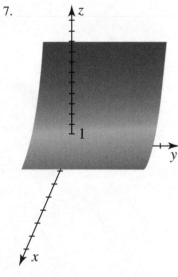

9.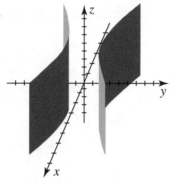

11.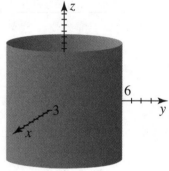

13.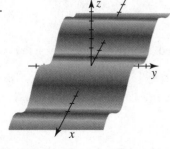

15.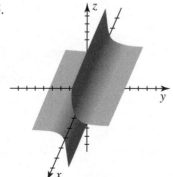

17.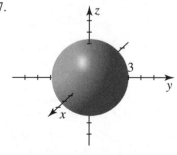

19.

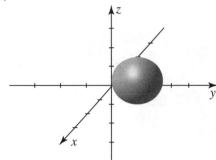

center: $(1,1,1)$
radius: 1

21. $(x^2+8x+16)+(y^2-6x+9)+(z^2-4z+4)=7+16+9+4$
$(x+4)^2+(y-3)^2+(z-2)^2=36;$ center: $(-4,3,2);$ radius: 6

23. $x^2+y^2+(z^2-16z=64)=64;$ center: $(0,0,8);$ radius: 8

25. $(x+1)^2+(y-4)^2+(z-6)^2=3$

27. $(x-1)^2+(y-1)^2+(z-4)^2=16$

29. There are two solutions: one sphere is inside the given sphere and the other is outside.
$x^2+(y-8)^2+z^2=4$ or $x^2+(y-4)^2+z^2=4.$

31. The center is at $(1,4,2)$ and the radius is $\sqrt{(1-0)^2+(4+4)^2(2-7)^2}=3\sqrt{10}.$ The equation is
$(x-1)^2+(y-4)^2+(z-2)^2=90.$

33. The upper half of the sphere $x^2+y^2+(z-1)^2=4;$ a hemisphere

35. All points on and outside the unit sphere centered at the origin

37. $x^2+y^2+z^2=1$ represents a sphere of radius 1 and $x^2+y^2+z^2=9$ represents a sphere of radius 3. Therefore $1\le x^2+y^2+z^2\le 9$ represents the set of points lying between these two spheres. Thus, the geometric object is a hollowed out ball with outer radius 3 and inner radius 1.

11.8 Quadric Surfaces

1. paraboloid

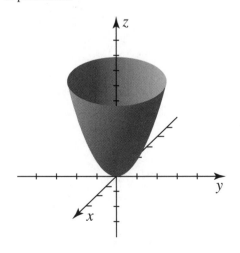

3. $x^2/4+y^2+z^2/9=1;$ ellipsoid

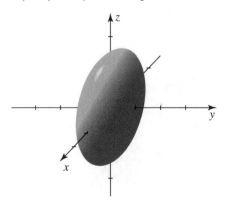

5. $x^2/4 - y^2/144 + z^2/16 = 1$
   hyperboloid of one sheet

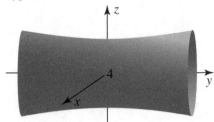

7. elliptical cone

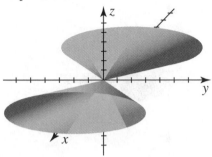

9. hyperbolic paraboloid

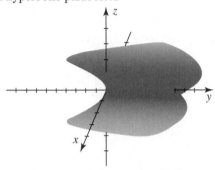

11. $x^2/4 - y^2/4 - z^2/4 = 1$
    hyperboloid of two sheets

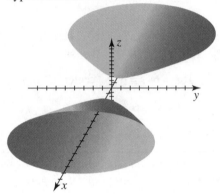

13. $y^2 + \dfrac{z^2}{1/4} = x$
    paraboloid

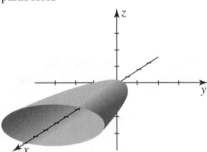

15.

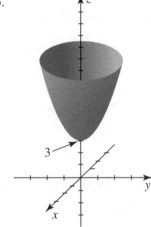

17.

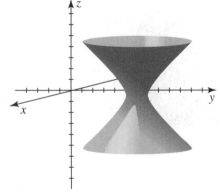

19. The equation can be written as $x^2 + (\pm\sqrt{y^2+z^2})^2 = 1$. The surface is generated by revolving the circles $x^2 + y^2 = 1$ or $x^2 + z^2 = 1$ about the x-axis. [Alternatively, the surface is generated by revolving the circles $x^2 + y^2 = 1$ or $y^2 + z^2 = 1$ about the y-axis, or the circles $x^2 + z^2 = 1$ or $y^2 + z^2 = 1$ about the z-axis.]

21. The equation can be written as $y = e^{\pm\sqrt{x^2+z^2})^2}$. The surface is generated by revolving the curves $y = e^{x^2}$ or $y = e^{z^2}$ about the y-axis.

23. Replacing x by $\pm\sqrt{x^2+y^2}$ we have $y = \pm 2\sqrt{x^2+z^2}$ or $y^2 = 4x^2 + 4z^2$.

25. Replacing z by $\pm\sqrt{y^2+z^2}$ we have $\pm\sqrt{y^2+z^2} = 9 - x^2$ or $y^2 + z^2 = (9-x^2)^2$, $x \geq 0$.

27. Replacing z by $\sqrt{y^2+z^2}$ we have $x^2 - (\pm\sqrt{y^2+z^2})^2 = 4$ or $x^2 - y^2 - z^2 = 4$.

29. Replacing y by $\sqrt{x^2+y^2}$ we have $z = \ln\sqrt{x^2+y^2}$.

31. The surface is Problem 11 is a surface of revolution about the x-axis. The surface in Problem 2 is a surface of revolution about the y-axis. The surface is Problems 1, 4, 6, 10, and 14 are surfaces of revolution about the z

33. The first equation is the lower nappe of the cone $(z+2)^2 = x^2 + y^2$ whose axis of revolution is the z-axis and whose vertex is at $(0,0,-2)$.

35. (a) Writing the equation of the ellipse in the form $x^2/(c-z)a^2 + y^2/(c-z)b^2 = 1$ we see that the area of a cross-section is $\pi a\sqrt{c-z}\, b\sqrt{c-z} = \pi ab(c-z)$.

    (b) $V = \int_0^c \pi ab(c-x)dz = \pi ab\left(-\frac{1}{2}\right)(c-z)^2\Big|_0^c = \frac{1}{2}\pi abc^2$

37. Expressing the line in the form $(x-2)/4 = (y+2)/(-6) = (z-6)/3$ we see that parametric equations for the line are $x = 2+4t$, $y = -2 = 6t$, $z = 6+3t$. Writing the equation of the ellipse as $36x^2 + 9y^2 + 4z^2 = 324$ and substituting, we obtain $36(2+4t)^2 + 9(-2-6t)^2 + 4(6+3t)^2 = 936t^2 + 936t + 324$ or $936t(t+1) = 0$. When $t = 0$ we obtain the point $(2,-2,6)$, and when $t = -1$ we obtain the point $(-2,4,3)$.

Chapter 11 in Review

A. True/False

1. True

3. False; since a normal to the plane is $\langle 2,3,-4\rangle$ which is not a multiple of the direction vector $\langle 5,-2,2\rangle$ of the line.

5. True

7. True

9. True

11. True. The normal vector of the first plane is $\langle 1,2,-1\rangle$ while the normal vector of the second plane is $\langle -2,-4,2\rangle$. Since the second vector is a scalar multiple of the first, the planes are parallel.

13. True. This is a parabolic cylinder similar to that shown in Figure 11.7.6.

15. False. Find the equation of the plane containing the first three points, $P_1(0,1,2)$, $P_2(1,-1,1)$, and $P_3(3,2,6)$. This plane must contain the vectors $\overrightarrow{P_1P_2} = \langle 1,-2,1\rangle$ and $\overrightarrow{P_1P_3} = \langle 3,1,4\rangle$. Define $\mathbf{n} = \overrightarrow{P_1P_2} \times \overrightarrow{P_1P_3} = \begin{vmatrix} \mathbf{i} & \mathbf{j} & \mathbf{k} \\ 1 & -2 & -1 \\ 3 & 1 & 4 \end{vmatrix} = \langle -7,-7,-7\rangle$. Then $\mathbf{n}$ must be normal to the plane. Using $\mathbf{n}$ and the point P_1, the equation of the plane becomes $-7x - 7y + 7z = 7$ or $z+y-z = -1$. The fourth point $P_4(2,1,2)$ does not lie on the plane since $(2) + (1) - (2) \neq -1$.

17. False. The trace in the yz-plane is described by the equation $9y^2 + z^2 = 1$ which represents an ellipse.

19. True. $|\mathbf{a} \times \mathbf{b}| = |\mathbf{a}||\mathbf{b}||\sin\theta| = |\mathbf{a}||\mathbf{b}|$ since $\theta = 90°$

B. Fill In the Blanks

1. $9\mathbf{i} + 2\mathbf{j} + 2\mathbf{k}$

3. $-5(\mathbf{k} \times \mathbf{j}) = -5(-\mathbf{i}) = 5\mathbf{i}$

5. $\sqrt{(-12)^2 + 4^2 + 6^2} = 14$

7. $\begin{vmatrix} 2 & -5 \\ 4 & 3 \end{vmatrix} = 2(3) - (-5)(4) = 6 + 20 = 26$

9. $-6\mathbf{i} + \mathbf{j} - 7\mathbf{k}$

11. Writing the line in parametric form, we have $x = 1 + t$, $y = -2 + 3t$ $z = -1 + 2t$. Substituting into the equation of the plane yields $(1+t) + 2(-2+3t) - (-1+2t) = 13$ or $t = 3$. Thus, the point of intersection is $x = 1 + 3$, $y = -2 + 3(3) = 7$, $z = -1 + 2(3) = 5$, or $(4, 7, 5)$.

13. $x_2 - 2 = 3$, $x_2 = 5$; $y_2 - 1 = 5$, $y_2 = 6$; $z_2 - 7 = -4$, $z_2 = 3$; $P_2 = (5, 6, 3)$

15. $(7.2)(10) \cos 135° = -36\sqrt{2}$

17. $12, -8, 6$

19. $A = \dfrac{1}{2}|5\mathbf{i} - 4\mathbf{j} - 7\mathbf{k}| = 3\sqrt{10}/2$

21. $|-5 - (-3)| = 2$

23. The equation can be transformed into something more recognizable by completing the square: $x^2 + 2y^2 + 2z^2 - 4y - 12z = 0$
$\implies x^2 + 2(y^2 - 2y) + 2(z^2 - 6z) = 0$
$\implies x^2 + 2(y^2 - 2y + 1) + 2(z^2 - 6z = 9) = 20$
$\implies x^2 + 2(y - 1)^2 + 2(z - 3)^2 = 20$
This is the equation of an ellipsoid centered at $(0, 1, 3)$.

C. Exercises

1. $\mathbf{a} \times \mathbf{b} = \begin{vmatrix} \mathbf{i} & \mathbf{j} & \mathbf{k} \\ 1 & 1 & 0 \\ 1 & -2 & 1 \end{vmatrix} = \begin{vmatrix} 1 & 0 \\ -2 & 1 \end{vmatrix}\mathbf{i} - \begin{vmatrix} 1 & 0 \\ 1 & 1 \end{vmatrix}\mathbf{i} + \begin{vmatrix} 1 & 1 \\ 1 & -2 \end{vmatrix}\mathbf{k} = \mathbf{i} - \mathbf{j} + 3\mathbf{k}$

   A unit vector perpendicular to both $\mathbf{a}$ and $\mathbf{b}$ is

$$\frac{\mathbf{a} \times \mathbf{b}}{|\mathbf{a} \times \mathbf{b}|} = \frac{1}{\sqrt{1+1+9}}(\mathbf{i} - \mathbf{j} - 3\mathbf{k} = \frac{1}{\sqrt{11}}\mathbf{i} - \frac{1}{\sqrt{11}}\mathbf{j} - \frac{3}{\sqrt{11}}\mathbf{k}.$$

3. $\text{comp}_\mathbf{b}\mathbf{a} = \mathbf{a} \cdot \mathbf{b}/|\mathbf{b}| = \langle 1, 2, -2 \rangle \cdot \langle 4, 3, 0 \rangle / 5 = 2$

5. First we compute $2\mathbf{a} = \langle 2, 4, -4 \rangle$, $|\mathbf{b}| = \sqrt{16+9} = 5$, and $2\mathbf{a} \cdot \mathbf{b} = 20$. So $\text{proj}_\mathbf{b} 2\mathbf{a} = \dfrac{2\mathbf{a} \cdot \mathbf{b}}{|\mathbf{b}|^2}\mathbf{b} = \dfrac{20}{25}\langle 4, 3, 0 \rangle = \langle \dfrac{16}{5}, \dfrac{12}{5}, 0 \rangle$.

7. $\dfrac{x^2}{16} + \dfrac{y^2}{4} = 1$; elliptical cylinder

9. $-\dfrac{x^2}{9} - \dfrac{y^2}{9/4} + \dfrac{z^2}{9} = 1$; hyperboloid of two sheets

11. $x^2 - y^2 = 9z$; hyperbolic paraboloid

13. Replacing x by $\pm\sqrt{x^2 + z^2}$ we have $(\pm\sqrt{x^2+z^2})^2 - y^2 = 1$ or $x^2 + z^2 - y^2 = 1$, which is a hyperboloid of one sheet. Replacing y by $(\pm\sqrt{y^2+z^2})$ we have $x^2 - (\pm\sqrt{y^2+z^2})^2 = 1$ or $x^2 - y^2 - z^2 = 1$, which is a hyperboloid of two sheets.

15. Let $\mathbf{a} = \langle a, b, c \rangle$ and $\mathbf{r} = \langle x, y, z \rangle$. Then

    (a) $(\mathbf{r} - \mathbf{a}) \cdot \mathbf{r} = \langle x - a, y - b, z - c \rangle \cdot \langle x, y, z \rangle = x^2 - ax + y^2 - by + z^2 - ac = 0$ implies $\left(x - \dfrac{a}{2} \right)^2 + \left(y - \dfrac{b}{x} \right)^2 + \left(z - \dfrac{c}{2} \right)^2 = $

    $\dfrac{a^2 + b^2 + c^2}{4}$. The surface is a sphere.

    (b) $(\mathbf{r} - \mathbf{a}) \cdot \mathbf{a} = \langle x - a, y - b, z - c \rangle \cdot \langle a, b, c \rangle = a(x - a) + b(y - b) + c(z - c) = 0$
    The surface is a plane.

17. A direction vector of the given line is $\langle 4, -2, 6 \rangle$. A parallel line containing $(7, 3, -5)$ is
$(x - 7)/4 = (y - 3)/(-2) = (z + 5)/6$.

19. The direction vectors are $\langle -2, 3, 1 \rangle$ and $\langle 2, 1, 1 \rangle$. Since $\langle -2, 3, 1 \rangle \cdot \langle 2, 1, 1 \rangle = 0$, the lines are orthogonal. Solving $1 - 2t = x = 1 + 2s$, $3t = y = -4 + s$, we obtain $t = -1$ and $s = 1$. The point $(3, -3, 0)$ obtained by letting $t = -1$ and $s = 1$ is common to the two lines, so they do intersect.

21. The lines are parallel with direction vector $\langle 1, 4, -2 \rangle$. Since $(0, 0, 0)$ is on the first line and $(1, 1, 3)$ is on the second line, the vector $\langle 1, 1, 3 \rangle$ is in the plane. A normal vector to the plane is thus $\langle 1, 4, -2 \rangle \times \langle 1, 1, 3 \rangle = \langle 14, -5, -3 \rangle$. An equation of the plane is $14x - 5y - 3z = 0$.

23. A normal vector is $(\mathbf{i} - 2\mathbf{j}) \times (2\mathbf{i} + 3\mathbf{k}) = -6\mathbf{i} - 3\mathbf{j} + 4\mathbf{k}$. Thus, an equation of the plane is $-6(z - 1) - 3(y + 1) + 4(z - 2) = 0$ or $6x + 3y - 4z = -5$.

25. We compute $(\mathbf{a} \times \mathbf{b}) \cdot \mathbf{c}$. First $\mathbf{a} \times \mathbf{b} = -3\mathbf{i} + 3\mathbf{j} - 3\mathbf{k}$. Then $(\mathbf{a} \times \mathbf{b}) \cdot \mathbf{c} = -3(4) + 3(5) - 3(1) = 0$, and the vectors are coplanar.

27.   (a) We have $\mathbf{v} = v\mathbf{j}$ and $\mathbf{B} = B\mathbf{i}$. Then $\mathbf{F} = q(\mathbf{v} \times \mathbf{B} = q(v\mathbf{j} \times B\mathbf{i}) = q(-vB\mathbf{k}) = -qvB\mathbf{k}$.

    (b) We first note that $\mathbf{L} = m\mathbf{r} \times \mathbf{v}$ and $\mathbf{r} \times \mathbf{v} = 0$. Then

$$\mathbf{r} \times \mathbf{L} = \mathbf{r} \times (m\mathbf{r} \times \mathbf{v}) = m[\mathbf{r} \times (\mathbf{r} \times \mathbf{v})] = m[(\mathbf{r} \cdot \mathbf{v})\mathbf{r} - (\mathbf{r} \cdot \mathbf{r})\mathbf{v}] = -m|\mathbf{r}|^2 \mathbf{v},$$

    and so $\mathbf{v} = -\dfrac{1}{m|\mathbf{r}|^2} (\mathbf{r} \times \mathbf{L}) = \dfrac{1}{m|\mathbf{r}|^2} (\mathbf{L} \times \mathbf{r})$.

29. $\mathbf{F} = 5\sqrt{2}\mathbf{i} + 5\sqrt{2}\mathbf{j} + 50\mathbf{i} = (5\sqrt{2} + 50)\mathbf{i} + 5\sqrt{2}\mathbf{j};$ $\mathbf{d} = 3\mathbf{i} + 3\mathbf{j}$

$W = 15\sqrt{2} + 150 + 15\sqrt{2} = 30\sqrt{2} + 150$ N-m ≈ 192.4 N-m

Vector-Valued Functions

12.1 Vector Functions

1. Since the square root function is only defined for nonnegative values, we must have $t^2 - 9 \geq 0$. So the domain is $(-\infty, -3) \cup [3, \infty)$.

3. Since the inverse sine function is only defined for values between -1 and 1, the domain is $[-1, 1]$.

5. $\mathbf{r}(t) = \sin \pi t \mathbf{i} + \cos \pi t \mathbf{j} - \cos^2 \pi t \mathbf{k}$

7. $\mathbf{r}(t) = e^{-t} \mathbf{i} + e^{2t} \mathbf{j} + e^{3t} \mathbf{k}$

9. $x = t^2, \quad y = \sin t, \quad z = \cos t$

11. $x = \ln t, \quad y = 1 + t, \quad z = t^3$

13.

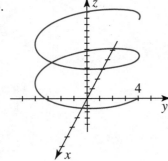

15.

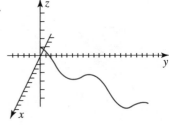

17.

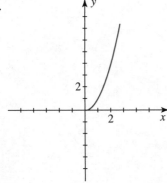

19.

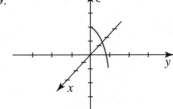

21.

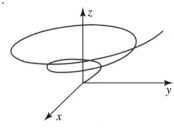

Note: the scale is distorted in this graph. For $t = 0$, the graph starts at $(1,0,1)$. The upper loop shown intersects the xz-plane at about $(286751, 0, 286751)$.

23.

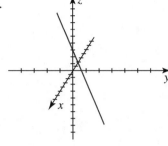

25. $\mathbf{r}(t) = \langle 4, 0 \rangle + \langle 0 - 4, 3 - 0 \rangle t = (4 - 4t)\mathbf{i} + 3t\mathbf{j}, \ 0 \leq t \leq 1$

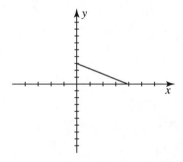

27. $x = t, \ y = t, \ z = t^2 + t^2 = 2t^2; \ \mathbf{r}(t) = t\mathbf{i} + t\mathbf{j} + 2t^2\mathbf{k}$

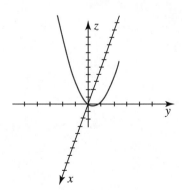

29. $x = 3\cos t, \; z = 9 - 9\cos^2 t = 0\sin^2 t; \; y = 3\sin t; \mathbf{r}(t) = 3\cos t \mathbf{i} + 3\sin t \mathbf{j} + 9\sin^2 t \mathbf{k}$

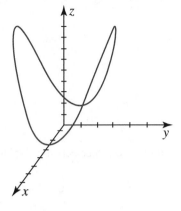

31. $x = t, \; y = t, \; z = 1 - 2t; \; \mathbf{r}(t) = t\mathbf{i} + t\mathbf{j} + (1 - 2t)\mathbf{k}$

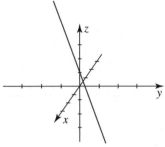

33. (b); Notice that the y and z values consistently increase while the x values oscillate rapidly between -1 and 1. The only vector fucntion that describes this behavior is (b).

35. (d); Notice that the z value is contant. The only vector function that satisfies this constraint is (d).

37. Letting $x = at\cos t, \; y = bt\sin t,$ and $z = ct$, we have
$$\frac{z^2}{c^2} = \frac{c^2 t^2}{c^2} = t^2 = t^2 \cos^2 t + t^2 \sin^2 t$$
$$= \frac{a^2 t^2 \cos^2 t}{a^2} + \frac{b^2 t^2 \sin^2}{b^2}$$
$$= \frac{x^2}{a^2} + \frac{y^2}{b^2}$$

39. Letting $x = ae^{kt}\cos t, \; y = be^{kt}\sin t,$ and $z = ce^{kt}$, we have
$$\frac{z^2}{c^2} = \frac{c^2 e^{kt}}{c^2} = e^{2kt} = e^{2kt} \cos^2 t + e^{2kt} \sin^2 t$$
$$= \frac{a^2 e^{2kt} \cos^2 t}{a^2} + \frac{b^2 e^{2kt} \sin^2 t}{b^2}$$
$$\frac{x^2}{a^2} + \frac{y^2}{b^2}$$

41. $x^2 + y^2 + z^2 = a^2 \sin^2 kt \cos^2 t + a^2 \sin^2 kt \sin^2 t + a^2 \cos^2 kt$
$$= a^2 \sin^2 kt + a^2 \cos^2 kt$$
$$= a^2$$

43. (a)

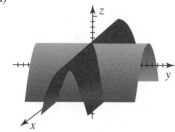

(b) $\mathbf{r}_1(t) = t\mathbf{i} + t\mathbf{j} + (4 - t^2)\mathbf{k}$
$\mathbf{r}_2(t) = t\mathbf{i} - t\mathbf{j} + (4 - t^2)\mathbf{k}$

(c)

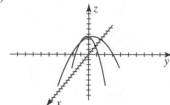

45.

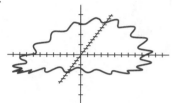

47. $k = 2$ $\qquad\qquad$ $k = 4$

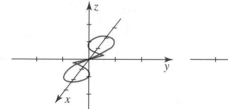 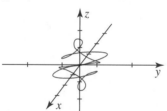

12.2 Calculus of Vector Functions

1. $\lim\limits_{t \to 2}[t^3\mathbf{i} + t^4\mathbf{j} + t^5\mathbf{k}] = 2^3\mathbf{i} + 2^4\mathbf{j} + 2^5\mathbf{k} = 8\mathbf{i} + 16\mathbf{j} + 32\mathbf{k}$

3. Using L'Hôpital's Rule, we have
$$\lim_{t \to 1}\left\langle \frac{t^2 - 1}{t - 1}, \frac{5t - 1}{t + 1}, \frac{2e^{t-1} - 2}{t - 1} \right\rangle = \lim_{t \to 1} = \left\langle \frac{2t}{1}, \frac{5t - 1}{t + 1}, \frac{2e^{t-1}}{1} \right\rangle = \langle 2, 2, 2 \rangle$$

5. $\lim\limits_{t \to a}[-4\mathbf{r}_1(t) + 3\mathbf{r}_2(t)] = -4(\mathbf{i} - 2\mathbf{j} + \mathbf{k}) + 3(2\mathbf{i} + 5\mathbf{j} + 7\mathbf{k}) = 2\mathbf{i} + 23\mathbf{j} + 17\mathbf{k}$

7. Notice that the $\mathbf{k}$ component $\ln(t - 1)$ is not defined at $t = 1$. Therefore, $\mathbf{r}(t)$ is not continuous at $t = 1$.

9. $\mathbf{r}'(t) = 3\mathbf{i} + 8t\mathbf{j} + (10t - 1)\mathbf{k}$
so $\mathbf{r}'(1) = 3\mathbf{i} + 8\mathbf{j} + 9\mathbf{k} = \langle 3, 8, 9 \rangle$
while $\dfrac{\mathbf{r}(1.1) - \mathbf{r}(1)}{0.1} = \dfrac{\langle 3(1.1) - 1, 4(1.1)^2, 5(1.1)^2 - (1.1) \rangle - \langle 3(1) - 1, 4(1)^2, 5(1)^2 - (1) \rangle}{0.1}$

$\qquad\qquad = \dfrac{\langle 2.3, 4.84, 4.95 \rangle - \langle 2, 4, 4 \rangle}{0.1}$

$\qquad\qquad = \dfrac{\langle 0.3, 0.84, 0.95 \rangle}{0.1} = \langle 3, 8.4, 9.5 \rangle$

11. $\mathbf{r}'(t) = \dfrac{1}{t}\mathbf{i} - \dfrac{1}{t^2}\mathbf{j}; \quad \mathbf{r}''(t) = -\dfrac{1}{t^2}\mathbf{i} + \dfrac{2}{t^3}\mathbf{j}$

13. $\mathbf{r}'(t) = \langle 2te^{2t} + e^{2t}, 3t^2, 8t - 1 \rangle; \quad \mathbf{r}''(t) = \langle 4te^{2t} + 4e^{2t}, 6t, 8 \rangle$

15. $\mathbf{r}'(t) = -2\sin t\,\mathbf{i} + 6\cos t\,\mathbf{j}$
    $\mathbf{r}'(\pi/6) = -\mathbf{i} + 3\sqrt{3}\,\mathbf{j}$

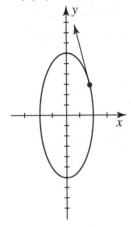

17. $\mathbf{r}'(t) = \mathbf{j} - \dfrac{8t}{(1+t^2)^2}\mathbf{k}$
    $\mathbf{r}'(-1) = \mathbf{j} - 2\mathbf{k}$

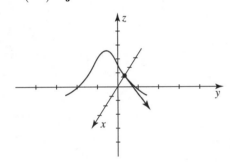

19. $\mathbf{r}(t) = t\mathbf{i} + \dfrac{1}{2}\mathbf{j} + \dfrac{1}{3}t^3\mathbf{k}; \quad \mathbf{r}(2) = 2\mathbf{i} + 2\mathbf{j} + \dfrac{8}{3}\mathbf{k}; \quad \mathbf{r}'(t) = \mathbf{i} + t\mathbf{j} + t^2\mathbf{k}; \quad \mathbf{r}'(2) = \mathbf{i} + 2\mathbf{j} + 4\mathbf{k}$
    Using the point $(2, 2, 8/3)$ and the direction vector $\mathbf{r}'(2)$, we have $x = 2 + t, \; y = 2 + 2t, \; z = 8/3 + 4t$.

21. $\mathbf{r}'(t) = \langle e^t + te^t, 2t + 2, 3t^2 - 1 \rangle$ so $\mathbf{r}'(0) = \langle 1, 2, -1 \rangle$ and $|\mathbf{r}'(0)| = \sqrt{1^2 + 2^2 + (-1)^2} = \sqrt{6}$

    The unit tangent vector at $t = 0$ is given by $\dfrac{\mathbf{r}'(0)}{|\mathbf{r}'(0)|} = \dfrac{\langle 1, 2, -1 \rangle}{\sqrt{6}} = \left\langle \dfrac{1}{\sqrt{6}}, \dfrac{2}{\sqrt{6}}, \dfrac{-1}{\sqrt{6}} \right\rangle$

    To find the parametric equations of the tangent line at $t = 0$, we first compute $\mathbf{r}(0) = \langle 0, 0, 0 \rangle$. The tangent line is then given in vector form as $\mathbf{p}(t) = \langle 0, 0, 0 \rangle + t\left\langle \dfrac{1}{\sqrt{6}}, \dfrac{2}{\sqrt{6}}, \dfrac{-1}{\sqrt{6}} \right\rangle = \left\langle \dfrac{1}{\sqrt{6}}t, \dfrac{2}{\sqrt{6}}t, \dfrac{-1}{\sqrt{6}}t \right\rangle$ or in parametric form as $x = \dfrac{1}{\sqrt{6}}t, \; y = \dfrac{2}{\sqrt{6}}t,$
    $z = \dfrac{-1}{\sqrt{6}}t.$

23. $\mathbf{r}(\pi/3) = \left\langle \dfrac{1}{2}, \dfrac{\sqrt{3}}{2}, \dfrac{\pi}{3} \right\rangle$
    $\mathbf{r}'(t) = \langle -\sin t, \cos t, 1 \rangle$
    $\mathbf{r}'(\pi/3) = \left\langle -\dfrac{\sqrt{3}}{2}, \dfrac{1}{2}, 1 \right\rangle$

so the tangent line is given by

$$\mathbf{p}(t) = \left\langle \frac{1}{2}, \frac{\sqrt{3}}{2}, \frac{\pi}{3} \right\rangle + t \left\langle -\frac{\sqrt{3}}{2}, \frac{1}{2}, 1 \right\rangle$$

$$= \left\langle \frac{1}{2} - \frac{\sqrt{3}}{2}t, \frac{\sqrt{3}}{2} + \frac{1}{2}t, \frac{\pi}{3} + t \right\rangle$$

25. $\dfrac{d}{dt}[\mathbf{r}(t) \times \mathbf{r}'(t)] = \mathbf{r}(t) \times \mathbf{r}''(t) + \mathbf{r}'(t) \times \mathbf{r}'(t) = \mathbf{r}(t) \times \mathbf{r}''(t)$

27. $\dfrac{d}{dt}[\mathbf{r}(t) \cdot (\mathbf{r}'(t) \times \mathbf{r}''(t))] = \mathbf{r}(t) \cdot \dfrac{d}{dt}(\mathbf{r}'(t) \times \mathbf{r}''(t)) + \mathbf{r}'(t) \cdot (\mathbf{r}'(t) \times \mathbf{r}''(t))$

$\qquad\qquad = \mathbf{r}(t) \cdot (\mathbf{r}'(t) \times \mathbf{r}'''(t) + \mathbf{r}''(t) \times \mathbf{r}''(t)) + \mathbf{r}'(t) \cdot (\mathbf{r}'(t) \times \mathbf{r}''(t))$

$\qquad\qquad = \mathbf{r}(t) \cdot (\mathbf{r}'(t) \times \mathbf{r}'''(t))$

29. $\dfrac{d}{dt}[\mathbf{r}_1(2t) + \mathbf{r}_2(\frac{1}{t})] = 2\mathbf{r}'(2t) - \dfrac{1}{t^2}\mathbf{r}_2'(\frac{1}{t})$

31. $\displaystyle\int_{-1}^{2} \mathbf{r}(t)dt = \left[\int_{-1}^{2} t\,dt\right]\mathbf{i} + \left[\int_{-1}^{2} 3t^2\,dt\right]\mathbf{j} + \left[\int_{-1}^{2} 4t^3\,dt\right]\mathbf{k} = \frac{1}{2}t^2\Big|_{-1}^{2}\mathbf{i} + t^3\Big|_{-1}^{2}\mathbf{j} + t^4\Big|_{-1}^{2}\mathbf{k} = \frac{3}{2}\mathbf{i} + 9\mathbf{j} + 15\mathbf{k}$

33. $\displaystyle\int \mathbf{r}(t)dt = \left[\int te^t\,dt\right]\mathbf{i} + \left[\int -e^{-2t}\,dt\right]\mathbf{j} + \left[\int te^{t^2}\,dt\right]\mathbf{k}$

$\qquad = [te^t - e^t + c_1]\mathbf{i} + \left[\frac{1}{2}e^{-2t} + c_2\right]\mathbf{j} + \left[\frac{1}{2}e^{t^2} + d_3\right]\mathbf{k} = e^t(t-1)\mathbf{i} + \frac{1}{2}e^{-2t}\mathbf{j} + \frac{1}{2}e^{t^2}\mathbf{k} + \mathbf{c},$

where $\mathbf{c} = c_1\mathbf{i} + c_2\mathbf{j} + c_3\mathbf{k}$.

35. $\mathbf{r}(t) = \int \mathbf{r}'(t)dt = [\int 6\,dt]\mathbf{i} + [\int 6t\,dt]\mathbf{j} + [\int 3t^2\,dt]\mathbf{k} = [6t + c_1]\mathbf{i} + [3t^2 + c_2]\mathbf{j} + [t^3 + c_3]\mathbf{k}$
Since $\mathbf{r}(0) = \mathbf{i} + 2\mathbf{j} + \mathbf{k} = c_1\mathbf{i} + c_2\mathbf{j} + c_3\mathbf{k}$, $c_1 - 1$, $c_2 = -2$, and $c_3 = 1$. Thus,
$$\mathbf{r}(t) = (6t + 1)\mathbf{i} + (3t^2 - 2)\mathbf{j} + (t^3 + 1)\mathbf{k}$$

37. $\mathbf{r}'(t) = \int \mathbf{r}''(t)dt = [\int 12t\,dt]\mathbf{i} + [\int -3t^{-1/2}\,dt]\mathbf{j} + [\int 2\,dt]\mathbf{k} = [6t^2 + c_1]\mathbf{i} + [-6t^{1/2} + c_2]\mathbf{j} + [2t + c_3]\mathbf{k}$
Since $\mathbf{r}'(1) = \mathbf{j} = (6 + c_1)\mathbf{i} + (-6 + c_2)\mathbf{j} + (2 + c_3)\mathbf{k}$, $c_1 = -6$, $c_2 = 7$, and $c_3 = -2$. Thus,
$$\mathbf{r}'(t) = (6t^2 - 6)\mathbf{i} + (-6t^{1/2} + 7)\mathbf{j} + (2t - 2)\mathbf{k}.$$

$\mathbf{r}(t) = \displaystyle\int \mathbf{r}'(t)dt = \left[\int (6t^2 - 6)dt\right]\mathbf{i} + \left[\int (-6t^{1/2} + 7)dt\right]\mathbf{j} + \left[\int (2t - 2)dt\right]\mathbf{k}$

$\qquad = [2t^3 - 6t + c_4]\mathbf{i} + [-4t^{3/2} + 7t + c_5]\mathbf{j} + [t^2 - 2t + c_6]\mathbf{k}.$
Since
$$\mathbf{r}(1) = 2\mathbf{i} - \mathbf{k} = (-4 + c_4)\mathbf{i} + (3 + c_5)\mathbf{j} + (-1 + c_6)\mathbf{k},$$
$c_4 = 6$, $c_5 = -3$, and $c_6 = 0$. Thus,
$$\mathbf{r}(t) = (2t^3 - 6t + 6)\mathbf{i} + (-4t^{3/2} + 7t - 3)\mathbf{j} + (t^2 - 2t)\mathbf{k}.$$

39. $\mathbf{r}'(t) = -a\sin t\,\mathbf{i} + a\cos t\,\mathbf{j} + c\mathbf{k}$; $\quad |\mathbf{r}'(t)| = \sqrt{(-a\sin t)^2 + (a\cos t)^2 + c^2} = \sqrt{a^2 + c^2}$
$s = \int_0^{2\pi} \sqrt{a^2 + c^2}\,dt = \sqrt{a^2 + c^2}\,t\Big|_0^{2\pi} = 2\pi\sqrt{a^2 + c^2}$

41. $\mathbf{r}'(t) = (-2e^t\sin 2t + e^t\cos 2t)\mathbf{i} + (2e^t\cos 2t + e^t\sin 2t)\mathbf{j} + e^t\mathbf{k}$
$|\mathbf{r}'(t)| = \sqrt{5e^{2t}\cos^2 2t + 5e^{2t}\sin^2 2t + e^{2t}} = \sqrt{6e^{2t}} = \sqrt{6}e^t$
$s = \int_0^{3\pi} \sqrt{6}e^t\,dt = \sqrt{6}e^t\Big|_0^{3\pi} = \sqrt{6}(e^{3\pi} - 1)$

43. From $\mathbf{r}'(t) = \langle 9\cos t, -9\sin t \rangle$, we find $|\mathbf{r}'(t)| = 9$. Therefore, $s = \int_0^t 9\,du = 9t$ so that $t = \dfrac{s}{9}$. By substituting for t in $\mathbf{r}(t)$, we

obtain $\mathbf{r}(s) = \left\langle 9\sin\dfrac{s}{9}, 9\cos\dfrac{s}{9} \right\rangle$. Note that $\mathbf{r}'(s) = \left\langle \sin\dfrac{s}{9}, \cos\dfrac{s}{9} \right\rangle$ so that $\left|\mathbf{r}'(s)\right| = \sqrt{\sin^2\dfrac{s}{9} + \cos\dfrac{s}{9}} = 1$.

45. From $\mathbf{r}'(t) = \langle 2, -3, 4 \rangle$, we find $|\mathbf{r}'(t)| = \sqrt{29}$. Therefore, $s = \int_0^t \sqrt{29} \, du = \sqrt{20}t$ so that $t = \dfrac{s}{\sqrt{29}}$. By substituting for t in $\mathbf{r}(t)$,

we obtain $\mathbf{r}(s) = \left\langle 1 + \dfrac{2}{\sqrt{29}} s, 5 - \dfrac{3}{\sqrt{29}} s, 2 + \dfrac{4}{\sqrt{29}} s \right\rangle$.

Note that $\mathbf{r}'(s) = \left\langle \dfrac{2}{\sqrt{29}}, -\dfrac{3}{\sqrt{29}}, \dfrac{4}{\sqrt{29}} \right\rangle$ so that $\mathbf{r}'(s) = \sqrt{\dfrac{4}{29} + \dfrac{9}{29} + \dfrac{16}{29}} = 1$.

47. Since $\dfrac{d}{dt}(\mathbf{r} \cdot \mathbf{r}) = \dfrac{d}{dt}|\mathbf{r}|^2 = \dfrac{d}{dt} c^2 = 0$ and $\dfrac{d}{dt}(\mathbf{r} \cdot \mathbf{r}) = \mathbf{r} \cdot \mathbf{r}' + \mathbf{r}' \cdot \mathbf{r} = 2\mathbf{r} \cdot \mathbf{r}'$, we have $\mathbf{r} \cdot \mathbf{r}' = 0$. Thus, $\mathbf{r}'$ is perpendicular to $\mathbf{r}$.

49. From $\mathbf{r}(t) = \mathbf{r}_0 + t\mathbf{v}$, we get $\mathbf{r}'(t) = \mathbf{v}$ so that $|\mathbf{r}'(t)| = |\mathbf{v}|$. Therfore $s = \int_0^t |\mathbf{r}'(t)| \, du = \int_0^t |\mathbf{v}| \, du = |\mathbf{v}|t$ which gives $t = \dfrac{s}{|\mathbf{v}|}$.

Substituting for t in $\mathbf{r}(t)$, we have $\mathbf{r}'(s) = \mathbf{r}_0 + \dfrac{s}{|\mathbf{v}|}\mathbf{v} = \mathbf{r}_0 + s\dfrac{\mathbf{v}}{|\mathbf{v}|}$. Note that $\mathbf{r}'(s) = \dfrac{\mathbf{v}}{|\mathbf{v}|}$ so that $|\mathbf{r}'(s)| = \dfrac{|\mathbf{v}|}{|\mathbf{v}|} = 1$.

12.3 Motion on a Curve

1. $\mathbf{v}(t) = 2t\mathbf{i} + t^3\mathbf{j}$; $\mathbf{v}(1) = 2\mathbf{i} + \mathbf{j}$; $|\mathbf{v}(1)| = \sqrt{4+1} = \sqrt{5}$;
   $\mathbf{a}(t) = 2\mathbf{i} + 3t^2\mathbf{j}$; $\mathbf{a}(1) = 2\mathbf{i} + 3\mathbf{j}$

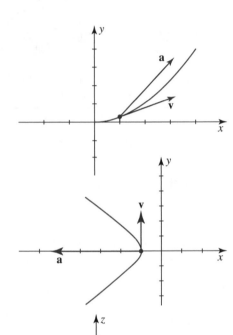

3. $\mathbf{v}(t) = -2\sinh 2t\mathbf{i} + 2\cosh 2t\mathbf{j}$; $\mathbf{v}(1) = 2\mathbf{j}$; $|\mathbf{v}(0)| = 2$;
   $\mathbf{a}(t) = -4\cosh 2t\mathbf{i} + +4\sinh 2t\mathbf{j}$; $\mathbf{a}(0) = -4\mathbf{i}$

5. $\mathbf{v}(t) = (2t - 2)\mathbf{j} + \mathbf{k}$; $\mathbf{v}(2) = 2\mathbf{j} + \mathbf{k}$; $|\mathbf{v}(2)| = \sqrt{4+1} = \sqrt{5}$;
   $\mathbf{a}(t) = 2\mathbf{j}$; $\mathbf{a}(2) = 2\mathbf{j}$

7. $\mathbf{v}(t) = \mathbf{i} + 2t\mathbf{j} + 3t^2\mathbf{k}$;
   $\mathbf{v}(1) = \mathbf{i} + 2\mathbf{j} + 3\mathbf{k}$; $|\mathbf{v}(1)| = \sqrt{1+1+9} = \sqrt{14}$;
   $\mathbf{a}(t) = 2\mathbf{j} + 6t\mathbf{k}$; $\mathbf{a}(1) = 2\mathbf{j} + 6\mathbf{k}$

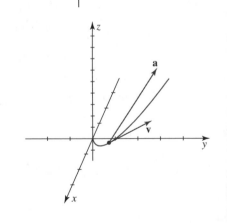

9. The particle passes through the xy-plane when $z(t) = t^2 - 5t = 0$ or $t = 0, 5$ which gives us the points $(0,0,0)$ and $(25,115,0)$.
$\mathbf{v}(t) = 2t\mathbf{i} + (3t^2 - 2)\mathbf{j} + (2t - 5)\mathbf{k}$; $\mathbf{v}(0) = -2\mathbf{j} - 5\mathbf{k}$, $\mathbf{v}(5) = 10\mathbf{i} + 73\mathbf{j} + 5\mathbf{k}$; $\mathbf{a}(t) = 2\mathbf{i} + 6t\mathbf{j} + 2\mathbf{k}$; $\mathbf{a}(0) = 2\mathbf{i} + 30\mathbf{j} + 2\mathbf{k}$

11. Initially we are given $s_0 = \mathbf{0}$ and $\mathbf{v}_0 = (480\cos 30°)\mathbf{i} + (480\cos 30°)\mathbf{j} = 240\sqrt{3} + 240\mathbf{j}$. Using $\mathbf{a}(t) = -32\mathbf{j}$ we find

$$\mathbf{v}(t) = \int \mathbf{a}(t)dt = -32t\mathbf{j} + \mathbf{c}$$

$$240\sqrt{3}\mathbf{i} + 240\mathbf{j} = \mathbf{v}(0) = \mathbf{c}$$

$$\mathbf{v}(t) = -32t\mathbf{j} + 240\sqrt{3}\mathbf{i} + 240\mathbf{j} = 240\sqrt{3}\mathbf{i} + (240 - 32t)\mathbf{j}$$

$$\mathbf{r}(t) = \int \mathbf{v}(t)dt = 240\sqrt{3}t\mathbf{i} + (240t - 16t^2)\mathbf{j} + \mathbf{b}$$

$$\mathbf{0} = \mathbf{r}(0) = \mathbf{b}.$$

(a) The shell's trajectory is given by $\mathbf{r}(t) = 240\sqrt{3}t\mathbf{i} + (240t - 16t^2)\mathbf{j}$ or $x = 240\sqrt{3}t$, $y = 240 - 16t^2$.

(b) Solving $dy/dt = 240 - 32t = 0$, we see that y is maximum when $t = 15/2$. The maximum altitude is $y(15/2) = 900$ ft.

(c) Solving $y(t) = 240t - 16t^2 = 16t(15 - t) = 0$, we see that the shell is at ground level when $t = 0$ and $t = 15$. The range of the shell is $s(15) = 3600\sqrt{3} \approx 6235$ ft.

(d) From (c), impact is when $t = 15$. The speed at impact is

$$|\mathbf{v}(15)| = |240\sqrt{3}\mathbf{i} + (240 - 32 \cdot 15)\mathbf{j}| = \sqrt{240^2 \cdot 3 + (-240)^2} = 480 \text{ ft/s}.$$

13. We are given $s_0 = 81\mathbf{j}$ and $\mathbf{v}_0 = 4\mathbf{i}$. Using $\mathbf{a}(t) = -32\mathbf{j}$, we have

$$\mathbf{v}(t) = \int \mathbf{a}(t)dt = -32t\mathbf{j} + \mathbf{c}$$

$$4\mathbf{i} = \mathbf{v}(0) = \mathbf{c}$$

$$\mathbf{v}(t) = 4\mathbf{i} - 32t\mathbf{j}$$

$$\mathbf{r}(t) = \int \mathbf{v}(t)dt = 4t\mathbf{i} - 16t^2\mathbf{j} + \mathbf{b}$$

$$81\mathbf{j} = \mathbf{r}(0) = \mathbf{b}$$

$$\mathbf{r}(t) = 4t\mathbf{i} + (81 - 16t^2)\mathbf{j}.$$

Solving $y(t) = 81 - 16t^2 = 0$, we see that the car hits the water when $t = 9/4$. Then

$$|v(9/4)| = |4\mathbf{i} - 32(9/4)\mathbf{j}| = \sqrt{4^2 + 72^2} = 20\sqrt{13} \approx 72.11 \text{ft/s}.$$

15. Let s be the initial speed. Then $\mathbf{v}(0) = s\cos 45°\mathbf{i} + s\sin 45°\mathbf{j} = \frac{s\sqrt{2}}{2}\mathbf{i} + \frac{s\sqrt{2}}{2}\mathbf{j}$. Using $\mathbf{a}(t) = -32\mathbf{j}$, we have

$$\mathbf{v}(t) = \int \mathbf{a}(t)dt = -32\mathbf{j} + \mathbf{c}$$

$$\frac{s\sqrt{2}}{2}\mathbf{i} + \frac{s\sqrt{2}}{2}\mathbf{j} = \mathbf{v}(0) = \mathbf{c}$$

$$\mathbf{v}(t) = \frac{s\sqrt{2}}{2}\mathbf{i} + \left(\frac{s\sqrt{2}}{2} - 32t\right)\mathbf{j}$$

$$\mathbf{r}(t) = \frac{s\sqrt{2}}{2}t\mathbf{i} + \left(\frac{s\sqrt{2}}{2}t - 16t^2\right)\mathbf{j} + \mathbf{b}.$$

Since $\mathbf{r}(0) = \mathbf{0}$, $\mathbf{b} = \mathbf{0}$ and

$$\mathbf{r}(t) = \frac{s\sqrt{2}}{2}t\mathbf{i} + \left(\frac{s\sqrt{2}}{2}t - 16t^2\right)\mathbf{j}.$$

Setting $y(t) = s\sqrt{2}t/2 - 16t^2 = t(2\sqrt{2}/2 - 16t) = 0$ we see that the ball hits the ground when $t = \sqrt{2}s/32$. Thus, using $x(t) = s\sqrt{2}t/2$ and the fact that 100 yd = 300 ft, $300 = x(t) = \frac{s\sqrt{2}}{2}(\sqrt{2}s/32) = \frac{s^2}{32}$ and $s = \sqrt{9600} \approx 97.98$ ft/s.

17. $\mathbf{r}'(t) = \mathbf{v}(t) = -r_0\omega\sin\omega t\,\mathbf{i} + r_0\omega\cos\omega t\,\mathbf{j};\quad v = |\mathbf{v}(t)| = \sqrt{r_0^2\omega^2\sin^2\omega t + r_0^2\omega^2\cos^2\omega t} = r_0\omega$

$\omega = v/r_0;\quad \mathbf{a}(t) = \mathbf{r}''(t) = -r_0\omega^2\cos\omega t\,\mathbf{i} - r_0\omega^2\sin\omega t\,\mathbf{j}$

$a = |\mathbf{a}(t)| = \sqrt{r_0^2\omega^4\cos^2\omega t + r_0^2\omega^4\sin^2\omega t} = r_0\omega^2 = r_0(v/r_0)^2 = v^2/r_0.$

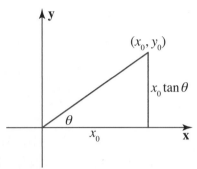

19. Let the initial speed of the projectile be s and let the target be at (x_0, y_0). Then $\mathbf{v}_p(0) = s\cos\theta\,\mathbf{i} + s\sin\theta\,\mathbf{j}$ and $\mathbf{v}_t(0) = \mathbf{0}$. Using $\mathbf{a}(t) = -32\mathbf{j}$, we have

$\mathbf{v}_p(t) = \int \mathbf{a}\,dt = -32t\mathbf{j} + \mathbf{c}$

$s\cos\theta\,\mathbf{i} + s\sin\theta\,\mathbf{j} = \mathbf{v}_p(0) = \mathbf{c}$

$\mathbf{v}_p(t) = s\cos\theta\,\mathbf{i} + (s\sin\theta - 32t)\mathbf{j}$

$\mathbf{r}_p(t) = st\cos\theta\,\mathbf{i} + (st\sin\theta - 16t^2)\mathbf{j} + \mathbf{b}.$

Since $\mathbf{r}_p(0) = \mathbf{0}$, $\mathbf{b} = \mathbf{0}$ and $\mathbf{r}_p(t) = st\cos\theta\,\mathbf{i} + (st\sin\theta - 16t^2)\mathbf{j}$. Also, $\mathbf{v}_t(t) = -32t\mathbf{j} + \mathbf{c}$ and since $\mathbf{v}_t(0) = \mathbf{0}$, $\mathbf{c} = \mathbf{0}$ and $\mathbf{v}_t(t) = -32t\mathbf{j}$. Then $\mathbf{r}_t(t) = -16t^2\mathbf{j} + \mathbf{b}$. Since $\mathbf{r}_t(0) = x_0\mathbf{i} + y_0\mathbf{j}$, $\mathbf{b}x_0\mathbf{i} + y_0\mathbf{j}$ and $\mathbf{r}_t(t) = x_0\mathbf{i} + (y_0 - 16t^2)\mathbf{j}$. Now, the horizontal component of $\mathbf{r}_p($ will be x_0 when $t = x_0/s\cos\theta$ at which time the vertical component of $\mathbf{r}_p(t)$ will be

$$(sx_0/s\cos\theta)\sin\theta - 16(x_0/s\cos\theta)^2 = x_0\tan\theta - 16(x_0/s\cos\theta)^2 = y-) - 16(x_0/s\cos\theta)^2.$$

Thus, $\mathbf{r}_p(x_0/s\cos\theta) = \mathbf{r}_t(x_0/s\cos\theta)$ and the projectile will strike the target as it falls.

21. By Problem 17, $a = v^2/v_0 = 1530^2/(4000\cdot 5280) \approx 0.1108$. We are given $mg = 192$, so $m = 192/32$ and $w_e = 1192 - (192/32)(0.1108) \approx 191.33$ lb.

23. Solving $x(t) = (v_0\cos\theta)t$ for t and substituting into $y(t) - \frac{1}{2}gt^2 + (v_0\sin\theta)t + s_0$ we obtain

$$y = -\frac{1}{2}g\left(\frac{x}{v_0\cos\theta}\right)^2 + (v_0\sin\theta)\frac{x}{v_0\cos\theta} + s = -\frac{g}{2v_0^2\cos^2\theta}x^2 + (\tan\theta)x + s_0,$$

which is the equation of a parabola.

25. Letting $\mathbf{r}(t) = x(t)\mathbf{i} + y(t)\mathbf{j} + z(t)\mathbf{k}$, the equation $d\mathbf{r}/dt = \mathbf{v}$ is equivalent to $dx/dt = 6t^2x$, $dy/dt = -4ty^2$, $dz/dt = 2t(z+1)$. Separating the variables and integrating, we obtain $x/x = 6t^2dt$, $dy/y^2 = -4tdt$, $dz/(z+1) = 2tdt$, and $\ln x = 2t^3 + c_1$, $-1/y = 2t^2 + c_2$, $\ln(z+1) + t^2 + c_3$. Thus,

$$\mathbf{r}(t) = k_1 e^{2t^3}\mathbf{i} + \frac{1}{2t^2 + k_2}\mathbf{j} + (k_3 e^{t^2} - 1)\mathbf{k}.$$

27. (a) Since $\mathbf{F}$ is directed along $\mathbf{r}$ we have $\mathbf{F} = c\mathbf{r}$ for some constant c. Then

$$\tau = \mathbf{r}\times\mathbf{F} = \mathbf{r}\times(c\mathbf{r}) = c(\mathbf{r}\times\mathbf{r}) = \mathbf{0}.$$

(b) If $\tau = \mathbf{0}$ then $d\mathbf{L}/dt = \mathbf{0}$ and $\mathbf{L}$ is constant.

12.4 Curvature and Acceleration

1. $\mathbf{r}'(t) = -t\sin t\,\mathbf{i} + t\cos t\,\mathbf{j} + 2t\mathbf{k};\quad |\mathbf{r}'(t)| = \sqrt{t^2\sin^2 t + t^2\cos^2 t + 4t^2} = \sqrt{5}t;\ \mathbf{T} = -\frac{\sin t}{\sqrt{5}}\mathbf{i} + \frac{\cos t}{\sqrt{5}}\mathbf{j} + \frac{2}{\sqrt{5}}\mathbf{k}$

3. We assume $a > 0$. $\mathbf{r}'(t) = -a\sin t\,\mathbf{i} + a\cos t\,\mathbf{j} + c\mathbf{k};\quad |\mathbf{r}'(t)| = \sqrt{a^2\sin^2 t + a^2\cos^2 t + c^2} = \sqrt{a^2 + c^2};$

$\mathbf{T}(t) - \frac{a\sin t}{\sqrt{a^2+c^2}}\mathbf{i} + \frac{a\cos t}{\sqrt{a^2+c^2}}\mathbf{j} + \frac{c}{\sqrt{a^2+c^2}}\mathbf{k};\quad \frac{d\mathbf{T}}{dt} = -\frac{a\cos t}{\sqrt{a^2+c^2}}\mathbf{i} - \frac{a\sin t}{\sqrt{a^2+c^2}}\mathbf{j},$

$\left|\frac{d\mathbf{T}}{dt}\right| = \sqrt{\frac{a^2\cos^2 t}{a^2+c^2} + \frac{a^2\sin^2 t}{a^2+c^2}} = \frac{a}{\sqrt{a^2+c^2}};\quad \mathbf{N} = -\cos t\,\mathbf{i} - \sin t\,\mathbf{j};$

$$\mathbf{B} = \mathbf{T} \times \mathbf{N} = \begin{vmatrix} \mathbf{i} & \mathbf{j} & \mathbf{k} \\ -\dfrac{a\sin t}{\sqrt{a^2+c^2}} & \dfrac{a\cos t}{\sqrt{a^2+c^2}} & \dfrac{c}{\sqrt{a^2+c^2}} \\ -\cos t & -\sin t & 0 \end{vmatrix} = \dfrac{c\sin t}{\sqrt{a^2+c^2}}\mathbf{i} - \dfrac{c\cos t}{\sqrt{a^2+c^2}} + \dfrac{a}{\sqrt{a^2+c^2}}\mathbf{k};$$

$$\kappa = \dfrac{|d\mathbf{T}/dt|}{\mathbf{r}'(t)} = \dfrac{a/\sqrt{a^2+c^2}}{\sqrt{a^2+c^2}} = \dfrac{a}{a^2+c^2}$$

5. From Example 1 in the text, a normal to the osculating plane is $\mathbf{B}(\pi/4) = \frac{1}{26}(3\mathbf{i} - 3\mathbf{j} + 2\sqrt{2}\mathbf{k})$. The point on the curve when $t = \pi/4$ is $(\sqrt{2}, \sqrt{2}, 3\pi/4)$. An equation of the plane is $3(x - \sqrt{2}) - 3(y - \sqrt{2}) + 2\sqrt{2}(z - 3\pi/4 (= 0, \; 3x - 3y + 2\sqrt{2}z = 3\pi/2,$ or $3\sqrt{2}x - 3\sqrt{2}y + 4z = 3\pi.$

7. $\mathbf{v}(t) = \mathbf{j} + 2t\mathbf{k}, \quad |\mathbf{v}(t)| = \sqrt{1 + 4t^2}; \quad \mathbf{a}(t) = 2\mathbf{k}; \quad \mathbf{v} \cdot \mathbf{a} = 4t, \quad \mathbf{v} \times \mathbf{a} = 2\mathbf{i}, \quad |\mathbf{v} \times \mathbf{a}| = 2; \; a_{\mathbf{T}} = \dfrac{4t}{\sqrt{1 + 4t^2}}, \quad a_{\mathbf{N}} = \dfrac{2}{\sqrt{1 + 4t^2}}$

9. $\mathbf{v}(t) = 2t\mathbf{i} + 2t\mathbf{j} + 4t\mathbf{k}, \quad |\mathbf{v}(t)| = 2\sqrt{6}t, \quad t > 0; \quad \mathbf{a}(t) = 2\mathbf{i} + 2\mathbf{j} + 4\mathbf{k}; \quad \mathbf{v} \cdot \mathbf{a} = 24t, \quad \mathbf{v} \times \mathbf{a} = \mathbf{0}; \; a_{\mathbf{T}} = \dfrac{24t}{2\sqrt{6}t} = 2\sqrt{6},$
$a_{\mathbf{N}} = 0, \; t > 0$

11. $\mathbf{v}(t) = 2\mathbf{i} + 2t\mathbf{j}, \quad |\mathbf{v}(t)| = 2\sqrt{1 + t^2}; \quad \mathbf{a}(t) = 2\mathbf{j}; \quad \mathbf{v} \times \mathbf{a} = 4\mathbf{k}, \quad |\mathbf{v} \times \mathbf{a}| = 4; a_{\mathbf{T}} = \dfrac{2t}{\sqrt{1 + t^2}}, \quad a_{\mathbf{N}} = \dfrac{2}{\sqrt{1 + t^2}}$

13. $\mathbf{v}(t) = -5\sin t\mathbf{i} + 5\cos t\mathbf{j}, \quad |\mathbf{v}(t)| = 5; \quad \mathbf{a}(t) = -5\cos t\mathbf{i} - 5\sin t\mathbf{j}; \quad \mathbf{v} \cdot \mathbf{a} = 0, \mathbf{v} \times \mathbf{a} = 25\mathbf{k}, \quad |\mathbf{v} \times \mathbf{a}| = 25; a_{\mathbf{T}} = 0, \quad a_{\mathbf{N}} = 5$

15. $\mathbf{v}(t) = e^t(\mathbf{i} + \mathbf{j} + \mathbf{k}), \quad |\mathbf{v}(t)| = \sqrt{3}e^{-t}; \quad \mathbf{a}(t) = e^{-t}(\mathbf{i} + \mathbf{j} + \mathbf{k}); \quad \mathbf{v} \cdot \mathbf{a} = -3e^{-2t}; \quad \mathbf{v} \times \mathbf{a} = \mathbf{0}, |\mathbf{v} \times \mathbf{a}| = 0; \quad a_{\mathbf{T}} = -\sqrt{3}e^{-t},$
$a_{\mathbf{N}} = 0$

17. $\mathbf{v}(t) = -a\sin t\mathbf{i} + b\cos t\mathbf{j} + c\mathbf{k}, \quad |\mathbf{v}(t)| = \sqrt{a^2\sin^2 t + b^2\cos^2 t + c^2}; \quad \mathbf{a}(t) = -a\cos t\mathbf{i} - b\sin t\mathbf{j}; \mathbf{v} \times \mathbf{a} = bc\sin t\mathbf{i} - ac\cos t\mathbf{j} +$
$ab\mathbf{k}, \quad |\mathbf{v} \times \mathbf{a}| = \sqrt{b^2c^2\sin^2 t + a^2c^2\cos^2 t + a^2b^2}\,\kappa = \dfrac{|\mathbf{v} \times \mathbf{a}|}{|\mathbf{v}|^3} = \dfrac{\sqrt{b^2c^2\sin^2 t + a^2c^2\cos^2 t + a^2b^2}}{(a^2\sin^2 t + b^2\cos^2 t + c^2)^{3/2}}$

19. The equation of a line is $\mathbf{v}(t) = \mathbf{b} + t\mathbf{c}$, when $\mathbf{b}$ and $\mathbf{c}$ are constant vectors. $\mathbf{v}(t) = \mathbf{c}, \quad |\mathbf{v}(t)| = |\mathbf{c}|; \quad \mathbf{a}(t) = \mathbf{0}; \quad \mathbf{v} \times \mathbf{a} = \mathbf{0};$
$\kappa = |\mathbf{v} \times \mathbf{a}|/|\mathbf{v}|^3 = 0$

21. $\mathbf{v}(t) = f'(t)\mathbf{i} + g'(t)\mathbf{j}, \quad |\mathbf{v}(t)| = \sqrt{[f'(t)]^2 + [g'(t)]^2}; \quad \mathbf{a}(t) = f''(t)\mathbf{i} + g''(t)\mathbf{j}; \mathbf{v} \times \mathbf{a} = [f'(t)g''(t) - g'(t)f''(t)]\mathbf{k},$
$|\mathbf{v} \times \mathbf{a}| = |f'(t)g''(t) - g'(t)f''(t)|; \kappa = \dfrac{|\mathbf{v} \times \mathbf{a}|}{|\mathbf{v}|^3} = \dfrac{|f'(t)g''(t) - g'(t)f''(t)|}{([f'(t)]^2 + [g'(t)]^2)^{3/2}}$

23. $F(x) = x^2, \quad F(0) = 0, \quad F(1) = 1; \quad F'(x) = 2x, \quad F'(0) = 0, \quad F'(1) = 2; F''(x) = 2, \quad F''(0) = 2, \quad F''(1) = 2;$
$\kappa(0) = \dfrac{2}{(1 + 0^2)^{3/2}} = 2; \quad \rho(0) = \dfrac{1}{2}; \kappa(1) = \dfrac{2}{(1 + 2^2)^{3/2}} = \dfrac{2}{5\sqrt{5}} \approx 0.18; \rho(1) = \dfrac{5\sqrt{5}}{2} \approx 5.59;$ Since $2 > 2/5\sqrt{5}$, the curve is "sharper" at $(0,0)$.

25. Letting $F(x) = x^2$, we can use Problem 22 to get $\kappa(x) = \dfrac{|F''(x)|}{|1 + (F'(x))^2|^{3/2}}$.

Now, $F'(x)2x, \quad F''(x) = 2$, and $(F'(x))^2 = 4x^2$ so that $\kappa = \dfrac{2}{(1 + 4x^2)^{3/2}}$.
As $x \to \pm\infty$, the denominator grows without bound. Therefore, $\kappa(x) \to 0$ as $x \to \pm\infty$.

27. Since $(c, F(c))$ is an inflection point and F'' exists on an interval containg c, we must have $F''(c) = 0$. Therefore, using the formula from Problem 22, we see that the curvature is zero.

Chapter 12 in Review

A. True/False

1. True; $|\mathbf{v}(t)| = \sqrt{2}$

3. True

5. True

7. True

9. False; consider $\mathbf{r}_1(t) = \mathbf{r}_2(t) = \mathbf{i}$.

B. Fill In the Blanks

1. $y = 4$

3. $\mathbf{r}'(t) = \langle 1, 2t, t^2 \rangle$ so $\mathbf{r}'(1) = \langle 1, 2, 1 \rangle$

5. $\mathbf{r}'(1) \times \mathbf{r}''(1) = \begin{vmatrix} \mathbf{i} & \mathbf{j} & \mathbf{j} \\ 1 & 2 & 1 \\ 0 & 2 & 2 \end{vmatrix} = \langle 2, -2, 2 \rangle$ so $\mathbf{r}'(1) \times \mathbf{r}''(1) = \sqrt{12}$.

Since $\mathbf{r}'(1)| = \sqrt{6}$, we have $\kappa(1) = \dfrac{\mathbf{r}'(1) \times \mathbf{r}''(1)}{|\mathbf{r}'(1)|^3} = \dfrac{\sqrt{12}}{6\sqrt{6}} = \dfrac{\sqrt{2}}{6}$.

7. $\mathbf{T}(t) = \dfrac{\mathbf{r}'(t)}{|\mathbf{r}'(t)|} = \dfrac{\langle 1, 2t, t^2 \rangle}{\sqrt{1 + 4t^2 + t^4}} = \left\langle \dfrac{1}{\sqrt{1 + 4t^2 + t^4}}, \dfrac{2t}{\sqrt{1 + 4t^2 + t^4}}, \dfrac{t^2}{\sqrt{1 + 4t^2 + t^4}} \right\rangle$

So $\mathbf{T}'(t) = \left\langle \dfrac{-2(t^2 + 2)}{(t^4 + 4t^2 + 1)^{3/2}}, \dfrac{-2(t^4 - 1)}{(t^4 + 4t^2 + 1)^{3/2}}, \dfrac{2t(2t^2 + 1)}{(t^4 + 4t^2 + 1)^{3/2}} \right\rangle$.

This gives $\mathbf{T}'(1) = \left\langle \dfrac{-6}{6^{3/2}}, 0, \dfrac{6}{6^{3/2}} \right\rangle = \left\langle \dfrac{-1}{\sqrt{6}}, 0, \dfrac{1}{\sqrt{6}} \right\rangle$ and $|\mathbf{T}'(1)| = \sqrt{\tfrac{1}{6} + \tfrac{1}{6}} = \dfrac{1}{\sqrt{3}}$.

Therefore $\mathbf{N}(1) = \dfrac{\mathbf{T}'(1)}{|\mathbf{T}'(1)|} = \dfrac{\left\langle \frac{-1}{\sqrt{6}}, 0, \frac{1}{\sqrt{6}} \right\rangle}{\left(\frac{1}{\sqrt{3}} \right)} = \langle \tfrac{-1}{\sqrt{2}}, 0, \tfrac{1}{\sqrt{2}} \rangle$.

9. A normal to the normal plane is $\mathbf{T}(1) = \left\langle \frac{1}{\sqrt{6}}, \frac{2}{\sqrt{6}}, \frac{1}{\sqrt{6}} \right\rangle$ so we can use $\mathbf{n} = \langle 1, 2, 1 \rangle$ as a vector normal to the plane. Since $\mathbf{r}(1) = \langle 1, 1, \tfrac{1}{3} \rangle$, the point $(1, 1, \tfrac{1}{3})$ lies on the normal plane at $t = 1$. Thus an equation of the normal plane is $(x - 1) + 2(y - 1) + (z - \tfrac{1}{3}) = 0$ or $x + 2y + z = \tfrac{1}{3}$ or $3x + 6y + 3z = 10$

C. Exercises

1. $\mathbf{r}'(t) = \cos t \, \mathbf{i} + \sin t \, \mathbf{j} + \mathbf{k};$ $s = \int_0^\pi \sqrt{\cos^2 t + \sin^2 + 1} \, dt = \int_0^\pi \sqrt{2} \, dt = \sqrt{2}\pi$

3. $\mathbf{r}(3) = -27\mathbf{i} + 8\mathbf{j} + \mathbf{k};$ $\mathbf{r}'(t) = -6t\mathbf{i} = \dfrac{2}{\sqrt{t+1}} + \mathbf{k};$ $\mathbf{r}'(2) = -18\mathbf{i} + \mathbf{j} + \mathbf{k}.$ The tangent line is $x = -27 - 18t$, $y = 8 + t$, $z = 1 + t$.

5.

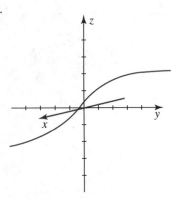

7. $\dfrac{d}{dt}[\mathbf{r}_1(t) \cdot \mathbf{r}_2(t)] = \mathbf{r}_1(t) \cdot \dfrac{d}{dt}\mathbf{r}_2(t) + \dfrac{d}{dt}\mathbf{r}_1(t) \cdot \mathbf{r}_2(t)$

$$= (\cos t\,\mathbf{i} - \sin t\,\mathbf{j} + 4t^3\mathbf{k}) \cdot (2t\,\mathbf{i} + \sin t\,\mathbf{j} + 2e^{2t}\mathbf{k})$$

$$(-\sin t\,\mathbf{i} - \cos t\,\mathbf{j} + 12t^2\mathbf{k}) \cdot (t^2\mathbf{i} + \sin t\,\mathbf{j} + e^{2t}\mathbf{k})$$

$$= 2t\cos t - \sin t\cos t + 8t^3 e^{2t} - t^2\sin t - \sin t\cos t + 12t^2 e^{2t}$$

$$= 2t\cos t - t^2\sin t - 2\sin t\cos t + 8t^3 e^{2t} + 12t^2 e^{2t}$$

$\dfrac{d}{dt}[\mathbf{r}_1(t) \cdot \mathbf{r}_2(t)] = \dfrac{d}{dt}[t^2\cos t - \sin^2 t + 4t^3 e^{2t}] = -t^2\sin t + 2t\cos t - 2\sin t\cos t + 8t^3 e^{2t} + 12t^2 e^{2t}$

9. We are given $\mathbf{F} = m\mathbf{a} = 2\mathbf{j}$; $\mathbf{v}(0) = \mathbf{i}+\mathbf{j}+\mathbf{k}$. and $\mathbf{r}(0) = \mathbf{i}+\mathbf{j}$. Then

$$\mathbf{v}(t) = \int \mathbf{a}(t)\,dt = \int \frac{2}{m}\mathbf{j}\,dt = \frac{2}{m}t\mathbf{j} + \mathbf{c}$$

$$\mathbf{i} = \mathbf{j} + \mathbf{k} = \mathbf{v}(0) = \mathbf{c}$$

$$\mathbf{v}(t) = \mathbf{i} + \left(\frac{2}{m}t + 1\right)\mathbf{j} + \mathbf{k}$$

$$\mathbf{r}(t) = t\mathbf{i} + \left(\frac{1}{m}t^2 + t\right)\mathbf{j} + t\mathbf{k} + \mathbf{b}$$

$$\mathbf{i} + \mathbf{j} = \mathbf{r}(0) = \mathbf{b}$$

$$\mathbf{r}(t) = (t+1)\mathbf{i} + \left(\frac{1}{m}t^2 + t + 1\right)\mathbf{j} + t\mathbf{k}$$

The parametric equations are $x = t$, $y = \dfrac{1}{m}t^2 + t + 1$, $z = t$.

11. $\mathbf{v}(t) = 6\mathbf{i} + \mathbf{j} + 2t\mathbf{k}$; $\mathbf{a}(t) = 2\mathbf{k}$. To find when the particle passes through the plane, we solve $-6t + t + t^2 = -4$ or $t^2 - 5t + 4 = 0$. This gives $t = 1$ and $t = 4$. $\mathbf{v}(1) = 6\mathbf{i} + \mathbf{j} + 2\mathbf{k}$, $\mathbf{a}(1) = 2\mathbf{k}$; $\mathbf{v}(4) = 6\mathbf{i} + \mathbf{j} + 8\mathbf{k}$, $\mathbf{a}(4) = 2\mathbf{k}$

13. $\mathbf{v}(t) = \int \mathbf{a}(t)\,dt = \int(\sqrt{2}\sin t\,\mathbf{i} + \sqrt{2}\cos t\,\mathbf{j})\,dt = -\sqrt{2}\cos t\,\mathbf{i} + \sqrt{2}\sin t\,\mathbf{j} + \mathbf{c}$;
$-\mathbf{i} + \mathbf{j} + \mathbf{k} = \mathbf{v}(\pi/4) = -\mathbf{i} + \mathbf{j} + \mathbf{c}$, $\mathbf{c} = \mathbf{k}$; $\mathbf{v}(t) = -\sqrt{2}\cos t\,\mathbf{i} + \sqrt{2}\sin t\,\mathbf{j} + \mathbf{k}$;
$\mathbf{r}(t) = -\sqrt{2}\sin t\,\mathbf{i} - \sqrt{2}\cos t\,\mathbf{j} + t\mathbf{k} + \mathbf{b}$; $\mathbf{i} + 2\mathbf{j} + (\pi/4)\mathbf{k} = \mathbf{r}(\pi/4) = -\mathbf{i} - \mathbf{j} + (\pi/4)\mathbf{k} + \mathbf{b}$, $\mathbf{b} = 2\mathbf{i} + 3\mathbf{j}$;
$\mathbf{r}(t) = (2 - 2\sqrt{2}\sin t)\mathbf{i} + (3 - \sqrt{2}\cos t)\mathbf{j} + t\mathbf{k}$; $\mathbf{r}(3\pi/4) = \mathbf{i} + 4\mathbf{j} + (3\pi/4)\mathbf{k}$

15. $\mathbf{r}'(t) = \sinh t\,\mathbf{i} + \cosh t\,\mathbf{j} + \mathbf{k}$, $\mathbf{r}'(1) = \sinh 1\,\mathbf{i} + \cosh 1\,\mathbf{j} + \mathbf{k}$;
$|\mathbf{r}'(t)| = \sqrt{\sinh^2 t + \cosh^2 t + 1} = \sqrt{2\cosh^2 t} = \sqrt{2}\cosh t$; $|\mathbf{r}'(1)| = \sqrt{2}\cosh 1$;
$\mathbf{T} = \dfrac{1}{\sqrt{2}}\tanh t\,\mathbf{i} + \dfrac{1}{\sqrt{2}}\mathbf{j} + \dfrac{1}{\sqrt{2}}\operatorname{sech} t\,\mathbf{k}$, $\mathbf{T}(1) = \dfrac{1}{\sqrt{2}}(\tanh 1\,\mathbf{i} + \mathbf{j} + \operatorname{sech} 1\,\mathbf{k})$;
$\dfrac{d\mathbf{T}}{dt} = \dfrac{1}{\sqrt{2}}\operatorname{sech}^2 t\,\mathbf{i} - \dfrac{1}{\sqrt{2}}\operatorname{sech} t\,\tanh t\,\mathbf{k}$; $\dfrac{d}{dt}\mathbf{T}(1) = \dfrac{1}{\sqrt{2}}\operatorname{sech}^2 1\,\mathbf{i} - \dfrac{1}{\sqrt{2}}\operatorname{sech} 1\,\tanh 1\,\mathbf{k}$,
$\left|\dfrac{d}{dt}\mathbf{T}(1)\right| = \dfrac{\operatorname{sech} 1}{\sqrt{2}}\sqrt{\operatorname{sech}^2 1 + \tanh^2 1} = \dfrac{1}{\sqrt{2}}\operatorname{sech} 1$; $\mathbf{N}(1) = \operatorname{sech} 1\,\mathbf{i} - \tanh 1\,\mathbf{k}$;
$\mathbf{B}(1) = \mathbf{T}(1) \times \mathbf{N}(1) = -\dfrac{1}{\sqrt{2}}\tanh 1\,\mathbf{i} + \dfrac{1}{\sqrt{2}}(\tanh^2 1 + \operatorname{sech}^2 1)\mathbf{j} - \dfrac{1}{\sqrt{2}}\operatorname{sech} 1\,\mathbf{k}$

$$= \dfrac{1}{\sqrt{2}}(-\tanh 1\,\mathbf{i} + \mathbf{j} - \operatorname{sech} 1\,\mathbf{k})$$

$\kappa = \left|\dfrac{d}{dt}\mathbf{T}(1)\right| / |\mathbf{r}'(1)| = \dfrac{(\operatorname{sech} 1)/\sqrt{2}}{\sqrt{2}\cosh 1} = \dfrac{1}{2}\operatorname{sech}^2 1$

Chapter 13

Partial Derivatives

13.1   Functions of Several Variables

1. $\{(x,y)|(x,y) \neq (0,0)\}$

3. $\{(t,Y)|y \neq x^2\}$

5. $\{(s,t)|s,\ t \text{ any real numbers}\}$

7. $\{(r,s)|\ |s| \geq 1\}$

9. $(u,v,w)|u^2 + v^2 + q^2 \geq 16$

11. (c); The domain of $f(x,y) = \sqrt{x} + \sqrt{y-x}$ is $\{(x,y)|x \geq 0,\ y-x \geq 0\} = \{(x,y)|x \geq 0,\ y \geq x\}$

13. (b); The domain of $f(x,y) = \ln(x-y^2)$ is $\{(x,y)|x-y^2 > 0\} = \{(x,y)|x > y^2\}$

15. (d); The domain of $f(x,y) = \sqrt{\frac{x}{y}-1}$ is $\left\{(x,y)|\frac{x}{y}-1 \geq 0\right\} = \left\{(x,y)|\frac{x}{y} \geq 1\right\}$

17. (f); The domain of $f(x,y) = \sin^{-1}(xy)$ is $\{(x,y)|\ |xy| \leq 1\}$

19. $\{(x,y)|x \geq 0 \text{ and } y \geq 0\}$

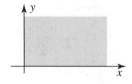

21. $\{(x,y)|y-x \geq 0\}$

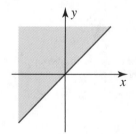

23. $\{z\,|\,z \geq 10\}$

25. $\{w\,||; -1 \leq w \leq 1\}$

27. $f(2,3) = \int_2^4 (2t-1)dt = (t^2-t)|_2^4 = 12-2 = 10$
    $f(-1,1) = \int_{-1}^1 (2t-1)dt = (t^2-t)|_{-1}^1 = 0-2 = -2$

29. $f(-1,1,-1) = (-2)^2 = 4; \quad f(2,3,-2) = 2^2 = 4$

31. A plane through the origin perpendicular to the xz-plane

33. The upper half of a cone lying above the xy-plane with axis along the positive z-axis

35. The upper half of an ellipsoid

37. $y = -\frac{1}{2}x + C$

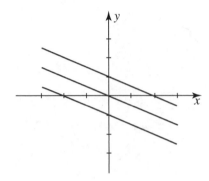

39. $x^2 - y^2 = 1 + c^2$

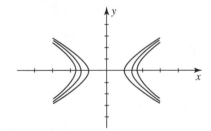

41. $y = x^2 + \ln c, \ c > 0$

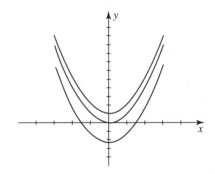

43. $x^2/9 + z^2/4 = c$; elliptical cylinder

45. $x^2 + 3y^2 + 6z^2 = c$; ellipsoid

47.

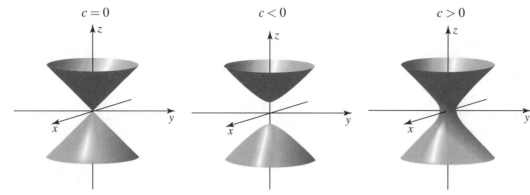

$c = 0$ $c < 0$ $c > 0$

49.

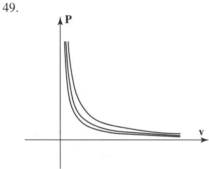

51. $C(r,h) = \pi r^2 (1.8) + \pi r^2 (1) + 2\pi h (2.3) = 2.8\pi r^2 + 4.6\pi rh$

53. $V + \pi r^2 g + \frac{1}{3}\pi r^2 \left(\frac{2}{3}h\right) = \frac{11}{9}\pi r^2 h$

55. $X = 2(156)(50) = 15,600$ sq cm

57. (a) The distance the water falls in time t is $s(t) = \frac{1}{2}gt^2 + vt$ where v is the velocity of the water at the top level ($t = 0$). The velocity of the water at time t is $v(t) = gt + v$. If t_1 is the time it takes a cross-section of water to fall from the top level to the bottom level, then $V = gt_1 + v$ and $t_1 = (V - v)/g$. The distance traveled in time t_1 is

$$h = \frac{1}{2}gt_1^2 + vt_1 = \frac{1}{2}g\left(\frac{V-v}{g}\right)^2 + v\left(\frac{V-v}{g}\right)$$

Simplifying the equation we obtain $2gh = V^2 - v^2$. Now the rates at the top and bottom levels are $Z = v\pi r^2$ and $Q = V\pi r^2$ (recall that the flow rate is constant). Solving for v and V and substituting into $2gh = V^2 - v^2$ we obtain $2gh = (Q/\pi r^2)^2 - (Q/\pi R^2)^2$. Solving for Q we find $Q = \dfrac{\pi r^2 R^2 \sqrt{2gh}}{\sqrt{R^4 - r^4}}$.

(b) When $r = 0.2$ cm, $R = 1$ cm, and $h = 10$, $Q \approx 7.61$ cm^3/s.

13.2 Limits and Continuity

1. $\displaystyle\lim_{(x,y)\to(5,-1)} (x^2 + y^2) = 25 + 1 = 26$

3. On $y = 0$, $\displaystyle\lim_{(x,y)\to(0,0)} \frac{5x^2 + y^2}{x^2 + y^2} = \lim_{(x,y)\to(0,0)} \frac{5x^2}{x^2} = 5$.

   On $x = 0$, $\displaystyle\lim_{(x,y)\to(0,0)} \frac{5x^2 + y^2}{x^2 + y^2} = \lim_{(x,y)\to(0,0)} \frac{y^2}{y^2} = 1$. The limit does not exist.

5. $\displaystyle\lim_{(x,y)\to(1,1)} \frac{4 - x^2 - y^2}{x^2 + y^2} = \frac{4 - 1 - 1}{1 + 1} = 1$

نوع (x,y) سین ... عبرو و راه نوع

path كين

7. On $y = x$, $\displaystyle\lim_{(x,y)\to(0,0)} \frac{x^2y}{x^4+y^2} = \lim_{(x,y)\to(0,0)} \frac{x^3}{x^4+x^2} = \lim_{(x,y)\to(0,0)} \frac{x}{x^2+1} = 0.$

On $y = x^2$, $\displaystyle\lim_{(x,y)\to(0,0)} \frac{x^2y}{x^4+y^2} = \lim_{(x,y)\to(0,0)} \frac{x^4}{x^4+x^4} = \frac{1}{2}.$ The limit does not exist.

9. $\displaystyle\lim_{(x,y)\to(1,2)} x^3y^2(x+y)^3 = 1(4)(27) = 108$

11. $\displaystyle\lim_{(x,y)\to(0,0)} \frac{e^{xy}}{x+y+1} = \frac{1}{1} = 1$

13. $\displaystyle\lim_{(x,y)\to(2,2)} \frac{xy}{x^3+y^2} = \frac{4}{8+4} = \frac{1}{3}$

15. $\displaystyle\lim_{(x,y)\to(0,0)} \frac{x^2-3y+1}{x+5y-3} = -\frac{1}{3}$

17. $\displaystyle\lim_{(x,y)\to(4,3)} xy^2\frac{x+2y}{x-y} = 4(9)\frac{4+6}{4-3} = 360$

19. $\displaystyle\lim_{(x,y)\to(1,1)} \frac{xy-x-y+1}{x^2+y^2-2x-2y+2} = \lim_{(x,y)\to(1,1)} \frac{(x-1)(y-1)}{(x-1)^2+m^2(x-1)^2}$

On $y - x = m(x-1)$,

$\displaystyle\lim_{(x,y)\to(1,1)} \frac{(x-1)(y-1)}{(x-1)^2+m^2(x-1)^2} = \lim_{(x,y)\to(1,1)} \frac{(x-1)m(x-1)}{(x-1)^2+m^2(x-1)^2} = \frac{m}{1+m^2}.$

The limit does not exist.

21. $\displaystyle\lim_{(x,y)\to(0,0)} \frac{x^3y+xy^3-3x^2-3y^2}{x^2+y^2} = \lim_{(x,y)\to(0,0)} \frac{xy(x^2+y^2)-3(x^2+y^2)}{x^2+y^2}$

$\displaystyle = \lim_{(x,y)\to(0,0)} (xy-3) = -3$

23. $\displaystyle\lim_{(x,y)\to(1,1)} \ln(2x^2-y^2) = \ln(2-1) = 0$

In Problems 25–30 let $x = r\cos\theta$ and $y = r\sin\theta$. Then $x^2+y^2 = r^2$ and $(x,y) \to (0,0)$ if and only if $r \to 0$. We also use the facts that $|\cos\theta| \leq 1$ and $|\sin\theta| \leq 1$ for all θ.

25. $\displaystyle\lim_{(x,y)\to(0,0)} \frac{(x^2-y^2)^2}{x^2+y^2} = \lim_{r\to0} \frac{(r^2\cos^2\theta - r^2\sin^2\theta)^2}{r^2} = \lim_{r\to0} \frac{r^4(\cos^2\theta-\sin^2\theta)^2}{r^2}$

$\displaystyle = \lim_{r\to0} r^2\cos^2 2\theta = 0$

27. $\displaystyle\lim_{(x,y)\to(0,0)} \frac{6xy}{\sqrt{x^2+y^2}} = \lim_{r\to0} \frac{6r^2\cos\theta\sin\theta}{\sqrt{r^2}} = \lim_{r\to0} 3|r|\sin2\theta = 0$

29. $\displaystyle\lim_{(x,y)\to(0,0)} \frac{x^3}{x^2+y^2} = \lim_{r\to0} \frac{r^3\cos^3\theta}{r^2} = \lim_{r\to0} r\cos^3\theta = 0$

31. $\{(x,y) \mid x \geq 0 \text{ and } y \geq -x\}$

33. $\{(x,y) \mid y \neq 0 \text{ and } x/y \neq \pi/2 + k\pi, k \text{ and integer}\}$

35. (a) For $x^2+y^2 < 1$, $f(x,y) = 0$ is continuous

(b) For $x \geq 0$, $f(x,y)$ is not continuous since it is discontinuous at $(2,0)$.

(c) For $y > x$, $f(x,y)$ is not continuous since it is discontinuous at $(2,3)$.

37. Since

$$\lim_{(x,y)\to(0,0)} f(x,y) = \lim_{(x,y)\to(0,0)} \frac{6x^2y^3}{(x^2+y^2)^2} = \lim_{r\to0} \frac{6r^5\cos^2\theta\sin^3\theta}{r^4} = \lim_{r\to0} 6r\cos^2\theta\sin^3\theta = 0 = f(0,0)$$

the function is continuous at $(0,0)$.

39. Choose $\varepsilon > 0$. Using $x = r\cos\theta$ and $y = r\sin\theta$ we have

$$\frac{3xy^2}{2x^2 + 2y^2} = \frac{3t\cos\theta\, r^2\sin^2\theta}{2r^2} \frac{3}{2} r\cos\theta\sin^2\theta.$$

Let $\delta = \frac{2\varepsilon}{3}$. Now, whenever $r = \sqrt{x^2 + y^2} < \delta$, we have

$$\left|\frac{3xy^2}{2x^2 + 2y^2}\right| = \frac{3}{2}|r\cos\theta\sin^2\theta| \le \frac{3}{2}|r| < \frac{3}{2}\delta = \frac{3}{2}\left(\frac{2\varepsilon}{3}\right) = \varepsilon.$$

Thus $\displaystyle\lim_{(x,y)\to(0,0)} \frac{3xy^2}{x^2 + y^2} = 0$.

41. Where $y \ne x$, we have

$$f(x,y) = \frac{x^3 - y^3}{x - y} = \frac{(x-y)(x^2 + xy + y^2)}{x - y} = x^2 + xy + y^2.$$

When $y = x$, we have

$$x^2 + xy + y^2 = x^2 + x^2 + x^2 = 3x^2 = f(x,y).$$

Therefore, $f(x,y) = x^2 + xy + y^2$ throughout the entire plane. Since $x^2 + xy + y^2$ is a polynomial, f must be continuous throughout the plane and thus has no discontinuities.

13.3 Partial Derivatives

1. $\dfrac{\partial z}{\partial x} = \lim\limits_{\triangle \to 0} \dfrac{7(x = \triangle x) + 8y^2 - 7x - 8y^2}{\triangle x} = \lim\limits_{\triangle \to 0} \dfrac{7\triangle x}{\triangle x} = 7$

$\dfrac{\partial z}{\partial y} = \lim\limits_{\triangle y \to 0} \dfrac{7x + 8(y + \triangle y)^2 - 7x - 8y^2}{\triangle y} = \lim\limits_{\triangle y \to 0} \dfrac{16y\triangle y + 8(\triangle y)^2}{\triangle y}$

$\qquad = \lim\limits_{\triangle y \to 0}(16y + 8\triangle y) = 16y$

3. $\dfrac{\partial z}{\partial x} = \lim\limits_{\triangle x \to 0} \dfrac{3(x + \triangle x)^2 y + 4x + \triangle x)y^2 - 3x^2 y - 4xy^2}{\triangle x}$

$\qquad = \lim\limits_{\triangle x \to 0} \dfrac{3x^2 y + 6x(\triangle x)y + 3(\triangle x)^2 y + 4xy^2 + 4(\triangle x)y^2 - 3x^2 y - 4xy^2}{\triangle x}$

$\qquad = \lim\limits_{\triangle x \to 0} \dfrac{6x(\triangle x)y + 3(\triangle x)^2 y + 4(\triangle x)y^2}{\triangle x} = \lim\limits_{\triangle x \to 0}(6xy + 3(\triangle x)y + 4y^2) = 6xy + 4y^2$

$\dfrac{\partial z}{\partial y} = \lim\limits_{\triangle y \to 0} \dfrac{3x^2(y + \triangle y) + 4x(y + \triangle y)^2 - 3x^2 y - 4xy^2}{\triangle y}$

$\qquad = \lim\limits_{\triangle y \to 0} \dfrac{3x^2 y + 3x^2\triangle y + 4xy^2 + 8xy\triangle y + 4x(\triangle y)^2 - 3x^2 y - 4xy^2}{\triangle y}$

$\qquad = \lim\limits_{\triangle y \to 0} \dfrac{3x^2\triangle y + 8xy\triangle y + 4x(\triangle y)^2}{\triangle y} = \lim\limits_{\triangle y \to 0}(3x^2 + 8xy + 4x\triangle y) = 3x^2 + 8xy$

5. $z_x = 2x - y^2$; $z_y = -2xy + 20y^4$

7. $z_x = 20x^3 y^3 - 2xy^6 + 30x^4$; $z_y = 15x^4 y^2 - 6x^2 y^5 - 4$

9. $z_x = \dfrac{2}{\sqrt{x}(3y^2 + 1)}$; $z_y = -\dfrac{24y\sqrt{x}}{(3y^2 + 1)^2}$

11. $z_x = -(x^3 - y^2)^{-2}(3x^2) = -3x^2(x^3 - y^2)^{-2}$; $z_y = -(x^3 - y^2)^{-2}(-2y) = 2y(x^3 - y^2)^{-2}$

13. $z_x = 2(\cos 5x)(-\sin 5x)(5) = -10\sin 5x\cos 5x$; $z_y = 2(\sin 5y)(\cos 5y)(5) = 10\sin 5y\cos 5y$

15. $f_x = x(3x^2 ye^{x^3 y}) + e^{x^3 y}$; $f_y = x^4 e^{x^3 y}$

17. $f_x = \dfrac{(x+2y)3 - (3x-y)}{(x+2y)^2} = \dfrac{7y}{(x+2y)^2};\ f_y = \dfrac{(x+2y)(-1) - (3x-y)(2)}{(x+2y)^2} = \dfrac{-7x}{(x+2y)^2}$

19. $g_u = \dfrac{8u}{4u^2 - 5v^3};\ g_v = \dfrac{15v^2}{4u^2 + 5v^3}$

21. $w_x = \dfrac{y}{\sqrt{x}};\ w_y = 2\sqrt{x} - y\left(\dfrac{1}{z}e^{y/z}\right) = 2\sqrt{x} - \left(\dfrac{y}{z} + 1\right)e^{y/z};\ w_z = -ye^{y/z}\left(-\dfrac{y}{z^2}\right) = \dfrac{y^2}{z^2}e^{y/z}$

23. $F_u = 2uw^2 - v^3 - vwt^2\sin(ut^2);\ F_v = -3uv^2 + w\cos(ut^2);$
    $F_x = 3(2x^2t)^3(4xt) = 16xt(2x^2t)^3 = 128x^7t^4;\ F_t = -2uvwt\sin(ut^2) + 64x^8t^3$

25. $z_y = 16x^3y^3,\ z_y(1,-1) = -16$

27. $f_y = \dfrac{(x+y)18x - 18xy}{(x+y)^2} = \dfrac{18x^2}{(x+y)^2},\ f_y(-1,4) = 2.$ An equation of the tangent line is given by $x = -1$ and $z + 24 = 2(y-4)$.
    Parametric equations of the line are $x = -1,\ y = 4 + t,\ z = -24 + 2t$.

29. $z_x = \dfrac{-x}{\sqrt{9 - x - y^2}},\ z_x(2,2) = -2$

31. $\dfrac{\partial z}{\partial x} = ye^{xy};\ \dfrac{\partial^2 z}{\partial x^2} = y^2e^{xy}$

33. $f_x = 10xy^2 - 2y^3;\ f_{xy} = 20xy - 6y^2$

35. $w_t = 3u^2v^3t^2,\ w_{tu}6uv^3t^2;\ w_{tuv}18uv^2t^2$

37. $F_r = 2re^{r^2}\cos\theta;\ F_{r\theta} - 2re^{r^2}\sin\theta;\ F_{r\theta r} - 2r(2re^{r^2})\sin\theta - 2e^{r^2}\sin\theta = -2e^{r^2}(2r^2 + 1)\sin\theta$

39. $\dfrac{\partial z}{\partial y} = -5x^4y^2 + 8xy;\ \dfrac{\partial^2 z}{\partial x\partial y} = -60x^3y^2 + 8y;\ \dfrac{\partial z}{\partial x} = 6x^5 - 20x^3y^3 + 4y^2;\ \dfrac{\partial^2 z}{\partial y\partial x} = -60x^3y^2 + 8y$

41. $w_u = 3u^2v^4 - 8uv^2t^3,\ w_{uv} = 12u^2v^3 - 16uvt^3,\ w_{uvt} = -48uvt^2;\ w_t = -12u^2v^2t^2 + v^2,$
    $w_{tv} = -24u^2vt^2 + 2v,\ w_{tvu} = -48uvt^2;\ w_v = 4u^3v^3 - 8u^2vt^3 + 2vt,\ w_{vu} = 12u^2v^3 - 16uvt^3,$
    $w_{vut} = -48uvt^2$

43. $2x + 2zz_x = 0,\ z_x = -x/z;\ 2y + 2zz_y = 0,\ z_y = -y/z$

45. $2zz_u + 2uv^3 - uvz_u - vz = 0 \Longrightarrow (2z - uv)z_u = vz - 2uv^3 \Longrightarrow z_u = \dfrac{vz - 2uv^3}{2z - uv};$

    $2zz_v + 3u^2v^2 - uvz_v - uz = 0 \Longrightarrow (2z - uv)z_v = uz - 3u^2v^2 \Longrightarrow z_v = \dfrac{uz - 3u^2v^2}{2z - uv}$

47. $a_x = y\sin\theta,\ A_y = x\sin\theta,\ A_\theta = xy\cos\theta$

49. $\dfrac{\partial u}{\partial x} = 2\pi(\cosh 2\pi y + \sinh 2\pi y)\cos 2\pi x;\ \dfrac{\partial^2 u}{\partial x^2} = -4\pi^2(\cosh 2\pi y + \sinh 2\pi y)\sin 2\pi x;$

    $\dfrac{\partial u}{\partial y} = (2\pi\sinh 2\pi y + 2\pi\cosh 2\pi y)\sin 2\pi x;\ \dfrac{\partial^2 u}{\partial y^2} = (4\pi^2\cosh 2\pi y + 4\pi^2\sinh 2\pi y)\sin 2\pi x;$

    $\dfrac{\partial^2 u}{\partial x^2} + \dfrac{\partial^2 u}{\partial y^2} = -4\pi^2(\cosh 2\pi y + \sinh 2\pi y)\sin 2\pi x + 4\pi^2(\cosh 2\pi y + \sinh 2\pi y)\sin 2\pi x = 0$

51. $\dfrac{\partial z}{\partial x} = \dfrac{2x}{x^2 + y^2},\ \dfrac{\partial^2 z}{\partial x^2} = \dfrac{(x^2 + y^2)2 - 2x(2x)}{(x^2 + y^2)^2} = \dfrac{2y^2 - 2x^2}{(x^2 + y^2)^2};\ \dfrac{\partial z}{\partial y} = \dfrac{2y}{x^2 + y^2},$
    $\dfrac{\partial^2 z}{\partial y^2} = \dfrac{(x^2 + y^2)2 - 2y(2y)}{(x^2 + y^2)^2} = \dfrac{2x^2 - 2y^2}{(x^2 + y^2)^2};\ \dfrac{\partial^2 z}{\partial x^2} + \dfrac{\partial^2 z}{\partial y^2} = \dfrac{2y^2 - 2x^2 + 2x^2 - 2y^2}{(x^2 + y^2)^2} = 0$

53. $\dfrac{\partial u}{\partial x} = -\dfrac{x}{(x^2+y^2+z^2)^{3/2}};$ $\dfrac{\partial u}{\partial y} = -\dfrac{y}{(x^2+y^2+z^2)^{3/2}};$ $\dfrac{\partial u}{\partial z} = -\dfrac{z}{(x^2+y^2+z^2)^{3/2}};$

$\dfrac{\partial^2 u}{\partial x^2} = \dfrac{2x^2-y^2-z^2}{(x^2+y^2+z^2)^{5/2}};$ $\dfrac{\partial^2 u}{\partial y^2} = \dfrac{-x^2+2y^2-z^2}{(x^2+y^2+z^2)^{5/2}};$ $\dfrac{\partial^2 u}{\partial z^2} = \dfrac{-x^2-y^2+2z^2}{(x^2+y^2+z^2)^{5/2}};$

$\dfrac{\partial^2 u}{\partial x^2} + \dfrac{\partial^2 u}{\partial y^2} + \dfrac{\partial^2 u}{\partial z^2} = \dfrac{2x^2-y^2-z^2-x^2+2y^2-z^2-x^2-y^2+2z^2}{(x^2+y^2+z^2)^{5/2}} = 0$

55. $\dfrac{\partial u}{\partial x} = \cos at \cos x,$ $\dfrac{\partial^2 u}{\partial x^2} = -\cos at \sin x;$ $\dfrac{\partial u}{\partial t} = -a \sin at \sin x,$ $\dfrac{\partial^2 u}{\partial t^2} = -a^1 \cos at \sin x;$

$a^2 \dfrac{\partial^2 u}{\partial x^2} = a^2(-\cos at \sin x) = \dfrac{\partial^2 u}{\partial t^2}$

57. $\dfrac{\partial C}{\partial x} = -\dfrac{2x}{kt} t^{-/12} e^{-x/kt},$ $\dfrac{\partial^2 C}{\partial x^2} = \dfrac{4x^2}{k^2t^2} t^{-1/2} e^{-x^2/kt} - \dfrac{2}{kt} t^{-1/2} e^{-x^2/kt};$

$\dfrac{\partial C}{\partial t} = t^{-1/2} \dfrac{x^2}{kt^2} e^{-x^2/kt} - \dfrac{t^{-3/2}}{2} e^{-x^2/kt};$ $\dfrac{k}{4} \dfrac{\partial^2 C}{\partial x^2} = \dfrac{x^2}{kt^2} t^{-1/2} e^{-x^2/kt} - \dfrac{t^{-1/2}}{2t} e^{-x^2/kt} = \dfrac{\partial C}{\partial t}$

59. (a) $\dfrac{\partial u}{\partial t} = \begin{cases} -gx/z, & 0 \le x \le at \\ -gt, & x > at \end{cases}$

For $x > at$, the motion is that of a freely falling body.

(b) For $x > at$, $\dfrac{\partial u}{\partial x} = 0$. For $x > at$, the string is horizontal.

61. (a) $\dfrac{\partial^2 z}{\partial x^2} = \lim\limits_{\Delta x \to 0} \dfrac{f_x(x+\Delta x, y) - f_x(x,y)}{\Delta x}$

(b) $\dfrac{\partial^2 z}{\partial y^2} = \lim\limits_{\Delta y \to 0} \dfrac{f_y(x, y+\Delta y) - f_x(x,y)}{\Delta y}$

(c) $\dfrac{\partial^2 z}{\partial x \partial y} = \lim\limits_{\Delta x \to 0} \dfrac{f_y(x+\Delta x, y) - f_y(x,y)}{\Delta x}$

63. Consider the mixed partials:

$$\dfrac{\partial^2 z}{\partial y \partial x} = \dfrac{\partial}{\partial y}\left(\dfrac{\partial z}{\partial x}\right) = 2y \quad \text{and} \quad \dfrac{\partial^2 z}{\partial x \partial y} = \dfrac{\partial}{\partial x}\left(\dfrac{\partial z}{\partial y}\right) = 2x.$$

Since $\dfrac{\partial z}{\partial x}, \dfrac{\partial z}{\partial y}, \dfrac{\partial^2 z}{\partial y \partial x},$ and $\dfrac{\partial^2 z}{\partial x \partial y}$ are all continuous on an open set, we should have $\dfrac{\partial^2 z}{\partial y \partial x} = \dfrac{\partial^2 z}{\partial x \partial y}$ on that set. But the mixed partials are equal only on the line $y = x$, which contains no open set in the plane. Therefore, such a function cannot exist.

65. (a) There slopes of the surface in the x and y directions are zero everywhere. This implies that the surface must have constant height everywhere. Therefore f must have the form $f(x,y) = c$.

(b) Since the mixed partials are both zero, we have

$$\dfrac{\partial}{\partial x}\left(\dfrac{\partial z}{\partial y}\right) = 0 \quad \text{and} \quad dfrac{\partial}{\partial y}\left(\dfrac{\partial z}{\partial x}\right) = 0$$

which implies $\dfrac{\partial z}{\partial y}$ is a function of y alone and $\dfrac{\partial z}{\partial x}$ is a function of x alone. Therefore, z has no term that depends on both x and y. Hence z is of the form $z = g(x) + h(y) + c$ where g and h are twice continuously differentiable functions of a single variable.

67. $\dfrac{\partial z}{\partial x}\Big|_{(0,0)} = \lim\limits_{\Delta x \to 0} \dfrac{f(0+\Delta x, 0) - f(0,0)}{\Delta x} = \lim\limits_{\Delta x \to 0} \dfrac{0/2(\Delta x)^2}{\Delta x} = 0;$

$\dfrac{\partial z}{\partial y}\Big|_{(0,0)} = \lim\limits_{\Delta y \to 0} \dfrac{f(0, 0+\Delta y) - f(0,0)}{\Delta y} = \lim\limits_{\Delta y \to 0} \dfrac{0/2(\Delta y)^2}{\Delta y} = 0$

13.4 Linearization and Differentials

1. $\dfrac{\partial f}{\partial x} = 4y^2 - 6x^2 y$ so $\dfrac{\partial f}{\partial x}(1,1) = -2$

   $\dfrac{\partial f}{\partial y} = 8zy - 2x^3$ so $\dfrac{\partial f}{\partial y}(1,1) = 6$

   $f(1,1) = 2$ The linearization is $L(x,y) = 2 - 2(x-1) + 6(y-1) = -2x + 6y - 2$

3. $\dfrac{\partial f}{\partial x} = \sqrt{x^2 + y^2} + \dfrac{x^2}{\sqrt{x^2 + y^2}}$ so $\dfrac{\partial f}{\partial x}(8,15) = \dfrac{353}{17}$

   $\dfrac{\partial f}{\partial y} = \dfrac{xy}{\sqrt{x^2 + y^2}}$ so $\dfrac{\partial f}{\partial y}(8,15) = \dfrac{120}{17}$

   $f(8,15) = 136$ The linearization is $L(x,y) = 136 + \frac{353}{17}(x-8) + \frac{120}{17}(y-15) = \frac{353}{17}x + \frac{120}{17}y - 136$

5. $\dfrac{\partial f}{\partial x} = \dfrac{2x}{x^2 + y^3}$ so $\dfrac{\partial f}{\partial x}(-1,1) = -1$

   $\dfrac{\partial f}{\partial y} = \dfrac{3y^2}{x^2 + y^3}$ so $\dfrac{\partial f}{\partial y}(-1,1) = \dfrac{3}{2}$

   $f(-1,1) = \ln(2)$ The linearization is $L(x,y) = \ln(2) - (x+1) + \frac{3}{2}(y-1) = -x + \dfrac{3}{2}y - \dfrac{5}{2} + \ln(2)$

7. Note that we are trying to approximate $f(102,80)$ where $f(x,y) = \sqrt{x} + \sqrt[4]{y}$. Since $(102,80)$ is reasonably close to $(100,81)$, we can use the linearization of f at $(100,81)$ to approximate the value at $(102,80)$. To do this, we compute

   $\dfrac{\partial f}{\partial x} = \dfrac{1}{2\sqrt{x}}$, $\dfrac{\partial f}{\partial x}(100,81) = \dfrac{1}{20}$, $\dfrac{\partial f}{\partial y} = \dfrac{1}{4y^{3/2}}$, $\dfrac{\partial f}{\partial y}(100,81) = \dfrac{1}{4(27)} = \dfrac{1}{108}$, and $f(100,81) = 13$

   The linearization is $L(x,y) = 13 + \frac{1}{20}(x-100) + \frac{1}{108}(y-81)$. For the approximation, we have $L9102,80) = 13 + \frac{1}{20}(102 - 100) + \frac{1}{108}(80-81) = 13 + \frac{1}{10} - \frac{1}{108} = \frac{7069}{540} \approx 13.0907$

9. First, linearize f at $(2,2)$. To do this, compute $\dfrac{\partial f}{\partial x} = 2(x^2 + y^2)2x$, $\dfrac{\partial f}{\partial x}(2,2) = 64$, $\dfrac{\partial f}{\partial y} = 2(x^2 + y^2)2y$, $\dfrac{\partial f}{\partial y}(2,2) = 64$, and

   $f(2,2,) = 64$.

   The linearizationis $L(x,y) = 64 + 64)x - 2) + 64)y - 2)$. For the approximation, we have $L(1.95, 2.01) = 64 + 64(-0.05) + 64(0.01) \approx 61.44$.

11. $dz = 2x \sin 4y\, dx + 4x^2 \cos 4y\, dy$

13. $dz = \dfrac{2x}{\sqrt{2x^2 - 4y^3}}dx - \dfrac{6y^2}{\sqrt{2x^2 - 4y^3}}dy$

15. $df = \dfrac{(s+3t)2 - (2s-t)}{(s+3t)^2}ds + \dfrac{(s+t)(-1) - (2s-t)3}{(s+3t)^2}dt = \dfrac{7t}{(s+3t)^2} - \dfrac{7s}{(s+3t)^2}dt$

17. $dw = 2xy^4 z^{-5}dx + 4x^2 y^3 z^{-5}dy - 5x^2 y^4 z^{-6}dz$

19. $dF = 3r^2 dr - 2s^{-3}ds - 2t^{-1/2}dt$

21. $w = \ln u + \ln v - \ln s - \ln t$; $dw = \dfrac{du}{u} + \dfrac{dv}{v} - \dfrac{ds}{s} - \dfrac{dt}{t}$

23. $\Delta z = z(2.2, 3.9) - z(2,4) = (6.6 + 15.6 + 8) - (6 + 16 + 8) = 0.1$; $dz = 3dx + 4dy$

    When $x = 2$, $y = 4$, $dx = 0.2$, and $dy = -0.1$, $dz = 3(0.2) + 4(-0.1) = 0.2$

25. $\Delta z = z(3.1, 0.8) - z(3,1) = (3.1 + 0.8)^2 - (3+1)^2 = 15.21 - 16 = -0.79$;

    $dz = 2(x+y)dx + 2(x+y)dy$. When $x = 3$, $y = 1$, $dx = 0.1$, and $dy = -0.2$,

    $dz = 2(3+1)(0.1) + 2(3+1)(0.2) = 0.8 - 1.6 = -0.8$

27. $\Delta z = 5(x + \Delta x)^2 + 3(y + \Delta y) - (x + \Delta x)(y + \Delta y) - (5x^2 + 3y - xy)$

    $= 10x\Delta x + 5(\Delta x)^2 + 3\Delta y - x\Delta y - y\Delta x - \Delta x\Delta y$

    $= (10x - y)\Delta x + (3 - z)\Delta y + (5\Delta x)\Delta x - (\Delta x)\Delta y$

    $\varepsilon_1 = 5\Delta x$, $\varepsilon_2 = -\Delta x$

29. $\Delta x = (x+\Delta x)^2(y+\Delta y)^2 - x^2 y^2 = [x^2 + 2x\Delta x + (\Delta x)^2][y^2 + 2y\Delta y + (\Delta y)^2] - x^2 y^2$

$= 2x^2 y\Delta y + x^2(\Delta y)^2 + 2xy^2\Delta x + 4xy(\Delta x)\Delta y + 2x(\Delta x)(\Delta y)^2 + y^2(\Delta x)^2 + 2y(\Delta x)^2\Delta y + (\Delta x)^2(\Delta y)^2$

$= 2xy^2\Delta x + 2x^2 y\Delta y + [4xy\Delta y + 2x(\Delta y)^2 + y^2 x]\Delta x + [x^2\Delta y + 2y(\Delta x)^2 + (\Delta x)^2\Delta y]\Delta y$

$\varepsilon_1 = 4xy\Delta y + 2x(\Delta y)^2 + y^2 x, \quad \varepsilon_2 = x^2\Delta y + 2y(\Delta x)^2 + (\Delta x)^2\Delta y$

(Several other choices of ε_1 and ε_2 are possible.)

31. $R = \dfrac{R_1 R_2 R_3}{R_2 R_3 + R_1 R_3 + R_1 R_2}; \quad \Delta R_1 = \pm 0.009 R_1, \quad \Delta R_! = \pm 0.009 R_2, \quad \Delta R_3 = \pm 0.0009 R_3$

$|\Delta R| \approx |dR| \leq \left| \dfrac{R_2^2 R_3^2}{(R_2 R_3 + R_1 R_3 + R_1 R_2)^2}(\pm 0.009 R_1) \right| + \left| \dfrac{R_1^2 R_3^2}{(R_2 R_3 + R_1 R_3 + R_1 R_2)^2}(\pm 0.009 R_2) \right|$

$+ \left| \dfrac{R_1^2 R_2^2}{(R_2 R_3 + R_1 R_3 + R_1 R_2)^2}(\pm 0.009 R_3) \right| = 0.009 R\left(\dfrac{R_2 R_3 + R_1 R_3 + R_1 R_2}{R_2 R_3 + R_1 R_3 + R_1 R_2} \right) = 0.009 R$

The maximum percentage error is approximately 0.9%.

33. $dT = mg\dfrac{(2r^2 + R^2 - R(2R)}{(2r^2 + R^2)^2}dR + mg\dfrac{-R(4r)}{2r^2 + R^2)^2}dr = mg\dfrac{2r^2 - R^2}{(2r + R^2)^2}dR - mg\dfrac{4rR}{(2r + R^2)^2}dr$

When $R = 4$, $r = 0.8$, $dR = 0.1$, and $dr = 0.1$,

$$\Delta T \approx dT = mg\left[\dfrac{2(0.8)^2 - 4^2}{[2(0.8)^2 + 4^2]^2}(0.1) - \dfrac{4(0.8)4}{[2(0.8)^2 + 4^2]^2}(0.1) \right]$$

$$= mg\left(\dfrac{-1.472 - 1.28}{298.598} \right) \approx -0.009 \text{ mg.}$$

The tension decreases.

35. $V = lwh$, $dV = whdl + lhdq + lwdh$. With $dl = \pm 0.02l$, $dw = \pm 0.05w$, and $dh = \pm 0.08h$,

$$|\Delta V| \approx |dV| = |wh(\pm 0.02l) + lh(\pm 0.05w) + lw(\pm 0.08h)| \leq lwh(0.02 + 0.5 + 0.8) = 0.15V.$$

The approximate percentage increase in volume is 15%.

37. $dS = 0.1091(0.425)w^{-0.575}h^{0.725}dw + 0.1091(0.725)w^{0.425}h^{-0.275}dh$

With $dw = \pm 0.03w$ and $dh = \pm 0.05h$,

$$|\Delta S| \approx |dS| = 0.1091|0.425w^{-0.575}h^{0.725}(\pm 0.03w) + 0.725w^{0.425}h^{-0.275}(\pm 0.05h)|$$

$$\leq 0.1091[0.425w^{0.425}h^{0.725}(0.03)] + 0.1091[0.725w^{0.425}h^{0.725}(0.05)]$$

$$= 0.1091w^{0.425}h^{0.725}(0.013 + 0.036) = 0.049S.$$

The approximate maximum percentage error is 4.9%.

39. (a) If a function $w = f(x,y,z)$ is differentiable at a point (x_0, y_0, z_0), then the function

$$L(x,y,z) = f(x_0,y_0,z_0) + f_x(x_0,y_0,z_0)(x - x_0) + f_y(x_0,y_0,z_0)(y - y_0) + f_z(x_0,y_0,z_0)(z - z_0)$$

is a linearization of f at (x_0, y_0, z_0).

(b) Let $f(x,y,z) = \sqrt{x^2 + y^2 + z^2}$. Then we wish to approximate $f(9.1, 11.75, 19.98)$. To do this, linearize f at $(9, 12, 20)$. Compute

$$\frac{\partial f}{\partial x} = \frac{x}{\sqrt{x^2 + y^2 + z^2}}, \quad \frac{\partial f}{\partial x}(9, 12, 20) = \frac{9}{25}$$

$$\frac{\partial f}{\partial y} = \frac{y}{\sqrt{x^2 + y^2 + z^2}}, \quad \frac{\partial f}{\partial y}(9, 12, 20) = \frac{12}{25}$$

$$\frac{\partial f}{\partial z} = \frac{z}{\sqrt{x^2 + y^2 + z^2}}, \quad \frac{\partial f}{\partial z}(9, 12, 20) = \frac{4}{5} \text{ and } f(9, 12, 20) = 25$$

The linearization is $L(x,y,z) = 25 + \dfrac{9}{25}(x - 9) + \dfrac{12}{25}(y - 12) + \dfrac{4}{5}(z - 20)$. For the approximation, we have $L(9.1, 11.75, 19.98) = 25 + \dfrac{9}{25}(0.1) + \dfrac{12}{25}(-0.25) + \dfrac{4}{5}(-0.02) = 24.9$

41. (a) The graph of $z = f(x,y)$ is an inverted cone with vertex at the origin. Since the graph comes to a sharp "point" at the origin, there is no possible increment formula for Δz that will work in every direction there.

    (b) We show that the partial derivative f_x does not exist at $(0,0)$. If $h > 0$,

$$\frac{f(0+h)-f(0,0)}{h} = \frac{\sqrt{h^2+0^2}-\sqrt{0^2+0^2}}{h}$$

$$= \frac{\sqrt{h^2}}{h} = \frac{|h|}{h} = 1$$

But if $h < 0$, then $\dfrac{f(0+h)-f(0,0)}{h} = -1$.

Therefore, $\displaystyle\lim_{h \to 0} \frac{f(0+h)-f(0,0)}{h}$ does not exist. But this means f_x does not exist at $(0,0)$ and thus f is not differentiable at $(0,0)$.

43. (a) From the figure we see that $\alpha = \pi - (\theta + \phi)$. Then
$$x_h = L\cos\theta + l\cos\alpha = L\cos\theta + l\cos(\pi - \theta - \phi)$$
$$= L\cos\theta - l\cos(\theta + \phi)$$
and $y_h = L\sin\theta - l\sin\alpha = L\sin\theta - l\sin(\pi - \theta - \phi)$
$$= L\sin\theta - l\sin(\theta + \phi).$$

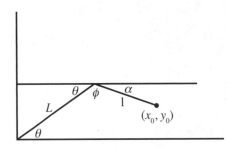

(b) Using $l\sin(\theta + \phi) = y_e - y_h$ and $l\cos(\theta + \phi) = x_e - x_h$, we have
$$dx_h = (-L\sin\theta + l\sin(\theta + \phi))d\theta + l\sin(\theta + \phi)d\phi = -y_h d\theta + (y_e - y_h)d\phi$$
$$dy_h = (L\cos\theta - l\cos(\theta + \phi))d\theta - l\cos(\theta + \phi)d\phi = x_h - (x_e - x_h)d\phi.$$

(c) One position has the lower arm reaching straight up, with the elbow on the x-axis, so that $\theta = 0$ and $\phi = 270°$. The other position has the lower arm reaching straight across, with the elbow on the y-axis, so that $\theta = \phi = 90°$. In both cases, $(x_h, y_h) = (L, L)$. In the first case, $(x_e, y_e) = (L, 0)$, and in the second case $(x_e, y_e) = (0, L)$. In general, the approximate maximum error in x_h is

$$|dx_h| = |-h_h d\theta + (y_e y_h)d\phi| \le L|d\theta| + |y_e - L||d\phi| = (L + |y_e - L|)\frac{\pi}{180}.$$

Thus, in the first case the approximate maximum error is $2\pi L/180$, while in the second case it is only $\pi L/180$.

13.5 Chain Rule

1. $\dfrac{dz}{dt} = \dfrac{\partial z}{\partial x}\dfrac{dx}{dt} + \dfrac{\partial z}{\partial y}\dfrac{dy}{dt} = \dfrac{2x}{x^2+y^2}(2t) + \dfrac{2y}{x^2+y^2}(-2t^{-3})$

$\qquad = \dfrac{4xt - 4yt^{-3}}{x^2+y^2}$

3. $\dfrac{dz}{dt} = \dfrac{\partial z}{\partial x}\dfrac{dx}{dt} + \dfrac{\partial z}{\partial y}\dfrac{dy}{dt} = -3\sin(3x+4y)(2) - 4\sin(3x+4y)(-1)$

At $t = \pi$, $x = \dfrac{5\pi}{2}$ and $y = \dfrac{-5\pi}{4}$

so $\left.\dfrac{dz}{dt}\right|_{t=\pi} = -6\sin\left(\dfrac{15\pi}{2} - 5\pi\right) + 4\sin\left(\dfrac{15\pi}{2} - 5\pi\right) = -6 + 4 = -2$

5. $\dfrac{dp}{du} = \dfrac{1}{2s+t}(2u) - \dfrac{2r}{(2s+t)^2}\left(-\dfrac{2}{u^3}\right) - \dfrac{r}{(2s+t)^2}\left(\dfrac{1}{2\sqrt{u}}\right) = \dfrac{2u}{2s+t} + \dfrac{4r}{u^3(2s+t)^2} - \dfrac{r}{2\sqrt{u}(2s+t)^2}$

7. $z_u = \dfrac{\partial z}{\partial x}\dfrac{\partial x}{\partial u} + \dfrac{\partial z}{\partial y}\dfrac{\partial y}{\partial u} = y^2 e^{xy^2}(3u^2) + 2xye^{xy^2}(1)$

$\quad = 3u^2 y^2 e^{xy^2} + 2xye^{xy^2}$

$\quad z_v = \dfrac{\partial z}{\partial x}\dfrac{\partial x}{\partial v} + \dfrac{\partial z}{\partial y}\dfrac{\partial y}{\partial v} = y^2 e^{xy^2}(0) + 2xye^{xy^2}(-2v)$

$\quad = -4vxye^{xy^2}$

9. $z_u = 4(4u^3) - 10y[2(2u-v)(2)] = 16u^3 - 40(2u-v)y$

$\quad z_v = 4(-24v^2) - 10y[2(2u-v)(-1)] = -96v^2 + 20(2u-v)y$

11. $w_t = \dfrac{3}{2}(u^2+v^2)^{1/2}(2u)(-e^{-t}\sin\theta) + \dfrac{3}{2}(u^2+v^2)^{1/2}(2v)(-e^{-t}\cos\theta)$

$\quad = -3u(u^2+v^2)^{1/2}e^{-t}\sin\theta - 3v(u^2+v^2)^{1/2}e^{-t}\cos\theta$

$\quad w_\theta = \dfrac{3}{2}(u^2+v^2)^{1/2}(2u)e^{-t}\cos\theta + \dfrac{3}{2}(u^2+v^2)^{1/2}(2v)(-e^{-t}\sin\theta)$

$\quad = 3u(u^2+v^2)^{1/2}e^{-t}\cos\theta - 3v(u^2+v^2)^{1/2}e^{-t}\sin\theta$

13. $R_u = s^2t^4(e^{v^2}) + 2rst^4(-2uve^{-u^2}) + 4rs^2t^3(2uv^2e^{u^2v^2}) = s^2t^4e^{v^2} - 4uvrst^4e^{-u^2} + 8uv^2rs^2t^3e^{u^2v^2}$

$\quad R_v = s^2t^4(2uve^{v^2}) + 2rst^4(e^{-u^2}) + 4rs^2t^3(2u^2ve^{u^2v^2}) = 2s^2t^4uve^{v^2} + 2rst^4e^{-u^2} + 8rs^2t^3u^2ve^{u^2v^2}$

15. $w_t = \dfrac{2x}{2\sqrt{x^2+y^2}}\dfrac{u}{rs+tu} + \dfrac{2y}{2\sqrt{x^2+y^2}}\dfrac{\cosh rs}{u} = \dfrac{xu}{\sqrt{x^2+y^2}(rs+tu)} + \dfrac{y\cosh rs}{u\sqrt{x^2+y^2}}$

$\quad w_r = \dfrac{2x}{2\sqrt{x^2+y^2}}\dfrac{s}{rs+tu} + \dfrac{2y}{2\sqrt{x^2+y^2}}\dfrac{st\sinh rs}{u} = \dfrac{xs}{\sqrt{x^2+y^2}(rs+tu)} - \dfrac{yst\sinh rs}{u\sqrt{x^2+y^2}}$

$\quad w_u = \dfrac{2x}{2\sqrt{x^2+y^2}}\dfrac{t}{rs+tu} + \dfrac{2y}{2\sqrt{x^2+y^2}}\dfrac{-t\cosh rs}{u^2} = \dfrac{xt}{\sqrt{x^2+y^2}(rs+tu)} - \dfrac{yt\cosh rs}{u^2\sqrt{x^2+y^2}}$

17. (a) $3x^2 - 2x^2(2yy') - 4xy^2 + y' = 0 \Longrightarrow (1-4x^2y)y' = 4xy^2 - 3x^2 \Longrightarrow y' = \dfrac{4xy^2 - 3x^2}{1-4x^2y}$

(b) $f_x = 3x^2 - 4xy^2,\quad f_y = -4x^2y + 1;\quad y' = -\dfrac{3x^2 - 4xy^2}{-4x^2y + 1} = \dfrac{4xy^2 - 3x^2}{1-4x^2y}$

19. (a) $y' = (\cos xy)(xy' + y) \Longrightarrow (1-x\cos xy)y' = y\cos xy \Longrightarrow y' = \dfrac{y\cos xy}{1-x\cos xy}$

(b) $f(x,y) = y - \sin xy;\quad f_x = -y\cos xy,\quad f_y = 1 - x\cos xy;$

$\quad y' = -\dfrac{-y\cos xy}{1-x\cos xy} = \dfrac{y\cos xy}{1-x\cos xy}$

21. $F_x = 2x,\ F_y = 2y,\ F_z = -2z;\quad \dfrac{\partial z}{\partial x} = -\dfrac{2x}{-2z} = \dfrac{x}{z};\quad \dfrac{\partial z}{\partial y} = -\dfrac{2y}{-2z} = \dfrac{y}{z}$

23. $F(x,y,z) = xy^2z^3 + x^2 - y^2 - 5z^2,\quad F_x = y^2z^3 + 2x,\quad F_y = 2xyz^3 - 2y,\quad F_z = 3xy^2z^2 - 10z$

$\quad \dfrac{\partial z}{\partial x} = -\dfrac{y^2z^3 + 2x}{3xy^2z^2 - 10z} = \dfrac{y^2z^3 + 2x}{10z - 3xy^2z^2};\quad \dfrac{\partial z}{\partial y} = -\dfrac{2xyz^3 - 2y}{3xy^2z^2 - 10z} = \dfrac{2xyz^3 - 2y}{10z - 3xy^2z^2}$

25. Let $y = x + at$ and $z = x - at$. Then $u(x,t) = F(y) + G(z)$ and

$$\frac{\partial u}{\partial x} = \frac{dF}{dy}\frac{\partial y}{\partial x} + \frac{dG}{dz}\frac{\partial z}{\partial x} = \frac{dF}{dy} + \frac{dG}{dz};\quad \frac{\partial^2 u}{\partial x^2} = \frac{d^2F}{dy^2}\frac{\partial y}{\partial x} + \frac{d^2G}{dz^2}\frac{\partial z}{\partial x} = \frac{d^2F}{dy^2} + \frac{d^2G}{dz^2};$$

$$\frac{\partial u}{\partial t} = \frac{dF}{dy}\frac{\partial y}{\partial t} + \frac{dG}{dz}\frac{\partial z}{\partial t} = a\frac{dF}{dy} - a\frac{dG}{dz};\quad \frac{\partial^2 u}{\partial xt^2} = a\frac{d^2F}{dy^2}\frac{\partial y}{\partial t} - a\frac{d^2G}{dz^2}\frac{\partial z}{\partial t} = a^2\frac{d^2F}{dy^2} + a^2\frac{d^2G}{dz^2}.$$

Thus, $a^2\dfrac{\partial^2 u}{\partial x^2} = a^2\dfrac{d^2F}{dy^2} + a^2\dfrac{d^2G}{dz^2} = \dfrac{\partial^2 u}{\partial t^2}.$

27. With $x = r\cos\theta$ and $y = r\sin\theta$

$$\frac{\partial u}{\partial r} = \frac{\partial u}{\partial x}\frac{\partial x}{\partial r} + \frac{\partial u}{\partial y}\frac{\partial y}{\partial r} = \frac{\partial u}{\partial x}\cos\theta + \frac{\partial u}{\partial y}\sin\theta$$

$$\frac{\partial^2 u}{\partial r^2} = \frac{\partial^2 u}{\partial x^2}\frac{\partial x}{\partial r}\cos\theta + \frac{\partial^2 u}{\partial y^2}\frac{\partial y}{\partial r}\sin\theta = \frac{\partial^2 u}{\partial x^2}\cos^2\theta + \frac{\partial^2 u}{\partial y^2}\sin^2\theta$$

$$\frac{\partial u}{\partial\theta} = \frac{\partial u}{\partial x}\frac{\partial x}{\partial\theta} + \frac{\partial u}{\partial y}\frac{\partial y}{\partial\theta} = \frac{\partial u}{\partial y}(r\sin\theta) + \frac{\partial u}{\partial y}(-r\sin\theta) + \frac{\partial^2 u}{\partial y^2}\frac{\partial y}{\partial\theta}(r\cos\theta)$$

$$= -r\frac{\partial u}{\partial x}\cos\theta + r^2\frac{\partial^2 u}{\partial x^2}\sin^2\theta - r\frac{\partial u}{\partial y}\sin\theta - r^2\frac{\partial^2 u}{\partial y^2}\cos^2\theta.$$

Using $\dfrac{\partial^2 u}{\partial x^2} + \dfrac{\partial^2 u}{\partial y^2} = 0$, we have

$$\frac{\partial^2 u}{\partial r^2} + \frac{1}{r}\frac{\partial u}{\partial r} + \frac{1}{r^2}\frac{\partial^2 u}{\partial\theta^2} = \frac{\partial^2 u}{\partial x^2}\cos^2\theta + \frac{\partial^2 u}{\partial y^2}\sin^2\theta + \frac{1}{r}\left(\frac{\partial u}{\partial x}\cos\theta + \frac{\partial u}{\partial y}\sin\theta\right)$$

$$+ \frac{1}{r^2}\left(-r\frac{\partial u}{\partial x} + r^2\frac{\partial^2 u}{\partial x^2}\sin^2\theta - r\frac{\partial u}{\partial y}\sin\theta + r^2\frac{\partial^2 u}{\partial y^2}\cos^2\theta\right)$$

$$= \frac{\partial^2 u}{\partial x^2}(\cos^2\theta + \sin^2\theta) + \frac{\partial^2 u}{\partial y^2}(\cos^2\theta + \sin^2\theta) + \frac{\partial u}{\partial x}\left(\frac{1}{r}\cos\theta - \frac{1}{r}\cos\theta\right)$$

$$+ \frac{\partial u}{\partial y}\left(\frac{1}{r}\sin\theta - \frac{1}{r}\sin\theta\right)$$

$$= \frac{\partial^2 u}{\partial x^2} + \frac{\partial^2 u}{\partial y^2} = 0$$

29. Letting $u = y/x$ in Problem 28, we have

$$x\frac{\partial z}{\partial x} + y\frac{\partial z}{\partial y} = x\frac{dz}{du}\frac{\partial u}{\partial x} + y\frac{dz}{du}\frac{\partial u}{\partial y} = x\frac{dz}{du}\left(-\frac{y}{x^2}\right) + y\frac{dz}{du}\left(\frac{1}{x}\right) = \frac{dz}{du}\left(\frac{-y}{x} + \frac{y}{x}\right) = 0.$$

31. We first compute

$$\frac{\partial u}{\partial x} = B\frac{\partial}{\partial x}erf\left(\frac{x}{\sqrt{4kt}}\right) = B\frac{\partial}{\partial x}\left(\frac{2}{\pi}\int_0^{x/\sqrt{4kt}}e^{-v^2}dv\right) = B\frac{2}{\sqrt{\pi}}\frac{1}{\sqrt{4kt}}e^{-x^2/4kt}$$

$$\frac{\partial^2 u}{\partial x^2} = B\frac{2}{\sqrt{\pi}}\frac{1}{\sqrt{4kt}}\left(-\frac{x}{2kt}\right)e^{-x^2/4kt} = -B\frac{x}{2k\sqrt{\pi}\sqrt{kt^3}}e^{-x^2/4kt}$$

$$\frac{\partial u}{\partial t} = B\frac{\partial}{\partial t}erf\left(\frac{x}{\sqrt{4kt}}\right) = B\frac{\partial}{\partial t}\left(\frac{2}{\pi}\int_0^{x/\sqrt{4kt}}e^{-v^2}dv\right) = B\frac{2}{\sqrt{\pi}}\left[-\frac{x}{2\sqrt{k}}\left(-\frac{1}{2}t^{-3/2}\right)\right]e^{-x/4kt}$$

$$= -B\frac{x}{2\sqrt{\pi}\sqrt{kt^3}}e^{-x^2/4kt}$$

Then $k\dfrac{\partial^2 u}{\partial x^2} = \dfrac{\partial u}{\partial t}$.

33. Since the height of the triangle is $x\sin\theta$, the area is given by $A = \frac{1}{2}xy\sin\theta$. Then

$$\frac{dA}{dt} = \frac{\partial A}{\partial x}\frac{dx}{dt} + \frac{\partial A}{\partial y}\frac{dy}{dt} + \frac{\partial A}{\partial\theta}\frac{d\theta}{dt} = \frac{1}{2}y\sin\theta\frac{dx}{dt} + \frac{1}{2}x\sin\theta\frac{dy}{dt} + \frac{1}{2}xy\cos\theta\frac{d\theta}{dt}.$$

When $x = 10$, $y = 8$, $\theta = \pi/6$, $dz/dt = 0.3$, $dy/dt = 0.5$, and $d\theta/dt = 0.1$,

$$\frac{dA}{dt} = \frac{1}{2}(8)\left(\frac{1}{2}\right)(0.3) + \frac{1}{2}(10)\left(\frac{1}{2}\right)(0.5) + \frac{1}{2}(10)(8)\left(\frac{\sqrt{3}}{2}\right)(0.1)$$

$$= 0.6 + 1.25 + 2\sqrt{3} = 1.85 + 2\sqrt{2} \approx 5.31 \text{ cm}^2/\text{s}.$$

35. $\dfrac{dS}{dt} = 0.1091\left(0.425w^{-0.575}h^{0.725}\dfrac{dw}{dt} + 0.725w^{0.425}h^{-0.275}\dfrac{dh}{dt}\right)$

When $w = 25$, $h = 29$, $dw/dt = 4.2$, and $dh/dt = 2$,

$$\dfrac{dS}{dt} = 0.1091[0.425(25)^{-0.575}(29^{(0.725}(4.2) + 0.725(25)^{0.425}(29)^{-0.275}(2)] \approx 0.5976\,\text{in}^2/\text{yr}.$$

37. Since $dT/dT = 1$ and $\dfrac{\partial P}{\partial T} = 0$,

$$0 = F_T = \dfrac{\partial F}{\partial P}\dfrac{\partial P}{\partial T} + \dfrac{\partial F}{\partial V}\dfrac{\partial V}{\partial T} + \dfrac{\partial F}{\partial T}\dfrac{\partial T}{\partial T} \implies \dfrac{\partial V}{\partial T} = -\dfrac{\partial F/\partial T}{\partial F/\partial V} = -\dfrac{1}{\partial T/\partial V}.$$

39. (a) Using $f = \pi$, $l = 6$, $V = 100$, and $c = 330,000$ we obtain $f \approx 380.04$ cycles per second.

(b) We first note that $\dfrac{\partial f}{\partial V} = \dfrac{c}{4\pi}\left(\dfrac{A}{lV}\right)^{-1/2}\left(-\dfrac{A}{lv^2}\right) = -\dfrac{c}{4\pi}\sqrt{\dfrac{A}{lV}}\dfrac{1}{V} = -\dfrac{1}{2V}f$ and

$\dfrac{\partial f}{\partial l} = -\dfrac{1}{2l}f$. Then $\dfrac{df}{dt} = \dfrac{\partial f}{\partial V}\dfrac{dV}{dt} + \dfrac{\partial f}{\partial l}\dfrac{dl}{dt} = -\dfrac{1}{2V}f\dfrac{dV}{dt} - \dfrac{1}{2l}f\dfrac{dl}{dt} = -\dfrac{f}{2}\left(\dfrac{1}{V}\dfrac{dV}{dt} + \dfrac{1}{l}\dfrac{dl}{dt}\right)$.

Using $dV/dt = -10$, $dl/dt = 1$, $V = 100$, and $l = 6$ we find

$$\dfrac{df}{dt} = -\dfrac{f}{2}\left[\dfrac{1}{100}(-10) + \dfrac{1}{6}(1)\right] = -\dfrac{f}{2}\left(\dfrac{1}{6} - \dfrac{1}{10}\right) < 0.$$

The frequency is decreasing.

41.

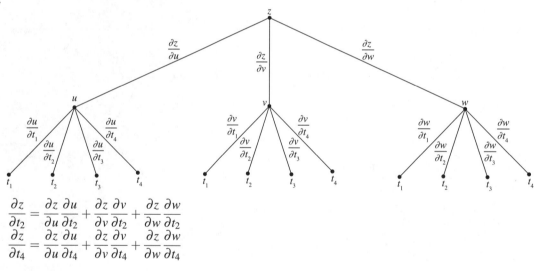

$$\dfrac{\partial z}{\partial t_2} = \dfrac{\partial z}{\partial u}\dfrac{\partial u}{\partial t_2} + \dfrac{\partial z}{\partial v}\dfrac{\partial v}{\partial t_2} + \dfrac{\partial z}{\partial w}\dfrac{\partial w}{\partial t_2}$$
$$\dfrac{\partial z}{\partial t_4} = \dfrac{\partial z}{\partial u}\dfrac{\partial u}{\partial t_4} + \dfrac{\partial z}{\partial v}\dfrac{\partial v}{\partial t_4} + \dfrac{\partial z}{\partial w}\dfrac{\partial w}{\partial t_4}$$

43. Letting $F(x,y,z,u) = -xyz + x^2yu + 2xy^3u - u^4 - 8$ we find $F_z = -yz + 2xyu + 2y^3u$,
$F_y = -xz + x^2u + 6xy^2u$, $F_z = -xy$, and $F_u = x^2y + 2xy^3 - 4u^3$. Then

$$\dfrac{\partial u}{\partial x} = -\dfrac{-yz + 2xyu + 2y^3u}{x^2y + 2xy^3 - 4u^3}, \quad \dfrac{\partial u}{\partial y} = -\dfrac{-xz + x^2y + 6xy^2u}{x^2y + 2xy^3 - 4u^3}, \quad \dfrac{\partial u}{\partial z} = \dfrac{xy}{x^2y + 2xy^3 - 4u^3}.$$

13.6 Directional Derivative

1. $\nabla f = (2x - 3x^2y^2)\mathbf{i} + (4y^3 - 2x^3y)\mathbf{j}$

3. $\nabla F = \dfrac{y^2}{z^3}\mathbf{i} + \dfrac{2xy}{z^3}\mathbf{j} - \dfrac{3xy^2}{z^4}\mathbf{k}$

5. $\nabla f = 2x\mathbf{i} - 8y\mathbf{j}$; $\nabla f(2,4) = 4\mathbf{i} - 32\mathbf{j}$

7. $\nabla F = 2xz^2 \sin 4y\mathbf{i} + 4x^2z^2 \cos 4y\mathbf{j} + 2x^2z \sin 4y\mathbf{k}$

$\nabla F(-2, \pi/3, 1) = -4\sin\dfrac{4\pi}{3}\mathbf{i} + 16\cos\dfrac{4\pi}{3}\mathbf{j} + 8\sin\dfrac{4\pi}{3}\mathbf{k} = 2\sqrt{3}\mathbf{i} - 8\mathbf{j} - 4\sqrt{3}\mathbf{k}$

9. $D_{\mathbf{u}}f(x,y) = \lim\limits_{h\to 0}\dfrac{f(x+h\sqrt{3}/2, y+h/2) - f(x,y)}{h} = \lim\limits_{h\to 0}\dfrac{(x+h\sqrt{3}/2)^2 + (y+h/2)^2)_x^2 - y^2}{h}$

$= \lim\limits_{h\to 0}\dfrac{h\sqrt{3}x + 3h^2/4 = hy + h^2/4}{h} = \lim\limits_{h\to 0}(\sqrt{3}x + 3h/4 + y + h/4) = \sqrt{3}x + y$

11. $\mathbf{u} = \dfrac{\sqrt{3}}{2}\mathbf{i}\dfrac{1}{2}\mathbf{j}$; $\nabla f = 15x^2y^6\mathbf{i} + 30x^3y^5\mathbf{j}$; $\nabla f(-1,1) = 15\mathbf{i} - 30\mathbf{j}$; $D_{\mathbf{u}}f(-1,1) = \dfrac{15\sqrt{3}}{2} - 15 = \dfrac{15}{2}(\sqrt{3}-2)$

13. $\mathbf{u} = \dfrac{\sqrt{10}}{10}\mathbf{i} - \dfrac{3\sqrt{10}}{10}\mathbf{j}$; $\nabla f = \dfrac{-y}{x^2+y^2}\mathbf{i} + \dfrac{x}{x^2+y^2}\mathbf{j}$; $\nabla f(2,-2) = \dfrac{1}{4}\mathbf{i} + \dfrac{1}{4}\mathbf{j}$; $D_{\mathbf{u}}f(2,-2) = \dfrac{\sqrt{10}}{40} - \dfrac{3\sqrt{10}}{40} = -\dfrac{\sqrt{10}}{20}$

15. $\mathbf{u} = (2\mathbf{i}+\mathbf{j})/\sqrt{5}$; $\nabla f = 2y(xy+1)\mathbf{i} + 2x(xy+1)\mathbf{j}$; $\nabla f(3,2) = 28\mathbf{i} + 42\mathbf{j}$; $D_{\mathbf{u}}f(3,2) = \dfrac{2(28)}{\sqrt{5}} + \dfrac{42}{\sqrt{5}} = \dfrac{98}{\sqrt{5}}$

17. $\mathbf{u} = \dfrac{1}{\sqrt{2}}\mathbf{j} + \dfrac{1}{\sqrt{2}}\mathbf{k}$; $\nabla f = 2xy^2(2z+1)^2\mathbf{i}2x^2y(2z+1)^2\mathbf{j} + 4x^2y^2(2z+1)\mathbf{k}$; $\nabla f(1,-1,1) = 18\mathbf{i} - 18\mathbf{j} + 12\mathbf{k}$;

$D_{\mathbf{u}}f(1,-1,1) = -\dfrac{18}{\sqrt{2}} + \dfrac{12}{\sqrt{2}} = -\dfrac{6}{\sqrt{2}} = -3\sqrt{2}$

19. $\mathbf{u} = -\mathbf{k}$; $\nabla f = \dfrac{xy}{\sqrt{x^2y+2y^2z}}\mathbf{i} + \dfrac{x^2+4z}{2\sqrt{x^2y+2y^2z}}\mathbf{j} + \dfrac{y^2}{\sqrt{x^2y+2y^2z}}\mathbf{k}$; $\nabla f(-2,2,1) = -\mathbf{i} + \mathbf{j} + \mathbf{k}$; $D_{\mathbf{u}}f(-2,2,1) = -1$

21. $\mathbf{u} = (-4\mathbf{i} - \mathbf{j}/\sqrt{17}$; $\nabla f = 2(x-y)\mathbf{i} - 2(x-y)\mathbf{j}$; $\nabla f(4,2) = 4\mathbf{i} - 4\mathbf{j}$; $D_{\mathbf{u}}f(4,2) = -\dfrac{16}{\sqrt{17}} + \dfrac{4}{\sqrt{17}} = -\dfrac{12}{\sqrt{17}}$

23. $\nabla f = 2e^{2x}\sin y\mathbf{i} + e^{2x}\cos y\mathbf{j}$; $\nabla f(0, \pi/4) = \sqrt{2}\mathbf{i} + \dfrac{\sqrt{2}}{2}\mathbf{j}$ The maximum $D_{\mathbf{u}}$ is $\left[(\sqrt{2})^2 + (\sqrt{2}/2)^2\right]^{1/2} = \sqrt{5/2}$ in the direction $\sqrt{2}\mathbf{i} + (\sqrt{2}/2)\mathbf{j}$.

25. $\nabla f = (2x+4z)\mathbf{i} + 2z^2\mathbf{j} + (4x+4yz)\mathbf{k}$; $\nabla f(1,2,-1) = -2\mathbf{i} + 2\mathbf{j} - 4\mathbf{k}$ The maximum $D_{\mathbf{u}}$ is $\left[(-2)^2 + (2)^2 + (-4)^2\right]^{1/2} = 2\sqrt{6}$ in the direction $-2\mathbf{i} + 2\mathbf{j} - 4\mathbf{k}$.

27. $\nabla f = 2x\sec^2(x^2+y^2)\mathbf{i} + 2y\sec^2(x^2+y^2)\mathbf{j}$; $\nabla f(\sqrt{\pi/6}, \sqrt{\pi/6}) = 2\sqrt{\pi/6}\sec^2(\pi/3)(\mathbf{i}+\mathbf{j}) = 8\sqrt{\pi/6}(\mathbf{i}+\mathbf{j})$ The minimum $D_{\mathbf{u}}$ is $-8\sqrt{\pi/6}(1^2+1^2)^{1/2} = -8\sqrt{\pi/3}$ in the direction $-(\mathbf{i}+\mathbf{j})$.

29. $\nabla f = \dfrac{\sqrt{z}e^y}{2\sqrt{x}}\mathbf{i} + \sqrt{xz}e^y\mathbf{j} + \dfrac{\sqrt{x}}{2\sqrt{z}}\mathbf{k}$; $\nabla f(16,0,9) = \dfrac{3}{8}\mathbf{i} + 12\mathbf{j} + \dfrac{2}{3}\mathbf{k}$. The minimum $D_{\mathbf{u}}$ is $-\left[(3/8)^2 + 12^2 + (2/3)^2\right]^{1/2} = -\sqrt{83281}/24$ in the direction $-\dfrac{3}{8}\mathbf{i} - 12\mathbf{j} - \dfrac{2}{3}\mathbf{k}$.

31. Using implicit differentiation on $2x^2 + y^2 = 9$ we find $y' = -2x/y$. At $(2,1)$ the slope of the tangent line is $-2(2)/1 = -4$. Thus, $\mathbf{u} = \pm(\mathbf{i} - 4\mathbf{j})/\sqrt{17}$. Now, $\nabla f = \mathbf{i} + 2y\mathbf{j}$ and $\nabla f(3,4) = \mathbf{i} + 8\mathbf{j}$. Thus, $D_{\mathbf{u}} = \pm(1/\sqrt{17} - 32\sqrt{17}) = \pm 31/\sqrt{17}$.

33. (a) Vectors perpendicular to $4\mathbf{i} + 3\mathbf{j}$ are $\pm(3\mathbf{i} - 4\mathbf{j})$. Take $\mathbf{u} = \pm\left(\dfrac{3}{5}\mathbf{i} - \dfrac{4}{5}\mathbf{j}\right)$.

(b) $\mathbf{u} = (4\mathbf{i} + 3\mathbf{j})/\sqrt{16+9} = \dfrac{4}{5}\mathbf{i} + \dfrac{3}{5}\mathbf{j}$

(c) $\mathbf{u} = -\dfrac{4}{5}\mathbf{i} - \dfrac{3}{5}\mathbf{j}$

35. (a) $\nabla f = (3x^2 - 6xy^2)\mathbf{i} + (-6x^2y + 3y^2)\mathbf{j}$

$D_{\mathbf{u}}f(x,y) = \dfrac{3(3x^2 - 6xy^2)}{\sqrt{10}} + \dfrac{-6x^2y + 3y^2}{\sqrt{10}} = \dfrac{9x^2 - 18xy^2 - 6x^2y + 3y^2}{\sqrt{10}}$

(b) $F(x,y) = \dfrac{3}{\sqrt{10}}(3x^2 - 3xy^2 - 2x^2y + y^2);\quad \nabla F = \dfrac{3}{\sqrt{10}}[(6x - 6y^2 - 4xy)\mathbf{i} + (-12xy - 2x^2 + 2y)\mathbf{j}]$

$D_{\mathbf{u}}F(x,y) = \left(\dfrac{3}{\sqrt{10}}\right)\left(\dfrac{3}{\sqrt{10}}\right)(6x - 6y^2 - 4xy) + \left(\dfrac{1}{\sqrt{10}}\right)\left(\dfrac{3}{\sqrt{10}}\right)(-12xy - 2x^2 + 2y)$

$= \dfrac{9}{5}(3x - 3y^2 - 2xy) + \dfrac{3}{5}(-6xy - x^2 + y) = \dfrac{1}{5}(27x - 27y^2 - 36xy - 3x^2 + 3y)$

37. $\nabla f = (3x^2 - 12)\mathbf{i} + (2y - 10)\mathbf{j}$. Setting $|\nabla f| = [(3x^2 - 12)^2 + (2y - 10)^2]^{1/2} = 0$, we obtain $3x^2 - 12 = 0$ and $2y - 10 = 0$. The points where $|\nabla f| = 0$ are $(2,5)$ and $(-2,5)$.

39. $\nabla T = 4x\mathbf{i} + 2y\mathbf{j};\quad \nabla T(4,2) = 16\mathbf{i} + 4\mathbf{j}$. The minimum change in temperature (that is, the maximum decrease in temperature) is in the direction $-\nabla T(4,3) = -16\mathbf{i} - 4\mathbf{j}$.

41. Let $x(t)\mathbf{i} + y(t)\mathbf{j}$ be the vector equation of the path. At (x,y) on this curve, the direction of a tangent vector is $x'(t)\mathbf{i} + y'(t)\mathbf{j}$. Since we want the direction of motion to be $\nabla T(x,y)$, we have $x'(t)\mathbf{i} + y'(t)\mathbf{j} = \nabla T(x,y) = -4x\mathbf{i} - 2y\mathbf{j}$. Separating variables in $dx/dt = -4x$, we obtain $dx/x = -4dt, \ln x = -4t + c_1$, and $x = C_1 e^{-4t}$. Separating variables in $dy/dt = -2y$, we obtain $dy/y = -2dt, \ln y = -2t + c_2$, and $y = C_2 e^{-2t}$. Since $x(0) = 3$ and $y(0) = 4$, we have $x = 3e^{-4t}$ and $y = 4e^{-2t}$. The equation of the path is $3e^{-4t}\mathbf{i} + 4e^{-2t}\mathbf{j}$, or eliminating the parameter, $16x = 3y^2, \ y \geq 0$.

43. $\nabla U = \dfrac{Gmx}{(x^2 + y^2)^{3/2}}\mathbf{i} + \dfrac{Gmy}{(x^2 + y^2)^{3/2}}\mathbf{j} = \dfrac{Gm}{(x^2 + y^2)^{3/2}}(x\mathbf{i} + y\mathbf{j})$

The maximum and minimum values of $D_{\mathbf{u}}U(x,y)$ are obtained when $\mathbf{u}$ is in the directions ∇U and $-\nabla U$, respectively. Thus, at a point (x,y), not $(0,0)$, the directions of maximum and minimum increase in U are $x\mathbf{i} + y\mathbf{j}$ and $-x\mathbf{i} - y\mathbf{j}$, respectively. A vector at (x,y) in the direction $\pm(x\mathbf{i} + y\mathbf{j})$ lies on a line through the origin.

45. $\nabla(cf) = \dfrac{\partial}{\partial x}(cf)\mathbf{i} + \dfrac{\partial}{\partial y}(cf)\mathbf{j} = cf_x\mathbf{i} + cf_y\mathbf{j} = c(f_x\mathbf{i} + f_y\mathbf{j}) = c\nabla f$

47. $\nabla(fg) = (fg_x + f_xg)\mathbf{i} + (fg_y + f_yg)\mathbf{j} = f(g_x\mathbf{i} + g_y\mathbf{j}) + g(f_x\mathbf{i} + f_y\mathbf{j}) = f\nabla g + g\nabla f$

49. $r(x,y) = \sqrt{x^2 + y^2}$ so $\dfrac{\partial r}{\partial x} = \dfrac{x}{\sqrt{x^2 + y^2}} = \dfrac{x}{r}$ and $\dfrac{\partial r}{\partial y} = \dfrac{y}{\sqrt{x^2 + y^2}} = \dfrac{y}{r}$

This gives $\nabla r = \left\langle \dfrac{x}{r}, \dfrac{y}{r} \right\rangle = \dfrac{1}{r}\langle x, y \rangle = \dfrac{\mathbf{r}}{r}$

51. Let $\mathbf{u} = u_1\mathbf{i} + u_2\mathbf{j}$ and $\mathbf{v} = v_1\mathbf{i} + v_2\mathbf{j}$.

$D_{\mathbf{v}}f = (f_x\mathbf{i} + f_y\mathbf{j}) \cdot \mathbf{v} = v_1 f_x + v_2 f_y$

$D_{\mathbf{u}}D_{\mathbf{v}}f = \left[\dfrac{\partial}{\partial x}(v_1 f_x + v_2 f_y)\mathbf{i} + \dfrac{\partial}{\partial y}(v_1 f_x + v_2 f_y)\mathbf{j}\right] \cdot \mathbf{u} = [(v_1 f_{xx} + v_2 f_{yz})\mathbf{i} + (v_1 f_{xy} + v_2 f_{yy})\mathbf{j}] \cdot \mathbf{u}$

$= u_1 v_1 f_{xx} + u_1 v_2 f_{yx} + u_2 v_1 f_{xy} + u_2 v_2 f_{yy}$

$D - uf = (f_x\mathbf{i} + f_y\mathbf{j}) \cdot \mathbf{u} = u_1 f_x + u_2 f_y$

$D_{\mathbf{v}}D_{\mathbf{u}}f = \left[\dfrac{\partial}{\partial x}(u_1 f_x + u_2 f_y)\mathbf{i} + \dfrac{\partial}{\partial y}(u_1 f_x + u_2 f_y)\mathbf{j}\right] \cdot \mathbf{v} = [(u_1 f_{xx} + u_2 f_{yx})\mathbf{i} + (u_1 f_{xy} + u_2 f_{yy})\mathbf{j}] \cdot \mathbf{v}$

$= u_1 v_1 f_{xx} + u_2 v_1 f_{yx} + u_1 v_2 f_{xy} + u_2 v_2 f_{yy}$

Since the second partial derivatives are continuous, $f_{xy} = f_{yx}$ and $D_{\mathbf{u}}D_{\mathbf{v}}f = D_{\mathbf{v}}D_{\mathbf{u}}f$. [Note that this result is a generalization $f_{xy} = f_{yx}$ since $D_iD_jf = f_{yx}$ and $D_jD_if = f_{xy}$]

13.7 Tangent Planes and Normal Lines

1. Since $f(6,1) = 4$, the level curve is $x - 2y = 4$. $\nabla f = \mathbf{i} - 2\mathbf{j};\ \nabla f(6,1) = \mathbf{i} - 2\mathbf{j}$

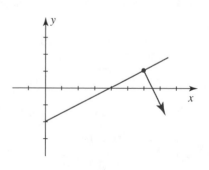

3. Since $f(2,5) = 1$, the level curve is $y = x^2 + 1$. $\nabla f = -2x\mathbf{i} + \mathbf{j}$; $\nabla f(2,5) = -4\mathbf{i} + \mathbf{j}$

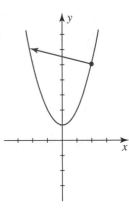

5. Since $f(-2,-3) = 2$, the level curve is $x^2/4 + y^2/0 = 2$ or
$x^2/8 + y^2/18 = 1$. $\nabla f = \dfrac{x}{2}\mathbf{i} + \dfrac{2y}{9}\mathbf{j}$; $\nabla f(-2,-3) = -\mathbf{i} - \dfrac{2}{3}\mathbf{j}$

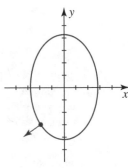

7. Since $f(1,1) = -1$, the level curve is $(x-1)^2 - y^2 = -1$ or $y^2 - (x-1)^2 = 1$.
$\nabla f = 2(x-1)\mathbf{i} - 2y\mathbf{j}$; $\nabla f(1,1) = -2\mathbf{j}$

9. Since $f(3,1,1) = 2$, the level curve is $y + z = 2$ $\nabla f = \mathbf{j} + \mathbf{k}$; $\nabla f(3,1,1) = \mathbf{j} + \mathbf{k}$

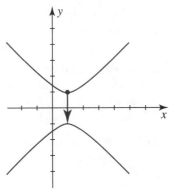

11. Since $F(3,4,0) = 5$, the level curve is $x^2 + y^2 + z^2 = 25$. $\nabla F = \dfrac{x}{\sqrt{x^2 + y^2 + z^2}}\mathbf{i} +$

$\dfrac{y}{\sqrt{x^2 + y^2 + z^2}}\mathbf{j} + \dfrac{z}{\sqrt{x^2 + y^2 + z^2}}\mathbf{k}$; $\nabla F(3,4,0) = \dfrac{3}{4}\mathbf{i} + \dfrac{4}{5}\mathbf{j}$

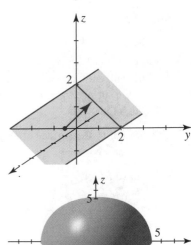

13. $F(x,y,z) = x^2 + y^2 - z$; $\nabla F = 2x\mathbf{i} + 2y\mathbf{j} - \mathbf{k}$. We want $\nabla F = c\left(4\mathbf{i} + \mathbf{j} + \frac{1}{2}\mathbf{k}\right)$ or $2x = 4c$, $2y = c$, $-1 = c/2$. From the third equation $c = -2$. Thus, $x = -4$ and $y = -1$. Since $z = x^2 + y^2 = 16 + 1 = 17$, the point on the surface is $(-4,-1,-17)$.

15. $F(x,y,z0 = x^2 + y^2 + z^2$; $\nabla F = 2x\mathbf{i} + 2y\mathbf{j} + 2z\mathbf{k}$. $\nabla F(-2,2,1) = -4\mathbf{i} + 4\mathbf{j} + 2\mathbf{k}$. The equation of the tangent plane is $-4(x + 2) + 4(y - 2) + 2(z - 1) = 0$ or $-2x + 2y + z = 9$.

17. $F(x,y,z) = x^2 - y^2 - 3z^2$; $\nabla F = 2x\mathbf{i} - 2y\mathbf{j} - 6z\mathbf{k}$; $\nabla F(6,2,3) = 12\mathbf{i} - 4\mathbf{j} - 18\mathbf{k}$. The equation of the tangent plane is $12(x-6) - 4(y-2) - 18(z-3) = 0$ or $6x - 2y - 9z = 5$.

19. $F(x,y,z) = x^2 + y^2 + z$; $\nabla F = 2x\mathbf{i} + 2y\mathbf{j} + \mathbf{k}$; $\nabla F(3,-4,0) = 6\mathbf{i} - 8\mathbf{j} + \mathbf{k}$. The equation of the tangent plane is $6(x-3) - 8(y+4) + z = 0$ or $6x - 8y + z = 50$.

21. $F(x,y,z) = \cos(2x+y) - z$; $\nabla F = -2\sin(2x+y)\mathbf{i} - \sin(2x+y)\mathbf{j} - \mathbf{k}$; $\nabla F(\pi/2, \pi/4, -1\sqrt{2}) = \sqrt{2}\mathbf{i} + \dfrac{\sqrt{2}}{2}\mathbf{j} - \mathbf{k}$. The equation of the tangent plane is $\sqrt{2}\left(x - \dfrac{\pi}{2}\right) + \dfrac{\sqrt{2}}{2}\left(y - \dfrac{\pi}{4}\right) - \left(z + \dfrac{1}{\sqrt{2}}\right) = 0$, $2\left(x - \dfrac{\pi}{2}\right) + \left(y - \dfrac{\pi}{4}\right) - \sqrt{2}\left(z + \dfrac{1}{\sqrt{2}}\right) = 0$, or $2x + y - \sqrt{2}z = \dfrac{5\pi}{4} + 1$.

23. $F(x,y,z) = \ln(x^2 + y^2) - z$; $\nabla F = \dfrac{2x}{x^2 + y^2}\mathbf{i} + \dfrac{2y}{x^2 + y^2}\mathbf{j} - \mathbf{k}$; $\nabla F(1/\sqrt{2}, 1/\sqrt{2}, 0) = \sqrt{2}\mathbf{i} + \sqrt{2}\mathbf{j} - \mathbf{k}$. The equation of the tangent plane is $\sqrt{2}\left(x - \dfrac{1}{\sqrt{2}}\right) + \sqrt{2}\left(y - \dfrac{1}{\sqrt{2}}\right) - (z-0) = 0$, $2\left(x - \dfrac{1}{\sqrt{2}}\right) + 2\left(y - \dfrac{1}{\sqrt{2}}\right) - \sqrt{2}z = 0$, or $2x + 2y - \sqrt{2}z = 2\sqrt{2}$.

25. The gradient of $F(x,y,z) = x^2 + y^2 + z^2$ is $\nabla F = 2x\mathbf{i} + 2y\mathbf{j} + 2z\mathbf{k}$, so the normal vector to the surface at (x_0, y_0, z_0) is $2x_0\mathbf{i} + 2y_0\mathbf{j} + 2z_0\mathbf{k}$. A normal vector to the plane $2x + 4y + 6z = 1$ is $2\mathbf{i} + 4\mathbf{j} + 6\mathbf{k}$. Since we want the tangent plane to be parallel to the given plane, we find c so that $2x_0 = 2c$, $2y_0 = 4c$, $2z_0 = 6c$ or $x_0 = c$, $y_0 = 2c$, $z_0 = 3c$. Now, (x_0, y_0, z_0) is on the surface, so $c^2 + (2c)^2 + (3c)^2 = 14c^2 = 7$ and $c = \pm 1/\sqrt{2}$. Thus, the points on the surface are $(\sqrt{2}/2, \sqrt{2}, 3\sqrt{2}/2)$ and $-\sqrt{2}/2, -\sqrt{2}, -3\sqrt{2}/2)$.

27. The gradient of $F(x,y,z) = x^2 + 4x + y^2 + z^2 - 2z$ is $\nabla F = (2x+4)\mathbf{i} + 2y\mathbf{j} + (2z-2)\mathbf{k}$, so a normal to the surface at (x_0, y_0, z_0) is $(2x_0 + 4)\mathbf{i} + 2y_0\mathbf{j} + (2z_0 - 2)\mathbf{k}$. A horizontal plane has normal $c\mathbf{k}$ for $c \neq 0$. Thus, we want $2x_0 + 4 = 0$, $2y_0 = 0$, $2z_0 - 2 = c$ or $x_0 = -2$, $y_0 = 0$, $z_0 = c + 1$. Since (x_0, y_0, z_0) is on the surface, $(-2)^2 + 4(-2) + (c+1)^2 - 2(c+1) = c^2 - 5 = 11$ and $c = \pm 4$. The points on the surface are $(-2, 0, 5)$ and $(-2, 0, -3)$.

29. If (x_0, y_0, z_0) is on $x^2/a^2 + y^2/b^2 + z^2/c^2 = 1$, then $x_0^2/a^2 + y_0^2/b^2 + z_0^2/c^2 = 1$ and $x_0, y_0, z_0)$ is on the plane $xx_0/a^2 + yy_0/b^2 + zz_0/c^2 = 1$. A normal to the surface at (x_0, y_0, z_0) is $\nabla F(x_0, y_0, z_0) = (2x - 0/a^2)\mathbf{i} + (2y_0/b^2)\mathbf{j} + (2z_0/c^2)\mathbf{k}$. A normal to the plane is $(x_0/a^2)\mathbf{i} + (y_0/b^2)\mathbf{j} + (z_0/c^2)\mathbf{k}$. Since the normal to the surface is a multiple of the normal to the plane, the normal vectors are parallel and the plane is tangent to the surface.

31. $F(x,y,z) = x^2 + 2y^2 + z^2$; $\nabla F = 2x\mathbf{i} + 4y\mathbf{j} + 2z\mathbf{k}$; $\nabla F(1,-1,1) = 2\mathbf{i} - 4\mathbf{j} + 2\mathbf{k}$. Parametric equations of the line are $x = 1 + 2t$, $y = -1 - 4t$, $z = 1 + 2t$.

33. $F(x,y,z) = 4x^2 + 9y^2 - z$; $\nabla F = 8x\mathbf{i} + 18y\mathbf{j} - \mathbf{k}$; $\nabla F(1/2, 1/3, 3) = 4\mathbf{i} + 6\mathbf{j} - \mathbf{k}$. Symmetric equations of the line are $\dfrac{x - 1/2}{4} = \dfrac{y - 1/3}{6} = \dfrac{z - 3}{-1}$.

35. Let $F(x,y,z) = x^2 + y^2 - z^2$. Then $\nabla F = 2x\mathbf{i} + 2y\mathbf{j} - 2z\mathbf{k}$ and a normal to the surface at (x_0, y_0, z_0) is $x_0\mathbf{i} + y_0\mathbf{j} - z_0\mathbf{k}$. An equation of the tangent plane at (x_0, y_0, z_0) is $x_0(x - x_0) + y_0(y - y_0) - z_0(z - z_0) = 0$ or $x_0 x + y_0 y - z_0 z = x_0^2 + y_0^2 - z_0^2$. Since (x_0, y_0, z_0) is on the surface, $z_0^2 = x_0^2 + y_0^2$ and $x_0^2 + y_0^2 - z_0^2 = 0$. Thus, the equation of the tangent plane is $x_0 x + y_0 y - z_0 z = 0$, which passes through the origin.

37. A normal to the surface at (x_0, y_0, z_0) is $\nabla F(x_0, y_0, z_0) = 2x_0\mathbf{i} + 2y_0\mathbf{j} + 2z_0\mathbf{k}$. Parametric equations of the normal line are $x = x_0 + 2x_0 t$, $y = y_0 + 2y_0 t$, $z = z_0 + 2z_0 t$. Letting $t = -1/2$, we see that the normal line passes through the origin.

39. We have $F(x,y,z) = x^2 + y^2 + z^2$ and $G(x,y,z) = x^2 + y^2 - z^2$.
$\nabla F = \langle 2x, 2y, 2z \rangle \neq 0$ except at the origin
$\nabla G = \langle 2x, 2y, -2z \rangle \neq 0$ except at the origin

Therefore, the gradient vectors are nonzero at each of the intersection points. Now

$$F_x G_x + F_y G_y + F_x G_z = (2x)(2x) + (2y)(2y) + (2z)(-2z)$$
$$= 4x^2 + 4y^2 - 4z^2$$
$$= 4(x^2 + y^2 + z^2) = 4(0) = 0$$

The second to last equality follows from the fact that the intersection points lie on both surfaces and hence satisfy the second equation $x^2 + y^2 - z^2 = 0$.

13.8 Extrema of Multivariable Functions

1. $f_x = 2x$; $f_{xx} = 2$; $f_{xy} = 0$; $f_y = 2y$; $f_{yy} = 2$; $D = 4$. Solving $f_x = 0$ and $f_y = 0$, we obtain the critical point $(0,0)$. Since $D(0,0) = 4 > 0$ and $f_{xx}(0,0) = 2 > 0$, $f(0,0) = 5$ is a relative minimum.

3. $f_x = -2x+8$; $f_{xx} = -2$; $f_{xy} = 0$; $f_y = -2y+6$; $f_{yy} = -2$; $D = 4$. Solving $f_x = 0$ and $f_y = 0$ we obtain the critical point $(4,3)$. Since $D(4,3) = 4 > 0$ and $f_{xx}(4,3) = -2 < 0$, $f(4,3,) = 25$ is a relative maximum.

5. $f_x = 10x+20$; $f_{xx} = 10$; $f_{xy} = 0$; $f_y = 10y-10$; $f_{yy} = 10$; $D = 100$. Solving $f_x = 0$ and $f_y = 0$, we obtain the critical point $(-2,1)$. Since $D(-2,1) = 100 > 0$ and $f_{xx}(-2,1) = 10 > 0$, $f(-2,1) = 15$ is a relative minimum.

7. $f_x = 12x^2 - 12$; $f_{xx} = 24x$; $f_{xy} = 0$; $f_y = 3y^2 - 3$; $f_{yy} = 6y$; $D = 144xy$. Solving $f_x = 0$ and $f_y = 0$, we obtain the critical points $(-1,-1)$, $(-1,1)$, $(1,-1)$, and $(1,1)$. Since $D(-1,1) = -144 < 0$ and $D(1,-1) = -144 < 0$, these points do not give relative extrema. Since $D(-1,-1) = 144 > 0$ and $f_{xx}(-1,-1) = -24 < 0$, $f(-1,-1) = 10$ is a relative maximum. Since $D(1,1) = 144 > 0$ and $f_{xx}(1,1) = 24 > 0$, $f(1,1) = -10$ is a relative minimum.

9. $f_x = 4x - 2y - 10$; $f_{xx} = 4$; $f_{xy} = -2$; $f_y = 8y - 2x - 2$; $f_{yy} = 8$; $D = 32 - (-2)^2 = 28$. Setting $f_x = 0$ and $f_y = 0$, we obtain $4x - 2y = 10$ and $8y - 2x = 2$ or $2x - y = 5$ and $4y - x = 1$. Solving, we obtain the critical point $(3,1)$. Since $D(3,1) = 28 > 0$ and $f_{xx}(3,1) = 4 > 0$, $f(3,1) = -14$ is a relative minimum.

11. $f_x = 2t - 8$; $f_{xx} = 0$; $f_{xy} = 2$; $f_y = 2x - 5$; $f_{yy} = 0$; $D = 0 - 2^2 = -4$. Since $D(x,y) = -4 < 0$ for all (x,y),   there are no relative extrema.

13. $f_x = -6x^2 + 6y$; $f_{xx} = -12x$; $f_{xy} = 6$; $f_y = -6y^2 + 6x$; $f_{yy} = -12y$; $D = 144xy - 36$. Setting $f_x = 0$ and $f_y = 0$, we obtain $-6x^2 + 6y = 0$ and $-6y^2 + 6x = 0$ or $y = x^2$ and $x = y^2$. Substituting $x = y^2$ into $y = x^2$, we obtain $y = y^4$ or $y(y^3 - 1) = 0$. Thus, $y = 0$ and $y = 1$. The critical points are $(0,0)$ and $(1,1)$. Since $D(0,0) = -36 < 0$, $(0,0)$ does not give a relative extremum. Since $D(1,1) = 108 > 0$ and $f_{xx}(1,1) = -12 < 0$, $f(1,1) = 12$ is a relative maximum.

15. $f_x = y + 2/x^2$; $f_{xx} = -4/x^3$; $f_{xy} = 1$; $f_y = x + 4/y^2$; $f_{yy} = -8/y^3$; $D = 32/x^3Y^3 - 1$. Setting $f_x = 0$ and $f_y = 0$ we obtain $y + 2/x^2 = 0$ and $x + 4/y^2 = 0$. Substituting $y = -2/x^2$ into $x + 4/y^2 = 0$ we obtain $x + x^4 = x(1 + x^3) = 0$. Since $x = 0$ is not in the domain of f, the only critical point is $(-1,-2)$. Since $D(-1,-2) = 3 > 0$ and $f_{xx}(-1,-2) = 4 > 0$, $f(-1,-2) = 14$ is a relative minimum.

17. $f_x = (xe^x + e^x)\sin y$; $f_{xx} = (xe^x + 2e^x)\sin y$; $f_{xy} = (xe^x + e^x)\cos y$; $f_y = xe^x \cos y$; $f_{yy} = -xe^x \sin y$; $D = -xe^{2x}(x+2)\sin^2 y - e^{2x}(x+1)^2 \cos^2 y$. Setting $f_x(x,y) = 0$ and $f_y(x,y) = 0$ we obtain $(xe^x + e^x)\sin y = 0$ and $xe^x \cos y = 0$. Since $e^x > 0$ for all x, we have $(x+1)\sin y = 0$ and $x\cos y = 0$. When $x = -1$, we must have $\cos y = 0$ or $y = \pi/2 + k\pi$, k an integer. When $x = 0$, we must have $\sin y = 0$ or $y = k\pi$, k an integer. Thus, the critical points are $(0, k\pi)$ and $(-1, \pi/2 + k\pi)$, k an integer. Since $D(0, k\pi) = 0 - \cos^2 k\pi < 0$, $(0, k\pi)$ does not give a relative extrema. Now, $D(-1, \pi/2 + k\pi) = e^{-2}\sin^2(\pi/2 + k\pi) - 0 > 0$ and $f_{xx}(-1, \pi/2 + k\pi) = e^{-1}\sin(\pi/2 + k\pi)$. Since $f_{xx}(-1, \pi/2 + k\pi)$ is positive for k even and negative for k odd, $f(-1, \pi/2 + k\pi) = -e^{-1}$ are relative minima for k even, and $f(-1, \pi/2 + k\pi) = e^{-1}$ are relative maxima for k odd.

19. $f_x = \cos x$; $f_{xx} = -\sin x$; $f_{xy} = 0$; $f_y = \cos y$; $f_{yy} = -\sin y$; $D = \sin x \sin y$. Solving $f_x = 0$ and $f_y = 0$, we obtain the critical points $(\pi/2 + m\pi, \pi/2 + n\pi)$ for m and n integers. For m even and n odd or m odd and n even, $D < 0$ and no relative extrema result. For m and n both even, $D > 0$ and $f_{xx} < 0$ and $f(\pi/2 + m\pi, \pi/2 + n\pi) = 2$ are relative maxima. For m and n both odd, $D > 0$ and $fxx > 0$ and $f(\pi/2 + m\pi, \pi/2 + n\pi) = -2$ are relative minima.

21. Let the numbers be x, y, and $21 - x - y$. We want to maximize $P(x,y) = xy(21 - x - y) = 21xy - x^2y - xy^2$. Now $P_x = 21y - 2xy - y^2$; $P_{xx} = -2y$; $P_{xy} = 21 - 2x - 2y$; $P_y = 21x - x^2 - 2xy$; $P_{yy} = -2x$; $D = 4xy - (21 - 2x - 2y)^2$. Setting $P_x = 0$ and $P_y = 0$, we obtain $y(21 - 2x - y) = 0$ and $x(21 - x - 2y) = 0$. Letting $x = 0$ and $y = 0$, we obtain the critical points $(0,0)$, $(0,21)$, and $(21,0)$. Each of these results in $P = 0$ which is clearly not a maximum. Solving $21 - 2x - y = 0$ and $21 - x - 2y = 0$, we obtain the critical point $(7,7)$. Since $D(7,7) = 147 > 0$ and $P_{xx}(7,7) = -14 < 0$, $P(7,7) = 343$ is a maximum. The three numbers are 7,7, and 7.

23. Let $(x, y, 1 - x - 2y)$ be a point on the plane $x + 2y + z = 1$. We want to minimize $f(x,y) = x^2 + y^2 + (1 - x - 2y)^2$. Now $f_x = 2x - 2(1 - x - 2y)$; $f_{xx} = 4$; $f_{xy} = 4$; $f_y = 2y - 4(1 - x - 2y)$; $f_{yy} = 10$; $D = 40 - 4^2 = 24$. Setting $f_x = 0$ and $f_y = 0$ we obtain $2x - 2(1 - x - 2y) = 0$ and $2y - 4(1 - x - 2y) = 0$ or $2x + 2y = 1$ and $2x + 5y = 2$. Thus, $(1/6, 1/3)$ is a critical point. Since $D = 24 > 0$ and $f_{xx} = 4 > 0$ for all (x,y), $f(1/6, 1/3) = 1/6$ is a minimum. Thus, the point on the plane closest to the origin is $(1/6, 1/3, 1/6)$.

25. Let $(x, y, 8/xy)$ be a point on the surface. We want to minimize the square of the distance to the origin or $f(x, y) = x^2 + y^2 + 64/x^2y^2$. Now $f_x = 2x - 128/x^3y^2$; $f_{xx} = 2 + 384/x^4y^2$; $f_{xy} = 256/x^3y^3$; $f_y = 2y - 128/x^2y^3$; $f_{yy} = 2 + 384/x^2y^4$; $D = (2 + 384/x^4y^2)(2 + 384/x^2y^4) - (256)^2/x^6y^6$. Setting $f_x = 0$ and $f_y = 0$ we obtain $2x - 128/x^3y^2 = 0$ and $2y - 128/x^2y^3 = 0$ or $x^4y^2 = 64$ and $x^2y^4 = 64$; $x \neq 0$, $y \neq 0$. This gives $x^4y^2 = x^2y^4$, $x^2y^2(x^2 - y^2) = 0$ or $x^2 = y^2$. Thus, $x^6 = 64$ and $x = \pm 2$. Similarly, $y = \pm 2$ and the critical points are $(-2, -2)$, $(-2, 2)$, $(2, -2)$, and $(2, 2)$. Since $D(\pm 2, \pm 2) = 48 > 0$ and $f_{xx}(\pm 2, \pm 2) = 8 > 0$, $f(\pm 2, \pm 2) = 12$ are minima. The points closest to the origin are $(-2, -2, 2)$, $(-2, 2, -2)$, $(2, -2, -2)$, and $(2, 2, 2)$. The minimum distance is $\sqrt{12} = 2\sqrt{3}$.

27. We will maximize the square of the volume of the box in the first octant,

$$V(x, y) = x^2 Y^2 z^2 = x^2 y^2 (c^2 - c^2 x^2/a^2 - c^2 y^2/b^2).$$

$V_x = 2c^2 xy^2 - 4c^2 x^3 y^2/a^2 - 2c^2 xy^4/b^2$; $V_{xx} = 2c^2 y^2 - 12c^2 x^2 y^2/a^2 - 2c^2 y^4/b^2$; $V_{xy} = 4c^2 xy - 8c^2 x^3 y/a^2 - 8c^2 xy^3/b^2$; $V_y = 2c^2 x^2 y - 2c^2 x^4/a^2 - 4c^2 x^2 y^3/b^2$; $V_{yy} = 2c^2 x^2 - 2c^2 x^4/a^2 - 12c^2 x^2 y^2/b^2$; $D = V_{xx} V_{yy} - V_{xy}^2$. Setting $V_x = 0$ and $V_y = 0$ we obtain $xy^2 - 2x^3 y^2/a^2 - xy^4/b^2 = 0$, $x^2 y - x^4 y/a^2 - 2x^2 y^3/b^2 = 0$, or, assuming $x > 0$ and $y > 0$, $2b^2 x^2 + a^2 y^2 = a^2 b^2$. Solving, we obtain $x^2 = a^2/3$ and $y^2 = b^2/3$. Thus, $(a/\sqrt{3}, b/\sqrt{3})$ is a critical point. Since

$$D(a/\sqrt{3}, b/\sqrt{3}) = (-\frac{14}{9}b^2 c^2)(-\frac{14}{9}a^2 c^2) - (-\frac{4}{9}abc^2)^2 = \frac{20}{9}a^2 b^2 c^4 > 0$$

and $v_{xx} = -\frac{14}{9}b^2 c^2 < 0$, $V(a/\sqrt{3}, b/\sqrt{3}) = a^2 b^2 c^2/27$. The maximum volume is

$$8\sqrt{V(a/\sqrt{3}, b/\sqrt{3})} = 8\sqrt{3}abc/9.$$

29. The perimeter is given by $P = 2x + 2y + 2x \sec\theta$ and the area is $a = 2xy + x^2 \tan\theta$. Solving P for $2y$ and substituting in A, we obtain $A = Px - 2x^2(1 + \sec\theta) + x^2 \tan\theta$. Now $A_x(x, \theta) = P - 4x(1 + \sec\theta)2x\tan\theta$; $A_{xx}(x\theta) = -4(1 + \sec\theta) + 2\tan\theta$; $A_{x\theta}(x, \theta) = -4x\sec\theta\tan\theta + 2x\sec^2\theta$; $A_\theta(x\theta) = x^2\sec\theta(\sec\theta - 2\tan\theta)$; $A_{\theta\theta}(x, \theta) = 2x^2\sec\theta(\tan\theta - 2\sec^2\theta + 1)$. We assume that $x > 0$ and $0 \leq \theta \leq \pi/2$.

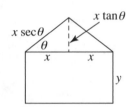

Setting $A_x = 0$ and $A_\theta = 0$, we obtain $P - 4x(1 + \sec\theta) + 2x\tan\theta = 0$ and $x^2\sec\theta(\sec\theta - 2\tan\theta) = 0$. We note from the second equation and the fact that $\sec\theta \neq 0$ for all θ that $\sec\theta - 2\tan\theta = 0$. Solving for θ, we obtain $\theta = 30°$ and solving $A_x = 0$ for x, we obtain $x_) = P/(4 + 2\sqrt{3})$. Since $D(x_0, 30°) = (-2\sqrt{3} + 2)(4x_0^2(\sqrt{3} - 5)/3\sqrt{3}) - 0^2 > 9$ and $A_{xx} = 2 - 2\sqrt{3} < 0$, $A(x_0, 30°)$ is a maximum. Letting $x = P/(4 + 2\sqrt{3})$ and $\theta = 30°$ in $P = 2x + 2y + 2x\sec\theta$, we obtain $P = 2y + P?\sqrt{3}$. Thus, the area is maximized for $x = P/(4 + 2\sqrt{3})$, $y = P(\sqrt{3} - 1)/2\sqrt{3}$, and $\theta = 30°$

31. $f_x = -\frac{2}{3}x^{-1/3}$, $f_y = -\frac{2}{3}y^{-1/3}$. Since $f_x = 0$ and $f_y = 0$ have no solutions, $f(x, y)$ has no critical points and Theorem 13.8.2 does not apply. However, for all $(x, y), f(0, 0) = 16 \geq 16 - (x^{1/3})^2 - (y^{1/3})^2 = f(x, y)$, and $f(0, 0) = 16$ is an absolute maximum.

33. $f_x = 10x$; $f_{xx} = 10$; $f_{xy} = 0$; $f_y = 4y^3$; $f_{yy} = 12y^2$; $D = 120y^2$. Solving $f_x = 0$ and $f_y = 0$ we obtain the critical point $(0, 0)$. Since $D(0, 0) = 0$, Theorem 13.8.2 does not apply. However, for any (x, y), $f(0, 0) = -8 \leq 5x^2 + y^4 - 8 = f(x, y)$ and $f(0, 0) = -8$ is an absolute minimum.

In Problems 35–38 we parameterize the boundary of R by letting $x = \cos t$ and $y = \sin t$; $0 \leq t \leq 2\pi$. Then, for $(x(t), y(t))$ on the boundary, we maximize or minimize $F(t) = f(\cos t, \sin t)$ on $[0, 2\pi]$.

35. $f_x = 1$; $f_y = \sqrt{3}$. There are no critical points on the interior of R. On the boundary we consider $F(t) = \cos t + \sqrt{3}\sin t$. Solving $F'(t) = -\sin t + \sqrt{3}\cos t = 0$, we obtain critical points at $t = \pi/3$ and $t = 4\pi/3$. Comparing $F(0) = 1, F(\pi/3) = 2$, and $F(4\pi/3) = -2$, we see that $f(1/2, \sqrt{3}/2) = 2$ is an absolute maximum and $f(-1/2, -\sqrt{3}/2) = -2$ is an absolute minimum.

37. $f_x = 2x + y$; $f_y = x + 2y$. Solving $f_x = 0$ and $f_y = 0$ we obtain the critical point $(0, 0)$ with corresponding function value $f(0, 0) = 0$. On the boundary we consider $F(t) = \cos^2 t + \cos t \sin t + \sin^2 t = 1 + \frac{1}{2}\sin 2t$. Solving $F'(t) = \cos 2t = 0$ we obtain critical points at $\pi/4$, $3\pi/4$, $5\pi/4$, and $7\pi/4$. Comparing $f(0, 0) = 0$, $F(0) = 1$, $F(\pi/4) = 3/2; F(3\pi/4) = 1/2$, $F(5\pi/4) = 3/2$, and $F(7\pi/4) = 1/2$, we see that $f(\sqrt{2}/2, \sqrt{w}/2) = f(-\sqrt{2}/2, -\sqrt{2}/2) = 3/2$ are absolute maxima and $f(0, 0) = 0$ is an absolute minimum.

39. $f_x = 4$; $f_y = -6$. There are no critical points over the region R, so absolute extrema must occur on the boundary. We parameterize the boundary by $x = 2\cos t$ and $y = \sin t$ for $0 \le t \le 2\pi$. Considering $F(t) = 8\cos t - 6\sin t$ we obtain $F'(t) = -8\sin t - 6\cos t$. Solving $F'(t) = 0$ we find $\tan t = -3/4$. Using $1 + \tan^2 t = \sec^2 t$ we see that $\sec^2 t = 25/16$ and $\cos t = -4/5$, t is in the second quadrant and $\sin t = 3/5$. The corresponding points on the boundary of R are $(8/5, -3/5)$ and $(-8/5, 3/5)$. Comparing $f(0) = F(2\pi) = f(2,0) = 8$, $f(8/5, -3/5) = 10$, and $f(-8/5, 3/5) = -10$ we see that the absolute minimum is $f(-8/5, 3/5) = -10$ and the absolute maximum is $f(8/5, -3/5) = 10$.

41. (a) $f_x = y\cos xy$; $f_y = x\cos xy$. Setting $f_x = 0$ and $f_y = 0$ we obtain $y\cos xy = 0$ and $x\cos xy = 0$. If $y = 0$ from the first equation, then necessarily $x = 0$ from the second equation. Thus, $(0,0)$ is a critical point. For $x \ne 0$ and $y \ne 0$ we have $\cos xy = 0$ or $xy = \pi/2$. Thus, all points $(x, \pi/2x)$ for $0 \le x \le \pi$ are also critical points.

    (b) Since $0 \le \sin xy \le 1$ for $0 \le x \le \pi$ and $0 \le y \le 1$, $f(x,y) = \sin xy$ has absolute minima at any points for which $\sin xy = 0$ and absolute maxima at any points for which $\sin xy = 1$. Thus, $f(x,y)$ has absolute minima when $xy = 0$ or $xy = \pi$, that is, at the points $(0,y)$, $(x,0)$, and $(\pi, 1)$ which are in the region. Absolute maxima occur when $xy = \pi/2$ or along the curve $y = \pi/2x$ in the region

    (c)

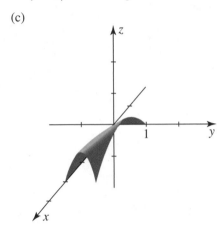

43. Since the volume of the box is 60, the height is $60/xy$. Then

$$C(x,y) = 10xy + 20xy + 2[2x60/xy + 2y60/xy] = 30xy + 240/y + 240/x.$$

$C_x = 30y - 240/x^2$; $C_{xx} = 480/x^3$, $C_{xy} = 30$; $C_y = 30x - 240/y^2$; $c_{yy} = 480/y^3$; $D = 480^2/x^3y^3 - 900$. Setting $C_x = 0$ and $C_y = 0$ we obtain $30y - 240/x^2 = 0$ and $30x - 240/y^2 = 0$ or $y = 8/x^2$ and $x = 8/y^2$. Substituting the first equation into the second, we have $x = x^4/8$ or $x(x^3 - 8) = 0$. Thus, $(2,2)$ is a critical point. Since $D(2,2) = 2700 > 0$ and $C_{xx}(2,2) = 60 > 0$, $C(2,2)$ is a minimum. Thus, the cost is minimized when the base of the box is 2 feet square and the height is 15 feet.

13.9 Method of Least Squares

1. $\displaystyle\sum_{i=1}^{4} x_i = 14$, $\displaystyle\sum_{i=1}^{4} y_i = 8$, $\displaystyle\sum_{i=1}^{4} x_i y_i = 30$, $\displaystyle\sum_{i=1}^{4} x_i^2 = 54$, $m = \dfrac{4(30) - 14(8)}{4(54) - (14)^2} = 0.4$,
$b = \dfrac{54(8) - 30(14)}{4(54) - (14)^2} = 0.6$, $y = 0.4x + 0.6$

3. $\displaystyle\sum_{i=1}^{5} x_i = 15$, $\displaystyle\sum_{i=1}^{5} y_i = 15$, $\displaystyle\sum_{i=1}^{5} x_i y_i = 56$, $\displaystyle\sum_{i=1}^{5} x_i^2 = 55$, $m = \dfrac{5(56) - 15(15)}{5(55) - (15)^2} = 1.1$,
$b = \dfrac{55(15) - 56(15)}{5(55) - (15)^2} = -0.3$, $y = 1.1x - 0.3$

5. $\displaystyle\sum_{i=1}^{7} x_i = 21$, $\displaystyle\sum_{i=1}^{7} y_i = 42$, $\displaystyle\sum_{i=1}^{7} x_i y_i = 164$, $\displaystyle\sum_{i=1}^{7} x_i^2 = 91$, $m = \dfrac{7(164) - 21(42)}{7(91) - (21)^2} \approx 1.35714$, $b = \dfrac{91(42) - 164(21)}{7(91) - (21)^2} \approx 1.92857$, $y \approx$
$1.35714x + 1.92857$

7. $\sum_{i=1}^{6} T_i = 420$, $\sum_{i=1}^{6} v_i = 1055$, $\sum_{i=1}^{6} T_i v_i = 68,000$, $\sum_{i=1}^{6} T_i^2 = 36,400$, $m = \dfrac{6(68,000) - 420(1055)}{6(36,400) - (420)^2} \approx -0.835714$,

$b = \dfrac{36,400(1055) - 68,000(420)}{6(36,400) - (420)^2} \approx 234.333$, $v \approx -0.835714T + 234.333$.

9. (a) least-squares line: $y = 0.5966x + 4.3665$
   least-squares quadratic: $y = -0.0232x^2 + 0.5618x + 4.5942$
   least-squares cubic: $y = 0.00079x^3 - 0.0212x^2 + 0.5498x + 4.5840$

   (b)   least-squares line   least-squares quadratic   least-squares cubic

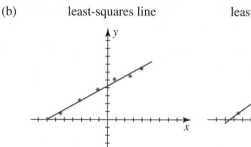 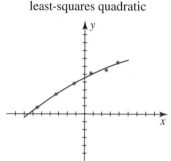 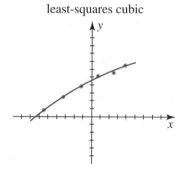

13.10   Lagrange Multipliers

1. f has constrained extrema where the
   level lines intersect the circle.

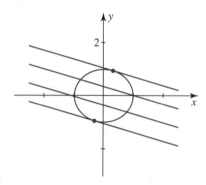

3. $f_x = 1$; $f_y = 3$; $g_x = 2x$; $g_y = 2y$. We need to solve $1 = 2\lambda x$, $3 = 2\lambda y$, $x^2 + y^2 - 1 = 0$. Dividing the second equation by the first, we obtain $3 = y/x$ or $y = 3x$. Substituting into the third equation, we have $x^2 + 9x^2 = 1$ or $x = \pm 1/\sqrt{10}$. For $x = 1/\sqrt{10}$, $y = 3/\sqrt{10}$ and for $x = -1/\sqrt{10}$, $y = -3/\sqrt{10}$. A constrained maximum is $f(1/\sqrt{10}, 3/\sqrt{10}) + \sqrt{10}$ and a constrained minimum is $f(-1/\sqrt{10}, -3/\sqrt{10}) = -\sqrt{10}$.

5. $f_x = y$; $f_y = x$; $g_x = 2x$; $g_y = 2y$. We need to solve $y = 2\lambda x$, $x = 2\lambda y$, $x^2 + y^2 - 2 = 0$. Substituting the second equation into the first, we obtain $y = 4\lambda^2 y$ or $y(4\lambda^2 - 1) = 0$. If $y = 0$, then from the second equation $x = 0$. Since $g(0,0) = -2 \neq 0$, $(0,0)$ does not satisfy the constraint. Thus, $\lambda = \pm 1/2$ and $y = \pm x$. Substituting into the third equation, we have $2x^2 = 2$ or $x = \pm 1$. Solutions of the system are $x = 1$, $y = 1$, $\lambda = 1/2$, $x = -1$, $y = -1$; $\lambda = 1/2$, $x = 1$, $y = -1$, $\lambda = -1/2$, and $x = -1$, $y = 1$, $\lambda = -1/2$. Thus, $f(1,1) = f(-1,-1) = 1$ are constrained maxima and $f(1,-1) = f(-1,1) = -1$ are constrained minima.

7. $f_x = 6x$; $f_y = 6y$; $g_x = 1$; $g_y = -1$. We need to solve $6x = \lambda$, $6y = -\lambda$; $x - y - 1 = 0$. From the first two equations, we obtain $x + y = 0$. Solving this with the third equation, we obtain $x = 1/2$, $y = -1/2$. Thus, $f(1/2, -1/2) = 13/2$ is a constrained extremum. Since $(1,0)$ satisfies the constraint and $f(1,0) = 8 > 13/2$, $f(1/2 - 1/2) = 13/2$ is a constrained minimum.

9. $f_x = 2x$; $f_y = 2y$; $g_x = 4x^3$; $g_y = 4y^3$. We need to solve $2x = 4\lambda x^3$, $2y = 4\lambda y^3$, $x^4 + y^4 - 1 = 0$ or $x(2\lambda x^2 - 1) = 0$, $y(2\lambda y^2 - 1) = 0$, $x^4 + y^4 = 1$. If $x = 0$, the from the third equation $y = \pm 1$. If $y = 0$, then $x = \pm 1$. From $2\lambda x^2 = 1 = 2\lambda y^2$ we have $x^2 = y^2$. Substituting into the third equation, we obtain $x = \pm 1/\sqrt[4]{2}$ and $y = \pm 1/\sqrt[4]{2}$. Solutions of the system are $(0, \pm 1)$, $(\pm 1, 0)$, and $(\pm 1/\sqrt[4]{2}, \pm 1/[4]2)$. Thus, $f(0, \pm 1) = f(\pm 1, 0) = 1$ are constrained minima and $f(\pm 1/\sqrt[4]{2}, \pm 1/\sqrt[4]{2}) = \sqrt{2}$ are constrained maxima.

11. $f_x = 3x^2y$; $f_y = x^3$; $g_x = 1/2\sqrt{x}$; $g_y = 1/2\sqrt{y}$. We need to solve $3x^2y = \lambda/2\sqrt{x}$, $x^3 = \lambda/2\sqrt{y}$, $\sqrt{x} + \sqrt{y} - 1 = 0$ or $6x^{5/2}y = \lambda$, $2x^3y^{1/2} = \lambda$, $\sqrt{x} + \sqrt{y} = 1$. From the first two equations, we obtain $3x^{5/2}y = x^3y^{1/2}$ and $3\sqrt{y} = \sqrt{x}$. Substituting into the third equation, we have $3\sqrt{y} + \sqrt{y} = 4\sqrt{y} = 1$. Then, $y = 1/16$ and $x = 9/16$. Since $(1/4, 1/4)$ satisfies the constraint and $f(1/4, 1/4) = 1/256$, $f(9/16, 1/16) + 729/65{,}536$ is a constrained maximum. We also consider $x = 0$, which requires $y = 1$; and $y = 0$, which requires $x = 1$. Since $x \ge 0$ and $y \ge 0$, $f(0,1) = f(1,0) = 0 \le x^3y = f(x,y)$ for all (x,y) which satisfy $\sqrt{x} + \sqrt{y} = 1$. Thus, $f(0,1) = 0$ and $f(1,0) = 0$ are constrained minima.

13. $F_x = 1$; $F_y = 2$; $F_z = 1$; $g_x = 2x$; $g_y = 2y$; $g_z = 2z$. We need to solve $1 = 2\lambda x$, $2 = 2\lambda y$; $1 = 2\lambda z$, $x^2 + y^2 + z^2 - 30 = 0$. From the first and second equations, we obtain $y = 2x$. From the first and third equations, we obtain $z = x$. Substituting into the fourth equation, we have $x^2 + 4x^2 + x^2 = 6x^2 = 30$. Thus, $x = \pm\sqrt{5}$, $y = \pm 2\sqrt{5}$, $z = \pm\sqrt{5}$. Then, $F(\sqrt{5}, 2\sqrt{5}, \sqrt{5}) = 6\sqrt{5}$ is a constrained maximum and $F(-\sqrt{5}, -2\sqrt{5}, -\sqrt{5}) = -6\sqrt{5}$ is a constrained minimum.

15. $F_x = yz$; $F_y = xz$; $F_z = xy$; $g_x = 2x$; $g_y = y/2$; $g_z = 2z/9$. We need to solve $yz = 2\lambda x$, $xz = \lambda y/2$, $xy = 2\lambda z/9$ or $xyz/2 = \lambda x^2$, $xyz/2 = \lambda y^2/4$, $xyz/2 = \lambda z^2/9$ along with $x^2 + y^2/4 + z^2/9 - 1 = 0$ or $x^2 + Y^2/4 + z^2/9 = 1$ for $x > 0$, $y > 0$, $z > 0$. From the first three equations and the fact that $\lambda \ne 0$, $x^2 = y^2/4 = z^2/9$. Substituting into the third equation, we obtain $x^2 + x^2 + x^2 = 3x^2 = 1$, so $x^2 = 1/3$, $y^2 = 4/3$, and $z^2 = 3$. Thus, $F(\sqrt{3}/3, 2\sqrt{3}/3, \sqrt{3}) = 2\sqrt{3}/3$ is a constrained extremum. Since $(\sqrt{23}/6, 1, 1)$ satisfies the constraint and $F(\sqrt{23}/6, 1, 1) = \sqrt{23}/6 < 2\sqrt{3}/3$, $F(\sqrt{3}/3, 2\sqrt{3}/3, \sqrt{3}) = 2\sqrt{3}/3$ is a constrained maximum.

17. $F_x = 3x^2$; $F_y = 3y^2$; $F_z = 3z^2$; $g_x = 1$; $g_y = 1$; $g_z = 1$. We need to solve $3x^2 = \lambda$, $3y^2 = \lambda$, $3z^2 = \lambda$, $x + y + z - 1 = 0$ for $x > 0$, $y > 0$, $z > 0$, and hence $\lambda > 0$. From the first three equations $x^2 = y^2 = z^2$, and since x, y, and z are positive, $x = y = z$. Then, from the fourth equation, $x = y = z = 1/3$ and $F(1/3, 1/3, 1/3) = 1/9$ is a constrained extremum. Since $(1/2, 1/4, 1/4)$ satisfies the constraint and $F(1/2, 1/4, 1/4) = 5/32 > 1/9$, $F(1/3, 1/3, 1/3) = 1/9$ is a constrained minimum.

19. $F_x = 2x$; $F_y = 2y$; $F_z = 2z$; $g_x = 2$; $g_y = 1$; $g_z = 1$; $h_x = -1$; $h_y = 2$; $h_z = -3$. We need to solve $2x = 2\lambda - \mu$, $2y = \lambda + 2\mu$, $2z = \lambda - 3\mu$ subject to $2x + y + z = 1$, $-x + 2y - 3z = 4$. Solving the first three equations for x, y, and z, respectively, and substituting into the constraint equations, we obtain $2\lambda - \mu + \lambda/2 - 3\mu/2 = 1$, $-\lambda + \mu/2 + \lambda + 2\mu - 3\lambda/2 + 9\mu/2 = 4$ or $6\lambda - 3\mu = 2$, $-3\lambda + 14\mu = 8$. From this, we obtain $\lambda = 52/75$ and $\mu = 54/75$. Then $x = 1/3$, $y = 16/15$, and $z = -11/15$. Thus, $F(1/3, 16/15, -11/15) = 134/75$ is a constrained minimum.

21. We want to maximize $A(x, y, xy/2)$ subject to $P(x, y) = x + y + \sqrt{x^2 + y^2} - 4 = 0$. $A_x = y/2$; $A_y = x/2$; $P_x = 1 + x/\sqrt{x^2 + y^2}$; $P_y = 1 + y/\sqrt{x^2 + y^2}$. We need to solve $y/2 = \lambda + \lambda x/\sqrt{x^2 + y^2}$, $x/2 = \lambda + \lambda y/\sqrt{x^2 + y^2}$, $x + y + \sqrt{x^2 + y^2} - 4 = 0$ for $x > 0, y > 0$, and hence $\lambda > 0$. Subtracting the second equation from the first, we have $(y - x)/2 = \lambda(x - y)/\sqrt{x^2 + y^2}$ or $(y - x) = (y - x)(-2\lambda/\sqrt{x^2 + y^2})$. Since $-2\lambda/\sqrt{x^2 + y^2}$ is negative, it cannot equal 1, and hence $y - x = 0$ or $y = x$. Substituting in the third equation gives $2x + \sqrt{2x^2} = (2 + \sqrt{2})x = 4$. Thus, $x = y = 4/(2 + \sqrt{2})$ and this maximum area is $A(4/(2 + \sqrt{2}), 4/(2 + \sqrt{2})) = 4/(3 + 2\sqrt{2})$.

23. We want to maximize $V(x, y) = 9\pi x + 3\pi y$ subject to $S(x, y) = 9\pi + 6\pi x + 3\pi\sqrt{9 + y^2} - 81\pi$. Now $V_x = 9\pi$; $V_y = 3\pi$; $S_x = 6\pi$; $S_y = 3\pi y/\sqrt{9 + y^2}$. We need to solve $9\pi = 6\pi\lambda$, $3\pi = 3\pi\lambda y/\sqrt{9 + y^2}$, $9\pi + 6\pi x + 3\sqrt{9 + y^2} - 81\pi = 0$ for $x > 0$ and $y > 0$. From the first equation, $\lambda = 3/2$. Using $\lambda = 3/2$ in the second equation gives $y = 6/\sqrt{5}$. From the third equation, we have $6x + 3\sqrt{9 + 36/5} = 72$ or $x = 12 - 9/2\sqrt{5}$. The volume is maximum when $x = 12 - 9/2\sqrt{5}m$ and $y = 6/\sqrt{5}m$.

25. We want to maximize $z(x, y) = P - x - y$ subject to $z^2/xy^3 = k$ or $(P - x - y)^2 - kxy^3 = 0$. Now $z_x = -1$; $z_y = -1$; $g_x = -2(P - x - y) - ky^3$; $g_y = -2(P - x - y) - 3kxy^2$. We need to solve $-1 = -2\lambda(P - x - y) - \lambda ky^3$, $-1 = -2\lambda(P - x - y) - 3\lambda kxy^2$, $(P - x - y)^2 = kxy^3$ for $x > 0$, $y > 0$, and $z > 0$. From the first two equations, we have $y = 3x$. Substituting into the third equation, we obtain $(P - 4x)^2 = 27kx^4$ or $\sqrt{27k}x^2 = P - 4x$. (Since $z > 0$, $z = P - x - y = P - 4x > 0$.) Using the quadratic formula and the fact that $x > 0$, we find

$$x = \frac{-4 + \sqrt{16 + 4P\sqrt{27k}}}{2\sqrt{27k}} = \frac{-2 + \sqrt{4 + P\sqrt{27k}}}{27k}.$$

Then the maximum value of z is $P - 4x = P + 4(2 - \sqrt{4 + P\sqrt{27k}}/\sqrt{27k}$.

27. $f(x, y)$ is the square of the distance from a point on the graph of $x^4 + y^4 = 1$ to the origin. The points $(0, \pm 1)$ and $(\pm 1, 0)$ are closest to the origin, while $(\pm 1/\sqrt[4]{2}, \pm 1/\sqrt[4]{2})$ are farthest from the origin.

29. F is the square of the distance of points on the intersection of the planes $2x + y + z = 1$ and $-x + 2y - 3z = 4$ from the origin. The point $(1/3, 16/15, -11/15)$ is closest to the origin.

31. We want to minimize $f(x,) = x^2 + y^2$ subject to $xy^2 = 1$. Now $f_x = 2x$; $f_y = 2y$; $g_x = y^2$; $g_y = 2xy$. We need to solve $2x = \lambda y^2$, $2y = 2\lambda xy$, $xy^2 - 1 = 0$ or $2xy = \lambda y^3$, $2xy = 2\lambda x^2 y$, $xy^2 = 1$ for $x > 0, y > 0$, and hence $\lambda > 0$. From the first two equations, we have $y^3 = 2x^2 y$ or $y^2 = 2x^2$. Substituting into the third equation gives $2x^3 = 1$ or $x = 2^{-1/3}$. Again, from the third equation we have $y = 1/(2^{-1/3})^{1/2} = 2^{1/6}$. Thus, the point closest to the origin is $(2^{-1/3}, 2^{1/6})$. Since the surface is $F(x,y,z) = xy^2 - 1 = 0$, $\nabla F = y^2 \mathbf{i} + 2xy\mathbf{j}$ is normal to the surface at (x,y,z). Thus, a normal to the surface at $(2^{-1/3}, 2^{1/6}, 0)$ is $\nabla F(2^{-1/3}, 2^{1/6}, 0)$ is $\nabla F(2^{-1/3}, \ 21/6, 0) = 2^{1/3}\mathbf{i} + 2(2^{-1/3})(2^{1/6})\mathbf{j} = 2^{1/3}\mathbf{i} + 2^{5/6}\mathbf{j} = 2^{2/3}(2^{-1/3}\mathbf{i} + 2^{1/6}\mathbf{j})$. Since $\nabla F(2^{-1/3}, 2^{1/6}, 0)$ is a multiple of the vector from the origin to $P(2^{-1/3}, 2^{1/6}, 0)$, this vector is perpendicular to the surface.

33. For any $x + y + z = k$, by Problem 32, $\sqrt[3]{xyz} \leq \dfrac{k}{3} = \dfrac{x+y+z}{3}$.

Chapter 13 in Review

A. True/False

1. False; see Example 3 in Section 13.2 in the text.

3. True

5. False; consider $z = y^2$.

7. False; ∇f is perpendicular to the level curve $f(x,y) = c$.

9. True

B. Fill In the Blanks

1. $\displaystyle\lim_{(x,y) \to (1,1)} \frac{3x^2 + xy^2 - 3xy - 2y^3}{5x^2 - y^2} = \frac{3+1-3-2}{5-1} = -\frac{1}{4}$

3. $3x^2 + y^2 = 3(2)^2 + (-4)^2 = 28$

5. $\dfrac{d}{dw} F(r,s) = \dfrac{\partial F}{\partial r}\dfrac{dr}{dw} + \dfrac{\partial F}{\partial s}\dfrac{ds}{dw} = F_r g'(w) + F_s h'(w)$

7. f_{yyzx}

9. Using the Fundamental Theorem of Calculus, we have $\dfrac{\partial f}{\partial y}(x,y) = \dfrac{\partial}{\partial y}[\int_x^y F(t)dt] = F(y)$

$\dfrac{\partial f}{\partial x}(x,y) = \dfrac{\partial}{\partial x}[\int_x^y F(t)dt] = \dfrac{\partial}{\partial x}\left[-\int_y^x F(t)dt\right] = -\dfrac{\partial}{\partial x}\left[\int_y^x F(t)dt\right] = -F(x)$

11. $F_{x,y,z} = \dfrac{\partial^2}{\partial z \partial y} f_x(x,y)g(y)h(z) = \dfrac{\partial}{\partial z}[f_x(x,y)g'(y)h(z) + f_{xy}(x,y)g(y)h(z)]$
$= f_x(x,y)g'(y)h'(z) + f_{xy}(x,y)g(y)h'(z)$

C. Exercises

1. $z_y = -x^3 y e^{-x^3 y} + e^{-x^3 y}$

3. $f_r = \dfrac{3}{2}r^2(r^3 + \theta^2)^{-1/2}$; $f_{r\theta} = -\dfrac{3}{2}r^2\theta(r^3 + \theta^2)^{-3/2}$

5. $\dfrac{\partial z}{\partial y} = 3x^2 y^2 \sinh x^2 y^3$; $\dfrac{\partial^2 z}{\partial y^2} = 9x^4 y^4 \cosh x^2 y^3 + 6x^2 y \sinh x^2 y^3$

7. $F_s = 3s^2 t^5 v^{-4}$; $F_{st} = 15s^2 t^4 v^{-4}$; $F_{stv} = -60s^2 t^4 v^{-5}$

9. $\nabla f = -\dfrac{y}{x^2}\dfrac{1}{1+y^2/x^2}\mathbf{i} + \dfrac{1}{x}\dfrac{1}{1+y^2/x^2}\mathbf{j} = -\dfrac{y}{x^2+y^2}\mathbf{i} + \dfrac{x}{x^2+y^2}\mathbf{j}$; $\nabla f(1,-1) = \dfrac{1}{2}\mathbf{i} + \dfrac{1}{2}\mathbf{j}$

11. $\nabla f = (2xy - y^2)\mathbf{i} + (x^2 - 2xy)\mathbf{j}$; $\mathbf{u} = \dfrac{2}{\sqrt{40}}\mathbf{i} + \dfrac{6}{\sqrt{40}}\mathbf{j} = \dfrac{1}{\sqrt{10}}(\mathbf{i} + 3\mathbf{j})$;

$D_{\mathbf{u}}f = \dfrac{1}{\sqrt{10}}(2xy - y^2 + 3x^2 - 6xy) = \dfrac{1}{\sqrt{10}}(3x^2 - 4xy - y^2)$

13. $\{(x,y) | (x+y)^2 \le 1\} = \{(x,y) | \ |x+y| \le 1\}$

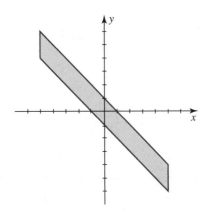

15. $\Delta z = 2(x+\Delta x)(y+\Delta y) - (y+\Delta y)^2 - (2xy - y^2) = 2x\Delta y + 2y\Delta x + 2\Delta x\Delta y - 2y\Delta y - (\Delta y)^2$

17. $z_x = \dfrac{4x + 3y - (x - 2y)4}{(4x+3y)^2} = \dfrac{11y}{(4x+3y)^2}$; $z_y = \dfrac{(4x+3y)(-2) - (x-2y)3}{(4x+3y)^2} = \dfrac{-11x}{(4x+3y)^2}$;

$dz = \dfrac{11y}{(4x+3y)^2}dx - \dfrac{11x}{(4x+3y)^2}dy$

19. $z_y = 4y/\sqrt{x^2 + 4y^2}$, $z_y(-\sqrt{5}, 1) = 4/3$, $z(-\sqrt{5}, 1) = 3$. The line is given by $x = -\sqrt{5}$ and $z - 3 = \dfrac{4}{3}(y - 1)$. Symmetric equations of the line are $x = -\sqrt{5}$, $\dfrac{z-3}{4} = \dfrac{y-1}{3}$.

21. $f_x = 2xy^4$, $f_y = 4x^2y^3$.

    (a) $\mathbf{u} = \mathbf{i}$, $D_{\mathbf{u}}(1,1) = f_x(1,1) = 2$

    (b) $\mathbf{u} = (\mathbf{i} - \mathbf{j}/\sqrt{2}$, $D_{\mathbf{u}}(1,1) = (2-4)/\sqrt{2} = -2/\sqrt{2}$

    (c) $\mathbf{u} = \mathbf{j}$, $D_{\mathbf{u}}(1,1) = f_y(1,1) = 4$

23. $F(x,y,z) = \sin xy - z$; $\nabla F = y\cos xy\,\mathbf{i} + x\cos xy\,\mathbf{j} - \mathbf{k}$; $\nabla F(1/2, 2\pi/3, \sqrt{3}/2) = \dfrac{\pi}{3}\mathbf{i} + \dfrac{1}{4}\mathbf{j} - \mathbf{k}$. The equation of the tangent plane is $\dfrac{\pi}{3}(x - \dfrac{1}{2}) + \dfrac{1}{4}(y - \dfrac{2\pi}{3}) - (z - \dfrac{\sqrt{3}}{2}) = 0$ or $4\pi x + 3y - 12z = 4\pi - 6\sqrt{3}$.

25. $\nabla F = 2x\mathbf{i} + 2y\mathbf{j}$; The equation of the tangent plane is $6(x - 3) + 8(y - 4) = 0$ or $3x + 4y = 25$.

27. We want to maximize $v(x,y,z) = xyz$ subject to $x + 2y + z = 6$. Now $V_x = yz$; $V_y = xz$; $V_z = xy$; $g_x = 1$; $g_y = 2$; $g_z = 1$. We need to solve $yz = \lambda$, $xz = 2\lambda$, $xy = \lambda$ or $xyz = \lambda x$, $xyz = 2\lambda y$, $xyz = \lambda z$ along with $x + 2y + z - 6 = 0$ or $x + 2y + z = 6$. From the first three equations, we have $x = 2y = z$. Substituting into the fourth equation gives $x + x + x = 3x = 6$ or $x = 2$. Then $y = 1$ and $z = 2$ and $V(2,1,2) = 4$ is the maximum volume.

29. We are given $v = 14\sqrt{5}ry^{-1/2}$, $dr = -1$, $dy = 1$, $r = 20$, and $y = 25$. Now, $dv = 14\sqrt{5}y^{-1/2}dr - 7\sqrt{5}ry^{-3/2}dy$ and the approximate change in volume is

$$\Delta v \approx 14\sqrt{5}(25)^{-1/2}(-1) - 7\sqrt{5}(20)(25)^{-3/2}(1) = -98\sqrt{5}/25 \approx -8.77\,\text{cm/s}.$$

31. Let $g(r,\theta) = \dfrac{R^2 - r^2}{R^2 - 2rR\cos(\theta - \phi) + r^2}$. Then, after a straightforward but lengthy computaiton, we find

$$g_r = \dfrac{(2r^2R + 2R^3)\cos(\theta - \phi) - 4rR^2}{(R^2 - 2rR\cos(\theta - \phi) + r^2)^2},$$

$$g_{rr} = \frac{8R^4\cos^2(\theta - \phi) + (-12rR^3 - 4r^3R)\cos(\theta - \phi) - 4R^4 + 12r^2R^2}{[R^2 - 2rR\cos(\theta - \phi) + r^2]^3},$$

$$g_{\theta\theta} = \frac{(4r^4R^2 - 4r^2R^4)\cos^2(\theta - \phi) + (2r^5R - 2rR^5)\cos(\theta - \phi) - 8r^4R^2 + 8r^2R^4}{(R^2 - 2rR\cos\theta + r^2)^3}$$

and $r^2 g_{rr} + rg_r + g_{\theta\theta}$. Then

$$r^2 U_{rr} + rU_r + U_{\theta\theta} = \frac{r^2}{2\pi}\int_{\pi}^{\pi} g(r,\theta)f(\phi)d\phi + \frac{r}{2\pi}\int_{\pi}^{\pi} g(r,\theta)f(\phi)d\phi + \frac{1}{2\pi}\int_{\pi}^{\pi} g(r,\theta)f(\phi)d\phi$$

$$= \frac{1}{2\pi}\int_{\pi}^{\pi} (r^2 g_{rr} + rg_r + g_{\theta\theta})d\phi = 0.$$

33. Since $D = 4(6) - 5^2 = -1 < 0$, $f(a,b)$ is not a relative extremum.

35. Since $D = (-5)(-9) - 6^2 = 9 > 0$ and $f_{xx} = -5 < 0$, $f(a,b)$ is a relative maximum.

37. Since $x = L\cos\theta$ and $y = L\sin\theta$,
$A = \frac{1}{2}xy = \frac{1}{2}L^2\sin\theta\cos\theta = \frac{1}{4}l^2\sin 2\theta.$

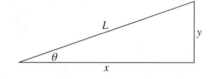

39. $A = xy - (y - 2z)(x - 2z) - z^2 = 2(x+y)z - 5z^2$

41. $V = (2x)(2y)z = 4xy\left(4 - \sqrt{x^2 - y^2}\right) = 16xy - 4xy\sqrt{x^2 + y^2}$

Multiple Integrals

14.1 The Double Integral

1. With $f(x,y) = x + 3y + 1$ and $\Delta A_k = 1$,

$$\int\int_R (x+3y+1)dA \approx f(1/2,1/2) + f(3/2,1/2) + f(5/2,1/2) + f(1/2,3/2)$$
$$+ f(3/2,3/2) + f(5/2,3/2) + f(1/2,5/2) + f(3/2,5/2)$$
$$= 3 + 4 + 5 + 6 + 7 + 8 + 9 + 10 = 52.$$

3. (a) With $f(x,y) = x + y$, and $\Delta A_k = 1$,

$$\int\int_R (x+y)\ dA \approx (-3/2+1/2) + (-1/2+1/2) + (1/2+1/2) + (3/2+1/2)$$
$$+ (-3/2+3/2) + (-1/2+3/2) + (1/2+3/2) + (3/2+3/2)$$
$$= \frac{1}{2}(-2+0+2+4+0+2+4+6) = \frac{16}{2} = 8.$$

(b) With $f(x,y) + y + 4$ and $\Delta A_k = 1$,

$$\int\int_R (x+y)\ dA \approx (-2+1) + (-1+1) + (0+1) + (1+1) + (-2+2) + (-1+2) + (0+2) + (1+2)$$
$$= 8.$$

5. $\int\int_R 10 dA = 10 \int\int_R dA = 10(6) = 60$

7. $\int\int_R 10 dA = 10 \int\int_R dA = 10\left[\frac{1}{4}\pi(2)^2\right] = 10\pi$

9. No, since $x + 5y$ is negative at $(3, -1)$ which is in R.

11. $\int\int_R 10 dA = 10 \int\int_R dA = 10(8) = 80$

13. $\int\int_R (2x+4y)dA = 2\int\int_R x dA + 4\int\int_R y dA = 2(3) + 4(7) = 34$

15. $\int\int_R (3x+7y+1)dA = 3\int\int_R x dA + 7\int\int_R y dA + \int\int_R dA = 3(3) + 7(7) + 8 = 66$

17. $\int\int_R f(x,y)dA = \int\int_{R_1} f(x,y)dA + \int\int_{R_2} f(x,y)dA = 4 + 14 = 18$

14.2 Iterated Integrals

1. $\int dy = y + c_1(x)$

3. By holding y fixed,
$$\int (6x^2 y - 3x\sqrt{y})dx = 6\left(\frac{x^3}{3}\right)y - 3\left(\frac{x^2}{2}\right)\sqrt{y} + c_2(y)$$
$$= 2x^3 y - \frac{3}{2}x^2\sqrt{y} + c_2(y)$$

5. By holding x fixed,
$$\int \frac{1}{x(y+1)}dy = \frac{\ln|y+1|}{x} + c_1(x)$$

7. By holding y fixed,
$$\int (12y\cos 4x - 3\sin y)dx = 12y\left(\frac{\sin 4x}{4}\right) - 3x\sin y + c_2(y)$$
$$= 3y\sin 4x - 3x\sin y + c_2(y)$$

9. By holding y fixed,
$$\int \frac{y}{\sqrt{2x+3y}}dy = y\sqrt{2x+3y} + c_2(y)$$

11. $\displaystyle\int_{-1}^{3} (6xy - 5e^y)\ dx = (3x^2 y - 5xe^y)\Big|_{-1}^{3} = (27y - 15e^y) - (3y + 5e^y) = 24y - 20e^y$

13. $\displaystyle\int_{1}^{3x} x^3 e^{xy} = x^2 e^{xy}\Big|_{1}^{3x} = x^2(e^{3x^2} - e^x)$

15. $\displaystyle\int_{0}^{2x} \frac{xy}{x^2 + y^2}\ dy = \frac{x}{2}\ln(x^2 + y^2)\Big|_{0}^{2x} = \frac{x}{2}[\ln(x^2 + 4x^2) - \ln x^2] = \frac{x}{2}\ln 5$

17. $\displaystyle\int_{\tan y}^{\sec y} (2x + \cos y)\ dx = (x^2 + x\cos y)\Big|_{\tan y}^{\sec y} = \sec^2 y + \sec y\cos y - \tan^2 y - \tan y\cos y$
$$= \sec^2 y + 1 - \tan^2 y - \sin y = 2 - \sin y$$

19. $\displaystyle\int_{x}^{\pi/2} \cos x\sin^3 y\,dy = \cos x\cos y\left(\frac{-\sin^2 y}{3} - \frac{2}{3}\right)\Big|_{x}^{\pi/2} = 0 - \cos^2 x\left(\frac{-\sin^2 x}{3} - \frac{2}{3}\right)$
$$= \frac{\cos^2 x\sin^3 x}{3} + \frac{2\cos^2 x}{3} = \frac{\cos^2(1 - \cos^2 x)}{3} + \frac{2\cos^2 x}{3}$$
$$= \frac{\cos^2 x}{3} - \frac{\cos^4 x}{3} + \frac{2\cos^2 x}{3} = \cos^2 x - \frac{1}{3}\cos^4 x$$

21. $\displaystyle\int_{1}^{2} \int_{-x}^{x^2} (8x - 10y + 2)\ dy\ dx = \int_{1}^{2} (8xy - 5y^2 + 2y)\Big|_{-x}^{x^2}\ dx$
$$= \int_{1}^{2} [(8x^3 - 5x^4 + 2x^2) - (-8x^2 - 5x^2 - 2x)]\ dx$$
$$= \int_{1}^{2} (8x^3 - 5x^4 + 15x^2 + 2x)\ dx = (2x^4 - x^5 + 5x^3 + x^2)\Big|_{1}^{2}$$
$$= 44 - 7 = 37$$

23. $\displaystyle\int_{0}^{\sqrt{2}} \int_{-\sqrt{2-y^2}}^{\sqrt{2-y^2}} (2x - y)\ dx\ dy = \int_{0}^{\sqrt{2}} (x^2 - xy)\Big|_{-\sqrt{2-y^2}}^{\sqrt{2-y^2}}$
$$= \int_{0}^{\sqrt{2}} \left[(2 - y^2 - y\sqrt{2-y^2}) - (2 - y^2 + y\sqrt{2-y^2})\right]\ dy$$
$$= \int_{0}^{\sqrt{2}} (-2\sqrt{2-y^2})\ dy = \frac{2}{3}(2 - y^2)^{3-2}\Big|_{0}^{\sqrt{2}} = \frac{2}{3}(0) - \frac{2}{3}2^{3/2} = -\frac{4}{3}\sqrt{2}$$

25. $\int_0^\pi \int_y^{3y} \cos(2x+y) \ dx \ dy \int_0^\pi \frac{1}{2}\sin(2x+y)\Big|_y^{3y} \ dy \frac{1}{2}\int_0^\pi (\sin 7y - \sin 3y) \ dy$

$$= \frac{1}{2}\left(-\frac{1}{7}\cos 7y + \frac{1}{3}\cos 3y\right)\Big|_0^\pi$$

$$= \frac{1}{2}\left[-\frac{1}{7}(-1) + \frac{1}{3}(-1) - \left(-\frac{1}{7} + \frac{1}{3}\right)\right] = -\frac{4}{21}$$

27. $\int_1^{\ln 3} \int_0^x 6e^{x+2y} \ dy \ dx = \int_1^{\ln 3} e^{x+2y}\Big|_0^x \ dx = \int_1^{\ln 3} (3e^{3x} - 3e^x) \ dx = (e^{3x} - 3e^x)\Big|_1^{\ln 3}$

$$= (27 - 9) - (e^3 - 3e) = 18 - e^3 + 3e$$

29. $\int_0^3 \int_{x+1}^{2x+1} \frac{1}{\sqrt{y-x}} \ dy \ dx = \int_0^3 2\sqrt{y-x}\Big|_{x+1}^{2x+1} \ dx = 2\int_0^3 (\sqrt{x+1} - 1) \ dx$

$$= 2\left[\frac{2}{3}(x+1)^{3/2}\right]\Big|_0^3 = 2\left[\left(\frac{16}{3} - 3\right) - \left(\frac{2}{3}\right)\right] = \frac{10}{3}$$

31. $\int_1^9 \int_0^x \frac{1}{x+y^2} \ dy \ dx = \int_1^9 \frac{1}{x}\tan^{-1}\frac{y}{x}\Big|_0^x = \int_1^9 \frac{\pi}{4x} \ dx = \frac{\pi}{4}\ln|x|\Big|_1^9 = \frac{\pi}{4}\ln 9$

33. $\int_1^e \int_1^y \frac{y}{x} \ dx \ dy = \int_1^e y\ln x\Big|_1^y \ dy = \int_1^e y\ln y \ dy$ $\boxed{\text{Integration by parts}}$

$$= \left(\frac{1}{2}y^2\ln y - \frac{1}{4}y^2\right)\Big|_1^e = \frac{1}{2}e^2 - \frac{1}{4}e^2 - \left(-\frac{1}{4}\right) = \frac{1}{4}(e^2 + 1)$$

35. $\int_0^6 \int_0^{\sqrt{25-y^2}/2} \frac{1}{\sqrt{(25-y^2)-x^2}} \ dx \ dy = \int_0^6 \left(\sin^{-1}\frac{x}{\sqrt{25-y^2}}\right)\Big|_0^{\sqrt{25-y^2}/2} \ dy$

$$= \int_0^6 \sin^{-1}\frac{1}{2} \ dy = \int_0^6 \frac{\pi}{6} \ dy = \pi$$

37. $\int_{\pi/2}^\pi \int_{\cos y}^0 e^x \sin y \ dx \ dy = \int_{\pi/2}^\pi e^x \sin y\Big|_{\cos y}^0 \ dy = \int_{\pi/2}^\pi (\sin y - e^{\cos y}\sin y) \ d$

$$= (-\cos y + e^{\cos y})|_{\pi/2}^\pi = (1 + e^{-1}) - (0 + 1) = e^{-1}$$

39. $\int_\pi^{2\pi} \int_0^x (\cos x - \sin y) \ dy \ dx = \int_\pi^{2\pi} (y\cos x + \cos y)\Big|_0^x \ dx = \int_\pi^{2\pi} (x\cos x + \cos x - 1) \ dx$

$$\boxed{\text{Integration by parts}}$$

$$= (\cos x + x\sin x + \sin x - x)|_\pi^{2\pi} = (1 - 2\pi) - (-1 - \pi) = 2 - \pi$$

41. $\int_{\pi/12}^{5\pi/12} \int_1^{\sqrt{2\sin 2\theta}} r \ dr \ d\theta = \int_{\pi/12}^{5\pi/12} \frac{1}{2}r^2\Big|_1^{\sqrt{2\sin 2\theta}} \ d\theta = \int_{\pi/12}^{5\pi/12} \left(\sin 2\theta - \frac{1}{2}\right) \ d\theta = -\frac{1}{2}(\cos 2\theta + \theta)\Big|_{\pi/12}^{5\pi/12}$

$$= -\frac{1}{2}\left[\left(-\frac{\sqrt{3}}{2} + \frac{5\pi}{12}\right) - \left(\frac{\sqrt{3}}{2} + \frac{5\pi}{12}\right)\right] = \frac{\sqrt{3}}{2} - \frac{\pi}{6}$$

43.

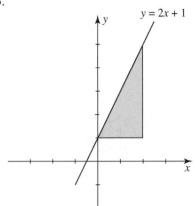

45.

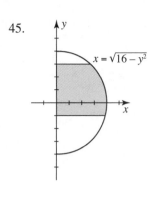

47. $\displaystyle \int_0^4 \int_{x/2}^{\sqrt{x}} x^2 y\, dy\, dx = \int_0^4 \frac{1}{2} x^2 y \Big|_{x/2}^{\sqrt{x}}\, dx = \int_0^4 \frac{1}{2} x^2 \left(x - \frac{x^2}{4} \right)\, dx$

$\displaystyle \qquad = \int_0^4 \left(\frac{1}{2} x^2 - \frac{1}{8} x^4 \right)\, dx = \left(\frac{1}{8} x^4 - \frac{1}{40} x^5 \right) \Big|_0^4$

$\displaystyle \qquad = 32 - \frac{128}{5} = \frac{32}{5}$

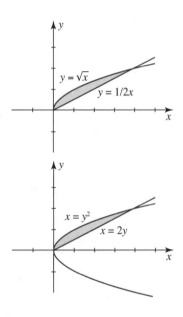

$\displaystyle \int_0^2 \int_{y^2}^{2y} x^2 y\, dx\, dy = \int_0^2 \frac{1}{3} x^3 y \Big|_{y^2}^{2y}\, dy = \int_0^2 \frac{1}{3} y(8y^3 - y^6)\, dy$

$\displaystyle \qquad = \int_0^2 \left(\frac{8}{3} y^4 - \frac{1}{3} y^7 \right) dy = \left(\frac{8}{15} y^5 - \frac{1}{24} y^8 \right) \Big|_0^2$

$\displaystyle \qquad = \frac{256}{15} - \frac{32}{3} = \frac{32}{5}$

Therefore $\displaystyle \int_0^4 \int_{x/2}^{\sqrt{x}} x^2 y\, dy\, dx = \int_0^2 \int_{y^2}^{2y} x^2 y\, dx\, dy$

49.

$$\int_{-1}^2 \int_0^3 x^2\, dy\, dx = \int_{-1}^2 x^2 y \Big|_0^3\, dx = \int_{-1}^2 3x^2\, dx = x^3 \Big|_{-1}^2 = 8 - (-1) = 9$$

$$\int_0^3 \int_{-1}^2 x^2\, dx\, dy = \int_0^3 \frac{1}{3} x^3 \Big|_{-1}^2\, dy = \int_0^3 \left(\frac{8}{3} + \frac{1}{3} \right)\, dy = \int_0^3 3\, dy = 3y \Big|_0^3 = 9$$

51. $\displaystyle \int_1^3 \int_0^\pi (3x^2 y - 4\sin y)\, dy\, dx = \int_1^3 \left(\frac{3}{2} x^2 y^2 - 4\cos y \right) \Big|_0^\pi\, dx = \int_1^3 \left[\left(\frac{3\pi^2}{2} x^2 - 4 \right) - (4) \right]\, dx$

$\displaystyle \qquad = \int_1^3 \left(\frac{3\pi^2}{2} x^2 - 8 \right)\, dx = \left(\frac{\pi^2}{2} x^3 - 8x \right) \Big|_1^3$

$\displaystyle \qquad = \left(\frac{27\pi^2}{2} - 24 \right) - \left(\frac{\pi^2}{2} - 8 \right) = 13\pi^2 - 16$

$\displaystyle \int_0^\pi \int_1^3 (3x^2 y - 4\sin y)\, dy\, dx = \int_0^\pi (x^3 y - 4x\sin y) \Big|_1^3\, dy = \int_0^\pi [(27y - 12\sin y) - (y - 4\sin y)]\, dy$

$\displaystyle \qquad = \int_0^\pi (26y - 8\sin y)\, dy = (13y^2 + 8\cos y) \Big|_0^\pi$

$\displaystyle \qquad = (13\pi^2 - 8) - (8) = 13\pi^2 - 16$

53. We use the fact that $\int_\alpha^\beta kF(t)\, dt = k \int_\alpha^\beta F(t)\, dt$. Then

$$\int_c^d \int_a^b f(x)g(y)\, dx\, dy = \int_c^d g(y) \left[\int_a^b f(x)\, dx \right]\, dy = \left[\int_a^b f(x)\, dx \right] \left[\int_c^d g(y)\, dy \right]$$

14.3 Evaluation of Double Integrals

1. $\displaystyle\int\int_R x^3y^2\,dA = \int_0^1\int_0^x x^3y^2\,dy\,dx = \int_0^1 \frac{1}{3}x^3y^3\Big|_0^x\,dx = \frac{1}{3}\int_0^1 x^6\,dx$

$\displaystyle\qquad = \frac{1}{21}x^7\Big|_0^1 = \frac{1}{21}$

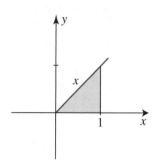

3. $\displaystyle\int\int_R (2x+4y+1)\,dA = \int_0^1\int_{x^3}^{x^2}(2x+4y+1)\,dy\,dx$

$\displaystyle\qquad = \int_0^1 (2xy+2y^2+y)\Big|_{x^3}^{x^2}\,dx$

$\displaystyle\qquad = \int_0^1 [(2x^3+2x^4+x^2)-(2x^4+2x^6+x^3)]\,dx$

$\displaystyle\qquad = \int_0^1 (x^3+x^2-2x^6)\,dx = \left(\frac{1}{4}x^4+\frac{1}{3}x^3-\frac{2}{7}x^7\right)\Big|_0^1$

$\displaystyle\qquad = \frac{1}{4}+\frac{1}{3}-\frac{2}{7} = \frac{25}{84}$

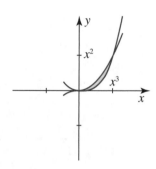

5. $\displaystyle\int\int_R 2xy\,dA = \int_0^2\int_{x^3}^8 2xy\,dy\,dx = \int_0^2 xy^2\Big|_{x^3}^8\,dx$

$\displaystyle\qquad = \int_0^2 (64x-x^7)\,dx = \left(32x^2-\frac{1}{8}x^8\right)\Big|_0^2 = 96$

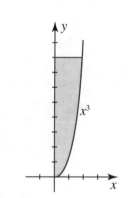

7. $\displaystyle\int\int_R \frac{y}{1+xy}\,dA = \int_0^1\int_0^1 \frac{y}{1+xy}\,dx\,dy = \int_0^1 \ln(1+xy)\Big|_0^1\,dy$

$\displaystyle\qquad = \int_0^1 1\ln(1+y)\,dy = [(1+y)\ln(1+y)-(1+y)]\big|_0^1$

$\displaystyle\qquad = (2\ln 2 - 2) - (-1) = 2\ln 2 - 1$

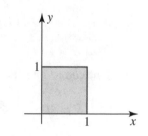

9. $\displaystyle\int\int_R \sqrt{x^2+1}\,dA = \int_0^{\sqrt{3}}\int_{-x}^x \sqrt{x^2+1}\,dy\,dx = \int_0^{\sqrt{3}} y\sqrt{x^2+1}\Big|_{-x}^x\,dx$

$\displaystyle\qquad = \int_0^{\sqrt{3}} (x\sqrt{x^2+1}+x\sqrt{x^2+1})\,dx$

$\displaystyle\qquad = \int_0^{\sqrt{3}} 2x\sqrt{x^2+1}\,dx = \frac{2}{3}(x^2+1)^{3/2}\Big|_0^{\sqrt{3}}$

$\displaystyle\qquad = \frac{2}{3}(4^{3/2}-1^{3/2}) = \frac{14}{3}$

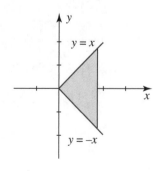

11. $\displaystyle\iint_R (x+y)\,dA = \int_0^4\int_0^2 (x+y)\,dx\,dy + \int_0^2\int_2^4 (x+y)\,dx\,dy$

$\displaystyle\qquad = \int_0^4\left(\tfrac{1}{2}x^2+xy\right)\Big|_0^2\,dy + \int_0^2\left(\tfrac{1}{2}x^2+xy\right)\Big|_2^4\,dy$

$\displaystyle\qquad = \int_0^4 (2+2y)\,dy + \int_0^2 [(8+4y)-(2+2y)]\,dy = (2y+y^2)\Big|_0^4 + (6y+y^2)\Big|_0^2$

$\displaystyle\qquad = 24 + 16 = 40$

13. $\displaystyle A = \int_0^3\int_{-x}^{2x-x^2} dy\,dx = \int_0^3 (2x-x^2+x)\,dx$

$\displaystyle\qquad\qquad = \left(\tfrac{3}{2}x^2-\tfrac{1}{3}x^3\right)\Big|_0^3 = \tfrac{9}{2}$

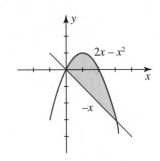

15. $\displaystyle A = \int_1^4\int_{\ln x}^{e^x} dy\,dx = \int_1^4 (e^x-\ln x)\,dx = (e^x-x\ln x+x)\Big|_1^4$

$\displaystyle\qquad = (e^4-4\ln 4+4)-(e+1) = e^4-e-4\ln 4+3$

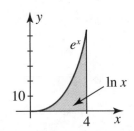

17. $\displaystyle A = \int_{-2}^1\int_{x^3}^{-2x+3} dy\,dx = \int_{-2}^1 (-2x+3-x^3)\,dx$

$\displaystyle\qquad\qquad = \left(-x^2+3x-\tfrac{1}{4}x^4\right)\Big|_{-2}^1 = \tfrac{7}{4}-(-14) = \tfrac{63}{4}$

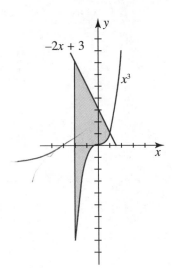

19. The correct integral is (**c**).

$\displaystyle V = 2\int_{-2}^2\int_0^{\sqrt{4-y^2}} (4-y)\,dx\,dy = 2\int_{-2}^2 (4-y)x\Big|_0^{\sqrt{4-y^2}}\,dy = 2\int_{-2}^2 (4-y)\sqrt{4-y^2}\,dy$

$\displaystyle\qquad = 2\left[2y\sqrt{4-y^2}+8\sin^{-1}\tfrac{y}{2}+\tfrac{1}{3}(4-y^2)^{3/2}\right]\Big|_{-2}^2 = 2(4\pi-(-4\pi)) = 16\pi$

21. Setting $z=0$ we have $y = 6-2x$.

$\displaystyle V = \int_0^3\int_0^{6-2x} (6-2x-y)\,dy\,dx = \int_0^3\left(6y-2xy-\tfrac{1}{2}y^2\right)\Big|_0^{6-2x}\,dx$

$\displaystyle\qquad = \int_0^3\left[6(6-2x)-2x(6-2x)-\tfrac{1}{2}(6-2x)^2\right]dx = \int_0^3 (18-12x+2x^2)\,dx$

$\displaystyle\qquad = \left(18x-6x^2+\tfrac{2}{3}x^3\right)\Big|_0^3 = 18$

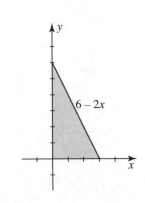

23. Solving for z, we have $x = 2 - \frac{1}{2}x + \frac{1}{2}y$. Setting $z = 0$, we see that this surface (plane) intersects the xy-plane in the line $y = x - 4$. Since $z(0,0) = 2 > 0$, the surface lies above the xy-plane over the quarter-circular region.

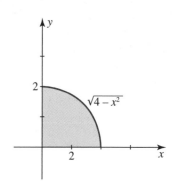

$$V = \int_0^2 \int_0^{\sqrt{4-x^2}} \left(2 - \frac{1}{2}x + \frac{1}{2}y\right) dy\,dx$$

$$= \int_0^2 \left(2y - \frac{1}{2}xy + \frac{1}{4}y^2\right)\Big|_0^{\sqrt{4-x^2}} dx$$

$$= \int_0^2 \left(2\sqrt{4-x^2} - \frac{1}{2}x\sqrt{4-x^2} + 1 - \frac{1}{4}x^2\right) dx$$

$$= \left[x\sqrt{4-x^2} + 4\sin^{-1}\frac{x}{2} + \frac{1}{6}(4-x^2)^{3/2} + x - \frac{1}{12}x^3\right]\Big|_0^2$$

$$= \left(2\pi + 2 - \frac{2}{3}\right) - \frac{4}{3} = 2\pi$$

25. Note that $z = 1 + x^2 + y^2$ is always positive. Then

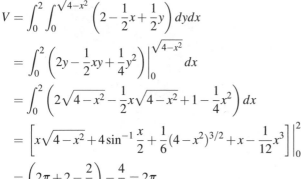

$$V = \int_0^1 \int_0^{3-3x} (1 + x^2 + y^2)\,dy\,dx = \int_0^1 \left(y + x^2 y + \frac{1}{3}y^3\right)\Big|_0^{3-3x} dx$$

$$= \int_0^1 [(3-3x) + x^2(3-3x) + 9(1-x)^3]\,dx$$

$$= \int_0^1 (12 - 30x + 30x^2 - 12x^3)\,dx$$

$$= (12x - 15x^2 + 10x^3 - 3x^4)\Big|_0^1 = 4.$$

27. In the first octant $z = 6/y$ is positive. Then

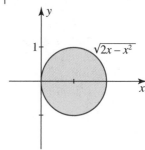

$$V = \int_1^6 \int_0^5 \frac{6}{y}\,dx\,dy = \int_1^6 \frac{6x}{y}\Big|_0^5 dy = 30 \int_1^6 \frac{dy}{y} = 30\ln y\Big|_1^6 = 30\ln 6.$$

29. Note that $z = 4 - y^2$ is positive for $|y| \leq 1$. Using symmetry,

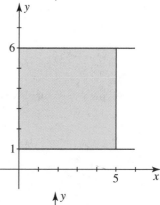

$$V = 2\int_0^2 \int_0^{\sqrt{2x-x^2}} (4 - y^2)\,dy\,dx = 2\int_0^2 \left(4y - \frac{1}{3}y^3\right)\Big|_0^{\sqrt{2x-x^2}} dx$$

$$= 2\int_0^2 \left[4\sqrt{2x-x^2} - \frac{1}{3}(2x-x^2)\sqrt{2x-x^2}\right] dx$$

$$= 2\int_0^2 \left(4\sqrt{1-(x-1)^2} - \frac{1}{3}[1-(x-1)^2]\sqrt{1-(x-1)^2}\right) dx$$

$$\boxed{u = x - 1, \ du = dx}$$

$$= 2\int_{-1}^1 \left[4\sqrt{1-u^2} - \frac{1}{3}(1-u^2)\sqrt{1-u^2}\right] du = 2\int_{-1}^1 \left(\frac{11}{3}\sqrt{1-u^2} + \frac{1}{3}u^2\sqrt{1-u^2}\right) du$$

$$\boxed{\text{Trig substitution}}$$

$$= 2\left[\frac{11}{6}u\sqrt{1-u^2} + \frac{11}{6}\sin u + \frac{1}{24}x(2x^2-1)\sqrt{1-u^2} + \frac{1}{24}\sin^{-1}u\right]\Big|_{-1}^1$$

$$= 2\left[\left(\frac{11}{6}\frac{\pi}{2} + \frac{1}{24}\frac{\pi}{2}\right) - \left(-\frac{11}{6}\frac{\pi}{2} - \frac{1}{24}\frac{\pi}{2}\right)\right] = \frac{15}{4}.$$

31. From $z = 4 - x - 2y$ and $z = x + y$, we have $4 - x - 2y = x + y$ or $x = 2 - \dfrac{3}{2}y$.

$$
\begin{aligned}
V &= \int_0^{4/3} \int_0^{2-3y/2} [(4 - x - 2y) - (x + y)]\,dx\,dy \\
&= \int_0^{4/3} \left(4x - x^2 - 3xy\right)\Big|_0^{2-3y/2}\,dy \\
&= \int_0^{4/3} \left[4\left(2 - \frac{3}{2}y\right) - \left(2 - \frac{3}{2}y\right)^2 - 3\left(2 - \frac{3}{2}y\right)y\right]dy \\
&= \int_0^{4/3} \left(4 - 6y + \frac{9}{4}y^2\right)dy \\
&= \left(4y - 3y^2 + \frac{3}{4}y^3\right)\Big|_0^{4/3} = \frac{16}{9}
\end{aligned}
$$

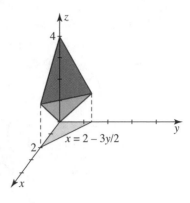

33. From $z = x^2$ and $z = -x + 2$ we have $x^2 = -x + 2$ or $x = 1$ (in the first octant). Then

$$
\begin{aligned}
V &= \int_0^5 \int_0^1 (-x + 2 - x^2)\,dx\,dy = \int_0^5 \left(-\frac{1}{2}x^2 + 2x - \frac{1}{3}x^3\right)\Big|_0^1\,dy \\
&= \int_0^5 \frac{7}{6}\,dy = \frac{35}{6}.
\end{aligned}
$$

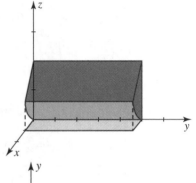

35. Solving $x = y^2$ for y, we obtain $y = \sqrt{x}$. Thus,

$$
\int_0^2 \int_0^{y^2} f(x, y)\,dx\,dy = \int_0^4 \int_{\sqrt{x}}^2 f(x, y)\,dy\,dx.
$$

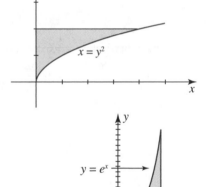

37. Solving $y = e^x$ for x, we obtain $x = \ln y$. Thus,

$$
\int_0^3 \int_1^{e^x} f(x, y)\,dy\,dx = \int_1^{e^3} \int_{\ln y}^3 f(x, y)\,dx\,dy.
$$

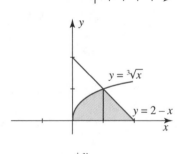

39. Solving $y = \sqrt[3]{x}$ and $y = 2 - x$ for x, we obtain $x = y^3$ and $x = 2 - y$. Thus,

$$
\int_0^1 \int_0^{\sqrt[3]{x}} f(x, y)\,dy\,dx + \int_1^2 \int_0^{2-x} f(x, y)\,dy\,dx = \int_0^1 \int_{y^3}^{2-y} f(x, y)\,dx\,dy.
$$

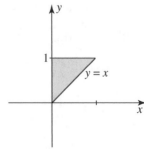

41. $\displaystyle\int_0^1 \int_x^1 x^2\sqrt{1+y^4}\,dy\,dx = \int_0^1 \int_0^y x^2\sqrt{1+y^4}\,dx\,dy = \int_0^1 \frac{1}{3}x^3\sqrt{1+y^4}\Big|_0^y\,dy$

$$
\begin{aligned}
&= \frac{1}{3}\int_0^1 y^3\sqrt{1+y^4}\,dy = \frac{1}{3}\left[\frac{1}{6}(1+y^4)^{3/2}\right]\Big|_0^1 \\
&= \frac{1}{18}(2\sqrt{2} - 1)
\end{aligned}
$$

43. $\displaystyle\int_0^2\int_{y^2}^4\cos x^{3/2}dxdy = \int_0^4\int_0^{\sqrt{x}}\cos x^{3/2}dydx = \int_0^4 y\cos x^{3/2}\Big|_0^{\sqrt{x}}dx$

$\displaystyle = \int_0^4\sqrt{x}\cos x^{3/2}dx = \frac{2}{3}\sin x^{3/2}\Big|_0^4 = \frac{2}{3}\sin 8$

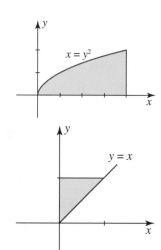

45. $\displaystyle\int_0^1\int_x^1\frac{1}{1+y^4}dydx = \int_0^1\int_0^y\frac{1}{1+y^4}dxdy = \int_0^1\frac{x}{1+y^4}\Big|_0^y dy$

$\displaystyle = \int_0^1\frac{y}{1+y^4}dy = \frac{1}{2}\tan^{-1}y^2\Big|_0^1 = \frac{\pi}{8}$

47. $\displaystyle f_{ave} = \frac{1}{A}\int_c^d\int_a^b xydxdy = \frac{1}{A}\int_c^d\frac{x^2y}{2}\Big|_a^b dy$

$\displaystyle = \frac{1}{A}\int_c^d\frac{(b^2-a^2)y}{2}dy = \frac{1}{A}\left[\frac{(b^2-a^2)y^2}{4}\right]\Big|_c^d$

$\displaystyle = \frac{1}{A}\frac{(b^2-a^2)(d^2-c^2)}{4}$

But $A = (b-a)(d-c)$, so

$$f_{ave} = \frac{(b^2-a^2)(d^2-c^2)}{4(b-a)(d-c)} = \frac{(b+a)(d+c)}{4}$$

49. Let S be the solid with base R and height described by the function $f(x,y)$. The volume of S is equal to the volume of the solid with base R and constant height f_{ave}.

51. By Problem 50(a) we have $\iint_{R_k}\cos 2\pi(x+y)dA = \iint_{R_k}\sin 2\pi(x+y)dA = 0$ for $k = 1, 2, \cdots, n$. Then

$$\iint_R\cos 2\pi(x+y)dA = \iint_{R_1}\cos 2\pi(x+y)dA + \cdots + \iint_{R_n}\cos 2\pi(x+y)dA = 0+\cdots+0 = 0$$

and

$$\iint_R\sin 2\pi(x+y)dA = \iint_{R_1}\sin 2\pi(x+y)dA + \cdots + \iint_{R_n}\sin 2\pi(x+y)dA = 0+\cdots+0 = 0.$$

Therefore by Problem 45(c), at least one of the two sides of R must have integer length.

14.4 Center of Mass and Moments

1. $\displaystyle m = \int_0^3\int_0^4 xydxdy = \int_0^3\frac{1}{2}x^2y\Big|_0^4 dy = \int_0^3 8ydy$

$\displaystyle = 4y^2\Big|_0^3 = 36$

$\displaystyle M_y = \int_0^3\int_0^4 x^2ydxdy = \int_0^3\frac{1}{3}x^3y\Big|_0^4 dy$

$\displaystyle = \int_0^3\frac{64}{3}ydy = \frac{32}{3}y^2\Big|_0^3 = 96$

$\displaystyle M_x = \int_0^3\int_0^4 xy^2dxdy = \int_0^3\frac{1}{2}x^2y^2\Big|_0^4$

$\displaystyle = \int_0^3 8y^2dy = \frac{8}{3}y^3\Big|_0^3 = 72$

$\bar{x} = M_y/m = 96/36 = 8/3$; $\bar{y} = M_x/m = 72/36 = 2$. The center of mass is $(8/3, 2)$.

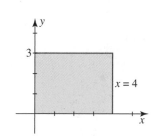

3. Since both the region and ρ are symmetric with respect to the line $x = 3$, $\bar{x} = 3$.

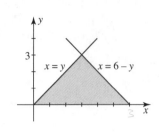

$$m = \int_0^3 \int_y^{6-y} 2y\,dx\,dy = \int_0^3 2xy \Big|_y^{6-y}$$

$$= \int_0^3 2y(6 - y - y)\,dy = \int_0^3 (12y - 4y^2)\,dy$$

$$= \left(6y^2 - \frac{4}{3}y^3\right)\Big|_0^3 = 18$$

$$M_x = \int_0^3 \int_y^{6-y} 2y^2\,dx\,dy = \int_0^3 2xy^2 \Big|_y^{6-y} dx\,dy = \int_0^3 2y^2(6 - y - y)\,dy = \int_0^3 (12y^2 - 4y^3)\,dy$$

$$= (4y^3 - y^4)\Big|_0^3 = 27$$

$\bar{y} = M_x/m = 27/18 = 3/2$. The center of mass is $(3, 3/2)$.

5. $m = \int_0^1 \int_0^{x^2} (x + y)\,dy\,dx = \int_0^1 \left(xy + \frac{1}{2}y^2\right)\Big|_0^{x^2} dx$

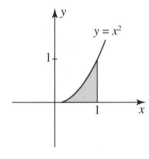

$$= \int_0^1 \left(x^3 + \frac{1}{2}x^4\right)dx = \left(\frac{1}{4}x^4 + \frac{1}{10}x^5\right)\Big|_0^1 = \frac{7}{20}$$

$$M_y = \int_0^1 \int_0^{x^2} (x^2 + xy)\,dy\,dx = \int_0^1 \left(x^2 y + \frac{1}{2}xy^2\right)\Big|_0^{x^2} dx$$

$$= \int_0^1 \left(x^4 + \frac{1}{2}x^5\right)dx = \left(\frac{1}{5}x^5 + \frac{1}{12}x^6\right)\Big|_0^1 = \frac{17}{60}$$

$$M_x = \int_0^1 \int_0^{x^2} (xy + y^2)\,dy\,dx = \int_0^1 \left(\frac{1}{2}xy^2 + \frac{1}{3}y^3\right)\Big|_0^{x^2} = \int_0^1 \left(\frac{1}{2}x^5 + \frac{1}{3}x^6\right)dx$$

$$= \left(\frac{1}{12}x^6 + \frac{1}{21}x^7\right)\Big|_0^1 = \frac{11}{84}$$

$\bar{x} = M_y = m = \dfrac{17/60}{7/20} = 17/21$; $\bar{y} + M_x/m = \dfrac{11/84}{7/20} = 55/147$.

The center of mass is $(17/21, 55/147)$.

7. The density is $\rho = ky$. Since both the region and ρ are symmetric with respect to the y-axis, $\bar{x} = 0$.

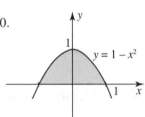

Using symmetry, $m = 2\int_0^1 \int_0^{1-x^2} ky\,dy\,dx = 2k\int_0^1 \frac{1}{2}y^2 \Big|_0^{1-x^2} dx = k\int 0^1 (1 - x^2)^2 dx$

$$= k\int_0^1 (1 - 2x^2 + x^4)\,dx = k\left(x - \frac{2}{3}x^3 + \frac{1}{5}x^5\right)\Big|_0^1$$

$$= k\left(1 - \frac{2}{3} + \frac{1}{5}\right) = \frac{8}{15}k$$

$$M_x = 2\int_0^1 \int_0^{1-x^2} ky^2\,dy\,dx = 2k\int_0^1 \frac{1}{3}y^3 \Big|_0^{1-x^2} dx = \frac{2}{3}k\int_0^1 (1 - x^2)^3 dx$$

$$= \frac{2}{3}k\int_0^1 (1 - 3x^2 + 3x^4 - x^6)\,dx = \frac{2}{3}k\left(x - x^3 + \frac{3}{5}x^5 - \frac{1}{7}x^7\right)\Big|_0^1$$

$$= \frac{2}{3}k\left(1 - 1 + \frac{3}{5} - \frac{1}{7}\right) = \frac{32}{105}k$$

$\bar{y} = M_x/m = \dfrac{32k/105}{8k/15} = 4/7$. The center of mass is $(0, 4/7)$.

9. $m = \int_0^1 \int_0^{e^x} y^3 \, dy \, dx = \int_0^1 \frac{1}{4} y^4 \Big|_0^{e^x} dx = \int_0^1 \frac{1}{4} e^{4x} dx$

$= \frac{1}{16} e^{4x} \Big|_0^1 = \frac{1}{16}(e^4 - 1)$

$M_y = \int_0^1 \int_0^{e^x} xy^3 \, dy \, dx = \int_0^1 \frac{1}{4} xy^4 \Big|_0^{e^x} dx$

$= \int_0^1 \frac{1}{4} xe^{4x} dx$ Integration by parts

$= \frac{1}{4}\left(\frac{1}{4} xe^{4x} - \frac{1}{16} e^{4x}\right)\Big|_0^1 = \frac{1}{4}\left(\frac{3}{16}e^4 + \frac{1}{16}\right) = \frac{1}{64}(3e^4 + 1)$

$M_x = \int_0^1 \int_0^{e^x} y^4 \, dy \, dx = \int_0^1 \frac{1}{5} y^5 \Big|_0^{e^x} dx = \int_0^1 \frac{1}{5} e^{5x} dx = \frac{1}{25} e^{5x}\Big|_0^1 = \frac{1}{25}(e^5 - 1)$

$\bar{x} = M_y/m = \frac{(3e^4+1)/64}{(e^4-1)/16} = \frac{3e^4+1}{4(e^4-1)}; \quad \bar{y} = M_x/m = \frac{(e^5-1)/25}{(e^4-1)/16} = \frac{16(e^5-1)}{25(e^4-1)}$

The center of mass is $\left(\dfrac{3e^4+1}{4(e^4-1)}, \dfrac{16(e^5-1)}{25(e^4-1)}\right) \approx (0.77, 1.76)$.

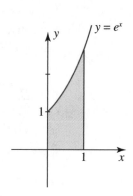

11. $I_x = \int_0^1 \int_0^{y-y^2} 2xy^2 \, dx \, dy = \int_0^1 x^2 y^2 \Big|_0^{y-y^2} dy = \int_0^1 (y-y^2)^2 y^2 \, dy$

$= \int_0^1 (y^4 - 2y^5 + y^6) \, dy = \left(\frac{1}{5} y^5 - \frac{1}{3} y^6 + \frac{1}{7} y^7\right)\Big|_0^1 = \frac{1}{105}$

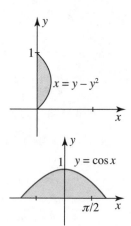

13. Using symmetry,

$I_x = 2\int_0^{\pi/2}\int_0^{\cos x} ky^2 \, dy \, dx = 2k\int_0^{\pi/2} \frac{1}{3} y^3 \Big|_0^{\cos x} dx = \frac{2}{3}k\int_0^{\pi/2}\cos^3 x \, dx$

$= \frac{2}{3}k\int_0^{\pi/2}\cos x(1-\sin^2 x)\, dx = \frac{2}{3}k\left(\sin x - \frac{1}{3}\sin^3 x\right)\Big|_0^{\pi/2} = \frac{4}{9}k.$

15. $I_y = \int_0^4 \int_0^{\sqrt{y}} x^2 y \, dx \, dy = \int_0^4 \frac{1}{3} x^3 y \Big|_0^{\sqrt{y}} dy = \frac{1}{3}\int_0^4 y^{3/2} y \, dy = \frac{1}{3}\int_0^4 y^{5/2} dy$

$= \frac{1}{3}\left(\frac{2}{7} y^{7/2}\right)\Big|_0^4 = \frac{2}{21}(4^{7/2}) = \frac{256}{21}$

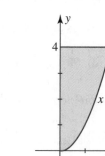

17. $I_y = \int_0^1 \int_y^3 (4x^3 + 3x^2 y) \, dx \, dy = \int_0^1 (x^4 + x^3 y)\Big|_y^3 dy$

$= \int_0^1 (81 + 27y - 2y^4) \, dy$

$= \left(81y + \frac{27}{2} y^2 - \frac{2}{5} y^5\right)\Big|_0^1 = \frac{941}{10}$

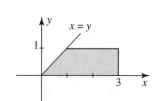

19. Using symmetry,

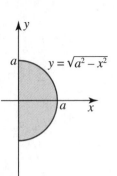

$$m = 2\int_0^a \int_0^{\sqrt{a^2-y^2}} x\,dx\,dy = 2\int_0^a \frac{1}{2}x^2 \Big|_0^{\sqrt{a^2-y^2}} dy = \int_0^a (a^2-y^2)\,dy$$

$$= \left(a^2 y - \frac{1}{3}y^3\right)\Big|_0^a = \frac{2}{3}a^3.$$

$$I_y = 2\int_0^a \int_0^{\sqrt{a^2 y^2}} x^3\,dx\,dy = 2\int_0^a \frac{1}{4}x^4 \Big|_0^{\sqrt{a^2-y^2}} dy = \frac{1}{2}\int_0^a (a^2-y^2)^2\,dy$$

$$= \frac{1}{2}\int_0^a (a^4 - 2a^2 y^2 + y^4)\,dy = \frac{1}{2}\left(a^4 y - \frac{2}{3}a^2 y^3 + \frac{1}{5}y^5\right)\Big|_0^a = \frac{4}{15}a^5$$

$$R_g = \sqrt{\frac{I_y}{m}} = \sqrt{\frac{4a^5/15}{2a^3/3}} = \sqrt{\frac{2}{5}}a$$

21. (a) Using symmetry,

$$I_x = 4\int_0^a \int_0^{b\sqrt{a^2-x^2}/a} y^2\,dy\,dx = \frac{4b^3}{3a^3}\int_0^a (a^2-x^2)^{3/2}\,dx \quad \boxed{x = a\sin\theta,\ dx = a\cos\theta\,d\theta}$$

$$= \frac{4}{3}ab^3 \int_0^{\pi/2} \cos^4\theta\,d\theta = \frac{4}{3}ab^3 \int_0^{\pi/2} \frac{1}{4}(1+\cos 2\theta)^2\,d\theta$$

$$= \frac{1}{3}ab^3 \int_0^{\pi/2} (1+\cos 2\theta + \frac{1}{2} + \frac{1}{2}\cos 4\theta)\,d\theta = \frac{1}{3}ab^3 \left(\frac{3}{2}\theta + \frac{1}{2}\sin 2\theta + \frac{1}{8}\sin 4\theta\right)\Big|_0^{\pi/2}$$

$$= \frac{ab^3 \pi}{4}.$$

(b) Using symmetry,

$$I_y = 4\int_0^a \int_0^{b\sqrt{a^2-x^2}/a} x^2\,dy\,dx = \frac{4b}{a}\int_0^a x^2 \sqrt{a^2-x^2}\,dx \quad \boxed{x = a\sin\theta,\ dx = a\cos\theta\,d\theta}$$

$$= 4a^3 b \int_0^{\pi/2} \sin^2\theta\cos^2\theta\,d\theta = 4a^3 b \int_0^{\pi/2} \frac{1}{4}(1-\cos^2 2\theta)\,d\theta$$

$$= a^3 b \int_0^{\pi/2} (1 - \frac{1}{2} - \frac{1}{2}\cos 4\theta)\,d\theta = a^3 b \left(\frac{1}{2}\theta - \frac{1}{8}\sin 4\theta\right)\Big|_0^{\pi/2} = \frac{a^3 b\pi}{4}.$$

(c) Using $m = \pi ab$, $R_g = \sqrt{I_x/m} = \frac{1}{2}\sqrt{ab^3\pi/\pi ab} = \frac{1}{2}b$.

(d) $R_g = \sqrt{I_y/m} = \frac{1}{2}\sqrt{a^3 b\pi/\pi ab} = \frac{1}{2}a$

23. From Problem 20, $m = \frac{1}{2}ka^2$ and $I_x = \frac{1}{12}ka^4$.

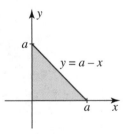

$$I_y = \int_0^a \int_0^{a-x} kx^2\,dy\,dx = \int_0^a kx^2 y\Big|_0^{a-x} dx = k\int_0^a x^2(a-x)\,dx$$

$$= k\left(\frac{1}{3}ax^3 - \frac{1}{4}x^4\right)\Big|_0^a = \frac{1}{12}ka^4$$

$$I_0 = I_x + I_y = \frac{1}{12}ka^4 + \frac{1}{12}ka^4 = \frac{1}{6}ka^4$$

25. The density is $\rho = k/(x^2+y^2)$. Using symmetry,

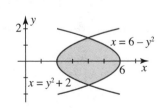

$$I_0 = 2\int_0^{\sqrt{2}} \int_{y^2+2}^{6-y^2} (x^2+y^2)\frac{k}{x^2+y^2}\,dx\,dy = 2\int_0^{\sqrt{2}} kx\Big|_{y^2+2}^{6-y^2} dy$$

$$= 2k\int_0^{\sqrt{2}} (6-y^2-y^2-2)\,dy = 2k\left(4y - \frac{2}{3}y^3\right)\Big|_0^{\sqrt{2}}$$

$$= 2k\left(\frac{8}{3}\sqrt{2}\right) = \frac{16\sqrt{2}}{3}k.$$

27. From Problem 20, $m = \dfrac{1}{2}ka^2$, and from Problem 21, $I_0 = \dfrac{1}{6}ka^4$.

Then $R_g = \sqrt{I_0/m} = \sqrt{\dfrac{ka^4/6}{ka^2/2}} = \sqrt{\dfrac{1}{3}}a$.

14.5 Double Integrals in Polar Coordinates

1. Using symmetry,
$$A = 2\int_{-\pi/2}^{\pi/2}\int_0^{3+3\sin\theta} r\,dr\,d\theta = 2\int_{-\pi/2}^{\pi/2}\frac{1}{2}r^2\Big|_0^{3+3\sin\theta}\,d\theta$$
$$= \int_{-\pi/2}^{\pi/2} 9(1+\sin\theta)^2\,d\theta = 9\int_{-\pi/x}^{\pi/x}(1+2\sin\theta+\sin^2\theta)\,d\theta$$
$$= 9\left(\theta - 2\cos\theta + \frac{1}{2}\theta - \frac{1}{4}\sin 2\theta\right)\Big|_{-\pi/2}^{\pi/2}$$
$$= 9\left[\frac{3}{2}\frac{\pi}{2} - \frac{3}{2}\left(-\frac{\pi}{2}\right)\right] = \frac{27\pi}{2}$$

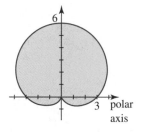

3. Solving $r = 2\sin\theta$ and $r = 1$, we obtain $\sin\theta = 1/2$ or $\theta = \pi/6$. Using symmetry,
$$A = 2\int_0^{\pi/6}\int_0^{2\sin\theta} r\,dr\,d\theta + 2\int_{\pi/6}^{\pi/2}\int_0^1 r\,dr\,d\theta$$
$$= 2\int_0^{\pi/6}\frac{1}{2}r^2\Big|_0^{2\sin\theta}\,d\theta + 2\int_{\pi/6}^{\pi/2}\frac{1}{2}r^2\Big|_0^1\,d\theta = \int_0^{\pi/6} 4\sin^2\theta\,d\theta + \int_{\pi/6}^{\pi/2}\,d\theta$$
$$= (2\theta - \sin 2\theta)\Big|_0^{\pi/6} + \left(\frac{\pi}{2} - \frac{\pi}{6}\right) = \frac{\pi}{3} - \frac{\sqrt{3}}{2} + \frac{\pi}{3} = \frac{4\pi - 3\sqrt{3}}{6}$$

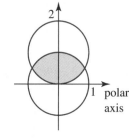

5. Using symmetry,
$$V = 2\int_0^{\pi/6}\int_0^{5\cos 3\theta} 4r\,dr\,d\theta = 4\int_0^{\pi/6} r^2\Big|_0^{5\cos 3\theta}\,d\theta = 4\int_0^{\pi/6} 25\cos^2 3\theta\,d\theta$$
$$= 100\left(\frac{1}{2}\theta + \frac{1}{12}\sin 6\theta\right)\Big|_0^{\pi/6} = \frac{25\pi}{3}$$

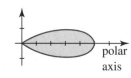

7. $V = \displaystyle\int_0^{2\pi}\int_1^3 \sqrt{16-r^2}\,r\,dr\,d\theta = \int_0^{2\pi} -\frac{1}{3}(16-r^2)^{3/2}\Big|_1^3\,d\theta$

$= -\dfrac{1}{3}\displaystyle\int_0^{2\pi}(7^{3/2} - 15^{3/2})\,d\theta = \dfrac{1}{3}(15^{3/2} - 7^{3/2})2\pi = \dfrac{2\pi(15\sqrt{15} - 7\sqrt{7})}{3}$

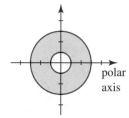

9. $V = \displaystyle\int_0^{\pi/2}\int_0^{1+\cos\theta}(r\sin\theta)r\,dr\,d\theta = \int_0^{\pi/2}\frac{1}{3}r^3\sin\theta\Big|_0^{1+\cos\theta}\,d\theta$

$= \dfrac{1}{3}\displaystyle\int_0^{\pi/2}(1+\cos\theta)^3\sin\theta\,d\theta = \frac{1}{3}\left[-\frac{1}{4}(1+\cos\theta)^4\right]\Big|_0^{\pi/2}$

$= -\dfrac{1}{12}(1 - 2^4) = \dfrac{5}{4}$

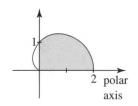

11. $m = \int_0^{\pi/2} \int_1^3 kr \, dr \, d\theta = k \int_0^{\pi/2} \frac{1}{2} r^2 \Big|_1^3 d\theta = \frac{1}{2} k \int_0^{\pi/2} 8 \, d\theta = 2k\pi$

$M_y = \int_0^{\pi/2} \int_1^3 kxr \, dr \, d\theta = k \int_0^{\pi/2} \int_1^3 r^2 \cos\theta \, dr \, d\theta = k \int_0^{\pi/2} \frac{1}{3} r^3 \cos\theta \Big|_1^3 d\theta$

$= \frac{1}{3} k \int_0^{\pi/2} 26 \cos\theta \, d\theta = \frac{26}{3} k \sin\theta \Big|_0^{\pi/2} = \frac{26}{3} k$

$\bar{x} = M_y/m = \frac{26k/3}{2k\pi} = \frac{13}{3\pi}$.

Since the region and density function are symmetric about the ray $\theta = \pi/4$, $\bar{y} = \bar{x} = 13/3\pi$ and the center of mass $(13/3\pi, \; 13/3\pi)$.

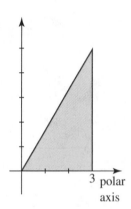

13. In polar coordinates the line $x = 3$ becomes $r \cos\theta = 3$ or $r = 3 \sec\theta$. The angle of inclination of the line $y = \sqrt{3}x$ is $\pi/3$.

$m = \int_0^{\pi/3} \int_0^{3\sec\theta} r^2 \, dr \, d\theta = \int_0^{\pi/3} \frac{1}{4} r^4 \Big|_0^{3\sec\theta} d\theta$

$= \frac{81}{4} \int_0^{\pi/3} \sec^4\theta \, d\theta = \frac{81}{4} \int_0^{\pi/3} (1 + \tan^2\theta) \sec^2\theta \, d\theta$

$= \frac{81}{4} \left(\tan\theta + \frac{1}{3}\tan^3\theta\right) \Big|_0^{\pi/3} = \frac{81}{4}(\sqrt{3} + \sqrt{3}) = \frac{81}{2}\sqrt{3}$

$M_y = \int_0^{\pi/3} \int_0^{3\sec\theta} xr^2 \, dr \, d\theta = \int_0^{\pi/3} \int_0^{3\sec\theta} r^4 \cos\theta \, dr \, d\theta$

$= \int_0^{\pi/3} \frac{1}{5} r^5 \cos\theta \Big|_0^{3\sec\theta} d\theta = \frac{243}{5} \int_0^{\pi/3} \sec^5\theta \cos\theta \, d\theta$

$= \frac{243}{5} \int_0^{\pi/3} \sec^4\theta \, d\theta = \frac{243}{5}(2\sqrt{3}) = \frac{486}{5}\sqrt{3}$

$M_x = \int_0^{\pi/3} \int_0^{3\sec\theta} yr^2 \, dr \, d\theta = \int_0^{\pi/3} \int_0^{3\sec\theta} r^4 \sin\theta \, d\theta = \int_0^{\pi/3} \frac{1}{5} r^5 \sin\theta \Big|_0^{3\sec\theta}$

$= \frac{243}{5} \int_0^{\pi/3} \sec^5\theta \sin\theta \, d\theta = \frac{243}{5} \int_0^{\pi/3} \tan\theta \sec^4\theta \, d\theta = \frac{243}{5} \int_0^{\pi/3} \tan\theta(1 + \tan^2\theta) \sec^2\theta \, d\theta$

$= \frac{243}{5} \int_0^{\pi/3} (\tan\theta + \tan^3\theta) \sec^2\theta \, d\theta = \frac{243}{5} \left(\frac{1}{2}\tan^2\theta + \frac{1}{4}\tan^4\theta\right) \Big|_0^{\pi/3} = \frac{243}{5}\left(\frac{3}{2} + \frac{9}{4}\right) = \frac{729}{4}$

$\bar{x} = M_y/m = \frac{486\sqrt{3}/5}{81\sqrt{3}/2} = 12/5$; $\bar{y} = M_x/m = \frac{729/4}{81\sqrt{3}/2} = 3\sqrt{3}/2$. The center of mass is $(12/5, 3\sqrt{3}/2)$.

15. The density is $\rho = k/r$.

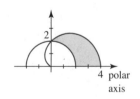

$m = \int_0^{\pi/2} \int_2^{2+2\cos\theta} \frac{k}{r} r \, dr \, d\theta = k \int_0^{\pi/2} \int_2^{2+2\cos\theta} dr \, d\theta$

$= k \int_0^{\pi/2} 2\cos\theta \, d\theta = 2k(\sin\theta) \Big|_0^{\pi/2} = 2k$

$M_y = \int_0^{\pi/2} \int_2^{2+2\cos\theta} x \frac{k}{r} r \, dr \, d\theta = k \int_0^{\pi/2} \int_2^{2+2\cos\theta} r \cos\theta \, dr \, d\theta = k \int_0^{\pi/2} \frac{1}{2} r^2 \Big|_2^{2+2\cos\theta} \cos\theta \, d\theta$

$= \frac{1}{2} k \int_0^{\pi/2} (8\cos\theta + 4\cos^2\theta) \cos\theta \, d\theta = 2k \int_0^{\pi/2} (2\cos^2\theta + \cos\theta - \sin^2\theta \cos\theta) \, d\theta$

$= 2k \left(\theta + \frac{1}{2}\sin 2\theta + \sin\theta - \frac{1}{3}\sin^3\theta\right) \Big|_0^{\pi/2} = 2k\left(\frac{\pi}{2} + \frac{2}{3}\right) = \frac{3\pi + 4}{3} k$

$$M_x = \int_0^{\pi/2} \int_2^{2+2\cos\theta} y\frac{k}{r}r\,dr\,d\theta = k\int_{\pi/2}\int_2^{2+2\cos\theta} r\sin\theta\,dr\,d\theta = k\int_0^{\pi/2} \frac{1}{2}r^2\Big|_2^{2+2\cos\theta}\sin\theta\,d\theta$$

$$= \frac{1}{2}k\int_0^{\pi/2}(8\cos\theta + 4\cos^2\theta)\sin\theta\,d\theta = \frac{1}{2}k\left(-4\cos^2\theta - \frac{4}{3}\cos^3\theta\right)\Big|_0^{\pi/2}$$

$$= \frac{1}{2}k\left[-\left(-4-\frac{4}{3}\right)\right] = \frac{8}{3}k$$

$$\bar{x} = M_y/m = \frac{(3\pi+4)k/3}{2k} = \frac{3\pi+4}{6}; \quad \bar{y} = M_x/m = \frac{8k/3}{2k} = \frac{4}{3}$$

The center of mass is $((3\pi+4)/6, 4/3)$.

17. $I_x = \int_0^{2\pi}\int_0^a y^2 kr\,dr\,d\theta = k\int_0^{2\pi}\int_0^a r^3\sin^2\theta\,dr\,d\theta = k\int_0^{2\pi}\frac{1}{4}r^4\sin^2\theta\Big|_0^a d\theta$

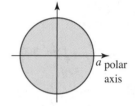

$$= \frac{ka^4}{4}\int_0^{2\pi}\sin^2\theta\,d\theta = \frac{ka^4}{4}\left(\frac{1}{2}\theta - \frac{1}{4}\sin 2\theta\right)\Big|_0^{2\pi} = \frac{k\pi a^4}{4}$$

19. Solving $a = 2a\cos\theta$, $\cos\theta = 1/2$ or $\theta = \pi/3$. The density is k/r^3. Using symmetry,

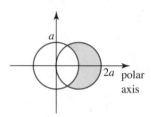

$$I_y = 2\int_0^{\pi/3}\int_a^{2a\cos\theta} x^2\frac{k}{r^3}r\,dr\,d\theta = 2k\int_0^{\pi/3}\int_a^{2a\cos\theta}\cos^2\theta\,dr\,d\theta$$

$$= 2k\int_0^{\pi/3}(2a\cos^3\theta - a\cos^2\theta)d\theta$$

$$= 2ak\left(2\sin\theta - \frac{2}{3}\sin^3\theta - \frac{1}{2}\theta - \frac{1}{4}\sin 2\theta\right)\Big|_0^{\pi/3}$$

$$= 2ak\left(\sqrt{3} - \frac{\sqrt{3}}{4} - \frac{\pi}{6} - \frac{\sqrt{3}}{8}\right) = \frac{5ak\sqrt{3}}{4} - \frac{ak\pi}{3}$$

21. From Problem 17, $I_x = k\pi a^4/4$. By symmetry, $I_y = I_x$. Thus $I_0 = k\pi a^4/2$.

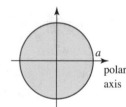

23. The density is $\rho = k/r$. $I_0 = \int_1^3\int_0^{1/r} r^2\frac{k}{r}r\,d\theta\,dr = k\int_1^3\int_0^{1/r} r^2\,d\theta\,dr$

$$= k\int_1^3 r^2\left(\frac{1}{r}\right)dr = k\left(\frac{1}{2}r^2\right)\Big|_1^3 = 4k$$

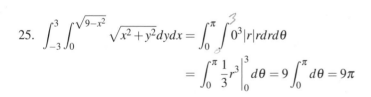

25. $\int_{-3}^3\int_0^{\sqrt{9-x^2}}\sqrt{x^2+y^2}\,dy\,dx = \int_0^\pi\int_0^3 r|r|\,dr\,d\theta$

$$= \int_0^\pi \frac{1}{3}r^3\Big|_0^3 d\theta = 9\int_0^\pi d\theta = 9\pi$$

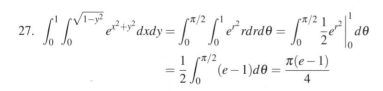

27. $\int_0^1\int_0^{\sqrt{1-y^2}} e^{x^2+y^2}\,dx\,dy = \int_0^{\pi/2}\int_0^1 e^{r^2}r\,dr\,d\theta = \int_0^{\pi/2}\frac{1}{2}e^{r^2}\Big|_0^1 d\theta$

$$= \frac{1}{2}\int_0^{\pi/2}(e-1)d\theta = \frac{\pi(e-1)}{4}$$

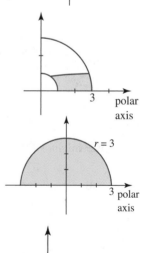

29. $$\int_0^1 \int_{\sqrt{1-x^2}}^{\sqrt{4-x^2}} \frac{x^2}{x^2+y^2}dydx = \int_1^2 \int_0^{\sqrt{4-x^2}} \frac{x^2}{x^2+y^2}dydx$$

$$= \int_0^{\pi/2} \int_1^2 \frac{r^2\cos^2\theta}{r^2}rdrd\theta$$

$$= \int_0^{\pi/2} \int_1^2 r\cos^2\theta drd\theta$$

$$= \int_0^{\pi/2} \frac{1}{2}r^2\Big|_1^2 \cos^2\theta d\theta = \frac{3}{2}\int_0^{\pi/2}\cos^2\theta d\theta$$

$$= \frac{3}{2}\left(\frac{1}{2}\theta + \frac{1}{4}\sin 2\theta\right)\Big|_0^{\pi/2} = \frac{3\pi}{8}$$

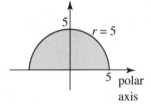

31. $$\int_{-5}^5 \int_0^{\sqrt{25-x^2}} (4x+3y)dydx = \int_0^\pi \int_0^5 (4r\cos\theta + 3r\sin\theta)rdrd\theta$$

$$= \int_0^\pi \int_0^5 (4r^2\cos\theta + 3r^2\sin\theta)drd\theta$$

$$= \int_0^\pi \left(\frac{4}{3}r^3\cos\theta + r^3\sin\theta\right)\Big|_0^5 d\theta$$

$$= \int_0^\pi \left(\frac{500}{3}\cos\theta + 125\sin\theta\right)d\theta = \left(\frac{500}{3}\sin\theta - 125\cos\theta\right)\Big|_0^\pi = 250$$

33. $$I^2 = \int_0^\infty \int_0^\infty e^{-(x^2+y^2)}dxdy = \int_0^{\pi/2} \int_0^\infty e^{-r^2}rdrd\theta = \int_0^{\pi/2} \lim_{t\to\infty} -\frac{1}{2}e^{-r^2}\Big|_0^t d\theta$$

$$= \int_0^{\pi/2} \lim_{t\to\infty}\left(-\frac{1}{2}e^{-t^2} + \frac{1}{2}\right)d\theta = \int_0^{\pi/2}\frac{1}{2}d\theta = \frac{\pi}{4}; \quad I = \frac{\sqrt{\pi}}{2}$$

35. The volume of the cylindrical portion of the tank is $V_c = \pi(4.2)^2 19.3 \approx 1069.56m^3$. We take the equation of the ellipsoid to be

$$\frac{x^2}{(4.2)^2} + \frac{x^2}{(5.15)^2} = 1 \text{ or } z = \pm\frac{5.15}{4.2}\sqrt{(4.2)^2 - x^2 - y^2}.$$

The volume of the ellipsoid is

$$V_e = 2\left(\frac{5.15}{4.2}\right)\int\int_R \sqrt{(4.2)^2 - x^2 - y^2}dxdy = \frac{10.3}{4.2}\int_0^{2\pi}\int_0^{4.2}[(4.2)^2 - r^2]^{1/2}rdrd\theta$$

$$= \frac{10.3}{4.2}\int_0^{2\pi}\left[\left(-\frac{1}{2}\right)\frac{2}{3}[(4.2)^2 - r^2]^{3/2}\Big|_0^{4.2}\right]d\theta = \frac{10.3}{4.2}\frac{1}{3}\int_0^{2\pi}(4.2)^3d\theta$$

$$= \frac{2\pi}{3}\frac{10.3}{4.2}(4.2)^3 \approx 380.53.$$

The volume of the tank is approximately $1069.56 + 380.53 = 1450.09m^3$.

37. (a) $$P = \int\int_C D(r)dA = \int_0^{2\pi}\int_0^R D_0 e^{-r/d}rdrd\theta = 2\pi D_0\int_0^R re^{-r/d}dr$$

$$= 2\pi D_0(-dre^{-r/d} - d^2 e^{r/d})\Big|_0^R = 2\pi dD_0[d - (R+d)e^{-R/d}]$$

(b) Using

$$\int\int_C rD(r)dA = \int_0^{2\pi}\int_0^R rD_0 e^{-r/d}rdrd\theta = 2\pi D_0\int_0^R r^2 e^{-r/d}dr$$

$$= 2\pi D_0(-2d^3 e^{-r/d} - 2d^2 re^{-r/d} - dr^2 e^{-r/d})\Big|_0^R$$

$$= 2\pi dD_0\left[2d^2 - (R^2 + 2dR + 2d^2)e^{-R/d}\right]$$

we have

$$\frac{\int \int_C rD(r)dA}{\int \int_C D(r)dA} = \frac{2d^2 - (R^2 + 2dR + 2d^2)e^{-R/d}}{d - (R+d)e^{-R/d}}$$

(c) Letting $R \longrightarrow \infty$ in the result of parts (a) and (b) we find that the total population is $2\pi d^2 D_0$ and the average commute for the total population is $2d^2/d = 2d$.

14.6 Surface Area

1. Letting $z = 0$, we have $2x + 3y = 12$. Using $f(x,y) = z = 3 - \frac{1}{2}x - \frac{3}{4}y$ we have
$f_x = -\frac{1}{2}$, $f_y = -\frac{3}{4}$, $1 + f_x^2 + f_y^2 = \frac{29}{16}$. Then

$$A = \int_0^6 \int_0^{4-2x/3} \sqrt{29/16}\,dy\,dx = \frac{\sqrt{29}}{4} \int_0^6 (4 - \frac{2}{3}x)dx$$

$$= \frac{\sqrt{29}}{4}(4x - \frac{1}{3}x^2)\Big|_0^6 = \frac{\sqrt{29}}{4}(24 - 12) = 3\sqrt{29}.$$

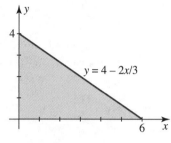

3. Using $f(x,y) = z = \sqrt{16 - x^2}$ we see that for $0 \le x \le 2$ and $0 \le y \le 5$, $z.0$.
Thus, the surface is entirely above the region. Now $f_x = -\frac{x}{\sqrt{16-x^2}}$, $f_y = 0$,

$$1 + f_x^2 + f_y^2 = 1 + \frac{x^2}{16 - x^2} = \frac{16}{16 - x^2} \text{ and}$$

$$A = \int_0^5 \int_0^2 \frac{4}{\sqrt{16 - x^2}}dxdy = 4\int_0^5 \sin^{-1}\frac{x}{4}\Big|_0^2 dy = 4\int_0^5 \frac{\pi}{6}dy = \frac{10\pi}{3}.$$

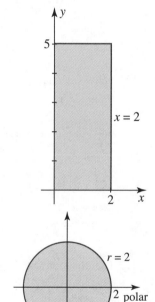

5. Letting $z = 0$ we have $x^2 + y^2 = 4$. Using $f(x,y) = z = 4 - (x^2 + y^2)$ we have
$f_x = -2x$, $f_y = -2y$, $1 + f_x^2 + f_y^2 = 1 + 4(x^2 + y^2)$. Then

$$A = \int_0^{2\pi} \int_0^2 \sqrt{1 + 4r^2}\,r\,dr\,d\theta = \int_0^{2\pi} \frac{1}{3}(1 + 4r^2)^{3/2}\Big|_0^2 d\theta$$

$$= \frac{1}{12}\int_0^{2\pi}(17^{3/2} - 1)d\theta = \frac{\pi}{6}(17^{3/2} - 1).$$

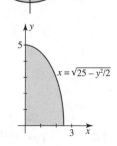

7. Using $f(x,y) = z = \sqrt{25 - x^2 - y^2}$ we have
$f_x = -\frac{x}{\sqrt{25 - x^2 - y^2}}$, $f_y = -\frac{y}{\sqrt{25 - x^2 - y^2}}$,
$1 + f_x^2 + f_y^2 = \frac{25}{25 - x^2 - y^2}$. Then

$$A = \int_0^5 \int_0^{\sqrt{25-y^2}/2} \frac{5}{\sqrt{25 - x^2 - y^2}}dxdy$$

$$= 5\int_0^5 \sin^{-1}\frac{x}{\sqrt{25 - y^2}}\Big|_0^{\sqrt{25-y^2}/2} dy = 5\int_0^5 \frac{\pi}{6}dy = \frac{25\pi}{6}.$$

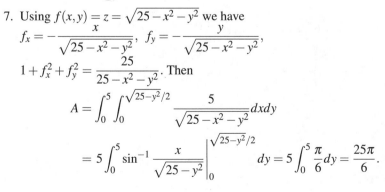

9. There are portions of the sphere within the cylinder both above and below the xy-plane. Using $f(x,y) = z = \sqrt{a^2 - x^2 - y^2}$ we have

$$f_x = -\frac{x}{\sqrt{a^2 - x^2 - y^2}}, \quad f_y = -\frac{y}{\sqrt{a^2 - x^2 - y^2}}, \quad 1 + f_x^2 + f_y^2 = \frac{a^2}{a^2 - x^2 - y^2}.$$

Then, using symmetry,

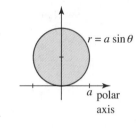

$$A = 2\left[2\int_0^{\pi/2}\int_0^{a\sin\theta}\frac{a}{\sqrt{a^2 - r^2}}r\,dr\,d\theta\right] = 4a\int_0^{\pi/2} -\sqrt{a^2 - r^2}\Big|_0^{a\sin\theta}\,d\theta$$

$$= 4a\int_0^{\pi/2}(a - a\sqrt{1 - \sin^2\theta})\,d\theta = 4a^2\int_0^{\pi/2}(1 - \cos\theta)\,d\theta$$

$$= 4a^2(\theta - \sin\theta)\Big|_0^{\pi/2} = 4a^2\left(\frac{\pi}{2} - 1\right) = 2a^2(\pi - 2).$$

11. There are portions of the surface in each octant with areas equal to the area of the portion in the first octant. Using $f(x,y) = z = \sqrt{a^2 - y^2}$ we have $f_x = 0$, $f_y = \frac{y}{\sqrt{a^2 - y^2}}$, $1 + f_x^2 + f_y^2 = \frac{a^2}{a^2 - y^2}$. Then

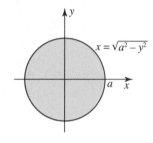

$$A = 8\int_0^a\int_0^{\sqrt{a^2 - y^2}}\frac{a}{\sqrt{a^2 - y^2}}\,dx\,dy$$

$$= 8a\int_0^a\frac{x}{\sqrt{a^2 - y^2}}\Big|_0^{\sqrt{a^2 - y^2}}\,dy = 8a\int_0^a dy = 8a^2.$$

13. The projection of the surface onto the xy-plane is shown in the graph. Using $f(x,z) = y = \sqrt{a^2 - x^2 - z^2}$ we have $f_x = -\frac{x}{\sqrt{a^2 - x^2 - z^2}}$, $f_z = -\frac{z}{\sqrt{a^2 - x^2 - z^2}}$, $1 + f_x^2 + f_z^2 = \frac{a^2}{a^2 - x^2 - z^2}$. Then

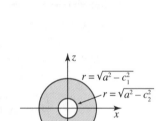

$$A = \int_0^{2\pi}\int_{\sqrt{a^2 - c_2^2}}^{\sqrt{a^2 - c_1^2}}\frac{a}{\sqrt{a^2 - r^2}}r\,dr\,d\theta = a\int_0^{2\pi} -\sqrt{a^2 - r^2}\Big|_{\sqrt{a^2 - c_2^2}}^{\sqrt{a^2 - c_1^2}}\,d\theta$$

$$= a\int_0^{2\pi}(c_2 - c_1)\,d\theta = 2\pi a(c_2 - c_1).$$

15. The equations of the spheres are $x^2 + y^2 + z^2 = a^2$ and $x^2 + y^2 + (z + a)^2 = 1$. Subtracting these equations, we obtain $(z - a)^2 - z^2 = 1 - a^2$ or $-2az + a^2 = 1 - a^2$. Thus, the spheres intersect on the plane $z = a - 1/2a$. The region of integration is $x^2 + y^2 + (a - 1/2a)^2 = a^2$ or $r^2 = 1 - 1/4a^2$. The area is

$$A = a\int_0^{2\pi}\int_0^{\sqrt{1 - 1/4a^2}}(a^2 - r^2)^{-1/r}r\,dr\,d\theta = 2\pi a[-(a^2 - r^2)^{1/2}]\Big|_0^{\sqrt{1 - 1/4a^2}}$$

$$= 2\pi a\left(a - \left[a^2 - \left(1 - \frac{1}{4a^2}\right)\right]^{1/2}\right) = 2\pi a\left(a - \left[\left(a - \frac{1}{2a}\right)^2\right]^{1/2}\right) = \pi.$$

14.7 The Triple Integral

1. $\int_2^4\int_{-2}^2\int_{-1}^1(x + y + z)\,dx\,dy\,dz = \int_2^4\int_{-2}^2\left(\frac{1}{2}x^2 + xy + xz\right)\Big|_{-1}^1\,dy\,dz$

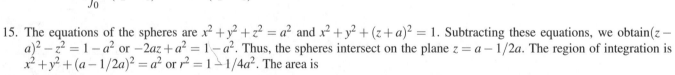

$= \int_2^4\int_{-2}^2(2y + 2z)\,dy\,dz = \int_2^4(y^2 + 2yz)\Big|_{-2}^2\,dz = \int_2^4 8z\,dz = 4z^2\Big|_2^4 = 48$

3. $\int_0^6 \int_0^{6-x} \int_0^{6-x-z} dy\,dz\,dx = \int_0^6 \int_0^{6-x} (6-x-z)dz\,dx = \int_0^6 (6z - xz - \frac{1}{2}z^2)\Big|_0^{6-x} dx$

$$= \int_0^6 \left[6(6-x) - x(6-x) - \frac{1}{2}(6-x)^2 \right] dx = \int_0^6 (18 - 6x + \frac{1}{2}x^2)dx$$

$$= \left(18x - 3x^2 + \frac{1}{6}x^3 \right)\Big|_0^6 = 36$$

5. $\int_0^{\pi/2} \int_0^{y^2} \int_0^y \cos\frac{x}{y}\,dz\,dx\,dy = \int_0^{\pi/2} \int_0^{y^2} y\cos\frac{x}{y}\,dx\,dy = \int_0^{\pi/2} y^2 \sin\frac{x}{y}\Big|_0^{y^2} dy$

$$= \int_0^{\pi/2} y^2 \sin y\,dy \quad \boxed{\text{Integration by parts}}$$

$$= (-y^2 \cos y + 2\cos y + 2y\sin y)\Big|_0^{\pi/2} = \pi - 2$$

7. $\int_0^1 \int_0^1 \int_0^{2-x^2-y^2} xye^z dz\,dx\,dy = \int_0^1 \int_0^1 xye^z \Big|_0^{2-x^2y^2} dx\,dy = \int_0^1 \int_0^1 (xye^{2-x^2-y^2} - xy)dx\,dy$

$$= \int_0^1 \left(-\frac{1}{2}ye^{2-x^2-y^2} - \frac{1}{2}x^2y \right)\Big|_0^1 dy = \int_0^1 \left(-\frac{1}{2}ye^{1-y^2} - \frac{1}{2}y + \frac{1}{2}ye^{2-y^2} \right) dy$$

$$= \left(\frac{1}{4}e^{1-y^2} - \frac{1}{4}y^2 - \frac{1}{4}e^{2-y^2} \right)\Big|_0^1 = \left(\frac{1}{4} - \frac{1}{4} - \frac{1}{4}e \right) - \left(\frac{1}{4}e - \frac{1}{4}e^2 \right) = \frac{1}{4}e^2 - \frac{1}{2}e$$

9. $\iiint_D z\,dV = \int_0^5 \int_1^3 \int_y^{y+2} z\,dx\,dy\,dz = \int_0^5 \int_1^3 2x\,dy\,dz$

$$= \int_0^5 2yz\Big|_1^3 dz = \int_0^5 4z\,dz = 2z^2\Big|_0^5 = 50$$

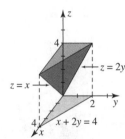

11. The other five integrals are

$\int_0^4 \int_0^{2-x/2} \int_{x+2y}^4 f(x,y,z)dz\,dy\,dx$, $\int_0^4 \int_0^z \int_0^{(z-x)/2} f(x,y,z)dy\,dx\,dz$,

$\int_0^4 \int_x^4 \int_0^{(z-x)/2} f(x,y,z)dy\,dz\,dx$, $\int_0^4 \int_0^{z/2} \int_0^{z-2y} f(x,y,z)dx\,dy\,dz$,

$\int_0^2 \int_{2y}^4 \int_0^{z-2y} f(x,y,z)dx\,dz\,dy$.

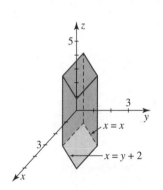

13. (a) $V = \int_0^2 \int_{x^3}^8 \int_0^4 dz\,dy\,dx$ (b) $V = \int_0^8 \int_0^4 \int_0^{y^{1/3}} dx\,dz$

(c) $V = \int_0^4 \int_0^2 \int_{x^2}^8 dy\,dx\,dz$

15.

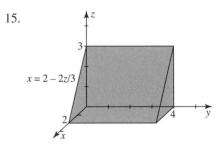

17.

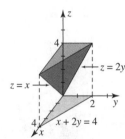

19.

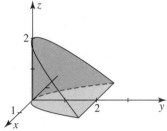

21. Solving $x = y^2$ and $4 - x = y^2$, we obtain $x = 2$, $y = \pm\sqrt{2}$. Using symmetry,

$$V = 2\int_0^3 \int_0^{\sqrt{2}} \int_{y^2}^{4-y^2} dx\,dy\,dz = 2\int_0^3 \int_0^{\sqrt{2}} (4 - 2y^2)\,dy\,dz$$

$$= 2\int_0^3 \left(4y - \frac{2}{3}y^3\right)\Big|_0^{\sqrt{2}} dz = 2\int_0^3 \frac{8\sqrt{2}}{3}\,dz = 16\sqrt{2}.$$

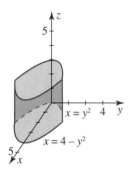

23. Adding the two equations, we obtain $2y = 8$. Thus, the paraboloids intersect in the plane $y = 4$. Their intersection is a circle of radius 2. Using symmetry,

$$V = 4\int_0^2 \int_0^{\sqrt{4-x^2}} \int_{x^2+z^2}^{8-x^2-z^2} dy\,dz\,dx$$

$$= 4\int_0^2 \int_0^{\sqrt{4-x^2}} (8 - 2x^2 - 2x^2)\,dz\,dx = 4\int_0^2 \left[2(4-x^2)z - \frac{2}{3}z^3\right]\Big|_0^{\sqrt{4-x^2}} dx$$

$$= 4\int_0^2 \frac{4}{3}(4-x^2)^{3/2}dx \quad \boxed{\text{Trig substitution}}$$

$$= \frac{16}{3}\left[-\frac{x}{8}(2x^2 - 20)\sqrt{4-x^2} + 6\sin^{-1}\frac{x}{2}\right]\Big|_0^2 = 16\pi.$$

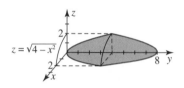

25. We are given $\rho(x,y,z) = kz$.

$$m = \int_0^8 \int_0^4 \int_0^{y^{1/3}} kz\,dx\,dz\,dy = k\int_0^8 \int_0^4 xz\Big|_0^{y^{1/3}} dz\,dy = k\int_0^8 \int_0^4 y^{1/3}z\,dz\,dy$$

$$= k\int_0^8 \frac{1}{2}y^{1/3}z^2\Big|_0^4 dy = 8k\int_0^8 y^{1/3}dy = 8k\left(\frac{3}{4}y^{4/3}\right)\Big|_0^8 = 96k$$

$$M_{xy} = \int_0^8 \int_0^4 \int_0^{y^{1/3}} kz^2\,dx\,dz\,dy = k\int_0^8 \int_0^4 xz^2\Big|_0^{y^{1/3}} dz\,dy = k\int_0^8 \int_0^4 y^{1/3}z^2\,dz\,dy$$

$$= k\int_0^8 \frac{1}{3}y^{1/3}z^3\Big|_0^4 dy = \frac{64}{3}k\int_0^8 y^{1/3}dy = \frac{64}{3}k\left(\frac{3}{4}y^{4/3}\right)\Big|_0^8 = 256k$$

$$M_{xz} = \int_0^8 \int_0^4 \int_0^{y^{1/3}} kyz\,dx\,dz\,dy = k\int_0^8 \int_0^4 xyz\Big|_0^{y^{1/3}} dz\,dy = i\int_0^8 \int_0^4 y^{4/3}z\,dz\,dy$$

$$= k\int_0^8 \frac{1}{2}y^{4/3}z^2\Big|_0^4 dy = 8k\int_0^8 y^{4/3}dy = 8k\left(\frac{3}{7}y^{7/3}\right)\Big|_0^8 = \frac{3072}{7}k$$

$$M_{yz} = \int_0^8 \int_0^4 \int_0^{y^{1/3}} kxz\,dx\,dz\,dy = k\int_0^8 \int_0^4 \frac{1}{2}x^2z\Big|_0^{y^{1/3}} dz\,dy = \frac{1}{2}k\int_0^8 \int_0^4 y^{2/3}z\,dz\,dy$$

$$= \frac{1}{2}k\int_0^8 \frac{1}{2}y^{2/3}z^2\Big|_0^4 dy = 4k\int_0^8 y^{2/3}dy = 4k\left(\frac{3}{5}y^{5/3}\right)\Big|_0^8 = \frac{384}{5}k$$

$$\bar{x} = M_{yz}/m = \frac{384k/5}{96k} = 4/5; \quad \bar{y} = M_{xz}/m = \frac{3072k/7}{96k} = 32/7; \quad \bar{z} = M_{xy}/m = \frac{256k}{96k} = 8/3$$

The center of mass is $(4/5, 32/7, 8/3)$.

27. The density is $\rho(x,y,z) = ky$. Since both the region and the density function are symmetric with respect to the xy- and yz-planes, $\bar{x} = \bar{z} = 0$. Using symmetry,

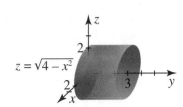

$$m = 4\int_0^3\int_0^2\int_0^{\sqrt{4-x^2}} ky\,dzdxdy = 4k\int_0^3\int_0^2 yz\Big|_0^{\sqrt{4-k^2}}\,dxdy$$

$$= 4k\int_0^3\int_0^2 y\sqrt{4-x^2}\,dxdy = 4k\int_0^3 y\left(\frac{x}{2}\sqrt{4-x^2}+2\sin^{-1}\frac{x}{2}\right)\Big|_0^2\,dy$$

$$= 4k\int_0^3 \pi y\,dy = 4\pi k\left(\frac{1}{2}y^2\right)\Big|_0^3 = 18\pi k$$

$$M_{xz} = 4\int_0^3\int_0^2\int_0^{\sqrt{4-x^2}} ky^2\,dzdxdy = 4k\int_0^3\int_0^2 y^2z\Big|_0^{\sqrt{4-x^2}}\,dxdy = 4k\int_0^3\int_0^2 y^2\sqrt{4-x^2}\,dxdy$$

$$= 4k\int_0^3 y^2\left(\frac{x}{2}\sqrt{4-x^2}+2\sin^{-1}\frac{x}{2}\right)\Big|_0^2\,dy = 4k\int_0^3 \pi y^2\,dy = 4\pi k\left(\frac{1}{3}y^3\right)\Big|_0^3 = 36\pi k.$$

$\bar{y} = M_{xz}/m = \dfrac{36\pi k}{18\pi k} = 2$. The center of mass is $(0,2,0)$.

29. $m = \displaystyle\int_{-1}^1\int_{-\sqrt{1-x^2}}^{\sqrt{1-x^2}}\int_{2+2y}^{8-y}(x+y+4)\,dzdydx$

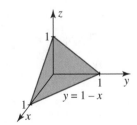

31. We are given $\rho(x,y,z) = kz$.

$$I_y = \int_0^8\int_0^4\int_0^{y^{1/3}} kz(x^2+z^2)\,dxdzdy = k\int_0^8\int_0^4\left(\frac{1}{3}x^3z+xz^3\right)\Big|_0^{y^{1/3}}\,dzdy$$

$$= k\int_0^8\int_0^4\left(\frac{1}{3}yz+y^{1/3}z^3\right)\,dzdy = k\int_0^8\left(\frac{1}{6}yz^2+\frac{1}{4}y^{1/3}z^4\right)\Big|_0^4\,dy$$

$$= k\int_0^8\left(\frac{8}{3}y+64y^{1/3}\right)\,dy = k\left(\frac{4}{3}y^2+48y^{4/3}\right)\Big|_0^8 = \frac{2560}{3}k$$

From Problem 25, $m = 96k$. Thus, $R_g = \sqrt{I_y/m} = \sqrt{\dfrac{2560k/3}{96k}} = \dfrac{4\sqrt{5}}{3}$.

33. $I_z = k\displaystyle\int_0^1\int_0^{1-x}\int_0^{1-x-y}(x^2+y^2)\,dzdydx$

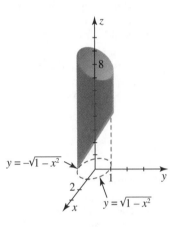

$$= k\int_0^1\int_0^{1-x}(x^2+y^2)(1-x-y)\,dydx$$

$$= k\int_0^1\int_0^{1-x}(x^2-X^3-x^2y+y^2-xy^2-y^3)\,dydx$$

$$= k\int_0^1\left[(x^2-x^3)y-\frac{1}{2}x^2y^2+\frac{1}{3}(1-x)y^3-\frac{1}{4}y^4\right]\Big|_0^{1-x}\,dx$$

$$= k\int_0^1[\frac{1}{2}x^2-x^3+\frac{1}{2}x^4+\frac{1}{12}(1-x)^4]\,dx = k\left[\frac{1}{6}x^6-\frac{1}{4}x^4+\frac{1}{10}x^5-\frac{1}{60}(1-x)^5\right]\Big|_0^1 = \frac{k}{30}$$

35. We are given $\rho(x,y,z) = k\sqrt{x^2+y^2+z^2}$. Both the region and the integrand are symmetric with respect to the yz- and xz-planes.

$$I_z = 4\int_0^5 \int_0^{\sqrt{25-x^2}} \int_{\sqrt{x^2+y^2}}^5 k(x^2+y^2)\sqrt{x^2+y^2+z^2}\,dz\,dy\,dx$$

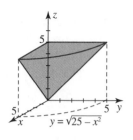

$y = \sqrt{25-x^2}$

14.8   Triple Integrals in Other Coordinate Systems

1. $x = 10\cos 3\pi/4 = -5\sqrt{2};\ \ y = 10\sin 3\pi/4 = 5\sqrt{2};\ \ (-5\sqrt{2}, 5\sqrt{2}, 5)$

3. $x = \sqrt{3}\cos\pi/3 = \sqrt{3}/2;\ \ y = \sqrt{3}\sin\pi/3 = 3/2;\ \ (\sqrt{3}/2, 3/2, -4)$

5. $x = 5\cos\pi/2 = 0;\ \ y = 5\sin\pi/2 = 5;\ \ (0, 5, 1)$

7. With $x = 1$ and $y = -1$ we have $r^2 = 2$ and $\tan\theta = -1$. The point is $(\sqrt{2}, -\pi/4, -9)$

9. With $x = -\sqrt{2}$ and $y = \sqrt{6}$ we have $r^2 = 8$ and $\tan\theta = -\sqrt{3}$. The point is $(2\sqrt{2}, 2\pi/3, 2)$.

11. With $x = 0$ and $y = -4$ we have $r^2 = 16$ and $\tan\theta$ undefined. The point is $(4, -\pi/2, 0)$.

13. $r^2 + z^2 = 25$

15. $r^2 - z^2 = 1$

17. $z = x^2 + y^2$

19. $r\cos\theta = 5,\ \ z = 5$

21. The equations are $r^2 = 4,\ \ r + z^2 = 16$, and $z = 0$.

$$V = \int_0^{2\pi}\int_0^2\int_0^{\sqrt{16-r^2}} r\,dz\,dr\,d\theta = \int_0^{2\pi}\int_0^2 r\sqrt{16-r^2}\,dr\,d\theta$$

$$= \int_0^{2\pi} -\frac{1}{3}(16-r^2)^{3/2}\Big|_0^2\,d\theta = \int_0^{2\pi}(64-24\sqrt{3})\,d\theta = \frac{2\pi}{3}(64-24\sqrt{3})$$

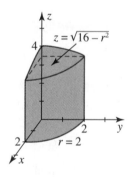

$z = \sqrt{16-r^2}$

$r = 2$

23. The equations are $z = r^2,\ \ r = 5$, and $z = 0$.

$$V = \int_0^{2\pi}\int_0^5\int_0^{r^2} r\,dz\,dr\,d\theta = \int_0^{2\pi}\int_0^5 r^3\,dr\,d\theta = \int_0^{2\pi}\frac{1}{4}r^4\Big|_0^5\,d\theta$$

$$= \int_0^{2\pi}\frac{625}{4}\,d\theta = \frac{625\pi}{2}$$

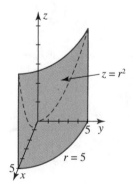

$z = r^2$

$r = 5$

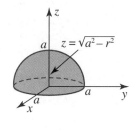

25. The equation is $z = \sqrt{a^2 - r^2}$. By symmetry, $\bar{x} = \bar{y} = 0$.

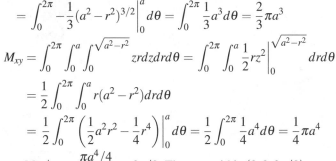

$$m = \int_0^{2\pi}\int_0^a\int_0^{\sqrt{a^2-r^2}} r\,dz\,dr\,d\theta = \int_0^{2\pi}\int_0^a r\sqrt{a^2-r^2}\,dr\,d\theta$$

$$= \int_0^{2\pi} -\frac{1}{3}(a^2-r^2)^{3/2}\Big|_0^a\,d\theta = \int_0^{2\pi}\frac{1}{3}a^3\,d\theta = \frac{2}{3}\pi a^3$$

$$M_{xy} = \int_0^{2\pi}\int_0^a\int_0^{\sqrt{a^2-r^2}} zr\,dz\,dr\,d\theta = \int_0^{2\pi}\int_0^a \frac{1}{2}rz^2\Big|_0^{\sqrt{a^2-r^2}}\,dr\,d\theta$$

$$= \frac{1}{2}\int_0^{2\pi}\int_0^a r(a^2-r^2)\,dr\,d\theta$$

$$= \frac{1}{2}\int_0^{2\pi}\left(\frac{1}{2}a^2r^2 - \frac{1}{4}r^4\right)\Big|_0^a\,d\theta = \frac{1}{2}\int_0^{2\pi}\frac{1}{4}a^4\,d\theta = \frac{1}{4}\pi a^4$$

$$\bar{z} = M_{xy}/m = \frac{\pi a^4/4}{2\pi a^3/3} = 3a/8. \text{ The centroid is } (0,0,3a/8).$$

27. The equation is $z = \sqrt{9-r^2}$ and the density is $\rho = k/r^2$. When $x = 2$, $r = \sqrt{5}$.

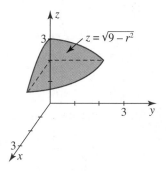

$$I_z = \int_0^{2\pi}\int_0^{\sqrt{5}}\int_2^{\sqrt{9-r^2}} r^2(k/r^2)r\,dz\,dr\,d\theta\,(k/r^2)$$

$$= k\int_0^{2\pi}\int_0^{\sqrt{5}} rz\Big|_2^{\sqrt{9-r^2}}\,dr\,d\theta$$

$$= k\int_0^{2\pi}\int_0^{\sqrt{5}} \left(r\sqrt{9-r^2} - 2r\right)dr\,d\theta$$

$$= k\int_0^{2\pi}\left[-\frac{1}{3}(9-r^2)^{3/2} - r^2\right]\Big|_0^{\sqrt{5}}\,d\theta$$

$$= k\int_0^{2\pi}\frac{4}{3}\,d\theta = \frac{8}{3}\pi k$$

29. (a) $x = (2/3)\sin(\pi/2)\sin(\pi/2)\cos(\pi/6) = \sqrt{3}/3$; $y = (2/3)\sin(\pi/2)\sin(\pi/6) = 1/3$;

    (b) With $x = \sqrt{3}/3$ and $y = 1/3$ we have $r^2 = 4/9$ and $\tan\theta = \sqrt{3}/3$. The point is $(2/3, \pi/6, 0)$.

31. (a) $x = 8\sin(\pi/4)\cos(3\pi/4) = -4$; $y = 8\sin(\pi/4)\sin(3\pi/4) = 4$; $z = 8\cos(\pi/4) = 4\sqrt{2}$;
    $(-4, 4, 4\sqrt{2})$

    (b) With $x = -4$ and $y = 4$ we have $r^2 = 32$ and $\tan\theta = -1$. The point is $(4\sqrt{2}, 3\pi/4, 4\sqrt{2})$.

33. (a) $x = 4\sin(3\pi/4)\cos 0 = 2\sqrt{2}$; $y = 4\sin(3\pi/4)\sin 0 = 0$; $z = 4\cos(3\pi/4) = -2\sqrt{2}$;
    $(2\sqrt{2}, 0, -2\sqrt{2})$

    (b) With $x = 2\sqrt{2}$ and $y = 0$ we have $r^2 = 8$ and $\tan\theta = 0$. The point is $(2\sqrt{2}, 0, -2\sqrt{2})$.

35. With $x = -5$, $y = -5$, and $z = 0$, we have $\rho^2 = 50$, $\tan\theta = 1$, and $\cos\phi = 0$.
    The point is $(5\sqrt{2}, \pi/2, 5\pi/4)$.

37. With $x = \sqrt{3}/2$, $y = 1/2$, and $z = 1$, we have $\rho^2 = 2$, $\tan\theta = 1/\sqrt{3}$, and $\cos\phi = 1/\sqrt{2}$.
    The point is $(\sqrt{2}, \pi/4, \pi/6)$.

39. With $x = 3$, $y = -3$, and $z = 3\sqrt{2}$, we have $\rho^2 = 36$, $\tan\theta = -1$, and $\cos\phi = -\sqrt{2}/2$.
    The point is $(6, \pi/4, -\pi/4)$

41. $\rho = 8$

43. $4z^2 = 3x^2 + 3y^2 + 3z^2$; $4\rho^2\cos^2\phi = 3\rho^2$; $\cos\phi = \pm\sqrt{3}/2$; $\phi = \pi/6,\ 5\pi/6$

45. $x^2 + y^2 + z^2 = 100$

47. $\rho\cos\phi = 2$; $z = 2$

49. The equations are $\phi = \pi/4$ and $\rho = 3$.

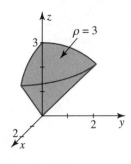

$$V = \int_0^{2\pi} \int_0^{\pi/4} \int_0^3 \rho^2 \sin\phi \, drhod\phi d\theta = \int_0^{2\pi} \int_0^{\pi/4} \frac{1}{3}\rho^3 \sin\phi \Big|_0^3 \, d\phi d\theta$$

$$= \int_0^{2\pi} \int_0^{\pi/4} 9\sin\phi \, d\phi d\theta = \int_0^{2\pi} -9\cos\phi \Big|_0^{\pi/4} \, d\theta$$

$$= -9 \int_0^{2\pi} \left(\frac{\sqrt{2}}{2} - 1 \right) d\theta = 9\pi(2 - \sqrt{2})$$

51. Using Problem 43, the equations are $\phi = \pi/6$, $\theta = \pi/2$, and $\rho\cos\phi = 2$.

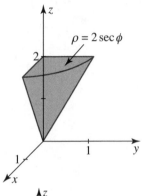

$$V = \int_0^{\pi/2} \int_0^{\pi/6} \int_0^{2\sec\phi} \rho^2 \sin\phi \, d\rho d\phi d\theta$$

$$= \int_0^{\pi/2} \int_0^{\pi/6} \frac{1}{3}\rho^2 \sin\phi \Big|_0^{2\sec\phi} \, d\phi d\theta$$

$$= \frac{8}{3} \int_0^{\pi/2} \int_0^{\pi/6} \sec^3\phi \sin\phi \, d\phi d\theta = \frac{8}{3} \int_0^{\pi/2} \int_0^{\pi/6} \sec^2\phi \tan\phi \, d\phi d\theta$$

$$= \frac{8}{3} \int_0^{\pi/2} \frac{1}{2}\tan^2\phi \Big|_0^{\pi/6} \, d\theta = \frac{4}{3} \int_0^{\pi/2} \frac{1}{3} d\theta = \frac{2}{9}\pi$$

53. By symmetry, $\bar{x} = \bar{y} = 0$. The equations are $\phi = \pi/4$ and $\rho = 2\cos\phi$.

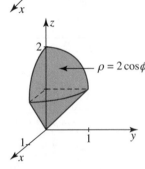

$$m = \int_0^{2\pi} \int_0^{\pi/4} \int_0^{2\cos\phi} \rho^2 \sin\phi \, d\rho d\phi d\theta = \int_0^{2\pi} \int_0^{\pi/4} \frac{1}{3}\rho^3 \sin\phi \Big|_0^{2\cos\phi} \, d\phi d\theta$$

$$= \frac{8}{3} \int_0^{2\pi} \int_0^{\pi/4} \sin\phi \cos^3\phi \, d\phi d\theta = \frac{8}{3} \int_0^{2\pi} -\frac{1}{4}\cos^4\phi \Big|_0^{\pi/4} \, d\theta$$

$$= -\frac{2}{3} \int_0^{2\pi} \left(\frac{1}{4} - 1 \right) d\theta = \pi$$

$$M_{xy} = \int_0^{2\pi} \int_0^{\pi/4} \int_0^{2\cos\phi} z\rho^2 \sin\phi \, d\rho d\phi d\theta \int_0^{2\pi} \int_0^{\pi/4} \int_0^{2\cos\phi} \rho^3 \sin\phi \cos\phi \, d\rho d\phi d\theta$$

$$= \int_0^{2\pi} \int_0^{\pi/4} \frac{1}{4}\rho^4 \sin\phi \cos\phi \Big|_0^{2\cos\phi} \, d\phi d\theta = 4 \int_0^{2\pi} \int_0^{\pi/4} \cos^5\phi \sin\phi \, d\phi d\theta$$

$$= 4 \int_0^{2\pi} -\frac{1}{6}\cos^6\phi \Big|_0^{\pi/4} = -\frac{2}{3} \int_0^{2\pi} \left(\frac{1}{8} - 1 \right) d\theta = \frac{7}{6}\pi$$

$\bar{z} = M_{xy}/m = \dfrac{7\pi/6}{\pi} = 7/6$. The centroid is $(0,0,7/6)$.

55. We are given density$= k/\rho$.

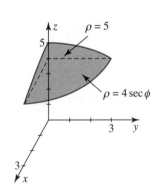

$$m = \int_0^{2\pi} \int_0^{\cos^{-1}4/5} \int_{4\sec\phi}^5 \frac{k}{\rho}\rho^2 \sin\phi \, d\rho d\phi d\theta$$

$$= k \int_0^{2\pi} \int_0^{\cos^{-1}4/5} \frac{1}{2}\rho^2 \sin\phi \Big|_{4\sec\phi}^5 \, d\phi d\theta$$

$$= \frac{1}{2}k \int_0^{2\pi} \int_0^{\cos^{-1}4/5} (25\sin\phi - 16\tan\phi \sec\phi) d\phi d\theta$$

$$\frac{1}{2}k \int_0^{2\pi} (-25\cos\phi - 16\sec\phi) \Big|_0^{\cos^{-1}4/5} \, d\theta$$

$$= \frac{1}{2}k \int_0^{2\pi} [-25(4/5) - 16(5/4) - (-25 - 16)]d\theta$$

$$= \frac{1}{2}k \int_0^{2\pi} d\theta = k\pi$$

14.9 Change of Variables in Multiple Integrals

1. $T:\ (0,0) \longrightarrow (0,0);\ (0,2) \longrightarrow (-2,8);\ (4,0) \longrightarrow (16,20);\ (4,2) \longrightarrow (14,28)$

3. The uv-corner points $(0,0),\ (2,0),\ (2,2)$ correspond to xy-points $(0,0),\ (4,2),\ (6,-4)$.

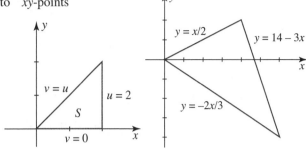

   $v = 0:\ x = 2u,\ y = u \Longrightarrow y = x/2$
   $u = 2:\ x = 4+v,\ y = 2-3v \Longrightarrow$
   $y = 2-3(x-4) = -3+14$
   $v = u:\ x = 3u,\ y = -2u \Longrightarrow y = -2x/3$

5. The uv-corner points $(0.0),$ $(1,0),\ (1,2),\ (0,2)$ correspond to the xy-points $(0,0),\ (1,0),$ $(-3,2),\ (-4,0)$.

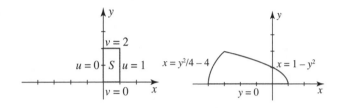

   $v = 0:\ x = u^2,\ y = 0 \Longrightarrow y = 0$ and $0 \le x \le 1$
   $u = 1:\ x = 1-v^2,\ y = v \Longrightarrow x = 1-y^2$
   $v = 2:\ x = u^2-4,\ y = 2u \Longrightarrow x = y^2/4-4$
   $u = 0:\ x = -v^2,\ y = 0 \Longrightarrow y = 0$ and $-4 \le x \le 0$

7. $\dfrac{\partial(x,y)}{\partial(u,v)} = \begin{vmatrix} -ve^{-u} & e^{-u} \\ ve^{u} & e^{u} \end{vmatrix} = -2v$

9. $\dfrac{\partial(u,v)}{\partial(x,y)} = \begin{vmatrix} -2y/x^3 & 1/x^2 \\ -y^2/x^2 & 2y/x \end{vmatrix} = -\dfrac{3y^2}{x^4} = -3\left(\dfrac{y}{x^2}\right)^2 = -3u^2;\ \dfrac{\partial(x,y)}{\partial(u,v)} = \dfrac{1}{-3u^2} = -\dfrac{1}{3u^2}$

11. (a) The uv-corner points $(0,0),\ (1,0),\ (1,1),\ (0,1)$ correspond to the xy-points $(0,0),\ (1,0),\ (0,1),\ (0,0)$.

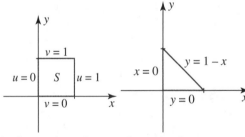

   $v = 0:\ x = u,\ y = 0 \Longrightarrow y = 0,\ 0 \le x \le 1$
   $u = 1:\ x = 1-v,\ y = v \Longrightarrow y = 1-x$
   $v = 1:\ x = 0,\ y = u \Longrightarrow x = 0,\ 0 \le y \le 1$
   $u = 0:\ x = 0,\ y = 0$

   (b) Since the segment $u = 0,\ 0 \le v \le 1$ in the uv-plane maps to the origin in the xy-plane, the transformation is not one-to-one.

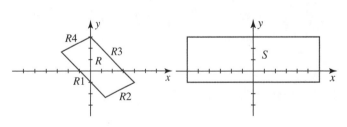

13. $R1:\ x+y = -1 \Longrightarrow v = -1$
    $R2:\ x-2y = 6 \Longrightarrow u = 6$
    $R3:\ x+y = 3 \Longrightarrow v = 3$
    $R4:\ x-2y = -6 \Longrightarrow u = -6$
    $\dfrac{\partial(u,v)}{\partial(x,y)} = \begin{vmatrix} 1 & -2 \\ 1 & 1 \end{vmatrix} = 3 \Longrightarrow \dfrac{\partial(x,y)}{\partial(u,v)} = \dfrac{1}{3}$
    $\int\int_R (x+y)dA = \int\int_S v(\frac{1}{3})dA' = \frac{1}{3}\int_{-1}^{3}\int_{-6}^{6} v\,du\,dv = \frac{1}{3}(12)\int_{-1}^{3} v\,dv = 4(\frac{1}{2})v^2 \Big|_{-1}^{3} = 16$

15. $R1: \quad y = x^2 \Longrightarrow u = 1$

$R2: \quad x = y^2 \Longrightarrow v = 1$

$R3: \quad y = \dfrac{1}{2}x^2 \Longrightarrow u = 2$

$R4: \quad x = \dfrac{1}{2}y^2 \Longrightarrow v = 2$

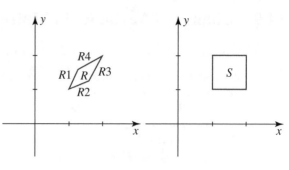

$\dfrac{\partial(u,v)}{\partial(x,y)} = \begin{vmatrix} 2x/y & -x^2/y^2 \\ -y^2/x^2 & 2y/x \end{vmatrix} = 3$

$\Longrightarrow \dfrac{\partial(x,y)}{\partial(u,v)} = \dfrac{1}{3}$

$\displaystyle\iint_R \dfrac{y^2}{x}\,dA = \iint_S v\left(\dfrac{1}{3}\right)dA' = \dfrac{1}{3}\int_1^2\int_1^2 v\,du\,dv = \dfrac{1}{3}\int_1^2 v\,dv = \dfrac{1}{6}v^2\Big|_1^2 = \dfrac{1}{2}$

17. $R1: \quad 2xy = c \Longrightarrow v = c$

$R2: \quad x^2 - y^2 = b \Longrightarrow u = b$

$R3: \quad 2xy = d \Longrightarrow v = d$

$R4: \quad x^2 - y^2 = a \Longrightarrow u = a$

$\dfrac{\partial(u,v)}{\partial(x,y)} = \begin{vmatrix} 2x & -2y \\ 2y & 2x \end{vmatrix} = 4(x^2 + y^2)$

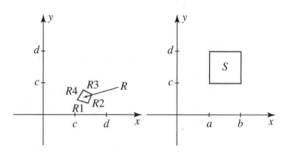

$\Longrightarrow \dfrac{\partial(x,y)}{\partial(u,v)} = \dfrac{1}{4(x^2+y^2)}$

$\displaystyle\iint_R (x^2+y^2)\,dA = \iint_S (x^2+y^2)\dfrac{1}{4(x^2+y^2)}\,dA' = \dfrac{1}{4}\int_c^d\int_a^b du\,dv = \dfrac{1}{4}(b-a)(d-c)$

19. $R1: \quad y = x^2 \Longrightarrow v + u = v - u \Longrightarrow u = 0$

$R2: \quad y = 4 - x^2 \Longrightarrow v + u = 4 - (v - u)$

$\qquad\qquad\qquad\qquad \Longrightarrow v + u = 4 - v + u \Longrightarrow v = 2$

$R3: \quad x = 1 \Longrightarrow v - u = 1 \Longrightarrow v = 1 + u$

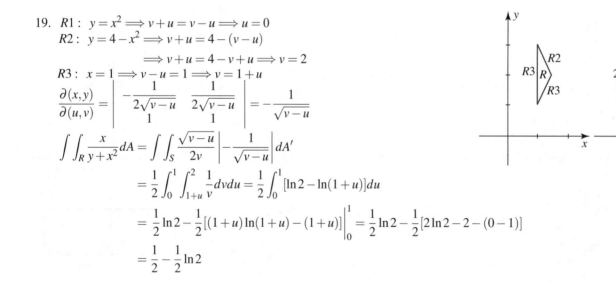

$\dfrac{\partial(x,y)}{\partial(u,v)} = \begin{vmatrix} -\dfrac{1}{2\sqrt{v-u}} & \dfrac{1}{2\sqrt{v-u}} \\ 1 & 1 \end{vmatrix} = -\dfrac{1}{\sqrt{v-u}}$

$\displaystyle\iint_R \dfrac{x}{y+x^2}\,dA = \iint_S \dfrac{\sqrt{v-u}}{2v}\left|-\dfrac{1}{\sqrt{v-u}}\right|dA'$

$\qquad = \dfrac{1}{2}\int_0^1\int_{1+u}^2 \dfrac{1}{v}\,dv\,du = \dfrac{1}{2}\int_0^1 [\ln 2 - \ln(1+u)]\,du$

$\qquad = \dfrac{1}{2}\ln 2 - \dfrac{1}{2}[(1+u)\ln(1+u) - (1+u)]\Big|_0^1 = \dfrac{1}{2}\ln 2 - \dfrac{1}{2}[2\ln 2 - 2 - (0-1)]$

$\qquad = \dfrac{1}{2} - \dfrac{1}{2}\ln 2$

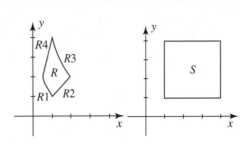

21. $R1: \; y = 1/x \Longrightarrow u = 1$
    $R2: \; y = x \Longrightarrow v = 1$
    $R3: \; y = 4/x \Longrightarrow u = 4$
    $R4: \; y = 4x \Longrightarrow v = 4$

$$\frac{\partial(u,v)}{\partial(x,y)} = \begin{vmatrix} y & x \\ -y/x^2 & 1/x \end{vmatrix} = \frac{2y}{x} \Longrightarrow \frac{\partial(x,y)}{\partial(u,v)} = \frac{x}{2y}$$

$$\int\int_R y^4 \, dA = \int\int_S u^2 v^2 \left(\frac{1}{2v}\right) du\,dv = \frac{1}{2}\int_1^4 u^2 v\,du\,dv = \frac{1}{2}\int_1^4 \frac{1}{3}u^3 v \Big|_1^4 dv = \frac{1}{6}\int_1^4 63v\,dv$$

$$= \frac{21}{4}v^2 \Big|_1^4 = \frac{315}{4}$$

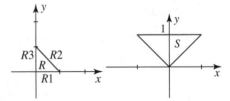

23. We let $u = y - x$ and $V = y + x$.
    $R1: \; y = 0 \Longrightarrow u = -x, \; v = x \Longrightarrow v = -u$
    $R2: \; x + y = 1 \Longrightarrow v = 1$
    $R3: \; x = 0 \Longrightarrow u = y, \; v = y, \; \Longrightarrow v = u$

$$\frac{\partial(u,v)}{\partial(x,y)} = \begin{vmatrix} -1 & 1 \\ 1 & 1 \end{vmatrix} = -2$$

$$\Longrightarrow \frac{\partial(x,y)}{\partial(u,v)} = -\frac{1}{2}$$

$$\int\int_R e^{(y-x)/(y+x)} \, dA = \int\int_S e^{u/v} \left|-\frac{1}{2}\right| dA'$$

$$= \frac{1}{2}\int_0^1 \int_{-v}^{v} e^{u/v} \, du\,dv = \frac{1}{2}\int_0^1 v e^{u/v}\Big|_{-v}^{v} dv$$

$$= \frac{1}{2}\int_0^1 v(e - e^{-1})\,dv = \frac{1}{2}(e - e^{-1})\frac{1}{2}v^2 \Big|_0^1 = \frac{1}{4}(e - e^{-1})$$

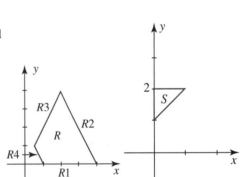

25. Noting that $R2$, $R3$, and $R4$ have equations $y + 2x = 8$, $y - 2x = 0$, and $y + 2x = 2$, we let $u = y/x$ and $v = y + 2x$.
    $R1: \; y = 0 \Longrightarrow u = 0, \; v = 2x \Longrightarrow u = 0, 2 \le v \le 8$
    $R2: \; y + 2x = 8 \Longrightarrow v = 8$
    $R3: \; y - 2x = 0 \Longrightarrow u = 2$
    $R4: \; y + 2x = 2 \Longrightarrow v = 2$

$$\frac{\partial(u,v)}{\partial(x,y)} = \begin{vmatrix} -y/x^2 & 1/x \\ 2 & 1 \end{vmatrix} = \frac{y + 2x}{x^2}$$

$$\Longrightarrow \frac{\partial(x,y)}{\partial(u,v)} = \frac{x^2}{y + 2x}$$

$$\int\int_R (6x + 3y)\,dA = 3\int\int_S (y + 2x)\left|-\frac{x^2}{y+2x}\right|dA' = 3\int\int_S x^2 \, dA'$$

From $y = ux$ we see that $v = ux + 2x$ and $x = v/(u+2)$. Then

$$3\int\int_S x^2 \, dA' = 3\int_0^2 \int_2^8 v^2(u+2)^2 \, dv\,du = \int_0^2 \frac{v^3}{(u+2)^2}\Big|_2^8 du = 504\int_0^2 \frac{du}{(u+2)^2} =$$

$$-\frac{504}{u+2}\Big|_0^2 = 126.$$

27. The image of the ellipse is the unit circle $x^2 + y^2 = 1$. From $\dfrac{\partial(x,y)}{\partial(u,v)} = \begin{vmatrix} 5 & 0 \\ 0 & 3 \end{vmatrix} = 15$ we obtain

$$\int\int_R \left(\frac{x^2}{25} + \frac{y^2}{9}\right)dA = \int\int_s (u^2 + v^2)15\,dA' = 15\int_0^{2\pi}\int_0^1 r^2 \cdot r\,dr\,d\theta = \frac{15}{4}\int_0^{2\pi} r^4\Big|_0^1 d\theta$$

$$= \frac{15}{4}\int_0^{2\pi} d\theta = \frac{15\pi}{2}.$$

29. The image of the ellipsoid $x^2/a^2 + y^2/b^2 + z^2/c^2 = 1$ under the transformation $u = x/a,\ v = y,\ w = z/c$, is the unit sphere $u^2 + v^2 + w^2 = 1$. The volume of this sphere is $\frac{4}{3}\pi$. Now

$$\frac{\partial(x,y,z)}{\partial(u,v,w)} = \begin{vmatrix} a & 0 & 0 \\ 0 & b & 0 \\ 0 & 0 & c \end{vmatrix} = abc$$

and

$$\iiint_D dV = \iiint E\,abc\,dV' = abc \iiint_E dV' = abc\left(\frac{4}{3}\pi\right) = \frac{4}{3}\pi abc.$$

Chapter 14 in Review

A. True/False

1. True; use $e^{x^2-y} = e^{x^2}e^{-y}$ and Problem 53 in Section 14.2

3. True

5. False; both the density function and the lamina must be symmetric about an axis.

B. Fill In the Blanks

1. $\displaystyle\int_{y^2+1}^5 \left(8y^3 - \frac{5y}{x}\right) dx = (8xy^3 - 5y\ln x)\big|_{y^2+1}^5 = 40y^3 - 5y\ln 5 - [8(y^2+1)y^3 - 5y\ln(y^2+1)]$

$$= -8y^5 + 32y^3 + 5y\ln\frac{y^2+1}{y}$$

3. square

5. $f(x,4) - f(x,2)$

7. $\int_0^4 \int_{x/2}^{\sqrt{x}} f(x,y)\,dy\,dx$

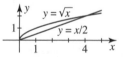

9. $r = 2\sin(\pi/4) = \sqrt{2};\ \theta = 2\pi/3;\ z = 2\cos(\pi/4) = \sqrt{2};\ (\sqrt{2}, 2\pi/3, \sqrt{2})$

11. $z = r^2;\ \rho = \cot\phi\csc\phi$

C. Exercises

1. Holding x fixed, $\displaystyle\int (12x^2 e^{-4xy} - 5x + 1)\,dy = \frac{12x^2 e^{-4xy}}{-4x} - 5xy + y + c_1(x)$

$$= -3xe^{-4xy} - 5xy + y + c_1(x)$$

3. $\displaystyle\int_{y^3}^y y^2 \sin xy\,dx = -y\cos xy\big|_{y^3}^y = y(\cos y^4 - \cos y^2)$

5. $\displaystyle\int_0^2 \int_0^{2x} ye^{y-x}\,dy\,dx = \int_0^2 (ye^{y-x} - e^{y-x})\Big|_0^{2x} dx \quad \boxed{\text{Integration by parts}}$

$$= \int_0^2 (2xe^x - e^x + e^{-x})\,dx \quad \boxed{\text{Integration by parts}}$$

$$= (2xe^x - 2e^x - e^x - e^{-x})\big|_0^2 = e^2 - e^{-2} + 4$$

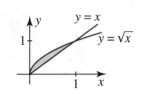

7. $\int_0^1 \int_x^{\sqrt{x}} \frac{\sin y}{y} dy dx = \int_0^1 \int_{y^2}^y \frac{\sin y}{y} dx dy = \int_0^1 \frac{\sin y}{y} x \Big|_{y^2}^y dy$

$\qquad = \int_0^1 (\sin y - y \sin y) dy$ $\boxed{\text{Integration by parts}}$

$\qquad = (-\cos y - s \in y + y \cos y)\Big|_0^1 = (-\cos 1 - \sin 1 + \cos 1) - (-1) = 1 - \sin 1$

9. $\int_0^5 \int_0^{\pi/2} \int_0^{\cos\theta} 3r^2 dr d\theta dz = \int_0^5 \int_0^{\pi/2} r^3 \Big|_0^{\cos\theta} d\theta dz = \int_0^5 \int_0^{\pi/2} \cos^3\theta \, d\theta dz$

$\qquad = \int_0^5 \int_0^{\pi/2} (1 - \sin^2\theta) \cos\theta \, d\theta dz = \int_0^5 \left(\sin\theta - \frac{1}{3}\sin^3\theta\right)\Big|_0^{\pi/2} dz$

$\qquad = \int_0^5 2/3 \, dz = 10/3$

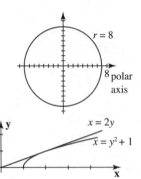

11. Using polar coorindates,

$\iint_R 5 dA = \int_0^{2\pi} \int_0^8 5r \, dr d\theta = 5 \int_0^{2\pi} \frac{1}{2} r^2 \Big|_0^8 d\theta = 5 \int_0^{2\pi} 32 \, d\theta = 320\pi.$

13. $\iint_R (2x + y) dA = \int_0^1 \int_{2y}^{y^2+1} (2x + y) dx dy = \int_0^1 (x^2 + xy)\Big|_{2y}^{y^2+1} dy$

$\qquad = \int_0^1 [(y^2 + 1)^2 + (y^2 + 1)y - (4y^2 + 2y^2)] dy$

$\qquad = (y^4 + y^3 - 4y^2 + y + 1) dy = \left(\frac{1}{5}y^5 + \frac{1}{4}y^4 - \frac{4}{3}y^3 + \frac{1}{2}y^2 + y\right)\Big|_0^1 = \frac{37}{60}$

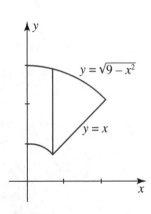

15. The circle $x^2 + y^2 = 1$ intersects $y = x$ at $x = 1/\sqrt{2}$. The circle $x^2 + y^2 = 9$ intersects $y = x$ at $x = 3/\sqrt{2}$.

$\iint_R \frac{1}{x^2 + y^2} dA = \int_0^{1/\sqrt{2}} \int_{\sqrt{1-x^2}}^{\sqrt{9-x^2}} \frac{1}{x^2 + y^2} dy dx$

$\qquad + \int_{1/\sqrt{2}}^{3/\sqrt{2}} \int_x^{\sqrt{9-x^2}} \frac{1}{x^2 + y^2} dy dx$

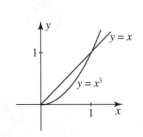

17.

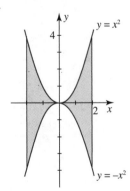

19. $\int_0^1 \int_y^{\sqrt[3]{y}} \cos x^2 dx dy = \int_0^1 \int_{x^3}^x \cos x^2 dy dx = \int_0^1 y \cos x^2 \Big|_{x^3}^x dx$

$\qquad = \int_0^1 (x \cos x^2 - x^3 \cos x^2) dx$

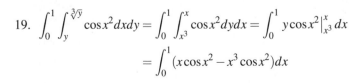

$$= \frac{1}{2}\sin x^2\Big|_0^1 = \int_0^1 x^2(x\cos x^2)dx$$

Integration by parts

$$= \frac{1}{2}\sin 1 - \left(\frac{1}{2}x^2\sin x^2 + \frac{1}{2}\cos x^2\right)\Big|_0^1$$

$$= \frac{1}{2}\sin 1 - \left(\frac{1}{2}\sin 1 + \frac{1}{2}\cos 1 - \frac{1}{2}\right)$$

$$= \frac{1-\cos 1}{2}$$

21. $\displaystyle\int_0^2\int_{1/2}^1\int_0^{\sqrt{x-x^2}}(4z+1)dydxdz = \int_0^2\int_{1/2}^1(4z-1)\sqrt{x-x^2}\,dxdz$

$$= \int_0^2\int_{1/2}^1(4z+1)\sqrt{\frac{1}{4}-\left(x-\frac{1}{2}\right)^2}\,dxdz \quad \boxed{\text{Trig substitution}}$$

$$= \int_0^2\left[(4z+1)\left(\frac{x-1/2}{2}\sqrt{x-x^2}+\frac{1}{8}\sin^{-1}\frac{x-1/2}{1/2}\right)\right]\Big|_{1/2}^1 dz$$

$$= \int_0^2(4z+1)\left(\frac{\pi}{16}-0\right)dz = \frac{\pi}{16}(2z^2+z)\Big|_0^2 = \frac{5\pi}{8}$$

23. $f_x = y$; $f_y = z$, $1+f_x^2+x_y^2 = 11+x^2+y^2$. Using cylindrical coordinates,

$$A = \int_0^{2\pi}\int_0^1\sqrt{1+r^2}\,rdrd\theta = \int_0^{2\pi}\frac{1}{3}(1+r^2)^{3/2}\Big|_0^1 d\theta = \frac{1}{3}\int_0^{2\pi}(2^{3/2}-1)d\theta = \frac{2\pi}{3}(2\sqrt{2}-1).$$

25. (a) $\displaystyle V = \int_0^1\int_x^{2x}\sqrt{1-x^2}\,dydx = \int_0^1 y\sqrt{1-x^2}\,\Big|_x^{2x}dx = \int_0^1 x\sqrt{1-x^2}\,dx$

$$= -\frac{1}{3}(1-x^2)^{3/2}\Big|_0^1 = \frac{1}{3}$$

(b) $\displaystyle V = \int_0^1\int_{y/2}^y\sqrt{1-x^2}\,dxdy + \int_1^2\int_{y/2}^1\sqrt{1-x^2}\,dxdy$

27. $\displaystyle I_y = \int_0^1\int_{x^3}^{x^2}k(x^4+x^2y^2)dydx = k\int_0^1\left(x^4y+\frac{1}{3}x^2y^3\right)\Big|_{x^3}^{x^2}dx$

$$= k\int_0^1\left(x^6+\frac{1}{3}x^8-x^7-\frac{1}{3}x^{11}\right)dx = k\left(\frac{1}{7}x^7+\frac{1}{27}x^9-\frac{1}{8}x^8-\frac{1}{36}x^{12}\right)\Big|_0^1 = \frac{41}{1512}k$$

29. We use spherical coordinates.
$$V = \int_0^{2\pi}\int_{\tan^{-1}1/2}^{\pi/4}\int_0^{3\sec\phi}\rho^2\sin\phi\,d\rho d\phi d\theta$$

$$= \int_0^{2\pi}\int_{\tan^{-1}1/2}^{\pi/4}\frac{1}{3}\rho^3\sin\phi\Big|_0^{3\sec\phi}d\phi d\theta$$

$$= \frac{1}{3}\int_0^{2\pi}\int_{\tan^{-1}1/2}^{\pi/4}27\sec^3\phi\sin\phi\,d\phi d\theta = 9\int_0^{2\pi}\int_{\tan^{-1}1/2}^{\pi/4}\tan\phi\sec^2\phi\,d\phi d\theta$$

$$= 9\int_0^{2\pi}\frac{1}{2}\tan^2\phi\Big|_{\tan^{-1}1/2}^{\pi/4}d\theta = \frac{9}{2}\int_0^{2\pi}\left(1-\frac{1}{9}\right)d\theta = 8\pi$$

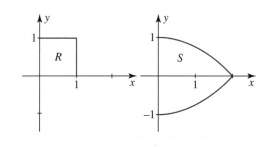

31. $x = 0 \Longrightarrow u = 0, \ v = -y^2 \Longrightarrow u = 0, \ -1 \le v \le 0$

$x = 1 \Longrightarrow u = 2y, \ v = 1 - y^2 = 1 - u^2/4$

$x = 1 \Longrightarrow u = 2y, \ v = 1 - y^2 = 1 - u^2/4$

$y = 0 \Longrightarrow u = 0, \ v = x^2 \Longrightarrow u = 0, \ 0 \le v \le 1$

$y = 1 \Longrightarrow u = 2x, \ v = x^2 - 1 = u^2/4 - 1$

$\dfrac{\partial(u,v)}{\partial(x,y)} = \begin{vmatrix} 2y & 2x \\ 2x & -2y \end{vmatrix} = -4(x^2 + y^2)$

$$\Longrightarrow \frac{\partial(x,y)}{\partial(u,v)} = -\frac{1}{4(x^2+y^2)}$$

$$\int\!\!\int_R (x^2+y^2)\sqrt[3]{x^2+y^2}\,dA = \int\!\!\int_S (x^2+y^2)\sqrt[3]{v}\left| -\frac{1}{4(x^2+y^2)} \right| dA' = \frac{1}{4}\int_0^2 \int_{u^2/4-1}^{1-u^2/4} v^{1/3}\,dv\,du$$

$$= \frac{1}{4}\int_0^2 \frac{3}{4} v^{4/3}\Big|_{u^2/4-1}^{1-u^2/4}\,du = \frac{3}{16}\int_0^2 \left[(1-u^2/4)^{4/3} - (u^2/4-1)^{4/3}\right] du$$

$$= \frac{3}{16}\int_0^2 \left[(1-u^2/4)^{4/3} - (1-u^2/4)^{4/3}\right] du = 0$$

Vector Integral Calculus

15.1 Line Integrals

1. $\displaystyle\int_C 2xy \ dx = \int_0^{\pi/4} 2(5\cos t)(5\sin t)(-5\sin t) \ dt = -250 \int_0^{\pi/4} \sin^2 t \cos t \ dt$

$$= -250 \left(\frac{1}{3}\sin^3 t\right)\Big|_0^{\pi/4} = -\frac{125\sqrt{2}}{6}$$

$\displaystyle\int_C 2xy \ dy = \int_0^{\pi/4} 2(5\cos t)(5\sin t)(5\cos t) \ dt = 250 \int_0^{\pi/4} \cos^2 t \sin t \ dt = 250 \left(-\frac{1}{3}\cos^3 t\right)\Big|_0^{\pi/4}$

$$= \frac{250}{3}\left(1 - \frac{\sqrt{2}}{4}\right) = \frac{125}{6}(4 - \sqrt{2})$$

$\displaystyle\int_C 2xy \ ds = \int_0^{\pi/4} 2(5\cos t)(5\sin t)\sqrt{25\sin^2 t + 25\cos^2 t} \ dt = 250 \int_0^{\pi/4} \sin t \cos t \ dt$

$$= 250 \left(\frac{1}{2}\sin^2 t\right)\Big|_0^{\pi/4} = \frac{125}{2}$$

3. $\displaystyle\int_C (3x^2 + 6y^2) \ dx = \int_{-1}^0 [3x^2 + 6(2x+1)^2] \ dx = \int_{-1}^0 (27x^2 + 24x + 6) \ dx = (9x^3 + 12x^2 + 6x)\Big|_{-1}^0$

$$= -(-9 + 12 - 6) = 3$$

$\displaystyle\int_C (3x^2 + 6y^2) \ dy = \int_{-1}^0 [3x^2 + 6(2x+1)^2]2 \ dx = 6$

$\displaystyle\int_C (3x^2 + 6y^2) \ ds = \int_{-1}^0 [3x^2 + 6(2x+1)^2]\sqrt{1+4} \ dx = 3\sqrt{5}$

5. $\displaystyle\int_C (x^2 + y^2) ds = \int_0^{2\pi} (25\cos^t - 25\sin^2 t)\sqrt{25\sin^2 t + 25\cos^2 t}\,dt = 125 \int_0^{2\pi} (\cos^2 t - \sin^2 t)dt$

$$= 125 \int_0^{2\pi} \cos 2t\,dt = \frac{125}{2}\sin 2t\Big|_0^{2\pi} = 0$$

7. $\displaystyle\int_C z \ dx = \int_0^{\pi/2} t(-\sin t) \ dt$ $\boxed{\text{Integration by parts}}$

$$= (t\cos t - \sin t)\Big|_0^{\pi/2} = -1$$

$\displaystyle\int_C z \ dy = \int_0^{\pi/2} t\cos t \ dt$ $\boxed{\text{Integration by parts}}$

$$= (t\sin t + \cos t)\Big|_0^{\pi/2} = \frac{\pi}{2} - 1$$

$$\int_C z\ dz = \int_0^{\pi/2} t\ dt = \frac{1}{2}t^2\Big|_0^{\pi/2} = \frac{\pi^2}{8}$$

$$\int_C z\ dx = \int_0^{\pi/2} t\sqrt{\sin^2 t + \cos^2 t + 1}\ dt = \sqrt{2}\int_0^{\pi/2} t\ dt = \frac{\pi^2\sqrt{2}}{8}$$

9. Using x as the parameter, $dy = dx$ and

$$\int_C = (2x+y)\ dx + xy\ dy = \int_{-1}^2 (2x + x + 3 + x^2 + 3x)\ dx = \int_{-1}^2 (x^2 + 6x + 3)\ dx$$

$$= \left(\frac{1}{3}x^3 + 3x^2 + 3x\right)\Big|_{-1}^2 = 21.$$

11. From $(-1,2)$ to $(2,2)$ we use x as a parameter with $y = 2$ and $dy = 0$. From $(2,2)$ to $(2,5)$ we use y as a parameter with $x = 2$ and $dx = 0$.

$$\int_C (2x+y)\ dx + xy\ dy = \int_{-1}^2 (2x+2)\ dx + \int_2^5 2y\ dy = (x^2+2x)\Big|_{-1}^2 + y^2\Big|_{-1}^2 = 9 + 21 = 30$$

13. Using x as a the parameter, $dy = 2x\,dx$.

$$\int_C y\ dx + x\ dy = \int_0^1 x^2\ dx + \int_0^1 x(2x)\ dx = \int_0^1 3x^2\ dx = x^3\Big|_0^1 = 1$$

15. From $(0,0)$ to $(0,1)$ we use y as a parameter with $x = dx = 0$. From $(0,1)$ to $(1,1)$ we use x as a parameter with $y = 1$ and $dy = 0$.

$$\int_C y\ dx + x\ dy = 0 + \int_0^1 1\ dx = 1$$

17. $$\int_C (6x^2 + 2y + 2)\ dx + 4xy\ dy = \int_4^9 (6t + 2t^2)\frac{1}{2}t^{-1/2}\ dt + \int_4^9 4\sqrt{t}t\ dt = \int_4^9 (3t^{1/2} + 5t^{3/2})\ dt$$

$$= (2t^{3/2} + 2t^{3/2})\Big|_4^9 = 460$$

19. $$\int_C 2x^3y\ dx + (3x+y)\ dy = \int_{-1}^1 2(y^6)y2y\ dy + \int_{-1}^1 (3y^2 + y)\ dy = \int_{-1}^1 (4y^8 + 3y^2 + y)\ dy$$

$$= \left(\frac{4}{9}y^9 + y^3 + \frac{1}{2}y^2\right)\Big|_{-1}^1 = \frac{26}{9}$$

21. From $(-2,0)$ to $(2,0)$ we use x as a parameter with $y = dy = 0$. From $(2,0)$ to $(-2,0)$ we parameterize the semicircle as $x = 2\cos\theta$ and $y = 2\sin\theta$ for $0 \le \theta \le \pi$.

$$\int_C (x^2 + y^2)\ dx - 2xy\ dy = \int_{-2}^2 x^2\ dx + \int_0^\pi 4(-2\sin\theta\ d\theta) - \int_0^\pi 8\cos\theta\sin\theta(2\cos\theta\ d\theta)$$

$$= \frac{1}{3}x^3\Big|_{-2}^2 - 8\int_0^\pi (\sin\theta + 2\cos^2\theta\sin\theta)\ d\theta$$

$$= \frac{16}{3} - 8\left(-\cos\theta - \frac{2}{3}\cos^3\theta\right)\Big|_0^\pi = \frac{16}{3} - \frac{80}{3} = -\frac{64}{3}$$

23. From $(1,1)$ to $(-1,1)$ and $(-1,-1)$ to $(1.-1)$ we use x as a parameter with $y = 1$ and $y = -1$, respectively, and $dy = 0$. From $(-1,1)$ to $(-1,-1)$ and $(1,-1)$ to $(1,1)$ we use y as a parameter with $x = -1$ and $z = 1$, respectively, and $dx = 0$.

$$\int_C x^2y^3\ dx - xy^2\ dy = \int_1^{-1} x^2(1)\ dx + \int_1^{-1} -(-1)y^2\ dy + \int_{-1}^1 x^2(-1)^3\ dx + \int_{-1}^1 -(1)y^2\ dy$$

$$= \frac{1}{3}x^3\Big|_1^{-1} + \frac{1}{3}y^3\Big|_{-1}^1 - \frac{1}{3}x^3\Big|_{-1}^1 - \frac{1}{3}y^3\Big|_{-1}^1 = -\frac{8}{3}$$

25. $\displaystyle\int_C y\,dx - x\,dy = \int_0^\pi 3\sin t(-2\sin t)\,dt - \int_0^\pi 2\cos t(3\cos t)\,dt = -6\int_0^\pi (\sin^2 t + \cos^2)\,dt$

$$= -6\int_0^\pi dt = -6\pi$$

Thus, $\displaystyle\int_{-C} y\,dx - x\,dy = 6\pi.$

27. We parameterize the line segment from $(0,0,0)$ to $(2,3,4)$ by $x = 2t,\ y = 3y,\ z = 4t$ for $0 \le t \le 1$. We parameterize the line segment from $(2,3,4)$ to $(6,8,5)$ by $x = 2+2t,\ y = 3+5t,\ z = 4+t,\ 0 \le t \le 1$.

$$\int_C y\,dx + z\,dy + x\,dz = \int_0^1 3t(2\,dt) + \int_0^1 4t(3\,dt) + \int_0^1 2t(4\,dt) + \int_0^1 (3+5t)(4\,dt)$$

$$\int_0^1 (4+t)(5\,dt) + \int_0^1 (2+4t)\,dt$$

$$= \int_0^1 (55t + 34)\,dt = \left(\frac{55}{2}t^2 + 34t\right)\Big|_0^1 = \frac{123}{2}$$

29. From $(0,0,0)$ to $(6,0,0)$ we use x as a parameter with $y = dy = 0$ and $z = dz = 0$. From $(6,0,0)$ to $(6,0,5)$ we use z as a parameter with $x = 6$ and $dx = 0$ and $y = dy = 0$. From $(6,0,5)$ to $(6,8,5)$ we use y as a parameter with $x = 6$ and $dz = 0$ and $z = 5$ and $dz = 0$.

$$\int_C y\,dx + z\,dy + z\,dz = \int_0^6 0\,dx + \int_0^5 6\,dz + \int_0^8 5\,dy = 70$$

31. $\displaystyle\int_C 10x\,dx - 2xy^2\,dy + 6xz\,dz = \int_0^1 10(t)\,dt - \int_0^1 2(t)(t^2)^2(2t)\,dt + \int_0^1 6(t)(t^3)(3t^2)\,dt$

$$= 5t^2\Big|_0^1 - 4t^6\Big|_0^1 + 18t^6\Big|_0^1$$

$$= 5 - 4 + 18 = 19$$

33. $\displaystyle\int_{C_1} y^2\,dx + xy\,dy = \int_0^1 (4t+2)^2 2\,dt + \int_0^1 (2t+1)(4t+2)4\,dt = \int_0^1 (64t^2 + 64t + 16)\,dt$

$$= \left(\frac{64}{3}t^3 + 32t^2 + 16t\right)\Big|_0^1 = \frac{64}{3} + 32 + 16 = \frac{208}{3}$$

$$\int_{C_2} y^2\,dx + xy\,dy = \int_1^{\sqrt{3}} 4y^4(2t)\,dt + \int_1^{\sqrt{3}} 2t^4(4t)\,dt = \int_1^{\sqrt{3}} 16t^5\,dt = \frac{8}{3}t^6\Big|_1^{\sqrt{3}} = 72 - \frac{8}{3} = \frac{208}{3}$$

$$\int_{C_3} y^2\,dx + xy\,dy = \int_e^{e^3} 4(\ln t)^2\frac{1}{t}\,dt + \int_e^{e^3} 2(\ln t)^2\frac{2}{t}\,dt = \int_e^{e^3} \frac{8}{t}(\ln t)^2\,dt = \frac{8}{3}(\ln t)^3\Big|_e^{e^3}$$

$$= \frac{8}{3}(27-1) = \frac{208}{3}$$

35. We are given $\rho = kx$. Then

$$m = \int_C \rho\,dx = \int_0^\pi kx\,ds = k\int_0^\pi (1+\cos t)\sqrt{\sin^2 t + \cos^2 t}\,dt = k\int_0^\pi (1+\cos t)\,dt$$

$$= k\,(t + \sin t)\big|_0^\pi = k\pi.$$

15.2 Line Integrals of Vector Fields

1.

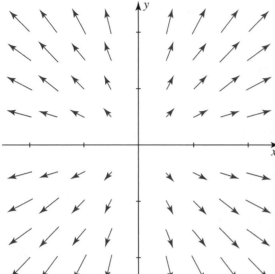

3.

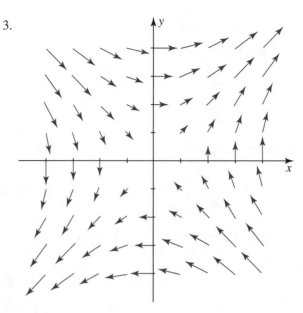

5.

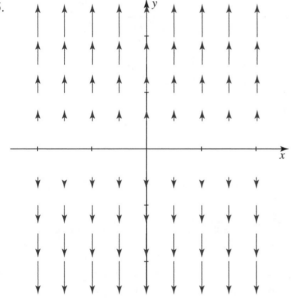

7. Since each vector points in a northeasterly direction, the vector field must have positive **i** and **j** components. Therefore, the answer is (b).

9. Since each vector points in a southwesterly direction, the vector field must have negative **i** and **j** components. Therefore, the answer is (d).

11. Note that the **k** component of each vector is always positive. Therefore, the answer is (d).

13. Note that each vector points directly away from the origin. Therefore, the answer is (a).

15.

$$\mathbf{F} = e^{3t}\mathbf{i} - (e^{-4t})e^{t}\mathbf{j} = e^{3t}\mathbf{i} - e^{-3t}\mathbf{j}; \quad d\mathbf{r} = (-2e^{-2t}\mathbf{i} + e^{t}\mathbf{j})dt; \quad \mathbf{F} \cdot d\mathbf{r} = (-2e^{t} - e^{-2t})dt;$$

$$\int_{c}\mathbf{F} \cdot d\mathbf{r} = \int_{0}^{\ln 2}(-2e^{t} - e^{-2t})dt = \left(-2e^{t} + \frac{1}{2}e^{-2t}\right)\Big|_{0}^{\ln 2} = -\frac{31}{8} - \left(-\frac{3}{2}\right) = -\frac{19}{8}$$

17. $\mathbf{F} = 2(2t-1)\mathbf{i} - 2(6t+1)\mathbf{j} = (4t-2)\mathbf{i} + (12t+2)\mathbf{j}; \quad d\mathbf{r} = (2\mathbf{i}+6\mathbf{j})\,dt; \mathbf{F}\cdot d\mathbf{r} = -64t - 16; \int_C \mathbf{F}\cdot d\mathbf{r} = \int_{-1}^{1}(-64t-16)$
$dt = -32t^2 - 16t\big|_{-1}^{1} = -32$

19. $\mathbf{F} = -3\sin t\mathbf{i} + 2\cos t\mathbf{j} + 6t\mathbf{k}; \quad d\mathbf{r} = (-2\sin t\mathbf{i} + 3\cos t\mathbf{j} + 3\mathbf{k})\,dt; \mathbf{F}\cdot d\mathbf{r} = (-6\sin^2 t + 6\cos^2 t + 18t)\,dt;$
$$\int_C \mathbf{F}\cdot d\mathbf{r} = \int_0^{\pi}(-6\sin^2 t + 6\cos^2 t + 18t)\,dt$$
$$= \int_0^{\pi} -6\left[\frac{1}{2}(1-\cos 2t)\right] + 6\left[\frac{1}{2}(1+\cos 2t)\right] + 18t\,dt \quad = 3\sin 2t + 9t^2\big|_0^{\pi}$$
$$= 9\pi^2 + 6\pi$$

21. Using x as a parameter, $\mathbf{r}(x) = x\mathbf{i} + \ln x\mathbf{j}$. Then $\mathbf{F} = \ln x\mathbf{i} + x\mathbf{j}, \quad d\mathbf{r} = (\mathbf{i} + \frac{1}{x}\mathbf{j})\,dx$, and
$$W = \int_C \mathbf{F}\cdot d\mathbf{r} = \int_1^{e}(\ln x + 1)\,dx = (x\ln x)\big|_1^{e} = e.$$

23. Let $\mathbf{r}_1 = (1+2t)\mathbf{i} + \mathbf{j}, \quad \mathbf{r}_2 = 3\mathbf{i} + (1+t)\mathbf{j}$, and $\mathbf{r}_3 = (3-2t)\mathbf{i} + (2-t)\mathbf{j}$ for $0 \le t \le 1$. Then

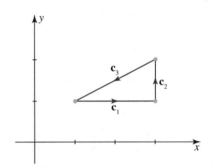

$$d\mathbf{r}_1 = 2\mathbf{i}, \quad d\mathbf{r}_2 = \mathbf{j}, \quad d\mathbf{r}_3 = -2\mathbf{i} - \mathbf{j},$$
$$\mathbf{F}_1 = (1+2t+2)\mathbf{i} + (6-2-4t)\mathbf{j} = (3+2t)\mathbf{i} + (4-4t)\mathbf{j},$$
$$\mathbf{F}_2 = (3+2+2t)\mathbf{i} + (6+6t-6)\mathbf{j} = (5+2t)\mathbf{i} + 6t\mathbf{j},$$
$$\mathbf{F}_3 = (3-2t+4-2t)\mathbf{i} + (12-6t-6+4t)\mathbf{j} = (7-4t)\mathbf{i} + (6-2t)\mathbf{j},$$

and
$$W = \int_{C_1} \mathbf{F}_1\cdot d\mathbf{r}_1 + \int_{C_2} \mathbf{F}_2\cdot d\mathbf{r}_2 + \int_{C_3} \mathbf{F}_3\cdot d\mathbf{r}_3$$
$$= \int_0^1 (6+4t)\,dt + \int_0^1 6t\,dt + \int_0^1 (-14+8t-6+2t)\,dt$$
$$= \int_0^1 (-14+20t)\,dt = (-14t+10t^2)\big|_0^1 = -4.$$

25. $\mathbf{r} = 3\cos t\mathbf{i} + 3\sin t\mathbf{j}, \quad 0 \le t \le 2\pi; \quad d\mathbf{r} = -3\sin t\mathbf{i} + 3\cos t\mathbf{j}; \quad \mathbf{F} = a\mathbf{i} + b\mathbf{j}; W = \int_C \mathbf{F}\cdot d\mathbf{r} = \int_0^{2\pi}(-3a\sin t + 3b\cos t)\,dt = (3a\cos t + 3b\sin t)\big|_0^{2\pi} = 0$

27. $\mathbf{F} = 10\cos t\mathbf{i} - 10\sin t\mathbf{j}; \quad d\mathbf{r} = (-5\sin t\mathbf{i} + 5\cos t\mathbf{j})\,dt; \mathbf{F}\cdot d\mathbf{r} = (-100\cos t\sin t)\,dt; \int_C \mathbf{F}\cdot d\mathbf{r} = \int_0^{2\pi}(-100\cos t\sin t)\,dt = 50\cos^2 t\big|_0^{2\pi} = 0$

29. On $C_1, \mathbf{T} = \mathbf{i}$ and $\mathbf{F}\cdot\mathbf{T} = \text{comp}_{\mathbf{T}}\mathbf{F} \approx 1$. On $C_2, \mathbf{T} = -\mathbf{j}$ and $\mathbf{F}\cdot\mathbf{T} = \text{comp}_{\mathbf{T}}\mathbf{F} \approx 2$. On $C_3, \mathbf{T} = -\mathbf{i}$ and $\mathbf{F}\cdot\mathbf{T} = \text{comp}_{\mathbf{T}}\mathbf{F} \approx 1.5$. Using the fact that the lengths of C_1, C_2, and C_3 are 4, 5, and 5, respectively, we have $W = \int_C \mathbf{F}\cdot\mathbf{T}ds = \int_{C_1}\mathbf{F}\cdot\mathbf{T}ds + \int_{C_2}\mathbf{F}\cdot\mathbf{T}ds + \int_{C_3}\mathbf{F}\cdot\mathbf{T}ds \approx 1(4)+2(5)+1.5(5)=21.5$ ft-lb.

31. $\nabla f(x,y) = \frac{1}{3}(3x-6y)3\mathbf{i} + \frac{1}{3}(3x-6y)(-6)\mathbf{j}$
$$= (3x-6y)\mathbf{i} + (-6x+12y)\mathbf{j}$$

33. $\nabla f(x,y,z) = \tan^{-1}zy\mathbf{i} + \frac{xz}{y^2z^2+1}\mathbf{j} + \frac{xy}{y^2z^2+1}\mathbf{k}$

35. $\nabla f(x,y,z) = e^{-y^2}\mathbf{i} + \left(1+2xye^{-y^2}\right)\mathbf{j} + \mathbf{k}$

37. $\nabla\left(x^2 + \frac{1}{2}y^2\right) = 2x\mathbf{i} + y\mathbf{j} = \mathbf{F}(x,y)$. Therefore, the answer is (b).

39. $\nabla\left(2x + \frac{1}{2}y^2 + 1\right) = 2\mathbf{i} + y\mathbf{j} = \mathbf{F}(x,y)$. Therefore, the answer is (d).

41. $\phi(x,y) = \sin x + y + \cos y$

43. $\phi(x,y) = x + y^2 - 4z^3$

45.

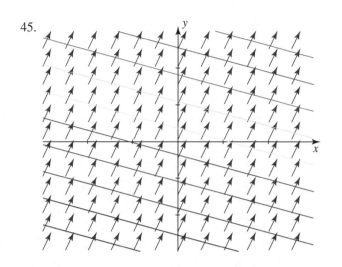

47.

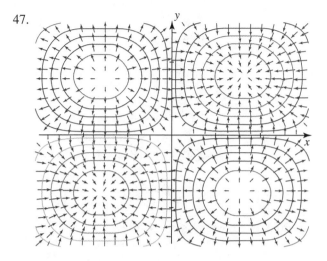

49.

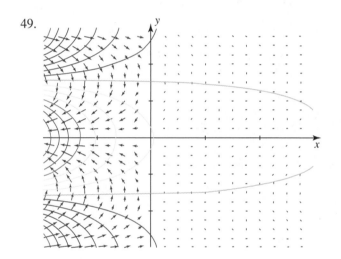

51. Let $\phi(x,y,z) = -c(x^2+y^2+z^2)^{-1/2}$. Then

$$\nabla\phi(x,y,z) = \frac{cx}{(x^2+y^2+z^2)^{3/2}}\mathbf{i} + \frac{cy}{(x^2+y^2+z^2)^{3/2}}\mathbf{j} + \frac{cz}{(x^2+y^2+z^2)^{3/2}}\mathbf{k}$$

$$= \frac{c(x\mathbf{i}+y\mathbf{j}+z\mathbf{k})}{(x^2+y^2+z^2)^{3/2}}$$

$$= \frac{c\mathbf{r}}{|\mathbf{r}|^3} = \mathbf{F}$$

15.3 Independence of the Path

1. (a) $P_y = 0 = Q_x$ and the integral is independent of path. $\phi_x = x^2$, $\phi = \frac{1}{3}x^3 + g(y)$, $\phi_y = g'(y) = y^2$, $g(y) = \frac{1}{3}y^3$,

$\phi = \frac{1}{3}x^3 + \frac{1}{3}y^3$, $\int_{(0,0)}^{(2,2)} x^2 dx + y^2 dy = \frac{1}{3}(x^3+y^3)\Big|_{(0,0)}^{(2,2)} = \frac{16}{3}$

(b) Use $y = x$ for $0 \le x \le 2$. $\int_{(0,0)}^{(2,2)} x^2 dx + y^2 dy = \int_0^2 (x^2+x^2)dx = \frac{2}{3}x^3\Big|_0^2 = \frac{16}{3}$

3. (a) $P_y = 2 = Q_x$ and the integral is independent of path. $\phi_x = x + 2y$, $\phi = \frac{1}{2}x^2 + 2xy + g(y)$, $\phi_y = 2x + g'(y) = 2x - y$,

$$g(y) = -\frac{1}{2}y^2, \quad \phi = \frac{1}{2}x^2 + 2xy - \frac{1}{2}y^2, \quad \int_{(1,0)}^{(3,2)}(x+2y)dx + (2x-y)dy = \left(\frac{1}{2}x^2 + 2xy - \frac{1}{2}y^2\right)\Big|_{(1,0)}^{(3,2)} = 14$$

(b) Use $y = x - 1$ for $1 \le x \le 3$.

$$\int_{(1,0)}^{(3,2)}(x+2y)dx + (2x-y)dy = \int_1^3 [x + 2(x-1) + 2x - (x-1)]dx$$

$$= \int_1^3 (4x-1)dx = (2x^2 - x)\Big|_1^3 = 14$$

5. (a) $P_y = 1/y^2 = Q_x$ and the integral is independent of path. $\phi_x = -\frac{1}{y}$, $\phi = -\frac{x}{y} + g(y)$, $\phi_y = \frac{x}{y^2} + g'(x) = \frac{x}{y^2}$, $g(y) = 0$,

$$\phi = -\frac{x}{y}, \quad \int_{(4,1)}^{(4,4)} -\frac{1}{y}dx + \frac{x}{y^2}dy = \left(-\frac{x}{y}\right)\Big|_{(4,1)}^{(4,4)} = 3$$

(b) Use $x = 4$ for $1 \le y \le 4$.

$$\int_{(4,1)}^{(4,4)} -\frac{1}{y}dx + \frac{x}{y^2}dy = \int_1^4 \frac{4}{y^2}dy = -\frac{4}{y}\Big|_1^4 = 3$$

7. (a) $P_y = 4xy = Q_x$ and the integral is independent of path. $\phi_x = 2y^2x - 3$, $\phi = x^2y^2 - 3x + g(y)$, $\phi_y = 2x^2y + g'(y) = 2x^2y + 4$, $g(y) = 4y$, $\phi = x^2y^2 - 3x + 4y$, $\int_{(1,2)}^{(3,6)}(2y^2x - 3)dx + (2yx^2 + 4)dy = (x^2y^2 - 3x + 4y)\Big|_{(1,2)}^{(3,6)} = 330$

(b) Use $y = 2x$ for $1 \le x \le 3$.

$$\int_{(1,2)}^{(3,6)}(2y^2x - 3)dx + (2yx^2 + 4)dy = \int_1^3 ([2(2x)^2x - 3] + [2(2x)x^2 + 4]2)dx$$

$$= \int_1^3 (16x^3 + 5)dx = (4x^4 + 5x)\Big|_1^3 = 330$$

9. (a) $P_y = 3y^2 + 3x^2 = Q_x$ and the integral is independent of path. $\phi_x = y^3 + 3x^2y$,
$\phi = xy^3 + x^3y + g(y)$, $\phi_y = 3xy^2 + x^3 + g'(y) = x^3 + 3y^2x + 1$, $g(y) = y$, $\phi = xy^3 + x^3y + y$,
$\int_{(0,0)}^{(2,8)}(y^3 + 3x^2y)dx + (x^3 + 3y^2x + 1)dy = .(xy^3 + x^3y + y)\Big|_{(0,0)}^{(2,8)} = 1096$

(b) Use $y = 4x$ for $0 \le x \le 2$.
$$\int_{(0,0)}^{(2,8)}(y^3 + 3x^2y)dx + (x^3 + 3y^2x + 1)dy = \int_0^2 [(64x^3 + 12x^3) + (x^3 + 48x^3 + 1)(4)]dx$$

$$= \int_0^2 (272x^3 + 4)dx = (68x^4 + 4x)\Big|_0^2 = 1096$$

11. $P_y = 12x^3y^2 = Q_x$ throughout the plane and the vector field is a conservative field. $\phi_x = 4x^3y^3 + 3$, $\phi = x^4y^3 + 3x + g(y)$, $\phi_y = 3x^4y^2 + g'(y) = 3x^4y^2 + 1$, $g(y) = y$, $\phi = x^4y^3 + 3x + y$

13. $P_y = -2xy^3 \sin xy^2 + 2y\cos xy^2$, $Q_x = -2xy^3 \cos xy^2 - 2y\sin xy^2$ throughout the plane and the vector is not a conservative field.

15. $P_y = 1 = Q_x$ throughout the plane and the vector field is a conservative field. $\phi_x = x^3 + y$, $\phi = \frac{1}{4}x^4 + xy + g(y)$, $\phi_y = x + g'(y) = x + y^3$, $g(y) = -\frac{1}{4}y^4$, $\phi = \frac{1}{4}x^4 + xy + \frac{1}{4}y^4$

17. $P_y = 0 = Q_x$, $P_x = 0 = R_x$, $Q_z = -1 = R_y$ throughout 3-space and the vector field is a conservative field.
$\phi_x = 2x$, $\phi = x^2 + g(y,z) \phi_y = \frac{\partial g}{\partial y} = 3y^2 - x$,
$g(y,z) = y^3 - yz + h(z), \phi = x^2 + y^3 - yz + h(z)$,
$\phi_z = -y + h'(z) = -y$, $h(z) = 0, \phi = x^2 + y^3 - yz$

19. Since $P_y = -e^{-y} = Q_x$, **F** is conservative and $\int_C \mathbf{F} \cdot d\mathbf{r}$ is independent of the path. Thus, instead of the given curve we may use the simpler curve $C_1 : y = x, \ 0 \leq x \leq 1$. Then

$$
\begin{aligned}
W &= \int_{C_1} (2x + e^{-y})dx + (4y - xe^{-y})dy \\
&= \int_0^1 (2x + e^{-x})dx + \int_0^1 (4x - e^{-x})dx \quad \boxed{\text{Integration by parts}} \\
&= (x^2 - e^{-x})\big|_0^1 + (2x^2 + xe^{-x})\big|_0^1 \\
&= [(1 - e^{-1}) - (-1)] + [(2 + e^{-1} + e^{-1}) - (1)] = 3 + e^{-1}.
\end{aligned}
$$

21. $P_y = z = Q_x, \ Q_z = x = R_y, \ R_x = y = P_z$, and the integral is independent of path. Parameterize the line segment between the points by $x = 1 + t, \ y = 1 + 3t, \ z = 1 + 7t$, for $0 \leq t \leq 1$. Then $dx = dt, \ dy = 3dt, \ dz = 7dt$ and

$$
\begin{aligned}
\int_{(1,1,1)}^{(2,4,8)} yz\,dx + xz\,dy + xy\,dz &= \int_0^1 [(1 + 3t)(1 + 7t) + (1 + t)(1 + 7t)(3) + (1 + t)(1 + 3t)(7)]dt \\
&= \int_0^1 (11 + 62t + 63t^2)dt = (11t + 31t^2 + 21t^3)\big|_0^1 = 63.
\end{aligned}
$$

23. $P_y = 2x \cos y = Q_x, \ Q_z = 0 = R_y, \ R_x = 3e^{3z} = P_z$, and the integral is independent of path. Integrating $\phi_x = 2x \sin y + e^{3z}$ we find $\phi = x^2 \sin y + xe^{3z} + g(y, z)$. Then $\phi_y = x^2 \cos y + g_y = Q = x^2 \cos y$, so $g_y = 0, \ g(y, z) = h(z)$, and $\phi = x^2 \sin y + xe^{3z} + h(z)$. Now $\phi_z = 3xe^{3z} + h'(z) = R = 3xe^{3z} + 5$, so $h'(z) = 5$ and $h(z) = 5z$. Thus $\phi = x^2 \sin y + xe^{3z} + 5z$ and

$$
\begin{aligned}
\int_{(1,0,0)}^{(2,\pi/2,1)} (2x \sin y + e^{3z})dx + x^2 \cos y\,dy + (3xe^{3z} + 5)dz \\
= (x^2 \sin y + xe^{3z} + 5z)\big|_{(1,0,0)}^{(2,\pi/2,1)} = [4(1) + 2e^3 + 5] - [0 + 1 + 0] = 8 + 2e^3.
\end{aligned}
$$

25. $P_y = 0 = Q_x; \ Q_z = 0 = R_y, \ R_x = 2e^{2z} = P_z$ and the integral is independent of path. Parameterize the line segment between the points by $x = 1 + t, \ y = 1 + t, \ z = \ln 3$, for $0 \leq t \leq 1$. Then $dx = dy = dt, \ dz = 0$ and

$$
\int_{(1,1,\ln 3)}^{(2,2,\ln 3)} e^{2z}dx + 3y^2 dy + 2xe^{2z}dz = \int_0^1 [e^{2\ln 3} + 3(1 + t)^2]dt = [9t + (1 + t)^3]\big|_0^1 = 16
$$

27. $P_y = 1 - z \sin x = Q_x, Q_z = \cos x = R_y, \ R_x = -y \sin x = P_z$ and the integral is independent of path. Integrating $\theta_x = y - yz \sin x$ we find $\theta = xy + yz \cos x + g(y, z)$. Then $\theta_y = x + z \cos x + g_y(y, z) = Q = x + z \cos x$, so $g_y = 0, \ g(y, z) = h(z)$, and $\theta = xy + yz \cos x + h(z)$. Now $\theta_z = y \cos x + h(z) = R = y \cos x$, so $h(z) = 0$ and $\theta = xy + yz \cos x$. Since $\mathbf{r}(0) = 4\mathbf{j}$ and $\mathbf{r}(\pi/2) = \pi \mathbf{i} + \mathbf{j} + 4\mathbf{k}$,

$$
\int_C \mathbf{F} \cdot d\mathbf{r} = (xy + yz \cos x)\big|_{(0,4,0)}^{(\pi,1,4)} = (\pi - 4) - (0 + 0) = \pi - 4.
$$

29. Since $P_y = Gm_1 m_2 (2xy/|r|^5) = Q_x, \ Q_z = Gm_1 m_2 (2yz/|r|^5) = R_y$, and $R_x = Gm_1 m_2 (2xz/|r|^5) = P_z$, the force field is conservative.

$$
\theta_x = -Gm_1 m_2 \frac{x}{(x^2 + y^2 + z^2)^{3/2}}, \quad \theta = Gm_1 m_2 (x^2 + y^2 + z^2)^{-1/2} + g(y, z),
$$

$$
\theta_y = -Gm_1 m_2 \frac{y}{(x^2 + y^2 + z^2)^{3/2}} + g_y(y, z) = -Gm_1 m_2 \frac{y}{(x^2 + y^2 + z^2)^{3/2}}, \quad g(y, z) = h(z),
$$

$$
\theta = Gm_1 m_2 (x^2 + y^2 + z^2)^{-1/2} + h(z),
$$

$$
\theta_z = -Gm_1 m_2 \frac{z}{(x^2 + y^2 + z^2)^{3/2}} + h'(z) = -Gm_1 m_2 \frac{z}{(x^2 + y^2 + z^2)^{3/2}},
$$

$$
h(z) = 0, \quad \theta = \frac{Gm_1 m_2}{\sqrt{x^2 + y^2 + z^2}} = \frac{Gm_1 m_2}{|r|}
$$

31. Since $\mathbf{F}$ is conservative, $\int_{C_1} \mathbf{F} \cdot d\mathbf{r} = \int_{-C_2} \mathbf{F} \cdot d\mathbf{r}$. Then, since the simply closed curve C is composed of C_1 and C_2,

$$\int_C \mathbf{F} \cdot d\mathbf{r} = \int_{C_1} \mathbf{F} \cdot d\mathbf{r} + \int_{C_2} \mathbf{F} \cdot d\mathbf{r} = \int_{C_1} \mathbf{F} \cdot d\mathbf{r} - \int_{-C_2} \mathbf{F} \cdot d\mathbf{r} = 0.$$

33. $P - y = -2x \sin y = Q_x$ throughout the plane and the vector field $\mathbf{F}$ is a conservative field. The path starts at point $(1, 0)$ and ends at point $(2, 1)$. Since $\mathbf{F}$ is conservative, the integral is path independent so we can use any path C starting at $(1, 0)$ and ending at $(2, 1)$. Use the path $y = x - 1$, $1 \le x \le 2$. Then

$$\int_C \mathbf{F} \cdot d\mathbf{r} = \int_1^2 (2x \cos(x - 1) - x^2 \sin(x - 1)) dx$$
$$= x^2 \cos(x - 1) \big|_1^2 = 4 \cos 1 - 1$$

35. $\mathbf{F}$ cannot be a conservative field in the region.

37. From Problem 45 in Exercises 15.2, $\dfrac{d\mathbf{v}}{dt} \cdot \dfrac{d\mathbf{r}}{dt} = \dfrac{d\mathbf{v}}{dt} \cdot \mathbf{v} = \dfrac{1}{2}\dfrac{d}{dt} v^2$. Then, using $\dfrac{dp}{dt} = \dfrac{\partial p}{\partial x}\dfrac{dx}{dt} + \dfrac{\partial p}{\partial y}\dfrac{dy}{dt} = \nabla p \cdot \dfrac{d\mathbf{r}}{dt}$, we have

$$\int m \frac{d\mathbf{v}}{dt} \cdot d\mathbf{r} dt dt + \int \nabla p \cdot \frac{d\mathbf{r}}{dt} = \int 0 dt$$
$$\frac{1}{2} m \int \frac{d}{dt} v^2 dt + \int \frac{dp}{dt} dt = \text{constant}$$
$$\frac{1}{2} mv^2 + p = \text{constant}.$$

15.4 Green's Theorem

1. The sides of the triangle are $C_1:\ y = 0,\ 0 \le x \le 1$; $C_2:\ x = 1, 0 \le y \le 3$; $C_3:\ y = 3x,\ 0 \le -x \le 1$.

$$\int_C (x - y)dx + xydy = \int_0^1 xdx + \int_0^3 ydy + \int_1^0 (x - 3x)dx + \int_1^0 x(3x)dx$$
$$= \left(\frac{1}{2}x^2\right)\bigg|_0^1 + \left(\frac{1}{2}y^2\right)\bigg|_0^3 + (-x^2)\big|_0^1 + (3x^2)\big|_1^0$$
$$= \frac{1}{2} + \frac{9}{2} + 1 - 3 = 3$$

$$\int\int_R (y + 1)dA = \int_0^1 \int_0^{3x} (y + 1)dydx = \int_0^1 \left(\frac{1}{2}y^2 + y\right)\bigg|_0^{3x} dx = \int_0^1 \left(\frac{9}{2}x^2 + 3x\right) dx$$
$$= \left(\frac{3}{2}x^3 + \frac{3}{2}x^2\right)\bigg|_0^1 = 3$$

3. $\displaystyle \int_C -y^2 dx + x^2 dy = \int_0^{2\pi} (-9\sin^2 t)(-3\sin t)dt + \int_0^{2\pi} 9\cos^2 t(3\cos t)dt$

$$= 27 \int_0^{2\pi} [(1 - \cos^2 t)\sin t + (1 - \sin^2 t)\cos t]dt$$
$$= 27 \left(-\cos t + \frac{1}{3}\cos^3 t + \sin t - \frac{1}{3}\sin^3 t\right)\bigg|_0^{2\pi} = 27(0) = 0$$

$$\int\int_R (2x + 2y)dA = 2\int_0^{2\pi}\int_0^3 (r\cos\theta + r\sin\theta)rdrd\theta = 2\int_0^{2\pi}\int_0^3 r^2(\cos\theta + \sin\theta)drd\theta$$
$$= 2\int_0^{2\pi} \left[\frac{1}{3}r^3(\cos\theta + \sin\theta)\right]\bigg|_0^3 d\theta = 18 \int_0^{2\pi} (\cos\theta + \sin\theta)d\theta$$
$$= 18(\sin\theta - \cos\theta)\big|_0^{2\pi} = 18(0) = 0$$

5. $P = 2y$, $P_y = 2$, $Q = 5x$, $Q_x = 5$

$$\int_C 2y\,dx + 5x\,dy = \int\int_R (5 - 2)dA$$
$$= 3\int\int_R dA = 3(25\pi) = 75\pi$$

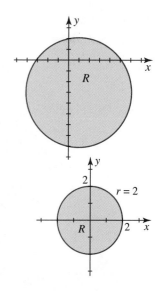

7. $P = x^4 - 2y^3$, $P_y = -6y^2$, $Q = 2x^3 - y^4$, $Q_x = 6x^2$. Using polar coordinates,

$$\int_C (x^4 - 2y^3)dx + (2x^3 - y^4)dy = \int\int_R (6x^2 + 6y^2)dA$$
$$= \int_0^{2\pi}\int_0^2 6r^2\,r\,dr\,d\theta$$
$$= \int_0^{2\pi}\left(\frac{3}{2}r^4\right)\Big|_0^2 d\theta = \int_0^{2\pi} 24\,d\theta = 48\pi.$$

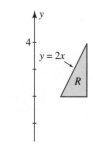

9. $P = 2xy$, $P_y = 2x$, $Q = 3xy^2$, $Q_x = 3y^2$

$$\int_C 2xy\,dx + 3xy^2\,dy = \int\int_R (3y^2 - 2x)dA = \int_1^2\int_2^{2x}(3y^2 - 2x)dy\,dx$$
$$= \int_1^2 (y^3 - 2xy)\Big|_2^{2x}dx = \int_1^2 (8x^3 - 4x^2 - 8 + 4x)dx$$
$$= \left(2x^4 - \frac{4}{3}x^3 - 8x + 2x^2\right)\Big|_1^2 = \frac{40}{3} - \left(-\frac{16}{3}\right) = \frac{56}{3}$$

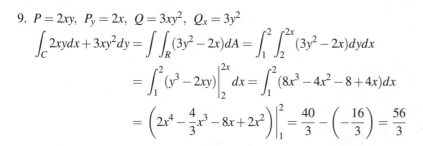

11. $P = xy$, $P_y = x$, $Q = x^2$, $Q_x = 2x$. Using polar coordinates,

$$\int_C xy\,dx + x^2\,dy = \int\int_R (2x - x)dA = \int_{-\pi/2}^{\pi/2}\int_0^1 r\cos\theta\,r\,dr\,d\theta$$
$$= \int_{-\pi/2}^{\pi/2}\left(\frac{1}{3}r^3\cos\theta\right)\Big|_0^1 d\theta = \int_{-\pi/2}^{\pi/2}\frac{1}{3}\cos\theta\,d\theta$$
$$= \frac{1}{3}\sin\theta\Big|_{-\pi/2}^{\pi/2} = \frac{2}{3}$$

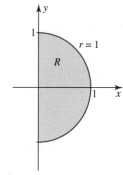

13. $P = \frac{1}{3}y^3$, $P_y = y^2$, $Q = xy + xy^2$, $Q_x = y + y^2$

$$\int_C \frac{1}{3}y^3\,dx + (xy + xy^2)dy = \int\int_R y\,dA = \int_0^{1/\sqrt{2}}\int_{y^2}^{1-y^2} y\,dx\,dy$$
$$= \int_0^{1/\sqrt{2}}(xy)\Big|_{y^2}^{1-y^2}dy$$
$$= \int_0^{1/\sqrt{2}}(y - y^2 - y^3)dy$$
$$= \left(\frac{1}{2}y^2 - \frac{1}{2}y^4\right)\Big|_0^{1/\sqrt{2}} = \frac{1}{4} - \frac{1}{8} = \frac{1}{8}$$

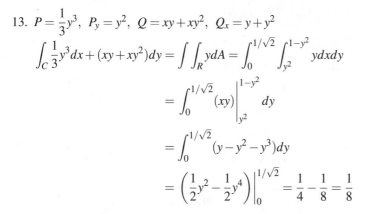

15. $P = ay$, $P_y = a$, $Q = bx$, $Q_x = b$.
$\int_C ay\,dx + bx\,dy = \int\int_R (b - a)dA = (b - a) \times$ (area bounded by C)

17. For the first integral: $P = 0$, $P_y = 0$, $Q = x$, $Q_x = 1$; $\int_C x\,dy = -\int\int_R 1\,dA =$ area of R. For the second integral: $P = y$, $P_y = 1$, $Q = 0$, $Q_x = 0$; $-\int_C y\,dx = -\int\int_R -dA =$ area of R. Thus, $\int_C x\,dy = -\int_C y\,dx$.

19. $A = \iint_R dA = \int_C x\,dy = \int_0^{2\pi} a\cos^3 t(3a\sin^2 t\cos t\,dt) = 3a^2\int_0^{2\pi}\sin^2 t\cos^4 t\,dt$

$$= 3a^2\left(\frac{1}{16}t - \frac{1}{64}\sin 4t + \frac{1}{48}\sin^3 2t\right)\Big|_0^{2\pi} = \frac{3}{8}\pi a^2$$

21. (a) Parameterize C by $x = x_1 + (x_2 - x_1)t$ and $y = y_1 + (y_2 - y_1)t$ for $0 \le t \le 1$. Then

$$\int_C -y\,dx + x\,dy = \int_0^1 -[y_1 + (y_2 - y_1)t](x_2 - x_1)\,dt + \int_0^1 [x_1 + (x_2 - x_1)t](y_2 - y_1)\,dt$$

$$= -(x_2 - x_1)[y_1 t + \frac{1}{2}(y_2 - y_1)t^2]\Big|_0^1 + (y_2 - y_1)[x_1 t + \frac{1}{2}(x_2 - x_1)t^2]\Big|_0^1$$

$$= -(x_2 - x_1)\left[y_1 + \frac{1}{2}(y_2 - y_1)\right] + (y_2 - y_1)\left[x_1 + \frac{1}{2}(x_2 - x_1)\right]$$

$$= x_1 y_2 - x_2 y_1.$$

(b) Let C_i be the line segment from (x_i, y_i) to (x_{i+1}, y_{i+1}) for $i = 1, 2, \cdots, n-1$, and C_2 the line segment from (x_n, y_n) to (x_1, y_1). Then

$$A = \frac{1}{2}\int_C -y\,dx + x\,dy \quad \boxed{\text{Problem 18}}$$

$$= \frac{1}{2}\left[\int_{C_1} -y\,dx + x\,dy + \int_{C_2} -y\,dx + x\,dy + \cdots + \int_{C_{n-1}} -y\,dx + x\,dy + \int_{C_n} -y\,dx + x\,dy\right]$$

$$= \frac{1}{2}(x_1 y_2 - x_2 y_1) + \frac{1}{2}(x_2 y_3 - x_3 y_2) + \frac{1}{2}(x_{n-1} y_n - x_n y_{n-1}) + \frac{1}{2}(x_n y_1 - x_1 y_n).$$

23. $P = 4x^2 - y^3$, $P_y = -3y^2$; $Q = x^3 + y^2$, $Q_x = 3x^2$.

$$\int_C (4x^2 - y^3)\,dx + (x^3 + y^2)\,dy = \iint_R (3x^2 + 3y^2)\,dA = \int_0^{2\pi}\int_1^2 3r^2(r\,dr\,d\theta) = \int_0^{2\pi}\left(\frac{3}{4}r^4\right)\Big|_1^2 d\theta$$

$$= \int_0^{2\pi}\frac{45}{4}\,d\theta = \frac{45\pi}{2}$$

25. We first observe that $P_y + (y^4 - 3x^2 y^2)/(x^2 + y^2)^3 = Q_x$. Letting C' be the circle $x^2 + y^2 = \frac{1}{4}$ we have

$$\int_C \frac{-y^3\,dx + xy^2\,dy}{(x^2 + y^2)^2} = \int_{C'} \frac{-y^3\,dx + xy^2\,dy}{(x_2 + y^2)^2}$$

$$\boxed{x = \frac{1}{4}\cos t,\ dx = -\frac{1}{4}\sin t\,dt,\ y = \frac{1}{4}\sin t,\ dy = \frac{1}{4}\cos t\,dt}$$

$$= \int_0^{2\pi} \frac{-\frac{1}{64}\sin^3 t\left(-\frac{1}{4}\sin t\,dt\right) + \frac{1}{4}\cos t\left(\frac{1}{16}\sin^2 t\right)\left(\frac{1}{4}\cos t\,dt\right)}{1/256}$$

$$= \int_0^{2\pi}(\sin^4 t + \sin^2 t\cos^2 t)\,dt = \int_0^{2\pi}(\sin^4 t + (\sin^2 t - \sin^4 t)\,dt$$

$$= \int_0^{2\pi}\sin^2 t\,dt = \left(\frac{1}{2}t - \frac{1}{4}\sin 2t\right)\Big|_0^{2\pi} = \pi$$

27. Writing $\iint_R x^2\,dA = \iint_R(Q_x - P_y)\,dA$ we identify $Q = 0$ and $P = -x^2 y$. Then, with $C:\ x = 3\cos t, y = 2\sin t,\ 0 \le t \le 2\pi$, we have

$$\iint_R x^2\,dA = \int_C P\,dx + Q\,dy = \int_C -x^2 y\,dx = -\int_0^{2\pi} 9\cos^2 t(2\sin t)(-3\sin t)\,dt$$

$$= \frac{54}{4}\int_0^{2\pi} 4\sin^2 t\cos^2 t\,dt = \frac{27}{2}\int_0^{2\pi}\sin^2 2t\,dt = \frac{27}{4}\int_0^{2\pi}(1 - \cos 4t)\,dt$$

$$= \frac{27}{4}\left(t - \frac{1}{4}\sin 4t\right)\Big|_0^{2\pi} = \frac{27\pi}{2}.$$

29. $P = x - y$, $P_y = -1$, $Q = x + y$, $Q_x = 1$;

$W = \int_C \mathbf{F} \cdot d\mathbf{r} = \int \int_R 2 \, dA = 2 \times \text{area} = 2(\frac{3\pi}{4}) = \frac{3}{2}\pi$

31. Let $P = 0$ and $W = x^2$. Then $Q_x - P_y = 2x$ and

$$\frac{1}{2A} \oint_C x^2 \, dy = \frac{1}{2A} \int \int_R 2x \, dA = \frac{\int \int_R x \, dA}{A} = \bar{x}.$$

Let $P = y^2$ and $Q = 0$. Then $Q_x - P_y = -2y$ and

$$-\frac{1}{2A} \oint_C y^2 \, dx = -\frac{1}{2A} \oint_C \int \int_R -2y \, dA = \frac{\int \int_R y \, dA}{A} = \bar{y}.$$

33. Since $\int_A^B P \, dx + Q \, dy$ is independent of path, $P_y = Q_x$ by Theorem 17.3. Then, by Green's Theorem

$$\int_C P \, dx + Q \, dy = \int \int_R (Q_x - P_y) \, dA = \int \int_R 0 \, dA = 0.$$

15.5 Parametric Surfaces and Area

1. $x = u$, $y = v$, $z = 4u + 3v - 2$, $-\infty < u < \infty$, $-\infty < v < \infty$

3. $x = u$, $y = -\sqrt{1 + u^2 + v^2}$, $z = v$, $-\infty < u < \infty$, $-\infty < v < \infty$

5. $\mathbf{r}(u, v) = u\mathbf{i} + v\mathbf{j} + (1 - v^2)\mathbf{k}$, $-2 \le u \le 2$, $-3 \le v \le 3$

7. $x^2 + y^2 = \cos^2 u + \sin^2 u = 1$, circular cylinder

9. $x = \sin u$, $y = \sin u \cos v$, $z = \sin u \sin v$
   $y^2 + z^2 = \sin^2 u \cos^2 v + \sin^2 u \sin^2 v = \sin^2(\cos^2 v + \sin^2 v) = \sin^2 u = z^2$,
   so $x^2 = y^2 + z^2$, portion of a circular cone

11. Surface is parameterzied by $x = u$, $y = \sin v$, $z = \cos v$ so R is defined by $0 \le u \le 4$, $0 \le v \le \frac{\pi}{2}$

13. Surface is parameterzied by $x = \sin \phi \cos \theta$, $y = \sin \phi \sin \theta$, $z = \cos \phi$ so R is defined by $0 \le \theta \le 2\pi, \frac{\pi}{2} \le \phi \le \pi$

15. At $u = \pi/6$, $v = 2$, we have $x = 5$, $y = 5\sqrt{3}$, $z = 2$.
    $\frac{\partial x}{\partial u}(\frac{\pi}{6}, 2) = 5\sqrt{3}$, $\frac{\partial y}{\partial u}(\frac{\pi}{6}, 2) = -5$, $\frac{\partial z}{\partial u}(\frac{\pi}{6}, 2) = 0$
    $\frac{\partial x}{\partial v}(\frac{\pi}{6}, 2) = 0$, $\frac{\partial y}{\partial v}(\frac{\pi}{6}, 2) = 0$, $\frac{\partial z}{\partial v}(\frac{\pi}{6}, 2) = 1$.

    A normal vector is given by $\mathbf{n} = \begin{vmatrix} \mathbf{i} & \mathbf{j} & \mathbf{k} \\ 5\sqrt{3} & -5 & 0 \\ 0 & 0 & 1 \end{vmatrix} = -5\mathbf{i} - 5\sqrt{3}\mathbf{j}$.

    The tangent plane is $-5(x - 5) - 5\sqrt{3}(y - 5\sqrt{3}) = 0$ or $x + \sqrt{3}y = 20$.

17. At $u = 1$, $v = 2$, we have $x = 3$, $y = 3$, $z = -3$.
    $\frac{\partial x}{\partial u}(1, 2) = 2$, $\frac{\partial y}{\partial u}(1, 2) = 1$, $\frac{\partial z}{\partial u}(1, 2) = 2$
    $\frac{\partial x}{\partial v}(1, 2) = 1$, $\frac{\partial y}{\partial v}(1, 2) = 1$, $\frac{\partial z}{\partial v}(1, 2) = -4$.

    A normal vector is given by $\mathbf{n} = \begin{vmatrix} \mathbf{i} & \mathbf{j} & \mathbf{k} \\ 2 & 1 & 2 \\ 1 & 1 & -4 \end{vmatrix} = -6\mathbf{i} + 10\mathbf{j} + \mathbf{k}$.

    The tangent plane is $-6(x - 3) + 10(y - 3) + (z + 3) = 0$ or $-6x + 10y + z = 9$.

19. At $u = 3$, $v = 3$, we have $x = 3$, $y = 3$, $z = 9$.
    $\frac{\partial x}{\partial u}(3, 3) = 1$, $\frac{\partial y}{\partial u}(3, 3) = 0$, $\frac{\partial z}{\partial u}(3, 3) = 3$
    $\frac{\partial x}{\partial v}(3, 3) = 0$, $\frac{\partial y}{\partial v}(3, 3) = 1$, $\frac{\partial z}{\partial v}(3, 3) = 3$.

A normal vector is given by $\mathbf{n} = \begin{vmatrix} \mathbf{i} & \mathbf{j} & \mathbf{k} \\ 1 & 0 & 3 \\ 0 & 1 & 3 \end{vmatrix} = -3\mathbf{i} - 3\mathbf{j} + \mathbf{k}$.

The tangent plane is $-3(x-3) - 3(y-3) + (z-9) = 0$ or $3x + 3y - z = 9$.

21. At $u = -2$, $v = 1$, we have $x = -1$, $y = 3$, $z = -2$.

$\dfrac{\partial x}{\partial u}(-2,1) = 1$, $\dfrac{\partial y}{\partial u}(-2,1) = -1$, $\dfrac{\partial z}{\partial u}(-2,1) = 1$

$\dfrac{\partial x}{\partial v}(-2,1) = 1$, $\dfrac{\partial y}{\partial v}(-2,1) = 1$, $\dfrac{\partial z}{\partial v}(-2,1) = -2$.

A normal vector is given by $\mathbf{n} = \begin{vmatrix} \mathbf{i} & \mathbf{j} & \mathbf{k} \\ 1 & -1 & 1 \\ 1 & 1 & -2 \end{vmatrix} = \mathbf{i} + 3\mathbf{j} + 2\mathbf{k}$.

The tangent plane is $(x+1) + 3(y-3) + 2(z+2) = 0$ or $x + 3y + 2z = 4$.

23. At $(1,7,5)$, we have $u = 2$, $v = 1$.

$\dfrac{\partial x}{\partial u}(2,1) = 1$, $\dfrac{\partial y}{\partial u}(2,1) = 2$, $\dfrac{\partial z}{\partial u}(2,1) = 4$

$\dfrac{\partial x}{\partial v}(2,1) = -1$, $\dfrac{\partial y}{\partial v}(2,1) = 3$, $\dfrac{\partial z}{\partial v}(2,1) = 2$.

A normal vector is given by $\mathbf{n} = \begin{vmatrix} \mathbf{i} & \mathbf{j} & \mathbf{k} \\ 1 & 2 & 4 \\ -1 & 3 & 2 \end{vmatrix} = -8\mathbf{i} - 6\ln 3\mathbf{j} + 5\mathbf{k}$.

The tangent plane is $-8(x-1) - 6(y-7) + 5(z-35) = 0$ or $8x + 6y - 5z = 25$.

25. $\dfrac{\partial \mathbf{r}}{\partial u} = \langle 2,1, \rangle$, $\dfrac{\partial \mathbf{r}}{\partial v} = \langle -1,1,0 \rangle$

$\dfrac{\partial \mathbf{r}}{\partial u} \times \dfrac{\partial \mathbf{r}}{\partial v} = \begin{vmatrix} \mathbf{i} & \mathbf{j} & \mathbf{k} \\ 2 & 1 & 1 \\ -1 & 1 & 0 \end{vmatrix} = \langle -1,-1,3 \rangle$

$\left| \dfrac{\partial \mathbf{r}}{\partial u} \times \dfrac{\partial \mathbf{r}}{\partial v} \right| = \sqrt{1+1+9} = \sqrt{11}$

$A = \int_0^2 \int_{-1}^1 \sqrt{11}\, dv\, du = 4\sqrt{11}$

27. $\dfrac{\partial \mathbf{r}}{\partial u} = \langle 1,0,2u \rangle$, $\dfrac{\partial \mathbf{r}}{\partial v} = \langle 0,1,2v \rangle$

$\dfrac{\partial \mathbf{r}}{\partial u} \times \dfrac{\partial \mathbf{r}}{\partial v} = \begin{vmatrix} \mathbf{i} & \mathbf{j} & \mathbf{k} \\ 1 & 0 & 2u \\ 0 & 1 & 2v \end{vmatrix} = \langle -2u,-2v,1 \rangle$

$\left| \dfrac{\partial \mathbf{r}}{\partial u} \times \dfrac{\partial \mathbf{r}}{\partial v} \right| = \sqrt{4u^2 + 4v^2 + 1}$

Since $0 \le z \le 4$, we have $0 \le u^2 + v^2 \le 4$. So

$A = \int_{-2}^2 \int_{-\sqrt{4-u^2}}^{\sqrt{4-u^2}} \sqrt{4u^2 + 4v^2 + 1}\, dv\, du$

$= \int_{-2}^2 \dfrac{v}{2}\sqrt{4u^2 + 4v^2 + 1} + \dfrac{(4u^2+1)}{4} \ln\left|2v + \sqrt{4u^2 + 4v^2 + 1}\right| \Big|_{-\sqrt{4-u^2}}^{\sqrt{4-u^2}}$

$= \dfrac{1}{4}\left[2(4u^2+1)\ln\left|2\sqrt{4-u^2} + \sqrt{17}\right| - (4u^2+1)\ln(4u^2+1) + 4\sqrt{-17(u^2-4)}\right]$

$= \int_0^{2\pi} \int_0^2 \sqrt{4r^2 + 1}\, r\, dr\, d\theta$ [polar transformation]

$= \int_0^{2\pi} \dfrac{(4r^2+1)^{3/2}}{12} \Big|_0^2\, d\theta = \int_0^{2\pi} \dfrac{17\sqrt{17}-1}{12}\, d\theta$

$= \dfrac{17\sqrt{17}-1}{12} \theta \Big|_0^{2\pi} = \dfrac{(17\sqrt{17}-1)\pi}{6}$

29. $\mathbf{r} = (r\cos\theta)\mathbf{i} + (r\sin\theta)\mathbf{j} + \theta\mathbf{k}$

$\dfrac{\partial \mathbf{r}}{\partial r} = \langle \cos\theta, \sin\theta, 0 \rangle, \quad \dfrac{\partial \mathbf{r}}{\partial \theta} = \langle -r\sin\theta, r\cos\theta, f \rangle$

$\dfrac{\partial \mathbf{r}}{\partial r} \times \dfrac{\partial \mathbf{r}}{\partial \theta} = \begin{vmatrix} \mathbf{i} & \mathbf{j} & \mathbf{k} \\ \cos\theta & \sin\theta & 0 \\ -r\sin\theta & r\cos\theta & 1 \end{vmatrix} = \langle \sin\theta, -\cos\theta, r \rangle$

$\left| \dfrac{\partial \mathbf{r}}{\partial r} \times \dfrac{\partial \mathbf{r}}{\partial \theta} \right| = \sqrt{\sin^2\theta + \cos^2\theta + r^2} = \sqrt{1 + r^2}.$

$A = \int_0^{2\pi} \int_0^2 \sqrt{1 + r^2} \, dr \, d\theta = 2\sqrt{5}\pi + \pi\ln(2 + \sqrt{5})$

31. We have $a = 2$, so $x = 2\sin\phi\cos\theta$, $y = 2\sin\phi\sin\theta$, $z = 2\cos\phi$, $\frac{\pi}{3} \le \phi \le \pi$, $0 \le \theta \le 2\pi$,

$A = \int_{\pi/3}^{\pi} \int_0^{2\pi} 4\sin\phi \, d\theta \, d\phi = 12\pi$

33. $x = 2\sin\phi\cos\theta$, $y = 2\sin\phi\sin\theta$, $z = 2\cos\phi$, $0 \le \phi \le \frac{\pi}{4}$, $0 \le \theta \le 2\pi$,

$A = \int_0^{\pi/4} \int_0^{2\pi} 4\sin\phi \, d\theta \, d\phi = 4\pi(2 - \sqrt{2})$

35.   (a)   (b)   (c)

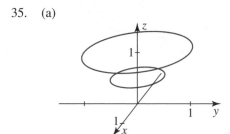 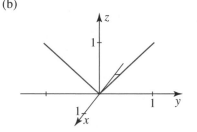 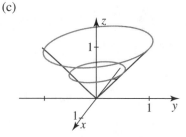

37. (f)

39. (d)

41. (c)

43.

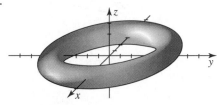

45. $x = 2u$, $y = 2v$, $z = 8u + 6v - 2$, $-\infty < u < \infty$, $-\infty < v < \infty$

47. We have $\mathbf{r} = u\mathbf{i} + f(u)\cos v\mathbf{j} + f(u)\sin v\mathbf{k}$.

$\dfrac{\partial r}{\partial u} = \langle 1, f'(u)\cos v, f'(u)\sin v \rangle, \quad \dfrac{\partial r}{\partial v} = \langle 0, -f(u)\sin v, f(u)\cos v \rangle,$

$\dfrac{\partial \mathbf{r}}{\partial u} \times \dfrac{\partial \mathbf{r}}{\partial v} = \begin{vmatrix} \mathbf{i} & \mathbf{j} & \mathbf{k} \\ 1 & f'(u)\cos v & f'(u)\sin v \\ 0 & -f(u)\sin v & f(u)\cos v \end{vmatrix}$

$\qquad = \langle f(u)f'(u), -f(u)\cos v \qquad -f(u)\sin v \rangle$

$\left| \dfrac{\partial \mathbf{r}}{\partial u} \times \dfrac{\partial \mathbf{r}}{\partial v} \right| = \sqrt{[f(u)f'(u)]^2 + [f(u)]^2\cos^2 v + [f(u)]^2\sin^2 v}$

$\qquad\qquad = \sqrt{[f(u)]^2[f'(u)]^2 + [f(u)]^2}$

$\qquad\qquad = f(u)\sqrt{1 + [f'(u)]^2}$

By (11), $A = \int_a^b \int_0^{2\pi} f(u)\sqrt{1 + [f'(u)]^2} \, dv \, du = 2\pi \int_a^b f(u)\sqrt{1 + [f'(u)]^2} \, du$

49. The surface is a plane passing through the point (x_0, y_0, z_0) with a normal vector $\mathbf{n} = \mathbf{v}_1 \times \mathbf{v}_2$.

15.6 Surface Integrals

1. $z_x = -2x$, $z_y = 0$; $dS = \sqrt{1+4x^2}\,dA$

$$\iint_S x\,dS = \int_0^4 \int_0^{\sqrt{2}} x\sqrt{1+4x^2}\,dx\,dy = \int_0^4 \frac{1}{12}(1+4x^2)^{3/2}\Big|_0^{\sqrt{2}}\,dy$$

$$= \int_0^4 \frac{13}{6}\,dy = \frac{26}{3}$$

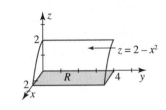

3. $z_x = \dfrac{x}{\sqrt{x^2+y^2}}$, $z_y = \dfrac{y}{\sqrt{x^2+y^2}}$; $dS = \sqrt{2}\,dA$.

Using polar coordinates,

$$\iint_S xz^3\,dS = \iint_R x(x^2+y^2)^{3/2}\sqrt{2}\,dA$$

$$= \sqrt{2}\int_0^{2\pi}\int_0^1 (r\cos\theta)r^{3/2}\,r\,dr\,d\theta$$

$$= \sqrt{2}\int_0^{2\pi}\int_0^1 r^{7/2}\cos\theta\,dr\,d\theta$$

$$= \sqrt{2}\int_0^{2\pi}\frac{2}{9}r^{9/2}\cos\theta\Big|_0^1\,d\theta$$

$$= \sqrt{2}\int_0^{2\pi}\frac{2}{9}\cos\theta\,d\theta = \frac{2\sqrt{2}}{9}\sin\theta\Big|_0^{2\pi} = 0.$$

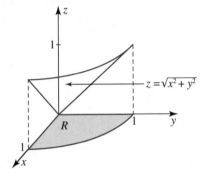

5. $z = \sqrt{36-x^2-y^2}$, $z_x = -\dfrac{x}{\sqrt{36-x^2-y^2}}$,

$z_y = -\dfrac{y}{36-x^2-y^2}$;

$$dS = \sqrt{1+\frac{x^2}{36-x^2-y^2}+\frac{y^2}{36-x^2-y^2}}\,dA$$

$$= \frac{6}{\sqrt{36-x^2-y^2}}\,dA.$$

Using polar coordinates,

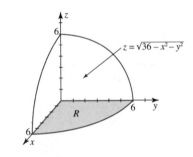

$$\iint_S (x^2+y^2)z\,dS = \iint_R (x^2+y^2)\sqrt{36-x^2-y^2}\frac{6}{\sqrt{36-x^2=y^2}}\,dA = 6\int_0^{2\pi}\int_0^6 r^2\,r\,dr\,d\theta$$

$$= 6\int_0^{2\pi}\frac{1}{4}r^4\Big|_0^6\,d\theta = 6\int_0^{2\pi}324\,d\theta = 972\pi.$$

7. $z_x = -x$, $z_y = -y$; $dS = \sqrt{1+x^2+y^2}\,dA$

$$\iint_S xy\,dS = \int_0^1\int_0^1 xy\sqrt{1+x^2+y^2}\,dx\,dy$$

$$= \int_0^1 \frac{1}{3}y(1+x^2+y^2)^{3/2}\Big|_0^1\,dy$$

$$= \int_0^1 \left[\frac{1}{3}y(2+y^2)^{3/2} - \frac{1}{3}y(1+y^2)^{3/2}\right]dy$$

$$= \left[\frac{1}{15}(2+y^2)^{5/2} - \frac{1}{15}(1+y^2)^{5/2}\right]\Big|_0^1$$

$$= \frac{1}{15}(3^{5/2} - 2^{7/2} + 1)$$

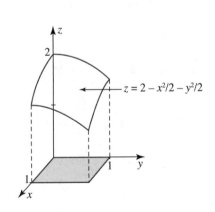

9. $y_x = 2x$, $y_z = 0$; $dS = \sqrt{1 + 4x^2}\, dA$

$$\int\int_S 24\sqrt{yz}\, dS = \int_0^3 \int_0^2 24xz\sqrt{1+4x^2}\, dx\, dz$$

$$= \int_0^3 2z(1+4x^2)^{3/2}\Big|_0^2\, dz$$

$$= 2(17^{3/2}-1)\int_0^3 z\, dz = 2(17^{3/2}-1)\left(\frac{1}{2}z^2\right)\Big|_0^3$$

$$= 9(17^{3/2}-1)$$

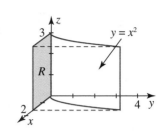

11. Write the equation of the surface as $y = \frac{1}{2}(6-x-3z)$.

$y_z = -\frac{1}{2}$, $y_z = -\frac{3}{2}$; $dS = \sqrt{1+1/4+9/4} = \frac{\sqrt{14}}{2}$.

$$\int\int_S (3z^2+4yz)\, dS = \int_0^2 \int_0^{6-3z}\left[3z^2+4z\frac{1}{2}(6-x-3z)\right]\frac{\sqrt{14}}{2}\, dx\, dz$$

$$= \frac{\sqrt{14}}{2}\int_0^2 \left[3z^2 x - z(6-x-3z)^2\right]\Big|_0^{6-3z}\, dz$$

$$= \frac{\sqrt{14}}{2}\int_0^2 [3z^2(6-3z)-0]-[0-z(6-3z)^2]\, dz$$

$$= \frac{\sqrt{14}}{2}\int_0^2 (36z-18z^2)\, dz = \frac{\sqrt{14}}{2}(18z^2-6z^3)\Big|_0^2 = \frac{\sqrt{14}}{2}(72-48) = 12\sqrt{14}$$

13. The density is $\rho = kx^2$. The surface is $z = 1-x-y$. Then $z_x = -1$, $z_y = -1$; $dS = \sqrt{3}\, dA$.

$$m = \int\int_S kx^2\, dS = k\int_0^1 \int_0^{1-x} x^2\sqrt{3}\, dy\, dx = \sqrt{3}k\int_0^1 \frac{1}{3}x^3\Big|_0^{1-x}\, dx$$

$$= \frac{\sqrt{3}}{3}k\int_0^1 (1-x)^3\, dx = \frac{\sqrt{3}}{3}k\left[-\frac{1}{4}(1-x)^4\right]\Big|_0^1 = \frac{\sqrt{3}}{12}k$$

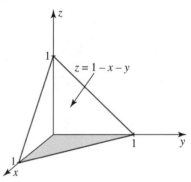

15. The surface is $g(x,y,z) = y^2 + z^2 - 4 = 0$. $\nabla g = 2y\mathbf{j} + 2z\mathbf{k}$,

$|\nabla g| = 2\sqrt{y^2+z^2}$; $\nabla \frac{y\mathbf{i}+z\mathbf{k}}{\sqrt{y^2+z^2}}$;

$$\mathbf{F}\cdot\nabla = \frac{2yz}{\sqrt{y^2+z^2}} + \frac{yz}{\sqrt{y^2+z^2}} = \frac{3yz}{\sqrt{y^2+z^2}};;\quad z = \sqrt{4-y^2},\ z_x = 0,$$

$$z_y = -\frac{y}{\sqrt{4-y^2}};\quad dS = \sqrt{1+\frac{y^2}{4-y^2}}\, dA = \frac{2}{\sqrt{4-y^2}}\, dA$$

$$\text{Flux} = \int\int_S \mathbf{F}\cdot\mathbf{n}\, dS = \int\int_R \frac{3yz}{\sqrt{y^2+z^2}}\frac{2}{\sqrt{4-y^2}}\, dA$$

$$= \int\int_R \frac{3y\sqrt{4-y^2}}{\sqrt{y^2+4-y^2}}\frac{2}{\sqrt{4-y^2}}\, dA$$

$$= \int_0^3 \int_0^2 3y\, dy\, dx = \int_0^3 \frac{3}{2}y^2\Big|_0^2\, dx = \int_0^3 6\, dx = 18$$

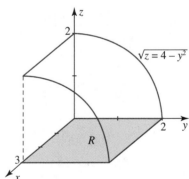

17. From Problem 16, $\mathbf{n} = \dfrac{2x\mathbf{i}+2y\mathbf{j}+\mathbf{k}}{\sqrt{1+4x^2+4y^2}}$. Then $\mathbf{F}\cdot\mathbf{n} = \dfrac{2x^2+2y^2+z}{\sqrt{1+4x^2+4y^2}}$. Also, from Problem 16, $dS = \sqrt{1+4x^2+4y^2}\, dA$. Using polar coordinates,

$$\text{Flux} = \int\int_S \mathbf{F} \cdot \mathbf{n}\,dS = \int\int_R \frac{2x^2 + 2y^2 + z}{\sqrt{1 + 4x^2 + 4y^2}}\sqrt{1 + 4x^2 + 4y^2}\,dA = \int\int_R (2x^2 + 2y^2 + 5 - x^2 - y^2)\,dA$$

$$= \int_0^{2\pi}\int_0^2 (r^2 + 5)r\,dr\,d\theta = \int_0^{2\pi}\left(\frac{1}{4}r^4 + \frac{5}{2}r^2\right)\Big|_0^2\,d\theta = \int_0^{2\pi} 14\,d\theta = 28\pi.$$

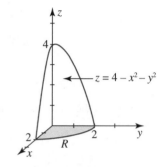

19. The surface is $g(x,y,z) = x^2 + y^2 + z - 4$. $\nabla g = 2x\mathbf{i} + 2y\mathbf{j} + \mathbf{k}$, $|\nabla g| = \sqrt{4x^2 + 4y^2 + 1}$;

$$\mathbf{n} = \frac{2x\mathbf{i} + 2y\mathbf{j} + \mathbf{k}}{\sqrt{4x^2 + 4y^2 + 1}}; \quad \mathbf{F}\cdot\mathbf{n} = \frac{x^3 + y^3 + z}{\sqrt{4x^2 + 4y^2 + 1}}; \quad z_x = -2x,\ z_y = -2y,$$

$dS = \sqrt{1 + 4x^2 + 4y^2}\,dA$. Using polar coordinates,

$$\text{Flux} = \int\int_S \mathbf{F}\cdot\mathbf{n}\,dS = \int\int_R (x^3 + y^3 + z)\,dA$$

$$= \int\int_R (4 - x^2 - y^2 + x^3 + y^3)\,dA$$

$$= \int_0^{2\pi}\int_0^2 (4 - r^2 + r^3\cos^3\theta + r^3\sin^3\theta)r\,dr\,d\theta$$

$$= \int_0^{2\pi}\left(2r^2 - \frac{1}{4}r^4 + \frac{1}{5}r^5\cos^3\theta + \frac{1}{5}r^5\sin^3\theta\right)\Big|_0^2\,d\theta$$

$$= \int_0^{2\pi}\left(4 + \frac{32}{5}\cos^3\theta + \frac{32}{5}\sin^3\theta\right)d\theta = 4\theta\Big|_0^{2\pi} + 0 + 0 = 8\pi.$$

21. For S_1: $g(x,y,z) = x^2 + y^2 - z$, $\nabla g = 2x\mathbf{i} + 2y\mathbf{j} - \mathbf{k}$, $|\nabla g| = \sqrt{4x^2 + 4y^2 + 1}$; $\mathbf{n}_1 = \frac{2x\mathbf{i} + 2y\mathbf{j} - \mathbf{k}}{\sqrt{4x^2 + 4y^2 + 1}}$;

$\mathbf{F}\cdot\mathbf{n}_1 = \frac{2xy^2 + 2x^2y - 5z}{\sqrt{4x^2 + 4y^2 + 1}}$; $z_x = 2x,\ z_y = 2y, dS_1 = \sqrt{1 + 4x^2 + 4y^2}\,dA$. For S_2: $g(x,y,z) = z - 1$, $\nabla g = \mathbf{k}$; $|\nabla g| = 1$;

$\mathbf{n}_2 = \mathbf{k}; \mathbf{F}\cdot\mathbf{n}_2 = 5z$; $z_x = 0, z_y = 0, dS_2 = dA$. Using polar coordinates and R: $x^2 + y^2 \le 1$ we have

$$\text{Flux} = \int\int_{S_1}\mathbf{F}\cdot\mathbf{n}_1\,dS_1 + \int\int_{S_2}\mathbf{F}\cdot\mathbf{n}_2\,dS_2 = \int\int_R (2xy^2 + 2x^2y - 5z)\,dA + \int\int_R 5z\,dA$$

$$= \int\int_R [2xy^2 + 2x^2y - 5(x^2 + y^2) + 5(1)]\,dA$$

$$= \int_0^{2\pi}\int_0^1 (2r^3\cos\theta\sin^2\theta + 2r^3\cos^2\theta\sin\theta - 5r^2 + 5)r\,dr\,d\theta$$

$$= \int_0^{2\pi}\left(\frac{2}{5}r^5\cos\theta\sin^2\theta + \frac{2}{5}r^5\cos^2\theta\sin\theta - \frac{5}{4}r^4 + \frac{5}{2}r^2\right)\Big|_0^1\,d\theta$$

$$= \int_0^{2\pi}\left[\frac{2}{5}(\cos\theta\sin^2\theta + \cos^2\theta\sin\theta) + \frac{5}{4}\right]d\theta = \frac{2}{5}\left(\frac{1}{3}\sin^3\theta - \frac{1}{3}\cos^3\theta\right)\Big|_0^{2\pi} + \frac{5}{4}\theta\Big|_0^{2\pi}$$

$$= \frac{2}{5}\left[-\frac{1}{3} - \left(-\frac{1}{3}\right)\right] + \frac{5}{2}\pi = \frac{5}{2}\pi.$$

23. The surface is $g(x,y,z) = x^2 + y^2 + z^2 - a^2 = 0$. $\nabla g = 2x\mathbf{i} + 2y\mathbf{j} + 2z\mathbf{k}$, $|\nabla g| = 2\sqrt{x^2 + y^2 + z^2}; \mathbf{n} = \frac{x\mathbf{i} + y\mathbf{j} + z\mathbf{k}}{\sqrt{x^2 + y^2 + z^2}}$;

$\mathbf{F}\cdot\mathbf{n} = -(2x\mathbf{i} + 2y\mathbf{j} + 2z\mathbf{k})\cdot\frac{x\mathbf{i} + y\mathbf{j} + z\mathbf{k}}{\sqrt{x^2 + y^2 + z^2}} = -\frac{2x^2 + 2y^2 + 2z^2}{\sqrt{x^2 + y^2 + z^2}} = -2\sqrt{x^2 + y^2 + z^2} = -2a$. Flux $= \int\int_S -2a\,dS = -2a \times \text{area} =$

$-2a(4\pi a^2) = -8\pi a^3$

25. Refering to the solution to Problem 23, we find $\mathbf{n} = \frac{x\mathbf{i} + y\mathbf{j} + z\mathbf{k}}{\sqrt{x^2 + y^2 + z^2}}$ and $dS = \frac{a}{\sqrt{a^2 - x^2 - y^2}}\,dA$.

$$\text{Now} \quad \mathbf{F}\cdot\mathbf{n} = kq\frac{\mathbf{r}}{|r|^3}\cdot\frac{\mathbf{r}}{|r|} = \frac{kq}{|r|^4}|r|^2 = \frac{kq}{|r|^2} = \frac{kq}{x^2 + y^2 + z^2} = \frac{kq}{a^2}$$

$$\text{and} \quad \text{Flux} = \int\int_S \mathbf{F}\cdot\mathbf{n}\,dS = \int\int_S \frac{kq}{a^2}\,dS = \frac{kq}{a^2} \times \text{area} = \frac{kq}{a^2}(4\pi a^2) = 4\pi kq.$$

27. The surface is $z = 6 - 2x - 3y$. Then $z_x = -2$, $z_y = -3$, $dS = \sqrt{1+4+9} = \sqrt{14}\,dA$. The area of the surface is

$$A(s) = \int\!\!\int_S dS = \int_0^3 \int_0^{2-2x/3} \sqrt{14}\,dy\,dx = \sqrt{14} \int_0^3 \left(2 - \frac{2}{3}x\right) dx$$

$$= \sqrt{14}\left(2x - \frac{1}{3}x^2\right)\Big|_0^3 = 3\sqrt{14}.$$

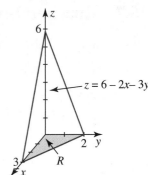

$$\bar{x} = \frac{1}{3\sqrt{14}}\int\!\!\int_S x\,dS = \frac{1}{3\sqrt{14}}\int_0^3 \int_0^{2-2x/3} \sqrt{14}\,x\,dy\,dx$$

$$= \frac{1}{3}\int_0^3 xy\Big|_0^{2-2x/3} dx = \frac{1}{3}\int_0^3 \left(2x - \frac{2}{3}x^2\right) dx$$

$$= \frac{1}{3}\left(x^2 - \frac{2}{9}x^3\right)\Big|_0^3 = 1$$

$$\bar{y} = \frac{1}{3\sqrt{14}}\int\!\!\int_S y\,dS = \frac{1}{3\sqrt{14}}\int_0^3 \int_0^{2-2x/3} \sqrt{14}\,y\,dy\,dx = \frac{1}{3}\int_0^3 \frac{1}{2}y^2\Big|_0^{2-2x/3} dx$$

$$= \frac{1}{6}\int_0^3 \left(2 - \frac{2}{3}x\right)^2 dx = \frac{1}{6}\left[-\frac{1}{2}(2 - \frac{2}{3}x)^3\right]\Big|_0^3 = \frac{2}{3}$$

$$\bar{z} = \frac{1}{3\sqrt{14}}\int\!\!\int_S z\,dS = \frac{1}{3\sqrt{14}}\int_0^3 \int_0^{2-2x/3} (6 - 2x - 3y)\sqrt{14}\,dy\,dx$$

$$= \frac{1}{3}\int_0^3 \left(6y - 2xy - \frac{3}{2}y^2\right)\Big|_0^{2-2x/3} dx = \frac{1}{3}\int_0^3 \left(6 - 4x + \frac{2}{3}x^2\right) dx = \frac{1}{3}\left(6x - 2x^2 + \frac{2}{9}x^3\right)\Big|_0^3 = 2$$

The centroid is $(1, 2/3, 2)$.

29. (a) The region in the xy-plane is $x^2 + y^2 \leq 16$. From $z_x = -x/\sqrt{x^2+y^2}$ and $z_y = -y/\sqrt{x^2+y^2}$ we see that

$$dS = \sqrt{1 + x^2/(x^2+y^2) + y^2/(x^2+y^2)}\,dA = \sqrt{2}\,da$$

and $\quad A(S) = \int\!\!\int_S dS = \int\!\!\int_R \sqrt{2}\,dA = \sqrt{2}\pi 4^2 = 16\sqrt{2}\pi.$

Then

$$\bar{x} = \frac{1}{16\sqrt{2}\pi}\int\!\!\int_S dS = \frac{1}{16\sqrt{2}\pi}\int\!\!\int_R \sqrt{2}\,x\,dA = \frac{1}{16\pi}\int_0^{2\pi}\int_0^4 r\cos\theta\,r\,dr\,d\theta$$

$$= \frac{1}{16\pi}\int_0^{2\pi} \frac{1}{3}r^3\cos\theta\Big|_0^4 d\theta = \frac{4}{3\pi}\int_0^{2\pi}\cos\theta\,d\theta = 0$$

$$\bar{y} = \frac{1}{16\sqrt{2}\pi}\int\!\!\int_S y\,dS = \frac{1}{16\sqrt{2}\pi}\int\!\!\int_R \sqrt{2}\,y\,dA = \frac{1}{16\pi}\int_0^{2\pi}\int_0^4 r\sin\theta\,r\,dr\,d\theta$$

$$= \frac{1}{16\pi}\int_0^{2\pi} \frac{1}{3}r^3\cos\theta\Big|_0^4 d\theta = \frac{4}{3\pi}\int_0^{2\pi}\sin\theta\,d\theta = 0$$

$$\bar{z} = \frac{1}{16\sqrt{2}\pi}\int\!\!\int_S z\,dS = \frac{1}{16\sqrt{2}\pi}\int\!\!\int_R \sqrt{2}(4 - \sqrt{x^2+y^2})\,dA = \frac{1}{16\pi}\int_0^{2\pi}\int_0^4 (4-r)r\,dr\,d\theta$$

$$= \frac{1}{16\pi}\int_0^{2\pi} \left(2r^2 - \frac{1}{3}r^3\right)\Big|_0^4 d\theta = \frac{2}{3\pi}\int_0^{2\pi} d\theta = \frac{4}{3}.$$

The centroid is $(0, 0, 4/3)$.

(b) $I_z = \int\!\!\int_S (x^2 + y^2)k\,dS = k\sqrt{2}\int\!\!\int_R (x^2+y^2)\,dA = k\sqrt{2}\int_0^{2\pi}\int_0^4 r^2\,r\,dr\,d\theta$

$$= \frac{k\sqrt{2}}{4}\int_0^{2\pi} r^4\Big|_0^4 d\theta = 64k\sqrt{2}\int_0^{2\pi} d\theta = 128k\pi\sqrt{2}$$

15.7   Curl and Divergence

1. $\text{curl}\mathbf{F} = (x-y)\mathbf{i} + (x-y)\mathbf{j}; \ \text{div}\mathbf{F} = 2z$

3. $\text{curl}\mathbf{F} = 0; \ \text{div}\mathbf{F} = 4y + 8z$ $\text{div}? \ 2z^3$

5. $\text{curl}\mathbf{F} = (4y^3 - 6xz^2)\mathbf{i} + (2x^3 - 3x^2)\mathbf{k}; \text{div}\mathbf{F} = 6xy$

7. $\text{curl}\mathbf{F} = (3e^{-z} - 8yz)\mathbf{i} - xe^{-z}\mathbf{j}; \ \text{div}\mathbf{F} = e^{-z} + 4z^2 - 3ye^{-z}$

9. $\text{curl}\mathbf{F} = (xy^2e^y + 2xye^y + x^3ye^z + x^3yze^z)\mathbf{i} - y^2e^y\mathbf{j} + (-3x^2yze^z - xe^x)\mathbf{k};$
   $\text{div}\mathbf{F} = xye^x + ye^x - x^3ze^z$

11. $\text{div}\,\mathbf{r} = 1 + 1 + 1 = 3$

13. $a \times \nabla = \begin{vmatrix} \mathbf{i} & \mathbf{j} & \mathbf{k} \\ a_1 & a_2 & a_3 \\ \partial/\partial x & \partial/\partial y & \partial/\partial z \end{vmatrix} = \left(a_2\dfrac{\partial}{\partial z} - a_3\dfrac{\partial}{\partial y}\right)\mathbf{i} + \left(a_3\dfrac{\partial}{\partial x} - a_1\dfrac{\partial}{\partial z}\right)\mathbf{j} + \left(a_1\dfrac{\partial}{\partial y} - a_2\dfrac{\partial}{\partial x}\right)\mathbf{k}$

$(a \times \nabla) \times \mathbf{r} = \begin{vmatrix} \mathbf{i} & \mathbf{j} & \mathbf{k} \\ a_2\dfrac{\partial}{\partial z} - a_3\dfrac{\partial}{\partial y} & a_3\dfrac{\partial}{\partial x} - a_1\dfrac{\partial}{\partial z} & a_1\dfrac{\partial}{\partial y} - a_2\dfrac{\partial}{\partial x} \\ x & y & z \end{vmatrix}$

$= (-a_1 - a_1)\mathbf{i} - (a_2 + a_2)\mathbf{j} + (-a_3 - a_3)\mathbf{k} = -2\mathbf{a}$

15. $\nabla \cdot (\mathbf{a} \times \mathbf{r}) = \begin{vmatrix} \partial/\partial x & \partial/\partial y & \partial/\partial z \\ a_1 & a_2 & a_3 \\ x & y & z \end{vmatrix} = \dfrac{\partial}{\partial x}(a_2 z - a_3 y) - \dfrac{\partial}{\partial y}(a_1 z - a_3 x) + \dfrac{\partial}{\partial z}(a_1 y - a_2 x) = 0$

17. $\mathbf{r} \times \mathbf{a} = \begin{vmatrix} \mathbf{i} & \mathbf{j} & \mathbf{k} \\ x & y & z \\ a_1 & a_2 & a_3 \end{vmatrix} = (a_3 y - a_2 z)\mathbf{i} - (a_3 x - a_1 z)\mathbf{j} + (a_2 x - a_1 y)\mathbf{k}; \ \mathbf{r} \cdot \mathbf{r} = x^2 + y^2 + z^2$

$\nabla \times [(\mathbf{r} \cdot \mathbf{r})\mathbf{a}] = \begin{vmatrix} \mathbf{i} & \mathbf{j} & \mathbf{k} \\ \partial/\partial x & \partial/\partial y & \partial/\partial z \\ (\mathbf{r} \cdot \mathbf{r})a_1 & (\mathbf{r} \cdot \mathbf{r})a_2 & (\mathbf{r} \cdot \mathbf{r})a_3 \end{vmatrix}$

$= (2ya_3 - 2za_2)\mathbf{i} - (2xa_3 - 2za_1)\mathbf{j} + (2xa_3 - 2ya_1)\mathbf{k} = 2(\mathbf{r} \times \mathbf{a})$

19. Let $\mathbf{F} = P(x,y,z)\mathbf{i} + Q(x,y,z)\mathbf{j} + R(x,y,z)\mathbf{k}$ and $\mathbf{G} = S(x,y,z)\mathbf{i} + T(x,y,z)\mathbf{j} + U(x,y,z)\mathbf{k}$.
   $\nabla \cdot (\mathbf{F} + \mathbf{G}) = \nabla \cdot [(P+S)\mathbf{i} + (Q+T)\mathbf{j} + (R+U)\mathbf{k}] = P_x + S_x + Q_y + T_y + R_x + U_z$
   $= (P_x + Q_y + R_z) + (S_x + T_y + U_z) = \nabla \cdot \mathbf{F} + \nabla \cdot \mathbf{G}$

21. $\nabla \cdot (f\mathbf{F}) = \nabla \cdot (fP\mathbf{i} + fQ\mathbf{j} + fR\mathbf{k}) = fP_x + Pf_x + fQ_y + Qf_y + fR_z + Rf_z$
   $= f(P_x + Q_y + R_z) + (Pf_x + Qf_y + Rf_z) = f(\nabla \cdot \mathbf{F}) + \mathbf{F} \cdot (\nabla f)$

23. Assuming continuous second partial derivatives,

$$\text{curl}(\text{grad}f) = \nabla \times (f_x\mathbf{i} + f_y\mathbf{j} + f_z\mathbf{k}) = \begin{vmatrix} \mathbf{j} & \mathbf{j} & \mathbf{k} \\ \partial/\partial x & \partial/\partial y & \partial/\partial z \\ f_x & f_y & f_z \end{vmatrix}$$

$$= (f_{zy} - f_{yz})\mathbf{i} - (f_{zx} - f_{xz})\mathbf{j} + (f_{yx} - f_{xy})\mathbf{k} = \mathbf{0}.$$

25. Let $\mathbf{F} = P(x,y,z)\mathbf{i} + Q(x,y,z)\mathbf{j} + R(x,y,z)\mathbf{k}$ and $\mathbf{G} = S(x,y,z)\mathbf{i} + Y(x,y,z)\mathbf{j} + U(x,y,z)\mathbf{k}$.
   $\mathbf{F} \times \mathbf{G} = \begin{vmatrix} \mathbf{i} & \mathbf{j} & \mathbf{k} \\ P & Q & R \\ S & T & U \end{vmatrix} = (QU - RT)\mathbf{i} - (PU - RS)\mathbf{j} + (PT - QS)\mathbf{k}$

$\text{div}\,(\mathbf{F} \times \mathbf{G}) = (QU_x + Q_xU - RT_x - R_xT) - (PU_y + P_yU - RS_y - R_yS)$

$+ (PT_z + P_zT - QS_z - Q_zS)$

$= S(R_y - Q_z) + T(P_z - R_x) + U(Q_x - P_y) - P(U - y - T_z) - Q(S_z - U_x)$

$- R(R_x - S_y)$

$= \mathbf{G} \cdot (\text{curl}\mathbf{F}) - \mathbf{F} \cdot (\text{curl}\mathbf{G})$

27. curl $\mathbf{F} = -8yz\mathbf{i} - 2z\mathbf{j} - x\mathbf{k}$; curl (curl $\mathbf{F}$) $= 2\mathbf{i} - (8y-1)\mathbf{j} + 8z\mathbf{k}$

29. $f_z = 6x + 4y - 9z$; $f_{xx} = 6$; $f_y = 10y + 4z$; $f_{yy} = 10$; $f_z = -9x - 16z$; $f_{zz} = -16$; $\nabla^2 f = f_{xx} + f_{yy} + f_{zz} = 6 + 10 - 16 = 0$

31. $f_x = 6x + 4y - 9z$; $f_{xx} = 6$; $f_y = 10y + 4x$; $f_{yy} = 10$; $f_z = -9x - 16z$; $f_{zz} = -16$; $\nabla^2 f + f_{xx} + f_{yy} + z_{zz} = 6 + 10 - 16 = 0$

33. $f_x = \dfrac{1}{1 + \dfrac{4y^2}{(x^2 + y^2 - 1)^2}}\left(-\dfrac{4xy}{(x^2 + y^2 - 1)^2}\right) = -\dfrac{4xy}{(x^2 + y^2 - 1)^2 + 4y^2}$

$f_{xx} = -\dfrac{[(x^2 + y^2 - 1)^2 + 4y^2]4y - 4xy[4x(x^2 + y^2 - 1)]}{[(x^2 + y^2 - 1) + 4y^2]^2} = \dfrac{12x^4 y - 4y^5 + 8x^2 y^3 - 8x^2 y - 8y^3 - 4y}{[(x^2 + y^2 - 1)^2 + 4y^2]^2}$

$f_y = \dfrac{1}{1 + \dfrac{4y^2}{(x^2 + y^2 - 1)^2}}\left[\dfrac{2(x^2 + y^2 - 1) - 4y^2}{(x^2 + y^2 - 1)^2}\right] = \dfrac{2(x^2 + y^2 - 1)^2}{(x^2 + y^2 - 1)^2 + 4y^2}$

$f_{yy} = \dfrac{[(x^2 + y^2 - 1)^2 + 4y^2](-4y) - 2(x^2 + y^2 - 1)^2[4y(x^2 + y^2 - 1)^2 + 8y]}{[(x^2 + y^2 - 1)^2 + 4y^2]^2}$

$= \dfrac{-12x^4 y + 4y^5 - 8x^2 y^3 + 8x^2 y + 8y^3 + 4y}{[(x^2 + y^2 - 1)^2 + 4y^2]^2}$

$\nabla^2 f = f_{xx} + f_{yy} = 0$

35. Using Problems 25 and 23,

$$\nabla \cdot \mathbf{F} = \text{div } (\nabla f \times \nabla g) = \nabla g \cdot (\text{curl } \nabla f) - \nabla f \cdot (\text{curl } g) = \nabla g \cdot \mathbf{0} - \nabla f \cdot \mathbf{0} = 0.$$

37. The surface is $g(x,y) = x^2 + y^2 + 4z^2 - 4 = 0$. $\nabla g = 2x\mathbf{i} + 2y\mathbf{j} + 8z\mathbf{k}$,

$|\nabla g| = 2\sqrt{x^2 + y^2 + 16z^2}$; $\mathbf{n} = \dfrac{2x\mathbf{i} + 2y\mathbf{j} + 8z\mathbf{k}}{2\sqrt{x^2 + y^2 + 16z^2}} = \dfrac{x\mathbf{i} + y\mathbf{j} + 4z\mathbf{k}}{\sqrt{x^2 + y^2 + 16z^2}}$;

$\nabla \times \mathbf{F} = (3x^2 - 3y^2)\mathbf{k}$, $(\nabla \times \mathbf{F}) \cdot \mathbf{n} = \dfrac{12z(x^2 + y^2)}{\sqrt{x^2 + y^2 + 16z^2}}$

Writing the equation of the surface as $z = \sqrt{1 - x^2/4 - y^2/4}$, we have

$z_x = -\dfrac{x}{4\sqrt{1 - x^2/4 - y^2/4}}$, $z_y = -\dfrac{y}{4\sqrt{1 - x^2/4 - y^2/4}}$, and $dS = \dfrac{\sqrt{16 - 3x^2 - 3y^2}}{2\sqrt{4 - x^2 - y^2}}dA$.

Then, using polar coordinates,

$\text{Flux} = \displaystyle\int\int_S (\nabla \times \mathbf{F}) \cdot \nabla dS = \int\int_R \dfrac{12z(x^2 - y^2)}{\sqrt{x^2 + y^2 + 16z^2}} \dfrac{\sqrt{16 - 3x^2 - 3y^2}}{2\sqrt{4 - x^2 - y^2}}dA$

$= \displaystyle\int\int_R \dfrac{6\sqrt{1 - x^2/4 - y^2/4}(x^2 - y^2)\sqrt{16 - 3x^2 - 3y^2}dA}{\sqrt{x^2 + y^2 + 16 - 4x^2 - 4y^2}\sqrt{4 - x^2 - y^2}}$

$= \displaystyle\int_0^{\pi/4}\int_0^2 \sqrt{1 - r^2/4}(r^2\cos^2\theta - r^2\sin^2\theta)r\,dr\,d\theta\sqrt{4 - r^2} = \int_0^{\pi/4}\int_0^2 3r^2\cos 2\theta\,dr\,d\theta$

$= \displaystyle\int_0^{\pi/4}\dfrac{3}{4}r^4\cos 2\theta\,\Big|_0^2\,d\theta = \int_0^{\pi/4} 12\cos 2\theta\,d\theta = 6\sin 2\theta\,\Big|_0^{\pi/4} = 6.$

39. curl $\mathbf{F} = -Gm_1 m_2 \begin{vmatrix} \mathbf{i} & \mathbf{j} & \mathbf{k} \\ \partial/\partial x & \partial/\partial y & \partial/\partial z \\ x/|\mathbf{r}|^3 & y/|\mathbf{r}|^3 & z/|\mathbf{r}|^3 \end{vmatrix}$

$= -Gm_1 m_2[(-3yz/|\mathbf{r}|^5 + 3yz/|\mathbf{r}|^5)\mathbf{i} - (-3xz/|\mathbf{r}|^5 + 3xz/|\mathbf{r}|^5)\mathbf{j} + (-3xy/|\mathbf{r}|^5 + 3xy/|\mathbf{r}|^5)\mathbf{k}]$

$= \mathbf{0}$

div $\mathbf{F} = -Gm_1 m_2\left[\dfrac{-2x^2 + y^2 + z^2}{|\mathbf{r}|^{5/2}} + \dfrac{x^2 - 2y^2 + z^2}{|\mathbf{r}|^{5/2}} + \dfrac{x^2 + y^2 - 2z^2}{|\mathbf{r}|^{5/2}}\right] = 0$

41. We first note that curl $(\partial\mathbf{H}/\partial t) = \partial(\text{curl } \mathbf{H})/\partial t$ and curl $(\partial\mathbf{E}/\partial t) = \partial(\text{curl } \mathbf{E})/\partial t$. Then, from Problem 30,

$$-\nabla^2\mathbf{E} = -\nabla^2\mathbf{E} + \mathbf{0} = -\nabla^2\mathbf{E} + \text{grad } 0 = -\nabla^2\mathbf{E} + \text{grad } (\text{div } \mathbf{E}) = \text{curl } (\text{curl } \mathbf{E})$$

$$= \text{curl } \left(-\dfrac{1}{c}\dfrac{\partial\mathbf{H}}{\partial t}\right) = -\dfrac{1}{c}\dfrac{\partial}{\partial t}\text{curl } \mathbf{H} = -\dfrac{1}{c}\dfrac{\partial}{\partial t}\left(\dfrac{1}{c}\dfrac{\partial\mathbf{E}}{\partial t}\right) = -\dfrac{1}{c^2}\dfrac{\partial^2\mathbf{E}}{\partial t}$$

and $\nabla^2 \mathbf{E} = \frac{1}{c^2} \partial^2 \mathbf{E}/\partial t^2$. Similarly,

$$-\nabla^2 \mathbf{H} = -\nabla^2 \mathbf{H} + \text{grad (div } \mathbf{H}) = \text{curl (curl } \mathbf{H}) = \text{curl}\left(\frac{1}{c}\frac{\partial \mathbf{E}}{\partial t}\right) = \frac{1}{c}\frac{\partial}{\partial t}\text{curl } \mathbf{E}$$

$$= \frac{1}{c}\frac{\partial}{\partial t}\left(-\frac{1}{c}\frac{\partial \mathbf{H}}{\partial t}\right) = -\frac{1}{c^2}\frac{\partial^2 \mathbf{H}}{\partial t^2}$$

and $\nabla^2 \mathbf{H} = \frac{1}{c^2}\partial^2 \mathbf{H}/\partial t^2$.

15.8 Stokes' Theorem

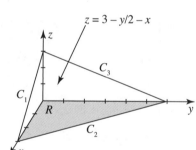

1. **Surface Integral:** $\text{curl}\mathbf{F} = -10\mathbf{k}$. Letting $g(x,y,z) = -1$, we have $\nabla g = \mathbf{k}$ and $\mathbf{n} = \mathbf{k}$. Then

$$\int\int_S (\text{curl}\mathbf{F}) \cdot \mathbf{n}\,dS = \int\int_S (-10)dS = -10 \times (\text{area of S}) = -10(4\pi) = -40\pi.$$

   Line Integral: Parameterize the curve C by $x = 2\cos t$, $y = 2\sin t$, $z = 1$, for $0 \le t \le 2\pi$. Then

$$\oint \mathbf{F}\cdot d\mathbf{r} = \oint 5y\,dx - 5x\,dy + 3\,dz = \int_0^{2\pi}[10\sin t(-2\sin t) - 10\cos t(2\cos t)]dt$$

$$= \int_0^{2\pi}(-20\sin^2 t - 20\cos^2 t)dt = \int_0^{2\pi} -20\,dt = -40\pi.$$

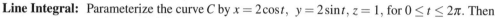

3. **Surface Integral:** $\text{curl}\mathbf{F} = \mathbf{i} + \mathbf{j} + \mathbf{k}$. Letting $g(x,y,z) = 2x + y + 2z - 6$, we have $\nabla g = 2\mathbf{i} + 2\mathbf{j} + 2\mathbf{k}$ and $\mathbf{n} = (2\mathbf{i} + \mathbf{j} + 2\mathbf{k})/3$. Then $\int\int_S(\text{curl}\mathbf{F})\cdot\mathbf{n}\,dS = \int\int_S \frac{5}{3}dS$.
   Letting the surface be $z = 3 - \frac{1}{2}y - x$ we have $z_x = -1$, $z_y = -\frac{1}{2}$, and
   $dS = \sqrt{1 + (-1)^2 + (-\frac{1}{2})^2}dA = \frac{3}{2}dA$. Then

$$\int\int_S(\text{curl}\mathbf{F})\cdot\mathbf{n}\,dS = \int\int_R \frac{5}{3}\left(\frac{3}{2}\right)dA = \frac{5}{2}\times(\text{area of }R) = \frac{5}{2}(9) = \frac{45}{2}.$$

   Line Integral: $C_1: z = 3 - x, 0 \le x \le 3, y = 0$; $C_2: y = 6 - 2x, 3 \ge x \ge 0, z = 0$; $C_3: z = 3 - y/2, 6 \ge y \ge 0, x = 0$.

$$\oint_C z\,dx + x\,dy + y\,dz = \int\int_{C_1} z\,dx + \int_{C_2} x\,dy + \int_{C_3} y\,dz$$

$$= \int_0^3 (3 - x)dx + \int_3^0 x(-2dx) + \int_6^0 y(-dy/2)$$

$$= \left(3x - \frac{1}{2}x^2\right)\Big|_0^3 - x^2\Big|_3^0 - \frac{1}{4}y^2\Big|_6^0 = \frac{9}{2} - (0 - 9) - \frac{1}{4}(0 - 36) = \frac{45}{2}$$

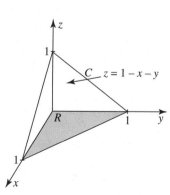

5. $\text{curl}\mathbf{F} = 2\mathbf{i} + \mathbf{j}$. A unit vector normal to the plane is $\mathbf{n} = (\mathbf{i} + \mathbf{j} + \mathbf{k})/\sqrt{3}$. Taking the equation of the plane to be $z = 1 - x - y$, we have $z_x = z_y = -1$. Thus, $dS = \sqrt{1 + 1 + 1}dA = \sqrt{3}dA$ and

$$\oint_C \mathbf{F}\cdot d\mathbf{r} = \int\int S(\text{curl}\mathbf{F})\cdot\mathbf{n}\,dS = \int\int_S \sqrt{3}dS = \sqrt{3}\int\int_R \sqrt{3}dA$$

$$= 3 \times (\text{area of }R) = 3(1/2) = 3/2.$$

7. $\text{curl}\mathbf{F} = -2y\mathbf{i} - z\mathbf{j} - x\mathbf{k}$. A unit vector normal to the plane is $\mathbf{n} = (\mathbf{j}+\mathbf{k})/\sqrt{2}$. From $z = 1-y$ we have $z_x = 0$ and $z_y = -1$. Then $dS = \sqrt{1+1}\,dA = \sqrt{2}\,dA$ and

$$\oint_c \mathbf{F} \cdot d\mathbf{r} = \int\int_S (\text{curl}\mathbf{F}) \cdot \mathbf{n}\,dS = \int\int_R \left[-\frac{1}{\sqrt{2}}(z+x) \right] \sqrt{2}\,dA = \int\int_R (y-x-1)\,dA$$

$$= \int_0^2 \int_0^1 (y-x-1)\,dy\,dx = \int_0^2 \left(\frac{1}{2}y^2 - xy - y \right)\bigg|_0^1 dx = \int_0^2 \left(-x - \frac{1}{2} \right) dx$$

$$= \left(-\frac{1}{2}x^2 - \frac{1}{2}x \right)\bigg|_0^2 = -3.$$

9. $\text{curl}\mathbf{F} = (-3x^2 - 3y^2)\mathbf{k}$. A unit vector normal to the plane is $\mathbf{n} = (\mathbf{i}+\mathbf{j}+\mathbf{k})/\sqrt{3}$. From $z = 1-x-y$, we have $z_x = z_y = -1$ and $dS = \sqrt{3}\,dA$. Then, using polar coordinates,

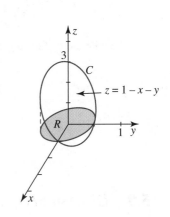

$$\oint_C \mathbf{F} \cdot d\mathbf{r} = \int\int_S (\text{curl}\mathbf{F}) \cdot \mathbf{n}\,dS = \int\int_R (-\sqrt{3}x^2 - \sqrt{3}y^2)\sqrt{3}\,dA$$

$$= 3\int\int_R (-x^2 - y^2)\,dA = 3\int_0^{2\pi} \int_0^1 (-r^2)r\,dr\,d\theta$$

$$= 3\int_0^{2\pi} -\frac{1}{4}r^4 \bigg|_0^1 d\theta = 3\int_0^{2\pi} -\frac{1}{4}d\theta = \frac{3\pi}{2}.$$

11. $\text{curl}\mathbf{F} = 3x^2y^2\mathbf{k}$. A unit vector normal to the surface is

$$\mathbf{n} = \frac{8x\mathbf{i} + 2y\mathbf{j} + 2z\mathbf{k}}{\sqrt{64x^2 + 4y^2 + 4z^2}} = \frac{4x\mathbf{i} + y\mathbf{j} + z\mathbf{k}}{\sqrt{16x^2 + y^2 + z^2}}.$$

From $z_x = -\dfrac{4x}{\sqrt{4 - 4x^2 - y^2}}$, $z_y = -\dfrac{y}{\sqrt{4 - 4x^2 - y^2}}$ we obtain $dS = 2\sqrt{\dfrac{1 + 3x^2}{4 - 4x^2 - y^2}}\,dA$. Then

$$\oint_C \mathbf{F} \cdot d\mathbf{r} = \int\int_S (\text{curl}\mathbf{F}) \cdot \mathbf{n}\,dS = \int\int_R \frac{3x^2y^2z}{\sqrt{16x^2 + y^2 + z^2}}\left(2\sqrt{\frac{1 + 3x^2}{4 - 4x^2 - y^2}}\right)dA$$

$$= \int\int_R 3x^2y^2\,dA \quad \boxed{\text{Using symmetry}}$$

$$= 12\int_0^1 \int_0^{2\sqrt{1-x^2}} x^2y^2\,dy\,dx = 12\int_0^1 \left(\frac{1}{3}x^2y^3 \right)\bigg|_0^{2\sqrt{1-x^2}} dx$$

$$= 32\int_0^1 x^2(1-x^2)^{3/2}dx \quad \boxed{x = \sin t, \ dx = \cos t\,dt}$$

$$= 32\int_0^{\pi/2} \sin^2 t \cos^4 t\,dt = \pi.$$

13. Parameterize C by $x = 4\cos t$, $y = 2\sin t, z = 4$, for $0 \le t \le 2\pi$. Then

$$\int\int_S (\text{curl}\mathbf{F}) \cdot \mathbf{n}\,dS = \oint_C \mathbf{F} \cdot d\mathbf{r} = \oint 6yz\,dx + 5x\,dy + yze^{x^2}dz$$

$$= \int_0^{2\pi} [6(2\sin t)(4)(-4\sin t) + 5(4\cos t)(2\cos t) + 0]\,dt$$

$$= 8\int_0^{2\pi} (-24\sin^2 t + 5\cos^2 t)\,dt = 8\int_0^{2\pi} (5 - 29\sin^2 t)\,dt = -152\pi.$$

15. Parameterize C by C_1: $x=0$, $z=0$, $2 \geq y \geq 0$; C_2: $z=x, y=0$, $0 \leq x \leq 2$;
C_3: $x=2$, $z=2$, $0 \leq y \leq 2$; C_4: $z=x, y=2$, $2 \geq x \geq 0$. Then

$$\int\int_S (\text{curl}\mathbf{F}) \cdot \mathbf{n}\,dS = \oint_C \mathbf{F} \cdot \mathbf{r} = \oint_C 3x^2\,dx + 8x^3y\,dy + 3x^2y\,dz$$

$$= \int_{C_1} 0\,dx + 0\,dy + 0\,dz + \int_{C_2} 3x^2\,dx$$

$$+ \int_{C_3} 64\,dy + \int_{C_4} 3x^2\,dx + 6x^2\,dx$$

$$= \int_0^2 3x^2\,dx + \int_0^2 64\,dy + \int_2^0 9x^2\,dx$$

$$= x^3\Big|_0^2 + 64y\Big|_0^2 + 3x^3\Big|_2^0 = 112.$$

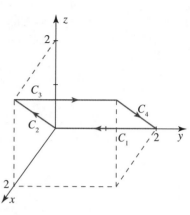

17. We take the surface to be $z=0$. Then $\mathbf{n}=\mathbf{k}$ and $dS=dA$. Since $\text{curl}\mathbf{F} = \dfrac{1}{1+y^2}\mathbf{i} + 2ze^{x^2}\mathbf{j} + y^2\mathbf{k}$,

$$\oint_C z^2e^{x^2}\,dx + xy\,dy + \tan^{-1}y\,dz = \int\int_S (\text{curl}\mathbf{F}) \cdot \mathbf{n}\,dS = \int\int_S y^2\,dS = \int\int_R y^2\,dA$$

$$= \int_0^{2\pi}\int_0^3 r^2\sin^2\theta\, r\,dr\,d\theta = \int_0^{2\pi} \frac{1}{4}r^4\sin^2\theta\Big|_0^3 d\theta$$

$$= \frac{81}{4}\int_0^{2\pi}\sin^2\theta\,d\theta = \frac{81\pi}{4}.$$

15.9 Divergence Theorem

1. $\text{div}\mathbf{F} = y+x+z$
   The Triple Integral:

$$\int\int\int_D \text{div}\mathbf{F}\,dV = \int_0^1\int_0^1\int_0^1 (x+y+z)\,dx\,dy\,dz$$

$$= \int_0^1\int_0^1 (\frac{1}{2}x^2 + xy + xz)\Big|_0^1 dy\,dz$$

$$= \int_0^1\int_0^1 (\frac{1}{2}+y+z)\,dy\,dz$$

$$= \int_0^1 (\frac{1}{2}y + \frac{1}{2}y^2 + yz)\Big|_0^1 dz$$

$$= \int_0^1 (1+z)\,dz = \frac{1}{2}(1+z^2)\Big|_0^1 = 2 - \frac{1}{2} = \frac{3}{2}$$

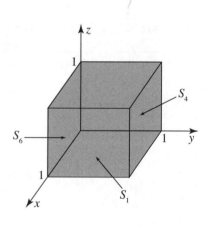

The Surface Integral: Let the surfaces be S_1 in $z=0$, S_2 in $z=1$, S_3 in $y=0$, S_4 in $y=1$, S_5 in $x=0$, and S_6 in $x=1$. The unit outward normal vectors are $-\mathbf{k}$, $\mathbf{k}$, $-\mathbf{j}$, $\mathbf{j}$, $-\mathbf{i}$ and $\mathbf{i}$, respectively. Then

$$\int\int_S \mathbf{F} \cdot \mathbf{n}\,dS = \int\int_{S_1} \mathbf{F} \cdot (-\mathbf{k})\,dS_1 + \int\int_{S_2} \mathbf{F} \cdot \mathbf{k}\,dS_2 + \int\int_{S_3} \mathbf{F} \cdot (-\mathbf{j})\,dS_3 + \int\int_{S_4} \mathbf{F} \cdot \mathbf{j}\,dS_4$$

$$+ \int\int_{S_5} \mathbf{F} \cdot (-\mathbf{i})\,dS_5 + \int\int_{S_6} \mathbf{F} \cdot \mathbf{i}\,dS_6$$

$$= \int\int_{S_1} (-xz)\,dS_1 + \int\int_{S_2} xz\,dS_2 + \int\int_{S_3} (-yz)\,dS_3 + \int\int_{S_4} yz\,dS_4$$

$$+ \int\int_{S_5} (-xy)\,dS^5 + \int\int_{S_6} xy\,dS_6$$

$$= \int\int_{S_2} x\,dS_2 + \int\int_{S_4} z\,dS_4 + \int\int_{S_6} y\,dS_6$$

$$= \int_0^1\int_0^1 x\,dx\,dy + \int_0^1\int_0^1 z\,dz\,dx + \int_0^1\int_0^1 y\,dy\,dz$$

$$= \int_0^1 \frac{1}{2}\,dy + \int_0^1 \frac{1}{2}\,dx + \int_0^1 \frac{1}{2}\,dz = \frac{3}{2}.$$

3. $\text{div}\mathbf{F} = 3x^2 + 3y^2 + 3z^2$. Using spherical coordinates,

$$\iint_S \mathbf{F} \cdot \mathbf{n}\, dS = \iiint_D 3(x^2 + y^2 + z^2)\, dV = \int_0^{2\pi} \int_0^\pi \int_0^a 3\rho^2 \rho^2 \sin\phi\, d\rho\, d\phi\, d\theta$$

$$= \int_0^{2\pi} \int_0^\pi \frac{3}{5}\rho^5 \sin\phi \Big|_0^a\, d\phi\, d\theta = \frac{3a^5}{5} \int_0^{2\pi} \int_0^\pi \sin\phi\, d\phi\, d\theta$$

$$= \frac{3a^5}{5} \int_0^{2\pi} -\cos\phi \Big|_0^\pi\, d\theta = \frac{6a^5}{5} \int_0^{2\pi} d\theta = \frac{12\pi a^5}{5}.$$

5. $\text{div}\mathbf{F} = 2(z - 1)$. Using cylindrical coordinates,

$$\iint_S \mathbf{F} \cdot \mathbf{n}\, dS = \iiint_D 2(z-1)V = \int_0^{2\pi} \int_0^4 \int_1^5 2(z-1)\, dz\, dr\, d\theta = \int_0^{2\pi} \int_0^4 (z-1)^2 \Big|_1^5 r\, dr\, d\theta$$

$$= \int_0^{2\pi} \int_0^4 16 r\, dr\, d\theta = \int_0^{2\pi} 8r^2 \Big|_0^4\, d\theta = 128 \int_0^{2\pi} d\theta = 256\pi.$$

7. $\text{div}\mathbf{F} = 3z^2$. Using cylindrical coordinates,

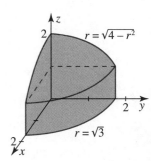

$$\iint_S \mathbf{F} \cdot \mathbf{n}\, dS = \iiint_D \text{div}\mathbf{F}\, dV = \int_0^{2\pi} \int_0^{\sqrt{3}} \int_0^{\sqrt{4-r^2}} 3z^2 r\, dz\, dr\, d\theta$$

$$= \int_0^{2\pi} \int_0^{\sqrt{3}} rz^3 \Big|_0^{\sqrt{4-r^2}}\, dr\, d\theta = \int_0^{2\pi} \int_0^{\sqrt{3}} r(4-r^2)^{3/2}\, dr\, d\theta$$

$$= \int_0^{2\pi} -\frac{1}{5}(4-r^2)^{5/2} \Big|_0^{\sqrt{3}}\, d\theta = \int_0^{2\pi} -\frac{1}{5}(1-32)\, d\theta$$

$$= \int_0^{2\pi} \frac{31}{5}\, d\theta = \frac{62\pi}{5}.$$

9. $\text{div}\mathbf{F} = \dfrac{1}{x^2 + y^2 + z^2}$. Using spherical coordinates,

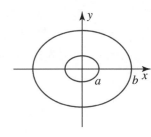

$$\iint_S \mathbf{F} \cdot \mathbf{n}\, dS = \iiint_D \text{div}\mathbf{F}\, dV = \int_0^{2\pi} \int_0^\pi \int_a^b \frac{1}{\rho^2}\rho^2 \sin\phi\, d\rho\, d\phi\, d\theta$$

$$= \int_0^{2\pi} \int_0^\pi (b-a)\sin\phi\, d\phi\, d\theta = (b-a) \int_0^{2\pi} -\cos\phi \Big|_0^\pi\, d\theta$$

$$= (b-a) \int_0^{2\pi} 2\, d\theta = 4\pi(b-a).$$

11. $\text{div}\mathbf{F} = 2z + 10y - 2z = 10y$.

$$\iint_S \mathbf{F} \cdot \mathbf{n}\, dS = \iiint_D 10y\, dV = \int_0^2 \int_0^{2-x^2/2} \int_z^{4-z} 10y\, dy\, dz\, dx$$

$$= \int_0^2 \int_0^{2-x^2/2} 5y^2 \Big|_z^{4-z}\, dz\, dx = \int_0^2 \int_0^{2-x^2/2} (80 - 40z)\, dz\, dx$$

$$= \int_0^2 (80z - 20z^2) \Big|_0^{2-x^2/2}\, dx = \int_0^2 (80 - 5x^4)\, dx = (80x - x^5) \Big|_0^2 = 128$$

13. $\text{div}\mathbf{F} = 6xy^2 + 1 - 6xy^2 = 1$. Using cylindrical coordinates,

$$\iint_S \mathbf{F} \cdot \mathbf{n}\, dS = \iiint_D dV = \int_0^\pi \int_0^{2\sin\theta} \int_{r^2}^{2r\sin\theta} dz\, r\, dr\, d\theta = \int_0^\pi \int_0^{2\sin\theta} (2r\sin\theta - r^2) r\, dr\, d\theta$$

$$= \int_0^\pi \left(\frac{2}{3}r^3 \sin\theta - \frac{1}{4}r^4\right) \Big|_0^{2\sin\theta}\, d\theta = \int_0^\pi \left(\frac{16}{3}\sin^4\theta - 4\sin^4\theta\right)\, d\theta$$

$$= \frac{4}{3} \int_0^\pi \sin^4\theta\, d\theta = \frac{4}{3}\left(\frac{3}{8}\theta - \frac{1}{4}\sin 2\theta + \frac{1}{32}\sin 4\theta\right) \Big|_0^\pi = \frac{\pi}{2}$$

15. Since div $\mathbf{a} = 0$, by the divergence Theorem

$$\iint_S (\mathbf{a} \cdot \mathbf{n})dS = \iiint_D \text{div } \mathbf{a}dV = \iiint_D 0dV = 0.$$

17. (a) $\text{div}\mathbf{E} = q\left[\dfrac{-2x^2+y^2+z^2}{(x^2+y^2+z^2)^{5/2}} + \dfrac{x^2-2y^2+z^2}{(x^2+y^2+z^2)^{5/2}} + \dfrac{x^2+y^2-2z^2}{(x^2+y^2+z^2)^{5/2}}\right] = 0$

$$\iint_{S \cup S_a} (\mathbf{E} \cdot \mathbf{n})dS = \iiint_D \text{div}\mathbf{E}dV = \iiint_D 0dV = 0$$

(b) From (a), $\int\int_S (\mathbf{E} \cdot \mathbf{n})dS + \int\int_{S_a} (\mathbf{E} \cdot \mathbf{n})dS = 0$ and $\int\int_S (\mathbf{E} \cdot \mathbf{n})dS = -\int\int_{S_a} (\mathbf{E} \cdot \mathbf{n})dS.$ on S_a, $|\mathbf{r}| = a$, $\mathbf{n} = -(x\mathbf{i}+y\mathbf{j}+z\mathbf{k})/a = -\mathbf{r}/a$ and $\mathbf{E} \cdot \mathbf{n} = (q\mathbf{r}/a^3) \cdot (-\mathbf{r}/a) = -qa^2/a^4 = -qa^2.$ Thus
$\int\int_S (\mathbf{E} \cdot \mathbf{n})dS = -\int\int_{S_a} (-\dfrac{q}{a^2})dS = \dfrac{q}{a^2}\int\int_{S_a} dS = \dfrac{q}{a^2} \times \text{(area of } S_a) = \dfrac{q}{a^2}(4\pi a^2) = 4\pi q.$

19. By the Divergence Theorem and Problem 21 in Section 15.7,

$$\iint_S (f\nabla g) \cdot \mathbf{n}dS = \iiint_D \text{div}(f\nabla g)dV = \iiint_D \nabla \cdot (f\nabla g)dV = \iiint_D [f(\nabla \cdot \nabla g) + \nabla g \cdot \nabla f]dV$$
$$= \iiint_D (f\nabla^2 g + \nabla g \cdot \nabla f)dV.$$

21. If $G(x,y,z)$ is a vector valued function then we define surface integrals and triple integrals of $\mathbf{G}$ component-wise. In this case, if $\mathbf{a}$ is a constant vector it is easily shown that

$$\iint_S \mathbf{a} \cdot \mathbf{G}dS = \mathbf{a} \cdot \iint_S \mathbf{G}dS \text{ and } \iiint_D \mathbf{a} \cdot \mathbf{G}dV = \mathbf{a} \cdot \iiint_D \mathbf{G}dV.$$

Now let $F = f\mathbf{a}$. Then

$$\iint_S \mathbf{F} \cdot \mathbf{n}dS = \iint_S (f\mathbf{a}) \cdot \mathbf{n}dS = \iint_S \mathbf{a} \cdot (f\mathbf{n})dS$$

and, using Problem 21 in Section 15.7 and the fact that $\nabla \cdot \mathbf{a} = 0$, we have

$$\iiint_D \text{div}\mathbf{F}dV = \iiint_D \nabla \cdot (f\mathbf{a})dV = \iiint_D [f(\nabla \cdot \mathbf{a}) + \mathbf{a} \cdot \nabla f]dV = \iiint_D \mathbf{a} \cdot \nabla f dV.$$

By the Divergence Theorem,

$$\iint_S \mathbf{a} \cdot (f\mathbf{n})dS = \iint_S \mathbf{F} \cdot \mathbf{n}dS = \iiint_D \text{div}\mathbf{F}dV = \iiint_D \mathbf{a} \cdot \nabla f dV$$

and

$$\mathbf{a} \cdot \left(\iint_S f\mathbf{n}dS\right) = \mathbf{a} \cdot \left(\iiint_D \nabla f dV\right) \text{ or } \mathbf{a} \cdot \left(\iint_S f\mathbf{n}dS - \iiint_D \nabla f dV\right) = 0.$$

Since $\mathbf{a}$ is arbitrary,

$$\iint_S f\mathbf{n}dS - \iiint_D \nabla f dV = 0 \text{ and } \iint_S f\mathbf{n}dS = \iiint_D \nabla f dV.$$

Chapter 15 in Review

A. True/False

1. True; the value is 4/3.

3. False; $\int_C xdx + x^2dy = 0$ from $(-1,0)$ to $(1,0)$ along the x-axis and along the semicircle $y = \sqrt{1-x^2}$, but since $xdx + x^2dy$ is not exact, the integral is not independent of path.

5. True; assuming that the first partial derivatives are continuous.

7. True

9. True

11. True

B. Fill In the Blanks

1. $\mathbf{F} = \nabla\phi = -x(x^2 + y^2)^{-3/2}\mathbf{i} - y(x^2 + y^2)^{-3/2}\mathbf{j}$

3. $2xy + 2xy + 2xy = 6xy$

5. $\dfrac{\partial}{\partial x}(2xz) - \dfrac{\partial}{\partial y}(2yz) + \dfrac{\partial}{\partial z}(y^2 - x^2) = 0$

7. 0; since $(y - 7e^{x^3})dx + (x + \ln\sqrt{y})dy$ is exact.

9. At $u = 1$, $v = 4$, we have $\mathbf{r} = \langle 1, 4, 4 \rangle$.

$\dfrac{\partial\mathbf{r}}{\partial u}(1,4) = \langle 1, 0, 2 \rangle$, $\dfrac{\partial\mathbf{r}}{\partial v}(1,4) = \langle 0, 1, 1/2 \rangle$

A normal vector is given by $\mathbf{n} = \begin{vmatrix} \mathbf{i} & \mathbf{j} & \mathbf{k} \\ 0 & 1 & 2 \\ 0 & 1 & 1/2 \end{vmatrix} = \langle -2, -1/2, 1 \rangle$.

The tangent plane is $-2(x-1) - \frac{1}{2}(y-4) + (z-4) = 0$ or $4x + y - 2z = 0$.

C. Exercises

1. $\displaystyle\int_C \frac{z^2}{x^2 + y^2}ds = \int_\pi^{2\pi} \frac{4t^2}{\cos^2 2t + \sin 2t}\sqrt{4\sin^2 2t + 4\cos^2 2t + 4}\,dt = \int_\pi^{2\pi} 8\sqrt{2}t^2\,dt$

$\qquad = \dfrac{8\sqrt{2}}{3}t^3\Big|_\pi^{2\pi} = \dfrac{56\sqrt{2}\pi^3}{3}$

3. Since $P - y = 6x^2y = Q_x$, the integral is independent of path.
$\phi_x = 3x^2y^2$, $\phi = x^3y^2 + g(y)$, $\phi_y = 2x^3y + g'(y) = 2x^3y - 3y^2$;
$g(y) = -y^3$; $\phi = x^3y^2 - y^3$;
$\displaystyle\int_{(0,0)}^{(-1,2)} 3x^2y^2\,dx + (2x^3y - 3y^2)\,dy = (x^3y^2 - y^3)\big|_{(0,0)}^{(-1,2)} = -12$

5. $\displaystyle\int_C y\sin\pi z\,dx + x^2e^y\,dy + 3xyz\,dz$

$\qquad = \displaystyle\int_0^1 [t^2\sin\pi t^3 + t^2e^{t^2}(2t) + 3tt^2t^3(3t^2)]\,dt = \int_0^1 (t^2\sin\pi t^3 + 2t^3e^{t^2} + 9t^8)\,dt$

$\qquad = \left(-\dfrac{1}{3\pi}\cos\pi t^3 + t^9\right)\Big|_0^1 + 2\displaystyle\int_0^1 t^3e^{t^2}\,dt \quad \boxed{\text{Integration by parts}}$

$\qquad = \dfrac{2}{3\pi} + 1 + (t^2e^{t^2} - e^{t^2})\Big|_0^1 = \dfrac{2}{3\pi} + 2$

7. Let $\mathbf{r}_1 = \dfrac{\pi}{2}t\mathbf{i}$ and $\mathbf{r}_2 = \dfrac{\pi}{2}\mathbf{i} + \pi t\mathbf{j}$ for $0 \le t \le 1$. Then $d\mathbf{r}_1 = \dfrac{\pi}{2}\mathbf{i}$, $d\mathbf{r}_2 = \pi\mathbf{j}$, $\mathbf{F}_1 = 0$,

$\qquad \mathbf{F}_2 = \dfrac{\pi}{2}\sin\pi t\mathbf{i} + \pi t\sin\dfrac{\pi}{2}\mathbf{j} = \dfrac{\pi}{2}\sin\pi t\mathbf{i} + \pi t\mathbf{j}$,

and

$\qquad W = \displaystyle\int_{C_1} \mathbf{F}_1 \cdot d\mathbf{r}_1 + \int_{C_2} \mathbf{F}_2 \cdot d\mathbf{r}_2 = \int_0^1 \pi^2 t\,dt = \dfrac{1}{2}\pi^2 t^2\Big|_0^1 = \dfrac{\pi^2}{2}.$

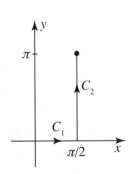

9. $P_y = 2x = Q_x$, $Q_z = 2y = R_y$, $R_x = 0 = P_z$ and the integral is independent of path. Parameterize the line segment between the points by $x = 1$, $y = 1$, $z = t$, $0 \le t \le \pi$. Then $dx = dy = 0$, $dz = dt$, and

$\qquad \displaystyle\int_{(1,1,0)}^{(1,1,\pi)} 2xy\,dx + (x^2 + 2yz)\,dy + (y^2 + 4)\,dz = \int_0^\pi [2(0) + (1 + 2t)(0) + (1 + 4)]\,dt = 5\pi.$

11. Using Green's Theorem,

$$\oint_C -4y\,dx + 8x\,dy = \int\int_R [8-(-4)]\,dA = 12\int\int_R dA = 12 \times (\text{area of } R)$$
$$= 12(16\pi - \pi) = 180\pi.$$

13. $z_x = 2x,\ z_y = 0;\ dS = \sqrt{1+4x^2}\,dA$

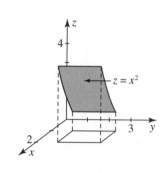

$$\int\int_S \frac{z}{xy}\,dS = \int_1^3 \int_1^2 \frac{x^2}{xy}\sqrt{1+4x^2}\,dx\,dy = \int_1^3 \frac{1}{y}\left[\frac{1}{12}(1+4x^2)^{3/2}\right]\Big|_1^2 dy$$

$$= \frac{1}{12}\int_1^3 \frac{17^{3/2}-5^{3/2}}{y}\,dy = \frac{17\sqrt{17}-5\sqrt{5}}{12}\ln y\Big|_1^3$$

$$= \frac{17\sqrt{17}-5\sqrt{5}}{12}\ln 3$$

15. The surface is $g(x,y,z) = y + e^{-x} - 2 = 0$. Then $\nabla g = -e^{-x}\mathbf{i}+\mathbf{j}$, $\mathbf{n} = (-e^{-x}\mathbf{i}+\mathbf{j})/\sqrt{e^{-2x}+1}$, and $dS = \sqrt{1+e^{-2x}}\,dA$.

$$\text{flux} = \int\int_S (\mathbf{F}\cdot\mathbf{n})\,dS = \int\int_R (-4e^{-x}+2-y)\,dA = \int_0^2\int_0^3 (-3e^{-x})\,dx\,dz = \int_0^2 3e^{-x}\Big|_0^3 dz$$

$$= \int_0^2 (3e^{-3}-3)\,dz = 6e^{-3}-6.$$

17. The surface is $g(x,y,z) = x^2+y^2+z^2-a^2 = 0$. $\nabla g = 2(x\mathbf{i}+y\mathbf{j}+z\mathbf{k}) = 2\mathbf{r}$ $\mathbf{n} = \mathbf{r}/|\mathbf{r}|$, $\mathbf{F} = c\nabla(1/|\mathbf{r}|) + c\nabla(x^2+y^2+z^2)^{-1/2} =$
$$c\frac{-x\mathbf{i}-y\mathbf{j}-z\mathbf{k}}{(x^2+y^2+z^2)^{3/2}} = c\mathbf{r}/|\mathbf{r}|^3$$

$$\mathbf{F}\cdot\mathbf{n} = -\frac{\mathbf{r}}{|\mathbf{r}|^3}\cdot\frac{\mathbf{r}}{|\mathbf{r}|} = -c\frac{\mathbf{r}\cdot\mathbf{r}}{|\mathbf{r}|^4} = -c\frac{|\mathbf{r}|^2}{|\mathbf{r}|^4} = -\frac{c}{|\mathbf{r}|^2} = -\frac{c}{a^2}$$

$$\text{flux} = \int\int_S \mathbf{F}\cdot\mathbf{n}\,dS = -\frac{c}{a^2}\int\int_S dS = -\frac{c}{a^2}\times(\text{area of } S) = -\frac{c}{a^2}(4\pi a^2) = -4\pi c$$

19. Since $\mathbf{F} = c\nabla(1/r)$, $\text{div}\mathbf{F} = \nabla\cdot(c\nabla(1/r)) = c\nabla^2(1/r) = c\nabla^2[(x^2+y^2+z^2)^{-1/2}] = 0$ by Problem 31 in Section 17.5. Then, by the Divergence Theorem,

$$\text{flux}\mathbf{F} = \int\int_S \mathbf{F}\cdot\mathbf{n}\,dS = \int\int\int_D \text{div}\mathbf{F}\,dV = \int\int\int_D 0\,dV = 0.$$

21. Identify $\mathbf{F} = -2y\mathbf{i}+3x\mathbf{j}+10z\mathbf{k}$. Then $\text{curl}\mathbf{F} = 5\mathbf{k}$. The curve C lies in the plane $z = 3$, so $\mathbf{n} = \mathbf{k}$ and $dS = dA$. Thus,

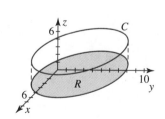

$$\oint_C \mathbf{F}\cdot d\mathbf{r} = \int\int_S (\text{curl}\mathbf{F})\cdot\mathbf{n}\,dS$$

$$= \int\int_R 5\,dA = 5\times(\text{area of } R) = 5(25\pi) = 125\pi.$$

23. $\text{div}\mathbf{F} = 1+1+1 = 3$;

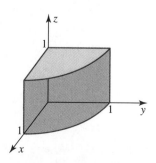

$$\int\int_S \mathbf{F}\cdot\mathbf{n}\,dS = \int\int\int_D \text{div}\mathbf{F}\,dV$$

$$= \int\int\int_D 3\,dV = 3\times(\text{volume of } D) = 3\pi$$

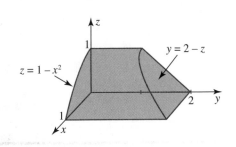

25. $\operatorname{div}\mathbf{F} = 2x + 2(x+y) - 2y = 4x$

$$\iint_S \mathbf{F} \cdot \mathbf{n}\,dS = \iiint_D \operatorname{div}\mathbf{F}\,dV = \iiint_D 4x\,dV$$

$$= \int_0^1 \int_0^{1-x^2} \int_0^{2-z} 4x\,dy\,dz\,dx$$

$$= \int_0^1 \int_0^{1-x^2} 4x(2-z)\,dz\,dx$$

$$= \int_0^1 \int_0^{1-x^2} (8x - 4xz)\,dz\,dx = \int_0^1 (8xz - 2xz^2)\Big|_0^{1-x^2}\,dx$$

$$= \int_0^1 [8x(1-x^2) - 2x(1-x^2)^2]\,dx$$

$$= \left[-2(1-x^2)^2 + \frac{1}{3}(1-x^2)^3\right]\Big|_0^1 = \frac{5}{3}$$

27. $x^2 - y^2 = u^2(\cosh v)^2 - u^2(\sinh v)^2$ hyperbolic paraboloid

$$= u^2\left[(\cosh v)^2 - (\sinh v)^2\right]$$

$$u^2 = z;$$

29. $y = x^2$; parabolic cylinder

Higher-Order Differential Equations

16.1 Exact First-Order Equations

1. Since $P_y = 0 = Q_x$, the equation is exact.

   $f_x = 2x + 4$, $f = x^2 + 4x + g(y)$, $f_y = g'(y) = 3y - 1$, $g(y) = \frac{3}{2}y^2 - y$

   The solution is $x^2 + 4x + \frac{3}{2}y^2 - y = C$.

3. Since $P_y = 4 = Q_x$, the equation is exact.

   $f_x = 5x + 4y$, $f = \frac{5}{2}x^2 + 4xy + g(y)$, $f_y = 4x + g'(y) = 4x - 8y^3$, $g(y) = -2y^4$

   The solution is $\frac{5}{2}x^2 + 4xy - 2y^4 = C$.

5. Since $P_y = 4xy = Q_x$, the equation is exact.
   $f_x = 2y^2x - 3$, $f = y^2x^2 - 3x + g(y)$, $f_y = 2yx^2 + g'(y) = 2yx^2 + 4$, $g(y) = 4y$
   The solution is $y^2x^2 - 3x + 4y = C$.

7. $(x^2 - y^2)dx + (x^2 - 2xy)dy = 0$. Since $P_y = -2y$ and $Q_x = 2x - 2y$, the equation is not exact.

9. $(y^3 - y^2\sin x - x)dx + (3xy^2 + 2y\cos x)dy = 0$. Since $P_y = 3y^2 - 2y\sin x = Q_x$, the equation is exact.
   $f_x = y^3 - y^2\sin x - x$, $f = xy^3 + y^2\cos x - \frac{1}{2}x^2 + g(y)$, $f_y = 3xy^2 + 2y\cos x + g'(y) = 3xy^2 + 2y\cos$, $g(y) = 0$
   The solution is $xy^3 + y^2\cos x - \frac{1}{2}x^2 = C$

11. Since $P_y = 1 + \ln y + xe^{-xy}$ and $Q_x = \ln y$, the equation is not exact.

13. $(2xe^x - y + 6x^2)dx - xdy = 0$. Since $P_y = -1 = Q_x$, the equation is exact. $f_x = 2xe^x - y + 6x^2$, $f = 2xe^x - 2e^x - yx + 2x^3 + g(y)$, $f_y = -x + g'(y) = -x$, $g(y) = 0$
    The solution is $2xe^x - 2e^x - yx + 2x^3 = C$.

15. Since $P_y = 3x^2y^2 = Q_x$, the equation is exact.

    $f_y = x^3y^2$, $f = \frac{1}{3}x^3y^3 + g(x)$, $f_x = x^2y^3 + g'(x) = x^2y^3 - \frac{1}{1=9x^2}$, $g'(x) = -\frac{1}{1+9x^2} = -\frac{1}{9}\frac{1}{1/9+x^2}$,

    $g(x) = -\frac{1}{9}\frac{1}{1/3}\tan^{-1}\frac{x}{1/3} = -\frac{1}{3}\tan^1 3x$

    The solution is $\frac{1}{3}x^3y^3 - \frac{1}{3}\tan^{-1}3x = C$ or $x^3y^3 = \tan - 13x = C_1$.

17. Since $P_y = \sin x\cos y = Q_x$, the equation is exact.
    $f_y = \cos x\cos y$, $f = \cos x\sin y + g(x)$, $f_x = -\sin x\sin y + g'(x) = \tan x - \sin x\sin y$, $g'(x) = \tan x$, $g(x) = \ln|\sec x|$
    The solution is $\cos x\sin y + \ln|\sec x| = C$ or $\cos x\sin y - \ln|\cos x| + C$.

19. Since $P_y = 4t^3 - 1 = Q_t$, the equation is exact.
    $f_t = 4t^3y - 15t^2 - y$, $f = t^4y - 5t^3 - yt + g(y)$, $f_y = t^4 - t + g'(y) = t^4 + 3y^2 - t$, $g'(y) - 3y^2$, $g(y) = y^3$.
    The solution is $t^4y - 5t^3 - yt + y^3 = C$.

21. Since $P_y = 2(x+y) = Q_x$, the equation is exact.

$f_x = (x+y)^2 = x^2 + 2xy + y^2$, $f = \frac{1}{3}x^3 + x^2y + xy^2 + g(y)$, $f_y = x^2 + 2xy + g'(y) = 2xy + x^2 - 1$

$g'(y) = -1$, $g(y) = -y$ A family of solutions is $\frac{1}{3}x^3 + x^2y + xy^2 - y = C$. Substituting $x = 1$ and $y = 1$ we obtain $\frac{1}{3} + 1 + 1 - 1 =$

$\frac{4}{3} = C$. The solution subject to the given condition is $\frac{1}{3}x^3 + x^2y + xy^2 - y = \frac{4}{3}$.

23. Since $P_y = 4 = Q_t$, the equation is exact.

$f_t = 4y + 2t - 5$, $f = 4ty + t^2 - 5t + g(y)$, $f_y = 4t + g'(y) = 6y + 4t - 1, g'(y) = 6y - 1$, $g(y) = 3y^2 - y$ A family of solutions is $4ty + t^2 - 5t + 3y^2 - y = C$. Substituting $t = -1$ and $y = 2$ we obtain $-8 + 1 + 5 + 12 - 2 = 8 = C$. The solution subject to the given condition is $4ty + t^2 - 5t + 3y^2 - y = 8$.

25. We want $P_y = Q_x$ or $3y^2 + 4kxy^3 = 3y^2 + 40xy^3$. Thus, $4k = 40$ and $k = 10$.

27. We need $P_y = Q_x$, so we must have $\frac{\partial M}{\partial y} = e^{xy} + xye^{xy} + 2y - \frac{1}{x^2}$. This gives $M(x,y) = \frac{1}{x}e^{xy} + \frac{(yx-1)e^{xy}}{x} + y^2 - \frac{y}{x^2} + g(x)$ for some function g.

29. Let $\mu(x,y = y^3$. Then $\frac{\partial}{\partial y}[\mu(x,y)M(x,y)] = \frac{\partial}{\partial y}[xy^4] = 4xy^3$

$\frac{\partial}{\partial z}[\mu(x,y)N(x,y)] = \frac{\partial}{\partial x}[2x^2y^3 + 3y^5 - 20y^3] = 4xy^3$

Therefore, $\mu(x,y)M(x,y)dx + \mu(x,y)N(x,y) = 0$ is exact, and $\mu(x,y)$ is an integrating factor.

Now, if $y^3[xydx + (2x^2 + 3y^2 - 20)dy] = 0$, then $xydx + (2x^2 + 3y^2 - 20)dy = 0$, provided $y \neq 0$. Therefore, to solve the original DE, we solve $xy^4dx + (2x^2y^3 + 3y^5 - 20y^3)dy = 0$.

$f_x = xy^4$, $f = \frac{1}{2}x^2y^4 + g(y)$, $f_y = 2x^2y + g'(y) = 2x^2y^3 + 3y^5 - 20y^3$,

$g'(y) = 3y^5 - 20y$, $g(y) = \frac{1}{2}y^6 - 5y^4$, $f = \frac{1}{2}x^2y^4 + \frac{1}{2}y^6 - 5y^4$.

The solution is therefore $\frac{1}{2}x^2y^4 + \frac{1}{2}y^6 - 5y^4 = C$.

16.2 Homogeneous Linear Equations

1. $3m^2 - m = 0 \Longrightarrow m(3m-1) = 0 \Longrightarrow m = 0, \ 1/3; \ y + C_1 + C_2e^{x/3}$

3. $m^2 - 16 = 0 \Longrightarrow m^2 = 16 \Longrightarrow m = -4, \ 4; \ y = C_1e^{-4x} + C_2e^{4x}$

5. $m^2 + 9 = 0 \Longrightarrow m^2 = -9 \Longrightarrow m = -3i, \ 3i; \ y = C_1\cos 3x + C_2\sin 3x$

7. $m^2 - 3m + 2 = 0 \Longrightarrow (m-1)(m-2) = 0 \Longrightarrow m = 1, \ 2; \ y = C_1e^x + C_2e^{2x}$

9. $m^2 + 8m + 16 = 0 \Longrightarrow (m+4)^2 = 0 \Longrightarrow m = -4, \ -4; \ y = C_1e^{-4x} + C_2xe^{-4x}$

11. $m^2 + 3m - 5 = 0 \Longrightarrow m = -3/2 \pm \sqrt{29}/2; \ y = C_1e^{(-3/2-\sqrt{29}/2)x} + C_2e^{(-3/2+\sqrt{29}/2)x}$

13. $12m^2 - 5m - 2 = 0 \Longrightarrow (3m-2)(4m+1) = 0 \Longrightarrow m = -1/4, \ 2/3; \ y = C_1e^{-x/4} + C_2e^{2x/3}$

15. $m^2 - 4m + 5 = 0 \Longrightarrow m = 2 \pm i; \ y = e^{2x}(C_1\cos x + C_2\sin x)$

17. $3m^2 + 2m + 1 = 0 \Longrightarrow m = -1/3 \pm (\sqrt{2}/3)i; \ y = e^{-x/3}\left(C_1\cos\frac{\sqrt{2}}{3}x + C_2\sin\frac{\sqrt{2}}{3}x\right)$

19. $9m^2 + 6m + 1 = 0 \Longrightarrow (3m+1)^2 = 0 \Longrightarrow m = -1/3, \ -1/3; \ y = C_1e^{-x/3} + C_2xe^{-x/3}$

21. $m^2 + 16 = 0 \Longrightarrow m^2 = -16 \Longrightarrow m = \pm 4i; \ y = C_1\cos 4x + C_2\sin 4x; \ y' = -4C_1\sin 4x + C_2\cos 4x$ Using $y(0) = 2$ we obtain $2 = C_1$. Using $y'(0) = -2$ we obtain $-2 = 4C_2$ or $C_2 = -1/2$. The solution is $y = 2\cos 4x - \frac{1}{2}\sin 4x$.

23. $m^2 + 6m + 5 = 0 \Longrightarrow (m+1)(m+5) = 0 \Longrightarrow m = -5, \ -1; \ y = C_1e^{-5x} + C_2e^{-x}; y' = -5C_1e^{-5x} - C_2e^{-x}$. Using $y(0) = 0$ and $y'(0) = 3$ we obtain the system $C_1 + C_2 = 0, -5C_1 - C_2 = 3$. Thus, $C_1 = -3/4$ and $c_2 = 3/4$. The solution is $y = -\frac{3}{4}e^{-5x} + \frac{3}{4}e^{-x}$.

25. $2m^2 - 2m + 1 = 0 \Longrightarrow m = 1/2 \pm (1/2)i$; $y = e^{x/2}(C_1 \cos \frac{1}{2}x + C_2 \sin \frac{1}{2}x)$; $y' = e^{x/2}[\frac{1}{2}(C_1 + C_2) \cos \frac{1}{2}x - \frac{1}{2}(C_1 - C_2) \sin \frac{1}{2}x]$.

Using $y(0) = -1$ and $y'(0) = 0$ we obtain the system $C_1 = -1, dfrac12 C_1 + \frac{1}{2}C_2 = 0$. Thus, $C_1 = -1$ and $C_2 = 1$. The solution is $y = e^{x/2}\left(\sin \frac{1}{2}x - \cos \frac{1}{2}x\right)$.

27. $m^2 + m + 2 = 0 \Longrightarrow m = -1/2 \pm (\sqrt{7}/2)i$; $y = e^{-x/2}(C_1 \cos \frac{\sqrt{7}}{2}x + C_2 \sin \frac{\sqrt{7}}{2}x)$;

$y' = e^{-x/2}\left[(-\frac{1}{2}C_1 + \frac{\sqrt{7}}{2}C_2) \cos \frac{\sqrt{7}}{2}x + (-\frac{\sqrt{7}}{2}C_1 - \frac{1}{2}C_2) \sin \frac{\sqrt{7}}{2}x\right]$.

Using $y(0) = y'(0) = 0$ we obtain the system $C_1 = 0$, $-\frac{1}{2}C_1 + \frac{\sqrt{7}}{2}C_2 = 0$. Thus, $C_1 = C_2 = 0$. The solution is $y = 0$.

29. $m^2 - 3m + 2 = 0 \Longrightarrow (m-1)(m-2) = 0 \Longrightarrow m = 1, \ 2$; $y = C_1 e^x + C_2 e^{2x}$; $y' = C_1 e^x + 2C_2 e^{2x}$. Using $y(1) = 0$ and $y'(1) = 1$ we obtain the system $eC_1 + e^2 C_2 = 0$, $eC_1 + 2e^2 C_2 = 1$. Thus, $C_1 = -e^{-1}$ and $C_2 = e-2$. The solution is $y = -e^{x-1} + e^{2x-2}$.

31. The auxiliary equation is $(m-4)(m+5) = m^2 + m - 20 = 0$. The differential equation is $y'' + y' - 20y = 0$.

33. The auxiliary equation is $m^2 + 1 = 0$, so $m = \pm i$. The general solution is $y = C_1 \cos x + C_2 \sin x$. The boundary conditions yield $y(0) = C_1 = 0$, $y(\pi) = -C_1 = 0$, so $y = C_2 \sin x$.

35. The general solution is $y = C_1 \cos x + C_2 \sin x$. The boundary conditions yield $y'(0) = C_2 = 0$, $y'\left(\frac{1}{2}\right) = -C_1 = 2$, so $y = -2 \cos x$.

37. The auxiliary equation is $m^2 - 2m + 2 = 0$, so $m = 1 \pm i$. The general solution is $y = e^x (C_1 \cos x + C_2 \sin x)$. The boundary conditions yield $y(0) = C_1 = 1$ and $y(\pi) = -e^\pi C_1 = -1$, which is a contradiction. No solution.

39. The auxiliary equation is $m^2 - 4m + 4 = 0$, so $m = 2$ is a repeated root. The general solution is $y = C_1 e^{2x} + C_2 x e^{2x}$. The boundary conditions yield $y(0) = C_1 = 0$ and $y(1) = C_2 e^2 = 1$, so $y = xe^{-2} e^{2x} = xe^{2(x-1)}$.

41. Assuming a solution of the form $y = e^{mx}$ we obtain the auxiliary equation $m^3 - 9m^2 + 25m - 17 = 0$. Since $y_1 = e^x$ is a solution we know that $m_1 = 1$ is a root of the auxiliary equation. The equation can then be written as $(m-1)(m^2 - 8m + 17) = 0$. The roots of this equation are 1 and $4 \pm i$. The general solution of the differential equation is $y = C_1 e^x + e^{4x}(C_2 \cos x + C_3 \sin x)$.

43. $y' = me^{mx}$, $y'' = m^2 e^{mx}$, $y''' = m^3 e^{mx}$; $m^3 e^{mx} - 4m^2 e^{mx} - 5me^{mx} = 0 \Longrightarrow (m^3 - 4m^2 - 5m)e^{mx} = 0$ $\Longrightarrow m^3 - 4m^2 - 5m = 0 \Longrightarrow m(m-5)(m+1) = 0 \Longrightarrow m = 0, \ -1, \ 5$; $y = C_1 + C_2 e^{-x} + C_3 e^{5x}$.

45. Case 1: $\lambda = -\alpha^2 < 0$
Auxiliary equation is $m^2 - \alpha^2 = 0$, so $m = \pm \alpha$ and general solution is $y = C_1 e^{\alpha x} + C_2 e^{-\alpha x}$. Boundary conditions yield $y(0) = C_1 + C_2 = 0$ and $y(1) = C_1 e^\alpha + C_2 e^{-\alpha} = 0$, or $C_1 = C_2 = 0$. So Case 1 yields no nonzero solutions.

Case 2: $\lambda = 0$
Auxiliary equation is $m^2 = 0$, so $m = 0$ is a repeated root and general solution is $y = C_1 + C_2 x$. Boundary conditions yield $y(0) = C_1 = 0$ and $y(1) = C_2 = 0$. So Case 2 yields no nonzero solutions.

Case 3: $\lambda = \alpha^2 > 0$
Auxiliary equation is $m^2 + \alpha^2 = 0$, so $m = \pm \alpha i$ and the general solution is $y = C_1 \cos \alpha x + C_2 \sin x$. Boundary conditions yield $y(0) = C_1 = 0$ and $y(1) = C_2 \sin \alpha = 0$. Hence, nonzero solutions exist only when $\sin \alpha = 0$, which implies $\alpha = \pm n\pi$ so that $\lambda = n^2 \pi^2$ for $n = 1, 2, 3, \ldots$ ($n = 0$ is excluded since that would give $\lambda = 0$).

16.3 Nonhomogeneous Linear Equations

1. $m^2 - 9 = 0 \Longrightarrow m = -3, \ 3$; $y_c = C_1 e^{-3x} + C_2 e^{3x}$; $y_p = A$, $y_p' = y_p'' = 0$; $-9A = 54 \Longrightarrow A = -6$; $y_p = -6$; $y = C_1 e^{-3x} + C_2 e^{3x} - 6$

3. $m^2 + 4m + 4 = 0 \Longrightarrow (m+2)^2 = 0 \Longrightarrow m = -2, \ -2$; $y_c = C_1 e^{-2x} + C_2 x e^{-2x}$; $y_p = Ax + B$, $y_p' = A$, $y_p'' = 0$; $4A + 4(Ax + B) = 2x + 6 \Longrightarrow 4Ax + 4(A+B) = 2x + 6$

Solving $4A = 2$, $4A + 4B = 6$, we obtain $A = 1/2$ and $B = 1$. Thus, $y = C_1 e^{-2x} + C_2 x e^{-2x} + \frac{1}{2}x + 1$.

5. $m^2 + 25 = 0 \implies m = \pm 5i; y_c = C_1 \cos 5x + C_2 \sin 5x; y_p = A \sin x + B \cos x, \ y_p' = A \cos x - B \sin x, \ y_p'' = -A \sin x - B \cos x;$
$-A \sin x - B \cos x + 25(A \sin x + B \cos x) = 6 \sin x \implies 24A \sin x + 24B \cos x = 6 \sin x; \quad A = 1/4, \quad B = 0; \quad y = C_1 \cos 5x + C_2 \sin 5x + \dfrac{1}{4} \sin x$

7. $m^2 - 2m - 3 = 0 \implies (m-3)(m+1) = 0 \implies m = -1, \ 3; \ y_c = C_1 e^{-x} + C_2 e^{3x} y_p = Ae^{2x} + Bx^3 + Cx^2 + Dx + E, \ y_p' = 2Ae^{2x} + 3Bx^2 + 2Cx + D, \ y_p'' = 4Ae^{2x} + 6Bx + 2C(4Ae^{2x} + 6Bx + 2C) - 2(2Ae^{2x} + 3Bx^2 + 2Cx + D) - 3(Ae^{2x} + Bx^3 + Cx^2 + Dx + E) = 4e^{2x} + 2x^3 \implies -3Ae^{2x}3Bx^3 + \ '(-6B - 3C)x^2 + (6B - 4C - 3D)x + (2C - 2D - 3E) = 4e^{2x} + 2x^3$ Solving $-3A = 4, \ -3B = 2, \ -6B - 3C = 0, \ 6B - 4C - 3D = 0, \ 2C - 2D - 3E = 0,$ we obtain $A = -4/3, \ B = -2/3, \ C = 4/3, \ D = -28/9,$ and $E = 80/27$. Thus,

$$y = C_1 e^{-x} + C_2 e^{3x} - \frac{4}{3}e^{2x} - \frac{2}{3}x^3 + \frac{4}{3}x^2 - \frac{28}{9}x + \frac{80}{27}.$$

9. $m^2 - 8m + 25 = 0 \implies m = 4 \pm 3i; \ y_c = e^{4x}(C_1 \cos 3x + C_2 \sin 3x); \ y_p = Ae^{3x} + B \sin 2x + C \cos 2x, \ y_p' = 3Ae^{3x} + 2B \cos 2x - 2C \sin 2x, \ y_p'' = 9Ae^{3x} - 4B \sin 2x - 4C \cos 2x(9Ae^{3x} - 4B \sin 2x - 4C \cos 2x) - 8(3Ae^{3x} + 2B \cos 2x - 2C \sin 2x) + 25(Ae^{3x} + B \sin 2x + C \cos 2x) = e^{3x} - 6 \cos 2x \implies 10Ae^{3x} + (21B + 16C) \sin 2x + (-16B + 21C) \cos 2x = e^{3x} - 6 \cos 2x$
Solving $10A = 1, \ 21B + 16C = 0, \ -16B + 21C = -6,$ we obtain $A = 1/10, \ B = 96/697,$ and $C = -126/697$. Thus,

$$y = e^{4x}(C_1 \cos 3x + C_2 \sin 3x) + \frac{1}{10}e^{3x} + \frac{96}{697} \sin 2x - \frac{126}{697} \cos 2x.$$

11. $m^2 - 64 = 0 \implies m = -8, \ 8; \ y_c = C_1 e^{-8x} + C_2 e^{8x}; \ y_p = A, \ y_p' = y_p'' = 0$
$-64A = 16 \implies A = -1/4; \ y = C_1 e^{-8x} + C_2 e^{8x} - \dfrac{1}{4}, \ y' = -8C_1 e^{-8x} + 8C_2 e^{8x}.$
Using $y(0) = 1$ and $y'(0) = 0$ we obtain $C_1 + C_2 - \dfrac{1}{4} = 1, \ -8C_1 + 8C_2 = 0,$ or $C_1 = C_2 = 5/8$. Thus, $y = \dfrac{5}{8}e^{-8x} + \dfrac{5}{8}e^{8x} - \dfrac{1}{4}.$

13. $m^2 + 1 = 0 \implies m = -i, \ i; \ y_c = C_1 \cos x + C_2 \sin x; \ W = \begin{vmatrix} \cos x & \sin x \\ -\sin x & \cos x \end{vmatrix} = 1$
$u_1' = -\sin x \sec x = -\tan x, \ u_1 = \ln|\cos x|; \ u_2' = \cos x \sec x = 1, u_2 = x$
$y_p = \cos x \ln|\cos x| + x \sin x; \ y = C_1 \cos x + C_2 \sin x + \cos x \ln|\cos x| + x \sin x$

15. $m^2 + 1 = 0 \implies m = -i, \ i; \ y_c = C_1 \cos x + C_2 \sin x; \ W = \begin{vmatrix} \cos x & \sin x \\ -\sin x & \cos x \end{vmatrix} = 1$
$u_1' = -\sin^2 x, \ u_1 = -\dfrac{1}{2}x + \dfrac{1}{2}\sin x \cos x; \ u_2' = \sin x \cos x, \ u_2 = \dfrac{1}{2}\sin^2 x$
$y_p = -\dfrac{1}{2}x \cos x + \dfrac{1}{2}\sin x \cos^2 x + \dfrac{1}{2}\sin^3 x$
$y = C_1 \cos x + C_2 \sin x - \dfrac{1}{2}x \cos x + \dfrac{1}{2}\sin x \cos^2 x + \dfrac{1}{2}\sin^3 x$
$= C_1 \cos x + C_2 \sin x - \dfrac{1}{2}x \cos x + \dfrac{1}{2}\sin x(\cos^2 x + \sin^2 x) = C_1 \cos x + C_3 \sin x - \dfrac{1}{2}x \cos x$

17. $m^2 + 1 = 0 \implies m = -i, \ i; \ y_c = C_1 \cos x + C_2 \sin x; \ W = \begin{vmatrix} \cos x & \sin x \\ -\sin x & \cos x \end{vmatrix} = 1$
$u_1' = -\sin x \cos^2 x \ u_1 = \frac{1}{3}\cos^3 x; \ u_2' = \cos x \cos^2 x = \cos x - \cos x \sin^2 x, \ u_2 = \sin x - \frac{1}{3}\sin^3 x$
$y_p = \frac{1}{3}\cos^4 x + \sin^2 x - \frac{1}{3}\sin^4 x = \sin^2 x + \frac{1}{3}(\cos^2 x - \sin^2 x) = \sin^2 x + \frac{1}{3}\cos 2x$
$y = C_1 \cos x + C_2 \sin x + \sin^2 x + \dfrac{1}{3}\cos 2x = C_1 \cos x + C_2 \sin x + \dfrac{1}{2} - \dfrac{1}{2}\cos 2x + \dfrac{1}{3}\cos 2x$
$= C_1 \cos x + C_2 \sin x + \dfrac{1}{2} - \dfrac{1}{6}\cos 2x$

19. $m^2 - 1 = 0 \implies m = -1, \ 1; \ y_c = C_1 e^{-x} + C_2 e^x; \ W = \begin{vmatrix} e^{-x} & e^x \\ -e^{-x} & e^x \end{vmatrix} = 2$
$u_1' = \dfrac{1}{2}e^x \cosh x = -\dfrac{1}{4}(e^{2x} + 1), \ u_1 = -\dfrac{1}{8}e^{2x} - \dfrac{1}{4}x;$
$u_2' = \dfrac{1}{2}e^{-x} \cosh x = \dfrac{1}{4}(1 + e^{-2x}) \ u_2 = \dfrac{1}{4}x - \dfrac{1}{8}e^{-2x},$

$$y_p = e^{-x}\left(-\frac{1}{8}e^{2x} - \frac{1}{4}x\right) + e^x\left(\frac{1}{4}x - \frac{1}{8}e^{-2x}\right) = -\frac{1}{8}e^x - \frac{1}{4}xe^{-x} + \frac{1}{4}xe^x - \frac{1}{8}d^{-x}$$

$$= -\frac{1}{8}e^x - \frac{1}{8}e^{-x} + \frac{1}{2}x\sinh x$$

$$y = C_1 e^{-x} + C_2 e^x - \frac{1}{8}e^x - \frac{1}{8}e^{-x} + \frac{1}{2}x\sinh x = C_3 e^{-x} + C_4 e^x + \frac{1}{2}x\sinh x$$

21. $m^2 - 4 = 0 \implies m = -2, 2$; $y_c = C_1 e^{-2x} + C_2 e^{2x}$; $W = \begin{vmatrix} e^{-2x} & e^{2x} \\ -2e^{-2x} & 2e^{2x} \end{vmatrix} = 4$

$$u_1' = -\frac{1}{4}e^{2x}\left(\frac{e^{2x}}{x}\right) = -\frac{1}{4}\frac{e^{4x}}{x}, \quad u_1 = -\frac{1}{4}\int_{x_0}^x \frac{e^{4t}}{t}dt; \quad u_2' = \frac{1}{4}e^{-2x}\left(\frac{e^{2x}}{x} = \frac{1}{4x}\right), \quad u_2 = \frac{1}{4}\ln|x|$$

$$y_p = -\frac{1}{4}e^{-2x}\int_{x_0}^x \frac{e^{4t}}{t}dt + \frac{1}{4}e^{2x}\ln|x|; \quad y = C_1 e^{-2x} + C_2 e^{2x} - \frac{1}{4}e^{2x}\int_{x_0}^x \frac{e^{4t}}{t}dt + \frac{1}{4}e^{2x}\ln|x|$$

23. $m^2 + 3m + 2 = 0 \implies (m+2)(m+1) = 0 \implies m = -2, -1$; $y_c = C_1 e^{-2x} + C_2 e^{-x}$

$$W = \begin{vmatrix} e^{-2x} & e^{-x} \\ -2e^{-2x} & -e^{-x} \end{vmatrix} = e^{-3x}; \quad u_1' = -\frac{1}{e^{-3x}}\frac{e^{-x}}{1+e^x} = -\frac{e^{2x}}{1+e^x}$$

$$u_1 = -\int \frac{e^{2x}}{1+e^x}dx \quad \boxed{v = 1 + e^x, \quad dv = e^x dx, \quad e^x = v - 1}$$

$$= -\int \frac{v-1}{v}dv = -v + \ln|v| = -1 - e^x + \ln(1+e^x)$$

$$y_p = e^{-2x}[-1 - e^x + \ln(1+e^x)] + e^{-x}\ln(1+e^x) = -e^{-2x} - e^{-x} + e^{-2x}\ln(1+e^x)$$
$$+ e^{-x}\ln(1+e^x)$$

$$y = C_1 e^{-2x} + C_2 e^{-x} - e^{-2x} - e^{-x} + e^{-2x}\ln(1+e^x) + e^{-x}\ln(1+e^x)$$

$$= C_3 e^{-2x} + C_4 e^{-x} + e^{-2x}\ln(1+e^x) + e^{-x}\ln(1+e^x)$$

25. $m^2 + 3m + 2 = 0 \implies (m+2)(m+1) = 0 \implies m = -2, -1$; $y_c = C_1 e^{-2x} + C_2 e^{-x}$

$$W = \begin{vmatrix} e^{-2x} & e^{-x} \\ -2e^{-2x} & -e^{-x} \end{vmatrix} = e^{-3x}; \quad u_1' = -\frac{1}{e^{-3x}}e^{-x}\sin e^x = -e^{2x}\sin e^x$$

$$u_1 = -\int e^{2x}\sin e^x dx \quad \boxed{\text{Integration by parts}}$$

$$= e^x\cos e^x - \sin e^x$$

$$u_2' = \frac{1}{e^{-3x}}e^{-2x}\sin e^x = e^x\sin e^x, \quad u_2 = -\cos e^x$$

$$y_p = e^{-2x}(e^x\cos e^x - \sin e^x) + e^{-x}(-\cos e^x) = -e^{-2x}\sin e^x; \quad y = C_1 e^{-2x} + C_2 e^{-x} - e^{-2x}\sin e^x$$

27. $m^2 - 2m + 1 = 0 \implies (m-1)^2 = 0 \implies m = 1, 1$; $y_c = C_1 e^x + C_2 xe^x$;

$$W = \begin{vmatrix} e^x & xe^x \\ e^x & xe^x + e^x \end{vmatrix} = e^{2x}; \quad u_1' = -\frac{1}{e^{2x}}xe^x\frac{e^x}{1+x^2} = -\frac{x}{1+x^2}, \quad u_1 = -\frac{1}{2}\ln(1+x^2);$$

$$u_2' = \frac{1}{e^{2x}}e^x\frac{e^x}{1+x^2} = \frac{1}{1+x^2}, \quad u_2 = \tan^{-1}x$$

$$y_p = -\frac{1}{2}e^x\ln(1+x^2) + xe^x\tan^{-1}x; \quad y = C_1 e^x + C_2 xe^x - \frac{1}{2}e^x\ln(1+x^2) + xe^x\tan^{-1}x$$

29. $m^2 + 2m + 1 = 0 \implies (m+1)^2 = 0 \implies m = -1, -1$; $y_c = C_1 e^{-x} + C_2 xe^{-x}$

$$W = \begin{vmatrix} e^{-x} & xe^{-x} \\ -e^{-x} & -xe^{-x} + e^{-x} \end{vmatrix} = e^{-2x}; \quad u_1' = -\frac{1}{e^{-2x}}xe^{-x}e^{-x}\ln x = -x\ln x$$

$$u_1 = -\int x\ln x dx \quad \boxed{\text{Integration by parts}}$$

$$= \frac{1}{4}x^2 - \frac{1}{2}x^2\ln x$$

$$u_2' = \frac{1}{e^{-2x}}e^{-x}e^{-x}\ln x = \ln x, \quad u_2 = x\ln x - x$$

$$y_p = e^{-x}\left(\frac{1}{4}x^2 - \frac{1}{2}x^2\ln x\right) + xe^{-x}(x\ln x - x) = \frac{1}{2}x^2 e^{-x}\ln x - \frac{3}{4}x^2 e^{-x}$$

$$y = C_1 e^{-x} + C_2 xe^{-x} + \frac{1}{2}x^2 e^{-x}\ln x - \frac{3}{4}x^2 e^{-x}$$

31. $4m^2 - 4m + 1 = 0 \implies (2m-1)^2 = 0 \implies m = 1/2, \ 1/2; \ y_c = C_1 e^{x/2} + C_2 x e^{x/2}$

$$W = \begin{vmatrix} e^{x/2} & xe^{x/2} \\ \frac{1}{2}e^{x/2} & \frac{1}{2}xe^{x/2} + e^{x/2} \end{vmatrix} = e^x; \ u_1' = -\frac{1}{e^x}xe^{x/2}(2e^{-x} + \frac{1}{4}x) = -2xe^{-3x/2} - \frac{1}{4}x^2 e^{-x/2}$$

$$u_1 = -2\int xe^{-3x/2}dx - \frac{1}{4}\int x^2 e^{-x/2}dx \quad \boxed{\text{Integration by parts}}$$

$$= \frac{4}{3}xe^{-3x/2} + \frac{8}{9}e^{-3x/2} + \frac{1}{2}x^2 e^{-x/2} + 2xe^{-x/2} + 4e^{-x/2}$$

$$u_2' = \frac{1}{e^x}e^{x/2}(2e^{-x} + \frac{1}{4}x) = 2e^{-3x/2} - \frac{1}{4}xe^{-x/2}$$

$$u_2 = 2\int e^{-3x/2}dx + \frac{1}{4}\int xe^{-x/2}dx \quad \boxed{\text{Integration by parts}}$$

$$= -\frac{4}{3}e^{-3x/2} - \frac{1}{2}xe^{-x/2} - e^{-x/2}$$

$$y_p = e^{x/2}(\frac{4}{3}xe^{-3x/2} + \frac{8}{9}e^{-3x/2} + \frac{1}{2}x^2 e^{-x/2} + 2xe^{-x/2} + 4e^{-x/2})$$

$$+ xe^{x/2}(-\frac{4}{3}e^{-3x/2} - \frac{1}{2}xe^{-x/2} - e^{-x/2}) = \frac{8}{9}e^{-x} + x + 4$$

$$y = C_1 e^{x/2} + C_2 x e^{x/2} + \frac{8}{9}e^{-x} + x + 4$$

33. $m^2 - 1 = 0 \implies m = -1, \ 1; \ y_c = C_1 e^{-x} + C_2 e^x; \ W = \begin{vmatrix} e^{-x} & e^x \\ -e^{-x} & e^x \end{vmatrix} = 2$

$$u_1' = \frac{1}{2}e^x x e^x = -\frac{1}{2}xe^{2x}$$

$$u_1 = -\frac{1}{2}\int xe^{2x}dx \quad \boxed{\text{Integration by parts}}$$

$$= \frac{1}{8}e^{2x} - \frac{1}{4}xe^{2x}$$

$$u_2' = \frac{1}{2}e^{-x}xe^x = \frac{1}{2}x, \ u_2 = \frac{1}{4}x^2; \ y_p = e^{-x}(\frac{1}{8}e^{2x} - \frac{1}{4}xe^{2x}) + e^x(\frac{1}{4}x^2) = \frac{1}{8}e^x - \frac{1}{4}xe^x + \frac{1}{4}x^2 e^x$$

$$y = C_1 e^{-x} + C_3 e^x - \frac{1}{4}xe^x + \frac{1}{4}x^2 e^x; \ y' = -C_1 e^{-x} + C_3 e^x - \frac{1}{4}e^x + \frac{1}{4}xe^x + \frac{1}{4}x^2 e^x$$

Using $y(0) = 1$ and $y'(0) = 0$ we have $C_1 + C_3 = 1, \ -C_1 + C_3 - \frac{1}{4} = 0$, or $C_1 = 3/8$ and $C_3 = 5/8$. Thus, $y = \frac{3}{8}e^{-x} + \frac{5}{8}e^x - \frac{1}{4}xe^x + \frac{1}{4}x^2 e^x$.

35. $y'' - \frac{1}{x}y' + \frac{1}{x^2}y = \frac{4}{x}\ln x; \ y_c = C_1 x + C_2 x \ln x; \ W = \begin{vmatrix} x & x\ln x \\ 1 & 1 + \ln x \end{vmatrix} = x$

$$u_1' = -\frac{1}{x}(x\ln x)(\frac{4}{x}\ln x) = -\frac{4}{x}(\ln x)^2, \ u_1 = -\frac{4}{3}(\ln x)^3; \ u_2' = \frac{1}{x}(x)(\frac{4}{x}\ln x) = \frac{4}{x}\ln x,$$

$$u_2 = 2(\ln x)^2; \ y_p = -\frac{4}{3}x(\ln x)^3 + 2x(\ln x)^3 = \frac{2}{3}x(\ln x)^3; \ y + C_1 x + C_2 x \ln x + \frac{2}{3}x(\ln x)^3$$

37. Writing the differential equation in the form $d^2 C/dx^2 - (1/\lambda^2)C = -C(\infty)/\lambda^2$ we see that the auxiliary equation is $m^2 - 1/\lambda^2 = 0$. Thus, $C_c = c_1 e^{x/\lambda} + c_2 e^{-x/\lambda}$. Using undetermined coefficients with $C_p = A$ we find that $A = C(\infty)$. Then $C(x) = c_1 e^{x/\lambda} + c_2 e^{-x/\lambda} + C(\infty)$. Since $C(0) = c_1 + c_2 + C(\infty) = 0$ and $\lim_{x\to\infty} C(x) = C(\infty)$ we see that $c_1 = 0$ and $c_2 = -C(\infty)$. Thus, $C(x) = C(\infty)(1 - e^{-x/\lambda})$.

39. (a) Substituting Ae^x in for y in the DE, we have $Ae^x + 2Ae^x - 3Ae^x = 10e^x$ or $0 = 10e^x$, which is a contradiction for any value of A.

    (b) Substituting Axe^x for y, we have
    $$A(x+2)e^x + A(2x+2)e^x - 3Axe^x = 10e^x.$$

    Equating coefficients of xe^x and coefficients of e^x, we get
    $$A + 2A - 3A = 0 \quad \text{and} \quad 2A + 2A = 10$$

    which gives $A = \frac{5}{2}$. Therefore, $y_p = \frac{5}{2}xe^x$.

(c) The auxiliary equation is $m^2 + 2m - 3 = 0$, so $m = -3$ or $m = 1$. This gives $y_c = C_1 e^{-3x} + C_2 e^x$. Therefore, the general solution is

$$y = y_c + y_p = C_1 e^{-3x} + C_2 e^x + \frac{5}{2} x e^x$$

16.4   Mathematical Models

1. A weight of 4 pounds is pushed up 3 feet above the equilibrium position. At $t = 0$ it is given an initial speed upward of 2 feet per second.

3. Using $m = W/g = 8/32 = 1/4$, the initial value problem is $\frac{1}{4} x'' + x = 0$; $x(0) = \frac{1}{2}$, $x'(0) = \frac{3}{2}$. The auxiliary equation is $\frac{1}{4} m^2 + 1 = 0$, so $m = \pm 2i$ and $x = C_1 \cos 2t + C_2 \sin 2t$, $x' = -2C_1 \sin 2t + 2C_2 \cos 2t$. Using the initial condition, we obtain $C_1 = 1/2$ and $C_2 = \frac{3}{4}$. The equation of motion is $x(t) = \frac{1}{2} \cos 2t + \frac{3}{4} \sin 2t$.

5. From Hooke's law we have $400 = k(2)$, so $k = 200$. The initial value problem is $50x'' + 200x = 0$; $x(0) = 0$, $x'(0) = -1 =$. The auxiliary equation is $50m^2 + 200 = 0$, so $m = \pm 2i$ and $x = C_1 \cos 2x + C_2 \sin 2x$, $x' = -2C_1 \sin 2x + 2C_2 \cos 2x$. Using the initial conditions, we obtain $C_1 = 0$ and $C_2 = -5$. Thus, $x(1) = -5 \sin 2x$.

7. A 2 pound weight is released from the equilibrium position with an upward speed of 1.5 ft/s. A damping force numerically equal to twice the instantaneous velocity acts on the system.

9. Using $m = W/g = 4/32 = 1/8$, the initial value problem is $\frac{1}{8} x'' + x' + 2x = 0$; $x(0) = -1$, $x'(0) = 8$. The auxiliary equation is $m^2/8 + m + 2 = 0$ or $(m + 4)^2 = 0$, so $m = -4$, -4 and $x = C_1 e^{-4t} + C_2 t e^{-4t}$, $x' = (C_2 - 4C_1)e^{-4t} - 4C_2 t e^{-4t}$. Using the initial conditions, we obtain $C_1 = -1$ and $C_2 = 4$. Thus, $x(t) = -e^{-4t} + 4t e^{-4t}$. Solving $x(t) = -e^{4t} + 4te^{-4t} = 0$, we see that the weight passes through the equilibrium position at $t = 1/4 s$. To find the maximum displacement we solve $x'(t) = 8e^{-4t} - 16te^{-4t} = 0$. This gives $t = 1/2$. Since $x(1/2) = e^{-2} \approx 0.14$, the maximum displacement is approximately 0.14 feet below the equilibrium position at $t = 1/2 s$.

11. From Hooke's law we have $10 = k(7 - 5)$, so $k = 5$. Using $m = W/g = 8/32 = 1/4$, the initial value problem is $\frac{1}{4} x'' + x' + 5x = 0$; $x(0) = \frac{1}{2}$; $x'(0) = 1$. The auxiliary equation is $m^2/4 + m + 5 = 0$ or $m^2 + 4m + 20 = 0$, so $m = -2 \pm 4i$. Thus, $x = e^{-2t}(C_1 \cos 4t + C_2 \sin 4t)$ and $x' = -2(C_1 - 2C_2)e^{-2t} \cos 4t - 2(2C_1 + C_2)e^{-2t} \sin 4t$. Using the initial conditions, we obtain $\frac{1}{2} = C_1$ and $1 = -2(\frac{1}{2} - 2C_2)$, so $C_1 = 1/2$ and $C_2 = 1/2$. Therefore $x(t) = \frac{1}{2} e^{-2t}(\cos 4t + \sin 4t)$.

13. From Hooke's law we have $10 = k(2)$, so $k = 5$. Using $m = W/g = 10/32 = 5/16$, the differential equation is $\frac{5}{16} x'' + \beta x' + 5 = 0$. The auxiliary is $\frac{5}{16} m^2 + \beta m + 5 = 0$ Using the quadratic formula, $m = (-\beta \pm \sqrt{\beta^2 - 25/4})/(5/8)$. For $\beta > 0$ the motion is

(a) overdamped when $\beta^2 - 25/4 > 0$ or $\beta > 5/2$

(b) critically damped when $\beta^2 - 25/4 = 0$ or $\beta = 5/2$

(c) underdamped when $\beta^2 - 25/4 < 0$ or $\beta < 5/2$.

15. The initial value problem is $x'' + 8x' + 16x = e^{-t} \sin 4t$; $x(0) = x'(0) = 0$. Using $x_p = Ae^{-t} \sin 4t + Be^{-t} \cos 4t$ we find $A = -7/625$ and $B = -24/625$. Thus,

$$x(t) = C_1 e^{-4t} + C_2 t e^{-4t} - \frac{7}{625} e^{-t} \sin 4t - \frac{24}{625} e^{-t} \cos 4t,$$

$$x'(t) = -4C_1 e^{-4t} - 4C_2 t e^{-4t} + C_2 e^{-4t} - \frac{28}{625} e^{-t} \cos 4t + \frac{7}{625} e^{-t} \sin 4t + \frac{96}{625} e^{-t} \sin 4t$$

$$+ \frac{24}{625} e^{-t} \cos 4t.$$

Using the initial conditions, we obtain $C_1 = 24/625$ and $C_2 = 100/625$. Thus,

$$x(t) = \frac{24}{625} e^{-4t} + \frac{100}{625} t e^{-4t} - \frac{7}{625} e^{-t} \sin 4t - \frac{24}{626} e^{-t} \cos 4t.$$

As $t \longrightarrow \infty$, $e^{-t} \longrightarrow 0$ and $x(t) \longrightarrow 0$.

17. The DE describing charge is $.05q'' + 2q' + 100q = 0$. The auxiliary equation is $0.5m^2 + 2m + 100 = 0$, so $m = -20 \pm 40i$. The general solution is $q = e^{-20t}(C_1 \cos 40t + C_2 \sin 40t)$. The initial conditions yield $q(0) = C_1 = 5$ and $i(0) = q'(0) = -20C_1 + 40C_2 = 0$, which gives $C_1 = 5$ and $C_2 = \frac{5}{2}$. Therefore $q(t) = e^{-20t}\left(5 \cos 40t + \frac{5}{2} \sin 40t\right)$, and $q(0.01) = 4.568C$. $q(t) = 0$ when $5 \cos 40t + \frac{5}{2} \sin 40t = 0$ which first occurs at $t = 0.0509$ s.

19. The DE is $\frac{5}{3}q'' + 10q' + 30q = 300$. The auxiliary equation is $\frac{5}{3}m^2 + 10m + 30 = 0$. so $m = -3 \pm 3i$. This gives $q_c = e^{-3t}(C_1 \cos 3t + C_2 \sin 3t)$. Assume a particular solution of the form $q_p = A$. Substituting into the DE, we have $30A = 300$ so that $A = 10$ and therefore $q_p = 10$. Thus, the general solution is $q = q_c + q_p = e^{-3t}(C_1 \cos 3t + C_2 \sin 3t) + 10$. The initial conditions yield $q(0) = C_1 + 10 = 0$ and $i(0) = q'(0) = -3(C_1 - C_2) = 0$, which gives $C_1 = -10$ and $C_2 = -10$. Therefore, $q(t) = e^{-3t}(-10 \cos 2t - 10 \sin 3t) + 10 = 10 - 10e^{-3t}(\cos 3t + \sin 3t)$, $i(t) = q'(t) = 60e^{-3t} \sin 3t$. The charge $q(t)$ attains a maximum of 10.432 C at $t = \frac{\pi}{3}$.

16.5 Power Series Solutions

1. $\displaystyle\underbrace{\sum_{n=2}^{\infty} n(n-1)c_n x^{n-2}}_{k=n-2} + \sum_{n=0}^{\infty} c_n x^n = \sum_{k=0}^{\infty} (k+2)(k+1)c_{k+2} x^k + \sum_{k=0}^{\infty} c_k x^k$

$$= \sum_{k=0}^{\infty} [(k+2)(k+1)c_{k+2} + c_k]x^k = 0$$

$(k+2)(k+1)c_{k+2} + c_k = 0; \quad c_{k+2} = -\dfrac{c_k}{(k+2)(k+1)}, \quad k = 0,1,2,\ldots$

$c_2 = -\dfrac{c_0}{2} = -\dfrac{c_0}{2!}, \quad c_3 = -\dfrac{c_1}{3 \cdot 2} = -\dfrac{c_1}{3!}, \quad c_4 = -\dfrac{c_2}{4 \cos 3} = \dfrac{c_0}{4 \cdot 3 \cdot 2!} = \dfrac{c_0}{4!},$

$c_5 = -\dfrac{c_3}{5 \cdot 4} = \dfrac{c_1}{5 \cdot 4 \cdot 3!} = \dfrac{c_1}{5!}, \quad c_6 = -\dfrac{c_4}{6 \cdot 5} = -\dfrac{c_0}{6 \cdot 5 \cdot 4!} = -\dfrac{c_0}{6!},$

$c_7 = -\dfrac{c_5}{7 \cdot 6} = -\dfrac{c_1}{7 \cdot 6 \cdot 5!} = -\dfrac{c_1}{7!}$

$y = c_0 \left[1 - \dfrac{1}{2!}x^2 + \dfrac{1}{4!}x^4 - \dfrac{1}{6!}x^6 + \cdots\right] + c_1 \left[x - \dfrac{1}{3!}x^3 + \dfrac{1}{5!}x^5 - \dfrac{1}{7!}x^7 + \cdots\right]$

$= c_0 \displaystyle\sum_{n=0}^{\infty} (-1)^n \dfrac{1}{(2n)!} x^{2n} + c_1 \sum_{n=0}^{\infty} (-1)^n \dfrac{1}{(2n+1)!} x^{2n+1}$

3. $\displaystyle\underbrace{\sum_{n=2}^{\infty} n(n-1)c_n x^{n-2}}_{k=n-2} - \underbrace{\sum_{n=1}^{\infty} nc_n x^{n-1}}_{k=n-1} = \sum_{k=0}^{\infty} (k+2)(k+1)c_{k+2} x^k - \sum_{k=0}^{\infty} (k+1)c_{k+1} x^k$

$$= \sum_{k=0}^{\infty} [(k+2)(k+1)c_{k+2} - (k+1)c_{k+1}]x^k = 0$$

$(k+2)(k+1)c_{k+2} - (k+1)c_{k+1} = 0; \quad c_{k+2} = \dfrac{c_{k+1}}{(k+2)}, \quad k = 0,1,2,\ldots; \quad c_2 = \dfrac{c_1}{2} = \dfrac{c_1}{2!},$

$c_3 = \dfrac{c_2}{3} = \dfrac{c_1}{3!}, \quad c_4 = \dfrac{c_3}{4} = \dfrac{c_1}{4!},$

$y = c_0 + c_1 \left[x + \dfrac{1}{2!}x^2 + \dfrac{1}{3!}x^3 + \cdots\right] = c_0 + c_1 \displaystyle\sum_{n=1}^{\infty} \dfrac{1}{n!} x^n$

5. $\underbrace{\sum_{n=2}^{\infty} n(n-1)c_n x^{n-2}}_{k=n-2} - x\underbrace{\sum_{n=0}^{\infty} c_n x^n}_{k=n+1} = \sum_{k=1}^{\infty}(k+2)(k+1)c_{k+2}x^k - \sum_{k=1}^{\infty} c_{k-1}x^k$

$$= 2c_2 + \sum_{k=0}^{\infty}[(k+2)(k+1)c_{k+2} - c_{k-1}]x^k = 0$$

$c_2 = 0; \ (k+2)(k+1)c_{k+2} - c_{k-1} = 0; \ c_{k+2} = \dfrac{c_{k-1}}{(k+2)(k+1)}, \ k = 1,2,3,\ldots;$

$c_3 = \dfrac{c_0}{3\cdot 2}, \ c_5 = \dfrac{c_2}{5\cdot 4} = 0, \ c_6 = \dfrac{c_3}{6\cdot 5} = \dfrac{c_0}{6\cdot 5\cdot 3\cdot 2}, \ c_7 = \dfrac{c_4}{7\cdot 6} = \dfrac{c_1}{7\cdot 6\cdot 4\cdot 3}$

$c_9 = \dfrac{c_5}{8\cdot 7} = 0, \ c_9 = \dfrac{c_6}{9\cdot 8} = c_9 = \dfrac{c_0}{9\cdot 8\cdot 6\cdot 5\cdot 3\cdot 2}, \ c_{10} = \dfrac{c_7}{10\cdot 9} = \dfrac{c_1}{10\cdot 9\cdot 7\cdot 6\cdot 4\cdot 3}$

$y = c_0\left[1 + \dfrac{1}{3\cdot 2}x^3 + \dfrac{1}{6\cdot 5\cdot 3\cdot 2}x^6 + \dfrac{1}{9\cdot 8\cdot 6\cdot 5\cdot 3\cdot 2}x^9 + \cdots\right]$

$\qquad + c_1\left[x + \dfrac{1}{4\cdot 3}x^4 + \dfrac{1}{7\cdot 6\cdot 4\cdot 3}x^7 + \dfrac{1}{10\cdot 9\cdot 7\cdot 6\cdot 4\cdot 3}x^{10} + \cdots\right]$

7. $\underbrace{\sum_{n=2}^{\infty} n(n-1)c_n x^{n-2}}_{k=n-2} - 2x\sum_{n=1}^{\infty} c_n x^{n-1} + \sum_{n=0}^{\infty} c_n x^n$

$$= \sum_{k=0}^{\infty}(k+2)(k+1)c_{k+2}x^k - \sum_{k=1}^{\infty} kc_k x^k + \sum_{k=0}^{\infty} c_k x^k$$

$$= c_0 + 2c_2 + \sum_{k=1}^{\infty}[(k+2)(k+1)c_{k+2} - (2k-1)c_k]x^k = 0$$

$c_0 + 2c_2 = 0; \ (k+2)(k+1)c_{k+2} - (2k-1)c_k = 0; \ c_2 = -\dfrac{c_0}{2}$

$c_{k+2} = \dfrac{(2k-1)c_k}{(k+2)(k+1)}, \ k = 1,2,3,\ldots; \ c_3 = \dfrac{c_1}{3\cdot 2} = \dfrac{c_1}{3!}, \ c_4 = \dfrac{3c_2}{4\cdot 3} = -\dfrac{3c_0}{4!},$

$c_5 = \dfrac{5c_3}{5\cdot 4} = \dfrac{5c_1}{5!}, \ c_6 = \dfrac{7c_4}{6\cdot 5} = \dfrac{7\cdot 3c_0}{6!}, \ c_7 = \dfrac{9c_5}{7\cdot 6} = \dfrac{9\cdot 5c_1}{7!}$

$y = c_0\left[1 - \dfrac{1}{2!}x^2 - \dfrac{3}{4!}x^4 - \dfrac{7\cdot 3}{6!}x^6 - \cdots\right] + c_1\left[x + \dfrac{1}{3!}x^3 + \dfrac{5}{5!}x^5 + \dfrac{9\cdot 5}{7!}x^7 + \cdots\right]$

9. $\underbrace{\sum_{n=2}^{\infty} n(n-1)c_n x^{n-2}}_{k=n-2} + x^2\underbrace{\sum_{n=1}^{\infty} nc_n x^{n-1}}_{k=n+1} + x\underbrace{\sum_{n=0}^{\infty} c_n x^n}_{k=n+1}$

$$= \sum_{k=0}^{\infty}(k+2)(k+1)c_{k+2}x^k + \sum_{k=2}^{\infty}(k-1)c_{k-1}x^k + \sum_{k=1}^{\infty} c_{k-1}x^k$$

$$= 2c_2 + (6c_3 + c_0)x + \sum_{k=2}^{\infty}[(k+2)(k+1)c_{k+2} + kc_{k-1}]x^k = 0$$

$2c_2 = 0, \ 6c_3 + c_0 = 0 \ (k+2)(k+1)c_{k+2} + kc_{k-1} = 0; \ c_2 = 0, \ c_3 = -\dfrac{c_0}{3\cdot 2}$

$c_{k+2} = -\dfrac{kc_{k-1}}{(k+2)(k+1)}, \ k = 2,3,4,\ldots; \ c_4 = -\dfrac{2c_1}{4\cdot 3}, \ c_5 = 0, \ c_6 = -\dfrac{4c_3}{6\cdot 5} = \dfrac{4c_0}{6\cdot 5\cdot 3\cdot 2}$

$c_7 = -\dfrac{5c_4}{7\cdot 6} = \dfrac{5\cdot 2c_1}{7\cdot 6\cdot 4\cdot 3}, \ c_8 = c_{11} = c_{14} = \cdots = 0, \ c_9 = -\dfrac{7c_6}{9\cdot 8} = -\dfrac{7\cdot 4c_0}{9\cdot 8\cdot 6\cdot 5\cdot 3\cdot 2}$

$c_{10} = -\dfrac{8c_7}{10\cdot 9} = -\dfrac{8\cdot 5\cdot 2c_1}{10\cdot 9\cdot 7\cdot 6\cdot 4\cdot 3}$

$y = c_0\left[1 - \dfrac{1}{3!}x^3 + \dfrac{4^2}{6!}x^6 - \dfrac{7^2\cdot 4^2}{9!}x^9\cdots\right] + c_1\left[x - \dfrac{2^2}{4}x^4 + \dfrac{5^2\cdot 2^2}{7!}x^7 - \dfrac{8^2\cdot 5^2\cdot 2^2}{10!}x^{10} + \cdots\right]$

11. $(x-1)\sum_{n=2}^{\infty} n(n-1)c_n x^{n-2} + \sum_{n=1}^{\infty} nc_n x^{n-1}$

$$= \sum_{n=2}^{\infty} n(n-1)c_n x^{n-1} - \underbrace{\sum_{n=2}^{\infty} n(n-1)c_n x^{n-2}}_{k=n-2} + \underbrace{\sum_{n=2}^{\infty} nc_n x^{n-1}}_{k=n-1}$$

with the first and third sums marked $k=n-1$.

$$= \sum_{k=1}^{\infty} (k+1)kc_{k+1}x^k - \sum_{k=0}^{\infty} (k+2)(k+1)c_{k+2}x^k + \sum_{k=0}^{\infty} (k+1)c_{k+1}x^k$$

$$= c_1 - 2c_2 + \sum_{k=1}^{\infty} [(k+1)kc_{k+1} - (k+2)(k+1)c_{k+2} + (k+1)c_{k+1}]x^k = 0$$

$c_1 - 2c_2 = 0$; $(k+1)kc_{k+1} - (k+2)(k+1)c_{k+2} + (k+1)c_{k+1} = 0$; $c_2 = \dfrac{c_1}{2}$

$c_{k+2} = \dfrac{(k+1)c_{k+1}}{k+2}$, $k = 1,2,3,\ldots$; $c_3 = \dfrac{2c_2}{3} = \dfrac{c_1}{3}$, $c_4 = \dfrac{3c_3}{4} = \dfrac{c_1}{4}$

$y = c_0 + c_1\left[x + \dfrac{1}{2}x^2 + \dfrac{1}{3}x^3 + \dfrac{1}{4}x^4 + \cdots\right] = c_0 + c_1 \sum_{n=1}^{\infty} \dfrac{1}{n}x^n$

13. $(x^2-1)\sum_{n=2}^{\infty} n(n-1)c_n x^{n-2} + 4x\sum_{n=1}^{\infty} nc_n x^{n-1} + 2\sum_{n=0}^{\infty} c_n x^n$

$$= \sum_{n=2}^{\infty} n(n-1)c_n x^n - \underbrace{\sum_{n=2}^{\infty} n(n-1)c_n x^{n-2}}_{k=n-2} + 4\sum_{n=1}^{\infty} nc_n x^n + 2\sum_{n=0}^{\infty} c_n x^n$$

$$= \sum_{k=2}^{\infty} k(k-1)c_k x^k - \sum_{k=0}^{\infty} (k+2)(k+1)c_{k+2}x^k + 4\sum_{k=1}^{\infty} kc_k x^k + 2\sum_{k=0}^{\infty} c_k x^k$$

$$= (2c_0 - 2c_2) + (2c_1 + 4c_1 - 6c_3)x + \sum_{k=2}^{\infty} [k(k-1)c_k - (k+2)(k+1)c_{k+2} + 4kc_k + 2c_k]x^k$$

$$= 0$$

$2c_0 - 2c_2 = 0$; $6c_1 - 6c_3 = 0$; $(k+2)(k+1)c_k - (k+2)(k+1)c_{k+2} = 0$; $c_2 = c_0$, $c_3 = c_1$;

$c_{k+2} = c_k$, $k = 2,3,4,\ldots$; $c_4 = c_2 = c_0$, $c_5 = c_3 = c_1$, $c_6 = c_4 = c_0$, $c_7 = c_5 = c_1$

$y = c_0\left[1 + x^2 + x^4 + \cdots\right] + c_1[x + x^3 + x^5 + \cdots] = c_0 \sum_{n=0}^{\infty} x^{2n} + c_1 \sum_{n=0}^{\infty} x^{2n+1}$

15. $(x^2+2)\sum_{n=2}^{\infty} n(n-1)c_n x^{n-2} + 3x\sum_{n=1}^{\infty} nc_n x^{n-1} - \sum_{n=0}^{\infty} c_n x^n$

$$= \sum_{n=2}^{\infty} n(n-1)c_n x^n + 2\underbrace{\sum_{n=2}^{\infty} n(n-1)c_n x^{n-2}}_{k=n-2} + 3\sum_{n=1}^{\infty} nc_n x^n - \sum_{n=0}^{\infty} c_n x^n$$

$$= \sum_{k=2}^{\infty} k(k-1)c_k x^k + 2\sum_{k=0}^{\infty} (k+2)(k+1)c_{k+2}x^k + 3\sum_{k=1}^{\infty} kc_k x^k - \sum_{k=0}^{\infty} c_k x^k$$

$$= (4c_2 - c_0) + (12c_3 + 3c_1 - c_1)x + \sum_{k=2}^{\infty} [k(k-1)c_k + 2(k+2)(k+1)c_{k+2} + 3kc_k - c_k]x^k$$

$$= 0$$

$4c_2 - c_0 = 0$; $12c_3 + 2c_1 = 0$; $2(k+2)(k+1)c_{k+2} + (k^2 + 2k - 1)c_k = 0$; $c_2 = \dfrac{c_0}{4}$, $c_3 = -\dfrac{c_1}{6}$;

$c_{k+2} = -\dfrac{(k^2 + 2k - 1)c_k}{2(k+2)(k+1)}$, $k = 2,3,4,\ldots$; $c_4 = -\dfrac{7c_2}{2\cdot4\cdot3} = -\dfrac{7}{4\cdot4!}c_0$, $c_5 = -\dfrac{14c_3}{2\cdot5\cdot4} = \dfrac{14}{2\cdot5!}c_1$

$c_6 = -\dfrac{23c_4}{2\cdot6\cdot5} = \dfrac{23\cdot7}{2^3\cdot6!}c_0$, $c_7 = -\dfrac{34c_5}{2\cdot7\cdot6} = -\dfrac{34\cdot14}{4\cdot7!}c_1$ $y = c_0\left[1 + \dfrac{1}{4}x^2 - \dfrac{7}{4\cdot4!}x^4 + \dfrac{23\cdot7}{8\cdot6!}x^6 - \cdots\right] +$

$c_1\left[x - \dfrac{1}{6}x^3 + \dfrac{14}{2\cdot5!}x^5 - \dfrac{34\cdot14}{4\cdot7!}x^7 + \cdots\right]$

17. $\displaystyle\sum_{n=2}^{\infty} n(n-1)c_n x^{n-2} - (x+1)\sum_{n=1}^{\infty} nc_n x^{n-1} - \sum_{n=0}^{\infty} c_n x^n$

$$= \underbrace{\sum_{n=2}^{\infty} n(n-1)c_n x^{n-2}}_{k=n-2} - \sum_{n=1}^{\infty} nc_n x^n - \underbrace{\sum_{n=1}^{\infty} nc_n x^{n-1}}_{k=n-1} - \sum_{n=0}^{\infty} c_n x^n$$

$$= \sum_{k=0}^{\infty} (k+2)(k+1)c_{k+2}x^k - \sum_{k=1}^{\infty} kc_k x^k - \sum_{k=0}^{\infty} (k+1)c_{k+1}x^k - \sum_{k=0}^{\infty} c_k x^k$$

$$= 2c_2 - c_1 - c_0 + \sum_{k=1}^{\infty}[(k+2)(k+1)c_{k+2} - kc_k - (k+1)c_{k+1} - c_k]x^k = 0$$

$2c_2 - c_1 - c_0 = 0$; $(k+2)(k+1)c_{k+2} - (k+1)c_{k+1} - (k+1)c_k = 0$; $c_2 = \dfrac{c_0 + c_1}{2}$

$c_{k+2} = \dfrac{c_k + c_{k+1}}{k+2}$, $k = 1, 2, 3, \ldots$; $c_3 = \dfrac{c_1 + c_2}{3} = \dfrac{c_1 + c_0/2 + c_1/2}{3} = \dfrac{c_0 + 3c_1}{6}$

$c_4 = \dfrac{c_2 + c_3}{4} = \dfrac{c_0/2 + c_1/2 + c_0/6 + c_1/2}{4} = \dfrac{2c_0 + 3c_1}{12}$

$c_5 = \dfrac{c_3 + c_4}{5} = \dfrac{c_0/6 + c_1/2 + c_0/6 + c_1/4}{5} = \dfrac{4c_0 + 9c_1}{60}$

$y = c_0\left[1 + \dfrac{1}{2}x^2 + \dfrac{1}{6}x^3 + \dfrac{1}{6}x^4 + \cdots\right] + c_1\left[x + \dfrac{1}{2}x^2 + \dfrac{1}{2}x^3 + \dfrac{1}{4}x^4 + \cdots\right]$

19. $(x-1)\displaystyle\sum_{n=2}^{\infty} n(n-1)c_n x^{n-2} - x\sum_{n=1}^{\infty} nc_n x^{n-1} + \sum_{n=0}^{\infty} c_n x^n$

$$= \underbrace{\sum_{n=2}^{\infty} n(n-1)c_n x^{n-1}}_{k=n-1} - \underbrace{\sum_{n=2}^{\infty} n(n-1)c_n x^{n-2}}_{k=n-2} - \sum_{n=1}^{\infty} nc_n x^n + \sum_{n=0}^{\infty} c_n x^n$$

$$= \sum_{k=1}^{\infty} (k+1)kc_{k+1}x^k - \sum_{k=0}^{\infty}(k+2)(k+1)c_{k+2}x^k - \sum_{k=1}^{\infty} kc_k x^k + \sum_{k=0}^{\infty} c_k x^k$$

$$= c_0 - 2c_2 + \sum_{k=1}^{\infty}[(k+1)kc_{k+1} - (k+2)(k+1)c_{k+2} - kc_k + c_k]x^k = 0$$

$c_0 - 2c_2 = 0$; $(k+1)kc_{k+1} - (k+2)(k+1)c_{k+2} - (k-1)c_k = 0$; $c_2 = \dfrac{1}{2}c_0$

$c_{k+2} = \dfrac{(k+1)kc_{k+1} - (k-1)c_k}{(k+2)(k+1)}$, $k = 1, 2, 3, \ldots$; $c_3 = \dfrac{2c_2}{3 \cdot 2} = \dfrac{c_0}{3 \cdot 2}$

$c_4 = \dfrac{3 \cdot 2c_3 - c_2}{4 \cdot 3} = \dfrac{c_0 - c_0/2}{4 \cdot 3} = \dfrac{c_0}{4 \cdot 3 \cdot 2}$

$y = c_0\left[1 + \dfrac{1}{2}x^2 + \dfrac{1}{3 \cdot 2}x^3 + \dfrac{1}{4 \cdot 3 \cdot 2}x^4 + \cdots\right] + c_1 x$; $y' = c_0\left[x + \dfrac{1}{2}x^2 + \cdots\right] + c_1$

Using the initial conditions, we obtain $-2 = y(0) = c_0$ and $6 = y'(0) = c_1$. The solution is

$$y = -2\left[1 + \dfrac{1}{2}x^2 + \dfrac{1}{3 \cdot 2}x^3 + \dfrac{1}{4 \cdot 3 \cdot 2}x^4 + \cdots\right] + 6x = -2 + 6x - x^2 - \dfrac{1}{3}x^3 - \dfrac{1}{4 \cdot 3}x^4 + \cdots$$

Chapter 16 in Review

A. True/False

1. True

3. False. y_2 is a constant multiple of y_1. Specifically, $y_2 = 0 \cdot y_1$.

5. True. Any constant function solves the DE.

7. True

B. Fill In the Blanks

1. By inspection, the constant function $y = 0$ solves the DE.

3. $10 = k(2.5) \Longrightarrow k = 4$ lb/ft;
   $32 = 4x \Longrightarrow x = 8$ ft

5. $y_p = Ax^2 + Bx + C + Dxe^{2x} + Ee^{2x}$

C. Exercises

1. $P_y = -6xy^2 \sin y^3 = Q_x$, and the equation is exact.
   $f_x = 2x \cos y^3$, $f = x^2 \cos y^3 + g(y)$, $f_y = -3x^2 y^2 \sin y^3 + g'(y) = -1 - 3x^2 y^2 \sin y^3$,
   $g'(y) = -1$, $g(y) = -y$, $f = x^2 \cos^3 y - y$.
   Therefore, the solution is $x^2 \cos y^3 - y = C$.

3. $P_y = -2xy^{-5} = Q_x$, and the equation is exact.
   $f_x = \frac{1}{2}xy^{-4}$, $f = \frac{1}{4}x^2 y^{-4} + g(y)$, $f_y = -x^2 y^{-5} + g'(y) = 3y^{-3} - x^2 y^{-5}$
   $g'(y) = 3y^{-3}$, $g(y) = -\frac{3}{2}y^{-2}$, $f = \frac{1}{4}x^2 y^{-4} - \frac{3}{2}y^{-2}$.
   Therefore, the general solution is $\frac{1}{4}x^2 y^{-4} - \frac{3}{2}y^{-2} = C$. Since $y(1) = 1$, we have $\frac{1}{4}(1)(1) - \frac{3}{2}(1) = C$ or $C = -\frac{5}{4}$. Thus, the solution is $\frac{1}{4}x^2 y^{-4} - \frac{3}{2}y^{-2} = -\frac{5}{4}$.

5. $m^2 - 2m - 2 = 0 \Longrightarrow m = 1 \pm \sqrt{3}$; $y = C_1 e^{(1-\sqrt{3})x} + C_2 e^{(1+\sqrt{3})x}$

7. $m^2 - 3m - 10 = 0 \Longrightarrow (m-5)(m+2) = 0 \Longrightarrow m = -2, 5$; $y = C_1 e^{-2x} + C_2 e^{5x}$

9. $9m^2 + 1 = 0 \Longrightarrow m = \pm\frac{1}{3}i$; $y = C_1 \cos\frac{x}{3} + C_2 \sin\frac{x}{3}$

11. Letting $y = ux$ we have

$$(x + uxe^u)dx - xe^u(udx + xdu) = 0 \Longrightarrow dx - xe^u du = 0 \Longrightarrow \frac{dx}{x} - e^u du = 0$$
$$\Longrightarrow \ln|x| - e^u = C_1 \Longrightarrow \ln|x| - e^{y/x} = C_1.$$

Using $y(1) = 0$ we find $C_1 = -1$. The solution of the initial-value problem is $\ln|x| = e^{y/x} - 1$.

13. $m^2 - m - 12 = 0 \Longrightarrow (m-4)(m+3) = 0 \Longrightarrow m = -3, 4$; $y_c = C_1 e^{-3x} + C_2 e^{4x}$
    $y_p = Axe^{2x} + Be^{2x}$, $y_p' = 2Axe^{2x} + (A + 2B)e^{2x}$; $y_p'' = 4Axe^{2x} + 4(A + B)e^{2x}$
    $[4Axe^{2x} + 4(A+B)e^{2x}] - [2Axe^{2x} + (A+2B)e^{2x}] - 12[Axe^{2x} + Be^{2x}]$
    $\qquad = -10Axe^{2x} + (3A - 10B)e^{2x} = xe^{2x} + e^x$
    Solving $-10A = 1$, $3A - 10B = 1$ we obtain $A = -1/10$ and $B = -13/100$. Thus,

$$y = C_1 e^{-3x} + C_2 e^{4x} - \frac{1}{10}xe^{2x} - \frac{13}{100}e^{2x}.$$

15. $m^2 - 2m + 2 = 0 \Longrightarrow m = 1 \pm i$; $y_c = e^x(C_1 \cos x + C_2 \sin x)$
    $$W = \begin{vmatrix} e^x \cos x & e^x \sin x \\ -e^x \sin x + e^x \cos x & e^x \cos x + e^x \sin x \end{vmatrix} = e^{2x}$$
    $u' = -\frac{1}{e^{2x}}e^x \sin x \, e^x \tan x = -\frac{\sin^2 x}{\cos x} = \frac{\cos^2 x - 1}{\cos x} = \cos x - \sec x$, $u = \sin x - \ln|\sec x + \tan x|$
    $v' = \frac{1}{e^{2x}}e^x \cos x \, e^x \tan x = \sin x$, $v = -\cos x$
    $y_p = e^x \cos x(\sin x - \ln|\sec x + \tan x|) - e^x \sin x \cos x = -e^x \cos x \ln|\sec x + \tan x|$
    $y = e^x(C_1 \cos x + C_2 \sin x) - e^x \cos x \ln|\sec x + \tan x|$

17. $m^2 + 1 = 0 \Longrightarrow m = \pm i$; $y_c = C_1 \cos x + C_2 \sin x$; $W = \begin{vmatrix} \cos x & \sin x \\ -\sin x & \cos x \end{vmatrix} = 1$
    $u' = -\sin x \sec^3 x = -\tan x \sec^2 x$, $u = -\frac{1}{2}\sec^2 x$; $v' = \cos x \sec^3 x = \sec^2 x$, $v = \tan x$

$$y_p = -\frac{1}{2}\cos x \sec^2 x + \sin x \tan x = \sin x \tan x - \frac{1}{2}\sec x = \frac{\sin^2 x}{\cos x} - \frac{1}{2\cos x}$$

$$= \frac{2\sin^2 x - 1}{2\cos x} = \frac{\sin^2 x - \cos^2 x}{2\cos x} = \frac{1}{2}\sin x \tan x - \frac{1}{2}\cos x$$

$$y = C_3 \cos x + C_2 \sin x + \frac{1}{2}\sin x \tan x, \quad y' = -C_3 \sin x + C_2 \cos x + \frac{1}{2}\sin x \sec^2 x + \sin x$$

Using the initial conditions, we obtain $C_3 = 1$ and $C_2 = 1/2$. Thus,

$$y = \cos x + \frac{1}{2}\sin x + \frac{1}{2}\sin x \tan x = \frac{2\cos^2 x}{2\cos x} + \frac{\sin^2 x}{2\cos x} + \frac{1}{2}\sin x$$

$$= \frac{\cos^2 x + 1}{2\cos x} + \frac{1}{2}\sin x = \frac{1}{2}(\sin x + \cos x + \sec x).$$

19. $\underbrace{\sum_{n=2}^{\infty} n(n-1)c_n x^{n-2}}_{k=n-2} + x\underbrace{\sum_{n=0}^{\infty} c_n x^n}_{k=n+1} = \sum_{k=0}^{\infty}(k+2)(k+1)c_{k+2}x^k + \sum_{k=1}^{\infty} c_{k-1}x^k$

$$= 2c_2 + \sum_{k=1}^{\infty}[(k+2)(k+1)c_{k+2} + c_{k-1}]x^k = 0$$

$c_2 = 0; \quad (k+2)(k+1)c_{k+2} + c_{k-1} = 0; \quad c_{k+2} = -\dfrac{c_{k-1}}{(k+2)(k+1)}, \quad k = 1,2,3,\ldots$

$c_3 = -\dfrac{c_0}{3\cdot 2}, \quad c_4 = -\dfrac{c_1}{4\cdot 3}, \quad c_5 = 0, \quad c_6 = -\dfrac{c_3}{6\cdot 5} = \dfrac{c_0}{6\cdot 5\cdot 3\cdot 2},$

$c_7 = -\dfrac{c_4}{7\cdot 6} = \dfrac{c_1}{7\cdot 6\cdot 4\cdot 3}$

$c_8 = 0. \quad c_9 = -\dfrac{c_6}{9\cdot 8} = -\dfrac{c_0}{9\cdot 8\cdot 6\cdot 5\cdot 3\cdot 2}, \quad c_{10} = -\dfrac{c_7}{10\cdot 9} = -\dfrac{c_1}{10\cdot 9\cdot 7\cdot 6\cdot 4\cdot 3}$

$y = c_0\left[1 - \dfrac{1}{3\cdot 2}x^3 + \dfrac{1}{6\cdot 5\cdot 3\cdot 2}x^6 - \dfrac{1}{9\cdot 8\cdot 6\cdot 5\cdot 3\cdot 2}x^9 + \cdots\right]$

$\quad + c_1\left[x - \dfrac{1}{4\cdot 3}x^4 + \dfrac{1}{7\cdot 6\cdot 4\cdot 3}x^7 - \dfrac{1}{10\cdot 9\cdot 7\cdot 6\cdot 4\cdot 3}x^{10} + \cdots\right]$

21. The differential equation is $mx'' + 4x' + 2x = 0$. The solutions of the auxiliary equation are

$$\frac{1}{2m}(-4 \pm \sqrt{16-8m}) = \frac{1}{m}(-2 \pm \sqrt{4-2m}).$$

The motion will be non-oscillatory when $4 - 2m \geq 0$ or $0 < m \leq 2$.

23. Using $m = W/g = 4/32 = 1/8$, the inital value problem is $\frac{1}{8}x'' + x' + 3x = e^{-t}$; $x(0) = 2$, $x'(0) = 0$. The auxiliary equation is $m^2/8 + m + 3 = 0$. Using the quadratic formula, $m = -4 \pm 2\sqrt{2}i$. Thus, $x_c = e^{-4t}(C_1 \cos 2\sqrt{2}t + C_2 \sin 2\sqrt{2}t)$. Using $x_p = Ae^{-t}$, we find $A = 8/17$. Thus,

$$x(t) = e^{-4t}(C_1 \cos 2\sqrt{2}t + C_2 \sin 2\sqrt{2}t) + \frac{8}{17}e^{-t}$$

$$\text{and} \quad x'(t) = e^{-4t}[(2\sqrt{2}C_2 - 4C_1)\cos 2\sqrt{2}t - (2\sqrt{2}C_1 - 4C_2)\sin 2\sqrt{2}t] - \frac{8}{17}e^{-t}.$$

Using the initial conditions, we obtain $2 = C_1 + 8/17$ and $0 = 2\sqrt{2}C_2 - 4C_1 - 8/17$. Then $C_1 = 26/17$ and $C_2 = 28\sqrt{2}/17$ and

$$x(t) = e^{-4t}\left(\frac{26}{17}\cos 2\sqrt{2}t + \frac{28}{17}\sqrt{2}\sin 2\sqrt{2}t\right) + \frac{8}{17}e^{-t}.$$

25. The auxiliary equation is $m^2/4 + m + 1 = 0$ or $(m+2)^2 = 0$, so $m = -2, -2$ and $x(t) = C_1 e^{-2t} + C_2 te^{-2t}$ and $x'(t) = -2C_1 e^{-2t} - 2C_2 te^{-2t} + C_2 e^{-2t}$. Using the initial conditions, we obtain $4 = C_1$ and $2 = -2C_1 + C_2$. Thus, $C_1 = 4$ and $C_2 = 10$. Therefore $x(t) = 4e^{-2t} + 10te^{-2t}$ and $x'(t) = 2e^{-2t} - 20te^{-2t}$. Setting $x'(t) = 0$ we obtain the critical point $t = 1/10$. The maximum vertical displacement is $x(1/10) = 5e^{-0.2} \approx 4.0937$.

Essay Exercises

1. Letting t_1 denote the time taken for the spiral to complete one turn, we have $\omega t_1 = 2\pi$. If the spiral is to reach the circle of radius R at the end of the first turn, we must have $v t_1 = R$ so that $t_1 = \frac{R}{v}$.

   Substituting this value for t_1 into the equation $\omega t_1 = 2\pi$, we get $\omega \left(\frac{R}{v}\right) = 2\pi$ or $\omega = \frac{2\pi v}{R}$.

   When the spiral intercepts the circle at time t_1, the circumferential velocity is given by $v t_1 \omega = 2\pi v$ so that the circumferential velocity is 2π times the radial velocity. Thus the ratio of the radial component of the velocity to the circumferential component is $1 : 2\pi$.

   Therefore, using the principle from Aristotle's Mechanics, we see that the diagonal of any parallelogram whose sides are in the ratio $1 : 2\pi$ will have the same slope as the tangent to the spiral at the desired point.

   On Figure 1, we can draw the rectangle with vertices $(0,0)$, $(R,0)$, $(R,-2\pi R)$, and $(0,-2\pi R)$. The sides of this rectangle are in the ratio $1 : 2\pi$. Therefore, the line containing the diagonal from $(0,-2\pi R)$ to $(R,0)$ intersects the spiral at the desired point $(R,0)$ and has the correct slope. Hence, it is the tangent.

2. (a) Since $DA = DG$, the cylinder which is formed by revolving $DEFG$ must have its center of gravity on the line AB. Since $AK = BK$, we also have that the center of gravity must lie on a disk through K and perpendicular to AB. Therefore, the center of gravity of the cylinder must be located at K.

   (b) Let the sections be taken at a distance h from the point B, and let dh represent the infinitesimal thickness of each section. Also, let r represent the radius of the sphere. Then the mass of the cylindrical section (assuming each figure has a mass density of 1) is $4\pi r^2 dh$. The mass of the spherical section is $(2\pi r h - \pi h^2)dh$ and the mass of the conical section is $\pi h^2 dh$. The torque produced by the cylindrical section is $(4\pi r^2 dh)(h) = 4\pi r^2 h dh$. Once they are removed to the point C, the magnitude of the torque produced by the sum of the spherical and conical sections is $[(2\pi r h - \pi h^2)dh + \pi h^2 dh](2r) = (2\pi r h dh)(2r) = 4\pi r^2 h dh$. Since the torques produced are equal in magnitude but opposite in direction, the pieces balance out.

   (c) Part (b) shows that the cylinder will balance the cone and sphere when they are concentrated at C. Part (a) shows that concentrating the cylinder at K will not alter the balance.

   (d) Since C is twice the distance from B as K, the balance from part (c) requires that the mass of the cylinder is twice the mass of the sum of the cone and the sphere. Since we have assumed a mass density of 1 for the three figures, we have that the volume of the cylinder is twice the volume of the sum of the sphere and the cone.

   (e) From part (d) we have

   $$V_{cylinder} = 2\left(V_{cone} + V_{sphere}\right)$$

   Since $V_{cone} = \frac{1}{3}V_{cylinder}$, this becomes

   $$V_{cylinder} = \frac{2}{3}V_{cylinder} + 2V_{sphere}$$

   or $\frac{1}{6}V_{cylinder} = V_{sphere}$.

   (f) Using part (e) we have

   $$V_{sphere} = \frac{1}{6}V_{cylinder} = \frac{1}{6}\left(8\pi r^3\right) = \frac{4}{3}\pi r^3$$

3. (a) Let C denote the center of the sphere, and let C_h denote the center of the circular section. Since the section is distance h below the center of the sphere, we can drop a segment of length h from C to C_h which is perpendicular to the section. We can then draw a radius from the sphere center C to any point P on the boundary of the section. This radius has length r.

Using the Pythagorean Theorem, we find that the length of the segment joining C_h and P is $\sqrt{r^2 - h^2}$. Thus, the radius of the section is $\sqrt{r^2 - h^2}$ and the diameter is $2\sqrt{r^2 - h^2}$.

(b) The square formed by the section of the double umbrella circumscribes the circular section. Therefore, its side length is $2\sqrt{r^2 - h^2}$ and its area is $4(r^2 - h^2)$. Since the section of the cube has area $4r^2$, the area between the section of the double umbrella is $4r^2 - 4(r^2 - h^2) = 4h^2$.

(c) The pyramid has altitude r and the base is a square with side length $2r$. Therefore, using the properties of similar triangles, we see that the section of the pyramid located a distance h below the sphere center C would be a square with side length $2h$. Thus, the area of the section is $4h^2$, which is equal to the area between the section of the cube and the section of the double umbrella. By symmetry, the sections of a mirror-image pyramid above the sphere center C would have the same property. Therefore, the volume of the sum of the two pyramids is equal to the volume between the cube and the double umbrella. Since the volume of the sum of the pyramids is one-third the volume of the cube, we see that the region between the cube and the double umbrella is one-third of the cube.

(d) From part (c), we see that the double umbrella has volume equal to two-thirds of the cube or $\frac{2}{3}(8r^3) = \frac{16}{3}r^3$.

(e) The area of the circular section of the sphere h units below the sphere center C is $\pi(r^2 - h^2)$. From part (b), the area of the corresponding square section of the double umbrella is $4(r^2 - h^2)$. Hence, the area of the section of the sphere is $\frac{\pi}{4}$ of the area of the section of the double umbrella.

(f) Taking the aggregate of the sections, we see that the volume of the sphere is $\pi/4$ of the volume of the double umbrella. Hence the volume of the sphere is $\frac{\pi}{4}\left(\frac{16}{3}r^3\right) = \frac{4}{3}\pi r^3$.

4. If we let R be the radius of the sphere, then each pyramid has altitude R and infinitesimally small base dA. The infinitesimal amount of volume filled by each pyramid is $dV = \frac{1}{3}(dA)R$ so that $\frac{3(dV)}{R} = dA$. Looking at the aggregate of all the pyramids, we see that $\frac{3V}{R} = A$ where V is the volume of the sphere and A is the surface area.